Table of Atomic Masses

ELEMENT	SYMBOL	ATOMIC NUMBER	ATOMIC MASS*	ELEMENT	SYMBOL	ATOMIC NUMBER	ATOMIC MASS*
Actinium	Ac	89	227.0	Mercury	Hg	80	200.6
Aluminum	Al	13	26.98	Molybdenum	Mo	42	95.94
Americium	Am	95	(243)	Neilsbohrium	Ns	107	(262)
Antimony	Sb	51	121.8	Neodymium	Nd	60	144.2
Argon	Ar	18	39.95	Neon	Ne	10	20.18
Arsenic	As	33	74.92	Neptunium	Np	93	237.0
Astatine	At	85	(210)	Nickel	Ni	28	58.69
Barium	Ba	56	137.3	Niobium	Nb	41	92.91
Berkelium	Bk	97	(247)	Nitrogen	N	7	14.01
Beryllium	Be	4	9.012	Nobelium	No	102	(259)
Bismuth	Bi	83	209.0	Osmium	Os	76	190.2
Boron	B	5	10.81	Oxygen	O	8	16.00
Bromine	Br	35	79.90	Palladium	Pd	46	106.4
Cadmium	Cd	48	112.4	Phosphorus	P	15	30.97
Calcium	Ca	20	40.08	Platinum	Pt	78	195.1
Californium	Cf	98	(251)	Plutonium	Pu	94	(244)
Carbon	C	6	12.01	Polonium	Po	84	(209)
Cerium	Ce	58	140.1	Potassium	K	19	39.10
Cesium	Cs	55	132.9	Praseodymium	Pr	59	140.9
Chlorine	Cl	17	35.45	Promethium	Pm	61	(145)
Chromium	Cr	24	52.00	Protactinium	Pa	91	231.0
Cobalt	Co	27	58.93	Radium	Ra	88	226.0
Copper	Cu	29	63.55	Radon	Rn	86	(222)
Curium	Cm	96	(247)	Rhenium	Re	75	186.2
Dysprosium	Dy	66	162.5	Rhodium	Rh	45	102.9
Einsteinium	Es	99	(252)	Rubidium	Rb	37	85.47
Erbium	Er	68	167.3	Ruthenium	Ru	44	101.1
Europium	Eu	63	152.0	Rutherfordium	Rf	104	(261)
Fermium	Fm	100	(257)	Samarium	Sm	62	150.4
Fluorine	F	9	19.00	Scandium	Sc	21	44.96
Francium	Fr	87	(223)	Selenium	Se	34	78.96
Gadolinium	Gd	64	157.3	Silicon	Si	14	28.09
Gallium	Ga	31	69.72	Silver	Ag	47	107.9
Germanium	Ge	32	72.61	Sodium	Na	11	22.99
Gold	Au	79	197.0	Strontium	Sr	38	87.62
Hafnium	Hf	72	178.5	Sulfur	S	16	32.07
Hahnium	Ha	105	(262)	Tantalum	Ta	73	180.9
Hassium	Hs	108	(265)	Technetium	Tc	43	(99)
Helium	He	2	4.003	Tellurium	Te	52	127.6
Holmium	Ho	67	164.9	Terbium	Tb	65	158.9
Hydrogen	H	1	1.008	Thallium	Tl	81	204.4
Indium	In	49	114.8	Thorium	Th	90	232.0
Iodine	I	53	126.9	Thulium	Tm	69	168.9
Iridium	Ir	77	192.2	Tin	Sn	50	118.7
Iron	Fe	26	55.85	Titanium	Ti	22	47.88
Krypton	Kr	36	83.80	Tungsten	W	74	183.9
Lanthanum	La	57	138.9	Unnilhexium	Unh	106	(263)
Lawrencium	Lr	103	(262)	Uranium	U	92	238.0
Lead	Pb	82	207.2	Vanadium	V	23	50.94
Lithium	Li	3	6.941	Xenon	Xe	54	131.3
Lutetium	Lu	71	175.0	Ytterbium	Yb	70	173.0
Magnesium	Mg	12	24.31	Yttrium	Y	39	88.91
Manganese	Mn	25	54.94	Zinc	Zn	30	65.39
Meitnerium	Mt	109	(266)	Zirconium	Zr	40	91.22
Mendelevium	Md	101	(258)				

* All molar masses rounded to four significant figures. A value given in parentheses indicates the mass number of the longest-lived or best known isotope.

Introduction to Chemical Principles

SIXTH EDITION

Introduction to Chemical Principles

SIXTH EDITION

Edward I. Peters
Emeritus
West Valley College

Robert C. Kowerski
College of San Mateo

Saunders Golden Sunburst Series

Saunders College Publishing
Harcourt Brace College Publishers

Fort Worth   Philadelphia   San Diego   New York   Orlando   Austin
San Antonio   Toronto   Montreal   London   Sydney   Tokyo

Text Typeface: Times Roman
Compositor: York Graphic Services
Publisher: John Vondeling
Developmental Editor: Beth Rosato
Managing Editor: Carol Field
Project Editors: Becca Gruliow, Anne Gibby
Copy Editor: Teresa Danielson
Manager of Art and Design: Carol Bleistine
Senior Art Director: Christine Schueler
Text Designer: Jeanne Calabrese
Cover Designer: Lawrence R. Didona
Text Artwork: J&R Technical Services; Rolin Graphics
Layout Artwork: Gene Harris
Photo Research: Sue C. Howard
Director of EDP: Tim Frelick
Production Managers: Charlene Squibb, Tim Frelick
Marketing Manager: Marjorie Waldron

Cover Credit: COMSTOCK, Inc.
Frontispiece: Charles C. Winters

Printed in the United States of America

Introduction to Chemical Principles, 6e

ISBN: 0-03-096818-6

Library of Congress Catalog Card Number: 93-083230

56 069 98765432

This book is dedicated to
Taylor C. Peters

Preface

As in all previous editions of this book, this edition is addressed to students whose experiences in chemistry are nonexistent, weak, or remote in time. Also, the text makes no assumption of mathematics skills beyond elementary algebra. After completing a one-term preparatory course using this text, students should be able to do the following:

- Read, write, and talk about chemistry, using a basic chemistry vocabulary;
- Write routine chemical formulas;
- Write routine chemical equations;
- Set up and solve chemistry problems;
- "Think" chemistry on an atomic or molecular level in fundamental theoretical areas—to visualize what happens in a chemical change.

Beginning students often enter a preparatory chemistry course with either or both of two weaknesses: (1) they don't know how to learn chemistry and/or (2) they don't know how to solve math problems. Sometimes the weaknesses are real; sometimes they are only imagined—feared. Either way, they are obstacles to learning. Either way, the obstacles can be removed. This book launches a frontal attack on these two obstacles. The pedagogical features by which this is done are described under the heading "Learning How to Learn."

NEW FEATURES

This book is now available in two editions. The soft-cover second edition of *Basic Chemical Principles* has 19 chapters. Three additional chapters, Nuclear Chemistry, Organic Chemistry, and Biochemistry, make up the hard-cover sixth edition of *Introduction to Chemical Principles*.

Chapter 22, **Biochemistry,** introduces the student to the chemistry of living systems.

Everyday Chemistry   Most chapters now have an "Everyday Chemistry" section that moves chemistry out of the textbook and classroom and into "life as we know it" in the closing years of the twentieth century.

Questions and Problems   Nearly all unanswered end-of-chapter questions and problems are new. These problems can be used for assignments.

Matching Sets   Matching Sets appear before the end-of-chapter questions. These replace "Terms and Concepts" and "Fill-in-the-Blank Summaries" in the previous edition, and they satisfy the same purpose more efficiently. Matching Set Answers are given after end-of-chapter Questions and Problems.

Structure of organic compounds   An optional section on the structure of the more common classes of organic compounds has been added to Chapter 12, the Structure and

Shape of Molecules. This gives the instructor the opportunity for a one-day assignment and lecture on organic chemistry if desired.

ORGANIZATIONAL CHANGES

Chemistry as an experimental science.   The introduction to chemistry and scientific thinking, previously given as a prologue, has been updated and incorporated into Chapter 1; but it doesn't end there. The experimental character of chemistry is extended into a new section in Chapter 4, brought to a conclusion in the discussion of Boyle's Law, and later referred to repeatedly throughout the book.

Problem solving.   The measurement and problem-solving topics that were formerly spread between two chapters are now combined in Chapter 3.

Nomenclature.   A full treatment of nomenclature now appears in Chapter 6, instead of being developed gradually as in prior editions.

Portable chapters.   Certain chapters or groups of chapters have been written so they are completely portable, i.e., they may be assigned and studied out of book sequence but with an order that is completely logical and supportive for the student. This gives the instructor options in the development of the course. These changes are described more fully below.

LEARNING HOW TO LEARN

In the fifth edition of *Introduction to Chemical Principles* we took a calculated risk by offering the student many more suggestions on how to learn chemistry—note, how to *learn,* not how to *study* (there is a difference)—than were found in any other chemistry book then available. These learning aids were couched in a five-chapter goal introduced in Chapter 1, practiced in detail in Chapter 2, and practiced in diminishing amounts in Chapters 3 through 5. The idea was to guide the student in learning how to learn the most chemistry in the least amount of time.

As work on this edition began, we asked reviewers of the previous edition to evaluate this specific feature. The responses were most encouraging. An unusually large majority of reviewers praised this feature. The most common recommendation was to leave the five-chapter goal just as it was. We've done almost that. It is shorter and more compact this time, but all of the general ideas have been retained. A full description of the five-chapter goal is in Section 1.3, page 6.

Section 1.3 also includes a comment about the relationship between instructor, student, and textbook. Before making our recommendations on how to learn, we state that the instructor may have additional suggestions. We then note that here and elsewhere, when the book says one thing and the instructor says another, "you will probably be better off following the instructor's suggestions rather than ours." This advice is recalled in more than a dozen places in the book where the instructor may prefer to teach a certain topic in a way that is different from ours. In some cases, both methods are described. In this and other ways, the book strives to support the instructor's choices among common alternatives.

LEARNING AIDS

In addition to the five-chapter goal, this text has all of the usual learning aids commonly found in beginning texts, plus a few that are not so common. These include:

Learning objectives Learning objectives (we call them performance goals) appear at the beginning of each section, where they are most valuable to the learner. At the end of the chapter all performance goals in the chapter are assembled as a ''Chapter in Review.'' In this location they are particularly helpful to students as they study for tests.

Example problems Most example problems are in a semiprogrammed format in which the student participates in solving the problem rather than just seeing how somebody else solved it. The success of this method is well established in over twenty years of classroom experience with prior editions of this text and other books, particularly in courses in which the text was the only source of instruction in problem solving. The semiprogrammed format also extends to topics not usually covered by examples, such as formula writing and equation writing.

Tear-out periodic tables The periodic table is emphasized as a source of elemental symbols, atomic numbers, and atomic masses. Equally important, it is emphasized as a guide in writing formulas and predicting properties. Tear-out periodic tables are provided for use as shields in solving programmed examples.

Flashbacks FLASHBACKS printed in the margin are brief reminders of material developed and used earlier in the text that is about to be used again. Cross references are given in case the student wants to review the item.

In-chapter summaries and procedures Clearly identified in-chapter summaries and procedures are given at the places where important concepts and techniques are presented.

Quick Checks Quick Check questions appear at the end of many sections to help the students determine if they have grasped the main idea of the section. Quick Check Answers are at the end of the chapter.

Study Hints and Pitfalls to Avoid This end-of-chapter feature has two main purposes: (1) to identify particularly important ideas and to offer suggestions on how they can be mastered; and (2) to alert students to some of the more common mistakes so that, forewarned, they will avoid that particular pitfall.

End-of-chapter problems This book contains a large number of end-of-chapter questions and problems that range in difficulty from very easy to challenging. Answers to most odd-numbered questions and problems are given in an answer section at the back of the book. The exceptions are those answers that are virtually direct quotes from the text. Answers to problem questions include complete calculation setups.

Removal of learning aids Beginning chemistry texts usually have more learning aids than general chemistry texts. This leads to the common complaint from general chemistry students that their new book is so much harder to understand than the book for the preparatory course. We therefore try to avoid this sudden change by making it occur *gradually within the beginning course,* rather than abruptly at the start of the general chemistry course. Beginning in Chapter 13, we gradually withdraw or reduce some of the learning aids, including semiprogrammed examples, performance goals, and Study Hints and Pitfalls to Avoid. With respect to performance goals, we encourage students to write their own, as the very act of writing produces learning. Thus the student becomes more self-reliant and less dependent on the text.

Review of mathematics   The appendix provides a review of algebra as it is used in this text. There are also instructions on how to use a calculator and procedures for estimating numerical answers. Calculator procedures and recommended algebraic procedures are also found in the text at the point where they are needed.

Glossary   A glossary at the back of the book provides a ready reference for scientific and technical terms discussed in the text, plus other terms that may have been used or that the student is apt to encounter in conversation and general reading.

PORTABLE CHAPTERS

We have written Chapters 4 and 10 so they are "portable chapters." They can be moved into different positions in the text to accommodate the sequence of topics preferred by the instructor. Whatever sequence the instructor selects, it will appear to students as a logical and continuous development in their study of chemistry.

Atomic theory   As in the previous edition, there are two chapters on atomic theory, but this time they are not consecutive. Chapter 5 introduces the atom and carries it through the Rutherford experiment, isotopes, atomic mass, and the periodic table and elemental symbols. Chapter 10 presents the quantum mechanical model, electron configuration and its correlation with the periodic table, and identifiable trends in the periodic table.

The advantage of the split between the chapters is that it postpones the highly abstract ideas of quantum mechanics until some of the more concrete and visible aspects of chemistry are mastered. The disadvantage is that the separation interrupts the logical sequence from the atom to bonds, compounds, nomenclature, and reactions. Both approaches have merit; the choice between them is one of personal preference. And both are accommodated in the book. Chapter 10 is completely portable and may be assigned after Chapter 5, restoring precisely the order of topics in the previous edition.

Gases   Unlike the previous edition, this book has two gas chapters. Chapter 4 covers the gas laws for a fixed quantity of gas—the P, V, T laws, including the "combined gas law." Chapter 13 extends the study of gases to the ideal gas law, partial pressure, and gas stoichiometry. This separation and placing the first gas chapter before the first chapter on atomic theory serves several purposes:

1. The gas chapter in the previous edition was very long. The topic is now broken into two more manageable chapters.
2. Chemistry is introduced more in the order in which it was discovered, through the observation of gas behavior. It is no accident that gases were studied before Dalton proposed his atomic theory; gases are evident to the senses and atoms are not. So today, visible, measurable, and demonstrable properties of gases are more easily comprehended by beginning chemistry students than the invisible, abstract concept of atoms.
3. Beginning courses that are accompanied by laboratories require experiments that are meaningful and supported by the text early in the semester. Gases meet that need; atoms do not.
4. With or without a laboratory, the experimental basis of the gas laws provides an excellent opportunity to emphasize early in the course that chemistry is an experimental science.

5. Measurements and problem-solving techniques usually appear in one of the first three chapters of an introductory text; they are in Chapter 3 in this book. Studying gas laws in Chapter 4 gives students an immediate opportunity to practice their newly acquired calculation skills. Without gases, there is in current introductory chemistry texts an average 161-page gap between the discussion of problem solving and its first serious application; in this book the math gap is 16 pages.

Chapter 4, "Introduction to Gases," is portable. Nothing between Chapters 4 and 13 depends on material in Chapter 4. Consequently, an instructor who wishes to cover gases as a single topic may postpone Chapter 4 until the chosen time and then assign Chapters 4 and 13 in sequence. This corresponds precisely with the arrangement of topics in prior editions.

CHEMICAL CALCULATIONS

A problem-solving strategy established in Chapter 3 begins by analyzing and summarizing the problem. (See Section 3.9 and Figure 3.6, page 71.) The analysis, or "plan" as it is referred to throughout the text, includes listing what is given, what is wanted, and any applicable equations and/or conversion factors. Then comes the decision: Do I solve the problem by algebra or by dimensional analysis? The criterion for that decision is precisely stated. Once the decision is made, calculating the answer becomes routine.

It is this strategy that helps students overcome difficulties in solving chemistry problems. It answers the question, "How do I begin?" We guide the students in using the strategy by the questions asked in semiprogrammed examples throughout the text, and encourage its use by using it ourselves in worked-out examples.

ANCILLARY PACKAGE

This textbook is supported with the following materials for professor and student: An **Instructor's Manual** provides answers to all the even-numbered end-of-chapter questions (the questions that are not answered in the text).

A **Study Guide** reinforces student learning skills and furnishes additional exercises to improve the student's understanding of chemistry.

Overhead Projection Transparencies of 100 four-color figures and illustrations from the text are available upon adoption.

A **Computerized Test Bank** features an extensive file of multiple-choice questions on both Macintosh and IBM (3.5″ and 5.25″) versions. A separate printed **Test Bank** is also available.

A **Laboratory Manual,** *Introduction to Chemical Principles: A Laboratory Approach* 4th edition, by Susan Weiner (West Valley College) and Edward Peters, provides 33 experiments, including the collection and analysis of experimental data. It is accompanied by an instructor's manual.

Shakashiri Chemical Demonstration Videotapes are also available. This set of 50, 4–7 minute classroom experiments brings to the classroom colorful and instructive illustrations of chemical principles described in the text. An instructor's manual accompanies the videotapes.

Saunders General Chemistry Videodisc (Version 2) enhances class and laboratory presentations with live-action footage of 44 chemical demonstrations and almost 2400 still images taken from five of Saunders' most popular chemistry texts.

Edward I. Peters
Robert C. Kowerski
September 1993

Acknowledgments

We wish to acknowledge and express our appreciation for the many helpful suggestions from those who reviewed the previous edition of this text and the early development of this edition. Some of the "improvements" we considered introducing did not exactly fit the definition of that word! The reviewers let us know, and we retreated. In other areas they were both supportive and encouraging, and their comments led to refinements in the approach to several topics. A significant reward for writing a text is the opportunity to receive fresh opinions on how best to present various concepts. This opportunity leads to conclusions that appear in print and that show up in our classrooms as well. We thank you.

Those who reviewed the text are:

Melvin T. Armold, Adams State College
Joe Asire, Cuesta College
Caroline Ayers, East Carolina University
Sharmaine Cady, East Stroudsburg State College
Pam Coffin, University of Michigan-Flint
Jerry A. Driscoll, University of Utah
Galen G. George, Santa Rosa Junior College
Jeffrey A. Hurlbut, Metropolitan State College
Jane V. Z. Krevor, California State University, San Francisco
Kenneth Miller, Milwaukee Area Technical College
Brian J. Pankuch, Union County College
Erwin W. Richter, University of Northern Iowa
John W. Singer, Alpena Community College
Linda Wilson, Middle Tennessee State University

A special word of thanks goes to Mark S. Cracolice of the University of Oklahoma. Mark read and checked the entire text at both the galley and page proof stages. He then wrote a computerized test bank for the book. The test bank is available through the publisher.

We wish to thank Beth Rosato, Becca Gruliow, and Christine Schueler for patiently working with us and making sure we didn't neglect any of the many details that contribute to a finished product. This also includes Anne Gibby, who joined the production team very late. She was suddenly submerged in a sea of these details, but she sorted them out in remarkably short time and guided the project to completion.

Edward I. Peters
Robert C. Kowerski

Contents Overview

1 Learning Chemistry—NOW 1

2 Matter and Energy 12

3 Measurement and Calculations 35

4 Introduction to Gases 87

5 Atomic Theory and the Periodic Table: The Beginning 114

6 The Language of Chemistry 136

7 Chemical Formula Problems 170

8 Chemical Reactions and Equations 198

9 Quantity Relationships in Chemical Reactions 222

10 Atomic Theory and the Periodic Table: A Modern View 253

11 Chemical Bonding: The Formation of Ionic Compounds and Molecules 288

12 The Structure and Shape of Molecules 307

13 The Ideal Gas Law and Gas Stoichiometry 335

14 Liquids and Solids 361

15 Solutions 397

16 Reactions That Occur in Water Solutions: Net Ionic Equations 434

17 Acid-Base (Proton Transfer) Reactions 459

18 Oxidation–Reduction (Electron Transfer) Reactions 486

19 Chemical Equilibrium 508

20 Nuclear Chemistry 537

21 Organic Chemistry 563

22 Biochemistry 607

Appendix I   Chemical Calculations A.1

Appendix II   The SI System of Units A.11

Answers to Questions and Problems A.13

Glossary G.1

Photo Credits C.1

Index I.1

Contents

1 Learning Chemistry—NOW 1

1.1 Why? 2

1.2 Bubbles 2

1.3 Learning How to Learn—or Learn Smarter, Not Harder 4

1.4 Meet Your Textbook 7

1.5 A Choice 11

2 Matter and Energy 12

2.1 Physical and Chemical Properties and Changes 13

2.2 Characteristics of Matter 15

2.3 The Electrical Character of Matter 22

2.4 Characteristics of a Chemical Change 23

2.5 Conservation Laws and Chemical Change 25

3 Measurement and Calculations 35

3.1 Introduction to Measurement 36

3.2 Exponential (Scientific) Notation 36

3.3 Dimensional Analysis 41

3.4 Metric Units 48

3.5 Significant Figures 54

3.6 Metric-English Conversions 63

3.7 Temperature 65

3.8 Density 68

3.9 A Strategy for Solving Problems 70

3.10 Practice, Practice, Practice 72

4 Introduction to Gases 87

4.1 Properties of Gases 88

4.2 The Kinetic Theory of Gases and the Ideal Gas Model 89

4.3 Gas Measurements 90

4.4 Interpretation of Experimental Data—Proportionality 94

4.5   Gay-Lussac's Law: Pressure and Temperature Variable, Volume and Amount Constant 96

4.6   Charles' Law: Volume and Temperature Variable, Pressure and Amount Constant 98

4.7   Boyle's Law: Pressure and Volume Variable, Temperature and Amount Constant 100

4.8   The Combined Gas Laws 104

5 Atomic Theory and the Periodic Table: The Beginning 114

5.1   Dalton's Atomic Theory 115

5.2   Subatomic Particles 117

5.3   The Nuclear Atom 118

5.4   Isotopes 119

5.5   Atomic Mass 121

5.6   The Periodic Table 123

5.7   Elemental Symbols and the Periodic Table 126

6 The Language of Chemistry 136

6.1   Introduction to Nomenclature 137

6.2   Formulas of Elements 137

6.3   Compounds Made from Two Nonmetals 138

6.4   Names and Formulas of Ions Formed by One Element 140

6.5   Names and Formulas of Acids and the Ions Derived from Acids 143

6.6   Names and Formulas of Acid Anions 151

6.7   Names and Formulas of Other Acids and Ions 152

6.8   Formulas of Ionic Compounds 154

6.9   Names of Ionic Compounds 157

6.10 Hydrates 159

6.11 Summary of the Nomenclature System 159

6.12 Common Names of Common Chemicals 159

7 Chemical Formula Problems 170

7.1   The Number of Atoms in a Formula 171

7.2   Molecular Mass; Formula Mass 172

7.3   The Mole Concept 175

7.4   Molar Mass 177

7.5   Conversion Between Mass, Number of Moles, and Number of Units 178

7.6   Mass Relationships Between Elements in a Compound: Percentage Composition 181

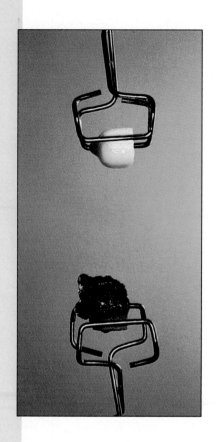

7.7   The Quantitative Meaning of a Chemical Formula: A Summary 184

7.8   Simplest (Empirical) Formula of a Compound 185

8 Chemical Reactions and Equations 198

8.1   Evolution of a Chemical Equation 199

8.2   Balancing Chemical Equations 201

8.3   Writing Chemical Equations 205

8.4   Combination Reactions 205

8.5   Decomposition Reactions 207

8.6   Complete Oxidation or Burning of Organic Compounds 208

8.7   Oxidation–Reduction (Redox) Reactions—Single Replacement Type 210

8.8   Precipitation Reactions 212

8.9   Neutralization Reactions 214

8.10 Summary of Reactions and Equations 216

9 Quantity Relationships in Chemical Reactions 222

9.1   Conversion Factors from a Chemical Equation 223

9.2   Mass Calculations 225

9.3   Percent Yield 229

9.4   Limiting Reactant Problems 233

9.5   Energy 242

9.6   Thermochemical Equations 243

9.7   Thermochemical Stoichiometry 244

10 Atomic Theory and the Periodic Table: A Modern View 253

10.1 The Bohr Model of the Hydrogen Atom 254

10.2 The Quantum Mechanical Model of the Atom 258

10.3 Electron Configuration 262

10.4 Valence Electrons 268

10.5 Trends in the Periodic Table 270

11 Chemical Bonding: The Formation of Ionic Compounds and Molecules 288

11.1 Monatomic Ions with Noble Gas Electron Configurations 289

11.2 Ionic Bonds 292

11.3 Covalent Bonds 293

11.4 Polar and Nonpolar Covalent Bonds 295

11.5 Multiple Bonds 298

11.6 Atoms That are Bonded to Two or More Other Atoms 299

11.7 Exceptions to the Octet Rule 300

12 The Structure and Shape of Molecules 307

12.1   Drawing Lewis Diagrams by Inspection 308

12.2   Drawing Complex Lewis Diagrams 310

12.3   Electron Pair Repulsion: Electron Pair Geometry 316

12.4   Molecular Geometry 319

12.5   The Geometry of Multiple Bonds 322

12.6   Polarity of Molecules 323

12.7   The Structures of Some Organic Compounds (Optional) 325

13 The Ideal Gas Law and Gas Stoichiometry 335

13.1   The Combined Gas Laws Revisited 336

13.2   Avogadro's Law: Volume and Amount Variable, Pressure and Temperature Constant 337

13.3   The Ideal Gas Equation 338

13.4   Applications of the Ideal Gas Equation 339

13.5   Gas Stoichiometry 344

13.6   Dalton's Law of Partial Pressures 351

13.7   Summary of Important Information about Gases 354

14 Liquids and Solids 361

14.1   Properties of Liquids 362

14.2   Types of Intermolecular Forces 365

14.3   Liquid-Vapor Equilibrium 369

14.4   The Boiling Process 372

14.5   Water—an "Unusual" Compound 373

14.6   The Nature of the Solid State 375

14.7   Types of Crystalline Solids 375

14.8   Energy and Change of State 378

14.9   Energy and Change of Temperature: Specific Heat 380

14.10   Change in Temperature Plus Change of State 382

15 Solutions 397

15.1   The Characteristics of a Solution 398

15.2   Solution Terminology 399

15.3   The Formation of a Solution 400

15.4   Factors That Determine Solubility 402

15.5   Solution Concentration: Percentage 403

15.6   Solution Concentration: Molarity 405

15.7   Solution Concentration: Molality (Optional) 408

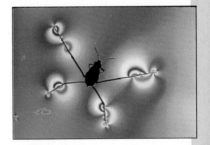

15.8   Solution Concentration: Normality (Optional) 409

15.9   Solution Concentration: A Summary 414

15.10 Dilution Problems 414

15.11 Solution Stoichiometry 415

15.12 Titration Using Molarity 417

15.13 Titration Using Normality (Optional) 419

15.14 Colligative Properties of Solutions (Optional) 421

16 Reactions That Occur in Water Solutions: Net Ionic Equations 434

16.1   Electrolytes and Solution Conductivity 435

16.2   Solution Inventories of Ionic Compounds 437

16.3   Strong Acids and Weak Acids 438

16.4   Net Ionic Equations: What They Are and How to Write Them 440

16.5   Redox Reactions That Are Described by ''Single Replacement'' Equations 441

16.6   Ion Combinations That Form Precipitates 444

16.7   Ion Combinations That Form Molecules 449

16.8   Ion Combinations That Form Unstable Products 451

16.9   Ion Combination Reactions with Undissolved Solutes 452

16.10 Summary of Net Ionic Equations 453

17 Acid-base (Proton Transfer) Reactions 459

17.1   Traditional Acids and Bases 460

17.2   The Arrhenius Theory of Acids and Bases 460

17.3   The Brönsted–Lowry Theory of Acids and Bases 461

17.4   The Lewis Theory of Acids and Bases (Optional) 464

17.5   Conjugate Acid-Base Pairs 465

17.6   Relative Strengths of Acids and Bases 467

17.7   Predicting Acid-Base Reactions 469

17.8   The Water Equilibrium 471

17.9   pH and pOH (Integer Values Only) 473

17.10 Noninteger pH–$[H^+]$ and pOH–$[OH^-]$ Conversions 477

18 Oxidation-Reduction (Electron Transfer) Reactions 486

18.1   Electrolytic and Voltaic Cells 487

18.2   Electron-Transfer Reactions 488

18.3   Oxidation Numbers and Redox Reactions 491

18.4   Oxidizing Agents (Oxidizers); Reducing Agents (Reducers) 495

18.5   Strengths of Oxidizing Agents and Reducing Agents 496

18.6   Predicting Redox Reactions 497

18.7   Redox Reactions and Acid-Base Reactions Compared 500

18.8   Writing Redox Equations 500

19 Chemical Equilibrium

19.1   The Character of an Equilibrium 508

19.2   The Collision Theory of Chemical Reactions 509

19.3   Energy Changes During a Molecular Collision 510

19.4   Conditions That Affect the Rate of a Chemical Reaction 512

19.5   The Development of a Chemical Equilibrium 515

19.6   Le Chatelier's Principle 516

19.7   The Equilibrium Constant 520

19.8   The Significance of the Value of K 522

19.9   Equilibrium Calculations (Optional) 523

20 Nuclear Chemistry 537

20.1   The Dawn of Nuclear Chemistry 538

20.2   Radioactivity 539

20.3   The Detection and Measurement of Radioactivity 539

20.4   The Effects of Radiation on Living Systems 541

20.5   Half-life 544

20.6   Nuclear Equations 547

20.7   Nuclear Reactions and Ordinary Chemical Reactions Compared 550

20.8   Nuclear Bombardment and Induced Radioactivity 550

20.9   Uses of Radionuclides 552

20.10   Nuclear Fission 553

20.11   Electrical Energy from Nuclear Fission 554

20.12   Nuclear Fusion 556

21 Organic Chemistry 563

21.1   The Nature of Organic Chemistry 564

21.2   The Molecular Structure of Organic Compounds 564

        HYDROCARBONS 565

21.3   Saturated Hydrocarbons: The Alkanes and Cycloalkanes 566

21.4   Unsaturated Hydrocarbons: The Alkenes and the Alkynes 572

21.5   Aromatic Hydrocarbons 575

21.6   Summary of the Hydrocarbons 577

21.7   Sources and Preparation of Hydrocarbons 577

21.8   Chemical Reactions of Hydrocarbons 578

21.9 Uses of Hydrocarbons 580

ORGANIC COMPOUNDS CONTAINING OXYGEN AND NITROGEN 582

21.10 Alcohols and Ethers 582

21.11 Aldehydes and Ketones 585

21.12 Carboxylic Acids and Esters 587

21.13 Amines and Amides 590

21.14 Summary of the Organic Compounds of Carbon, Hydrogen, Oxygen, and Nitrogen 591

POLYMERS 592

21.15 Addition Polymers 592

21.16 Condensation Polymers 595

22 Biochemistry 607

22.1 Amino Acids and Proteins 608

22.2 Globular Proteins: Enzymes 613

22.3 Carbohydrates 615

22.4 Lipids 620

22.5 Nucleic Acids 624

Appendix I Chemical Calculations A.1

A The Hand Calculator

B Arithmetic and Algebra

C Logarithms

D Estimating Calculation Results

Appendix II The SI System of Units A.11

Answers to Questions and Problems A.13

Glossary G.1

Photo Credits C.1

Index I.1

Learning Chemistry—NOW!

How many of these students are chemistry majors? Very few. How many need chemistry for their major? A lot of them. How many need chemistry to be an intelligent member of society? All of them; in a technological society, chemistry is everywhere. In this chapter we'll show you how this chemistry course can help you to "Learn smarter, not harder," now, and in the future.

1.1 WHY?

Why is "Why?" the title of this section? Because "Why?" is the fundamental question of chemistry. You have just opened a book and you have just started a course that begins to answer that question, which mere mortals will never answer fully. But the partial answers that have already been found underlie the vast creature comforts we in the United States, Canada, and other developed countries enjoy and take for granted. Such are the fruits of the physical scientists—chemists, in particular. Unfortunately, some of those partial answers have been abused to create a lot more than just discomfort for a large part of the world's population. Such are the challenges of the social scientists! We'll leave that topic to other books, other courses.

"Why?" The question is posed to the parents of toddlers everywhere. It expresses natural childlike curiosity about the world youngsters are just beginning to know. A decade or so later, when the toddler has become a teenager and "knows" more about the world than do the parents, "Why?" is often followed by "not?" Ah, the joys of parenthood!

What's important, though, is that the questions "Why?" and "Why not?" are good questions. It might be said that if we find the answers to enough *why's* we will become very *wise*. Not exactly. No more is a pun the height of humor than knowledge the ultimate of wisdom. Again, we are in the domain of the social scientist. But the social scientist who lacks knowledge of the physical world is no closer to wisdom than the physical scientist who ignores the interests of humanity. More personally, the individual, like you, who lacks knowledge of the physical world is ill equipped to exercise wisdom in his or her own daily life. And that can hurt.

So, welcome to the wonderful world of "Why?" (That's another way of describing the world of chemistry.)

1.2 BUBBLES

Imagine you're the diver in Figure 1.1. You look up and see a rising stream of bubbles—cute little things. After the dive, you talk to your friend in the boat about those tiny bubbles. Your friend, however, contradicts you: "No way! Those bubbles were big."

Because you were *curious* about bubbles, you and your friend have combined to *observe* that small bubbles in deep water expand as they rise. Why? (There's that question again. You will see it often in this book.) You sit down and *think* about this question. The bubble is a gas that is surrounded by liquid water. As long as the bubble doesn't break into two bubbles or join another bubble, you reason that the gas inside the bubble should remain the same. The water temperature is roughly the same all the way to the surface, so temperature should not be a factor. But, when you dive, you feel pressure on your eardrums that increases as you go deeper. Maybe that has something to do with bubble size.

Finally, you develop a *hypothesis:* The volume of a bubble is smaller in deep water because there is more water on top of the bubble to "push" it into a smaller size. As the bubble rises there is less pressure, or "push," over its surface, so it gets bigger.

Hypothesis is a fancy word for guess. It's not worth much in the sciences unless the hypothesis can be *tested by experiment.* You ask some friends, also divers, to help you conduct an experiment. You want to lower a long rope into the water and have each friend stay in the water at a given depth. As you exhale in the water at the bottom of

Figure 1.1

The "bubble" experiment. The deeper the water, the greater the pressure it exerts. As the pressure increases, the volume of the bubble becomes smaller.

the rope, your friends will measure your bubbles as the bubbles rise and then combine your data to see if the bubble expands.

A stunned silence follows; then everybody talks at once: "Just how are you going to exhale only one bubble?" "How can we tell which bubble is which, as we're all breathing out, too, and all the bubbles are going up?" "The water's murky; we won't be able to see the bubble very well." "Get a life . . ."

Your friends have stumbled upon the truth; there are too many chance factors in your first experimental design. Being practical as well as curious, you decide to build a single bubble. You take a drinking glass with straight sides and lower it vertically into the water, with its open end down. The volume of air trapped within the glass is your bubble. No air goes out of the bubble and no water gets in, but the bubble's volume can change depending how hard the water "presses" on it.

So, gleefully you take your glass down to the water and experiment with it. You dive, and your bubble in the glass shrinks; you surface, and the bubble grows. You hurriedly get back on shore and *communicate* your findings to your friends.

Your friends listen politely, but it's obvious they are not as curious about bubbles as you are. But one friend is curious, and also unsure.

"I don't believe you. Convince me," she says. So you *repeat your experiment,* this time with the *skeptical* friend. You both agree that, indeed, as you dived deeper in the water, the bubble trapped in the glass got smaller. The experiment gave *reproducible results.* Being *intellectually honest* as well as skeptical, your friend admits she was wrong when she didn't believe your results. She is, however, still not satisfied with your bubble hypothesis.

"Big deal," she says. "We learned that as pressure increases, the volume of a bubble decreases. As one goes up, the other goes down, an inverse relationship. Is that all there is?"

"Well, what do you want to do?"

"Before the next dive, let's calibrate the glass so we can measure the bubble. . . ."

Aha! She wants to know "How much?" This is a question that goes right along with "Why?"

Stand by for the exciting conclusion to this story—in Chapter 4.

Environmental study of water requires knowing the chemistry of water, such as pH.

Do you think those bubble-chasing divers are scientists? They are, for they are looking at the world around them and they are asking questions. Why? How much? Do you ask questions? We were all curious as children; we were all scientists. Now is just the right time to reclaim your gift of curiosity.

In this little story we have tried to provide a small glimpse of the character of chemistry. The observing, hypothesizing, experimenting, testing and retesting, theorizing, and finally, reaching conclusions have been going on for centuries. And they continue today more actively than ever before. Collectively they are often called the *scientific method.*

There is really no rigid order to the scientific method. Looking back over the history of science, though, the preceding features always seem to be present. They are the outcome of the day-to-day thinking of the scientifically curious people as they continually ask themselves, "What do I already know that can be applied here? What is the next logical step I can take?" Out of such questions come new hypotheses, new experiments, new theories.

Our story also lists in italicized words some of the qualities and actions of scientists. They surely are *observers* and they are *curious* about and *think* about what they see. They *develop a hypothesis* as a tentative explanation of their observations. They *conduct experiments* to test the hypothesis. If they find something new, they *communicate* with their fellow scientists, usually through scientific journals. They combine the qualities of *skepticism* with *intellectual honesty,* both of which leave scientists free to receive and evaluate new information that reaches them through many sources.

Our story furnishes one more important insight into the nature of chemistry. Notice that the first diver was concerned with *What* was taking place and *How.* Answers to these questions are considered to be the *qualitative* part of chemistry. The skeptical friend recognized that *What* and *How* furnish only some of the answers, but they cannot be used to make reliable predictions about the extent of chemical activity. She added the vital question, *How much?* We see, then, that the study of chemistry is both qualitative and *quantitative.* In this book we will consider both of these essential areas.

1.3 LEARNING HOW TO LEARN— OR LEARN SMARTER, NOT HARDER

Here is your first chemistry "test" question:

Which of the following is your most important goal in your introductory chemistry course?

A) To learn all the chemistry that I can learn in the coming term.
B) To spend as little time as possible studying chemistry.
C) To get a good grade in chemistry.
D) All of the above.

If you answered A, you have the ideal motive for studying chemistry—and any other course for which you have the same goal. Nevertheless, this is not the best answer.

If you answered B, we have a simple suggestion: Drop the course. Mission accomplished.

If you answered C, you have acknowledged the greatest short-term motivator of many college students. Fortunately, most students have a more honorable purpose for taking a course, although sometimes they hide it quite well.

If you answered D, you have chosen the best answer.

Let's examine answers A, B, and C in reverse:

C: There is nothing wrong in striving for a good grade in any course, just so it is not the major objective for taking the course. A student who has developed a high level of skill in cramming for and taking tests can get a good grade even though not much has really been learned. That helps the grade point average, but it foretells of trouble in the next course of a sequence. It is better to regard a good grade as a reward earned for good work. You have every right to expect to be rewarded in accord with your achievements in school and in everything that you do.

B: There is nothing wrong with spending "as little time as possible studying chemistry," as long as you *learn* the needed amount of chemistry in the same time. Reducing the time required to complete any task satisfactorily is a worthy objective. It even has a name; it is called *efficiency.*

A: There is nothing wrong with learning all the chemistry you can learn in the coming term, as long as it doesn't interfere with the rest of your school work and the rest of your life during the term. The more time you spend studying chemistry, the more you will learn. But maintain some balance. Mix some of answer B in your endeavor to learn. Again, the key is efficiency.

To summarize, the best goal for this chemistry course—and for all courses—is to learn as much as you can possibly learn in the smallest *reasonable* amount of time.

The rest of this chapter is devoted to suggestions on how to reach this goal.

Analysis of water, or air, or blood samples requires knowledge of chemistry.

Three Essential Ingredients

Learning chemistry—and most other subjects—requires three things:

1) *Time.* You must spend the time required. To keep time to a minimum, however, it must be spent regularly, doing each assignment each day. Unlike many subjects, chemistry must be learned a bit at a time, because what you learn in today's lecture depends on what you learned in doing yesterday's assignment. Falling behind is the biggest problem when it comes to learning chemistry.

2) *Concentration.* This means studying without distractions—without sounds, sights, people, or thoughts that take your attention away from chemistry. Every minute your mind wanders while you study must be added to your total study time. If your time is limited, that minute is lost forever.

3) *Good learning techniques.* A good place to begin developing good learning habits is to try the suggestions in the rest of this section. Our methods may not be perfect for you. However, unless you already have something better, they are a good place to start. Your instructor may offer additional suggestions. Unless you find something in them that is totally contrary to your learning style, you will probably be better off following your instructor's suggestions rather than ours. In fact, that advice goes for any time we recommend one procedure and your instructor recommends another. The most important thing is to begin. Nothing will happen unless you make it happen, beginning NOW.

Operating a DNA synthesizer, this scientist studies human genetics.

A Five-Chapter Goal

We invite you to join us in setting a goal to develop efficient and effective learning procedures by the end of Chapter 5 in this text. We will suggest several things for you to do. We will even do some of them with you.

What happens after Chapter 5? That's up to you. We'll continue dropping learning hints at different times, but, for the most part, you are on your own. As you progress, we will gradually withdraw some of these learning devices. Our aim is that the last chapters you read in this book have about the same level of learning aids that will be in the textbook for your next chemistry course.

Learn It NOW

Efficient learning means learning NOW. It doesn't mean ''studying'' now and learning later. It takes a little longer to *learn* now than it does to *study* now; but the payoff comes in all the time you save by not having to learn later.

There are two primary learning sources in a chemistry class—three, if the course includes a lab. The two are the lecture and the textbook. A brief word about lectures: Studies have shown that about one half of what is presented in lecture is *not learned* if you wait more than 24 hours to study your lecture notes. If you wait a week, figure on about 35% retention. The same studies show that about 90% of the lecture material is retained if you review the lecture the same day. The textbook material that goes with the lecture should be studied at the same time the notes are reviewed, if possible. But the text ideas will not vanish like lecture concepts do. It is a huge waste of study time to postpone learning from your lectures.

LEARN IT NOW is a major theme in our learning recommendations in the next four chapters. Learning suggestions are printed in blue. In addition, as a reminder that what follows is a ''Learn it NOW'' item, it begins with ''LEARN IT NOW'' in capital letters. Most of the LEARN IT NOW entries are in the margin, but longer entries run the full width of the page.

LEARN IT NOW This is what a LEARN IT NOW entry looks like when it is printed in the margin.

When you come to a LEARN IT NOW, stop. Do what it says to do. Think about it. Make a conscious effort to understand, learn, and, if necessary, memorize what is being presented. Write it down. When you are satisfied that this idea is firmly fixed in your mental storehouse, then go on. In short, learn it—NOW! Tomorrow is too late.

Write It Down

Three words in the foregoing paragraph are important: Write it down. Why? For two reasons: First, the very act of writing something down forces some level of learning. Second, because a few written pages of good notes are better than 50 to 100 pages of textbook when the time comes to prepare for a test. At that time your notes should replace the text as your main source of information. The text becomes more of a reference book.

Writing something down doesn't mean copying it directly from the book, although that's better than not writing anything. To learn by writing, write it in your own words. Generally, shorten it. You've got to understand something before you can summarize it in your own words. That's when learning occurs.

Occasionally you may expand something while writing it down. If there's something in the text that gave you trouble but you finally understood it, probably the text did not explain it clearly. Therefore, record your own explanation for future reference.

Highlighting a textbook is not a substitute for writing something down. A high-lighter hinders learning more than it helps. A heavily highlighted textbook is nothing more than a date book, a brightly colored engagement calendar. It lists all the things you recognize as being important and the dates you make to learn them later. ''Later'' usually means right before a test. And when later comes, there are so many dates to keep that it is impossible to do justice to more than a few.

Write something rather than highlighting it. That way you will learn it—NOW.

Your Chemistry Notebook

Question: Where do you write your notes? *Answer:* In your notebook. *Question:* Where do you *not* write your notes? *Answer:* On scraps or loose sheets of paper. *Question:* Why? *Answer:* You lose them. Or you can't separate the chemistry notes you've written on one side of the page from the history notes on the other side.

The essence of the preceding paragraph is to get organized. A spiral notebook *for each subject* keeps things together in the order in which they occur. But that order may not always be the best. It allows you no opportunity to rearrange and combine lecture notes and reading notes. It also makes no provision for lecture handouts. These needs are best met by a loose-leaf binder. You will have to decide which is best for you. But do decide. Then hold to that organization. It will save time and increase learning.

What form should your notes take? Whatever form works best for you. That is probably some kind of an *informal* outline. An informal outline may or may not have a I—A—1—a—(1)—(a) format. Either way, don't spend time being sure that every A has a B and every 1 has a 2, or that all entries are full sentences or all entries are fragments. Do try for *some* regular form, but don't let it cost you time. Be flexible enough to change for any good reason.

We do offer two concrete suggestions about your outline. First, rather than using I, II, and so on for your main topics, use section numbers from the text. That ties your notes directly to the book. Second, as you build your outline, include page number references so you can go back to the text easily if you wish to recheck the source of an entry.

We assume you will outline your text reading in the next four chapters. In Chapter 2 we will even tell you where we think an outline entry should be made. We'll even suggest what to write. This is also done in Chapter 3, but at the end of the section or subsection rather than within the section. In Chapter 4 we'll continue suggesting *where* an outline entry should be made, but leave it to you to compose the entry. However, we still offer a complete outline at the end of the chapter so you may compare your outline with ours. Chapter 5 has no outline suggestions within the chapter, but again, there is a full chapter outline at the end.

By the time you reach Chapter 6, your outlining methods should be well established. We therefore drop all outline suggestions for the rest of the book.

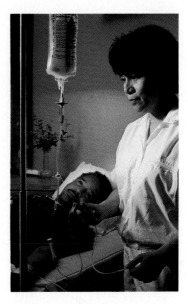

In many states, the nurse is legally responsible for assuring that every patient receives the correct amount of medication.

1.4 MEET YOUR TEXTBOOK

The most important tool in most college courses is the textbook. It is worth taking a few minutes to examine the book and find out what learning aids the author has provided. In this section we'll show you the features of this book that are designed specifically to help you learn chemistry.

Performance Goals

> **PG 1A** Read, write, and talk about chemistry, using a basic chemical vocabulary.
> **1B** Write routine chemical formulas and write names of chemicals when their formulas are given.
> **1C** Write and balance ordinary chemical equations.
> **1D** Set up and solve elementary chemical problems.
> **1E** "Think" chemistry in some of the simpler theoretical areas and visualize what happens on the atomic or molecular level.

As you approach most sections in this text, you will find one or more "performance goals" (PG), as you did here. They tell you exactly what you should learn as you study the section. If you focus your attention on learning what is in the performance goal, you will learn more in less time. All of the performance goals in a chapter are assembled as a Chapter in Review section at the end of the chapter.

The performance goals listed here are not for a section, but for this entire book and the course in which it will be used. They tell you exactly what you will be able to do when you have completed the course.

Few general chemistry textbooks include performance goals, although they sometimes appear in study guides that accompany those books. Consistent with our plan to make the latter chapters of this book like a general chemistry text, we discontinue performance goals at the beginning of the sections after Chapter 14. They will continue to appear as a Chapter in Review through Chapter 16. Thereafter they are dropped completely. It then becomes your responsibility to write the performance goals for yourself—to figure out what understanding or ability you are expected to gain in your study. Literally writing your own performance goals, incidentally, is an excellent way to prepare for a test.

Examples

As you study from this book, you will acquire certain "chemical skills." These include writing chemical formulas, writing chemical equations, and solving chemical problems—the things listed previously as Performance Goals 1B, 1C, and 1D. You will develop these skills by studying and working the examples in the text.

EXAMPLE 1.1

This is not an example, but this sentence and the following paragraph are written in the form of the examples throughout the book.

All examples begin with the word "Example," followed by the example number, and a solid line that runs across the page, printed in red. The end of the example is signaled by a line across the page, also in red.

There are two kinds of examples in this book. In the early chapters, most examples take you through a series of questions and answers in which you actually write the formula or equation or solve the problem yourself. The first such example appears in Section 3.2. You will find detailed instructions for working this kind of example problem at that point.

In the second kind of examples the problems are worked out for you. Comments are added as needed to explain the steps in the procedure. You learn from these examples by studying the solution given, being sure that you understand each step. Examples in general chemistry textbooks are written in this form. Accordingly, you will find more worked-out examples in the chapters toward the end of the book.

If you are to learn from examples, you need to work through each one as you come to it. Never postpone an example and read ahead. Learn it *NOW*. Quite often what you learn in an example is used immediately in the next section. You will not be able to understand that next section without understanding the earlier example.

Quick Checks

Many sections in this book end with one or more questions we call ''Quick Checks.'' They are identified by the heading

✔ QUICK CHECK 1.1

Chemical principles and theories are introduced with words and illustrations. Ideally you will learn and understand these ideas as you study the text and figures. A ''Quick Check'' is used to find out if you have caught on to the main ideas or methods immediately after they appear in the text. Quick Checks are also used to test your understanding of worked-out examples.

Most Quick Checks are relatively easy questions. Like examples, Quick Checks should be completed as you reach them. Immediate feedback is important in learning. Naturally you must know that your answers are correct. You will find the answers to Quick Checks at the end of the chapter.

Flashbacks

Quite often in the study of chemistry you meet some term or concept that was introduced in an earlier chapter. You may not recall the word or idea exactly. We use a

FLASHBACK to help you. We will denote a FLASHBACK by the symbol 💡; that light bulb should trigger a memory or idea. This is a brief reminder that usually appears in the margin. It always includes the number of the section where the term or concept first appeared so you may review it in detail, if necessary.

💡 FLASHBACK This is what a FLASHBACK looks like when it is printed in the margin.

In-Chapter Summaries and Procedures

SUMMARY

Throughout this book you will find summaries and step-by-step procedures printed within red lines, as this paragraph is printed. These give you, in relatively few words, the main ideas and/or methods you should learn from a more general discussion nearby. They should help you to clinch your understanding of the topic.

Occasionally summaries are in the form of a table or an illustration; some even combine the two. These forms are particularly helpful in reviewing for a test. Not only do they review the topic briefly, but they also create a mental image that is readily recalled during the exam.

Key Words and Matching Sets

Following each Chapter in Review, there are one or more Matching Sets. Members of the left-hand column begin with a blank space, followed by a phrase, definition, or sentence. Members of the right-hand column begin with a number, followed by a key word or phrase introduced in that chapter. When you match the numbers from the right column with the definitions in the left column, you will strengthen your understanding of the new ideas in the chapter. Answers to the Matching Sets are at the end of the chapter.

Study Hints and Pitfalls to Avoid

Following the Matching Sets you will find a brief section suggesting study methods that should make learning easier or more efficient. Some of these are ''Remember'' statements. Their purpose is to remind you of the importance of some word or method or concept that is often overlooked. ''Pitfalls'' identify the most common errors made by students in tests. The idea is that, if you are forewarned of a common mistake, you are less likely to make that mistake.

End-of-Chapter Questions and Problems

At the end of all chapters except this one, you will find study questions and problems. Generally they are grouped in pairs that are separated by a thin line. The questions in each pair involve similar reasoning and, in the case of problems, calculations. Some questions are easy, like the examples and quick checks. Others are more demanding. You may have to analyze a situation, apply a chemical principle, and then explain or predict some event or calculate some result. The more difficult questions are marked with an asterisk (*).

Generally, the answer for the first question or problem in each pair—an odd-numbered question—is given in the back of the book. Answers to problems include calculation setups. The numbers of answered questions are printed in blue, such as 23. If the first question of a pair is not answered, this indicates that the answer is a direct quotation from the text, or nearly so. No second questions—even-numbered questions —are answered in the book.

As you solve problems in the textbook, remember that your *main* objective is to understand the problem, not to get a correct answer. So, even when your answer is correct, stop and think about it for a moment. Don't leave the problem until you feel confident that you will recognize any new problem that is worded differently but requires similar—or even not so similar!—reasoning based on the same principle. Then be confident that you can solve such a problem.

Even more important is what you do when you do *not* get the correct answer to a problem. You will be tempted to return to the examples and quick checks, find one that matches your problem, and then solve the assigned problem step-by-step as in the example. *This temptation is to be resisted.* If you get stuck on a problem, it means you did not understand the earlier examples. Leave the problem. Turn back to the example. Study it again, by itself, until you understand it thoroughly. Then return to the assigned problem with a fresh start and work it to the end without further reference to the example.

Appendixes

The Appendix of this book has four parts. They are:

1) *Chemical Calculations.*   Here you will find suggestions on how to use a calculator specifically to solve chemistry problems. There is also a general review of arithmetic and algebraic operations that are used in this book. You will find these quite helpful if your math skills need dusting off before you can use them.
2) *The SI System of Units.*   This explains the units in which quantities are measured and expressed in current textbooks.
3) *Glossary.*   Like other fields of study, chemistry has its own special language, in which common words have very specialized and specific meanings. The Glossary lists these words in alphabetic order so you can find them easily and learn to use them correctly. The glossary contains nearly all the boldfaced words and words in the Matching Sets, as well as many others. Use the Glossary regularly; it's a real time-saver.
4) *Answers to Questions and Problems.*

Inside Front and Back Covers

Some reference items should be available quickly, without searching through pages in the book. Two of these are the periodic table, introduced in Section 5.6, and a list of elements. You will find these items inside the front cover of the textbook. Inside the back cover there is a list of important values, equations, and other items that we think you will find handy.

1.5 A CHOICE

You have a choice to make. You can choose to continue learning as before, or you can choose to improve your learning skills. Even if those skills are already good, they can be improved. This chapter gives you some specific suggestions on how to do this. It also invites you to adopt a five-chapter goal for upgrading your study habits, beginning here and ending in Chapter 5. We urge you to accept that invitation.

If you ever begin to feel that chemistry is a difficult subject, read this chapter again. Then ask yourself, and give an honest answer: Do I have trouble because the subject is difficult, or is it because I did not choose to improve my learning skills? Your honest answer will tell you what to do next.

At all stages of our lives we make choices. We then live with the consequences of those choices. Choose wisely—and enjoy chemistry!

2 Matter and Energy

A fireworks display includes many levels of chemistry. These include the physical and chemical properties of the substances used; the physical, chemical, and state changes that occur; the elements and compounds that react and are formed; and the transformation of chemical energy to light, heat, and sound energy. All of these are discussed in this chapter.

LEARN IT NOW   Begin the study of a new chapter with a brief preview. For each section, look at its title; glance quickly at the performance goals; scan the text for terms given in boldface. Look at the illustrations and tables, particularly if they are identified as summaries. Make a mental note of all summaries; they can shorten your chapter outline. Chapter 2 in Review at the end of the chapter lets you check your preview at a glance.

Recall what was said in Chapter 1 about writing an outline as you study a chapter rather than using a highlighter. Don't make a date to learn something later. Learn it—NOW!

Let's talk about the outline. In this chapter we "write" the outline along with you. Wherever the text contains material that should be in your outline, we write a suggestion for that outline entry. *Note:* It is a *suggestion,* not a statement of what the entry *should be.* *Your* words are better because they express *your* way of thinking about the topic; but your words and ours should express the same idea. In some cases we include a comment about the entry, explaining why we think it should appear at a particular place or in a particular form.

Keep your outline informal. Begin with the chapter number and title. Use textbook section numbers instead of I, II, III, and so on for your main headings. Page references help the outline, even though we use them only a few times in this chapter. Be brief, but not so brief that you must check the text to find out what the entry means.

Our outline is printed in the same typeface as this sentence so that you can distinguish between the text and the outline.

Chemistry is the study of matter and the energy associated with physical and chemical change. Rightfully, then, we begin by explaining some familiar but not widely understood features of matter and energy, the two major components of the universe.

2.1 PHYSICAL AND CHEMICAL PROPERTIES AND CHANGES

2.1 Physical and chemical properties and changes

PG 2A	Distinguish between physical and chemical properties.
2B	Distinguish between physical and chemical changes.

LEARN IT NOW   The two performance goals (PG's) tell you exactly what to look for as your study begins. In Chapters 2 and 3 we will place PG ALERT in the margin to help you find places where a performance goal is discussed. Always focus your study on the PG's. When you complete this section you should know what physical and chemical properties are and what physical and chemical changes are.

Usually, something printed in boldface type should be in your outline. The suggested outline entry for the first paragraph is:

A. Matter—has mass and occupies space

Matter is anything that has both mass (sometimes given as "weight") and takes up space. These two characteristics combine to define matter.

If you were asked to describe a bit of matter, you would also list its **physical properties.** What color is it? Charcoal is black, sulfur is yellow. Be cautious about touch: Glass is hard, bread dough is soft, and concentrated sulfuric acid hurts! Smell things cautiously, too: Enjoy the odor of baking bread, or of a rose, but avoid ammonia. Taste is *definitely not recommended* for describing laboratory chemicals, but for food chemists there is no substitute for distinguishing between things that are salty or sweet.

If you previewed the chapter, you know that the three concepts in this section are summarized in a table. We will wait until then and use a form of the table as our outline entry.

Other physical properties are measured in the laboratory. (Remember the bubble?) For example, we determine the temperature at which a substance boils or melts, called the *boiling point* or *melting point*. The relative "heaviness" of two substances, like lead versus aluminum, compares their *densities*.

Changes that alter the physical form of matter without changing its chemical identity are called **physical changes.** The melting of ice is a physical change, as is the freezing of liquid water. The substance is water both after the melt and after it is refrozen. Dissolving sugar in water is another physical change. The sugar seems to disappear, but if you taste the water, you'll know the sugar is there. The dissolved sugar can be recovered by evaporating the water, another physical change.

PG ALERT

A **chemical change** occurs when the chemical identity of a substance is destroyed and a new substance is formed. A chemical change is also called a **chemical reaction.** As a group, all of the chemical changes that are possible for a substance are its **chemical properties.**

PG ALERT

Chemical changes can usually be detected by one or more of the five physical senses. A change of color almost always indicates a chemical change, as in toasting bread. You can feel the heat and see the light given off as a match burns. You can smell milk that becomes sour. You can taste it, too. Explosions usually give off sound.

Table 2.1 summarizes chemical and physical changes and properties.

LEARN IT NOW Always pause at summaries. Master the ideas in a summary before proceeding. Learn it—NOW!

SUMMARY

Table 2.1

Chemical and Physical Properties and Changes

	Chemical	**Physical**
Change	Old substances destroyed New substances formed	New form of old substance No new substances formed
Properties	List of chemical changes possible	Description by senses— color, shape, odor, etc. Measurable properties— density, boiling point, etc.

B. Summary of chemical and physical changes and properties from page 14.

	Chemical	Physical
Change		
Properties		

LEARN IT NOW Table 2.1 is a tabular summary of the entire section. A table is a particularly good form of summary. The mental picture it creates is more easily recalled than just words. Study the table; clinch each concept in your mind. There is not much you can do to shorten the summary for your notebook. But you still want the learning advantage of writing in your own words the meaning of the four concepts in the section. So, summarize the section in your notebook in a table. But just *copying* the table will not help learning. We recommend instead that you draw a large skeleton of the table, skipping the outside lines, as if you were going to play a big game of tic-tac-toe. Don't fill in the spaces now, except for the column and row headings; the others come later. (This same procedure can be used for any tabular summary you encounter in the book.)

QUICK CHECK 2.1

Identify the true statements, and rewrite the false statements to make them true.

a) Baking bread is a chemical change.
b) The flammability of gasoline is a physical property.
c) Ethyl alcohol boils at 78°C. This is a chemical property.
d) Grinding sugar into a powder is a physical change.

LEARN IT NOW   Always complete a quick check immediately after studying a section. Check the answer at the end of the chapter. If any answer is wrong, find out why NOW! Reread the text, ask a classmate, and if necessary, talk to your instructor at the next class meeting. Learn it—NOW!

2.2 CHARACTERISTICS OF MATTER

States of Matter: Gases, Liquids, Solids

2.2 Characteristics of matter

> **PG 2C**   Identify and explain the differences between gases, liquids, and solids in terms of (a) visible properties and (b) particle movement.

A. States of matter: gases, liquids, solids

The air you breathe, the water you drink, and the food you eat are examples of the **states of matter** called **gases, liquids,** and **solids.** Water is the only substance that is familiar in all three states, as suggested in Figure 2.1. The differences among gases, liquids, and solids can be explained in terms of the **kinetic molecular theory.** According to this theory, all matter consists of extremely tiny particles that are in constant motion. **Kinetic** refers to motion; "molecular" comes from **molecule,** the smallest individual particle that is present in one kind of matter. Water, oxygen, and sugar are three common molecular substances.

1. Kinetic molecular theory. Particles of matter always moving. Kinetic means motion.

Under the kinetic theory, the "amount" of motion, or the speed at which the particles move, is faster at high temperatures and slower at low temperatures. As the temperature of a sample rises, the particles move faster and tend to separate, to "fly apart" from each other. When that happens, the sample exists as a **gas.** A gas must be held in a closed container to prevent the particles from escaping into the surrounding air. The particles move in a random fashion inside the container. They fill it completely, occupying its full volume.

PG ALERT

There is an attraction between the particles in any sample of matter, but it has almost no effect at the vigorous movement that goes with high temperatures. If the temperature is reduced, the particles slow down. The attractions become more important and the particles clump together to form a **liquid** drop. The drops fall to the bottom of the container, where the particles move freely among themselves, taking on the shape of the container. The volume of a liquid is almost constant, varying only slightly with changes in temperature.

PG ALERT

As temperature is reduced further, the particle movement becomes more and more sluggish. Eventually the particles no longer move among each other. Their movement is reduced to vibrating, or shaking, in fixed position relative to each other. This is the **solid** state. Like a liquid, a solid effectively has a fixed volume. But unlike a liquid, a solid has its own shape that remains the same wherever the sample may be placed.

PG ALERT

The table in Figure 2.1 summarizes the properties of gases, liquids, and solids.

	GAS	LIQUID	SOLID
WATER AS AN EXAMPLE	Gaseous water (steam)	Liquid water	Solid water (ice)
SHAPE	Variable—Same as a closed container	Variable—Same as the bottom of the container	Constant—Rigid, fixed
VOLUME	Variable—Same as a closed container	Constant	Constant
PARTICLE MOVEMENT	Completely independent (random); each particle may go anyplace in a closed container	Independent beneath the surface, limited to the volume of the liquid and the shape of the bottom of the container	Vibration in fixed position

Figure 2.1

The three states of matter illustrated by water.

A solid has both constant shape and constant volume; a liquid can assume any shape for its constant volume.

This is another tabular summary that may be reproduced in your notebook in skeleton form, to be completed later.

2. Summary from page 16.

	Gas	Liquid	Solid
Shape			
Volume			
Particle movement			

Note: With one exception, we will no longer show space for page numbers in our suggested outline. Your outline should include them for all entries that you feel you might like to consult later—particularly summaries. Otherwise the section number, already in your outline, gives you an approximate text location for all outline entries in the section.

✔ **QUICK CHECK 2.2**

Identify the true statements, and rewrite the false statements to make them true.

a) Particles move more freely in a gas than in a liquid.

b) The volume of a liquid may change, but its shape cannot.

Homogeneous and Heterogeneous Matter

B. Homogeneous and heterogeneous matter

| PG 2D | Distinguish between homogeneous matter and heterogeneous matter. |

Pure water has a uniform appearance. It is, in fact, uniform in more than appearance. If you were to take two samples of water from any place in a container, they would have exactly the same composition and properties. This is what is meant by **homogeneous: If a sample has a uniform appearance and composition throughout, it is said to be homogeneous.** The prefix *homo-* means "same."

PG ALERT

1. Homogeneous: same appearance, composition, and properties throughout

When water and alcohol are mixed, they dissolve in each other and form a **solution.** A solution also has a uniform appearance, and once properly stirred, it has a uniform composition too. It is homogeneous; in fact, a solution is sometimes defined as a homogeneous mixture. Clean air is a solution of nitrogen, oxygen, water vapor, carbon dioxide, and other gases.

2. Solution: a homogeneous mixture

When cooking oil and water are mixed, they quickly separate into two distinct layers or **phases,** forming a **heterogeneous** mixture. The prefix *hetero-* is used to describe things that are "different." The different phases in a heterogeneous sample of matter are usually visible to the naked eye.

PG ALERT

3. Heterogeneous: different phases visible, variable properties in different parts of sample

✔ QUICK CHECK 2.3

Classify the following as heterogeneous or homogeneous: (a) oil and vinegar dressing, (b) real lemonade, (c) beach sand, (d) gasoline.

Pure Substances and Mixtures

C. Pure substances and mixtures

| PG 2E | Distinguish between a pure substance and a mixture. |

At normal pressure, pure water boils at 100°C. As boiling continues, the temperature remains at 100° until all of the liquid has been changed to a gas. Pure water cannot be separated into separate parts by a physical change. Water from the ocean—salt water—is different. Not only does it boil at a higher temperature, but the boiling temperature continually increases as the boiling proceeds. If boiled long enough, the water boils off as a gas and the salt is left behind as a solid. (See Figs. 2.2 and 2.3.)

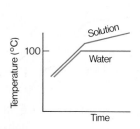

Figure 2.2

Comparison between boiling temperatures of a pure liquid and an impure liquid (solution). As water boils off the solution, what remains becomes more concentrated. This change in concentration causes the boiling temperature to increase.

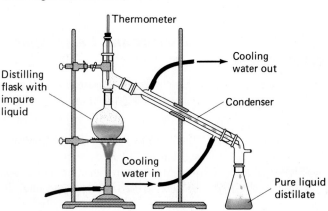

Figure 2.3

Laboratory distillation apparatus. When saltwater is heated, the water boils off, is cooled, is condensed (changed back into a liquid), and is collected as pure water.

Left, mixed iron and sulfur form a heterogeneous mixture. *Right,* the physical property magnetism allows separation of only iron, leaving behind pure sulfur.

1. Pure substance   **PG ALERT**
   a. One chemical with distinct set of physical and chemical properties
   b. Cannot be separated by physical changes
       PG ALERT

2. Mixture
   a. Two or more pure substances
   b. Properties vary, depending on relative amounts of pure substances
   c. Components can be separated by physical changes

The properties of pure water and ocean water illustrate the difference between a **pure substance** and a **mixture.** A pure substance* is a single chemical, one kind of matter. It has its own set of physical and chemical properties, not exactly the same as the properties of any other pure substance. These properties may be used to identify the substance. A pure substance cannot be separated into parts by physical means.

The word *mixture* has already been used to describe a sample of matter that consists of two or more different chemicals. The properties of a mixture depend on the substances in it. These properties vary as the relative amounts of the different parts change. The pure substances in a mixture may be separated by physical changes, as salt and water are separated by boiling off the water.

✔ QUICK CHECK 2.4

Specific gravity is a physical property. Three clear, colorless liquids are in beakers A, B, and C. The specific gravity of liquid A is 1.08; of liquid B, 1.00; and of liquid C, 1.12. The beakers are placed in a freezer until a solid crust forms across the surface of each. The crusts are removed and the liquids warmed to room temperature once again. Their specific gravities are now 1.10 for A, 1.00 for B, and 1.15 for C. Which beaker(s) contains a pure substance, and which contains a mixture? Explain your answers.

D. Elements and compounds

Elements and Compounds

PG 2F	Distinguish between elements and compounds.
2G	Distinguish between elemental symbols and the formulas of chemical compounds.

There are two kinds of pure substances. Silver represents one of them. Like all pure substances, it has its own unique set of physical and chemical properties, unlike the properties of any other substance. Among its chemical properties is that silver cannot be decomposed or separated into other stable pure substances, either chemically or physically. This identifies silver as an **element.**

1. Element: pure substance, cannot be decomposed chemically into other pure substances

---

*Technically, a ''substance'' is pure by definition. The word is so commonly used for any sample of matter, however, that we include the adjective ''pure'' when referring to a single kind of matter.

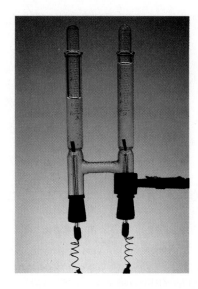

Figure 2.4
Decomposition of a compound. When electricity is
passed through certain water solutions, the water
decomposes into its elements, hydrogen and oxygen. The volume of hydrogen on the left is twice
the volume of oxygen on the right. This matches
the chemical formula of water, H_2O.

Water represents the second kind of pure substance. Unlike silver, it can be decomposed into other pure substances (Fig. 2.4). Any pure substance that can be decomposed by a chemical change into two or more other pure substances is a **compound.** Be
sure to catch the distinction between separating a pure compound into simpler pure
substances, elements or compounds, and separating an impure mixture into pure substances. This is shown in Figure 2.5.

Nature provides us with at least 88 elements. Most of the earth's crust is made up
of compounds containing a relatively small number of these elements (Table 2.2).
Copper, sulfur, silver, and gold are among the few well-known solid elements that
occur uncombined in nature. The atmosphere contains gaseous uncombined elements.
Nitrogen, oxygen, and argon make up about 98% of the air at the surface of the earth.
At 20°C at sea level, 75 elements are solids, 11 are gases, and 2 (mercury and bromine)
are liquids.

2. Compound: pure substance that can be
decomposed chemically into other pure
substances

Table 2.2
Composition of Earth's Crust*

Element	Percent by Weight
Oxygen	49.2
Silicon	25.7
Aluminum	7.5
Iron	4.7
Calcium	3.4
Sodium	2.6
Potassium	2.4
Magnesium	1.9
Hydrogen	0.9
All others	1.7

*The earth's "crust" includes the atmos-
phere and surface waters.

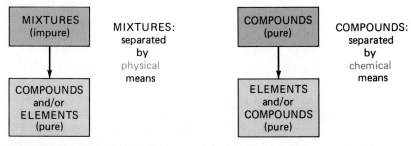

Figure 2.5
Separations. A mixture—an *impure* substance—is separated
into *pure* substances by *physical* means, using *physical*
changes. A compound—a *pure* substance—is separated into
other pure substances by *chemical* means, using *chemical*
changes.

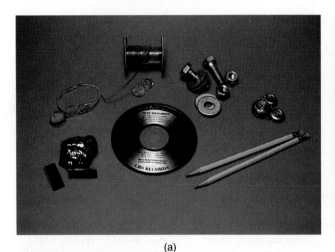

(a)

(b)

Figure 2.6

Common elements and compounds. (a) Familiar objects that are nearly pure elements: copper wire and the copper coat on pennies; iron nuts and bolts coated with zinc for corrosion protection; lead sinkers used by fishermen; graphite, a form of carbon, which is the ''lead'' in lead pencils; aluminum in a compact disk; a piece of silicon, used in the computer chips shown next to it; and silver in a bracelet and ring.
(b) Familiar substances that are compounds: drain cleaner,

sodium hydroxide (made up of the elements sodium, hydrogen, and oxygen); photographic fixer, sodium thiosulfate (sodium, sulfur, oxygen); water (hydrogen and oxygen); boric acid (hydrogen, boron, oxygen); milk of magnesia tablets, magnesium hydroxide (magnesium, hydrogen, oxygen); quartz crystal, silicon dioxide (silicon, oxygen); baking soda, sodium hydrogen carbonate (sodium, hydrogen, carbon, oxygen); chalk and an antacid, calcium carbonate (calcium, carbon, oxygen).

The name of an element is always a single word, like oxygen or iron. The chemical names of nearly all common compounds have two words, like sodium chloride (table salt) and calcium carbonate (limestone). A few familiar compounds have one-word names, like water and ammonia. At present, you may use the number of words in the name of a chemical to predict whether it is an element or a compound. Figure 2.6 shows some well-known elements and compounds.

PG ALERT

Chemists represent the elements by **elemental symbols.** The first letter of the name of the element, written as a capital, is often its symbol. If more than one element begins with the same letter, a second letter written in lowercase (a ''small'' letter) is added. Thus, the symbol for hydrogen is H, for oxygen, O, for carbon, C, and for chlorine, Cl. The symbols of some elements are derived from Latin names, such as Na for sodium (from *natrium*) and Fe for iron (from *ferrum*).

3. Elemental symbol: capital letter, sometimes followed by small letter

The names and symbols of all elements are listed inside the front cover of this book. The symbols are also in the periodic table inside the front cover. In Chapter 6 you will use the periodic table as an aid in learning the symbols of some of the elements.

PG ALERT

The ''symbol'' of a pure substance as it appears in nature is its **chemical formula.** A formula is a combination of the symbols of all the elements in the substance. The formula of most elements is the same as the symbol of the element. This indicates that the element is stable as a single **atom,** the smallest unit particle of the element. Some elements are not stable as single atoms but form stable units called **molecules** in which two or more atoms combine. Hydrogen and oxygen are two such elements. Their symbols are H and O, respectively, but their formulas as molecules are H_2 and O_2. The subscript 2 indicates, in each case, that a molecule of the element has two atoms.

4. Formula of compound: elemental symbols of elements in compound. Subscripts show number of atoms of each element

The formula of a compound also uses subscript numbers to show the number of atoms of each element in a "formula unit." (Some formula units are molecules, but some are not. You will learn to distinguish between them later.) If there is only one atom of an element in a unit, the subscript is omitted. One compound of iron and oxygen has the formula FeO. This indicates that there is a one-to-one atom ratio between iron and oxygen atoms. The formula of water is H_2O, which shows that there are two atoms of hydrogen for each atom of oxygen.

The precise ratio of atoms of different elements in a compound is responsible for the **Law of Definite Composition,** also called the **Law of Constant Composition.** This law states that any compound is always made up of elements in the same proportion by mass (weight). The source of the compound does not matter. For example, 100 grams of pure water always contain 11.1 grams hydrogen and 88.9 grams oxygen. It makes no difference if the water comes from a pond in Europe, a river in South America, a lake in the Alps, or the product of a chemical reaction in Asia. If 100 grams of the pure material contain 11.1 grams hydrogen and 88.9 grams oxygen, it's water.

The properties of compounds are always different from the properties of the elements from which they are formed. Sodium is a shiny metal that reacts vigorously when exposed to air or water; chlorine is a yellow-green gas that was the first poison gas used in World War I. Neither element is pleasant to work with. Yet the compound formed from these two elements, sodium chloride, is essential in the diet of many animals, including humans.

Figure 2.7 summarizes the classification system for matter.

While both sodium metal and elemental chlorine are deadly, sodium chloride is needed by animals and humans alike.

5. Law of Definite Composition: percentage by mass of different elements in a compound is *always* the same

6. CHECK SUMMARY, page 21

LEARN IT NOW Give particular attention to textbook summaries that combine several topics in one place. The top three-box portion summarizes Section 2.1. The middle section summarizes Section 2.2, as do the lower pairs of boxes. It's a good idea to record the location of summaries like this in your notebook as an outline entry.

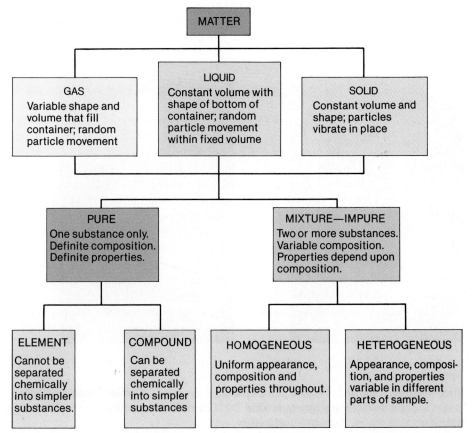

Figure 2.7

Summary of classification system for matter.

2.3 Electrical character of matter

✔ QUICK CHECK 2.5

a) Which of the following are compounds and which are elements: Na_2S, Br_2, C, potassium hydroxide, carbon dioxide, and fluorine?
b) Is a compound a pure substance or a mixture? On what differences between compounds and mixtures is your answer based?

2.3 THE ELECTRICAL CHARACTER OF MATTER

> **PG 2H**   Match electrostatic forces of attraction and repulsion with combinations of positive and negative charge.

If you release an object held above the floor, it falls to the floor. This is the result of gravity, an invisible attractive force between the object and the earth. There are two other invisible forces, both capable of repulsion as well as attraction. They are magnetic and **electrostatic forces.** The region in space where one of these forces is effective is called a **force field,** or simply a **field,** as the gravitational field of the earth.

A. Electrostatic forces attract or repel. Field is region where forces are effective

Electrostatic forces exist between objects that carry an **electrical charge.** A hard rubber rod that is rubbed with fur acquires a ''positive'' charge. If a glass rod is rubbed with silk, the rod gains a ''negative'' charge. These charges are like those you develop if you scrape your feet across a rug on a dry day. You can ''discharge'' yourself by touching another person, each receiving a mild shock in the process.

If a pith ball, a small spongy ball made of plant fiber, is touched with a positively charged rod, the pith ball itself becomes positively charged. When two pith balls that

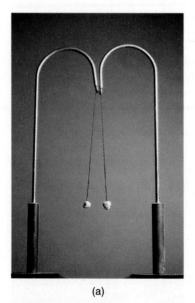
(a)

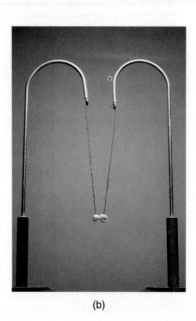
(b)

Figure 2.8
Electrostatic attraction and repulsion. (a) If both pith balls have a positive charge, or if both have a negative charge, they repel each other. (b) If one ball has a positive charge and the other a negative charge, they attract each other.

are positively charged are suspended close to each other, they repel each other (Fig. 2.8a). Similarly, two pith balls that have negative charges repel each other. However, if a positively charged pith ball is placed near a negatively charged pith ball, each one attracts the other (Fig. 2.8b). These experiments show that

Two objects having the same charge, both positive or both negative, repel each other.

Two objects having unlike charges, one positive and one negative, attract each other.

Electrostatic forces show that matter has electrical properties. These forces are responsible for the energy absorbed or released in chemical changes.

2.4 CHARACTERISTICS OF A CHEMICAL CHANGE

Chemical Equations

> **PG 2I**   Distinguish between reactants and products in a chemical equation.

In Section 2.1 we said that a chemical change occurs when a starting substance is destroyed and a new substance is formed. Chemists describe such a change by writing a **chemical equation.** The formulas of the beginning substances, called **reactants,** are written to the left of an arrow that points to the formulas of the substances formed, or **products.** The equation for the reaction of the element iron with the element sulfur to form the compound iron sulfide is

$$Fe + S \longrightarrow FeS$$

Notice how the equation represents the essence of a chemical change. The Fe, iron, and the S, sulfur, present at the beginning are destroyed, and a new substance, FeS, iron sulfide, is formed.

The decomposition of water (Fig. 2.4) is another example of a chemical change. The starting compound, water, H_2O, is broken into its elements, hydrogen, H_2, and oxygen, O_2. The equation is

$$2\,H_2O \longrightarrow 2\,H_2 + O_2$$

Energy in Chemical Change

> **PG 2J**   Distinguish between exothermic and endothermic changes.
> **2K**   Distinguish between potential energy and kinetic energy.

If you strike a match and hold your finger in the flame, you learn very quickly that the chemical change in burning wood or paper is releasing energy, which we call ''heat.'' Heat is only one form of energy, and it's the one that will concern us the most. A chemical change that releases energy to its surroundings is called an **exothermic reaction.**

PG ALERT

B. Like charges repel; unlike charges attract

2.4 Characteristics of a chemical change

A. Chemical Equations

PG ALERT

1. Equation: reactants $\rightarrow$ products

B. Energy in Chemical Change

1. Exothermic: energy released; endothermic: energy absorbed

Sometimes it takes energy to cause a reaction to occur. Electrical energy must be put into the system to force water to decompose according to the equation above. The chemical change absorbs energy from its surroundings. This is called an **endothermic change.**

Sometimes energy terms are included in chemical equations. The reaction between iron and sulfur is exothermic, releasing energy to the surroundings. This can be indicated in this way:

$$Fe + S \longrightarrow FeS + energy$$

Electrical energy is needed to decompose water. This is an endothermic change whose equation can be written

$$2\,H_2O + energy \longrightarrow 2\,H_2 + O_2$$

PG ALERT

Energy is closely associated with the physical concept of work. Work is the application of a force over a distance. If you lift a book from the floor to a table, you do work. You exert the force of raising the book against the attraction of gravity, and you exert this force over the distance from the floor to the table. The energy has been transferred from you to the book. On the table the book has a higher **potential energy** than it had on the floor.

2. Potential energy: energy because of position in force field

The potential energy of an object depends on its position in a field where forces of attraction and/or repulsion are present. With the book, the force is attraction in the earth's gravitational field. In chemistry, there are electrostatic forces between charged particles. Just as the book has a higher potential energy when separated farther from the earth that attracts it, so oppositely charged particles have greater potential energy when they are farther apart. Conversely, the closer the particles are, the lower the potential energy in the system. The relationships are reversed for two particles with the same charge because they repel each other. Their potential energy is greater when they are close than when they are far apart.

What is loosely called "chemical energy" comes largely from the rearrangement of charged particles in an electrostatic field.

3. Tendency toward minimization of energy a driving force for chemical reactions

If you push your book off the table, it falls to the floor. Its potential energy is reduced. Physical and chemical systems tend to change in a way that reduces their total energy. In fact, minimization of energy is one of the driving forces that cause chemical reactions to occur. We will mention this from time to time in this text.

PG ALERT

A moving automobile, an airplane in flight, and a falling book all possess another kind of energy called **kinetic energy.** We have already noted that the word *kinetic* refers to motion; and motion is the common feature of the automobile, the plane, and the book. Any moving object has kinetic energy. Most of what we call "mechanical energy" is kinetic energy. In a later chapter you will see that the temperature of an object is related to the average kinetic energy of its particles.

4. Kinetic energy: energy of motion

✔ QUICK CHECK 2.6

a)   Is the process of boiling water exothermic or endothermic? Explain.

b)   A charged object is moved closer to another object that has the same charge. This is an energy change. Is it a change in kinetic energy or potential energy? Is the energy change an increase or a decrease?

2.5 CONSERVATION LAWS AND CHEMICAL CHANGE

A "law" of science is not like a governmental law. A scientific "law" is simply a summary of past experimental results. Because the law worked well in the past, we expect it to work well in the future, also. The real experiments that matched the bubble experiment described in Chapter 1 led to one of the best known laws of chemistry and physics. You will meet that law in Chapter 4. In the meantime, we'll look at two other laws on which much of today's engineering and technology are based and see what happened to those laws when new ideas were explored.

The Law of Conservation of Mass

> **PG 2L** State the meaning of, or draw conclusions based on, the Law of Conservation of Mass.

Early chemists who studied burning, one of the most familiar of chemical changes, concluded that since the ash remaining was so much lighter than the object burned, something was "lost" in the reaction. Their reasoning was faulty. They did not realize that oxygen in the air, which they could not see, was a reactant, and carbon dioxide and water vapor, also invisible, were products. If we take account of those three gases we find that

$$\text{total mass of reactants} = \text{total mass of products}$$
$$\text{(wood + oxygen)} \qquad \text{(ash + carbon dioxide + water vapor)}$$

This equation is the **Law of Conservation of Mass: In a chemical change, mass is conserved; it is neither created nor destroyed.** The word *mass* refers to quantity of matter. It is closely related to the more familiar term, *weight*. The difference between the words is explained in Section 3.4.

PG ALERT

A. Law of Conservation of Mass: Mass conserved in chemical change; neither created nor destroyed

The Law of Conservation of Energy

> **PG 2M** State the meaning of, or draw conclusions based on, the Law of Conservation of Energy.

Energy changes take place all around us—and within us—all of the time. Driving an automobile starts with the chemical energy of a battery and fuel. It changes into kinetic, potential, sound, and thermal (heat) energy as the car moves up and down hills—and add light when the brake lights go on. The electric alarm clock that moves silently by your bed converts electrical energy to mechanical energy, and then to sound energy when the night is passed. Even as you sleep, metabolic processes in your body are processing the food you ate into the thermal energy that maintains body temperature. Other energy conversions are shown in Figure 2.9.

Careful study of energy conversions shows that the energy "lost" or "used" in one form is always exactly equal to the energy "gained" in another form. This leads to another conservation law, the **Law of Conservation of Energy: In any ordinary (non-nuclear) change, energy is conserved. It is neither created nor destroyed.**

PG ALERT

B. Law of Conservation of Energy: Energy is conserved, neither created nor destroyed, in ordinary change

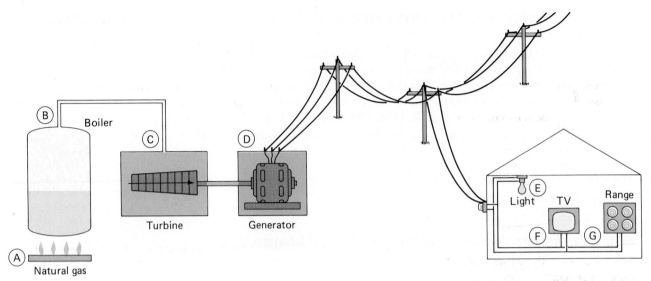

Figure 2.9

Energy changes. Common events in which energy is changed from one form to another. (A) Chemical energy of fuel is changed to heat energy. (B) Heat energy is changed to higher kinetic and potential energy of steam compared to water. (C) Kinetic energy changes to rotating mechanical energy in turbine. (D) Mechanical energy transmitted to generator, where it is changed to electrical energy. (E) Electrical energy changed to heat and light energy. (F) Electrical energy changed to light, sound, and heat energy. (G) Electrical energy changed to heat energy.

The Modified Conservation Law

C. Modified conservation law: Mass can be changed to energy. Combined laws say total mass + energy is constant.

Early in this century, Albert Einstein recognized a "sameness" between mass and energy. He suggested that it should be possible to convert one of these into the other. He proposed that mass and energy are related by the equation

$$\Delta E = \Delta mc^2$$

where ΔE is the energy change, Δm is the mass change, and c is the speed of light. Conversions between mass and energy occur primarily in the nuclei of atoms.

The fact that matter can be converted to energy and vice versa does not necessarily repeal the laws of conservation of mass and energy. For all non-nuclear changes the laws remain valid within our ability to measure such changes. If we include nuclear changes, the law must be modified by stating that the total of all the mass and energy in a change is conserved. Put another way, we now believe that the total of all mass plus energy in the universe is constant.

The amount of energy that can be produced from the conversion of mass is enormous. If it were possible to convert all of a given mass of coal to energy, that energy would be about 2.5 billion times as great as the energy derived from burning that same amount of coal (Fig. 2.10). This is why the explosion of nuclear devices is so destructive. It is also why nuclear energy is such an attractive alternative to traditional sources of energy, although clouded with serious questions of safety and cost overruns.

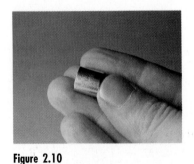

Figure 2.10

Nuclear fuel. A uranium fuel pellet of the size of the cylinder shown produces energy equal to the energy that would be produced by about one ton of coal.

EVERYDAY CHEMISTRY

The Ultimate Physical Property?

Chemists engaged in crime analysis are called forensic chemists. Forensic chemists and detectives have a lot in common. They both examine physical evidence in the hope that they can identify some fact, some object, even some person.

The Federal Bureau of Investigation has on file the fingerprints of approximately 24,000,000 people. This is almost a quarter of a billion fingerprints. If two sets of fingerprints share 16 characteristics, they are almost certain to come from the same person. Finding the identical set of fingerprints in the existing catalog is difficult. Matching fingerprint patterns is more quickly done by computers than by people. In the 1980's, the Tokyo Police pioneered a computer-based automatic fingerprint identification system that is now in use around the world.

What if there were no fingerprints? What if the suspect's fingerprints are not in the current catalogs? Is there another way to reach positive identification?

The physical characteristics of each individual are governed by the genetic information in that person's DNA, **d**eoxyribo**n**ucleic **a**cid. Because every human's DNA is believed to be unique, DNA analysis can, in theory, provide positive identification. As a result, DNA profiling, also called DNA "fingerprinting," has rapidly moved into courtrooms. (Before Operation Desert Storm, the United States began taking DNA samples from military personnel, to serve as an alternative to identification via the traditional dogtags.)

A DNA sample from blood, saliva, or semen is taken at a crime scene. At the laboratory, enzymes cut the DNA into smaller fragments. This is a chemical change, and the resulting products are analyzed by their changed physical properties. These fragments are sorted by size, until an identifying pattern is formed. Pattern matching to a suspect's DNA (also analyzed) is then done, usually by computer. The FBI is working on software that would allow mass computer storage of DNA data, much like their fingerprint database.

The chemistry behind DNA profiling depends on different DNA fragments having different chemical properties that allow separation into bands. The process is exacting and prone to inaccuracies unless carried out very precisely. As a result, the statistical reliability of DNA profiling is currently under question both in the scientific literature and in court.

In 1893, the British Home Secretary received a report from the Troup Committee that no two individuals had the same fingerprints; fingerprint evidence was not accepted in an English court until 1902. We are now in a similar probationary period for DNA fingerprinting. Some courts have accepted DNA fingerprinting as reliable; other courts have held it unreliable. If repeatable techniques of DNA analysis become widely available, we will have taken a long step towards identifying everyone by their "Ultimate Physical Property."

CHAPTER 2 IN REVIEW

2.1 Physical and Chemical Properties and Changes
2A Distinguish between physical and chemical properties.

2B Distinguish between physical and chemical changes.

2.2 Characteristics of Matter
2C Identify and explain the differences between gases, liquids, and solids in terms of (a) visible properties and (b) particle movement.

2D Distinguish between homogeneous matter and heterogeneous matter.

2E Distinguish between a pure substance and a mixture.

2F Distinguish between elements and compounds.

2G Distinguish between elemental symbols and the formulas of chemical compounds.

2.3 The Electrical Character of Matter
2H Match electrostatic forces of attraction and repulsion with combinations of positive and negative charge.

2.4 Characteristics of a Chemical Change
2I Distinguish between reactants and products in a chemical equation.

2J Distinguish between exothermic and endothermic changes.

2K Distinguish between potential energy and kinetic energy.

2.5 Conservation Laws and Chemical Change

2L State the meaning of, or draw conclusions based on, the Law of Conservation of Mass.

2M State the meaning of, or draw conclusions based on, the Law of Conservation of Energy.

CHAPTER 2 KEY WORDS AND MATCHING SETS

Set 1

10	Combination of elemental symbols that identifies a compound.	1. Solution
3	Has mass and occupies space.	2. Modified conservation law
7	Depends on position in force field.	3. Matter
4	Every compound is made up of elements in the same proportion by mass.	4. Law of Constant Composition
2	Total mass and energy in the universe is constant.	5. Element
1	Homogeneous mixture.	6. Endothermic
5	Cannot be broken down into simpler pure substance.	7. Potential energy
9	Variable volume, variable shape.	8. Physical change
11	Change that gives off heat.	9. Gas
8	Change in form without change in identity.	10. Chemical formula
6	Absorbs heat in physical or chemical change.	11. Exothermic

Set 2

10	All matter consists of tiny particles in constant motion.	1. Elemental symbol
1	One- or two-letter abbreviation that identifies an element.	2. Chemical properties
4	Mass of reactants is equal to mass of products in chemical change.	3. Molecule
5	Different phases and nonuniform composition.	4. Law of Conservation of Mass
6	Can be positive, negative, or neutral.	5. Heterogeneous
7	Has definite and constant physical and chemical properties.	6. Electrostatic charge
9	Starting substance in a chemical change.	7. Pure substance
8	Has constant volume but variable shape.	8. Liquid
11	Smallest particle of an element.	9. Reactant
2	Chemical capabilities of a substance.	10. Kinetic molecular theory
3	Term for smallest stable particle of elements whose formulas are N_2, F_2, and I_2.	11. Atom

Set 3

_____1_____ Starting substances destroyed, new substances formed.

_____3_____ Energy cannot be created or destroyed.

_____5_____ May be attraction or repulsion.

_____10_____ Characteristics observed with physical senses.

_____4_____ Uniform appearance and composition throughout.

_____7_____ Formula appears on right side of a chemical equation.

_____8_____ Has definite shape and volume.

_____9_____ Has to do with moving objects.

_____6_____ Parts can be separated by physical means.

_____2_____ Consists of two or more elements.

1. ~~Chemical change~~
2. ~~Compound~~
3. ~~Law of Conservation of Energy~~
4. ~~Homogeneous~~
5. ~~Electrostatic force~~
6. ~~Mixture~~
7. ~~Product~~
8. ~~Solid~~
9. ~~Kinetic~~
10. ~~Physical properties~~

STUDY HINTS AND PITFALLS TO AVOID

In determining whether a change is physical or chemical, remember that the starting substances are destroyed in a chemical change and new substances are formed. If there is no destruction and nothing new, the change must be physical.

A common pitfall in this chapter is not recognizing that both elements and compounds are pure substances. A compound is not a mixture. It takes a chemical change to break a compound into its elements. By contrast, a mixture is separated into its components, elements or compounds, by one or more physical changes.

CHAPTER 2 QUESTIONS AND PROBLEMS

Recall from Chapter 1 that end-of-chapter Questions and Problems are in pairs separated by a thin line. Generally the questions in each pair are matched; that is, they involve the same or related ideas, comparable calculations, or other similarities. Odd-numbered questions whose numbers are printed in blue are answered in the back of the book.

SECTION 2.1

1) Classify each of the following properties as chemical or physical: (a) hardness of a diamond; (b) corrosive character of an acid; (c) taste of salt; (d) elasticity of a rubber band; (e) combustibility of paper.

2) Which of the following properties are physical and which are chemical: (a) electrical resistance of rubber; (b) the ability of a cantaloupe to ripen; (c) odor of ammonia; (d) color of paint; (e) the ability of green leaves to change color in autumn?

3) Which among the following are physical changes: (a) fermenting grapes; (b) "evaporation" of dry ice; (c) decomposing a substance by heating it; (d) forming a snowflake; (e) blowing glass?

4) Describe each of the following changes as chemical or physical: (a) development of mold on stale bread; (b) carving a statue; (c) dissolving sugar in water; (d) digesting your dinner; (e) taking a photograph.

5) Will you use a physical change or a chemical change to boil an egg? On what do you base your answer?

6) If you wanted to obtain chlorine from sodium chloride (table salt), would you use a physical or a chemical change?

7) Golf balls and table tennis balls are about the same size. There are at least two ways they can be separated if many of both are mixed in a box. Describe the ways and identify the property on which your method is based.

8) Suppose somebody emptied some buckshot into a sugar bowl. Could you separate them? How? Would your method use a physical change or a chemical change?

SECTION 2.2

9) List gases, liquids, and solids in order of increasing freedom of movement of individual particles.

10) How do the volumes and shapes of samples of gases, liquids, and solids depend on the volume and shape of the container that holds them?

11) Which of the three states of matter is most easily compressed? Suggest a reason for this.

12) Which of the three states of matter is most likely to expand if opened to the atmosphere? Suggest a reason for this.

13) Which is the most rigid of the three states of matter? Suggest a reason for this.

14) Explain in terms of the properties of gases, liquids, and solids why a snowman disappears when warm weather comes.

---

15) The word "pour" is commonly used in reference to liquids, but not to solids or gases. Can a solid or a gas be poured? Why, or why not? If either answer is yes, can you give an example?

16) The slogan, "When it rains, it pours," has been associated with a brand of table salt for decades. How can salt, a solid, be poured? What unique feature—unique at one time, but probably not today—do you suppose was being emphasized by the slogan? In other words, under what circumstances would their salt "pour" while another brand would not, and why?

---

17) Is the air in a closed room homogeneous or heterogeneous? Justify your answer.

18) Is the air outside of your home homogeneous or heterogeneous? Justify your answer.

---

19) Is a one-liter sample of water from your kitchen faucet homogeneous or heterogeneous? What about a one-liter sample of "drinking water" that is sold in supermarkets?

20) Does the size of a sample of a substance influence the homogeneity or heterogeneity of the sample? Explain.

---

21) Identify the heterogeneous substances among the following: (a) a fresh egg (within the shell); (b) solid brass; (c) milk; (d) cooking oil; (e) linoleum.

22) Which are homogeneous among the following: (a) baking soda; (b) soda in an open can; (c) water in a stream; (d) cup of fresh coffee, just poured; (e) "lead" in a pencil?

---

23) An alloy is a metal that consists of two or more substances, at least one of which is also a metal. Most metal products are alloys. If an alloy that is completely uniform in appearance is etched with an acid and then examined under a microscope, individual crystals can be seen. Is an alloy homogeneous or heterogeneous? Justify your answer.

24) Some ice cubes are homogeneous and some are heterogeneous. In which group would you classify ice cubes from your home refrigerator? What is responsible for the heterogeneity of some ice cubes? How do you suppose water can be frozen so it forms homogeneous ice cubes?

---

25) Today you will see, feel, smell, or otherwise come into contact with the element carbon many times. Can you list five such occasions? For each, state whether the carbon is in elemental form or a part of a compound.

26) There are two elements that we make fresh contact with continuously. Can you identify them? (This question would be easier if we asked for only one element—but the second always comes right along with the first, although you may not be aware of that fact at this time.)

---

27) Silicon is the second most common element in the crust of the earth. In its natural form, it is part of a compound. Where would you go to find a sample of this compound?

28) How far are you from something that contains hydrogen (outside of your own body)? Can you reach it?

---

29) Which of the following are elements and which are compounds: (a) silver bromide (used in photography); (b) calcium carbonate (limestone); (c) sodium hydroxide (lye); (d) uranium; (e) tin? Identify the feature, on which your answer is based, that distinguishes an element from a compound.

30) Which of the following are elements and which are compounds: (a) KOH; (b) $CaCl_2$; (c) Ne; (d) S; (e) Al_2O_3? Identify the feature that distinguishes an element from a compound on which your answer is based.

---

31) Which of the materials listed in Question 29 are pure substances and which are mixtures? Identify the feature that distinguishes a pure substance from a mixture on which your answer is based.

32) Which of the materials listed in Question 30 are pure substances and which are mixtures? Identify the feature that distinguishes a pure substance from a mixture on which your answer is based.

---

33) If hydrogen and oxygen are placed in the same container and there is no chemical change, is the result a compound or a mixture? If a compound, identify it by name. If a mixture, identify its components.

34) If hydrogen and oxygen are placed in the same container and there is a chemical change, is the result a compound or a mixture? If a compound, identify it by name. If a mixture, identify its components. Explain your answer.

---

35) How can you tell if a substance is pure or a mixture? How can you tell if a substance is a compound or an element?

36) Can a compound be decomposed into two elements? Into an element and a compound? Into two compounds?

---

37) Why do chemists believe that water is not an element?

38) Why do chemists believe that hydrogen and oxygen are not compounds?

---

39) When heated, a liquid begins to boil at a certain temperature. It continues to boil at that temperature until it

is all changed to a gas. From this information alone, can you determine if the liquid is (a) a pure substance or a mixture; (b) an element or a compound? Explain why in each case.

40)* The liquid in Question 39 is placed in a sealed container in which there is a partial vacuum. It is again heated until it boils. Again it boils at a constant temperature until it is converted completely to a gas, but the temperature is lower than when the liquid is open to the atmosphere. Does this additional information change your answers to Question 39? Explain.

---

41) The text states that salt and water in a sample of saltwater can be separated by boiling off all the water and condensing it back to a liquid. How could you test the correctness of this statement?

42) If you freeze saltwater, ice will form and float on top of the water that is not yet frozen. What do you suppose happens to the salt? Does some of it freeze along with the ice, or does it remain dissolved in the water? If you don't know—and there is no reason why you should know from the material in this chapter—how could you find out?

---

43) Diamonds and graphite are two forms of the element carbon—they are called *allotropic* forms of carbon. If chunks of graphite are sprinkled among the diamonds on a jeweler's display tray, is the sample on the tray a pure substance or a mixture? Is it homogeneous or heterogeneous?

44)* Sulfur and phosphorus are two other elements that exist in allotropic forms. Do you suppose that the properties of sodium sulfate, a compound containing sulfur, made from one allotrope are the same as or different from the properties of sodium sulfate made from a different allotrope? Why or why not?

---

45) Table sugar is a specific compound. A manufacturer of sugar for sale in supermarkets emphasizes that its product is "cane sugar" from Hawaii. Another sugar manufacturer may process a product identified simply as "sugar," which is made from sugar beets grown anywhere in the United States. Why do you suppose the maker of cane sugar identifies the source of its product, while the maker of beet sugar does not?

46) Suppose on a grocery store shelf a generic brand of sugar, probably beet sugar, in a plain carton is placed next to a higher-price cane sugar (see Question 45) from a known and recognizable manufacturer. Which sugar would you buy? Why? Which kind of sugar, cane or beet, do you suppose is more widely used in the soft drink industry? Why?

---

47) In the text, the reaction between iron and sulfur to form iron sulfide was described by the equation Fe + S → FeS. Is the combination of iron and sulfur *before* the reaction a pure substance or a mixture? What properties,

chemical or physical, would be required to separate the iron from the sulfur? Can you suggest specifically how this separation might be accomplished?

48) Suppose that after the reaction described in Question 47, any unreacted iron or sulfur is removed, leaving only the product, iron sulfide. Is this a pure substance or a mixture? What properties, chemical or physical, would you use to separate the iron from the sulfur?

---

49) There are literally millions of compounds made from carbon and hydrogen. You probably use or benefit from some of these compounds every day. Can you name any of these compounds, either by their chemical names or by the names by which they are more commonly known? How are the properties of these compounds dependent upon the properties of hydrogen and carbon?

50) When carbon reacts with oxygen it can form two compounds. To what extent, if any, are the properties of the compounds related to the properties of the separate elements? How, if at all, are the properties of the compounds related to each other? Support your answers with facts about the elements and compounds that are well known.

---

51) A white crystalline solid that looks like table salt is heated. There is no visible change, but the weight of the material is reduced. Another sample of the original solid is then dissolved in water. The solution is clear and colorless. The solution is divided into two parts. When a clear solution of lead nitrate is added to one part, a yellow powder forms. When the other part is heated, the water evaporates, leaving a solid that looks just like the original solid. From the information given, can you tell if the original substance was a compound or an element? Explain your answer. Classify each change described as a physical or chemical change.

52) A clean piece of a reddish metal is carefully weighed. The color darkens after prolonged exposure to air. Foul smelling red-brown vapors are given off when the metal is placed in nitric acid. The resulting solution is green. A piece of zinc is placed in the solution. Some of the zinc dissolves and the remainder becomes coated with a dark powder. The powder is carefully scraped off the zinc, dried, and weighed. The weight of the powder is the same as the weight of the original metal. From the information given, can you tell if the original metal was an element or a compound? Explain your answer. Classify each change described as a physical or chemical change.

---

53) Which of the following are symbols of elements and which are formulas of compounds? State the number of elements in each compound formula.
Ba NaF P NH_3 $CuSO_4$ Cl C LiOH B $CaBr_2$.

54) How can you tell if the formula of a substance is the symbol or formula of an element or the formula of a compound?

Questions 55 through 58: Consider the following classifications as shown in the table below: gas, liquid, or solid (G, L, S); homogeneous or heterogeneous (Hom, Het); pure substance or mixture (P, M); element or compound (E, C). For each substance listed, place in the table the symbol, such as G, L, or S, that best describes the substance. (The first box is filled in as an example.)

55)	G, L, S	P, M	HOM, HET	E, C
Table salt	S			
Mercury				
Air in a closed jar				
Sand on a beach				
Automobile exhaust				

56)	G, L, S	P, M	HOM, HET	E, C
Butter in the refrigerator				
Distilled water				
Freshly squeezed orange juice				
Aluminum				
Carpet				

57)	G, L, S	P, M	HOM, HET	E, C
Contents of balloon				
Ice				
Baking soda				
Ethyl alcohol				
Wood				

58)	G, L, S	P, M	HOM, HET	E, C
Gasoline				
Copper				
Carbon dioxide				
Household ammonia				
Rock				

SECTION 2.3

59) Among electrostatic, gravitational, and magnetic forces, which one has a major difference between it and the other two? Explain. Can all three of these forces be exerted between two objects at the same time? If not, explain why not.

60) Identify the net electrostatic force (attraction, repulsion, or none) between the following pairs of substances: (a) a small, negatively charged marshmallow and a small, positively charged marshmallow; (b) two negatively charged dust particles; (c) a negatively charged chloride ion and a positively charged silver ion.

SECTION 2.4

61) Identify the reactants and products in the equation $Na_2SO_4 + BaCl_2 \rightarrow BaSO_4 + 2\,NaCl$.

62) What are the formulas of the reactants and the formulas of the products in the reaction $C_3H_5OH + 4\,O_2 \rightarrow 3\,CO_2 + 3\,H_2O$?

63) Identify the product(s) that are elements and the reactant(s) that are compounds in the equation $Mg + NiSO_4 \rightarrow MgSO_4 + Ni$.

64) Write the formula(s) of all elements that are reactants and the formula(s) of all compounds that are products in the equation $Cu + 2\,AgNO_3 \rightarrow 2\,Ag + Cu(NO_3)_2$.

65) As an object or substance experiences each of the following, is the energy change for the object exothermic or endothermic? (a) melting; (b) feeling cool when standing in front of a fan; (c) a bat strikes a ball; (d) a ball is struck by a bat; (e) water falls over Niagara Falls.

66) Which object undergoes an exothermic change, and which undergoes an endothermic change: (a) a hand that touches a hot piece of metal (assume 120°F); (b) a cold piece of metal (40°F) that is touched by a hand; (c) moisture in exhaled breath that can be "seen" on a cold day; (d) chemicals that heat a glass beaker when they react; (e) the body as it digests food?

67) A stone is dropped from a cliff into a stream below. At what time is its potential energy the greatest? At what time is its kinetic energy the greatest? Does its total energy ever change? If so, at what time is its total energy the least? Account for your answers in terms of the Law of Conservation of Energy.

68) A ball is thrown into the air and caught as it comes down. At what time is its (a) potential energy at a maximum; (b) potential energy at a minimum; (c) kinetic energy at a maximum; (d) kinetic energy at a minimum; (e) kinetic energy equal to its potential energy?

69) A positively charged object is held in place near a negatively charged object. If the positively charged object is released, will it move toward the negatively charged object, will it move away, or will it remain where it is? If it moves, will the potential energy decrease or increase as a result of the change in separation?

70) Two positively charged objects are held close to each other. If they are released, will they separate or come together? Will their potential energy decrease or increase as a result of their change in separation?

SECTION 2.5

71) Under very high pressure, the cheap graphite form of carbon can be converted to useful industrial diamonds. What relationship do you expect between the mass of graphite put in and the mass of diamonds obtained? Explain.

72) How would you expect the mass (weight) of a flashbulb before use to compare with the mass after use? Explain.

73) If solid limestone is heated, the rock that remains weighs less than the original limestone. How can this be?

74) If magnesium metal is burned, the mass (weight) of the ash formed is greater than the mass of the metal burned. How can this be explained?

75) If a baseball is thrown onto a field, it possesses a certain amount of kinetic energy as it leaves the hand of the thrower. Eventually it lands on the grass and rolls to a stop. It no longer has kinetic energy; it has even lost the potential energy above the ground that it had when thrown. Is this a violation of the Law of Conservation of Energy? Explain.

76) Trace the energy conversions that occur as a breakfast of orange juice, bacon, eggs, and toast is taken from the refrigerator or off the shelf, prepared for eating, eaten, the dishes are washed and dried, and everything is put away.

77) Identify several energy conversions that occur regularly in your home. State whether each is good (useful), bad (wasteful), or sometimes good, sometimes bad.

78) List several energy conversions that occur in your home between the time you get out of bed and leave the house. State what kind of energy is changed to a new kind of energy.

GENERAL QUESTIONS

79) Distinguish precisely and in scientific terms the differences between items in the following groups:
a) Physical change, physical property, chemical change, chemical property.
b) Gases, liquids, solids.
c) Element, compound.
d) Pure substance, mixture.
e) Homogeneous matter, heterogeneous matter.
f) Exothermic change, endothermic change.
g) Potential energy, kinetic energy.

80) Determine whether each statement that follows is true or false:
a) The fact that paper burns is a physical property.
b) Particles of matter are moving in gases and liquids, but not in solids.
c) A heterogeneous substance has a uniform appearance throughout.
d) Compounds are impure substances.
e) If one sample of sulfur dioxide is 50% sulfur and 50% oxygen, then all samples of sulfur dioxide are 50% sulfur and 50% oxygen.
f) A solution is a homogeneous mixture.
g) Two positively charged objects attract each other, but two negatively charged objects repel each other.
h) Mass is conserved in an ordinary (non-nuclear) endothermic chemical change, but not in an ordinary exothermic chemical change.
i) Potential energy can be related to positions in an electric force field.
j) Chemical energy can be converted to kinetic energy.
k) Potential energy is more powerful than kinetic energy.
l) A chemical change always destroys something and always makes something.

81) A "natural food" store advertises that "no chemicals" are present in any food sold in the store. If their ad is true, what do you expect to find in the store?

82) Which among the following can be pure substances: mercury, milk, water, a tree, ink, iced tea, ice, carbon?

83) Can you have a mixture of two elements and a compound of the same two elements?

84) Can you have more than one compound made of the same two elements? If yes, try to give an example.

85) Rainwater comes from the oceans. Is rainwater more pure, less pure, or of the same purity as ocean water? Explain.

86) A large box contains a white powder of uniform appearance. A sample is taken from the top of the box and another sample is taken from the bottom. By analysis it is found that the percentage of oxygen in the sample from the top is 48.2%, whereas the sample from the bottom is 45.3% oxygen. Answer each question below independently and give a reason that supports your answer.
a) Is the powder an element or a compound?
b) Are the contents of the box homogeneous or heterogeneous?
c) Can you be certain that the contents of the box are either a pure substance or a mixture?

87) If energy can neither be created nor destroyed, as the Law of Conservation of Energy states, why are we so concerned about wasting our energy resources?

88) When two objects are held in place in a force field—gravitational, magnetic, or electrical—and one object is released, it moves. Sometimes it moves toward the other object, sometimes away. In each case, does the potential energy increase or decrease as a result of the move? Explain your answer.

89) Imagine three samples of matter, one a gas, one a liquid, and one a solid. Imagine they are all enclosed in a balloon, and that they have identical shapes and volumes— in other words, a regular balloon, a water balloon, and a "solid" balloon, all the same size and shape. Describe what will happen to both the shape and volume as you press your thumb against the balloon surface.

90) Is it possible for two pure substances to have exactly the same physical and chemical properties? Why, or why not?

91) Explain how a nuclear reaction appears to contradict the laws of conservation of energy and conservation of mass. Does this contradiction mean we can no longer use or rely upon these laws in ordinary day-by-day events? *Do* we continue to rely upon them? If yes, can you cite an example?

92) Is it possible to alter the properties of iron? How about the properties of steel? Explain.

93)* Is the following statement true or false? The oxygen gas and the hydrogen gas produced from water have less energy than an equal mass of liquid water at the same temperature. Explain.

94) Carbon dioxide is a gaseous compound that is formed by burning most substances. "Dry ice" is also carbon dioxide, but in a solid form, and it is very cold. Why is dry ice colder than gaseous carbon dioxide? Have you ever seen or heard of liquid carbon dioxide?

MATCHING SET ANSWERS

Set 1: 10–3–7–4–2–1–5–9–11–8–6

Set 2: 10–1–4–5–6–7–9–8–11–2–3

Set 3: 1–3–5–10–4–7–8–9–6–2

QUICK CHECK ANSWERS

2.1 a and d: True. b: The flammability of gasoline is a chemical property. c: Ethyl alcohol boils at 78°F. This is a physical property.

2.2 a: True. b: The volume of a liquid is constant, but its shape is the same as the bottom of the container that holds it.

2.3 a–c: Heterogeneous. d: Homogeneous.

2.4 Beaker B holds a pure substance because its specific gravity, a physical property, is constant. Beakers A and C hold mixtures because their specific gravities are variable.

2.5 a: Na_2S, potassium hydroxide, and carbon dioxide are compounds; Br_2, C, and fluorine are elements. b: A compound is a pure substance because it has definite physical and chemical properties.

2.6 a: Boiling water is endothermic. The water must absorb energy to boil. b: The change is an increase in potential energy.

Measurement and Calculations

3

Why does one full can float while the other sinks? The obvious answer is density, but *why*
is the liquid in one can more dense than in the other? We'll learn about density in this
chapter, and about solutions in Chapter 15.

LEARN IT NOW In Chapter 3 we use a different form for our outline suggestions. Instead of showing a full outline entry at the place in the text that it would normally appear, we simply put the pencil symbol in the margin. The outline entries are accumulated and printed at the end of each section. Write your outline entry each time you encounter , and then compare your outline with ours when the section is completed.

Don't forget the importance of a chapter preview. If your instructor assigned this chapter before Chapter 2, look back to the beginning of Chapter 2 for a discussion of the preview.

Chemistry is both qualitative and quantitative. In its qualitative role it explains *how* and *why* chemical changes occur. Quantitatively it considers *how much* of a substance is used or produced. *How much* means measurements, calculations, and problem solving.

You will no doubt use a calculator to find the numerical answers to problems. The Appendix includes a discussion of calculators and how to use them in solving chemistry problems. Even if you have a calculator and know how to use it, you might find it helpful to check the section on "chain calculations." Many beginning chemistry students are not familiar with the more efficient ways to use a calculator for a series of multiplications and divisions.

We must assume your arithmetic and first-year algebra skills are ready for use. Even so, we include brief reviews of some operations in the text at the place where they are needed. If you require more than this, please check the mathematics review section in Appendix I Part B.

use of calculator (margin handwritten note)

3.1 INTRODUCTION TO MEASUREMENT

There is nothing new about measurement. You make measurements every day. What time is it? How tall are you? What do you weigh? What is the temperature? How many quarts of milk do you want from the store?

What may be new to you in this chapter, if you live in the United States, are the units in which most measurements are made. Scientific measurements are made in the **metric system.** Modern scientists use **SI units,** which are included in the metric system. SI is an abbreviation for the French name for the International System of Units.

The SI system defines seven **base units.** Three of these, mass, length, and temperature, are described in this chapter. We also refer to time, which is another base unit. Other quantities are made up of combinations of base units; the combinations are called **derived units.** Two of these, volume and density, appear in this chapter. A summary of the SI system appears in Appendix II.

Chapter 3—Measurement and Calculations
3.1 Introduction
  A. SI units are metric units
  B. Seven base units, including mass, length, time, temperature
  C. Units made up of base units are derived units
  D. SI details in Appendix II.

3.2 EXPONENTIAL (SCIENTIFIC) NOTATION

> **PG 3A** Write in exponential notation a number given in ordinary decimal form; write in ordinary decimal form a number given in exponential notation.
> **3B** Add, subtract, multiply, and divide numbers expressed in exponential notation.

Larger and smaller units for the same measurement in the metric system differ by multiples of 10, such as 10 (10^1), 100 (10^2), 10,000 (10^4), or 0.001 (10^{-3}). It is often convenient to express these numbers as **exponentials,** as shown.* Furthermore, chemistry problems sometimes involve very large or very small numbers. For example, the mass of a helium atom is 0.0000000000000000000000664 gram. And in one liter of helium at 0°C and 1 atmosphere of pressure there are 26,880,000,000,000,000,000,000 helium atoms. These are two very good reasons to use **exponential notation,** also known as **scientific notation,** for very large and very small numbers. It is also a good reason to devote a section to reviewing calculation methods with these numbers.

A number, n, may be written in exponential notation as follows:

$$n = C \times 10^e$$

where C is the **coefficient** and 10^e is an **exponential.** The exponent, e, is an integer (whole number); it may be positive or negative. In **standard exponential notation** the coefficient is always equal to or greater than 1, but less than 10. Unless there is reason for doing otherwise, C is written in that range.

One way to change an ordinary number to standard exponential notation uses a "larger/smaller" approach:

Scientists deal with large numbers; this picture was taken by *Voyager 2,* when it was 3.7×10^9 m from Saturn's surface.

PROCEDURE

1) Beginning with the first nonzero digit, rewrite the number, placing the decimal after the first digit. Then write "× 10."
2) Count the number of places the decimal in the original number moved to its new place in the coefficient. Write that number as the exponent of 10.
3) Compare the size of the original number with the coefficient in step 1.
   a) If the coefficient in step 1 is *smaller* than the original number (C < n), the exponential is larger than 1 *and* the exponent is *larger* than 0 (e > 0); it has a positive value. It is not necessary to write the + sign.
   b) If the coefficient in step 1 is *larger* than the original number (C > n), the exponential is *smaller* than 1 *and* the exponent is less than 0 (e < 0); it has a negative value. Insert a minus sign in front of the exponent.

PG ALERT

Recall from Section 2.1 that the words "PG ALERT" point to text that is directly related to one or more performance goals for the section.

It is common, in step 3, to say that the exponent is positive if the decimal moves left, and negative if it moves right. This rule is easily learned, but just as easily reversed in one's memory. The larger/smaller approach works no matter which way the decimal moves. Also, it can be used for moving the decimal of a number already in exponential notation, as you will see shortly.

Instructions   Most of the examples in this book are written in a self-teaching style, a series of questions and answers that guide you to understanding a problem. To reach that understanding, you should answer each question *before* looking at the answer that is printed on the next line of the page. This requires a shield to cover that answer while you consider the question. Two tear-out shields are provided in the book. Thumb the pages and you will find them. On one side you will find instructions on how to use the shield, copied from this section. On the other side is a periodic table that you will use for reference purposes later.

---

*An exponential is a number, called the **base,** raised to some power, called the **exponent.** The mathematics of exponentials is described in Appendix I, Part B. You may wish to review it before studying this section.

The procedure for solving a question-and-answer example is:

1) When you come to an example, locate a set of red lines on each side of the page:

———   ———

Use the shield to cover the page below the lines.
2) Read the example question. Write in the open space above the shield, or on separate paper, any answers or calculations needed.
3) Move the shield down to the next set of lines, if any.
4) Compare your answer with the one you can now read in the book. Be sure you
understand the example up to that point before going on.
5) Repeat the procedure until you finish the example.

We will guide you through this procedure as you work the next example.

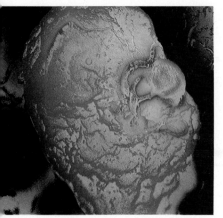

Scientists deal with small numbers.
The AIDS virus HIV-3 is budding
from the surface of an infected
lymphocyte. The lymphocyte is
about 1.5×10^{-5} m across.

EXAMPLE 3.1 ————————————————

Write each of the following in exponential notation:

$3,672,199 =$ $0.00461 =$

$0.000098 =$ $198.75 =$

Locate the red lines below. Place your shield over everything after the lines. Write
your answers to the questions. When you are finished, move the shield down to the next
set of lines, if any.

———   ———

$3,672,199 = 3.672199 \times 10^6$ C < n   e > 0   6 places   e = +6
$0.000098 = 9.8 \times 10^{-5}$ C > n   e < 0   5 places   e = -5
$0.00461 = 4.61 \times 10^{-3}$ C > n   e < 0   3 places   e = -3
$198.75 = 1.9875 \times 10^2$ C < n   e > 0   2 places   e = +2

There are no more lines, so this is the end of the problem. Check your answers against
those above. If any answer is different, find out why before proceeding.

PG ALERT   To change a number written in exponential notation to ordinary decimal form,
simply perform the indicated multiplication. The size of the exponent tells you how
many places to move the decimal point. A positive exponent indicates a large number,
so the decimal is moved to the right; a negative exponent says the number is small, so
the decimal is moved to the left. Thus, the positive exponent in 7.89×10^5 says the
number is large, so the decimal is moved five places to the right: 789,000. The negative
exponent in 5.37×10^{-4} indicates a small number, so the decimal moves four places to
the left: 0.000537.

E X A M P L E 3 . 2

Write each of the following numbers in ordinary decimal form:

$$3.49 \times 10^{-11} = \qquad\qquad 3.75 \times 10^{-1} =$$

$$5.16 \times 10^4 = \qquad\qquad 43.71 \times 10^{-4} =$$

_____

$$3.49 \times 10^{-11} = 0.0000000000349 \qquad 3.75 \times 10^{-1} = 0.375$$

$$5.16 \times 10^4 = 51,600 \qquad\qquad 43.71 \times 10^{-4} = 0.004371$$

Table 3.1

Calculator Procedure

AOS Logic	
Press	Display
3.96	*3.96*
EE	*3.96 00*
4	*3.96 04*
×	*3.96 04*
5.19	*5.19*
EE	*5.19 00*
7	*5.19 07*
+/−	*5.19−07*
=	*2.0552−02*

RPN Logic	
Press	Display
3.96	*3.96*
EEX	*3.96 00*
4	*3.96 04*
ENTER	*3.9600 04*
5.19	*5.19*
EEX	*5.19 00*
7	*5.19 07*
CHS	*5.19−07*
×	*2.0552−02*

If two exponentials with the same base are multiplied, the product is the same base raised to a power equal to the sum of the exponents. When exponentials are divided, the denominator exponent is subtracted from the numerator exponent:

$$10^a \times 10^b = 10^{a+b} \qquad 10^c \div 10^d = \frac{10^c}{10^d} = 10^{c-d} \qquad (3.1)$$

In multiplying or dividing numbers in exponential notation, we rearrange the factors. All coefficients are placed in one group and all exponentials are placed in a second group. The two groups are evaluated separately. The decimal result and the exponential result are then combined. For example,

$$(3.96 \times 10^4)(5.19 \times 10^{-7}) = 3.96 \times 10^4 \times 5.19 \times 10^{-7} \qquad \text{Remove parentheses}$$

$$= (3.96 \times 5.19)(10^4 \times 10^{-7}) \qquad \text{Rearrange/regroup}$$

$$= 20.6 \times 10^{-3} \qquad \text{Evaluate separately}$$

$$= 2.06 \times 10^{-2} \qquad \text{Relocate decimal}$$

PG ALERT

$$\frac{3.96 \times 10^4}{5.19 \times 10^{-7}} = \frac{3.96}{5.19} \times \frac{10^4}{10^{-7}} = 0.763 \times 10^{11} = 7.63 \times 10^{10}$$

regroup   evaluate separately   relocate decimal

Note: The preceding answers have been rounded off to three digits. We will consider *how* to round off in Section 3.5.

This is where the larger/smaller idea for coefficients and exponentials is applied to relocating the decimal point. In the multiplication example the coefficient became smaller as the decimal moved 1 place from 20.6 to 2.06. The exponent therefore was made larger—more positive—by 1, from 10^{-3} to 10^{-2}. In the division example the decimal moved 1 place while the coefficient became larger, so the exponent was reduced by 1, from 11 to 10. If 543×10^6 were to be changed to standard form, it would become 5.43×10^8. In this case the exponent would be raised by 2 because the coefficient became smaller with the decimal moving 2 places.

If your calculator works in exponential notation, you will no doubt use it to solve problems like the previous examples. The factors are entered as described in Appendix I Part A. The calculator operations for $(3.96 \times 10^4) \, (5.19 \times 10^{-7})$ are shown in Table 3.1. The decimal automatically appears after the first digit in the coefficient in the displayed result.

EXAMPLE 3.3 _____

Perform each of the following calculations. Our answers are rounded off to three digits, beginning with the first nonzero digit:

$$(3.26 \times 10^4)(1.54 \times 10^6) = \qquad (8.39 \times 10^{-7})(4.53 \times 10^9) =$$

$$(6.73 \times 10^{-3})(9.11 \times 10^{-3}) = \qquad (2.93 \times 10^5)(4.85 \times 10^6)(5.58 \times 10^{-3}) =$$

$$\frac{8.94 \times 10^6}{4.35 \times 10^4} = \qquad \frac{5.08 \times 10^{-3}}{7.23 \times 10^{-9}} =$$

$$\frac{(3.05 \times 10^{-6})(2.19 \times 10^{-3})}{5.48 \times 10^{-5}} =$$

_____

$$(3.26 \times 10^4)(1.54 \times 10^6) = 5.02 \times 10^{10}$$

$$(8.39 \times 10^{-7})(4.53 \times 10^9) = 38.0 \times 10^2 = 3.80 \times 10^3$$

$$(6.73 \times 10^{-3})(9.11 \times 10^{-3}) = 61.3 \times 10^{-6} = 6.13 \times 10^{-5}$$

$$(2.93 \times 10^5)(4.85 \times 10^6)(5.58 \times 10^{-3}) = 79.3 \times 10^8 = 7.93 \times 10^9$$

$$\frac{8.94 \times 10^6}{4.35 \times 10^4} = 2.06 \times 10^2 \qquad \frac{5.08 \times 10^{-3}}{7.23 \times 10^{-9}} = 0.703 \times 10^6 = 7.03 \times 10^5$$

$$\frac{(3.05 \times 10^{-6})(2.19 \times 10^{-3})}{5.48 \times 10^{-5}} = 1.22 \times 10^{-4}$$

PG ALERT   To add or subtract exponential numbers, we need to align digit values (hundreds, units, tenths, etc.) vertically. This is done by adjusting coefficients and exponents so all exponentials are 10 raised to the same power. The coefficients are then added or subtracted in the usual way. This adjustment is automatic on calculators. Shown here is the addition of 6.44×10^{-7} to 1.3900×10^{-5} as ordinary decimal numbers and as exponentials to 10^{-5} and 10^{-7}.

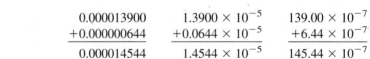

0.000013900	1.3900×10^{-5}	139.00×10^{-7}
+0.000000644	$+0.0644 \times 10^{-5}$	$+6.44 \times 10^{-7}$
0.000014544	1.4544×10^{-5}	145.44×10^{-7}

EXAMPLE 3.4 _____

Add or subtract the following numbers:

$$3.971 \times 10^2 + 1.98 \times 10^{-1} =$$

$$1.05 \times 10^{-4} - 9.7 \times 10^{-5} =$$

_____

3.971 $\times 10^2$	10.5×10^{-5}
+0.00198 $\times 10^2$	$- 9.7 \times 10^{-5}$
3.97298 $\times 10^2$	$0.8 \times 10^{-5} = 8 \times 10^{-6}$

3.2 Exponential notation (scientific notation)

A. Standard form

$$C \times 10^e \leftarrow \text{exponent or power}$$

coefficient   exponential

B. See procedure for writing numbers in exponential notation, page 37.
1) Write coefficient with decimal after first nonzero digit. Follow with "×10."
2) Exponent is number of places decimal moved.
3) If coefficient is smaller than number ($C < n$), exponent is larger than 0 ($e > 0$, a positive number); if coefficient is larger ($C > n$), exponent is smaller than 0 ($e < 0$, a negative number).

C. To change from exponential notation to decimal form, perform multiplication. Positive exponent means large number, so decimal moves right same number of places as exponent. Negative exponent, small number, move decimal left.

D. To multiply or divide in exponential notation, work with coefficients and exponentials separately and then combine.

E. Use larger/smaller changes in coefficient and exponential when moving decimal in exponential numbers.

F. How to use calculator for multiplication and division with exponential notation, page 39.

G. To add or subtract in exponential notation, adjust all exponentials to same power, then add or subtract coefficients.

3.3 DIMENSIONAL ANALYSIS

PG 3C Identify given and wanted quantities in a problem that are related by a "per" expression. Set up and solve the problem by dimensional analysis.

Most problems in beginning chemistry can be solved in either or both of two ways. One way uses algebra, with which you are already familiar. The second way uses **dimensional analysis.** We will use the decimal-based monetary system in the United States and Canada to introduce dimensional analysis. Money calculations are just like calculations with metric units, which are also decimal-based.

How many dimes are equal to 4 dollars?

Already you have probably figured out the answer: 40 dimes. How did you get it? You probably reasoned that there are 10 dimes in each dollar, so there must be 4×10 dimes in 4 dollars: $4 \times 10 = 40$. If you thought along these lines, congratulations! You have just solved your first problem—first in this book, at least—by dimensional analysis.

Let's examine this basic dimes-in-4-dollars problem in detail. The "10 dimes in each dollar" relationship between these two units can be stated as "10 dimes per dollar." We call this a **"per" expression.** This "per" expression can also be written   **PG ALERT** as a fraction, 10 dimes/dollar, or an equality, 10 dimes = 1 dollar. Similarly, there are 10 pennies per dime (10 pennies/dime or 10 pennies = 1 dime), and there are 4 quarters per dollar (4 quarters/dollar or 4 quarters = 1 dollar). Any "per" expression can be written as a fraction or an equality.

The mathematical requirement for a per expression between two quantities is that they are **directly proportional** to each other. What you pay for hamburger, for example, is directly proportional to the amount you buy. Two pounds cost twice as much as one pound. If hamburger is priced at 2.25 dollars per pound (2.25 dollars/pound), three pounds—three times as many pounds—costs 6.75 dollars—three times as many dollars.

A per expression can be used as a **conversion factor,** which we will identify in examples by the symbol FACTOR. In a problem setup, a conversion factor is written as a fraction. Each per expression yields two conversion factors, one the reciprocal of the other. For example, 10 dimes per dollar produces

$$\frac{10 \text{ dimes}}{1 \text{ dollar}} \quad \text{and} \quad \frac{1 \text{ dollar}}{10 \text{ dimes}}$$

A conversion factor is used to convert a quantity of either unit to an equivalent amount of the other unit. The conversion follows a **unit path** (PATH) from the given quantity (GIVEN) to the wanted quantity (WANTED). A unit path may have any number of steps, but you must know the conversion factor for each step in the path.

In the dollars-to-dimes example, the one-step unit path is dollars to dimes, which may be written dollars → dimes. Mathematically you multiply the given quantity, 4 dollars, by the conversion factor, 10 dimes/1 dollar, to get the number of dimes that has the same value as 4 dollars. The calculation setup is

PG ALERT

$$4 \text{ dollars} \times \frac{10 \text{ dimes}}{1 \text{ dollar}} = 40 \text{ dimes} \qquad 4\,\$ \times \frac{10 \text{ d}}{4\,\$} = 40 \text{ d}$$

Notice that units are always included in the calculation setup. Moreover, the units are treated in exactly the same way that variables are treated in algebra. Specifically, they are canceled, as dollars and $ are canceled in the preceding setup. In fact, one way to write a calculation setup is to write the given quantity and then multiply it by the conversion fraction that causes the unwanted unit to cancel out. In time, you will probably do this. It is important, though, to understand *why* this method works, just as you understood why you had to multiply 4 by 10 to get the number of dimes in 4 dollars *before* you saw the calculation setup.

What would have happened if you had selected the wrong conversion factor, the reciprocal of 10 dimes/dollars? Let's see:

$$4 \text{ dollars} \times \frac{1 \text{ dollar}}{10 \text{ dimes}} = 0.4 \text{ dollar}^2/\text{dime}$$

It wouldn't take long to recognize that this answer is wrong—even if you saw the numerical answer on your calculator! (Of course, you wouldn't use a calculator to move a decimal point when dividing by 10, would you?) First of all, you know that the number of dimes in 4 dollars can't be 0.4. There's no such thing as 4/10 of a dime. Second, what is a "dollar^2/dime?" This unit makes no sense. This is what we call a "nonsense unit." In the incorrect setup above, the dollars don't cancel to leave only the wanted dimes, as they should. Any time your calculation setup yields nonsense units, you can be sure that the answer is wrong in *both numbers and units.*

This is one of the valuable features of dimensional analysis. If you get an answer with nonsense units, you *know* you have made a mistake. Units are your friends in solving chemistry problems. Always include them in your calculation setups.

Let's try a nickel-and-dime dimensional analysis problem:

When you "make change" for a dollar, you are doing dimensional analysis. Dimensional analysis changes the *form* of the quantity, but not its *value*. All these stacks have a value of one dollar.

EXAMPLE 3.5

How many nickels are equal to 6 dimes?

It is helpful to plan how to solve a dimensional analysis problem by identifying the given quantity and the wanted quantity, writing a unit path, and then writing the conversion factor you will use. Begin by identifying the given and wanted quantities, and also write the unit path.

_____   _____

GIVEN: 6 dimes.   WANTED: nickels.   PATH: dimes → nickels.

The given quantity is almost always a number of something, a number of some kind of units. The wanted quantity is the number of whatever it is you are asked to calculate, a number of some other units. The unit path begins with the given quantity and ends with the wanted quantity.

Now write the conversion factor between nickels and dimes. Think, "_____ nickels = _____ dimes." Fill in the blanks, and then write the conversion factor as a per expression.

_____   _____

FACTOR: 2 nickels per dime, or 2 nickels/dime.

Now write the setup for converting six dimes to a number of nickels, using 2 nickels/dime as the conversion factor. Calculate the answer.

_____   _____

$$6 \text{ dimes} \times \frac{2 \text{ nickels}}{\text{dime}} = 12 \text{ nickels}$$

The important final step in solving any problem is to be sure the answer makes sense. With something as familiar as nickels and dimes, you know almost without thinking about it that 6 dimes are equal to 12 nickels. But let's use this simple example to establish a method of checking conversions between units that are not so familiar. Notice that a dime is a larger unit than a nickel. Notice also that 6 is a smaller number than 12. The two equal quantities, 6 dimes and 12 nickels, are made up of a smaller number of larger units and a larger number of smaller units. This is the same "larger/ smaller" reasoning we used when moving a decimal point in working with exponential notation in the last section.

The next example is a little bit tricky. See if you can figure out how to handle it.

EXAMPLE 3.6

How many quarters are equal to 75 pennies?

Plan how you will solve the problem by writing the given quantity, the wanted quantity, and the unit path and conversion factor that connects them.

_____   _____

GIVEN: 75 pennies.   WANTED: quarters.   PATH: pennies → quarters.
FACTOR: 25 pennies/quarter.

Now set up the problem, as before. Be careful about how you use the conversion factor.

_____   _____

$$75 \text{ pennies} \times \frac{1 \text{ quarter}}{25 \text{ pennies}} = 3 \text{ quarters}$$

Note two things: First, in order to have the pennies cancel, you had to select 1 quarter/ 25 pennies for the conversion factor rather than 25 pennies/quarter. Always use the conversion factor that leads to the cancellation of an unwanted unit. Second, notice the larger number (75) of smaller units (pennies) and the smaller number (3) of larger units (quarters). By this standard, the answer is reasonable.

In Example 3.6 you probably recognized quickly that you could get the answer by dividing 75 by 25. The same reasoning holds for dimensional analysis: You divide 75 pennies by 25 pennies per quarter. The details of this operation are

$$75 \text{ pennies} \div \frac{25 \text{ pennies}}{\text{quarter}} = 75 \text{ pennies} \times \frac{1 \text{ quarter}}{25 \text{ pennies}} = 3 \text{ quarters}$$

Does this ring a memory bell for you? The rule for dividing by a fraction is to invert the divisor and multiply: "invert and multiply," as you probably first learned it. In more sophisticated language, to divide by a fraction, multiply by its inverse. This rule works for units just as it works for numbers. Later you are more apt to encounter such a setup as a complex fraction, that is, a fraction in which either the numerator or denominator, or both, are fractions. In this instance,

$$\frac{75 \text{ pennies}}{25 \text{ pennies/quarter}} = 75 \text{ pennies} \times \frac{1 \text{ quarter}}{25 \text{ pennies}} = 3 \text{ quarters}$$

Most of the time you will be able to "see" what must be done in order for the units to cancel, just as we did in the first setup of this problem. Other times, particularly with problems solved by algebra rather than by dimensional analysis, you will find "invert and multiply" to be a handy tool in solving the problem.

EXAMPLE 3.7

How many dimes are equal to 6 quarters?

This time the conversion factor is not so obvious. To be sure we get off on the right start, write the given quantity, the wanted quantity, and the unit path.

_____   _____

GIVEN: 6 quarters.   WANTED: dimes.   PATH: quarters → dimes.

Next, the conversion factor. Do you know an equality that relates dimes and quarters? Think, "_____ quarters = _____ dimes," as you did for dimes and nickels in Example 3.5. Once you have numbers in the blanks, you can write the conversion factor.

————   ————

FACTOR: 2 quarters/5 dimes.

2 quarters/5 dimes is only one of any number of acceptable answers. It is the most likely one for you to write if you thought, "2 quarters = 5 dimes." You would have reached this conclusion on realizing that both quantities are equal to 50 cents, so they are equal to each other. You might also have written 5 dimes/2 quarters, which is the inverse, or reciprocal, of 2 quarters/5 dimes.

Now you are ready to multiply the given quantity, 6 quarters, by the conversion factor to get dimes. Which conversion factor will you use, 2 quarters/5 dimes, or 5 dimes/2 quarters? It makes no difference, really, except that in one case you will have to invert before multiplying.

————   ————

$$6 \text{ quarters} \times \frac{5 \text{ dimes}}{2 \text{ quarters}} = 15 \text{ dimes}$$

You had to multiply by 5 dimes/2 quarters in order to get the quarters to cancel and leave dimes as the only surviving unit. There is a small number (6) of larger units (quarters), and vice versa, so the answer is reasonable.

The foregoing example can be used to illustrate another technique of dimensional analysis. Sometimes you cannot get from the given quantity to the wanted quantity in one step. However, if you can develop a multiple-step unit path for which you know the conversion factor for each step in the path, the setup can be written as a series of steps. Suppose you had not recognized that 2 quarters = 5 dimes, but knew that there are 10 dimes/dollar and 4 quarters/dollar. Then you could have followed this unit path: quarters → dollars → dimes. The complete setup is

$$6 \text{ quarters} \times \frac{1 \text{ dollar}}{4 \text{ quarters}} \times \frac{10 \text{ dimes}}{\text{dollar}} = 15 \text{ dimes}$$

EXAMPLE 3.8 _____

Use a two-step unit path through dollars to find the number of quarters that are equal to 675 pennies.

Plan your solution identifying the given and wanted quantities, writing the unit path, and listing the conversion factors you will use.

————   ————

GIVEN: 675 pennies.   WANTED: quarters.
PATH: pennies → dollars → quarters.
FACTORS: 100 pennies/dollar; 4 quarters/dollar.

Now set up the problem and calculate the answer.

$$675 \text{ pennies} \times \frac{1 \text{ dollar}}{100 \text{ pennies}} \times \frac{4 \text{ quarters}}{\text{dollar}} = 27 \text{ quarters}$$

More small units (pennies) than large units (quarters). Answer makes sense.

Notice that, in Example 3.8, it was not necessary to calculate the "intermediate answer," the number of dollars that is equal to 675 pennies. Unless there is a specific reason for calculating intermediate answers, we recommend that you avoid them. Fewer errors are made when the complete setup is written from the GIVEN quantity to the WANTED quantity and all calculations are performed at the same time.

EXAMPLE 3.9

If you went into a bank and obtained 325 dollars worth of dimes, how many dimes would you receive?

Even if you already know the answer to this question, set it up by dimensional analysis.

GIVEN: 325 dollars.   WANTED: dimes.   PATH: dollars → dimes.
FACTOR: 10 dimes/dollar.

$$325 \text{ dollars} \times \frac{10 \text{ dimes}}{\text{dollar}} = 3250 \text{ dimes} \qquad \text{Larger/smaller check: OK.}$$

That was easy! All you had to do to find the answer was move the decimal point. That's the way it is with a decimal system of units, such as the metric system. But notice how dimensional analysis told you which way to move the decimal. You didn't need that here with dollars and dimes, but with unfamiliar metrics, you might.

EXAMPLE 3.10

A little girl broke open her piggy bank. Inside she found 2628 pennies. How many dollars did she have?

Again, even though you probably know the answer, solve the problem by dimensional analysis. There's a reason. Trust us.

GIVEN: 2628 pennies.   WANTED: dollars.   PATH: pennies → dollars.
FACTOR: 100 pennies/dollar

$$2628 \; \cancel{\text{pennies}} \times \frac{1 \text{ dollar}}{100 \; \cancel{\text{pennies}}} = 26.28 \text{ dollars} \qquad \text{Larger/smaller check: OK.}$$

How easy it is to move a decimal point!

SUMMARY / How to Solve a Problem by Dimensional Analysis

In this section, we have followed a pattern for solving problems by dimensional analysis. This pattern will be used throughout the book. You may wish to adopt it too. The steps are:

1) Identify and write down the given quantity. Include units.
2) Identify and write down the wanted quantity. Include units.
3) Write down the unit path and the conversion factors for each step of the path.
4) Write the calculation setup. Include units.
5) Calculate the answer.
6) Check the answer to be sure both the number and the units make sense.

3.3 Dimensional analysis (DA)

A. "Per" expressions
   1. 10 dimes per dollar, 10 dimes/dollar, and 10 dimes = 1 dollar all have the same meaning.
   2. Per expressions can be used when two quantities are directly proportional to each other.

B. FACTOR, PATH, GIVEN, WANTED
   1. FACTOR in examples is a conversion factor that is written as a fraction from the per expression. Every conversion factor can be written in two ways, as two fractions that are reciprocals of each other.
   2. PATH is a unit path, a series of steps from the units of the given quantity (GIVEN) to the wanted quantity (WANTED). Written as GIVEN → WANTED through as many steps as are needed. Conversion factor must be known for each step.

C. Calculations
   1. Start with given quantity.
   2. Multiply by conversion factors in unit path.
   3. Include units in all setups.
   4. If setup yields "nonsense units"—units that don't make sense—the setup and answer are wrong.
   5. Be sure numerical answer makes sense. Use larger/smaller reasoning when possible.
   6. To divide by a fraction, multiply by its inverse; "invert and multiply."
   7. When there are two or more steps, do not calculate intermediate answers unless required.

D. See summary of dimensional analysis, page 47.

3.4 METRIC UNITS

PG 3D   Distinguish between mass and weight.
3E   Identify the metric units of mass, length, and volume.

Mass and Weight

Consider a tool carried to the moon by astronauts. Suppose that tool weighs 6 ounces on earth. On the surface of the moon it will weigh about 1 ounce. But halfway between the earth and moon it would be essentially weightless. Released in "midair," it would remain there, "floating," until moved by an astronaut to some other location. Yet in all three locations it would be the same tool, having a constant quantity of matter.

PG ALERT   **Mass is a measure of quantity of matter. Weight is a measure of the force of gravitational attraction.** Weight is proportional to mass, but the ratio between them depends on where in the universe you happen to be. Fortunately this proportionality is essentially constant over the surface of the earth. Therefore, when you "weigh" something—measure the force of gravity on the object—you can express this weight in terms of mass. In effect, "weighing" an object is one way of measuring its mass. In the laboratory, mass is measured on a balance (Fig. 3.1).

PG ALERT   The SI unit of mass is the **kilogram, kg.** It is defined as the mass of a platinum-iridium cylinder that is stored in a vault in Sevres, France. A kilogram weighs about 2.2 pounds, which is too large for most small-scale work in the laboratory. The basic *metric* mass unit is used instead: the **gram, g.** One gram is 1/1000 kilogram, or 0.001 kg. In reverse, we can say that 1 kg is 1000 g.

The names and symbols of the mass units in the foregoing paragraph are an example of how units are handled in the metric system. Units that are larger than the basic unit are larger by multiples of 10, that is, 10 times larger, 100 times larger, 1000 times larger, and so on. Similarly, smaller units are $\frac{1}{10}$ as large, $\frac{1}{100}$ as large, and so forth. This is what makes the metric system so easy to work with. To convert from one unit to another, all you have to do is move the decimal point.

PG ALERT   Larger and smaller metric units are identified by metric symbols, or prefixes. The prefix for the unit 1000 times larger than the base unit is *kilo-*, and its symbol is k. When the *kilo-* symbol, k, is combined with the unit symbol for grams, g, you have the symbol for *kilo*gram, kg. Similarly, *milli-*, symbol m, is the prefix for the unit that is

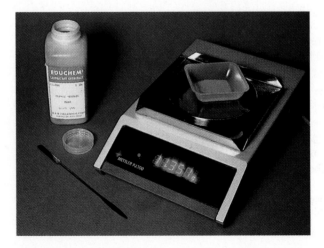

Figure 3.1

A laboratory balance. The balance is being used to measure the mass of a sample of copper sulfate.

Table 3.2

Metric Prefixes*

Large Units			Small Units		
Metric Prefix	Metric Symbol	Multiple	Metric Prefix	Metric Symbol	Submultiple
tera-	T	10^{12}	Unit (gram, meter, liter)		$1 = 10^0$
giga-	G	10^9	deci-	d	$0.1 = 10^{-1}$
mega-	M	$1,000,000 = 10^6$	**centi-**	**c**	**$0.01 = 10^{-2}$**
kilo-	**k**	**$1,000 = 10^3$**	**milli-**	**m**	**$0.001 = 10^{-3}$**
hecto-	h	$100 = 10^2$	micro-	μ	$0.000001 = 10^{-6}$
deca-	da	$10 = 10^1$	nano-	n	10^{-9}
Unit (gram, meter, liter) $1 = 10^0$			pico-	p	10^{-12}

*The most important prefixes are printed in boldface.

$\frac{1}{1000}$ as large as the unit. Thus $\frac{1}{1000}$ of a gram (0.001 g) is 1 milligram, mg. The unit $\frac{1}{100}$ as large as the base unit is a *centi*unit. The symbol for *centi-* is c. It follows that 1 cg (1 centigram) is 0.01 g.

Table 3.2 lists many of the metric prefixes and their symbols. Entries for the kilo-, centi-, and milli- units are shown in boldface. These should be memorized, and you should be able to apply them to any metric unit. We will have fewer occasions to use the prefixes and symbols for other larger and smaller units, but, with reference to the table, you should be able to work with them, too.

PG ALERT

Length

The SI unit of length is the **meter;*** its abbreviation is **m.** The meter has a very precise but awesome definition: the distance light travels in a vacuum in 1/299,792,468 second. It is not obvious why we need so precise a definition, but it is a fact that modern technology requires it. The meter is 39.37 inches long—about three inches longer than a yard.

PG ALERT

The common longer length unit, the kilometer (1000 meters) is about 0.6 mile. Both the centimeter and millimeter are used for small distances. A centimeter (cm) is about the width of a fingernail; a millimeter (mm) is roughly the thickness of a dime. Small metric and English length units are compared in Figure 3.2.

Volume

The SI volume unit is the cubic meter, m^3. This is a **derived unit** because it consists of three base units, all meters, multiplied by each other. A cubic meter is too large a volume—larger than a cube whose sides are 3 feet long—to use in the laboratory. A more practical unit is the **cubic centimeter, cm^3.** It is the volume of a cube with an edge of 1 cm. There are about 5 cm^3 in a teaspoon.

PG ALERT

*Outside the United states the length unit is spelled *metre,* and the liter, the volume unit that will appear shortly, is spelled *litre.* These spellings correspond with the French pronunciations of the words, and it was the French who introduced them. In this book we use the American spellings, which match the English pronunciations. (The English use the French spellings!)

centimeters

little lines are the millimeters

Figure 3.2
Length measurements: inches, centimeters, and millimeters. This illustration is very close to full
scale. One inch is equal to 2.54
centimeters (numbered line on the
metric scale) or 25.4 millimeters
(unnumbered lines).

Liquids and gases are not easily weighed, so we usually measure them in terms of
the volumes they occupy. The common unit for expressing their volumes is the **liter, L,
which is defined as exactly 1000 cubic centimeters.** Thus, there are 1000 cm³/L. This
volume is equal to 1.06 U.S. quarts. Smaller volumes are given in **milliliters, mL.**
Notice that there are 1000 mL in 1 liter (there are always 1000 milliunits in a unit), and
1 liter is 1000 cm³. This makes 1 mL and 1 cm³ exactly the same volume:

PG ALERT

$$1 \text{ mL} = 0.001 \text{ L} = 1 \text{ cm}^3 \tag{3.2}$$

Figure 3.3 shows some laboratory devices for measuring volume.

LEARN IT NOW This simple relationship is often missed. There is 1 mL in
1 cm³, not 1000.

Unit Conversions Within the Metric System

PG 3F	State and write with appropriate metric prefixes the relationship between any metric unit and its corresponding kilounit, centiunit, and milliunit.
3G	Using Table 3.2, state and write with appropriate metric prefixes the relationship between any metric unit and other larger and smaller metric units.
3H	Given a mass, length, or volume expressed in metric units, kilounits, centiunits, or milliunits, express that quantity in the other three units.

Conversions from one metric unit to another are straightforward applications of
dimensional analysis. Performance Goal 3H says you should be able to make these
conversions between the unit, kilounit, centiunit, and milliunit. In this sense, "unit" (u)
may be gram (g), meter (m), or liter (L). These relationships are summarized here as
per expressions and their resulting conversion factors:

1000 units per kilounit	1000 units/kilounit	1000 u/ku
100 centiunits per unit	100 centiunits/unit	100 cu/u
1000 milliunits per unit	1000 milliunits/unit	1000 mu/u

Your instructor may add other units to those you are required to know from memory. If
you can look at Table 3.2, you should be able to convert from any unit in the table to
any other unit.

Figure 3.3

Volumetric glassware used in the laboratory. The beaker is suitable only for estimating volumes. The tall graduated cylinders are used to measure volumes. The flask with the tall neck (volumetric flask) and the pipet are used to obtain samples of fixed but precisely measured volumes. The buret is used to measure volumes with high precision.

EXAMPLE 3.11

How many meters are in 2628 centimeters?

You can solve this problem by following the steps given at the end of the previous section. Identify the given and wanted quantities, the unit path that you will use, and write the conversion factor.

GIVEN: 2628 cm.   WANTED: m.   PATH: cm → m.   FACTOR: 100 cm/m.

Now you have all the information before you. Set up the problem and calculate the answer.

$$2628 \text{ cm} \times \frac{1 \text{ m}}{100 \text{ cm}} = 26.28 \text{ m} \qquad 2628 \text{ cm} \times \frac{1 \text{ m}}{10^2 \text{ cm}} = 26.28 \text{ m}$$

Larger/smaller check: OK. We've written the conversion factor both as 100 cm/m and 10^2 cm/m. Some students find it easier to work with exponentials.

When did you first *know* the answer to the preceding example? Was it as soon as you read the question? It might well have been. You probably knew the answer to Example 3.10 that quickly. In essence, that problem was, "How many dollars are in 2628 pennies?" (We told you there was a reason to write that problem out—to trust us.) Aside from the units, Examples 3.10 and 3.11 are exactly the same problem. In both examples, all you had to do was to move the decimal point.

(The closest English unit to the centimeter is the inch, and the closest English unit to the meter is the yard. You can compare unit conversions in the metric system with similar conversions in the English system by calculating the number of yards in 2628 inches. Again, it is a straightforward dimensional analysis problem. The setup and the answer are given at the end of this chapter.)

EXAMPLE 3.12 _____

Calculate the number of meters in 1.64 kilometers.

The purpose of this example is (1) to give you a chance to try for the answer simply by moving the decimal point, (2) to give you practice in solving the problem by dimensional analysis, and (3) to help you decide which method is best for you. First, what do you think the answer is?

————   ————

1640 m   Larger/smaller check: OK.

There are 1000 meters in each kilometer—1000 times as many meters as the number of kilometers. $1.64 \times 1000 = 1640$. The decimal moves three places to the right. If you answered 0.00164 m, you will be particularly interested in the discussion after the example.

Now set up and solve the problem by dimensional analysis.

————   ————

GIVEN: 1.64 km.   WANTED: m.   PATH: km $\rightarrow$ m.   FACTOR: 1000 m/km.

$$1.64 \ \text{km} \times \frac{1000 \ \text{m}}{\text{km}} = 1640 \ \text{m} \qquad\qquad 1.64 \ \text{km} \times \frac{10^3 \ \text{m}}{\text{km}} = 1640 \ \text{m}$$

Larger/smaller check: OK.

_____

The most common error made by students in metric conversions is moving the decimal the wrong way. The best protection against that mistake is to set up the problem by dimensional analysis, including all units. Always check your result with the larger/smaller rule: If your givens and wanteds have a large number of small units and a small number of large units, and if you've moved the decimal the right number of places, the answer should be correct.

EXAMPLE 3.13

How many millimeters are in 3.04 cm?

If, from the previous unit-centiunit-milliunit relationships, you can see how many milliunits there are in one centiunit, you can solve this problem in one step. Through meters, it takes two steps. (Remember the one-step and two-step solutions to the quarters-and-dimes problem in Example 3.7.) Plan how you will solve the problem, set it up, and calculate the answer. Be sure to check the answer.

Two steps
GIVEN: 3.04 cm.   WANTED: mm.
PATH: cm $\longrightarrow$ m $\longrightarrow$ mm
FACTORS: 100 cm/m, 1000 mm/m

$$3.04 \; \cancel{cm} \times \frac{1 \; \cancel{m}}{100 \; \cancel{cm}} \times \frac{1000 \; mm}{1 \; \cancel{m}} = 30.4 \; mm$$

Larger/smaller check: OK

One step
GIVEN: 3.04 cm.   WANTED: mm.
PATH: cm $\longrightarrow$ mm
FACTOR: 1000 mm/100 cm or 10 mm/cm

$$3.04 \; \cancel{cm} \times \frac{1000 \; mm}{100 \; \cancel{cm}} = 30.4 \; mm$$

or

$$3.04 \; \cancel{cm} \times \frac{10 \; mm}{\cancel{cm}} = 30.4 \; mm$$

The 1000 mm/100 cm conversion factor comes from the fact that both 1000 mm and 100 cm are equal to 1 m. If there are 1000 mm/m and 100 cm = 1 m, then there are 1000 mm/100 cm. This can be reduced to 10 mm/cm.

EXAMPLE 3.14

How many grams are in 0.528 kg?

Complete the problem.

GIVEN: 0.528 kg.   WANTED: g.   PATH: kg $\rightarrow$ g.   FACTOR: 1000 g/kg.

$$0.528 \; \cancel{kg} \times \frac{1000 \; g}{\cancel{kg}} = 528 \; g \qquad \text{Larger/smaller check: OK}$$

EXAMPLE 3.15

A fruit drink is sold in bottles that contain 1892 mL. Express the volume in cubic centimeters and in liters.

GIVEN: 1892 mL.   WANTED: cm^3 and L.   PATHS: mL $\rightarrow$ cm^3 and mL $\rightarrow$ L.
FACTORS: 1 cm^3/mL, 1000 mL/L.

$$1892 \text{ mL} = 1892 \text{ cm}^3 \qquad 1892 \text{ mL} \times \frac{1 \text{ L}}{1000 \text{ mL}} = 1.892 \text{ L} \qquad \text{Larger/smaller check OK.}$$

Recall that 1 mL and 1 cm^3 are the same volume.

3.4 Metric units
  A. Mass
   1. Weight measures gravitational attraction; mass measures amount of matter. Weight depends on where in universe object is; mass always the same anywhere.
   2. Kilogram, kg, is official mass unit, about 2.2 pounds. Gram, g, more common
   3. Memorize prefixes and symbols: *kilo-*, k, 1000; *centi-*, c, 1/100, or 0.01; *milli-*, m, 1/1000, or 0.001. Others in Table 3.2, page 49.
   4. Combine metric symbol with unit abbreviation, as kg = kilogram, cg = centigram, mg = milligram.
  B. Length
   1. Base unit is meter—3 inches more than 1 yard.
   2. Kilometer, 0.6 mile; centimeter, width of fingernail; millimeter, thickness of dime
  C. Volume
   1. Cubic meter, m^3, a derived unit (a combination of base units)
   2. Cubic centimeter, cm^3, more common
   3. Liquid volume: liter (L) and milliliter (mL)
   4. 1 mL = 1 cm^3
  D. Metric conversions
   1. Use dimensional analysis to convert metric units.
   2. Most metric conversions completed by moving the decimal point.
   3. Use larger/smaller rule to be sure decimal moved in correct direction.

3.5 SIGNIFICANT FIGURES

Uncertainty in Measurement

No physical measurement is exact. Every measurement has some **uncertainty.** This is illustrated in Figure 3.4, in which the length of a board is measured by a series of meter sticks graduated in progressively smaller distances. Study the caption carefully. Notice how the uncertainty is shown. The uncertain digit in a measurement is also called the **doubtful digit.**

LEARN IT NOW   Textbook authors sometimes use illustrations to introduce major concepts. If the text that accompanies the figure is long, it may be written into the caption. This keeps the text and the picture together when the book is assembled. When this is done, *the caption becomes the text.* Figure 3.4 is such an illustration.

It is important in scientific work to make accurate measurements and to record them correctly. That record should include some indication of the size of the uncertainty. Attaching a ± uncertainty to the measurement is one way. Another way is to use

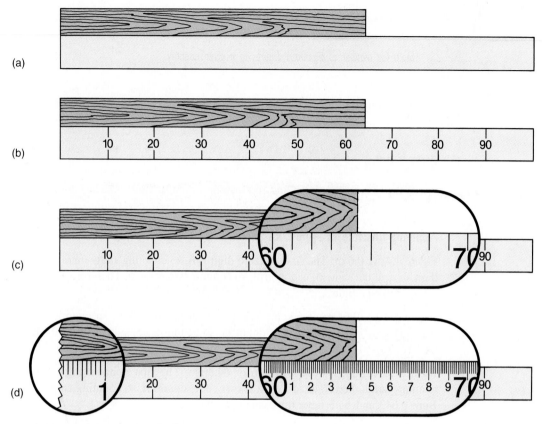

Figure 3.4

Uncertainty in measurement. The length of a board is "measured" (estimated) by comparing it with meter sticks with different graduation marks.

(a) There are no marks. The board is definitely more than half a meter long, probably close to two-thirds. In decimal numbers this is between 0.6 and 0.7 meter (m). The number of tenths is uncertain. Uncertainty is often added to a measurement as a "plus or minus" (±) value. In this case the length might be recorded as 0.6 ± 0.1 m or 0.7 ± 0.1 m.

(b) Graduation marks appear at every 0.1 m, but they are numbered at every 10 cm. The board is less than halfway between 60 and 70 cm (0.6 and 0.7 m). The length might be closer to 64 cm (0.64 m) than 65 cm (0.65 m), but it is hard to tell. Either 64 ± 1 cm or 0.64 ± 0.01 m are reasonable estimates.

(c) Now the centimeter lines are added to the graduations shown in the magnified view. The board's length is clearly closer to 64 cm than 65 cm. It also appears to be about one-fourth to one-third of the way between 64 and 65. Estimating the closest tenth of a centimeter (0.001 m) gives 64.2 ± 0.1 cm (0.642 ± 0.001 m) or 64.3 ± 0.1 cm (0.643 ± 0.001 m) as reasonable estimates.

(d) When millimeter lines are added, the board is clearly closest to, but a little less than, 64.3 cm (0.643 m). Do we estimate the next decimal? Usually you can estimate between the smallest graduation marks, but in this case wear or roughness at the end of the meter stick can introduce errors as much as several 0.01 cm (0.0001 m). It is best to accept 64.3 ± 0.1 cm or 0.643 ± 0.001 m as the most reliable measurement that can be made.

significant figures. Significant figures are related to measurements and to quantities that are calculated from measurements. They do not apply to exact numbers.

The number of significant figures in a quantity is **the number of digits that are known accurately plus the first uncertain digit—the doubtful digit.** It follows that, if a quantity is recorded correctly in terms of significant figures, *the doubtful digit is the last digit written.*

LEARN IT NOW "... the doubtful digit is the last digit written." If this rule is applied consistently, it is not necessary to show a ± uncertainty.

Counting Significant Figures

> **PG 31**   State the number of significant figures in a given quantity.

PG ALERT   From the preceding discussion we can state the rule for counting the number of significant figures in any quantity:

PROCEDURE

 Begin with the first nonzero digit and end with the doubtful digit—the last digit shown.

Applying this rule to Figure 3.4 shows that

Figure 3.4(a), 0.6 m: The first nonzero digit is 6, so counting starts there. The last digit shown, the same 6, is doubtful, so counting stops there. Therefore, 0.6 m has one significant figure.

Figure 3.4(b), 0.64 m or 64 cm: The first nonzero digit is 6, so counting starts there. The last digit shown, 4, is doubtful, so counting stops there. Both 0.64 and 64 have two significant figures.

Figures 3.4(c) and 3.4(d), 0.643 m or 64.3 cm: The first nonzero digit is 6, so counting starts there. The last digit shown, 3, is doubtful, so counting stops there. Both 0.643 and 64.3 have three significant figures.

It should be no surprise that both 0.643 m and 64.3 cm in Figure 3.4(d) have three significant figures. Both quantities came from the same measurement. Therefore, they should have the same uncertainty, the same doubtful digit. Only because the units are different is the decimal point in different places. Hence, the important conclusion: *The measurement process, not the units in which a result is expressed, determines the number of significant figures in a quantity. THEREFORE, THE LOCATION OF THE DECIMAL POINT HAS NOTHING TO DO WITH SIGNIFICANT FIGURES.*

LEARN IT NOW When an author emphasizes a point so strongly, there must be a reason. Many errors are made because students tend to relate significant figures to the location of the decimal point.

If 0.643 m is written in kilometers, it is 0.000643 km. This also is a three-significant figure number. The first three zeros after the decimal point are not significant, but they are required to locate the decimal point. Counting still begins with the first nonzero digit, 6. Specifically, *do not begin counting at the decimal point.*

If the 0.643 m is written in nanometers (nm), $(1 \text{ m} = 10^9 \text{ nm})$, it is 643,000,000 nm. This is still a three-significant figure number, but the uncertainty is 1,000,000 nm. This time we have six zeros that are not significant, but they are required to locate the decimal point. How do we end the recorded value with the doubtful digit, 3? The answer is to write it in exponential notation: 6.43×10^8 nm. The coefficient shows clearly the number of significant figures in the quantity. Exponential notation works with very small numbers too: 6.43×10^{-4} km.

The last two examples show that, in very large and very small numbers, zeros whose only purpose is to locate the decimal point are not significant.

Sometimes—one time in ten, on the average—the doubtful digit is a zero. If so, it still must be the last digit recorded. Suppose, for example, the length of a board is 75 centimeters plus or minus 0.1 cm. To record this as 75 cm is incorrect; it implies that uncertainty is ± 1 cm. The correct way to write this number is 75.0 cm. If the measurement has been uncertain to 0.01 cm, it would be recorded as 75.00 cm. Always, *the doubtful digit is the last digit written, even if it is a zero to the right of the decimal point.*

As a matter of fact, if the doubtful digit is a zero, it is best to have it to the right of the decimal point. If 75.0 cm were to be written in the next smaller decimal unit, it would be 750 mm. The reader of this measurement is faced with the questions, "Is the zero significant, or is it a place holder for the decimal point?" With 75.0 cm or 7.50×10^2 mm, there is no question; the zero is significant.

At the beginning of this section it was stated that significant figures do not apply to **exact numbers.** An exact number has no uncertainty; it is infinitely significant. Counting numbers are exact. A bicycle has exactly two wheels. Numbers fixed by definition are exact. There are exactly 16 ounces in a pound and exactly 12 eggs in 1 dozen eggs.

SUMMARY / The Number of Significant Figures in a Measurement

1) Significant figures are applied to measurements and quantities calculated from measurements. They do not apply to exact numbers.
2) The number of significant figures in a quantity is the number of digits that are known accurately plus one that is uncertain—the doubtful digit.
3) The measurement process, not the units in which the result is expressed, determines the number of significant figures in a quantity.
4) The location of the decimal point has nothing to do with significant figures.
5) The doubtful digit is the last digit written. If the doubtful digit is a zero to the right of the decimal point, that zero must be written.
6) Begin counting significant figures with the first nonzero digit.
7) End counting with the doubtful digit, the last digit written.
8) Exponential notation must be used for very large numbers to show if final zeros are significant.

		We Sell At	We Buy At
CANADA	CAD	.928	.839
UNITED KINGDOM	GBP	1.1884	1.703
JAPAN	YEN	.00812	.00733
SWITZERLAND	SFR	.796	.665
FRANCE	FRF	.191	.173
W. GERMANY	DEM	.653	.590
AUSTRALIA	AS	.8120	.7550
SPAIN	PTS	.01026	.00950
ITALY	LIRE	.000855	.000801

Currency conversions are exact, counted quantities; significant figures do not apply to these.

EXAMPLE 3.16

How many significant figures are in each of the following quantities?

| 45.26 ft____ | 0.109 in.____ | 0.00025 kg____ | 2.3659×10^{-8} cm____ |
| 163 mL____ | 0.60 ft. ____ | 62,700 cm ____ | 5.890×10^5 L ____ |

_____

| 45.26 ft _4_ | 0.109 in. _3_ | 0.00025 kg _2_ | 2.3659×10^{-8} cm _5_ |
| 163 mL _3_ | 0.60 ft. _2_ | 62,700 cm _?_ | 5.890×10^5 L _4_ |

Notes—(a) 0.00025 kg: begin counting significant figures with the first nonzero digit, the 2. (b) 0.60 ft and 5.890×10^5 L: the final zeros identify them as the doubtful digits, so they are significant. (c) 62,700 cm: exponential notations must be used to show if the 7, the first 0, or the second 0, is the doubtful digit:

6.27 $\times 10^4$ is doubtful in hundreds   (three significant figures)

6.270 $\times 10^4$ is doubtful in tens   (four significant figures)

6.2700 $\times 10^4$ is doubtful in ones   (five significant figures)

Rounding Off

PG ALERT   Sometimes when experimentally measured quantities are added, subtracted, multiplied, or divided, the answer given by a calculator contains figures that are not significant. When this happens, the result must be **rounded off.** Rules for rounding off are:

PROCEDURE

1) If the first digit to be dropped is less than 5, leave the digit before it unchanged.
2) If the first digit to be dropped is 5 or more, increase the digit before it by 1.

Other rules for rounding off vary if the first digit to be dropped is exactly 5. For any individual round off by any method, every rule has a 50% chance of being "more correct." Even when "wrong," the rounded off result is acceptable because only the doubtful digit is affected.

EXAMPLE 3.17

Round off each of the following quantities to three significant figures:

a)   1.42752 cm^3 e)   45853 cm
b)   643.349 cm^2 f)   0.03944498 m
c)   0.0074562 kg   g) 3.605×10^{-7} cm
d) 2.103×10^4 mm   h)   3.5000 g

a)   1.43 cm^3 e) 4.59×10^4 cm
b)   643 cm^2 f)   0.0394 m or 3.94×10^{-2} m
c)   0.00746 kg or 7.46×10^{-3} kg   g) 3.61×10^{-7} cm
d) 2.10×10^4 mm   h)   3.50 g

Addition and Subtraction

> **PG 3K** Add or subtract given quantities, and express the result in the proper number of significant figures.

The **significant figure rule for addition and subtraction** can be stated as follows: **PG ALERT**

PROCEDURE

Round off the answer to the first column that has a doubtful digit.

Example 3.18 shows how this rule is applied:

EXAMPLE 3.18

A student weighs four different chemicals into a preweighed beaker. The individual weights and their sum are as follows:

Beaker	319.542 g
Chemical A	20.460 g
Chemical B	0.0639 g
Chemical C	38.2 g
Chemical D	4.173 g
Total	382.4389 g

Express the sum to the proper number of significant figures.

This sum is to be rounded off to the first column that has a doubtful digit. What column is this: hundreds, tens, units, tenths, hundredths, thousandths, or ten thousandths?

———   ———

Tenths. The doubtful digit in 38.2 is in the tenths column. In all other numbers the doubtful digit is in the hundredths column or smaller.

According to the rule, the answer must now be rounded off to the nearest number of tenths. What answer should be reported?

———   ———

382.4 g

This example may be used to justify the rule for addition and subtraction. A sum or difference digit must be doubtful if any number entering into that sum or difference is doubtful or unknown. In the left addition that follows, all doubtful digits are shown in color, and all digits to the right of a colored digit are simply unknown:

319.542	319.5⦙42
20.460	20.4⦙60
0.0639	0.0⦙639
38.2	38.2⦙
4.173	4.1⦙73
382.4389 = 382.4	382.4⦙389 = 382.4

In the left addition the 4 in the tenths column is clearly the first doubtful digit.

The addition at the right shows a mechanical way to locate the first doubtful digit in a sum. Draw a vertical line after the last column in which every space is occupied. The doubtful digit in the sum will be just left of that line. The result must be rounded off to the line.

The same rule, procedure, and rationalization hold for subtraction.

EXAMPLE 3.19 _____

In an experiment in which oxygen is produced by heating potassium chlorate in the presence of a catalyst, a student assembled and weighed a test tube, test tube holder, and catalyst. He then added potassium chlorate and weighed the assembly again. The data were as follows:

Test tube, test tube holder, catalyst,
and potassium chlorate   26.255 g
Test tube, test tube holder, and catalyst   24.05 g

The weight of potassium chlorate is the difference between these numbers. Express this weight in the proper number of significant figures.

——————   ——————

    26.255 g
− 24.05 g
   2.205 g = 2.21 g

The 24.05 is doubtful in the hundredths column, so the difference is rounded off to hundredths.

_____

Multiplication and Division

PG 3L Multiply or divide given measurements and express the result in the proper number of significant figures.

PG ALERT The **significant figure rule for multiplication and division** is:

PROCEDURE

Round off the answer to the same number of significant figures as the smallest number of significant figures in any factor.

Again, application will be illustrated by an example.

EXAMPLE 3.20

If the mass of 1.000 L of a gas is 1.436 g, what is the mass of 0.0573 L of the gas?

This problem can be solved by dimensional analysis. If 1.436 grams of the gas occupies 1.000 liters, we can state this relationship in a per expression: 1.436 g/1.000 L. Can you now identify the given and wanted quantities, plot a unit path between them, and write the conversion factor?

GIVEN: 0.0573 L.   WANTED: g.   PATH: L → g.
FACTOR: 1.436 g/1.000 L.

The factor is written 1.436 g/1.000 L rather than 1.436 g/L because the purpose of this example is to emphasize significant figures in calculations.

Write a dimensional analysis setup for the problem. Calculate the answer and write it down just as it appears in your calculator.

$$0.0573 \text{ L} \times \frac{1.436 \text{ g}}{1.000 \text{ L}} = 0.0822828 \text{ g}$$

Three numbers entered into this multiplication/division problem: 0.0573, 1.436, and 1.000. The answer is to be rounded off to the smallest number of significant figures in any of these numbers. What is that smallest number of significant figures?

There are three significant figures in 0.0573. The other numbers have four significant figures.

Round off the calculated 0.0822828 g to three significant figures.

0.0823 g.

If you wrote 0.082 g, you forgot that counting significant figures begins at the first nonzero digit.

Example 3.20 may be used to justify the rule for multiplication and division. If both quantities have final digits that are one number higher than their true values, the true answer would be $0.0572 \times 1.435 = 0.0821$. If both are too low by 1, the problem is $0.0574 \times 1.437 = 0.0825$. Uncertainty appears in the third significant figure, just as predicted. Alternately, each product number reached with a doubtful multiplier must itself be doubtful. Colored numbers indicate the doubtful digits in the detailed multiplication:

$$
\begin{array}{r}
1.436 \\
\times 0.0573 \\
\hline
4308 \\
10052 \\
7180 \\
\hline
0.0822828
\end{array}
$$

EXAMPLE 3.21 _____

Assuming the numbers are derived from experimental measurements, solve

$$\frac{(2.86 \times 10^4)(3.163 \times 10^{-2})}{1.8} =$$

and express the answer in the correct number of significant figures.

_____   _____

5.0×10^2.

   The answer should not be shown as 500, as the number of significant figures could be read as one, two, or three. Two significant figures are set by the 1.8.

EXAMPLE 3.22 _____

How many inches (in.) are in 6.294 feet (ft)? Express the answer in the proper number of significant figures.

   This is another problem that can be worked by dimensional analysis. We assume you already know the conversion factor. Set up and solve completely. Round off the answer to the correct number of significant figures. Careful!

[NOTE: In order to avoid confusions with the word *in*, the symbol "in." for inches includes a period. The is the only unit symbol in this book that has a period.]

_____   _____

   GIVEN: 6.294 ft   WANTED: in.   PATH: ft → in.   FACTOR: 12 in./ft

$$6.294 \, \cancel{ft} \times \frac{12 \text{ in.}}{\cancel{ft}} = 75.53 \text{ in.} \qquad \text{Larger/smaller check: OK.}$$

By definition, there are *exactly* 12 in. in 1 ft. Exact numbers are infinitely significant. They never limit the number of significant figures in a calculated result.

Significant Figures and This Book

Modern calculators report answers in all the digits they are able to display, usually eight or more. Such answers are unrealistic; they should never be used. Calculations in this book have generally been made using all digits given, and the final answer has been rounded off. If, in a problem with several steps, you round off at each step, your answers may differ slightly from those in the book.

3.5 Significant figures (sig figs)
   A. See caption to Figure 3.4, page 55. Uncertainty is the ± value of a measurement. The "doubtful digit" is the digit value in which the uncertainty appears, such as hundreds, units, tenths, and so on.
   B. Significant figures (sig figs) ≡ number of digits in a quantity that are known for sure plus 1, the doubtful digit.
   C. Doubtful digit is last digit written.
   D. Counting sig figs: Begin with first nonzero digit, end with doubtful digit, the last digit shown.
   E. *Location of decimal point has nothing to do with sig figs.*
   F. In very small numbers, zeros between decimal point and first nonzero digit are not significant. Begin counting at first nonzero digit, not at decimal point.
   G. In very large numbers, zeros before decimal point usually are not significant. Write large numbers in exponential notation (exno) to put doubtful digit to right of decimal point.
   H. Zeros to locate decimal are never significant.
   I. If doubtful digit is a zero to right of decimal, it must be written. Doubtful digit must be the last digit shown. If doubtful digit is a zero, use exno if necessary to put it to right of decimal.
   J. Counting numbers and numbers fixed by definitions are exact. They have no uncertainty. They are infinitely significant.
   K. Summary of sig figs on page 57.
   L. Rules for rounding off
      1. If the first digit to be dropped is less than 5, leave the digit before it unchanged
      2. If the first digit to be dropped is 5 or more, increase the digit before it by 1
   M. Sig fig rule for addition and subtraction: Round off to first column with a doubtful digit. See how to round off addition/subtraction on page 60.
   N. Sig fig rule for multiplication and division: Answer has same *number* of sig figs as the smallest number of sig figs in any factor.

3.6 METRIC–ENGLISH CONVERSIONS

PG 3M   Given a metric–English conversion table and a quantity expressed in any unit in the table, express that quantity in corresponding units in the other system.

The more common conversion relationships between the English and metric systems are given in Table 3.3. The mass, length, and volume sections are for the units we are using in this chapter. The pressure and energy units will appear later in the book. Notice the ≡ symbol instead of the usual = sign between 1 in. and 2.54 cm in the

Table 3.3

Metric–English Conversion Factors

Mass	Length	Volume
1 lb = 454 g 1 oz = 28.3 g 2.20 lb = 1 kg	1 in. ≡ 2.54 cm (definition) 1 ft = 30.5 cm 39.4 in. = 1 m 3.28 ft = 1 m 1.09 yd = 1 m 1 mile = 1.61 km	1.06 qt = 1 L 1 gal = 3.785 L 1 gal (imp) = 4.546 L $1 \text{ in.}^3 = 16.39 \text{ cm}^3$ $1 \text{ ft}^3 = 2.832 \times 10^4 \text{ cm}^3$
Pressure		**Energy**
$14.69 \text{ lb/in.}^2 = 1 \text{ atm} \equiv 760 \text{ torr}$ (definition) 29.92 in. mercury = 1 atm ≡ 760 mm mercury (definition) = 101.3 kPa		1 calorie ≡ 4.184 J (definition) 1 Btu = 1.05 kJ

length section. That symbol is used to identify a definition. Thus exactly 1 in. is equal to exactly 2.54 cm, and the numbers are infinitely significant. There are three other definitions in the table. All other numbers are rounded, correct to the number of significant figures shown.

EXAMPLE 3.23

Find the mass in grams of an object that weighs 13.4 ounces.

Table 3.3 states that 28.3 g = 1 oz. Complete the problem.

GIVEN: 13.4 oz.   WANTED: g.   PATH: oz → g.   FACTOR: 28.3 g/oz.

$$13.4 \text{ oz} \times \frac{28.3 \text{ g}}{\text{oz}} = 379 \text{ g}$$

An ounce is larger than a gram. Larger/smaller check: OK.

EXAMPLE 3.24

An American tourist planning a vacation in Canada learns from an AAA tour map that the distance between Toronto and Montreal is 555 km. How many miles is this? Find the conversion factor in Table 3.3.

GIVEN: 555 km.   WANTED: miles.
PATH: km → miles.   FACTOR: 1.61 km/mile.

$$555 \text{ km} \times \frac{1 \text{ mile}}{1.61 \text{ km}} = 345 \text{ miles}$$

A mile is larger than a kilometer. Larger/smaller check: OK.

Does 89 km/hour seem faster than 55 miles/hour?

EXAMPLE 3.25

Table 3.3 shows that 1 in. ≡ 2.54 cm and 1 ft = 30.5 cm.

a) Calculate the number of cm in 54.00 in.
b) Calculate the number of cm in 4.500 ft.

Explain why the answers are not the same.

The distances 54.00 in. and 4.500 ft are the same distance. They should have the same number of centimeters. But the answers to (a) and (b) are not the same. Make both calculations and explain the difference.

GIVEN: 54.00 in.   WANTED: cm.
PATH: in. → cm.   FACTOR: 2.54 cm/in.

$$54.00 \; \text{in.} \times \frac{2.54 \; \text{cm}}{\text{in.}} = 137.2 \; \text{cm}$$

Larger/smaller check: OK.

GIVEN: 4.500 ft   WANTED: cm.
PATH: ft → cm.   FACTOR: 30.5 cm/ft.

$$4.500 \; \text{ft} \times \frac{30.5 \; \text{cm}}{\text{ft}} = 137 \; \text{cm}$$

The conversion factor 2.54 cm/in. is based on a definition, so 2.54 is an exact number. The answer therefore has four significant figures, the same as the number in 54.00 in. The conversion factor 30.5 cm/ft is a three-significant-figure conversion factor, so the second answer can have only three significant figures.

3.6 Metric–English conversions—see Table 3.3, page 64.

3.7 TEMPERATURE

PG 3N Given a temperature in either Celsius or Fahrenheit degrees, convert it to the other scale.
   30 Given a temperature in Celsius degrees or kelvins, convert it to the other scale.

The familiar temperature scale in the United States is the **Fahrenheit** scale. Most of the rest of the world uses the **Celsius** scale. Both scales are based on the temperature

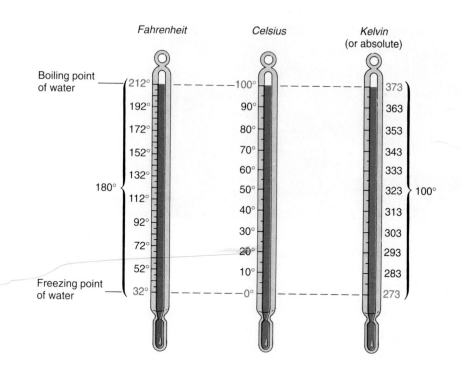

Figure 3.5

A comparison of Fahrenheit, Celsius, and Kelvin temperature scales. The reference or starting point for the Kelvin scale is *absolute zero* (0 K = −273.15°C), the lowest temperature theoretically and experimentally obtainable. Note that the abbreviation K for the kelvin unit is used without the degree sign (°). Also note that 1°C = 1 K = (9/5)°F.

at which water freezes and the temperature at which water boils, both at standard atmospheric pressure (see Section 4.8). The Fahrenheit scale divides the range between freezing and boiling into 180 degrees, starting at 32°F for freezing and 212°F for boiling. The Celsius scale divides the range into 100 degrees, from 0° to 100°. These scales are compared in Figure 3.5.

The way the Fahrenheit and Celsius temperature scales are defined leads to the following relationship between them:

$$T_{°F} - 32 = 1.8 T_{°C} \tag{3.3}$$

$T_{°F}$ is the Fahrenheit temperature and $T_{°C}$ is the Celsius temperature.

SI units include a third temperature scale known as the **kelvin** or **absolute** temperature scale. The "degree" on the kelvin scale is the same size as a Celsius degree. The origin of the kelvin scale is discussed in the next chapter. It is based on zero at the lowest temperature possible, which is 273.15° below zero on the Celsius scale. The two scales are therefore related by the equation

$$T_K = T_{°C} + 273.15 \tag{3.4}$$

K in this equation represents **kelvins,** the actual temperature unit. The degree sign, °, is not used for kelvin temperatures.

PG ALERT

Converting a temperature from one scale to another is done by algebra. In this book we will always solve algebra problems by first solving the equation *algebraically* for the wanted quantity. When the unknown is the only term on one side of the equation, given values are substituted on the other side and the result calculated. The advantage of this procedure is not so apparent in temperature conversions, but it will become clear near the end of the chapter.

EXAMPLE 3.26

What is the Celsius temperature on a comfortable 72°F day?

Begin by identifying the given and wanted quantities, and then writing the equation (Equation 3.3), solved for the wanted quantity.

GIVEN: 72°F.   WANTED: $T_{°C}$.   EQUATION: $T_{°C} = \dfrac{T_{°F} - 32}{1.8}$.

Equation 3.3 is solved for $T_{°C}$ by dividing both sides of the equation by the coefficient of $T_{°C}$, 1.8.

Substitute the given value and calculate the answer.

$$T_{°C} = \frac{T_{°F} - 32}{1.8} = \frac{72 - 32}{1.8} = 22°C.$$

A larger/smaller check of the answer is not readily done for Example 3.26, or for most problems solved algebraically. It is true, however, that Celsius and Fahrenheit temperatures are the same at −40°. At all temperatures above −40°, the Fahrenheit temperature is the larger number, and the Celsius temperature is the lower number. The number 75 is larger than the number 22, so they meet that standard. Beyond that, checking the answer for reasonableness involves some mental arithmetic. The numerator is 72 − 32, which is 40. The denominator is about 2. Dividing 40 by 2 gives 20, which is close to the calculated 22. You will find more about estimating answers in this way in Appendix I, Part D.

EXAMPLE 3.27

At −25°C, it's a cold day! Calculate the corresponding Fahrenheit temperature.

Write the given and wanted quantities, solve Equation 3.3 for $T_{°F}$, substitute, and solve.

GIVEN: −25°C.   WANTED: $T_{°F}$.

EQUATION: $T_{°F} = 1.8T_{°C} + 32 = 1.8(-25) + 32 = -13°F$

This time $T_{°F}$ is isolated by adding 32 to each side of Equation 3.3. Answer check: $2 \times (-25)$ is −50. Adding about 30 gives −20, which is close to −13. The −13 on the Fahrenheit scale is above −25 on the Celsius scale, which is as it should be.

EXAMPLE 3.28

Calculate the kelvin temperature that matches 23°C.

GIVEN: 23°C.   WANTED: T_K.

EQUATION: $T_K = T_{°C} + 273.15 = 23 + 273.15 = 296$ K

As you use the kelvin scale you will recognize that all kelvin temperatures are positive and 273 higher than the corresponding Celsius temperature. The answer is reasonable.

3.7 Temperature
   A. Three scales, Celsius, Fahrenheit, kelvin
      1. $T_{°F} - 32 = 1.8 \, T_{°C}$
      2. $T_K = T_{°C} + 273.15$
      3. Degree sign not used for kelvins
   B. Temperature conversions done by algebra
      1. First solve equation algebraically for the unknown.
      2. Then substitute given values and calculate answer.

3.8 DENSITY

PG 3P   Given two of the following for a pure substance, calculate the third: mass of a sample; volume of the sample; density.

Density is an example of the combination of base units to define a physical property of a pure substance. The formal definition of density is **mass per unit volume.** Expressed mathematically, the **defining equation** of density is

$$\text{density} \equiv \frac{\text{mass}}{\text{volume}} \qquad D \equiv \frac{m}{V} \tag{3.5}$$

We can think of density as a measure of the relative "heaviness" of a substance, in the sense that a block of iron is heavier than a block of aluminum of the same size. Table 3.4 lists the densities of some common materials.

The definition of density establishes its units. Mass is commonly measured in grams (g); volume is measured in cubic centimeters (cm^3). Therefore, according to Equation 3.5 and the "mass per unit volume" definition, the units of density are grams per cubic centimeter, g/cm^3. Liquid densities are often given in grams/milliliter, g/mL. (Recall that one milliliter is exactly 1 cubic centimeter by definition.) There are, of course, other units in which density can be expressed, but they all must reflect the definition in terms of mass/volume. Examples are grams/liter, usually used for gases because their densities are so low, and the English pounds/cubic foot.

PG ALERT   In order to find the density of a substance, it is necessary to know both the mass and volume of a sample of the substance. Dividing mass by volume yields density.

Table 3.4

Approximate Densities of Some Common Substances
(g/cm^3 at 20°C and 1 atm)

Helium	0.00017	Aluminum	2.7
Air	0.0012	Iron	7.8
Lumber		Copper	9.0
Pine	0.5	Silver	10.5
Maple	0.6	Lead	11.4
Oak	0.8	Mercury	13.6
Water	1.0	Gold	19.3
Glass	2.5		

E X A M P L E   3 . 2 9

A 12.0-cm^3 piece of magnesium weighs 20.9 grams. Find the density of magnesium.

GIVEN: 12.0 cm^3; 20.9 g.   WANTED: density.

EQUATION: $D = \dfrac{m}{V} = \dfrac{20.9 \text{ g}}{12.0 \text{ cm}^3} = 1.74 \text{ g/cm}^3$

Answer check: Units OK; approximately 20/10 = 2, which is close to 1.74.

Density problems are particularly useful at the beginning of a chemistry course because they illustrate both methods of solving problems, by algebra and by dimensional analysis. Notice that the definition of density is a per expression: mass per unit volume, m/V. It is therefore a conversion factor for m → V or V → m conversions. Both algebra and dimensional analysis are used in the next example.

E X A M P L E   3 . 3 0

The density of a certain cooking oil is 0.862 g/mL. Find the volume occupied by 196 grams of that oil.

Let's try dimensional analysis first. List the GIVEN, WANTED, PATH, and FACTOR. Set up and solve.

GIVEN: 196 g.   WANTED: mL.   PATH: g → mL.   FACTOR: 0.862 g/mL.

$196 \text{ g} \times \dfrac{\text{mL}}{0.862 \text{ g}} = 227 \text{ mL}$

GIVEN: 196 g
WANTED: mL

Now for algebra. Solve the density equation for volume, substitute the given quantities, and calculate the answer. *Include units when you substitute the given quantities.*

GIVEN: 196 g.   WANTED: mL.

$$\text{EQUATION: } V = \frac{m}{D} = \frac{196\ g}{0.862\ g/mL} = 196\ \cancel{g} \times \frac{mL}{0.862\ \cancel{g}} = 227\ mL$$

Notice that the resulting setups are the same, whether the problem is solved by dimensional analysis or by algebra.

Why was it so strongly suggested that units be included in the algebraic setup of the problem? Answer: An estimated one out of four students solving Example 3.30 without units would have come up with 169 mL for the answer. Starting with a correct substitution of numbers only into the original equation,

$$0.862 = \frac{196}{V}$$

they would have calculated the answer by multiplying 196 by 0.862 instead of dividing. The difference between a correct 227 and an incorrect 169 is not enough to give you an instinctive "feeling" that 169 is wrong, as is the case in nickel-and-dime conversions, so the numerical answer appears reasonable. If units were included in the multiplication, however, the result would have been

$$196\ g \times 0.862\ g/mL = 169\ g^2/mL$$

g^2/mL are nonsense units, and they would have signaled an error in the setup.

3.8 Density
  A. Density ≡ mass per unit volume. ≡ means "is defined as"
  B. Word definition can be written as a defining equation:

$$\text{density} \equiv \frac{mass}{volume} \qquad D \equiv \frac{m}{V}$$

  C. Units set by definition and defining equation: mass units over volume units. Examples: kg/m^3, g/cm^3, g/mL, g/L

3.9 A STRATEGY FOR SOLVING PROBLEMS

At this point we have solved example problems by the two methods that serve for about 90% of the problems in this text. It is time to organize them into a form that will be useful to you. This is done as a flow chart in Figure 3.6. We call this a six-step strategy for solving chemistry problems. In the first three steps you "*plan*" how to solve the problem; in the last three steps you execute your plan by solving the problem. In summary, the six steps are listed:

HOW TO SOLVE A CHEMISTRY PROBLEM

Circled numbers correspond to the six steps for solving a problem.

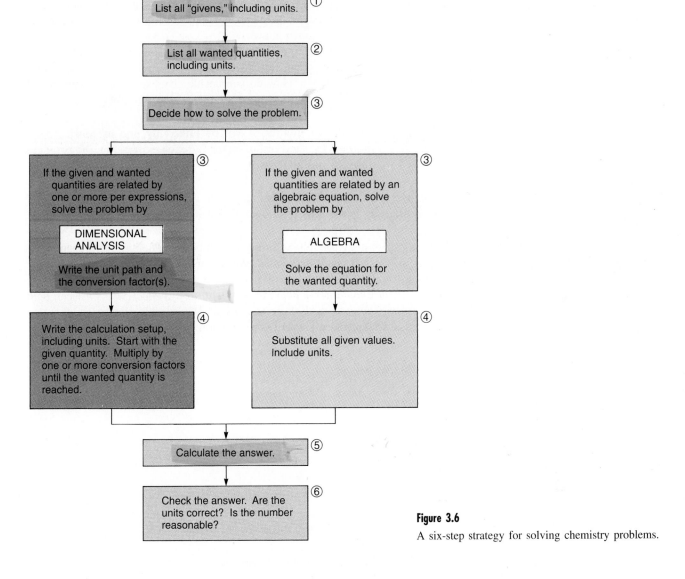

Figure 3.6
A six-step strategy for solving chemistry problems.

PROCEDURE

THE PLAN	THE EXECUTION
1) List everything that is given. Include units.	4) Write the calculation setup for the problem. Include units.
2) List everything that is wanted. Include units.	5) Calculate the answer. Include units.
3) Identify the relationship between the given and wanted quantities. Decide how to solve the problem.	6) Challenge the answer. Be sure that the number is reasonable and that the units are correct (make sense).

In examples throughout this book, we will ask you to "plan" your solution to a problem. "Plan," printed in quotation marks, is a one-word instruction that means to complete the first three steps given above.

Step 3 is the heart of solving a problem. This is where you think and decide what to do. Once the decision is made—once the plan is completed—the rest is usually routine. Deciding is the crucial step. How do you decide? By examining the given and wanted quantities and determining how they are related to each other:

If the given and wanted quantities can be linked by one or more "per" expressions and you know the conversion factor for each expression, the problem can be solved by dimensional analysis.

If the given and wanted quantities appear in an algebraic equation in which the wanted quantity is the only unknown, the problem can be solved by algebra.

LEARN IT NOW *This is a major point.* If you can decide which of these two methods works on a problem, you are well on your way to solving the problem.

The six-step strategy for solving problems has been applied in every example in this chapter to which it could be applied. We will use it consistently throughout the book. We may omit the last step later. It is largely a mental operation, rather than one that is written. *You should never omit the last step. Always check your answers, both numbers and units, to be sure they are reasonable.* We strongly recommend that you read the section in Appendix I, Part D on estimating answers for this purpose.

3.9 Strategy for Solving problems
- A. Six steps
    1. List GIVENs.
    2. List WANTEDs.
    3. Decide whether to use DA or algebra.
    4. Write calculation setup.
    5. Calculate answer.
    6. Check answer, number, and units.
- B. Include units everywhere.
- C. "Plan" in quotation marks means to complete the first three steps.
- D. How to decide—Step 3
    1. If GIVENs and WANTED are linked by per expressions, use DA.
    2. If GIVENs and WANTED are in an algebraic equation, and WANTED is only unknown, use algebra.

3.10 PRACTICE, PRACTICE, PRACTICE

The only way to learn how to solve problems is to solve them yourself. If you have followed the question-and-answer approach to the examples in this chapter—if you have covered the answers to the steps in the examples until you have figured out the answer for yourself—you have solved problems and not merely looked at how we solved them. But you still need more practice.

The end-of-chapter problems with answers give you lots of opportunities to solve problems with immediate feedback on the correctness of your methods. Be sure to keep in mind your reason for solving problems. It is not to get the answer we got, but to *learn how to solve the problem.*

"Solving" a problem with one finger at its solution in the answer section so you can check your progress is neither *solving* the problem nor *learning how to solve it.* When you tackle a problem, solve it completely. If you get stuck at any point, put the problem aside and check the part of the chapter that covers that point. Learn there what you need to learn. *Do not check the answer section.* Return to your solution of the problem and complete it. Then compare your solution to the one in the back of the book. If they do not agree, find out why. Even if they do agree, be sure you *understand* the problem before going to the next one.

We close this chapter with a few more examples with which you can practice while we guide you in following the above strategy.

EXAMPLE 3.31

Calculate the mass of 7.04 cm^3 of silver.

At first glance, it might seem that you don't have enough information to solve this problem. Sometimes you must look elsewhere. Try Table 3.4. When you have what you need, "plan" how you will solve the problem. You can choose between dimension analysis or algebra—or both, for practice. We will show both solutions. Solve the problem completely.

Dimensional analysis
GIVEN: 7.04 cm^3.   WANTED: g.   PATH cm$^3 \rightarrow$ g.   FACTOR: 10.5 g/cm^3.

$$7.04 \text{ cm}^3 \times \frac{10.5 \text{ g}}{\text{cm}^3} = 73.9 \text{ g}$$

Algebra
GIVEN: 7.04 cm^3.   WANTED: g.

$$\text{EQUATION: } m = V \times D = 7.04 \text{ cm}^3 \times \frac{10.5 \text{ g}}{\text{cm}^3} = 73.9 \text{ g}$$

EXAMPLE 3.32

The SI unit of heat energy is the joule (J). The more familiar heat unit in the United States is the calorie (cal). Calculate the number of calories in 6.56×10^5 J.

The English–metric conversion factor table (Table 3.3) gives you additional information. To "plan" your solution to this problem, you must decide if you will solve it by algebra or by dimensional analysis. Which do you choose, and why?

How many calories in
6.56×10^5 J?

Solve the problem by dimensional analysis. There is no algebraic equation you can use, but there is a conversion factor available from Table 3.3.

Complete the problem.

_____

_____

GIVEN: 6.56×10^5 J.   WANTED: cal.   PATH: J $\rightarrow$ cal.   FACTOR: 4.184 J/cal.

$$6.56 \times 10^5 \ \cancel{J} \times \frac{cal}{4.184 \ \cancel{J}} = 1.57 \times 10^5 \ cal$$

EXAMPLE 3.33 _____

The specific heat of a substance is the amount of heat energy needed to raise the temperature of 1 gram of the substance 1°C. Its units are joules per gram degree, which are written J/g · °C. The raised dot in the denominator is a "times" dot, signifying that the gram unit is multiplied by the °C unit. The equation for finding how much heat, Q, is needed to raise the temperature of m grams of the substance by ΔT degrees is $Q = m \times c \times \Delta T$ in which c is the specific heat. Calculate the amount of heat needed to raise the temperature of 43.6 grams of aluminum by 22°C. The specific heat of aluminum is 0.88 J/g · °C.

Believe it or not, this is an easy problem! Don't let the words and unfamiliar terms bother you. By reading the problem, you should be able to decide *how* to solve it, and from there on, it is routine. Plan your approach by writing the given and wanted quantities and state your decision.

_____

_____

GIVEN: m = 43.6 g; ΔT = 22°C; c = 0.88 J/g · °C.   WANTED: Q.

Solve by algebra because the given and wanted quantities are related by an equation in which the wanted quantity is the only unknown.

Substitute both numbers and units into the equation and complete the solution.

_____

_____

EQUATION: $Q = m \times c \times \Delta T = 43.6 \ \cancel{g} \times \dfrac{0.88 \ J}{\cancel{g} \cdot °\cancel{C}} \times 22°\cancel{C} = 8.4 \times 10^2 \ J$

The answer is given in two significant figures because two of the factors have only two significant figures. Units check. 0.88×43.6 is close to 1×40, and 40×20 is 800, which is close to the 840 calculated.

EXAMPLE 3.34 _____

The number of grams of an element (m) that can be deposited in an electroplating bath is calculated by the equation $m = a \times h \times Z$ where a is current measured in amperes (amp), h is time, measured in hours (hr), and Z is a constant that is a property for each element. For copper, Z = 1.18 g/amp · hr. (As in Example 3.33, the raised dot is a "times" dot.) Calculate the current, a, needed to deposit 7.43 g of copper in 0.75 hr.

Again, the words and ideas are unfamiliar, but the math is straightforward. "Plan" how you will solve the problem.

———————   ———————

GIVEN: m = 7.43 g; h = 0.75 hr; Z = 1.18 g/amp · hr.   WANTED: a.

Solve by algebra because given and wanted quantities are related by an equation in which the wanted quantity is the only unknown.

Solve the equation for the wanted quantity, substitute numbers and units, and calculate the answer.

———————   ———————

EQUATION: $a = \dfrac{m}{h \times Z} = \dfrac{m}{h} \times \dfrac{1}{Z} = \dfrac{7.43\ \cancel{g}}{0.75\ \cancel{hr}} \times \dfrac{amp \cdot \cancel{hr}}{1.18\ \cancel{g}} = 8.4$ amps

Units check. Dividing 7.43 by 0.75 yields about 10. If 10 is divided by a little more than 1, the answer is a little less than 10. The calculated 8.4 is reasonable.

EXAMPLE 3.35 ——————————————————————————

College chemistry departments buy sulfuric acid by the case. A case contains six bottles, each of which holds 2.5 L of acid. If the storeroom manager decides to buy a minimum of 125 L of sulfuric acid, how many cases should he order?

Can you "plan" your solution to this problem without leading questions? Carry the plan as far as you can.

———————   ———————

GIVEN: 125 L.   WANTED: cases.

You may have gone farther with your plan than we have in this first step. If you haven't seen *how* to solve the problem, ask yourself a few questions. Are the given and wanted quantities related by an equation? Can the given and wanted quantities be linked by one or more per expressions and conversion factors? The answers should tell you whether to use algebra or dimensional analysis. Complete the "plan" and solve the problem.

———————   ———————

GIVEN: 125 L.   WANTED: cases.   PATH: L → bottles → cases.
FACTORS: 2.5 L/bottle; 6 bottles/case.

$125\ \cancel{L} \times \dfrac{1\ \cancel{bottle}}{2.5\ \cancel{L}} \times \dfrac{1\ case}{6\ \cancel{bottles}} = 8.3$ cases = 9 cases

The 125 L is a minimum quantity, 2.5 L is a two-significant figure number, and 6 is a counting number, infinitely significant. The answer in cases must be a counting number, an integer. To round off to two significant figures (8.3) means buying a fraction of a case. The normal round off of 8.3 to 8 would yield $8 \times 6 \times 2.5 = 120$ L, which is short of the 125 L minimum. Therefore the answer is rounded up to the next integer, 9.

Example 3.35 shows that dimensional analysis can be applied to everyday problems as well as those associated with chemistry. The final example is one you may be able to identify with personally.

EXAMPLE 3.36

Suppose that you have just landed a part-time job that pays $6.25 an hour. You will work five shifts each week, and the shifts are 4 hours long. You plan to save all of your earnings to pay cash for a stereo system that costs $724.26, tax included. You are paid weekly. How many weeks must you work in order to save enough money to buy the stereo system? You might also be interested in knowing how much change you will have left for cassettes, CD's, or other goodies.

This is a long example, and we ask you to solve it completely and without help. Develop your "plan," set up the problem, and calculate the answer.

GIVEN: 724.26 dollars (dol).   WANTED: weeks.   PATH: dol → hr → shifts → week
FACTORS: 6.25 dol/hr; 4 hr/shift; 5 shifts/week.

$$724.26 \text{ dol} \times \frac{1 \text{ hr}}{6.25 \text{ dol}} \times \frac{1 \text{ shift}}{4 \text{ hr}} \times \frac{1 \text{ week}}{5 \text{ shifts}} = 5.79408 \text{ weeks} = 6 \text{ weeks}$$

All of the numbers in the calculation are exact numbers, but it will take your sixth paycheck to get the $724.26 you need.

What will be your total pay at the end of the sixth week? And how much "change" will there be?

GIVEN: 6 weeks.   WANTED: dollars.   PATH: weeks → shifts → hr → dollars.
FACTORS: 5 shifts/week; 4 hr/shift; 6.25 dol/hr.

$$6 \text{ weeks} \times \frac{5 \text{ shifts}}{\text{week}} \times \frac{4 \text{ hr}}{\text{shift}} \times \frac{6.25 \text{ dol}}{\text{hr}} = 750.00 \text{ dollars}$$

$750.00 earned − $724.26 = $25.74 "change"

Again, all the numbers are exact numbers, to the penny.

Oh, oh. We forgot about withholding. . . .

Should the United States Convert to Metric Units?—An Editorial

In the 1970's the four major English-speaking nations of the world—the United States, Great Britain, Canada, and Australia—took action to replace the English system of units with the Metric system. Three of these great nations have made the change successfully. Anyone living in the United States knows which country failed. But we're not alone in the world. Liberia and Burma still use English units. It's nice to have company!

Why does the United States still cling to English units? Simply because of resistance to change. We've grown up with English units. For the vast majority of Americans, there is no advantage to metrics. One kilogram of potatoes is no more convenient a quantity than 2.2 pounds of potatoes, so why bother to change? But Americans can get used to metric quantities. Example: What is the quantity in the large plastic bottles of soft drinks sold in grocery stores? You probably know. The containers are 2-liter bottles. Is 2 liters less convenient than 2.11 quarts, the equivalent English volume? The number is not important once you get used to it . . .

. . . until you get to calculating. If you must work with measured quantities—calculate as we must in the sciences, or buy, sell, or build as they do in commerce and industry—metrics are so much easier and so much more logical that the choice is obvious. If, in solving the problems in this book, you had to work with quantities expressed in English units, your chemistry course would be much more difficult than it is. And this is not because of the chemistry, but because of the English units!

To illustrate, let's restate and solve Example 3.30 with identical quantities given in English units. The original problem was to find the volume occupied by 196 grams of a cooking oil whose density is 0.862 g/mL. Now we will find the volume of 6.91 ounces of the oil, and the density in English units is 53.8 lb/ft^3. We will repeat the metric calculation, too, so you may compare the two. By equation,

$$V = \frac{m}{D} = 196 \; \cancel{g} \times \frac{1 \; mL}{0.862 \; \cancel{g}} = 227 \; mL$$

$$V = \frac{m}{D} = \frac{6.91 \; oz}{53.8 \; lb/ft^3} = 0.128 \; oz \; ft^3/lb$$

Already the difficulty with English units appears. The usual units of mass and density in the English system yield an answer with a bundle of units that must be "sorted out," as the English might say. This is more easily done by a dimensional analysis setup of the problem:

$$6.91 \; \cancel{oz} \times \frac{1 \; \cancel{lb}}{16 \; \cancel{oz}} \times \frac{1 \; ft^3}{53.8 \; \cancel{lb}} = 0.00803 \; ft^3$$

Now, despite your familiarity with the English system, you probably don't have much of a "feel" for how big 0.00803 ft^3 is. Let's extend the setup to cubic inches:

$$6.91 \; \cancel{oz} \times \frac{1 \; \cancel{lb}}{16 \; \cancel{oz}} \times \frac{1 \; \cancel{ft^3}}{53.8 \; \cancel{lb}} \times \frac{12^3 \; in.^3}{\cancel{ft^3}} = 13.9 \; in.^3$$

Now 13.9 cubic inches is a little easier to visualize—a little larger than $3'' \times 2'' \times 2''$. But we don't usually measure liquid volumes in cubic inches. Fluid ounces (these are not the same as the avoirdupois ounces we've already used) would be the most appropriate unit for this volume. Did you know that there are 1.80 cubic inches in a fluid ounce? In fact, are you "familiar" with fluid ounces at all? Probably not, unless you use measuring cups in the kitchen. Maybe pints are more familiar. The calculation setups for both units are

$$6.91 \; \cancel{oz} \times \frac{1 \; \cancel{lb}}{16 \; \cancel{oz}} \times \frac{1 \; \cancel{ft^3}}{53.8 \; \cancel{lb}} \times \frac{12^3 \; \cancel{in.^3}}{\cancel{ft^3}} \times \frac{1 \; fl \; oz}{1.80 \; \cancel{in.^3}}$$
$$= 7.71 \; fl \; oz$$

$$6.91 \; \cancel{oz} \times \frac{1 \; \cancel{lb}}{16 \; \cancel{oz}} \times \frac{1 \; \cancel{ft^3}}{53.8 \; \cancel{lb}} \times \frac{12^3 \; \cancel{in.^3}}{\cancel{ft^3}} \times \frac{1 \; pt}{28.9 \; \cancel{in.^3}}$$
$$= 0.480 \; pt$$

This is nearly half a pint, just about the same as a small-size milk carton.

continued on next page

In summary, and adding a couple of other common volume units in both systems, we have

$$6.91 \text{ av oz} = 0.00802 \text{ ft}^3 = 13.9 \text{ in.}^3 = 7.71 \text{ fl oz}$$
$$= 0.480 \text{ pt} = 0.240 \text{ qt} = 0.0600 \text{ gal}$$

$$196 \text{ g} = 227 \text{ mL} = 0.227 \text{ L} = 227 \text{ cm}^3$$
$$= 0.227 \text{ dm}^3 = 0.000227 \text{ m}^3$$

In the English system we now have six distinctly different numbers by which the same volume is expressed. In metrics the volume is expressed in three different numbers having identical digits, but differing only in the location of the decimal point.

We leave it to you to compare and decide which system of units is easier, English or metric.

The fact is that the general populace has little day-to-day need to calculate, beyond the simple arithmetic at the checkout counter—and even that has been largely replaced by bar codes and cash registers that figure sales tax and the customer's change. And it is the general public in the United States that has the political clout that can vote out of office any politician who would be so bold as to propose that the change to metrics be mandatory.

Changing to metric units is not nearly as difficult as the opponents of change would have us think. The Canadians, English, and Australians did it. Are Americans less capable than they? More to the point, consider all the immigrants to the United States who have changed from their native metric system to the English system—a much more difficult change than from English to metric. If they can do that, Americans are surely able to make the easier change in the other direction.

So has the conversion to metrics been lost in the United States? Not really. A driving force for the conversion that is stronger than public opposition to change is money. If, by adopting metrics and paying the one-time costs that go along with that adoption, a company can (a) make more money, (b) save money, or (c) avoid losing money, you can be quite sure that metrics will be adopted. Here are a few examples.

One major manufacturing firm delivered a large number of appliances to a middle Eastern country only to have them rejected because the connecting cord was six feet long instead of two meters.

The European Economic Community (EEC) has served notice on the United States that it may, as a matter of policy, refuse U.S. imports that fail to meet metric standards. In regard to mechanical equipment, the EEC's position is, "You may build it with English dimensions, but the tools we use to fix it are metric. If we can't fix it, we won't buy it." It's hard to argue against that.

All equipment purchased under our own Pentagon contracts must meet metric specifications. Also, the conversion to metrics is at or close to 100% in the U.S. automobile and drug industries.

One major U.S. manufacturer has reduced the number of screw sizes it holds in inventory from 70 to 15 by using only metric screw products. Another large exporter reports "saving tens of millions of dollars by avoiding double inventory costs and operating all our 32 domestic and foreign plants on one system."

Many U.S. companies that have converted to metrics report the cost of conversion to be much lower than expected—less than half in some cases—and that those costs have been recovered quickly and often unexpectedly.

In the ten years after an American firm introduced metrics, their number of employees doubled to meet their new export demand. In a two-year slack domestic market, those exports were the only thing that kept the company afloat.

An American company adopted metrics in order to hold a few Canadian customers. The change opened so many new markets that they now ship 28% of their output abroad.

Will the United States ever become a metric nation? As these examples attest, it is already a metric nation where it really counts. Industry and government could no longer wait around for a reluctant populace to do what had to be done. So we are, in effect, a nation of two systems of units right now. Maybe that's the way it's supposed to be. But metrics are so much easier, so logical, and have so many advantages. . . . Oh, well!

CHAPTER 3 IN REVIEW

3.1 Introduction to Measurement

3.2 Exponential (Scientific) Notation
3A Write in exponential notation a number given in ordinary decimal form; write in ordinary decimal form a number given in exponential notation.
3B Add, subtract, multiply, and divide numbers expressed in exponential notation.

3.3 Dimensional Analysis
3C Identify given and wanted quantities in a problem that are related by a ''per'' expression. Set up and solve the problem by dimensional analysis.

3.4 Metric Units
3D Distinguish between mass and weight.
3E Identify the metric units of mass, length, and volume.
3F State and write with appropriate metric prefixes the relationship between any metric unit and its corresponding kilounit, centiunit, and milliunit.
3G Using Table 3.2, state and write with appropriate metric prefixes the relationship between any metric unit and other larger units and smaller units.
3H Given a mass, length, or volume expressed in metric units, kilounits, centiunits, or milliunits, express that quantity in the other three units.

3.5 Significant Figures
3I State the number of significant figures in a given quantity.
3J Round off given numbers to a specified number of significant figures.
3K Add or subtract given quantities, and express the result in the proper number of significant figures.
3L Multiply or divide given measurements and express the results in the proper number of significant figures.

3.6 Metric–English Conversions
3M Given a metric–English conversion table and a quantity expressed in any unit on the table, express that quantity in corresponding units in the other system.

3.7 Temperature
3N Given a temperature in either Celsius or Fahrenheit degrees, convert it to the other scale.
3O Given a temperature in Celsius degrees or kelvins, convert it to the other scale.

3.8 Density
3P Given two of the following for a pure substance, calculate the third: mass of a sample; volume of the sample, density.

3.9 A Strategy for Solving Problems

3.10 Practice, Practice, Practice

CHAPTER 3 MATCHING SETS

Set 1

_____ Problem solving method that involves cancellation of units.

_____ Temperature unit on absolute temperature scale.

_____ Unit of length.

_____ Relationship between two quantities when 50% reduction in one yields 50% reduction in the other.

_____ Mass per unit volume.

_____ A measure of gravitational attraction.

_____ 1000.

_____ GIVEN, WANTED, and EQUATION or PATH and FACTOR.

_____ Includes metric units.

_____ A method by which the uncertainty of a quantity can be indicated.

_____ Base number raised to a power.

1. Exponential
2. Weight
3. *kilo-*
4. Meter
5. Significant figures
6. Kelvin
7. Density
8. SI system
9. Dimensional analysis
10. Directly proportional
11. ''Plan''

Set 2

_____ A way to state how many of one kind of unit there are in a fixed number of other units.	1. Base unit
_____ Temperature scale used almost every place on earth except the United States.	2. "Per" expression
_____ Unit of mass.	3. Unit path
_____ The largest digit in which a quantity is uncertain.	4. Gram
_____ Counting number or defined number.	5. *milli-*
_____ Meter, kilogram, second, and kelvin are four of them.	6. Doubtful digit
_____ A number expressed as the product of a coefficient and an exponential.	7. Celsius
_____ 1/1000.	8. Defining equation
_____ The order of unit changes in a dimensional analysis calculation.	9. Exact number
_____ Mathematical expression of a definition.	10. Exponential notation

Set 3

_____ A number that is used to change the units in which a quantity is expressed to different units.	1. Round off
_____ A meaningless label that signals an incorrect answer.	2. *centi-*
_____ To reduce the number of digits in which a quantity is expressed.	3. Exponent
_____ A measure of the quantity of matter.	4. Conversion factor
_____ Unit of volume.	5. Mass
_____ Power to which a base is raised.	6. Coefficient
_____ An amount that is known in a problem.	7. Nonsense units
_____ Temperature scale that is used in the United States.	8. Liter
_____ Decimal factor in scientific notation.	9. Given quantity
_____ 1/100.	10. Fahrenheit

STUDY HINTS AND PITFALLS TO AVOID

Exponential notation is not usually a problem, except for careless errors when relocating the decimal in the coefficient and adjusting the exponent. These errors are not apt to occur if you make sure the exponent and the coefficient move in opposite directions, one larger and one smaller. It sometimes helps even to think about an ordinary decimal number as being written in exponential notation in which the exponent is 10^0. Thus 0.0024 becomes 0.0024×10^0, and the larger/smaller changes in the coefficient and exponent are quite clear when changing to 2.4×10^{-3}.

Include units in every problem you solve. This is the best advice that can be given in a chapter on how to solve chemistry problems.

Challenging every problem answer in both size and units is important. Many errors would never be seen by a test grader if the test taker had simply checked the reasonableness of an answer. Part D in Appendix I offers some suggestions on how to estimate the numerical result in a problem. Please, read this section and put it into practice with every problem you solve.

A common error in metric conversions is moving the decimal point the wrong way. To avoid this, write out fully the dimensional analysis setup. Then challenge your answer. Use the "larger/smaller" rule. For a given amount of anything, the number of larger units is small, and the number of smaller units is larger.

Significant figures can indeed be troublesome, but they need not be if you learn to follow a few basic rules. There are four common errors to watch out for:

1) Starting to count significant figures at the decimal point of a very small number instead of at the first nonzero digit.
2) Using the significant figure rule for multiplication/division when rounding off an addition or subtraction result.

3) Failing to show a doubtful tail-end zero on the right-hand side of the decimal.
4) Failing to use exponential notation when writing large numbers, thereby causing the last digit shown to be other than the doubtful digit.

Most of the arithmetic operations you will perform are multiplications and/or divisions. Students often learn the rule for those operations well, but then apply that rule also to the occasional addition or subtraction problem they meet. Products and quotients have the same number of significant figures as the smallest number in any factor. Sums can have more significant figures than the largest number in any number added. Example: $68 + 61 = 129$. Differences can have fewer significant figures than the smallest number in either number in the subtraction. Example: $68 - 61 = 7$.

CHAPTER 3 QUESTIONS AND PROBLEMS

Recall from Chapter 1 that end-of-chapter Questions and Problems are in pairs separated by space. Generally the questions in each pair are matched; that is, they involve the same or related ideas, comparable calculations, or other similarities. Odd-numbered questions whose numbers are printed in blue are answered in the back of the book.

SECTION 3.2

1)   Write the following numbers in exponential notation:
a)   0.0826
b)   2,630,000
c)   924,000

2)   Express the following numbers in exponential notation:
a)   1,760,000,000
b)   0.000000378
c)   0.0103

3)   Write the ordinary decimal form of the following exponential numbers:
a) 1.23×10^{-4}
b) 7.51×10^5
c) 3.82×10^{-3}

4)   Write the following exponential numbers in the usual decimal form:
a) 2.91×10^7
b) 4.37×10^6
c) 6.02×10^{23}

5)   Complete the following operations:
a) $(8.16 \times 10^6)(5.71 \times 10^5) =$
b) $(4.12 \times 10^{-3})(9.22 \times 10^{-11}) =$
c) $(3.82 \times 10^4)(3.56 \times 10^{-8}) =$
d) $(2.73 \times 10^4)(4.29 \times 10^7)(7.99 \times 10^{-7}) =$

6)   Complete the following operations:
a) $(1.26 \times 10^3)(4.75 \times 10^{-7}) =$
b) $(9.01 \times 10^{-5})(3.61 \times 10^{-4}) =$
c) $(2.81 \times 10^6)(5.86 \times 10^{12}) =$
d) $(7.28 \times 10^{-1})(6.42 \times 10^2)(2.63 \times 10^{-15}) =$

7)   Complete the following operations:
a) $\dfrac{6.82 \times 10^5}{1.37 \times 10^3} =$

b) $\dfrac{2.71 \times 10^6}{2.94 \times 10^{-4}} =$

c) $\dfrac{4.83 \times 10^{-3}}{8.51 \times 10^{8}} =$

d) $\dfrac{7.27 \times 10^{-9}}{6.19 \times 10^{-7}} =$

8) Complete the following operations:

a) $\dfrac{3.57 \times 10^{3}}{6.91 \times 10^{2}} =$

b) $\dfrac{5.28 \times 10^{-7}}{7.73 \times 10^{5}} =$

c) $\dfrac{1.37 \times 10^{9}}{6.10 \times 10^{-4}} =$

d) $\dfrac{1.12 \times 10^{-3}}{1.01 \times 10^{-8}} =$

_____

9) Complete the following operations:

a) $\dfrac{(1.23 \times 10^{2})(1.99 \times 10^{6})}{(5.73 \times 10^{3})} =$

b) $\dfrac{378(8.84 \times 10^{6})}{(2.79 \times 10^{8})(6.36 \times 10^{2})} =$

10) Complete the following operations:

a) $\dfrac{1.16 \times 10^{-4}}{(4.40 \times 10^{1})(3.65 \times 10^{-2})} =$

b) $\dfrac{(8.76 \times 10^{9})(5.93 \times 10^{3})}{0.0404(8.21 \times 10^{-6})} =$

_____

11) Complete the following operations:

a) $8.71 \times 10^{-6} + 7.99 \times 10^{-5} =$

b) $9.27 \times 10^{14} - 6.40 \times 10^{13} =$

12) Complete the following operations:

a) $1.97 \times 10^{9} + 1.56 \times 10^{10} =$

b) $7.39 \times 10^{-1} - 4.52 \times 10^{-2} =$

SECTION 3.3

13) What is the distance in feet between two points exactly 4 miles apart? There are 5280 feet per mile.

14) How long will it take to travel the 591 miles between El Paso and Dallas at an average speed of 55 miles per hour?

_____

15) If sound travels 1.09×10^{3} feet per second (ft/sec), how many seconds are needed for sound to travel 0.375 mile? There are exactly 5280 feet per mile.

16) If light travels 1.86×10^{5} miles per second, how many seconds does it take for light to travel the 93,000,000 miles from the sun to the earth?

_____

17) A walker travels 9.31 miles in 3.00 hours. At this pace, how far will the walker travel in 8.33 hours?

18) The marathon run is approximately 26.2 miles. What is your average speed if you complete this run in 3.75 hours?

_____

19) If the rate of exchange is 1.21 Canadian dollars per American dollar, what is the cost in American dollars of a book that costs $29.95 in Canadian dollars?

20) If the rate of exchange is 1.76 American dollars per English pound (£), what is the cost in dollars of an umbrella that costs 6.75 pounds (£)?

_____

21) An English shilling is equal to 12 old pence. An old half-crown coin was equal to 2 shillings, 6 pence. How many old pence are in a half-crown?

22) There were 20 old shillings to the English pound (£). If there are 12 old pence per shilling, how many old pence are there in 2.50 £?

SECTION 3.4

23) Television broadcasts from the space shuttle often show astronauts floating in the shuttle's cabin. Television announcers describe the astronauts as being "weightless" as they float around. Are the astronauts also massless?

24) Russian cosmonauts spend extended periods in space, months rather than days. A major physiological problem from extended times spent in low gravity is loss of muscle mass. Would loss of muscle weight in space be a physical problem?

_____

25) What is the metric unit of length?
26) What is the metric unit of volume?

_____

27) One centiliter is equal to how many liters?
28) One kilometer is equal to how many meters?

_____

29) Some microcomputers can address 16 megabytes of physical memory. How many bytes of physical memory is this?

30) Modern microcomputers can address 4 gigabytes of physical memory. How many bytes of physical memory is this?

_____

31) What is the name of the unit whose symbol is μm? Is this a long distance or a short distance? Just how long or how short is the μm?

32) Would centigrams or kilograms be better suited for expressing your mass?

Questions 33–38: Make each conversion indicated. Use exponential notation where needed. Write your answers without looking at a conversion table.

33) 503 g = _____ kg
197 g = _____ mg
592 mg = _____ kg

34) 4.08 cg = _____ g
1.89 kg = _____ g
4.84×10^{3} cg = _____ mg

_____

35) 5.49 km = _____ m
5.04 mm = _____ m
6.44 cm = _____ mm

36) 14.3 m = _____ cm
563 m = _____ km
0.467 km = _____ cm

37) 10.3 cL = _____ L
3.41 L = _____ mL
5.60 mL = _____ L
9.68 mL = _____ cm^3

38) 7.62 L = _____ mL
5.16 cm^3 = _____ mL
1.99 L = _____ cm^3
4.11 L = _____ cL

Questions 39 and 40: Refer to Table 3.2 for less common prefixes in these metric conversions.

39) 2.96 Mm = _____ m
2.72 ng = _____ g
9.27 ML = _____ mL

40) 5.95 hg = _____ g
8.49 × 10^{-2} m = _____ mm
2.72 × 10^{14} Gm = _____ km

SECTION 3.5

Questions 41 and 42: To how many significant figures is each quantity expressed?

41) 14.2107 g sand
0.78 mm diameter glass fiber
9.96 × 10^{-5} kg silver
6110 mL water
0.0298 ft wire
8.8440 × 10^6 L seawater
327 cm^3 gasoline
4300.0 cg salt

42) 56.3 g seashells
79.563 mL vinegar
0.0444 cm diameter wire
27,000 ft telephone cable
0.70 m fabric
0.369 kg flounder
21.277 × 10^2 cm^3 string
2.640 × 10^{-9} L oil

Questions 43 and 44: Round off each quantity to three significant figures:

43) 80.40 mL neon
65.716 g gold

11.79 mg aspirin
57,641,000 μg platinum
0.0055874 kg mercury
44) 7.524 × 10^{-3} km fiberoptic cable
0.0173 g sugar
67,000 m chain
81,960 pounds nitric acid
$106.67

45) A student added 4.98 grams of potassium bromide, 8.0 grams of lithium sulfate, and 0.939 gram of sodium chloride in 144 grams of water. Calculate the total mass of this solution and express the sum in the proper number of significant figures.

46) In some states, drivers convicted of driving under the influence of alcohol can perform as part of their court sentence cleaning of freeway shoulders. One such crew picked up on a glorious Saturday 154 pounds of plastic, a couch that weighed 72.3 pounds, a broken trailer that weighed 1.3 × 10^2 pounds, and 498.69 pounds of newspaper and cardboard. Calculate and express in the correct number of significant figures the total weight of this load.

47) An empty beaker has a mass of 91.27 grams. After some potassium chromate is added, the total mass is 102.189 grams. What is the mass of the potassium chromate in the beaker?

48) A buret contains 32.17 mL sodium hdyroxide solution. Because of a leak, the volume drops down to 29.0 mL a few minutes later. How many milliliters of solution have drained from the buret?

49) If exactly 1 liter of solution contains 34.3 grams of sugar, how many grams of sugar are contained in exactly 2 liters? How many grams are in 2.74 liters? Express the results in the correct number of significant figures.

50) The mole is the SI standard for amount of a substance. The mass of one mole of pure table salt is 58.45 grams. How many grams of table salt are in exactly 2 moles? What is the mass of 0.947 mole?

SECTION 3.6

Questions 51–60: You may consult the table of Metric–English Conversions, Table 3.3, while answering these questions.

51) A subcompact car weighs 1689 pounds. Calculate the mass of this car in kilograms and express the results to the correct number of significant figures.

52) The driver of the subcompact car in the previous question weighs 153 pounds. Calculate the mass of the car and its driver. Round your answer to the correct number of significant figures.

53)   The distance between Calgary and Edmonton in the Canadian province of Alberta is 294 kilometers. How many miles is this?

54)   When rotating, the blades of a window fan form a circle 18.0 inches in diameter. If 2.54 cm = 1 inch, what is the diameter of this circle in cm?

---

55)   The mass of aspirin in one aspirin tablet is 325 mg. Calculate the weight in ounces of aspirin in one aspirin tablet.

56)   The mass of ibuprofen in one ibuprofen tablet is 2.00×10^2 mg. Calculate the weight in pounds of ibuprofen in one ibuprofen tablet.

---

57)   In jewelry, 1 carat = 0.200 gram. Calculate the mass of a 6.24-carat diamond in grams and then in ounces.

58)   The payload of a compact car is 665 pounds. How many kilograms is this?

---

59)   The largest recorded difference of weight between spouses is 922 lb. The husband weighed 1020 lb, and his dear wife 98 lb. Express this difference in kilograms.

60)   There are 9.0 grams of protein in a 227-g serving of yogurt. How many ounces of protein is this?

---

61)*   The weight of a new baby is listed as 7 lb, 7 oz. How would this baby's mass be described in a metric hospital?

62)   A boxer reads 71.1 kg when he steps on a scale in his home gymnasium in Germany. Should he be classified in the United States as a welterweight (136 to 147 lb) or as a middleweight (148 to 160 lb)?

---

63)   The shortest gymnast in the 1992 Olympics was 132 cm tall. What is this height in feet and inches?

64)   This shortest gymnast had a mass of 31 kg. What is this weight in lb?

---

65)   The Amazon River is 3.9×10^2 miles long. How long is this in km?

66)   The height of Niagara Falls (the Horseshoe Falls) on the Canadian side is 48 m. How high is this in (a) feet; (b) yards?

---

67)   Standard-size paper in the United States is 8.5 in. wide. Metric size A4 paper is 210 mm wide. Which paper is wider?

68)   Metric size A4 paper is 297 mm long. Standard size paper in the United States is 11.0 inches long. Which paper is longer?

---

69)   An automobile has a 40.0-L fuel tank. Express this volume in U.S. gallons.

70)   An air-cooled car has a 10.4-L oil capacity. How many quarts of oil are needed for a complete oil change?

---

71)*   How many grams of milk are in 8.00 fluid ounces? The density of milk is 64.4 lb/ft^3. There are 7.48 gal/ft^3; by definition, there are 4 qt/gal and 32 fluid ounces/quart.

72)*   The fuel tank in an automobile has a capacity of 18.5 gal. There are 7.48 gal/ft^3, and the density of gasoline is 42.0 lb/ft^3. What is the mass of fuel when this tank is full in (a) lb; (b) kilograms?

SECTION 3.7

Questions 73 and 74: Fill in the spaces in the tables below so that each temperature is expressed in all three scales. Round off answers to the nearest degree.

73)

Celsius	Fahrenheit	Kelvin
	35	
		486
−19		
		142
821		
	−52	

74)

Celsius	Fahrenheit	Kelvin
65		
	−17	
		104
	41	
		397
−192		

---

75)   The boiling point of rubbing alcohol is 82.5°C. What is this temperature in °F?

76)   The melting point of rubbing alcohol is −88.5°C. What is this temperature in °F?

---

77)   To save heating fuel in the winter, it is recommended that thermostats be set no higher than 68°F. What temperature is this in °C?

78)   To save electrical power in summer, power companies suggest that air conditioners be set so they do not turn on until the inside temperature exceeds 78°F. What is this temperature in °C?

---

79) Normal body temperature is 37.0°C. What is this temperature in °F?

80) A person has a body temperature of 100.6°F. What is this temperature in °C?

SECTION 3.8

81) Commercial airplanes are made of aluminum because aluminum has a very low density for a metal. Calculate the density of aluminum if a 415-cm³ block has a mass of 1121 grams.

82) MTBE, methyl tert-butyl ether, is used as an octane booster in gasoline. 114 g of MTBE fills a graduated cylinder to the 154-mL mark. What is the density of MTBE?

83) A rectangular block of iron 7.45 cm × 19.7 cm × 9.70 cm has a mass of 11.2 kg. What is the density of iron?

84) Gases have very low densities compared to liquids or solids. Calculate the density of air, in g/L, if the mass of 12.7 L is 15.0 g.

85) Find the mass of 14.9 mL of alcohol if its density is 0.810 g/mL.

86) Halothane, an inhalation anesthetic, has a density of 1.87 g/mL. What is the volume of 672 g of halothane?

87) The newly popular canola oil has a density of 0.915 g/mL. What is the volume of one pound (454 grams) of canola oil?

88) A cookie recipe calls for one half cup of butter (121 cm³). What is the mass of the butter if its density is 0.86 g/cm³?

GENERAL QUESTIONS

89) Distinguish precisely and in scientific terms the differences between items in each of the following groups:
a) Coefficient, exponent, exponential
b) "Per" expression, conversion factor, unit path
c) Mass, weight
d) Unit, kilounit, centiunit, milliunit
e) Significant figures, doubtful digit
f) Uncertainty, exact number
g) Fahrenheit, Celsius, kelvin

90) Determine whether each statement that follows is true or false:
a) The SI system includes metric units.
b) The exponential notation form of a number smaller than 1 has a positive exponent.
c) In changing an exponential notation number whose coefficient is not between 1 and 10 to standard exponential notation, the exponent becomes smaller if the decimal in the coefficient is moved to the right.

d) Dimensional analysis can be used to change from one unit to another only when the quantities are related by a "per" expression.
e) In a direct proportionality, when one value increases, the other decreases.
f) A unit path begins with the units of the given quantity and ends with the units in which the answer is to be expressed.
g) The mass of an object is independent of its location in the universe.
h) There are 1000 mL in a cubic centimeter.
i) There are 1000 kilounits in a unit.
j) There are 10 milliunits in a centiunit.
k) The doubtful digit is the last digit written when a number is expressed properly in significant figures.
l) 76.2 g means the same as 76.200 g.
m) The number of significant figures in a sum may be more than the number of significant figures in any of the quantities added.
n) The number of significant figures in a difference may be fewer than the number of significant figures in any of the quantities subtracted.
o) The number of significant figures in a product may be more than the number of significant figures in any of the quantities multiplied.
p) The density of a substance can be used as a conversion factor between mass and volume.
q) There is no advantage to using units in a problem that is solved by algebra.
r) If the quantity in the answer to a problem is familiar, it is not necessary to check to make sure the answer is reasonable.

91) How tall are you in (a) meters; (b) decimeters; (c) centimeters; (d) millimeters? Which of the four units do you think would be most useful in expressing a person's height without resorting to decimal factors?

92) What do you weigh in (a) milligrams; (b) grams; (c) kilograms? Which of these units do you think is best for expressing a person's weight? Why?

93) The density of aluminum is 2.7 g/cm³. An ecology-minded student has gathered 126 empty aluminum cans for recycling. If there are 21 cans per pound, how many grams of aluminum does the student have? What is the volume of metal in cubic centimeters?

94) In Example 3.36 you calculated that you would have to work 6 weeks to earn enough money to buy a $724.26 stereo system. You would be working five shifts of 4 hours each at $6.25/hr. But, alas, when you received your first paycheck, you would find that exactly 23% of your earnings had been withheld for social security, federal and state income tax, and workmen's compensation insurance. Taking these costs into account, how many weeks are needed to earn the $724.26?

MATCHING SET ANSWERS

Group 1: 9-6-4-10-7-2-3-11-8-5-1

Group 2: 2-7-4-6-9-1-10-5-3-8

Group 3: 4-7-1-5-8-3-9-10-6-2

ANSWER TO PROBLEM FOLLOWING EXAMPLE 3.11

$$2628 \text{ inches} \times \frac{1 \text{ yard}}{36 \text{ inches}} = 73 \text{ yards}$$

Introduction to Gases

Scuba divers and others who explore underwater regions must provide for their breathing requirements. This leads to variations in pressure, temperature, volume, and amount of air. Three of the four variables appear in the gas laws you will study in this chapter.

Life on earth exists because of a precious envelope of gases we call the atmosphere. Near sea level the atmosphere contains about 80% nitrogen and 20% oxygen. (Star Trek fans will recognize this mixture as defining "a class M planet," like earth.) We live by "swimming" in this sea of gases; our very lives depend on the behavior of these gases as they interact with our bodies, especially the lungs. The abundance of manufactured goods in our lives evolved from mass-production techniques of the Industrial Revolution. The Industrial Revolution itself was made possible by the invention of the steam engine, an application of gas properties. Because gas behavior is so important, we'll start with general properties of all gases.

4.1 PROPERTIES OF GASES

Some of the familiar characteristics of air—in fact, of all gases—are the following:

1) *Gases may be compressed.* A fixed quantity of air may be made to occupy a smaller volume by applying pressure. Figure 4.1a shows a quantity of air in a cylinder having a leak-proof piston that can be moved to change the volume occupied by the air. Push the piston down by applying more force, and the volume of air is reduced (Fig. 4.1b).

2) *Gases expand to fill their containers uniformly.* If less force were applied to the piston, as shown in Figure 4.1c, air would respond immediately, pushing the piston upward, expanding to fill the larger volume uniformly. If the piston were pulled up (Fig. 4.1d), air would again expand to fill the additional space.

Figure 4.1

Compression and expansion properties of gases. The piston and cylinder show that gases may be compressed and that they expand to fill the volume available to them.

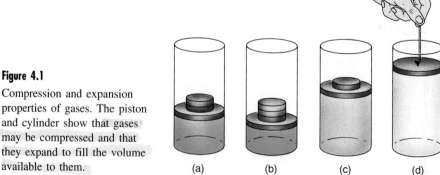

(a)   (b)   (c)   (d)

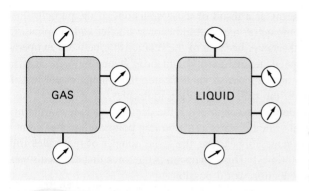

Figure 4.2
Pressures in gases and liquids. Gas pressures are exerted uniformly in all directions; liquid pressures depend on the depth of the liquid.

3) *All gases have low density.* The density of air is 0.0013 g/cm³. The density of water is 770 times greater than the density of air; and iron is 6000 times more dense than air.

4) *Gases may be mixed.* "There's always room for more," is a phrase that may be applied to gases. You may add the same or a different gas to that gas already occupying a rigid container of fixed volume, provided there is no chemical reaction between them.

5) *A confined gas exerts contant pressure on the walls of its container uniformly in all directions.* This pressure, illustrated in Figure 4.2, is a unique property of a gas, independent of external factors such as gravitational forces.

4.2 THE KINETIC THEORY OF GASES AND THE IDEAL GAS MODEL

PG 4A Explain or predict physical phenomena relating to gases in terms of the ideal gas model.

In trying to account for the properties of gases, scientists have devised the **kinetic molecular theory.** The theory describes an **ideal gas model** by which we visualize the nature of the gas by comparing it with a physical system that can be seen or at least readily imagined. The main features of the ideal gas model are:

FLASHBACK The kinetic molecular theory proposes that all matter consists of molecules in constant motion. See Section 2.2. (*See* description of FLASHBACKS on page 9.)

SUMMARY

1) Gases consist of molecular particles moving at any given instant in straight lines.
2) Molecules collide with each other and with the container walls without loss of energy.
3) Gas molecules behave as independent particles; attractive forces between them are negligible.
4) Gas molecules are very widely spaced.
5) The actual volume of molecules is negligible compared to the space they occupy.

Particle motion explains why gases fill their containers. It also suggests how they exert pressure. When an individual particle strikes a container wall, it exerts a force at the point of collision. When this is added to billions upon billions of similar collisions occurring continuously, the total effect is the steady force that is responsible for gas pressure.

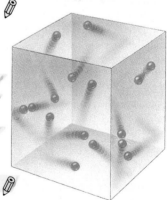

Representation of particle motion in a gas. Particles collide with each other and with the walls of the container, the latter being responsible for the pressure exerted by the gas.

There can be no loss of energy as a result of these collisions. If the particles lost energy or slowed down, the combined forces would become smaller and the pressure would gradually decrease. Furthermore, because of the relationship between temperature and average molecular speed, temperature would drop if energy were lost in collisions. But these things do not happen, so we conclude that energy is not lost in molecular collisions, either with the walls or between molecules.

Gas molecules must be widely spaced; otherwise the densities of gases would not be as low as they are. One gram of liquid water at the boiling point occupies 1.04 cm^3. When changed to steam at the same temperature, the same number of molecules fills 1670 cm^3, an expansion of 1600 times! (This expansion is the force that drives a steam engine that made our mass-production world possible.)

If the water molecules were touching each other in the liquid state, they must be widely separated in the gas state. The compressibility and mixing ability of gases are also attributable to the open spaces between gas molecules. Finally, the large distances between gas particles ensure us that attractions between these molecules are negligible.

4.3 GAS MEASUREMENTS

Experiments with a gas usually involve measuring or controlling its quantity, volume, pressure, and temperature. Finding quantity by weighing a gas is different from weighing a liquid or a solid, but it is readily done. In this chapter we will consider only fixed amounts of gas; quantity will be constant. Unlike liquids and solids, a gas always fills its container, so its volume (V) is the same as the volume of the container. The other two measurements, pressure (P) and temperature (T), deserve closer examination.

Pressure

> **PG 4B**   Given a gas pressure in atmospheres, torr, millimeters of mercury, centimeters of mercury, inches of mercury, pascals, kilopascals, or pounds per square inch, express that pressure in each of the other units.

By definition **pressure** is the force exerted on a unit area:

$$\text{pressure} \equiv \frac{\text{force}}{\text{area}} \qquad \text{or} \qquad P \equiv \frac{F}{A} \tag{4.1}$$

Units of pressure come from the definition. In the English system, if force is measured in pounds and area in square inches, the pressure unit is pounds per square inch (psi). The SI unit of pressure is the **pascal (Pa),** which is one newton per square meter. (The *newton* is the SI unit of force.) One pascal is a very small pressure; the **kilopascal (kPa)** is a more practical unit. The **millimeter of mercury,** or its equivalent, the **torr,** and the **atmosphere** are the common units for expressing pressure. The millimeter of mercury is usually abbreviated mm Hg. (Hg is the elemental symbol for mercury.)

Weather bureaus generally report *barometric* pressure, the pressure exerted by the atmosphere at a given weather station, in inches of mercury or in kilopascals. Atmospheric pressure is often measured by a **barometer.** This device was developed by Evangelista Torricelli in the seventeenth century (Fig. 4.3). On a day when the mer-

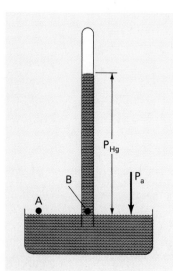

Figure 4.3

Mercury barometer. Two operational principles govern the mercury barometer. (1) The total pressure at any point in a liquid system is the sum of the pressures of each gas or liquid phase above that point. (2) The total pressures at any two points at the same level in a liquid system are always equal. Point A at the liquid surface outside the tube is at the same level as Point B inside the tube. The only thing exerting downward pressure at A is the atmosphere; P_a represents atmospheric pressure. The only thing exerting downward pressure at Point B is the mercury above that point, designated P_{Hg}. A and B being at the same level, the pressures at these points are equal: $P_a = P_{Hg}$.

FLASHBACK   Equation 4.1 is a defining equation for pressure, like the defining equation for density in Section 3.8.

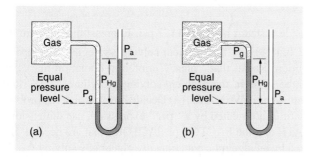

Figure 4.4

Open-end manometers. Open-end manometers are governed by the same principles as mercury barometers (Fig. 4.3). The pressure of the gas, P_g, is exerted on the mercury surface in the closed (left) leg of the manometer. Atmospheric pressure, P_a, is exerted on the mercury surface in the open (right) leg. Using a meter stick, the difference between these two pressures, P_{Hg}, may be measured directly in millimeters of mercury (torr). Gas pressure is determined by equating the total pressures at the lower liquid level. In (a), the pressure in the left leg is the gas pressure, P_g. Total pressure at the same level in the right leg is the pressure of the atmosphere, P_a, plus the pressure difference, P_{Hg}. Equating the pressures, $P_g = P_a + P_{Hg}$. In (b) total pressure in the closed leg is $P_g + P_{Hg}$, which is equal to the atmospheric pressure, P_a. Equating and solving for P_g yields $P_g = P_a - P_{Hg}$, P_a. Equating and solving for P_g yields $P_g = P_a - P_{Hg}$. In effect, the pressure of a gas, as measured by a manometer, may be found by adding the pressure difference to, or subtracting the pressure difference from, atmospheric pressure; $P_g = P_a \pm P_{Hg}$.

This wall barometer is calibrated to measure atmospheric pressure in inches of mercury.

cury column in a barometer is 752 mm high, we say that atmospheric pressure is 752 mm Hg.

One standard atmosphere of pressure is defined as 760 mm Hg. This is a typical barometric pressure at sea level. The atmosphere unit is particularly useful in referring to very high pressures. Other common pressure units and their relationships to each other are:

$$1 \text{ atm} \equiv 760 \text{ mm Hg} = 760 \text{ torr} = 76.0 \text{ cm Hg} = 1.013 \times 10^5 \text{ Pa}$$

$$= 101.3 \text{ kPa} = 29.92 \text{ in. Hg} = 14.69 \text{ psi} \quad (4.2)$$

The *torr*, which honors the work of Torricelli, and the *millimeter of mercury* are identical pressure units. Both terms are widely used, and the choice between them is one of personal preference. The advantage of the *millimeter of mercury* is that it has physical meaning; it may be read by direct observation of an open-end **manometer,** the instrument most commonly used to measure pressure in the laboratory (Fig. 4.4). *Torr,* however, is easier to say and write. We will use *torr* hereafter in this text.

Outside the laboratory, mechanical gauges are used to measure gas pressure. A typical tire gauge is probably the most familiar. Mechanical gauges show the pressure *above* atmospheric pressure, rather than the absolute pressure measured by a manometer. Even a flat tire contains air that exerts pressure. If it did not, the entire tire would collapse, not just the bottom. The pressure of the gas remaining in a flat tire is equal to atmospheric pressure. If a tire gauge shows 25 psi, that is the **gauge pressure** of the gas (air) in the tire. The absolute pressure is nearly 40 psi—the 25 psi shown by the gauge plus about 15 psi from the atmosphere.

EXAMPLE 4.1 _____

The pressure inside a steam boiler is 1127 psi. Express this pressure in atmospheres.

Now is the time to use the problem-solving strategies developed in Chapter 3. "Plan" the solution. Recall that "plan," printed in quotation marks, means to complete the first three steps in the problem-solving procedure. (1) Write down what is GIVEN. (2) Write what is WANTED. (3) Decide how to solve the problem. If the given and wanted quantities are related by a "per" expression, use dimensional analysis and write the unit PATH and the FACTOR or FACTORS that are needed. If the givens and wanted are related by an algebraic equation, use algebra by solving the EQUATION for the wanted quantity. "Plan" your solution now.

_____   _____

GIVEN: 1127 psi   WANTED: atm   PATH: psi → atm   FACTOR: 14.69 psi/atm

In this case, Equation 4.2 gives us a "per" relationship between the given and wanted quantities, 14.69 psi/atm. Therefore, the problem is solved by dimensional analysis.

Now "execute" your "plan" by completing the last three steps of the strategy: (4) Write the calculation setup, including units. (5) Calculate the answer, including units. (6) Check the answer; be sure it is reasonable in numbers and correct in units.

_____   _____

$$1127 \, \text{psi} \times \frac{1 \, \text{atm}}{14.69 \, \text{psi}} = 76.72 \, \text{atm}$$

An atmosphere is a larger unit than a psi, so there should be a smaller number of atm in a given pressure. The number is reasonable, and the unit is what was wanted.

✔ **QUICK CHECK 4.1**

Express 746 torr in atmospheres and kilopascals.

EXAMPLE 4.2 _____

Taking blood pressure requires two measurements expressed in torr, or mm Hg, which are written on one line such as 125/82. The first number indicates peak pressure on blood vessel walls when the heart's ventricles contract; the second number presents the pressure when the ventricles relax. Given that 1 atmosphere = 33.9 ft of water, express a blood pressure of 125 torr in feet of water.

"Plan" and complete the solution of the problem as in Example 4.1.

_____   _____

GIVEN: 125 mm Hg   WANTED: ft water
PATH: mm Hg → atmospheres → ft water   FACTOR: 760 torr/atm; 33.9 ft/atm

$$125 \, \text{torr} \times \frac{1 \, \text{atm}}{760 \, \text{torr}} \times \frac{33.9 \, \text{ft}}{1 \, \text{atm}} = 5.58 \, \text{ft}$$

The product of the given quantity and the numerator is about 100×35, or 3500; the denominator is about 700, so the answer should be about 3500/700, or about 5. There are fewer feet of water than torr in an atmosphere, so the larger/smaller check is OK.

But why feet of water? Well, blood has about the same density as water, so a blood pressure of 5.56 feet of water means the blood can still flow to your head when you stand up.

✔ QUICK CHECK 4.2

How do you think the blood pressure of a giraffe compares to the blood pressure of a human?

Temperature

Gas temperatures are ordinarily measured with a thermometer and expressed in Celsius degrees (°C). In solving gas problems, however, we must use absolute temperature, expressed in kelvins (K). As noted in Section 3.7, the kelvin is the SI unit of temperature. It is related to the Celsius degree by the equation

$$T_K = T_{°C} + 273* \tag{4.3}$$

To change Celsius degrees to kelvins, all you must do is add 273 to the Celsius temperature. Section 4.5 describes an experiment that shows the source of the 273.

In order to understand the behavior of gases as it depends on temperature, we need to know just what temperature measures. Experiments indicate that the temperature of a substance is a measure of the *average* kinetic energy of the particles in the sample. Kinetic energy is the energy of motion as a particle goes from one place to another. It is expressed mathematically as $\frac{1}{2}\,mv^2$, where m is the mass of the particle and v is its velocity, or speed.

Since the mass of a particle is constant, the particle speed must be higher at high temperatures and lower at low temperatures. If the speed reaches zero—if the particle stops moving—the absolute temperature becomes zero, or 0 K. This is referred to as **absolute zero,** the lowest possible temperature.

Notice the word *average* in the phrase "average kinetic energy." It suggests correctly that not all of the particles in a sample of matter have the same kinetic energy. Some have more, some have less, and as a whole they have an average energy that is

---

*Actually the equation is $T_K = T_{°C} + 273.15$, and it appears in that form in Equation 3.4. In this book the Celsius temperatures to which 273.15 is added will never have a doubtful digit smaller than units. According to the rules of significant figures, the ".15" drops out.

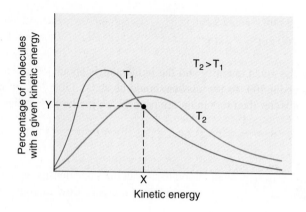

Figure 4.5

Kinetic energy distribution curve for a sample of matter at two temperatures. At temperature T_1 (blue) Y% of the molecules in the sample have a kinetic energy equal to X. The area beneath the curve represents all (100%) of the molecules in the sample. If the sample is heated to T_2 (red), the average molecular kinetic energy increases, flattening the curve and shifting it to the right. Because the size of the sample remains the same, the area beneath the blue curve remains equal to the area beneath the red curve.

proportional to absolute temperature. This is illustrated in Figure 4.5, which is a graph of the fraction, or percentage of a sample, plotted vertically, that has a given amount of kinetic energy, plotted horizontally. There is no occasion to use this graph in this chapter, but we will return to it in the chapters ahead.

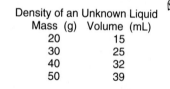

Density of an Unknown Liquid	
Mass (g)	Volume (mL)
20	15
30	25
40	32
50	39

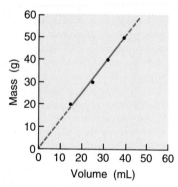

Figure 4.6

Interpretation of data from a laboratory experiment as recorded in a student's notebook.

4.4 INTERPRETATION OF EXPERIMENTAL DATA— PROPORTIONALITY

In Sections 4.5 through 4.7 we will be describing experiments that illustrate three of the so-called "gas laws" on which so much of our understanding of gases is based. Better than anything else in a beginning chemistry course, these laws show how laboratory observations become scientific knowledge. Each gas law is expressed in a mathematical conclusion. In this section we will describe an imaginary experiment and analyze the results. Our purpose is to help you understand how measurements in the laboratory produce mathematical conclusions.

Suppose a student is to determine the relationship between mass and volume of an unknown liquid. She measures the mass and volume of several samples of the liquid. Her results are tabulated at the top of Figure 4.6. A table of numbers doesn't really mean much unless we can find how the numbers are related to each other. Also, she recognizes that her measurements are subject to experimental error.

One way to find out how data are related and to "average out" experimental error is to graph the results. She plots each point in a graph of mass (vertical axis) versus volume (horizontal axis), as in the bottom part of Figure 4.6. Notice that the plotted points (dots) are almost in a straight line. This is good. A straight-line graph is the easiest to analyze. So she draws a solid line between the points that were measured and plotted. She also extends the line beyond the measured values with a dashed line.

The extension of a graph beyond measured values is called an *extrapolation*. It assumes that the data would fit into the same pattern if measurements had been made over a wider range. Sometimes extrapolations are valid, and sometimes they are not. If there is any doubt, they should be confirmed in the laboratory before they are depended upon. In this case, the extrapolation passes through the origin, 0, 0. This is just what we would expect: The mass of zero mL is zero grams.

A straight-line graph that passes through the origin is the graph of a direct proportionality. If one variable in a direct proportionality increases, so does the other. If the volume of the liquid increases, so does its mass. The relationship is quantitative. If volume is doubled, mass is doubled; two half-gallon cartons of milk are twice as heavy as one half-gallon carton. If volume is reduced by 1/3, so is mass reduced by 1/3.

A direct proportionality between two variables, such as mass (m) and volume (V), is indicated by m $\propto$ V, where the symbol $\propto$ means "is proportional to." The proportionality can be changed into an equation by inserting a multiplier called a **proportionality constant.** If we let D be the proportionality constant,

$$m = D \times V$$

Look what happens if the equation is solved for D:

$$D \equiv \frac{m}{V}$$

Does this look familiar? It is the defining equation for density that appeared in Section 3.8. The value of D can be determined from the graph of the data. The slope of the line—the "rise over the run," or change in mass per change in volume between any two points—is equal to the density. In other words, in the straight-line graph of a direct proportionality, the slope of the line is the proportionality constant. In some cases, as with density, the proportionality constant is a physical property of a substance.

In our imaginary experiment, then, the student turns to her graph to find the quantitative relationship between mass and volume of the liquid she analyzed. Selecting coordinates from any two points on the line, she calculates the slope, the difference in vertical values (mass) divided by the difference in horizontal values (volume). Choosing 0, 0 as one point makes the calculation easy; the difference to any point is simply the value of the coordinate. The line on the original larger-scale graph passes through 30.0, 38.0. From that point,

$$\frac{(38.0 - 0)\ \text{g}}{(30.0 - 0)\ \text{mL}} = \frac{38.0\ \text{g}}{30.0\ \text{mL}} = 1.27\ \text{g/mL}$$

One final observation: We said earlier that a graph "averages out" experimental errors. Assuming values in Figure 4.6 are good to three significant figures, the calculated densities from each individual measurement are 1.33 g/mL, 1.20 g/mL, 1.25 g/mL, and 1.28 g/mL. The average is 1.27 g/mL, which is the same as the slope. This is an example of why scientists use graphs to express the results of their experiments.

Density is only one of many proportionalities that emerge from laboratory experiments. Proportionalities produce the "per" expressions that we use to solve chemistry problems by dimensional analysis. They also lead to algebraic equations with which other chemistry problems are solved. We will see how this happens in the remaining sections in this chapter.

$$0.97 \text{ atm} \times \frac{304 \text{ K}}{291 \text{ K}} = 1.01 \text{ atm}$$

Arithmetically, the answer appears to be reasonable. The fraction 304/291 is somewhat more than 1, so the value of the number it multiplies should increase accordingly, which it does. But, is the fraction itself correct? We'll consider that next.

There are two ways to solve gas law problems: by algebra, as shown above, and by "reasoning." Reasoning is based on the proportionality between the variables. In this case the variables are temperatures and pressure, which are directly proportional to each other. They move in the same direction; if one goes up, the other goes up, and vice versa.

Equation 4.7:

$$P_2 = P_1 \times \frac{T_2}{T_1}$$

Now notice the nature of Equation 4.7. The initial pressure is multiplied by a ratio of temperatures—a temperature fraction. In this example there are two possible temperature ratios, 304 K/291 K and 291 K/304 K. Without thinking of the equation, which temperature ratio is correct? Here's the thought process:

Temperature increased from 291 K to 304 K.

Pressure is directly proportional to temperature; they move in the same direction.

Therefore pressure must increase.

If pressure is to increase, the initial pressure must be multiplied by a temperature fraction greater than 1.

In a fraction greater than 1, the numerator is larger than the denominator.

304 > 291, so 304/291 is the correct fraction.

If $T_2 > T_1$, then $P_2 > P_1$ Fraction > 1

The reasoning approach leads to exactly the same calculation setup as the equation, but it is reached by a different thought process.

We suggest that you solve gas law problems by algebra, and then check your setup by reasoning. The check is important. Inverting the fraction by which the beginning quantity is multiplied is the most common error made by students when solving gas law problems. Some instructors prefer that their students solve the problem by reasoning rather than by algebra. If yours is among them, we recommend that you follow the instructor's lead rather than the book. If you do it correctly, you will always get the right answer with either method.

If $T_2 < T_1$, then $P_2 < P_1$ Fraction < 1

4.6 CHARLES' LAW: VOLUME AND TEMPERATURE VARIABLE, PRESSURE AND AMOUNT CONSTANT

PG 4D Given the initial volume (or temperature) and the initial and final temperatures (or volumes) of a fixed quantity of gas at constant pressure, calculate the final volume (or temperature).

Many experiments have been performed in which the volume of a fixed quantity of gas has been measured at constant pressure but different temperatures. The results are remarkably similar to those in the last section. They even predict absolute zero at −273°C. When volume is plotted against absolute temperature, the result is a straight line passing through the origin—just like Figure 4.7c, except that the vertical axis is volume instead of pressure. This leads to the conclusion known as **Charles' Law: The volume of a fixed quantity of gas at constant pressure is directly proportional to absolute temperature.** Finally, the set of equations developed for pressure and temperature, Equations 4.4 to 4.7, is duplicated for volume and temperature, ending with

$$V_2 = V_1 \times \frac{T_2}{T_1} \qquad (4.8)$$

Either the algebraic approach or a reasoning approach may be used to solve volume-temperature problems, exactly as you solved pressure-temperature problems. Both Gay-Lussac's and Charles' Laws describe a direct proportional relationship between a physical property of a gas and its temperature.

EXAMPLE 4.4

1.67 liters of a gas, measured at 32°C, are heated to 55°C at constant pressure. What is the new volume of the gas?

As before, start by setting up a table to organize the variables.

Jacques Charles (1746–1823)

	Volume	Temperature	Pressure	Amount
Initial Value (1)	1.67 L	32°C; 305 K	Constant	Constant
Final Value (2)	V_2	55°C; 328 K	Constant	Constant

Now you are ready to set up the equation, substitute known values, and calculate the answer.

$$V_2 = V_1 \times \frac{T_2}{T_1} = 1.67 \text{ L} \times \frac{328 \text{ K}}{305 \text{ K}} = 1.80 \text{ L}$$

Now to check the answer. First, is the temperature increasing or decreasing?

Increasing.

Will the volume increase or decrease as the temperature increases?

Because volume varies directly with temperature, the volume will increase.

The multiplier is a ratio of temperatures. Should this ratio be more than 1 or less than 1, if the final volume is larger than the initial?

If $T_2 > T_1$,
then $V_2 > V_1$ Fraction > 1

If $T_2 < T_1$,
then $V_2 < V_1$ Fraction < 1

The ratio should be larger than 1, so the larger temperature is on top.

How about the number? Does 1.80 look like a reasonable answer when 1.67 is multiplied by 328/305?

Yes. The fraction is roughly 33/30, or about 1.1. Therefore the result should be somewhat more than 1.67. The calculated 1.80 fits.

✔ QUICK CHECK 4.3

If the final temperature is less than the initial temperature, how will the size of the final volume (pressure and amount constant) compare to the initial volume? How must the ratio of temperatures compare to 1?

4.7 BOYLE'S LAW: PRESSURE AND VOLUME VARIABLE, TEMPERATURE AND AMOUNT CONSTANT

> **4E**   Given the initial volume (or pressure) and initial and final pressures (or volumes) of a fixed quantity of gas at constant temperature, calculate the final volume (or pressure).

Do you recall the two friends who were examining bubbles back in Chapter 1? They had observed that bubbles became larger as they rose to the surface of a lake. They developed the hypothesis that the volume increased as the pressure decreased. They recognized this as an inverse relationship. They weren't completely satisfied, however. The observation was qualitative. Are volume and pressure related quantitatively, that is, mathematically? We left them about to conduct another experiment in which they would measure the bubble at different depths.

They performed their experiment. They collected some numbers. They then prepared a graph of their results. No straight line passes through the origin this time. The shape of their graph is shown in the margin. It shows that one value goes up while the other goes down, just as the two friends observed. Could it be that the relationship between pressure and volume is an **inverse proportionality?** An inverse *proportionality* is more precise than an inverse *relationship;* it is quantitative.

Robert Boyle, in the seventeenth century, investigated the quantitative relationship between pressure and volume of a fixed amount of gas at constant temperature. A modern laboratory experiment finds this relationship with a mercury-filled manometer such as that shown in Figure 4.8a. A graph (Fig. 4.8c) of pressure versus volume measured in the experiment suggests an inverse proportionality between the variables. Expressed mathematically

$$\text{pressure} \propto \frac{1}{\text{volume}} \quad or \quad P \propto \frac{1}{V} \tag{4.9}$$

Introducing a proportionality constant gives

$$P = k_b \frac{1}{V} \tag{4.10}$$

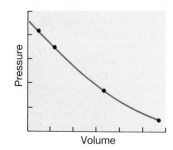
Pressure vs. Volume

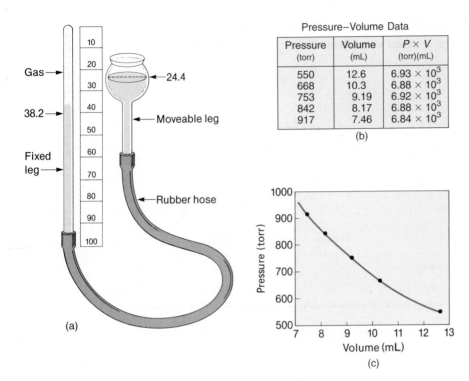

Pressure–Volume Data

Pressure (torr)	Volume (mL)	$P \times V$ (torr)(mL)
550	12.6	6.93×10^3
668	10.3	6.88×10^3
753	9.19	6.92×10^3
842	8.17	6.88×10^3
917	7.46	6.84×10^3

(b)

(a)

(c)

Figure 4.8

Boyle's Law. These are data from a student's experiment to find the relationship between pressure and volume of a fixed amount of gas at constant temperature. By raising or lowering the movable leg of the apparatus (a), the volume of the trapped gas was found at different pressures. The first two columns of the table (b) are the student's data. A graph of pressure vs. volume (c) indicates that the variables are inversely proportional to each other. This is confirmed by the constant product of pressure × volume (within experimental error, ±0.7% from the average value), shown in the third column of the table.

Robert Boyle (1627–1691)

Multiplying both sides of the equation by V yields

$$PV = k_b \tag{4.11}$$

Within experimental error, PV is indeed a constant (Fig. 4.8b).

Boyle's Law, which this experiment illustrates, states that **for a fixed quantity of gas at constant temperature, pressure is inversely proportional to volume.** Equation 4.11 is the usual mathematical statement of Boyle's Law. Since the product of P and V is constant, when either factor increases the other must decrease, and vice versa. This is what is meant by an inverse proportionality. Notice the differences between an inverse proportionality and the direct proportionalities of the two previous sections.

From Equation 4.11 we see that $P_1V_1 = k_b = P_2V_2$, or

$$P_1V_1 = P_2V_2 \tag{4.12}$$

where subscripts 1 and 2 refer to first and second measurements of pressure and volume at constant temperature. Solving for V_2, we have

$$V_2 = V_1 \times \frac{P_1}{P_2} \qquad (4.13)$$

EXAMPLE 4.5

A certain gas sample occupies 5.18 liters at 776 torr. Find the volume of the gas sample if the pressure is changed to 827 torr. Temperature remains constant.

Begin by setting up the table of initial and final values.

	Volume	Temperature	Pressure	Amount
Initial value (1)	5.18 L	Constant	776 torr	Constant
Final value (2)	V_2	Constant	827 torr	Constant

Now write the equation, insert values, and solve the problem.

$$V_2 = V_1 \times \frac{P_1}{P_2} = 5.18\ \text{L} \times \frac{776\ \text{torr}}{827\ \text{torr}} = 4.86\ \text{L}$$

Now to check the answer. First, is the pressure increasing or decreasing?

Increasing from 776 torr to 827 torr.

Will the volume increase or decrease as the pressure increases? Careful!

Boyle's Law works on marshmallows too! Place some marshmallows in a flask (top) then use a vacuum pump to lower pressure in the flask (bottom). The air trapped inside the marshmallows expands as the pressure is decreased. Unfortunately, the process is reversible. Open the flask to get to the giant marshmallows, and they shrink back to normal (top) size.

If $P_2 > P_1$,
then $V_2 < V_1$ Ratio < 1

If $P_2 < P_1$,
then $V_2 > V_1$ Ratio > 1

Because volume varies inversely with pressure, the volume will decrease.

In an inverse proportion the variables go in opposite ways; one goes up and the other goes down.

The multiplier is a ratio of pressures. Should this ratio be more than or less than 1, if the final pressure is larger than the initial?

The ratio should be smaller than 1, so the lower pressure is on top.

Is 4.86 a reasonable result when 5.18 is multiplied by 776/827?

Yes. The fraction is somewhat smaller than 1, and 4.86 is somewhat smaller than 5.18.

Did you know that you've been using Boyle's Law all your life? You breathe 10–12 times a minute, taking in about 500 cm^3 of air with each breath. How do you think the air gets into and out of your lungs? When you inhale, you lower your diaphragm and/or expand your rib cage. Either motion increases the volume of the chest cavity that surrounds your lungs. As volume increases, the pressure of air inside your lungs decreases, and atmospheric pressure forces air into your lungs. When you exhale, your diaphragm moves up and/or your rib cage contracts. Either motion decreases the volume of your chest cavity and increases the gas pressure inside your lungs. The gases are exhaled against the lower atmospheric pressure outside.

Let's do another Boyle's Law problem, to make sure you've got the inverse relationship right.

EXAMPLE 4.6

1.61 liters of a gas at 0.912 atmosphere are compressed to 0.207 liter, at constant temperature. What is the new pressure of the gas?

As before, start by setting up a table to organize the variables.

	Volume	Temperature	Pressure	Amount
Initial Value (1)	1.61 L	Constant	0.912 atm	Constant
Final Value (2)	0.207 L	Constant	P_2	Constant

Now the equation and the calculation.

$$P_2 = P_1 \times \frac{V_1}{V_2} = 0.912 \text{ atm} \times \frac{1.61 \text{ L}}{0.207 \text{ L}} = 7.09 \text{ atm}$$

Let's check. The fraction is more than 1. Is this as it should be? And how about 7.09? Is this a reasonable number?

If the volume is decreasing, pressure should increase because pressure and volume are inversely related. The fraction should be larger than 1. Roughly, $1.61/0.207 \approx 16/2 = 8$. Multiplying, $0.9 \times 8 = 7.2$. Close enough to 7.09.

✔ QUICK CHECK 4.4

Assume constant temperature and amount of gas. (a) If the final pressure is less than the initial pressure, how will the final volume compare to the initial volume? (b) If the final volume is greater than the initial volume, how will the final pressure compare to the initial volume?

Let's take one last look at the bubble experiment in Chapter 1. At the beginning of this section we showed the shape of the graph the data produced. It was not a straight line through the origin. Scientists often look for a different way to express their data mathemati-

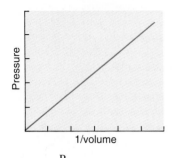

Slope $= \dfrac{P}{1/V} = PV = k$

Compare to Equation 4.11

cally so they plot as a direct proportionality. If the bubble chasers had plotted pressure versus the *reciprocal* of volume, they would have had their straight line. They would have had a graph of Equation 4.10, $P = k_b \times 1/V$, which is the equation of a direct proportionality. Compare that equation with $m = D \times V$ from Section 4.4, which gave the straight-line graph in Figure 4.6. They both have the form $y = mx$. Back in beginning algebra you learned that $y = mx$ is the equation for a straight line through the origin and that $m = $ slope. Remember?

 ## 4.8 THE COMBINED GAS LAWS

> **PG 4F** For a fixed quantity of a confined gas, given the initial volume, pressure, and temperature and the final values of any two variables, calculate the final value of the third. (Initial or final conditions may be standard temperature and pressure, STP.)

We have seen that $P \propto T$ (Section 4.5) and $V \propto T$ (Section 4.6). Whenever the same quantity (T) is proportional to each of two other quantities (P and V), it is proportional to the product of those quantities: $T \propto PV$. This becomes an equation when a proportionality constant is introduced: $k_c T = PV$. Rearranging, we get

$$\frac{PV}{T} = k_c \tag{4.14}$$

Again, using subscripts 1 and 2 for initial and final values of all variables, we obtain

$$\frac{P_1 V_1}{T_1} = k_c = \frac{P_2 V_2}{T_2} \quad or \quad \frac{P_1 V_1}{T_1} = \frac{P_2 V_2}{T_2} \tag{4.15}$$

If five of the six variables in Equation 4.15 are known, the value of the remaining variable can be calculated. If the unknown is the final volume, V_2, Equation 4.15 can be solved for that unknown:

$$V_2 = V_1 \times \frac{P_1}{P_2} \times \frac{T_2}{T_1} \tag{4.16}$$

EXAMPLE 4.7

A cylinder in an automobile engine has a volume of 352 cm³. This engine takes in air at 20°C and at 0.945 atm pressure. The compression stroke squeezes this gas until the temperature is 95°C and the pressure is 4.95 atm. What is the final volume in this cylinder?

Set up a table as before.

	Volume	Temperature	Pressure	Amount
Initial Value (1)	352 cm³	20°C; 293 K	0.945 atm	Constant
Final Value (2)	V_2	95°C; 368 K	4.95 atm	Constant

Write the equation, substitute values, and solve.

_____   _____

$$V_2 = V_1 \times \frac{P_1}{P_2} \times \frac{T_2}{T_1} = 352 \text{ cm}^3 \times \frac{0.945 \text{ atm}}{4.95 \text{ atm}} \times \frac{368 \text{ K}}{293 \text{ K}} = 84.4 \text{ cm}^3$$

Once again, to check our problem setup we'll use a reasoning approach, but this time we'll use it twice. First, how does increasing pressure affect volume?

_____   _____

Pressure is inversely proportional to volume, so as pressure increases, volume decreases.

Will the ratio of pressures be greater than or less than 1?

_____   _____

Less than 1.

Now the temperature effect: Will the volume increase or decrease as the temperature increases? What, then, for the temperature ratio: larger than 1 or smaller than 1?

_____   _____

Volume varies directly with temperature, so volume will increase. The temperature ratio must be larger than 1.

The setup looks all right. How about the numerical answer?

_____   _____

We start with about 350. The first fraction is close to 1/5, and one-fifth of 350 is 70. Then $368/293 \approx 360/300 = 1.2$, and $1.2 \times 70 = 84$. The calculated 84.4 looks good.

By the way, the ratio of volumes, $352 \text{ cm}^3/84.4 \text{ cm}^3$, is called the compression ratio of this engine.

The volume of a fixed quantity of gas depends on its temperature and pressure. It is therefore not possible to state the amount of gas in volume units without also specifying the temperature and pressure. These are often given as 0°C (273 K) and 1 atmosphere (760 torr), which are known as **standard temperature and pressure (STP).** Many gas law problems require changing volume to or from STP. The procedure is just like that in Example 4.7.

EXAMPLE 4.8 _____

What would be the volume at STP of 3.62 liters of nitrogen gas, measured at 649 torr and 16°C?

_____   _____

	Volume	Temperature	Pressure	Amount
Initial Value (1)	3.62 L	16°C; 289 K	649 torr	Constant
Final Value (2)	V_2	0°C; 273 K	760 torr	Constant

$$V_2 = V_1 \times \frac{P_1}{P_2} \times \frac{T_2}{T_1} = 3.62\ L \times \frac{649\ \cancel{torr}}{760\ \cancel{torr}} \times \frac{273\ \cancel{K}}{289\ \cancel{K}} = 2.92\ L$$

CHAPTER 4 OUTLINE

4.1   Properties of gases
  A.   May be compressed
  B.   Fill containers
  C.   Low density
  D.   Can be mixed
  E.   Uniform pressure in all directions
4.2   Kinetic molecular theory and ideal gas model
  A.   Theory describes model that can be "seen"
    1.   Consists of molecules that move in straight lines
    2.   Molecules collide with each other and with walls without losing energy
    3.   Molecules independent; negligible attractions
    4.   Molecules widely spaced
    5.   Volume of molecules small compared to space occupied
  B.   Use model to explain properties in 4.1
4.3   Gas measurements
  A.   Four measurements: quantity, volume (V), pressure (P), temperature (T)
    1.   Quantity constant in this chapter
    2.   Temperature measured with thermometer; volume is volume of container
  B.   Pressure
    1.   Pressure ≡ force per unit area, F/A
    2.   Units: pascal (Pa) or kilopascal (kPa); millimeter of mercury (mm Hg—Hg is mercury); torr (= to mm Hg); atmosphere
    3.   Measured with barometer (atmospheric pressure) or manometer
    4.   Standard atmosphere: 1 atm ≡ 760 mm Hg by definition
    5.   Values compared in Equation 4.2, page 91; note torr and mm Hg are the same thing
    6.   Gauge pressure is pressure of gas *above* pressure of atmosphere
  C.   Temperature
    1.   Must use absolute temperature—kelvins—in gas problems; $T_K = T_{°C} + 273$; add 273 to °C to get kelvins
    2.   Temperature a measure of average kinetic energy of gas molecules; note *average*; molecules move at different speeds
    3.   Absolute zero is when molecules stop moving
4.4   Experimental data—proportionality
  A.   Analyze data by graphing results; draw solid line within measured ranges, dashed line outside; dashed line is extrapolation, which may or may not be valid
  B.   Straight line through origin is direct proportionality; Significance: when one value increases, other increases; when one value decreases, other value decreases; they move together
  C.   ∝ is proportionality symbol
  D.   Proportionality changed to equation with a proportionality constant
4.5   Gay-Lussac's Law
  A.   Pressure proportional to absolute temperature
  B. $P_2 = P_1 \times (T_2/T_1)$; subscripts 1 and 2 are first and second values
  C.   IMPORTANT: TEMPERATURE MUST BE IN KELVINS
4.6   Charles' Law
  A.   Volume proportional to absolute temperature
  B. $V_2 = V_1 \times (T_2/T_1)$
4.7   Boyle's Law
  A.   Pressure *inversely* proportional to volume; significance: when one value increases, other value decreases; they move in opposite directions
  B. $V_2 = V_1 \times (P_1/P_2)$

4.8 Combined gas laws
 A. $V_2 = V_1 \times (P_1/P_2) \times (T_2/T_1)$
 B. Volume of fixed quantity of gas depends on temperature and pressure;

volume doesn't tell amount unless T and P are specified
 C. Gas volumes often related to standard temperature and pressure (STP), 0°C or 273 K and 1 atm or 760 torr

CHAPTER 4 IN REVIEW

4.1 Properties of Gases

4.2 The Kinetic Theory of Gases and the Ideal Gas Model
4A Explain or predict physical phenomena relating to gases in terms of the ideal gas model.

4.3 Gas Measurements
4B Given a gas pressure in atmospheres, torr, millimeters of mercury, centimeters of mercury, inches of mercury, pascals, kilopascals, or pounds per square inch, express that pressure in each of the other units.

4.4 Interpretation of Experimental Data—Proportionality

4.5 Gay-Lussac's Law: Pressure and Temperature Variable, Volume and Amount Constant
4C Given the initial pressure (or temperature) and initial and final temperatures (or pressures) of a fixed quantity of gas at constant volume, calculate the final pressure (or temperature).

4.6 Charles' Law: Volume and Temperature Variable, Pressure and Amount Constant
4D Given the initial volume (or temperature) and initial and final temperatures (or volumes) of a fixed quantity of gas at constant pressure, calculate the final volume (or temperature).

4.7 Boyle's Law: Pressure and Volume Variable, Temperature and Amount Constant
4E Given the initial volume (or pressure) and initial and final pressures (or volumes) of a fixed quantity of gas at constant temperature, calculate the final volume (or pressure).

4.8 The Combined Gas Laws
4F For a fixed quantity of a confined gas, given the initial volume, pressure, and temperature and the final values of any two variables, calculate the final value of the third. (Initial or final conditions may be standard temperature and pressure, STP.)

CHAPTER 4 KEY WORDS AND MATCHING SETS

Set 1

_____ Pressure is inversely proportional to volume.

_____ 1 atm less than gas pressure.

_____ Matter is made up of particles in constant motion.

_____ Force per unit area.

_____ Measure of average kinetic energy.

_____ Equal to 760 torr.

_____ Pressure directly proportional to absolute temperature.

_____ Instrument for measuring gas pressure.

_____ Unit of pressure as measured by above instrument.

_____ Volume directly proportional to absolute temperature.

1. mm Hg
2. Kinetic molecular theory
3. Gauge pressure
4. Temperature
5. Manometer
6. Boyle's Law
7. Pressure
8. Charles' Law
9. 1 atm
10. Gay-Lussac's Law

Set 2

_____ Quantities increase together and decrease together.

_____ Absolute temperature unit.

_____ Temperature at which there is no particle motion.

_____ Gas consists of widely separated molecules in constant motion that collide with each other with no loss of energy.

_____ 0°C, 1 atm.

_____ Instrument for measuring atmospheric pressure.

_____ SI pressure unit.

_____ PV/ T = k.

_____ Same as 1 mm Hg.

_____ One goes up, other goes down.

1. Absolute zero
2. Kelvin
3. Combined gas law
4. Barometer
5. STP
6. Inverse proportionality
7. Torr
8. Pascal
9. Ideal gas model
10. Direct proportionality

STUDY HINTS AND PITFALLS TO AVOID

There are two major pitfalls in this chapter. The first is failing to change Celsius temperatures to kelvins. If you make that change when analyzing the problem, even before thinking about how to solve it, you will not forget that critical step.

The second pitfall is inverting one or both correction ratios in combined gas law problems. Making a table of initial (1) and final (2) conditions will help you avoid that mistake. Set up the problem fully, showing all the values substituted in the algebraic equation. Then check the ratios to determine if they should be larger than 1 or smaller than 1. In using the ratios, remember that pressure and volume make up the only inverse proportion you will encounter. The others are direct proportions.

CHAPTER 4 QUESTIONS AND PROBLEMS

Recall from Chapter 1 that end-of-chapter Questions and Problems are in pairs separated by space. Generally the questions in each pair are matched; that is, they involve the same or related ideas, comparable calculations, or other similarities. Questions whose numbers are printed in color are answered in the back of the book.

SECTION 4.2
1)   According to the kinetic molecular theory of an ideal gas,
a)   What do we assume about the size of gas particles?
b)   Why doesn't friction slow down the gas particles and cause them to stop moving?
c)   Why can gases be compressed to a smaller volume easily?
2)   According to the kinetic molecular theory of an ideal gas,
a)   What do we assume about the speed of gas particles?
b)   What causes pressure on the walls of a container of gas?

c)   Why do gases quickly diffuse throughout a room?

_____

3)   If cigarette smoke is observed under a microscope, the solid particles are seen to be moving constantly in a random fashion known as _Brownian motion_. Explain this effect.

4)* Does gravity ever have an effect on gas properties (a) in a small room; (b) in the earth's atmosphere?

Questions 5 through 12: Explain how the physical phenomenon described is related to one or more of the features of the ideal gas model.

5)   Very small dust particles, seen in a beam of light passing through a darkened room, appear to be moving about erratically.

6)   A dust particle falls through the air more slowly than a tiny piece of paper.

_____

7) Gas bubbles always rise through a liquid.

8) Any container, regardless of size, will be completely filled by one gram of helium.

---

9) Pressure is exerted on the top of a tank holding a gas as well as on the sides and bottom.

10) At a given pressure, a fixed amount of a liquid occupies a much larger volume when it is changed to a gas.

---

11) Even though an automobile tire is "filled" with air, you can add more air without increasing the size of the tire significantly.

12) Properties of gases are less "ideal"—the substance behaves in ways not typical of a gas—at very high pressures when the particles are close to each other.

SECTION 4.3

13) Which of the following physical properties of a gas sample can be measured directly: density, volume, particle speed, pressure?

14) Which of the following physical properties of a gas sample can be measured directly: particle size, color, temperature, mass?

---

15) Identify the instruments that are used to measure the properties of gases that can be measured directly.

16)* Suggest a method by which a gas sample might be weighed.

---

17) What is the difference between a barometer and a manometer? Explain how each works.

18) Two basic principles underlie the function of both barometers and manometers. What are they?

Questions 19 and 20: Complete the table by converting the given pressure to each of the other pressures:

19)

atm	5.52			
psi		60.9		
in. Hg			21.4	
cm Hg				47.6
mm Hg				
torr				
Pa				
kPa				

20)

atm				
psi				
in. Hg				
cm Hg				
mm Hg	819			
torr		410		
Pa			2.64×10^4	
kPa				358

---

21) A mercury manometer is used to measure pressure in the container illustrated. Calculate the pressure exerted by the gas if atmospheric pressure is 731 torr.

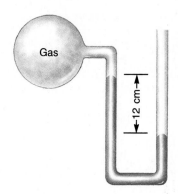

22) Find the pressure of the gas in the top of the mercury-filled J-tube shown in the sketch if atmospheric pressure is 693 mm Hg.

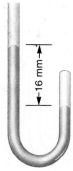

23)* Special kinds of glass reaction flasks can withstand a high inside vacuum without collapsing. Do you think such a flask could tolerate a 1.00 atm pressure inside and a vacuum outside? Explain.

24)* Suppose we know that a flask is able to withstand an inside vacuum of 10^{-2} torr. What are its chances of withstanding the best vacuum we can get in a modern laboratory, about 10^{-10} torr? Explain.

25) Precisely what does temperature measure? Do all the particles in a gas move at the same speed? Does each individual particle move at a constant speed? If not, why not? What is constant at a given temperature?

26) What does *absolute zero* mean? What physical condition is presumed to exist at absolute zero? Why is it theoretically impossible to have a negative value for temperature on the Kelvin scale?

27) Can you ice skate on a river if the temperature is 6°C? What about 260 K?

28) What kelvin temperature corresponds with 6°C? What is the Celsius temperature at 229 K?

29) Would you enjoy a picnic if the temperature were 295 K?

30) Among the following kelvin temperatures, select the one that is best suited for swimming and the one that is best for skiing:

22   81   98   166   263   301   415   550

SECTION 4.5

31) Why does the pressure of air in automobile tires increase when the car is driven at high speeds for a considerable period of time?

32) Explain the relationship between temperature and pressure in terms of the kinetic molecular theory.

33) A gas is confined in a rigid, constant-volume container. What happens inside the container that causes the gas to increase its pressure when it is heated?

34) Use the kinetic molecular theory and the model of an ideal gas to explain Gay-Lussac's Law.

35) How does a graph of pressure vs. temperature show the value of absolute zero?

36)* Why do you suppose all lines in Figure 4.7 converge at zero kelvins rather than 0°C?

37) Why must Kelvin temperatures be used in making calculations based on Gay-Lussac's Law? Support your answer by relating it to Figure 4.7.

38)* The caption in Figure 4.7 indicates that the two blue lines are for different quantities of gas than the black line. Do you suppose the sample that produced the blue line above the black line had a larger quantity of gas or a smaller quantity than the sample that produced the black line? Explain.

39) A gas collected in a laboratory experiment exerts a pressure of 912 torr at 68°C. What will the pressure be after cooling to 27°C, assuming no change in volume?

40) Air in a steel cylinder is heated from 14°C to 56°C. If the initial pressure was 3.55 atm, what is the final pressure?

41)* A gas storage tank is designed to hold a fixed volume and quantity of gas at 2.74 atm and 25°C. To prevent excessive pressure due to overheating, the tank is fitted with a relief valve that opens at 3.00 atm. To what temperature must the gas rise in order to open the valve?

42)* A gas in a heavy metal cylinder shows a gauge pressure of 217 kPa while sitting on a loading dock on a cold Canadian night when the temperature is −29°C. What will the pressure gauge show when the tank is brought inside and its contents warm up to 19°C?

SECTION 4.6

43) A variable-volume container holds 26.0 L of nitrogen at 39°C. If the pressure remains constant, what will the volume be if the temperature drops to 14°C?

44) 22.1 L of a gas are cooled from 74°C to 26°C. What will the volume be at the lower temperature?

45) A spring-loaded closure maintains constant pressure on a gas system that holds a fixed quantity of gas, but a bellows allows the volume to adjust for temperature changes. From a starting point of 16.0 L and 41°C, to what volume will the system change if the temperature rises to 75°C?

46) A large industrial gas-storage tank delivers fuel to a boiler at constant pressure. The pressure is maintained by a piston that rises or falls to adjust gas volume. The volume of the tank is 5.57 m^3 on a Monday morning before the heat is turned on and the temperature is 14°C. What will the volume be when the gas is heated to the temperature at which it is used, 35°C?

47) A gas is held in a container whose volume can be varied by adjusting the temperature. The volume is 38.5 L at 26°C. To what must the temperature be changed to expand the gas to 42.1 L?

48) A 5.83-liter sample of stack gases is collected at 96°C. To what temperature must the gas be cooled to reduce its volume to 4.71 L so it can be analyzed in the laboratory?

49)* A party balloon—the shiny metallic-looking type that doesn't stretch—is "filled" with helium in a store. The hostess of a party buys the balloon and puts it in her car in a parking lot on a cold winter day as she proceeds with other shopping. When she returns to her car, she is depressed to find the balloon appears to have "lost" some of the gas and is partially deflated. She complains to the store manager, who calmly assures her the balloon will be quite satisfactory when she takes it into her house and the gas warms to room temperature. She takes it home, and sure enough, it fills out completely, and even a bit more firmly than when she bought it. Explain what happened (a) in the car and (b) when the balloon was taken inside. What can you say about the relative temperatures of the store and the house?

50* Referring to Question 49, the balloon is filled to a volume of 6.12×10^3 cm^3 when the temperature is 19°C. What volume does the balloon occupy when it is in the cold car? What is the volume of the balloon when it has warmed to room temperature, as described in Question 49? In terms of the ideal gas law, how can you explain the balloon's having filled out, "even a bit more firmly than when she bought it"?

SECTION 4.7

51) If you squeeze the bulb of a dropping pipet (eye dropper) when the tip is below the surface of a liquid, bubbles appear. Why does this occur? On releasing the bulb, liquid flows into the pipet. Why? Explain both observations in terms of Boyle's Law.

52) A mountain climber was surprised to find his bag of potato chips popped open when he reached the peak. Why did it break?

---

53) A party balloon (see Question 49) contains helium at 753 torr and occupies a volume of 4.47 L. Calculate the volume the balloon will occupy if it is submerged in water to a depth that increases the total pressure to 977 torr.

54) Calculate the volume that will be occupied at 687 torr if a gas fills 8.18 L at 212 torr.

---

55) A gas in a variable-volume tank exerts a pressure of 804 cm of mercury when the volume is 39.2 liters. To what must the volume be changed to make the same gas exert a pressure of 664 cm of mercury?

56) The relief valve on a cylinder is set to open when the enclosed gas reaches a pressure of 8.23 atmospheres. A piston in the cylinder can be operated to increase or decrease the volume of the gas in the cylinder. If the volume is 2.01 liters at 3.55 atmospheres, calculate the gas volume at which the valve will open.

---

57)* An air compressor delivers air to a 59.8-liter storage tank at 9.50 atmospheres. A 3.16-liter portable compressed air tank is filled by connecting it to the larger tank and opening the valve between them. The air distributes itself between the tanks. If the compressor delivers no additional air to the system during the process, what will be the pressure in the portable tank when they are disconnected? in the storage tank? (Disregard the air initially in the smaller tank.)

58) If a light bulb contained air, the tungsten filament in it would oxidize and break; if the bulb held a vacuum, the filament would vaporize. Bulbs therefore contain argon, an unreactive gas, to prevent both events. 130 mL argon at 1.0 atm are introduced to an evacuated 150-mL bulb, which is then sealed. What is the pressure of the argon in the bulb?

---

59)* The volume of the chamber of a bicycle pump is 0.26 L. The volume of a bicycle tire, including the hose between the pump and the tire, is 2.40 L. If the air in both the tire and the pump chamber begins at 712 torr, what will be the pressure in the tire after a single stroke of the pump?

60)* A 7.15-liter gas chamber contains nitrogen at 1.34 atm. It is connected through a closed valve to another chamber whose volume is 2.06 L. The smaller chamber is evacuated to negligible pressure. What pressure will be reached in both chambers if the valve between them is opened and the nitrogen occupies the total volume?

SECTION 4.8

61) Why is the volume of a gas not a suitable measure of the amount of gas?

62) Under what conditions might volume be used to compare the amount of gas in two different containers?

---

63) What is the purpose of identifying "standard temperature and pressure" (STP)?

64) What values are associated with standard temperature and pressure? Are they values that are practical for laboratory work? Explain.

---

65) The gas in a 0.768-liter cylinder of a diesel engine has a pressure of 1.01 atm at 31°C. What will be the volume of the gas when the piston compresses it to 46.9 atm at a temperature of 498°C?

66) A gas occupies 44.7 L at 65°C at a pressure of 0.853 atm. Calculate the volume if the gas is heated to 82°C while the pressure is increased to 2.27 atm.

---

67) 9.75 L of a gas at 76°C and 4.01 atm are expanded to 18.2 L and cooled to 34°C. What is the final pressure exerted by the gas?

68) A gas has a pressure of 0.454 atm in a 7.04-L container at 25°C. What will the pressure be if the volume is reduced to 6.34 L and the temperature dropped to 6°C?

---

69)   If one cubic foot—28.4 L—of air at typical room conditions of 20°C and 755 torr were adjusted to STP, what would be the final volume?

70)   An industrial process yields 173 L of nitric oxide at 30°C and 889 torr. What would this volume become at STP?

———————

71)   The STP volume of a sample of neon is 9.59 L. What would be the volume of the sample at 0.231 atm and 41°C?

72)   Calculate the volume of a gas at 778 torr and 34°C if its STP volume is 248 mL.

GENERAL QUESTIONS

73)   Distinguish precisely and in scientific terms the differences between items in each of the following groups:

   Kinetic theory of gases, kinetic molecular theory
   Pascal, mm Hg, torr, atmosphere, psi
   Barometer, manometer
   Pressure, gauge pressure
   Gay-Lussac's Law, Charles' Law, Boyle's Law
   Celsius and Kelvin temperature scales
   0°C, 0 K
   Temperature, pressure, standard temperature and pressure (STP)

74)   Determine whether each statement that follows is true or false:
a)   Gas molecules change speed when they collide with each other.
b)   A gas molecule may gain or lose kinetic energy in colliding with another molecule.
c)   The total kinetic energy of two molecules is the same before and after they collide with each other.
d)   Gas molecules are strongly attracted to each other.
e)   Gauge pressure is always greater than absolute pressure except in a vacuum.

f)   For a fixed amount of gas at constant temperature, if volume increases, pressure decreases.
g)   For a fixed amount of gas at constant pressure, if temperature increases, volume decreases.
h)   For a fixed amount of gas at constant volume, if temperature increases, pressure increases.
i)   At a given temperature, the number of degrees Celsius is larger than the number of kelvins.
j)   Both temperature and pressure ratios are larger than 1 when calculating the gas volume as conditions change from STP to 15°C and 0.984 atm.

75)   Answer each question below from the following values of pressure: (a) 1.06 atm; (b) 743 mm Hg; (c) 20.0 psi; (d) 620 torr; (e) 76.0 cm Hg

   Identify the highest pressure.   _____

   Identify the lowest pressure.   _____

   Identify the middle pressure in the group.   _____

   Which one is standard pressure?   _____

76)   What do we mean by an "ideal" gas? Do common real gases behave as an ideal gas would behave? Under what conditions do you suppose a gas would be most apt to depart from ideal behavior?

77)   If gas to your home is metered, would you reduce your gas bill by warming or by cooling the gas before it passes through the meter? (This procedure is *not* recommended!)

78)   If you press your finger into a ball of clay and then remove it, the depression from your finger remains. If you press your finger into an inflated balloon and then remove it, the depression disappears. Explain the difference.

MATCHING SET ANSWERS

Set 1: 6–3–2–7–4–9–10–5–1–8

Set 2: 10–2–1–9–5–4–8–3–7–6

QUICK CHECK ANSWERS

4.1 $746 \text{ torr} \times \dfrac{1 \text{ atm}}{760.0 \text{ torr}} = 0.982 \text{ atm}$;

$746 \text{ torr} \times \dfrac{101.3 \text{ kPa}}{760.0 \text{ torr}} = 99.4 \text{ kPa}$

4.2 The blood pressure of a giraffe is *much* higher than the blood pressure of a human being. Although giraffes have an extremely high systolic (first number) blood pressure compared to humans, giraffes do not seem prone to the effects, such as stroke, of high blood pressure in humans. Medical researchers are working to determine why this is so.

4.3 If $T_2 < T_1$, $V_2 < V_1$, because $V \propto T$. The ratio of temperatures must be <1.

4.4 Remember that Boyle's Law is an *inverse* proportion: $V \propto 1/P$. Because $P_2 < P_1$, $V_2 > V_1$. If $V_2 > V_1$, $P_2 < P_1$.

5 Atomic Theory and the Periodic Table: The Beginning

This early variation of a gas discharge tube played an important role in learning about the structure of atoms. It is also a forerunner of neon signs, the devices by which we can measure the masses of atoms, and even the picture tube of a television set.

John Dalton

LEARN IT NOW Suggestions for where outline entries might be made are no longer given. However, there is a full chapter outline at the end of the chapter. Again we recommend that you complete your own outline without looking at ours.

This is the last chapter that is outlined for you. By now you should have developed skill in reducing many pages of text into a few pages of brief summaries. These are great time-savers when you prepare for a test. But that is only the reward. Your purpose for making an outline is to learn. The very act of condensing and writing in your own words what you have read produces the learning you seek. We encourage you to continue this practice.

As early as 400 BC Greek philosophers had proposed that matter consisted of tiny, indivisible particles, which they called **atoms.** In 1808 John Dalton, an English chemist and schoolteacher, revived the concept. We now know that the atom consists of even smaller particles. Today some of the most sophisticated research methods ever developed continue to seek an understanding of how atoms are put together. But it all started with the vision of John Dalton.

In this chapter we begin the study of the atom. We learn about three of the parts of the atom. We also see that different combinations of these parts account for the different elements. The arrangement of elements into groups that have similar properties is introduced.

5.1 DALTON'S ATOMIC THEORY

PG 5A Identify the main features of Dalton's atomic theory.

Dalton knew about the Law of Definite Composition: The percentage by mass of the elements in a compound is always the same (Section 2.2). He was also familiar with the Law of Conservation of Mass: In a chemical change, mass is conserved; it is neither created nor destroyed (Section 2.5). **Dalton's atomic theory** explained these observations. The main features of his theory are (Fig. 5.1):

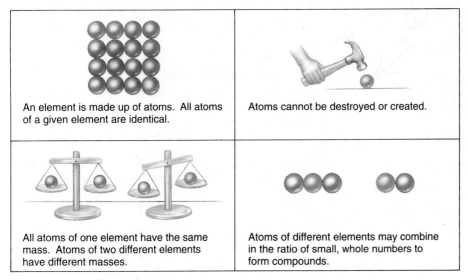

An element is made up of atoms. All atoms of a given element are identical.

Atoms cannot be destroyed or created.

All atoms of one element have the same mass. Atoms of two different elements have different masses.

Atoms of different elements may combine in the ratio of small, whole numbers to form compounds.

Figure 5.1

Atoms according to Dalton's atomic theory.

1) Each element is made up of tiny, individual particles called atoms.
2) Atoms are indivisible; they cannot be created or destroyed.
3) All atoms of each element are identical in every respect.
4) Atoms of one element are different from atoms of any other element.
5) Atoms of one element may combine with atoms of another element, usually in the ratio of small, whole numbers, to form chemical compounds.

Dalton's theory accounts for chemical reactions in this way: Before the reaction, the reactants contain a certain number of atoms of the different elements in those reactants. As the reaction proceeds, the atoms are rearranged to form the products. The atoms are neither created nor destroyed, but simply arranged differently. The starting *arrangement* is destroyed (reactants are destroyed in a chemical change) and a new arrangement is formed (new substances are formed).

As with many new ideas, Dalton's theory was not immediately accepted. However, it led to a prediction that *must* be true if the theory is correct. This is now known as the **Law of Multiple Proportions.** It states that when two elements combine to form more than one compound, the different weights of one element that combine with the same weight of the other element are in a simple ratio of whole numbers (Fig. 5.2). This is like threading one, two, or three identical nuts onto the same bolt. The mass of the bolt is constant. The mass of two nuts is twice the mass of one; of three nuts, three times the mass of one. The masses of nuts are in a simple ratio of whole numbers, 1:2:3.

The multiple proportion prediction can be confirmed by experiment. Using a theory to predict something unknown, and having the prediction confirmed, is convincing proof that the theory is correct. With supporting evidence such as this, Dalton's atomic theory was accepted.

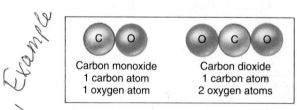

Carbon monoxide
1 carbon atom
1 oxygen atom

Carbon dioxide
1 carbon atom
2 oxygen atoms

Figure 5.2

Explanation of the Law of Multiple Proportions. Carbon and oxygen combine to form two compounds, carbon monoxide, CO, and carbon dioxide, CO_2. A CO molecule consists of one carbon atom and one oxygen atom. A CO_2 molecule has one carbon atom and two oxygen atoms. Considering both molecules, for a fixed number of carbon atoms—one in each molecule—the ratio of oxygen atoms is 1 to 2, or $\frac{1}{2}$. If all oxygen atoms have the same mass, M, the mass ratio is also $\frac{1}{2}$:

$$\frac{M \text{ grams (1 atom)}}{2 M \text{ grams (2 atoms)}} = \frac{1}{2}$$

Michael Faraday

Table 5.1

Subatomic Particles

Subatomic Particle	Symbol	Fundamental Charge	Mass		Location	Discovered
			Grams	amu* $^{12}C = 12.00000$		
Electron	e^-	−1	9.107×10^{-28}	$0.000549 \approx 0$	Outside nucleus	1897 Thomson
Proton	p or p^+	+1	1.672×10^{-24}	$1.00728 \approx 1$	Inside nucleus	1919 Rutherford
Neutron	n or n^0	0	1.675×10^{-24}	$1.00867 \approx 1$	Inside nucleus	1932 Chadwick

*An *amu* is a very small unit of mass used for atomic-sized particles. It is defined in Section 5.5.

5.2 SUBATOMIC PARTICLES

PG 5B Identify the features of Dalton's atomic theory that are no longer considered valid, and explain why.

5C Identify the three major subatomic particles by charge and approximate atomic mass, expressed in atomic mass units.

Despite the general acceptance of the atomic theory, it was soon challenged. As early as the 1820's, laboratory experiments suggested that the atom contains even smaller parts, or **subatomic particles.** The brilliant works of Michael Faraday and William Crookes, among others, led to the discovery of the **electron;** but it was not until 1897 that J. J. Thomson described some of its important properties. The electrical charge on an electron has been assigned a value of −1. The mass of an electron is extremely small, 9.107×10^{-28} grams (g).

The second subatomic particle, the **proton,** was isolated and identified in 1919 by Ernest Rutherford. Its mass is about 1837 times greater than the mass of an electron. The proton carries a +1 charge, equal in size but opposite in sign to the negative charge of the electron. The third particle, the **neutron,** was discovered by James Chadwick in 1932. As its name suggests, it is electrically neutral. The mass of a neutron is slightly more than the mass of a proton. Masses of atoms and parts of atoms are often expressed in **atomic mass units** (Section 5.5).

Today we know that atoms of all elements are made up of different combinations of some of more than 100 subatomic particles. Only the 3 described here are important in this course. The properties of the electron, proton, and neutron are summarized in Table 5.1.

✔ **QUICK CHECK 5.1**

Identify the true statements, and rewrite the false statements to make them true.

a) An atom is made up of electrons, protons, and neutrons.
b) The mass of an electron is less than the mass of a proton.

FLASHBACK Some of the electrical properties of matter (Section 2.3) were known in Dalton's day, but there was no explanation for them. Dalton's theory did not account for them. Faraday and Crookes opened the door that led to understanding electricity in terms of parts of atoms.

LEARN IT NOW If you noticed Table 5.1 in your chapter preview, you probably recognized it as a summary of this section and therefore postponed outline notes until now. Recall the open tabular summaries you used in Chapter 2. Table 5.1 can be used in the same way. You might wish to limit your table to what is needed for PG 5C, but leave space to add another column.

J.J. Thomson

LEARN IT NOW   Recall that in Section 3.4 the caption of an illustration became the text by which uncertainty in measurement was introduced. Figures 5.3 and 5.4 are similar illustrations. Their captions are the only place you will find what you need to satisfy PG 5D.

LEARN IT NOW   If you set up an open table from Table 5.1, you probably did not include the column showing where atomic particles are located. Now this location becomes important, it is part of the interpretation of Rutherford's experiment (PG 5D). You may wish to add this column.

c)   The mass of a proton is about 1 g.
d)   Electrons, protons, and neutrons are electrically charged.

5.3 THE NUCLEAR ATOM

PG 5D   Describe and/or interpret the Rutherford scattering experiment and the nuclear model of the atom.

In 1911 Ernest Rutherford and his students performed a series of experiments that are described in Figures 5.3 and 5.4. The results of these experiments led to the following conclusions:

1)   Every atom contains an extremely small, extremely dense **nucleus.**
2)   All of the positive charge and nearly all of the mass of an atom are concentrated in the nucleus.
3)   The nucleus is surrounded by a much larger volume of nearly empty space that makes up the rest of the atom.
4)   The space outside the nucleus is very thinly populated by electrons, the total charge of which exactly balances the positive charge of the nucleus.

This description of the atom is called the **nuclear model of the atom.**

The "emptiness" of the atom is difficult to visualize. If the nucleus of the atom were the size of a pea, the distance to its closest neighbor would be about 0.6 mile, or 1 kilometer (km). Between them would be almost nothing—only a small number of electrons of negligible size and mass. If it were possible to eliminate all of this nearly empty space and pack nothing but nuclei into a sphere the size of a period on this page, that sphere could, for some elements, weigh as much as a million tons!

When protons and neutrons were later discovered, it was concluded that these relatively massive particles make up the nucleus of the atom. But electrons were already known in 1911, and it was natural to wonder what they did in the vast open space they occupied. The most widely held opinion was that they traveled in circular orbits around the nucleus, much as planets move in orbits around the sun. The atom would then have the character of a miniature solar system. This is called the **planetary model of the atom.** In Chapter 10 we will examine this model more closely—and find out why it is wrong.

Figure 5.3

Rutherford scattering experiment. A narrow beam of alpha particles (helium atoms stripped of their electrons) from a radioactive source was directed at a very thin gold foil. Most of the particles passed right through the foil, striking a fluorescent screen and causing it to glow. Some particles were deflected through moderate angles (red lines). The larger deflections were surprises, but the 0.001% of the total that were reflected at acute angles (blue line) were totally unexpected. Similar results were observed using other foils.

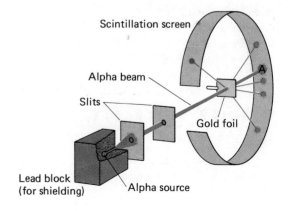

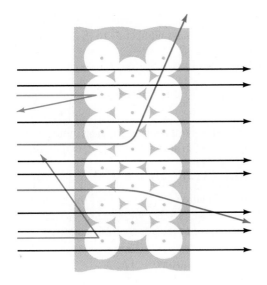

Figure 5.4

Interpretation of the Rutherford scattering experiment. The atom is pictured as consisting mostly of "open" space. At the center is a tiny and extremely dense nucleus that contains all of the atom's positive charge and nearly all of its mass. The electrons are thinly distributed throughout the open space. Most of the positively charged alpha particles (black) pass through the open space undeflected, not coming near any gold nuclei. The few that pass fairly close to a nucleus (red paths) are repelled by electrostatic force and thereby deflected. The very few particles that are on a "collision course" with gold nuclei (blue paths) are repelled backward at acute angles. Calculations based on the results of the experiment indicated that the diameter of the open-space portion of the atom is from 10,000 to 100,000 times greater than the diameter of the nucleus.

✔ QUICK CHECK 5.2

Identify the true statements, and rewrite the false statements to make them true.

a) Atoms are like small, hard spheres.
b) An atom is electrically neutral.
c) An atom consists mostly of empty space.

5.4 ISOTOPES

> **PG 5E** Explain what isotopes of an element are and how they differ from each other.
> **5F** For an isotope of any element whose chemical symbol is known, given one of the following, state the other two: (a) nuclear symbol, (b) number of protons and neutrons in the nucleus, (c) atomic number and mass number.

More than 100 years after Dalton's atomic theory was first suggested, another of its features was shown to be incorrect. All atoms of an element are not identical. Some atoms have more mass than other atoms of the same element.

We now know that every atom of a particular element has the same number of protons. This number is called the **atomic number** of the element. It is represented by the symbol **Z**. Atoms are electrically neutral, so the number of electrons must be the same as the number of protons. It follows that the total contribution to the mass of an atom from protons and electrons is the same for every atom of the element. That leaves neutrons. We conclude that the mass differences must be caused by different numbers of neutrons. **Atoms of the same element that have different masses—different numbers of neutrons—are called isotopes.**

An isotope is identified by its **mass number, A, the total number of protons and neutrons in the nucleus:**

$$\text{mass number} = \text{number of protons} + \text{number of neutrons}$$
$$A \quad = \quad Z \quad + \text{number of neutrons}$$

(5.1)

The name of an isotope is its elemental name followed by its mass number. Thus, an oxygen atom that has 8 protons and 8 neutrons has a mass number of 16 (8 + 8), and its name is "oxygen sixteen." It is written "oxygen-16."

An isotope is represented by a **nuclear symbol** that has the form

$$\frac{\text{number of protons}+\text{number of neutrons}}{\text{number of protons}}\text{Sy} \quad \text{or} \quad \frac{\text{mass number}}{\text{atomic number}}\text{Sy} \quad \text{or} \quad {}_{Z}^{A}\text{Sy}$$

Sy is the chemical symbol of the element. The symbol and the mass number are actually all that is needed to identify an isotope, so the atomic number, Z, is sometimes omitted. The symbol for oxygen-16 is ${}_{8}^{16}\text{O}$ or ${}^{16}\text{O}$.

Two natural isotopes of carbon are ${}_{6}^{12}\text{C}$ and ${}_{6}^{13}\text{C}$, carbon-12 and carbon-13. From the name and symbol of the isotopes and from Equation 5.1 you can find the number of neutrons in each nucleus. In carbon-12, if you subtract the atomic number (protons) from the mass number (protons + neutrons), you get the number of neutrons:

$$
\begin{array}{lr}
\text{mass number} = \text{protons} + \text{neutrons} & 12 \\
\text{atomic number} = \underline{\text{protons}} & = \underline{-6} \\
\text{neutrons} = & 6
\end{array}
$$

In carbon-13 there are 7 neutrons: $13 - 6 = 7$.

You can find the mass number and nuclear symbol of an isotope from the number of protons and neutrons. A nucleus with 12 protons and 14 neutrons has the atomic number 12, the same as the number of protons. From Equation 5.1, the mass number is $12 + 14 = 26$. The symbol of the element may be found by searching the atomic number column in the list of elements inside the front cover for 12. It is more easily found from the periodic table on the facing page of the front cover. The number at the top of each box is the atomic number. The elemental symbol corresponding to Z = 12 is Mg for magnesium. The isotope is therefore magnesium-26, and its nuclear symbol is ${}_{12}^{26}\text{Mg}$.

FLASHBACK The chemical symbols of the elements (Section 2.2) are shown in the alphabetical list of elements and the periodic table, both of which are inside the front cover of this book.

EXAMPLE 5.1

Fill in all of the blanks in the following table. Use the table of elements inside the front cover and Equation 5.1 for needed information. The number at the top of each box in the periodic table is the atomic number of the element whose symbol is in the middle of the box.

Name of Element	Elemental Symbol	Atomic No., Z	Number of Protons	Number of Neutrons	Mass Number, A	Nuclear Symbol	Name of Isotope
barium						${}_{56}^{138}\text{Ba}$	
oxygen				10			
		82			206		
							zinc-66

Name of Element	Elemental Symbol	Atomic No., Z	Number of Protons	Number of Neutrons	Mass Number, A	Nuclear Symbol	Name of Isotope
barium	Ba	56	56	82	138	$^{138}_{56}\text{Ba}$	barium-138
oxygen	O	8	8	10	18	$^{18}_{8}\text{O}$	oxygen-18
lead	Pb	82	82	124	206	$^{206}_{82}\text{Pb}$	lead-206
zinc	Zn	30	30	36	66	$^{66}_{30}\text{Zn}$	zinc-66

The atomic number, Z, and the number of protons are the same in each case, by definition. Also by definition, the mass number, A, is equal to the sum of the number of protons and the number of neutrons. For Z = 82 you must search for 82 in the atomic number column in the table of elements to identify the element as lead.

✔ QUICK CHECK 5.3

Identify the true statements, and rewrite the false statements to make them true.

a) The atomic number of an element is the number of protons in its nucleus.
b) All atoms of a specific element have the same number of protons.
c) The difference between isotopes of an element is a difference in the number of neutrons in the nucleus.
d) The mass number of an atom is always equal to or larger than the atomic number.

5.5 ATOMIC MASS

> PG 5G Define and use the atomic mass unit, amu.
> 5H Given the relative abundances of the natural isotopes of an element and the atomic mass of each isotope, calculate the atomic mass of the element.

The mass of an atom is very small—much too small to be measured on a balance. Nevertheless, early chemists did find ways to isolate samples of elements that contained the same number of atoms. These samples were weighed and compared. The ratio of the masses of equal numbers of atoms of different elements is the same as the ratio of the masses of individual atoms. From this, a scale of relative atomic masses was developed. They didn't know about isotopes at that time, so the idea of "atomic weight" applied to all of the natural isotopes of an element.

Today we recognize that a sample of a pure element contains atoms that have different masses. By worldwide agreement, these masses are expressed in **atomic mass units (amu),** which are **exactly 1/12 of the mass of a carbon-12 atom.** Both protons and neutrons have atomic masses very close to 1 amu (see Table 5.1). We use the amu to define the **atomic mass of an element: the average mass of the atoms of an element compared to an atom of carbon-12 at exactly 12 amu.**

If a is the mass of one atom of A, b is the mass of one atom of B, and N is any number of atoms, then $N \times a =$ mass of N atoms of A and $N \times b =$ mass of N atoms of B.

$$\frac{N \times a}{N \times b} = \frac{N}{N} \times \frac{a}{b} = \frac{a}{b}$$

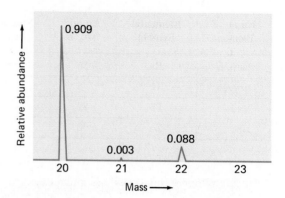

Figure 5.5

Mass spectrum of neon (1+ ions only). Neon contains three isotopes, of which neon-20 is by far the most abundant (90.9%). The mass of that isotope, to five decimal places, is 19.99244 amu.

To find the atomic mass of an element, you must know the atomic mass of each isotope and the fraction of each isotope in a sample. (See Fig. 5.5.) Fortunately, that fraction is constant for all elements as they occur in nature. Table 5.2 gives the percent abundance of the natural isotopes of some common elements. The following example shows you how to calculate the atomic mass of an element from these data.

Table 5.2

Percent Abundance of Some Natural Isotopes

Symbol	Mass (amu)	Percent	Symbol	Mass (amu)	Percent
$^{1}_{1}H$	1.007825	99.985	$^{19}_{9}F$	18.99840	100
$^{2}_{1}H$	2.0140	0.015	$^{32}_{16}S$	31.97207	95.0
$^{3}_{2}He$	3.01603	0.00013	$^{33}_{16}S$	32.97146	0.76
$^{4}_{2}He$	4.00260	100	$^{34}_{16}S$	33.96786	4.22
$^{12}_{6}C$	12.00000	98.89	$^{36}_{16}S$	35.96709	0.014
$^{13}_{6}C$	13.00335	1.11	$^{35}_{17}Cl$	34.96885	75.53
$^{14}_{7}N$	14.00307	99.63	$^{37}_{17}Cl$	36.96590	24.47
$^{15}_{7}N$	15.00011	0.37	$^{39}_{19}K$	38.96371	93.1
$^{16}_{8}O$	15.99491	99.759	$^{40}_{19}K$	39.974	0.00118
$^{17}_{8}O$	16.99474	0.037	$^{41}_{19}K$	40.96184	6.88
$^{18}_{8}O$	17.99477	0.204			

EXAMPLE 5.2

The natural distribution of isotopes of boron is 19.6% $^{10}_{5}B$ at a mass of 10.01294 amu, and 80.4% $^{11}_{5}B$ at 11.00931 amu. Calculate the atomic mass of boron.

SOLUTION

The "average" boron atom consists of 19.6% of an atom with mass 10.01294 amu. Therefore, it contributes $0.196 \times 10.01294 = 1.96$ amu to the mass of the average boron atom. A similar calculation for the other isotope is added.

$$0.196 \times 10.01294 \text{ amu} = 1.96 \text{ amu}$$
$$\underline{0.804 \times 11.00931 \text{ amu} = 8.85 \text{ amu}}$$
$$1.000 \text{ average atom} \qquad 10.81 \text{ amu}$$

The presently accepted value of the atomic weight of boron is 10.81, which matches our calculated value to four significant figures. (Note that we gained a significant figure in the addition process, because we ended with two digits to the left of the decimal point in the sum.)

We recommend that you identify the contribution of each isotope to an atomic mass, as in Example 5.2. When checking a result, it is always desirable to use a different sequence to avoid repeating a mechanical error. You can do this here if your calculator automatically completes multiplications and divisions before additions and subtractions. Exact sequences may vary on different calculators, but Table 5.3 shows one typical sequence for Example 5.2. Notice that the calculator offers no help in telling you the column to which the result should be rounded off.

Now you try an atomic mass calculation.

Table 5.3

Typical Calculator Sequence

AOS logic	
Press	**Display**
.196	0.196
×	0.196
10.01294	10.01294
+	1.96253624
.804	0.804
×	0.804
11.00931	11.00931
=	10.81402148

EXAMPLE 5.3

Calculate the atomic mass of potassium (symbol K), using data from Table 5.2.

$$
\begin{aligned}
0.931 &\times 38.96371\ \text{amu} = 36.3 &\text{amu}\\
0.0000118 &\times 39.974\ \ \text{amu} = 0.000472\ \text{amu}\\
0.0688 &\times 40.96184\ \text{amu} = \underline{2.82} &\text{amu}\\
& \qquad\qquad\qquad\qquad 39.1 &\text{amu}
\end{aligned}
$$

The accepted value of the atomic mass of potassium is 39.0983 amu.

5.6 THE PERIODIC TABLE

PG 5I Distinguish between groups and periods in a periodic table and identify them by number.
5J Given the atomic number of an element, use a periodic table to find the symbol and atomic mass of that element, and identify the period and group in which it is found.

During the period of research on the atom, even before any subatomic particles were identified, other chemists searched for an order among the elements. In 1869 it was found independently by two men at the same time. Dmitri Mendeleev and Lothar Meyer observed that when elements are arranged according to their atomic masses, certain properties repeat at regular intervals.

Mendeleev and Meyer arranged the elements in tables so that the elements with similar properties were in the same column or row. These were the first **periodic tables** of the elements. The arrangements were not perfect. In order for all elements to fall into the proper groups, it was necessary to switch a few of the elements. This interrupted the orderly increase in atomic masses. Of the two reasons for this, one was anticipated at that time: There were errors in atomic weights as they were known in 1869. The second was more important. About 50 years later it was found that the correct ordering property is the atomic number, Z, rather than the atomic mass.

Dmitri Mendeleev

Table 5.4

The Predicted and Observed Properties of Germanium

Property	Predicted by Mendeleev	Observed
Atomic weight	72	72.60
Density of metal	5.5 g/cm^3	5.36 g/cm^3
Color of metal	Dark gray	Gray
Formula of oxide	GeO_2	GeO_2
Density of oxide	4.7 g/cm^3	4.703 g/cm^3
Formula of chloride	$GeCl_4$	$GeCl_4$
Density of chloride	1.9 g/cm^3	1.887 g/cm^3
Boiling point of chloride	Below 100°C	86°C
Formula of ethyl compound	$Ge(C_2H_5)_4$	$Ge(C_2H_5)_4$
Boiling point of ethyl compound	160°C	160°C
Density of ethyl compound	0.96 g/cm^3	Slightly less than 1.0 g/cm^3

We have seen how Dalton used his atomic theory to predict the Law of Multiple Proportions, and thus gain acceptance for his theory. Mendeleev did the same with the periodic table. He noticed that there were blank spaces in the table. He reasoned that the blank spaces belonged to elements that had not yet been discovered. By averaging the properties of the elements above and below or on each side of the blanks, he predicted the properties of the unknown elements. Germanium is one of the elements about which he made these predictions. Table 5.4 summarizes the predicted properties and their presently accepted values.

The amazing accuracy of Mendeleev's predictions showed that the periodic table "made sense," but nobody knew why. That came later; and for us, it comes later, too. The reason for the strange shape of the table is explained in Chapter 10.

Figure 5.6 is a modern periodic table. It also appears inside the front cover of this book. You will find yourself referring to the periodic table throughout your study of chemistry. This is why a partial periodic table is printed on the opaque shields provided for working examples.

The number at the top of each box in our periodic tables is the atomic number of the element (Fig. 5.7). The chemical symbol is in the middle, and the atomic mass, rounded to four significant figures, is at the bottom. The boxes are arranged in horizontal rows called **periods.** Periods are numbered from top to bottom, but the numbers are not usually printed. Periods vary in length. The first period has 2 elements; the second and third have 8 elements each; and the fourth and fifth have 18. Period 6 has 32 elements, including atomic numbers 58 to 71, which are printed separately at the bottom to keep the table from becoming too wide. Period 7 also has 32 elements, but elements beyond $Z = 109$ are not yet known.

Elements with similar properties are placed in vertical columns called **groups** or **chemical families.** Groups are identified by two rows of numbers across the top of the table. The top row is the group numbers that are commonly used in the United States.* European chemists use the same numbers, but a different arrangement of A's and B's. The International Union of Pure and Applied Chemistry (IUPAC) has recently approved a compromise that simply numbers the columns in order from left to right. This is the second row of numbers at the top of Figure 5.6.

*Roman numerals are frequently used, as IIIA instead of 3A.

Periods

Figure 5.6

1	2A / 2	← Current American usage → New notation →	3B / 3	4B / 4	5B / 5	6B / 6	7B / 7	← 8B →			1B / 11	2B / 12	3A / 13	4A / 14	5A / 15	6A / 16	7A / 17	0 / 18

1 — H (1, 1.008); H (1, 1.008), He (2, 4.003)

2 — Li (3, 6.941), Be (4, 9.012); B (5, 10.81), C (6, 12.01), N (7, 14.01), O (8, 16.00), F (9, 19.00), Ne (10, 20.18)

3 — Na (11, 22.99), Mg (12, 24.31); Al (13, 26.98), Si (14, 28.09), P (15, 30.97), S (16, 32.07), Cl (17, 35.45), Ar (18, 39.95)

4 — K (19, 39.10), Ca (20, 40.08), Sc (21, 44.96), Ti (22, 47.88), V (23, 50.94), Cr (24, 52.00), Mn (25, 54.94), Fe (26, 55.85), Co (27, 58.93), Ni (28, 58.69), Cu (29, 63.55), Zn (30, 65.39), Ga (31, 69.72), Ge (32, 72.61), As (33, 74.92), Se (34, 78.96), Br (35, 79.90), Kr (36, 83.80)

5 — Rb (37, 85.47), Sr (38, 87.62), Y (39, 88.91), Zr (40, 91.22), Nb (41, 92.91), Mo (42, 95.94), Tc (43, (98)), Ru (44, 101.1), Rh (45, 102.9), Pd (46, 106.4), Ag (47, 107.9), Cd (48, 112.4), In (49, 114.8), Sn (50, 118.7), Sb (51, 121.8), Te (52, 127.6), I (53, 126.9), Xe (54, 131.3)

6 — Cs (55, 132.9), Ba (56, 137.3), *La (57, 138.9), Hf (72, 178.5), Ta (73, 180.9), W (74, 183.9), Re (75, 186.2), Os (76, 190.2), Ir (77, 192.2), Pt (78, 195.1), Au (79, 197.0), Hg (80, 200.6), Tl (81, 204.4), Pb (82, 207.2), Bi (83, 209.0), Po (84, (209)), At (85, (210)), Rn (86, (222))

7 — Fr (87, (223)), Ra (88, 226.0), †Ac (89, 227.0), Rf (104, (261)), Ha (105, (262)), φ (106, (263)), Ns (107, (262)), Hs (108, (265)), Mt (109, (266))

	58	59	60	61	62	63	64	65	66	67	68	69	70	71
*Lanthanide series	Ce 140.1	Pr 140.9	Nd 144.2	Pm (145)	Sm 150.4	Eu 152.0	Gd 157.3	Tb 158.9	Dy 162.5	Ho 164.9	Er 167.3	Tm 168.9	Yb 173.0	Lu 175.0
	90	91	92	93	94	95	96	97	98	99	100	101	102	103
†Actinide series	Th 232.0	Pa 231.0	U 238.0	Np 237.0	Pu (244)	Am (243)	Cm (247)	Bk (247)	Cf (251)	Es (252)	Fm (257)	Md (258)	No (259)	Lr (260)

All atomic weights have been rounded off to four significant figures.

φ The International Union for Pure and Applied Chemistry has not adopted official names or symbols for these elements. All atomic weights have been rounded off to four significant figures.

Periodic table of the elements.

We will use both sets of numbers, leaving it to your instructor to recommend which you should use. When we have occasion to refer to a group number, we will give the American number first, followed by the IUPAC number in parentheses. Thus, the column headed by carbon, $Z = 6$, is Group 4A (14).

Two other regions in the periodic table separate the elements into special classifications. Elements in the "A" groups (1, 2, 13 to 18) are called **representative elements.** Similarly, elements in the "B" groups (3 to 12) are known as **transition elements,** or **transition metals.** The stair-step line that begins between atomic numbers 4 and 5 in Period 2 and ends between 84 and 85 in Period 6 separates the **metals** on the left from the **nonmetals** on the right. Chemical reasons for these classifications appear in Chapter 10.

The location of an element in the periodic table is given by its period and group numbers.

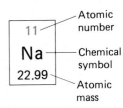

Figure 5.7

Sample box from the periodic table. The box is for sodium.

EXAMPLE 5.4

List the atomic number, chemical symbol, and atomic mass of the third period element in Group 6A (16).

$Z = 16$; symbol, S; atomic mass, 32.07 amu

In Group 6A (16), the third column from the right side of the table, you find $Z = 8$ in period 2, and $Z = 16$ in period 3. The element is sulfur.

EXAMPLE 5.5 _____

List the group, period, symbol, and atomic mass for lead (Z = 82).

—————   —————

Group 4A (14); period 6; symbol, Pb; atomic mass, 207.2 amu

The first element in Group 4A (14), Z = 6, is in the second period. Counting down from there, Z = 82 in Period 6.

5.7 ELEMENTAL SYMBOLS AND THE PERIODIC TABLE

PG 5K   Given the name or the symbol of an element in Figure 5.8, write the other.

The periodic table contains a large amount of information, far more than the three numbers in each box indicate. Its usefulness will unfold gradually in the next four chapters, climaxing in Chapter 10 when you learn how to use the table to describe the behavior of electrons in atoms. This behavior is the basis for our understanding of chemical properties, chemical changes, and the energy that accompanies chemical changes.

In this section we will use the periodic table to learn the names and symbols of 35 elements. Learning symbols and names is much easier if you learn the location of the elements in the periodic table at the same time. Here's how to do it:

Part (a) of Figure 5.8, the table at the top, gives the name, symbol, and atomic number of the elements whose names and symbols are to be learned.* Part (b) is a partial periodic table showing the atomic numbers and symbols of the same elements. Their names are listed in alphabetical order beneath the periodic table.

Study Part (a) briefly. Try to learn the symbol that goes with each element, but don't spend more than a few minutes doing this. Then cover Part (a) and look at the periodic table in Part (b). Run through the symbols mentally and see how many elements you can name. If you can't name one, glance through the alphabetical list below and see if it jogs your memory. If you still can't get it, note the atomic number and check Part (a) for the elemental name. Do this a few times until you become fairly quick in naming most of the elements from the symbols.

Next, reverse the process with Part (a) still covered. Look at the alphabetical list beneath the periodic table. For each name, mentally "write"—in other words, _think_— the symbol. Glance up to the periodic table and find the element. Again use Part (a) as a temporary help only if necessary. Repeat the procedure several times, taking the elements in random order. Move in both directions, from name to symbol in the periodic table and from symbol to name.

When you feel reasonably sure of yourself, try the following example. Do not refer to Figure 5.8. Instead, use only the more complete periodic table that is on one side of the tear-out sheet provided for this purpose.

———

*Your instructor may require you to learn different names and symbols, or perhaps more or fewer. If so, follow your instructor's directions. If they include elements not among our 35, we recommend that you add them to all three places in Figure 5.8 where names and symbols of other elements appear.

Common Elements

Atomic Number	Symbol	Element	Atomic Number	Symbol	Element	Atomic Number	Symbol	Element
1	H	Hydrogen	13	Al	Aluminum	28	Ni	Nickel
2	He	Helium	14	Si	Silicon	29	Cu	Copper
3	Li	Lithium	15	P	Phosphorus	30	Zn	Zinc
4	Be	Beryllium	16	S	Sulfur	35	Br	Bromine
5	B	Boron	17	Cl	Chlorine	36	Kr	Krypton
6	C	Carbon	18	Ar	Argon	47	Ag	Silver
7	N	Nitrogen	19	K	Potassium	50	Sn	Tin
8	O	Oxygen	20	Ca	Calcium	53	I	Iodine
9	F	Fluorine	24	Cr	Chromium	56	Ba	Barium
10	Ne	Neon	25	Mn	Manganese	80	Hg	Mercury
11	Na	Sodium	26	Fe	Iron	82	Pb	Lead
12	Mg	Magnesium	27	Co	Cobalt			

(a)

1A 1	2A 2	3B 3	4B 4	5B 5	6B 6	7B 7	8	8B 9	10	1B 11	2B 12	3A 13	4A 14	5A 15	6A 16	7A 17	0 18
1 H																1 H	2 He
3 Li	4 Be											5 B	6 C	7 N	8 O	9 F	10 Ne
11 Na	12 Mg											13 Al	14 Si	15 P	16 S	17 Cl	18 Ar
19 K	20 Ca				24 Cr	25 Mn	26 Fe	27 Co	28 Ni	29 Cu	30 Zn					35 Br	36 Kr
										47 Ag			50 Sn			53 I	
	56 Ba										80 Hg		82 Pb				

(b)

Figure 5.8

Partial periodic table showing the symbols and locations of the more common elements. The symbols above and the list that follows identify the elements you should be able to recognize or write, referring only to a complete periodic table. Associating the names and symbols with the table makes learning them much easier. The elemental names are:

aluminum	bromine	chromium	iodine	magnesium	nitrogen	silver
argon	calcium	copper	iron	manganese	oxygen	sodium
barium	carbon	fluorine	krypton	mercury	phosphorus	sulfur
beryllium	chlorine	helium	lead	neon	potassium	tin
boron	cobalt	hydrogen	lithium	nickel	silicon	zinc

EXAMPLE 5.6 _____

For each element below, write the name; for each name, write the symbol.

N_____   P _____   carbon _____   potassium_____

F _____   Cl_____   aluminum_____   zinc _____

I _____   Fe_____   copper _____   bromine _____

_____

N, nitrogen	P, phosphorus	carbon, C	potassium, K
F, fluorine	Cl, chlorine	aluminum, Al	zinc, Zn
I, iodine	Fe, iron	copper, Cu	bromine, Br

Look closely at your symbols for aluminum, zinc, copper, and bromine. If you wrote AL, ZN, CU, or BR, the symbol is wrong. The letters are right, but the symbol is not. Whenever writing a symbol that has two letters, the first letter is always capitalized, but THE SECOND LETTER IS ALWAYS WRITTEN IN LOWERCASE, or as a small letter. You can enjoy a long and happy life with a pile of Co in your house, but the day that you take a few stiff whiffs of CO will be your last!*

*Co is the metal, cobalt. CO is the deadly gas, carbon monoxide, that is present in the exhaust of an automobile.

CHAPTER 5 OUTLINE

Chapter 5 Atomic Theory and the Periodic Table: The Beginning

5.1 Dalton's atomic theory
   A. Main ideas
      1. Element made up of atoms
      2. Atoms indivisible, can't be made or destroyed
      3. Atoms of given element identical
      4. Atoms of different elements are different
      5. Atoms of different elements combine to form compounds
   B. Multiple proportions: two or more compounds from same elements; for fixed mass of one element, masses of second element are in ratio of whole numbers

5.2 Subatomic particles
   A. Atom made up of parts; can be destroyed
   B. Summary

Particle	Charge	Mass (amu)	Location

5.3 The nuclear atom
   A. Rutherford experiment: particles fired into thin foil; most pass through; some deflected; some reflected back
   B. Nuclear model of atom
      1. Small, dense nucleus contains all + charge and most of the mass of an atom

2. Huge space around nucleus holds electrons

C. Planetary model says electrons travel around nucleus in orbits; model is wrong

5.4 Isotopes

A. Atomic number ≡ number of protons in nucleus. Symbol: Z.

B. Isotopes ≡ atoms of same element that have different masses because of different numbers of neutrons

C. Mass number ≡ number of protons + number of neutrons; symbol: A. $A = Z$ + number of neutrons

D. Name of isotope is elemental name followed by mass number; example, carbon-12

E. Symbol of isotope

$$\frac{\text{mass number}}{\text{atomic number}}Sy \quad \text{or} \quad {}_{Z}^{A}Sy \quad \text{or} \quad {}^{A}Sy$$

5.5 Atomic mass

A. 1 atomic mass unit (amu) ≡ exactly $1/12$ mass of one atom of ${}^{12}C$

B. Atomic mass of element ≡ average mass of atoms of element compared to mass of one atom of ${}^{12}C$ at 12 amu

5.6 Periodic Table

A. Each box in periodic table contains atomic number, Z, of element; elemental symbol; atomic mass of element.

B. Horizontal rows are periods, numbered top to bottom

C. Columns have elements with similar properties; called groups or chemical families; two numbering systems; we use _____

D. Elements in (A groups) (Groups 1, 2, 13 to 18—the taller columns) are representative elements; elements in (B groups) (Groups 3 to 12—the valley) are transition elements or transition metals

E. Elements left of stair-step line are metals; right, nonmetals

5.7 Elemental symbols

A. For each element in Fig. 5.8, page 127, learn the name, symbol, and location in the periodic table

B. For our class:
1. Add:
2. Omit:

CHAPTER 5 IN REVIEW

5.1 Dalton's Atomic Theory

5A Identify the main features of Dalton's atomic theory

5.2 Subatomic Particles

5B Identify the features of Dalton's atomic theory that are no longer considered valid, and explain why.

5C Identify the three major subatomic particles by charge and approximate atomic mass, expressed in atomic mass units.

5.3 The Nuclear Atom

5D Describe and/or interpret the Rutherford scattering experiment and the nuclear model of the atom.

5.4 Isotopes

5E Explain what isotopes of an element are and how they differ from each other.

5F For an isotope of any element whose chemical symbol is known, given one of the following, state the other two: (a) nuclear symbol, (b) number of protons and neutrons in the nucleus, (c) atomic number and mass number.

5.5 Atomic Mass

5G Define and use the atomic mass unit, amu.

5H Given the relative abundances of the natural isotopes of an element and the atomic mass of each isotope, calculate the atomic mass of the element.

5.6 The Periodic Table

5I Distinguish between groups and periods in a periodic table and identify them by number.

5J Given the atomic number of an element, use a periodic table to find the symbol and atomic mass of that element, and identify the period and group in which it is found.

5.7 Elemental Symbols and the Periodic Table

5K Given the name or the symbol of an element in Figure 5.8, write the other.

MATCHING SETS

Set 1

___2___ Atoms of one element that have different masses.

___7___ Has a mass of 1 amu and no charge.

___6___ Average mass of atoms of an element compared to mass of one atom of carbon-12.

___4___ Vertical column in periodic table.

___3___ Element in Groups 1B–8B (3–12).

___8___ Protons, neutrons, and electrons.

___1___ Smallest particle of an element.

___5___ Atom having a very dense center and electrons occupying large space outside.

1. Atom
2. Isotopes
3. Transition element (metal)
4. Group
5. Nuclear model of atom
6. Atomic mass
7. Neutron
8. Subatomic particles

Set 2

___3___ Elements in a column in periodic table that have similar chemical properties.

___7___ 1/12 of mass of a carbon-12 atom.

___1___ Number of protons in all atoms of a given element.

___5___ Elements made up of tiny, identical, indestructible particles unlike particles of any other element.

___2___ Atom having dense nucleus with orbiting electrons.

___6___ Arrangement of elemental symbols in order of recurring properties.

___4___ Negatively charged subatomic particle.

1. Atomic number
2. Planetary model of atom
3. Chemical family
4. Electron
5. Dalton's atomic theory
6. Periodic table
7. Atomic mass unit

Set 3

___4___ Dense center of an atom.

___7___ In two compounds of same two elements, masses of one element that combine with fixed mass of other are in ratio of integers.

___1___ Element from A group of periodic table (Groups 1, 2, 13–18).

___2___ Number of protons plus neutrons in nucleus of an atom.

___8___ Elements left of stair-step line in periodic table.

___5___ Positively charged subatomic particle.

___3___ Horizontal row in periodic table.

___6___ Elements right of stair-step line in periodic table.

1. Representative element
2. Mass number
3. Period
4. Nucleus
5. Proton
6. Nonmetals
7. Law of Multiple Proportions
8. Metals

STUDY HINTS AND PITFALLS TO AVOID

The most common student error involving the symbols of elements is writing both letters in a two-letter elemental symbol as capitals. The first letter is always a capital letter. If a second letter is present, it is always written in lower-case. Chemistry is a very precise foreign language, and correctly written symbols are a large part of the vocabulary of that language.

CHAPTER 5 QUESTIONS AND PROBLEMS

SECTION 5.1

1) Summarize the main ideas in Dalton's atomic theory.

2) Hydrogen and oxygen form two different compounds, the familiar bleach, hydrogen peroxide, and the even more familiar water. How does Dalton's atomic theory explain this fact?

3) The mass of ash left by a burned log is less than the mass of the original log. How does Dalton's atomic theory account for this apparent contradiction of the Law of Conservation of Mass?

4) When alcohol burns, the number of carbon atoms produced as carbon dioxide is the same as the number of carbon atoms in the original fuel. Explain this in terms of Dalton's atomic theory.

5) Show that the Law of Definite Composition is explained by Dalton's atomic theory.

6) Question 2 states that hydrogen and oxygen form two compounds—water, H_2O, and hydrogen peroxide, H_2O_2. Explain how these compounds can be used as an example of the Law of Multiple Proportions.

7)* If hydrochloric acid, HCl, reacts with potassium hydroxide, KOH, water and a compound sometimes used as a substitute for ordinary table salt are formed. From this description, (a) write the formulas of the products, one of which is water; (b) guess the name of the new compound; and (c) describe what the atoms of the four elements did as they rearranged themselves into the products.

8) If milk of magnesia were filtered, a compound called magnesium hydroxide would be collected. It has three elements, magnesium, hydrogen, and oxygen. If magnesium hydroxide is heated, water vapor is driven off and solid magnesium oxide is left behind. From this description, tell where the atoms of magnesium, hydrogen, and oxygen in the original magnesium hydroxide may be found at the end of the reaction.

SECTION 5.2

9) What parts of Dalton's atomic theory are no longer regarded as being correct? Why?

10) What is an electron?

11) Describe a neutron in terms of its charge, relative mass, and position in the atom.

12) Describe a proton in terms of its charge, relative mass, and position in the atom.

SECTION 5.3

13) In the Rutherford scattering experiment, what caused some of the alpha particles fired into the foil to be deflected from their paths, and even bounced back toward their source?

14) What conclusions were drawn from the fact that most of the alpha particles fired into a sheet of gold passed right through the sheet in straight lines?

15) What is the center of an atom called?

16) What major subatomic particles are in the center of an atom?

17) According to the planetary model of the atom, what does the electron do?

18)* The electron was the only subatomic particle that had been isolated and identified when the planetary model of the atom was first proposed. Why, when the next two major subatomic particles were discovered, did scientists believe they were in the center of the atom?

SECTION 5.4

19) What does the term *atomic number* mean?

20) What is the mass number of an atom?

21) How do isotopes of one element differ from each other?

22) In what ways are isotopes of one element the same?

23) Can two different isotopes of one element have the same mass number? Can atoms of two different elements have the same mass number? Explain.

24) Helium, as found in nature, consists of two isotopes. Most of the atoms have a mass number 4, but a few have a mass number 3. For each isotope, indicate the (a) atomic number; (b) number of protons; (c) number of neutrons; (d) mass number; (e) nuclear charge.

Questions 25 and 26: From the information given in the table, complete as many blanks as you can without looking at any reference. If there are any unfilled spaces, continue referring to your periodic table. As a last resort, check the table of elements inside the front cover.

25)

Name of Element	Nuclear Symbol	Atomic Number	Mass Number	Protons	Number of Neutrons	Electrons
	$^{13}_{6}C$					
			29	14		
		20			24	

26)

Name of Element	Nuclear Symbol	Atomic Number	Mass Number	Protons	Number of Neutrons	Electrons
	$^{79}_{35}Br$					
Silver			107			
					79	56

SECTION 5.5

While this set of questions is based on material in Section 5.5, some parts of some problems assume that you have also studied Section 5.6 and can use the periodic table as a source of atomic masses.

27) What is an atomic mass unit? Why do chemists express the masses of atoms in atomic mass units rather than in grams?

28) The mass equal to 1 amu is 1.66×10^{-24} grams. Calculate the mass of one atom of carbon-12.

29) Why is it necessary to define the atomic mass of an element as the "average mass" of atoms of the element instead of saying simply that atomic mass is the mass of an atom of the element?

30) The "average mass" of atoms of a certain element is $\frac{3}{4}$ of the mass of an atom of carbon-12. What is the atomic mass of that element? Locate the element in the periodic table. What are its atomic number, elemental symbol, and name?

31) Tellurium ($Z = 52$) has an atomic mass of 127.6 amu, while the atomic mass of iodine ($Z = 53$) is 126.9. Isn't this strange? Explain how this can be.

32) There are two other pairs of elements in the periodic table in which the atomic number increases by one, but the atomic mass becomes smaller. Can you find those pairs of elements?

33) The mass of 60.4% of the atoms of an element is 68.9257 amu. The element has only one other natural isotope, and its mass is 70.9249 amu. Calculate the average atomic mass of the element. Using the periodic table and/or the table of elements inside the front cover of this book, write the symbol and name of the element.

34) An element has two natural isotopes. If 50.69% of the atoms of the element have a mass of 78.9183 amu and the remainder have a mass of 80.9163 amu, calculate the atomic mass of the element. Use any other source of information available to you to identify the element.

Questions 35 through 40: Atomic masses (amu) and percentage abundances of the natural isotopes of an element are given. (1) Calculate the atomic mass of each element from these data. (2) Using other information that is available to you, identify the element.

	Percentage Abundance	Atomic Mass (amu)
35)	69.09	62.9298
	30.91	64.9278
36)	69.2	62.9296
	30.8	64.9278
37)	37.07	184.9530
	62.93	186.9560
38)	92.23	27.97693
	4.67	28.97649
	3.10	29.97376
39)	67.88	57.9353
	26.23	59.9332
	1.19	60.9310
	3.66	61.9283
	1.08	63.9280

40)

8.0	45.95263
7.5	46.9518
73.7	47.98236
5.5	48.94787
5.3	49.9448

41) The CRC Handbook, a large reference book of chemical and physical data from which many of the values in this book are taken, lists two isotopes of chlorine. One isotope makes up 75.77% of all naturally occurring chlorine, and its mass is 34.96885 amu. The mass of the other isotope is not listed. The atomic mass of chlorine is 35.45 amu. Calculate the mass of the other isotope. Express the result in the number of significant figures justified by the data given.

42) The atomic mass of rubidium ($Z = 37$) is 85.47 amu. Rubidium has two naturally occurring isotopes. Their atomic masses are 84.9117 amu and 86.9178 amu. Calculate the percentage distribution of these isotopes.

SECTION 5.6

43) How many elements are in Period 3 of the periodic table? Write the atomic numbers of the elements in Group 2A (2) of the periodic table.

44) Write the symbols of the elements in Group 4A (14) of the periodic table. Write the atomic numbers of the elements in Period 5.

---

45) Locate on a periodic table each element whose atomic number is given and identify first the number of the period it is in, and then the number of the group: (a) 21; (b) 9; (c) 55.

46) Locate on the periodic table each pair of elements whose atomic numbers are given below. Do the elements in each pair belong to the same period or the same family? (a) 19 and 30; (b) 8 and 52; (c) 24 and 74; (d) 3 and 10.

---

47) Using only a periodic table for reference, list the atomic masses of the elements whose atomic numbers are 22, 45, and 54.

48) Referring only to a periodic table, list the atomic masses of the elements having atomic numbers 5, 29, and 53.

---

49) What are the atomic masses of phosphorus and neon?

50) Write the atomic masses of cobalt and boron.

SECTION 5.7

51) The names, atomic numbers, or symbols of some of the elements in Figure 5.8 are entered into Table 5.5. Fill in the open spaces, referring only to a periodic table for any information you need.

Table 5.5

Table of Elements

Name of Element	Atomic Number	Symbol of Element
Sodium		
		Pb
Aluminum		
	26	
		F
Boron		
	18	
Silver		
	6	
Copper		
		Be
Krypton		
Chlorine		
	1	
		Mn
	24	
Cobalt		
	80	

52) The names, atomic numbers, or symbols of some of the elements in Figure 5.8 are entered into Table 5.6 (turn to page 134). Fill in the open spaces, referring only to a periodic table for any information you need.

GENERAL QUESTIONS

53) Distinguish precisely and in scientific terms the differences between items in each of the following groups:
a) Atom, subatomic particle
b) Electron, proton, neutron
c) Nuclear model of the atom, planetary model of the atom
d) Atomic number, mass number
e) Chemical symbol of an element, nuclear symbol
f) Atom, isotope
g) Atomic mass, atomic mass unit
h) Atomic mass of an element, atomic mass of an isotope

Table 5.6

Name of Element	Atomic Number	Symbol of Element
Neon		
	56	
		Sn
Potassium		
	28	
		He
Lithium		
	35	
		I
Magnesium		
	20	
		O
Sulfur		
	15	
		Si
	7	
		Zn

i) Period, group or family (in the periodic table)
j) Representative element, transition element

54) Determine whether each statement that follows is true or false:
a) Dalton proposed that atoms of different elements always combine on a one-to-one basis.
b) According to Dalton, all oxygen atoms have the same diameter.
c) The mass of an electron is about the same as the mass of a proton.
d) There are subatomic particles in addition to the electron, proton, and neutron.
e) The mass of an atom is uniformly distributed throughout the atom.
f) Most of the particles fired into the gold foil in the Rutherford experiment were not deflected.
g) The masses of the proton and electron are equal but opposite in sign.
h) Isotopes of an element have different electrical charges.
i) The atomic number of an element is proportional to its atomic mass.

j) An oxygen-16 atom has the same number of protons as an oxygen-17 atom.
k) The nuclei of nitrogen atoms have a different number of protons than the nuclei of any other element.
l) Neutral atoms of sulfur have a different number of electrons than neutral atoms of any other element.
m) Isotopes of different elements that have the same mass number have similar chemical behavior.
n) The mass of a carbon-12 atom is exactly 12 g.
o) Periods are arranged vertically in the periodic table.
p) The atomic mass of the second element in the right column of the periodic table is 10 amu.
q) Nb is the symbol of the element for which $Z = 41$.
r) Elements in the same column of the periodic table have similar properties.
s) The element for which $Z = 38$ is in both Group 2A and the fifth period.

55) The first experiments to suggest that an atom consisted of smaller particles showed that one particle had a negative charge. From that fact, what could be said about the charge of other particles that might be present?

56)* When Thomson identified the electron, he found that the ratio of its charge to its mass (the e/m ratio) was the same regardless of the element from which the electron came. Positively charged particles found at about the same time did not all have the same e/m ratio. What does that suggest about the mass, particle charge, and minimum number of particles present in the positive particles from different elements?

57)* Why were scientists inclined to think of an atom as a miniature solar system in the planetary model of the atom? What are the similarities and differences between electrons in orbit around a nucleus and planets in orbit around the sun?

58)* The existence of isotopes did not appear until nearly a century after Dalton proposed the atomic theory, and then it appeared in experiments more closely associated with physics than with chemistry. What does this suggest about the chemical properties of isotopes?

59)* A carbon-12 atom contains six electrons, six protons, and six neutrons. Assuming the mass of the atom is the sum of the masses of those parts as given in Table 5.1, calculate the mass of the atom. Why is it not exactly 12 amu, as the definition of atomic mass unit would suggest?

60) Using the figures in Question 59, calculate the percentage each kind of subatomic particle contributes to the mass of a carbon-12 atom.

61) The element carbon occurs in two crystal forms, diamond and graphite. The density of the diamond form is

3.51 g/cm^3, and of graphite, 2.25 g/cm^3. The volume of a carbon atom is 1.9×10^{-24} cm^3. As stated in Section 5.5, one atomic mass unit is 1.66×10^{-24} g.

a)   Calculate the density of a carbon atom.

b)   Suggest a reason for the density of the atom being so much larger than the density of either form of carbon.

c)   The radius of a carbon atom is roughly 1×10^5 times larger than the radius of the nucleus. What is the volume of that nucleus? (*Hint:* Volume is proportional to the cube of the radius.)

d)   Calculate the density of the nucleus.

e)   The radius of a period on this page is about 0.02 cm. The volume of a sphere that size is 4×10^{-5} cm^3. Calculate the mass of that sphere if it were completely filled with carbon nuclei. Express the mass in tons.

MATCHING SET ANSWERS

Set 1: 2–7–6–4–3–8–1–5

Set 2: 3–7–1–5–2–6–4

Set 3: 4–7–1–2–8–5–3–6

QUICK CHECK ANSWERS

5.1 b: True. a: An atom is made up of electrons, protons, and other particles. c: The mass of a proton is about 1 amu. d: Electrons and protons are electrically charged, but neutrons have no charge, or zero charge.

5.2 b and c: True. a: Atoms are mostly empty space.

5.3 All true.

6 The Language of Chemistry

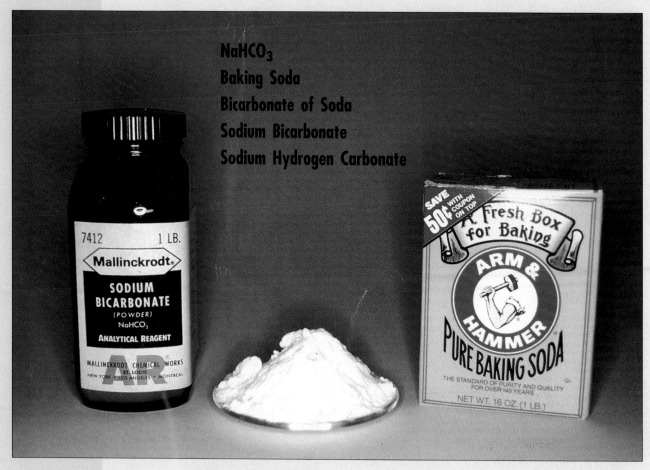

NaHCO₃
Baking Soda
Bicarbonate of Soda
Sodium Bicarbonate
Sodium Hydrogen Carbonate

The same substance may be known by several names, but it can have only one chemical formula. "Baking soda" is the common name of $NaHCO_3$ because of its widespread use in baking. Its old chemical name, "bicarbonate of soda," was replaced by a similar "sodium bicarbonate." Its official name today is "sodium hydrogen carbonate." The current name and formula are part of a *system* of nomenclature that you will learn in this chapter.

6.1 INTRODUCTION TO NOMENCLATURE

Have you ever been invited to a party and introduced to a large number of people you have never seen before, and then spent the rest of the evening trying to remember their names? It's not easy, is it? If we were to tell you we could introduce you to several hundred people and within an hour you would be able to identify them all by name, would you believe it? Probably not. We wouldn't believe it either. Unless . . .

Unless we could give you, along with the introductions, a *system* by which you could *learn* the names instead of just memorizing them. You'd still have to memorize the system, of course, but that's much easier than memorizing hundreds of individual names. In fact, 1440 chemical names and formulas. That's the number that fall within the performance goals of this chapter. Furthermore, by applying the system you will be able to write names and formulas of hundreds of other compounds that extend beyond the performance goals.

But you will be able to do all these wonderful things only if you LEARN THE SYSTEM!

Within the present chapter you will not see the chemical logic that lies behind the nomenclature system. This will appear in Chapters 10 and 11. In the meantime, our immediate purpose is to learn the language of chemistry so we can use it intelligently in Chapters 7 through 9. In fact, you'll find Chapters 7 through 9—and all those that follow—quite difficult if you can't speak the language.

Is there anything else we can add that can strengthen the advice in these three words?: LEARN THE SYSTEM!

6.2 FORMULAS OF ELEMENTS

PG 6A Given the name or the formula of an element in Figure 5.8, write the other.

In Chapter 5 you learned the symbols of 35 elements and their locations in a partial periodic table that contained only those symbols. That periodic table was Figure 5.8. In this section you will learn the chemical formulas of those elements as they are written in chemical equations.

The formula of most elements is simply the elemental symbol. In a chemical reaction the element behaves as if it was a bunch of independent atoms. This indicates that the tiniest individual unit in a sample of the element is one atom. We call this individual unit a **formula unit.**

The formula units of seven elements are not single atoms at normal temperatures and pressures.* Instead, two atoms of these elements are chemically bonded to each other to form a **molecule.** A molecule is the tiniest independent particle of a pure substance, element or compound, in a sample of that substance. The formula unit of these elements contains two atoms of the element. Their chemical formulas are therefore the elemental symbols followed by a subscript 2. Because there are two atoms in the molecule, they are referred to as **diatomic** molecules. *Di-* is a prefix that means "two."

---

*At normal temperatures, two other elements, sulfur and phosphorus, form molecules having more than one atom. Because this polyatomic (many-atom) structure does not affect equation calculations, their formulas are usually written simply as S and P, respectively.

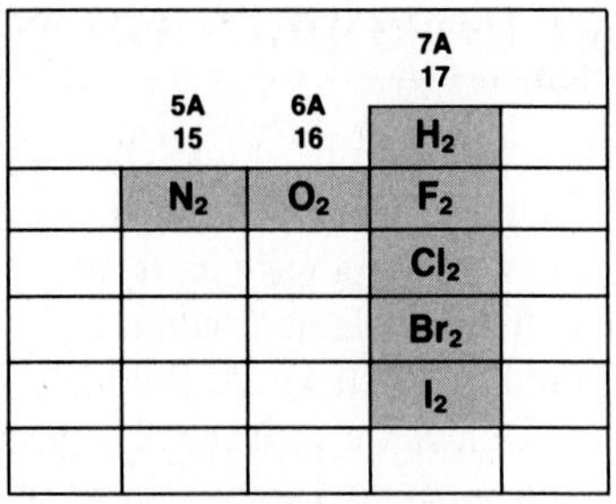

			7A 17	
	5A 15	6A 16	H_2	
	N_2	O_2	F_2	
			Cl_2	
			Br_2	
			I_2	

Figure 6.1

Elements that form stable diatomic molecules.

You must be able to recognize the elements that form diatomic molecules and to write their formulas correctly. Listed in a way that will help you remember them, they are:

ELEMENTS	FORMULAS
Nitrogen and oxygen, the two elements that make up about 97% of the atmosphere	N_2, O_2
Hydrogen, fluorine, chlorine, bromine, and iodine, the first five elements in Group 7A (17).	H_2, F_2, Cl_2, Br_2, I_2

The formulas of these elements are shown in their positions in the periodic table in Figure 6.1.

When you write chemical equations, it is *absolutely essential* that you write the formulas of these elements as diatomic molecules. *Failure to do so is probably the most common mistake made by beginning chemistry students.*

EXAMPLE 6.1 _____

Write the formulas of the following elements as they would be written in a chemical equation: potassium, fluorine, hydrogen, nitrogen, calcium.

———   ———

Potassium, K; fluorine, F_2; hydrogen, H_2; nitrogen, N_2; calcium, Ca.

6.3 COMPOUNDS MADE FROM TWO NONMETALS

PG 6B   Given the name or the formula of a binary molecular compound, write the other.
6C   Given the name or the formula of water, write the other; given the name or the formula of ammonia, write the other.

Recall from Section 5.6 the purpose of the stair-step line in the periodic table that begins between atomic numbers 4 and 5 and ends between 84 and 85. This line separates elements that are **metals,** such as iron, copper, and lead, from elements that are **nonmetals,** such as hydrogen, oxygen, and nitrogen. Many compounds are formed by two nonmetal elements. These compounds are all molecular; their tiniest individual unit is a molecule. Compounds formed by two nonmetals are called **binary** (two-element) molecular compounds. *Bi-* is another prefix that means "two."

The name of a molecular compound formed by two nonmetals has two words:

SUMMARY

1) The first word is the name of the element appearing first in the chemical formula, including a prefix to indicate the number of atoms of that element in the molecule.
2) The second word is the name of the element appearing second in the chemical formula, changed to end in -*ide*, and also including a prefix to indicate the number of atoms of that element in the molecule.

The same two nonmetals often form more than one binary compound. Their names are distinguished by the prefixes mentioned in the preceding rules. Silicon and chlorine form silicon tetrachloride, $SiCl_4$, and disilicon hexachloride, Si_2Cl_6. The prefix *tetra-* identifies four chlorine atoms in a molecule of $SiCl_4$. In Si_2Cl_6 *di-* indicates two silicon atoms and *hexa-* shows six chlorine atoms in the molecule. Technically, $SiCl_4$ should be monosilicon tetrachloride, but the prefix *mono-* for one is usually omitted. If an element has no prefix in the name of a binary molecular compound, you may assume that there is only one atom of that element in the molecule.

Table 6.1 gives the first ten number prefixes. The letter "o" in *mono*, and the letter "a" in prefixes for 4 to 10, are omitted if the resulting word "sounds better." This usually occurs when the next letter is a vowel. For example, a compound whose formula ends in O_5 is a *pentoxide* rather than a *pentaoxide*.

The oxides of nitrogen are ideal for practicing the nomenclature of binary molecular compounds.

Table 6.1
Numerical Prefixes Used in Chemical Names

Number	Prefix
1	*mono-*
2	*di-*
3	*tri-*
4	*tetra-*
5	*penta-*
6	*hexa-*
7	*hepta-*
8	*octa-*
9	*nona-*
10	*deca-*

EXAMPLE 6.2

For each name below, write the formula; for each formula, write the name.

nitrogen monoxide	NO_2
dinitrogen oxide	N_2O_3
dinitrogen pentoxide	N_2O_4

nitrogen monoxide, NO	NO_2, nitrogen dioxide
dinitrogen oxide, N_2O	N_2O_3, dinitrogen trioxide
dinitrogen pentoxide, N_2O_5	N_2O_4, dinitrogen tetroxide

Nitrogen dioxide could be correctly identified as mononitrogen dioxide. Two of the above compounds continue to be called by their older names: N_2O is nitrous oxide and NO is nitric oxide.

In "Study Hints and Pitfalls to Avoid" at the end of this chapter there are some suggestions that will help you to memorize the prefixes in Table 6.1.

Two compounds are so common they are always called by their traditional names rather than their chemical names. H_2O is water rather than dihydrogen oxide, and NH_3 is ammonia rather than nitrogen trihydride. The name and formula of ammonia are important; they should be memorized.

6.4 NAMES AND FORMULAS OF IONS FORMED BY ONE ELEMENT

PG 6D Given the name or the formula of an ion in Figure 6.2, write the other.

An atom contains the same number of protons (positive charges) and electrons (negative charges). That makes the atom electrically neutral; there is no net charge. But an atom can gain or lose one or more electrons. When it does, the balance between positive and negative charges is upset. Thus, the particle acquires a net electrical charge that is equal to the number of electrons gained or lost. If electrons are lost, the charge is positive; if electrons are gained, the charge is negative.

The charged particle formed when an atom has gained or lost electrons is an **ion.** If the ion has a positive charge, it is a **cation** (pronounced cat'-ion, not ca'-shun). If the ion has a negative charge, it is an **anion** (an'-ion). An ion that is formed from a single atom is a **monatomic ion.**

The formation of a monatomic ion from an atom can be expressed in a chemical equation. For example, a magnesium atom forms a magnesium ion by losing two electrons:

$$\text{Group 2A (2), magnesium ion: } Mg \longrightarrow Mg^{2+} + 2\,e^-$$

This equation illustrates two important rules about monatomic ions:

The formula of a monatomic ion is the symbol of the element followed by its electrical charge, written in superscript. The size of the charge (2) is written in front of the + or − sign. (If the charge is 1+ or 1−, the number is omitted.)

The name of a monatomic cation is the name of the element followed by the word "ion."

Notice the *essential difference* between the formula of an element and the formula of an ion formed by that element. The formula of most elements is the elemental symbol. The formula of the ion is the elemental symbol *followed by the ionic charge, written in superscript.* An elemental symbol, without a charge, is *never* the formula of an ion.

The thing that makes the periodic table so helpful in writing chemical formulas is that *all* of the elements in Group 2A (2) that form monatomic ions do so by losing two electrons. It is this loss of two electrons that gives the elements of Group 2A (2) their similar chemical properties.

As you might expect, this idea applies to other groups in the periodic table too. Group 1A (1) elements form monatomic ions by losing one electron per atom, and Group 3A (13) elements frequently lose three electrons per atom. From Period 3

Group 1A (1), sodium ion: $Na \rightarrow Na^+ + e^-$

Group 3A (13), aluminum ion: $Al \rightarrow Al^{3+} + 3\,e^-$

Atoms of the nonmetals in Groups 5A, 6A, and 7A (15, 16, and 17) of the periodic table form monatomic ions by gaining electrons. These ions have negative charges; they are anions. The equations for the ions formed by atoms of the Period 3 elements are:

Group 5A (15), phosphide ion: $P + 3\,e^- \rightarrow P^{3-}$

Group 6A (16), sulfide ion: $S + 2\,e^- \rightarrow S^{2-}$

Group 7A (17), chloride ion: $Cl + e^- \rightarrow Cl^-$

Notice that the name of a monatomic anion comes from the name of the parent element, but it is not exactly the same:

> *The name of a monatomic anion is the name of the element changed to end in -ide, followed by the word "ion."*

Thus phosph*orus* becomes phosph*ide*, sulf*ur* becomes sulf*ide*, and chlorine becomes chlor*ide*.*

EXAMPLE 6.3

Look only at a complete periodic table as you write (a) the names of Br^- and Ba^{2+} and (b) the formulas of the potassium and oxide ions.

—————   —————

a) bromide ion and barium ion; (b) K^+ and O^{2-}

Figure 6.2 places on a periodic table all of the monatomic ions formed by the elements whose symbols you learned in Section 5.7. Thus far we have examined the monatomic ions formed by elements in Groups 1A, 2A, 3A, and 5A–7A (1, 2, 13, 15–17). Notice that all ions formed by nonmetals are anions (they have negative charges), and all ions formed by metals are cations (positive charges). This is one of the features that distinguish metals from nonmetals.

Some transition elements—elements in the B groups (Groups 3 to 12) of the periodic table—are able to form two different monatomic ions that have different charges. Iron is one example. If a neutral atom loses two electrons, the ion has a 2+ charge, Fe^{2+}. A neutral atom can also lose three electrons, resulting in an ion with a 3+ charge: Fe^{3+}. To distinguish between the two ions, we include the size of the charge, but not its sign, when naming the ion. Thus, Fe^{2+} is the "iron two ion," and Fe^{3+} is the "iron three ion." In writing, the ion charge appears in roman numerals and enclosed in parentheses: Fe^{2+} is the iron(II) ion, and Fe^{3+} is the iron(III) ion. Note that there is no space between the last letter of the elemental name and the opening parenthesis.

---

*As elements, phosphorus, sulfur, and chlorine exist as polyatomic molecules that must be separated into atoms before ions can form.

The charge on a monatomic ion is formally referred to as the **oxidation state,** or **oxidation number,** of the element. Thus the oxidation state of Na^{+} is 1+; of Fe^{2+}, 2+; and of Fe^{3+}, 3+. These oxidation states are a part of a broader scheme that assigns oxidation numbers to all elements in a compound or ion, but we do not need these additional features in this chapter. Nevertheless, many instructors choose to include oxidation state as a part of nomenclature. If your instructor is among them, by all means, learn the concept of oxidation numbers now.

There is an older, but still widely used, way to distinguish between the Fe^{2+} and Fe^{3+} ions. Fe^{2+} is called the *ferrous* ion and Fe^{3+} is the *ferric* ion. The general rule is that two common ion charges are distinguished by using an *-ous* ending for the lower charge and an *-ic* ending for the higher charge. The endings are often applied to the Latin name of the element, which, for iron, is *ferrum.*

Notice that including the charge in the name of an ion is done *only when an element exhibits two common charges.* The student who refers to a "sodium one ion" or writes "calcium(II) ion" will attract attention—and not very favorable attention, at that.

Figure 6.2 entries for oxygen, mercury, tin, and lead deserve special comment. Oxygen and mercury each form a monotomic ion. These monotomic ions cause no naming problems. However, these two elements also form diatomic ions. The $O_2{}^{2-}$ is called a peroxide ion. (You will see the prefix *per-* used to indicate an extra oxygen in naming acids and polyatomic ions shortly.) The $Hg_2{}^{2+}$ has its 2+ charge shared by two mercury atoms, just as if each atom was contributing 1+ to the total charge of the ion. Accordingly, its name is mercury(I) ion.

1+	2+											3+		3−	2−	1−	
H^{+}																H^{-}	
Li^{+}	Be^{2+}													N^{3-}	O^{2-} O_2^{2-}	F^{-}	
Na^{+}	Mg^{2+}											Al^{3+}		P^{3-}	S^{2-}	Cl^{-}	
K^{+}	Ca^{2+}				Cr^{2+} Cr^{3+}	Mn^{2+} Mn^{3+}	Fe^{2+} Fe^{3+}	Co^{2+} Co^{3+}	Ni^{2+}	Cu^{+} Cu^{2+}	Zn^{2+}					Br^{-}	
										Ag^{+}			Sn^{2+} Sn^{4+}			I^{-}	
	Ba^{2+}										Hg_2^{2+} Hg^{2+}		Pb^{2+} Pb^{4+}				

Figure 6.2

Partial periodic table of common ions. Notes: (1) $O_2{}^{2-}$ is a diatomic elemental ion. Its name is peroxide ion. (2) $Hg_2{}^{2+}$ is a diatomic elemental ion. Its name is mercury(I), indicating a 1+ charge from each atom in the ion.

The tin(II) and lead(II) ions, Sn^{2+} and Pb^{2+}, behave as other monatomic ions. Tin and lead in compounds in which the elements appear to have 4+ charges have properties that differ from other compounds made up of ions. This notwithstanding, the compounds are named as if the ions were true ions.

6.5 NAMES AND FORMULAS OF ACIDS AND THE IONS DERIVED FROM ACIDS

PG 6E Given the name (or formula) of an acid of a Group 4A–7A (14–17) element, or of an ion derived from such an acid, write its formula (or name).

Acids: What They Are and What They Do

An **acid** is a hydrogen-bearing molecular compound that, when dissolved in water, loses one or more hydrogen ions, H^+. The particle that remains is an anion having a negative charge that is equal to the number of hydrogen ions lost. Hydrogen chloride, HCl, a molecular binary compound, behaves in this way. In water, the HCl molecule appears to **ionize,** or separate into a hydrogen ion and a chloride ion. The equation that describes this process is

$$HCl \longrightarrow H^+ + Cl^-$$

The acid formed by hydrogen chloride is hydrochloric acid. Its formula is the same as the formula of the parent compound, HCl.*

A hydrogen atom consists of one proton and one electron. To form a hydrogen ion, H^+, the neutral atom must lose its electron. That leaves only the proton; a hydrogen ion is simply a proton. Acids are sometimes classified by the number of hydrogen ions, or protons, that can be released by a single molecule. Hydrochloric acid is a **monoprotic acid** because it releases only one proton. If an acid has two ionizable hydrogens, it is **diprotic;** and a **triprotic** acid has three ionizable hydrogens. **Polyprotic** is a general term that may be applied to any acid having two or more ionizable hydrogens.

Binary Acids and the Ions Derived from Them

A **binary acid** has only two elements, hydrogen and another nonmetal. Hydrochloric acid, HCl, is a binary acid. The name *hydrochloric* shows how all binary acids are named. The name begins with the prefix, *hydro-*. This is followed by the name of the other element, changed so it ends in *-ic*. Hence, *hydro*-chlor-*ic*.

One of the characteristics of a chemical family is that its members usually form similar compounds. This is true of the binary acids of the Group 7A (17) elements. Rather than simply supplying the names and formulas of these acids, we will give you the opportunity to find them yourself, and thereby learn the system. We'll even include a binary acid that contains a Group 6A (16) element.

---

*The actual process by which a molecule separates into ions is described more fully in later chapters. You will also learn how to show whether a formula describes an acid or its parent molecular compound.

EXAMPLE 6.4

For each of the following names, write the formula; for each formula, write the name.

hydrofluoric acid   HI

hydrobromic acid H_2S

Chlorine, fluorine, bromine, and iodine are all from the same family. If you know the formula of hydrochloric acid, you should be able to find the formulas of hydrofluoric and hydrobromic acids by substitution of the elemental symbols. Then, if you reverse the thought process, you should be able to write the names of HI and H_2S.

hydrofluoric acid, HF   HI, hydroiodic acid

hydrobromic acid, HBr H_2S, hydrosulfuric acid

The names are established by the rule for naming binary acids: the prefix *hydro-* followed by the elemental name changed to end in *-ic*.

In regard to H_2S in Example 6.5, you may wonder how you might have predicted the formula if given the name. In other words, why is the formula H_2S rather than HS? First, sulfur is in Group 6A (16), not 7A (17), as are the other elements in the example. This should cause you to expect a formula that is different from the other acids. Is there another element in Group 6A (16) that forms a compound with hydrogen whose formula you know? How about oxygen and its famous oxide, H_2O, also known as water? If the compound of hydrogen and oxygen is H_2O, then the compound of hydrogen and sulfur should be H_2S. This kind of reasoning shows you how you can use the periodic table to predict names or formulas of compounds. Extending this example, can you name H_2Se (Se is the symbol for selenium, Z = 34) or write the formula of hydrotelluric acid? Try it, and then look at the answers at the bottom of the page.*

When a binary acid loses a hydrogen ion, it leaves behind a monatomic anion. This is indicated in the earlier equation, $HCl \rightarrow H^+ + Cl^-$. The anion is named by the rule for monatomic anions: Change the name of the element to end in *-ide*. Cl^- is the chloride ion.

Let us summarize the nomenclature rules for binary acids and the ions they produce:

SUMMARY

1) A binary acid consists of hydrogen and one nonmetal element.
2) The name of a binary acid is *hydro-* followed by the name of the nonmetal, changed to end in *-ic*.
3) The anion formed by the removal of all hydrogen from the original acid is a monatomic ion, which is named by the rule for monatomic anions: the name of the element, changed to end in *-ide*.
4) The negative charge on the anion formed by the removal of all hydrogen from the original acid is equal to the number of hydrogens removed.

---

*H_2Se, hydroselenic acid; hydrotelluric acid, H_2Te.

Oxyacids That End in -ic and the Oxyanions Derived from Them

An acid that contains oxygen in addition to hydrogen and another nonmetal is an **oxyacid.*** When a hydrogen ion is removed from an oxyacid, the oxygen stays with the nonmetal as part of an **oxyanion.*** The ionization of nitric acid, HNO_3, shows this:

$$HNO_3 \longrightarrow H^+ + NO_3^-$$

The name of the anion, NO_3^-, is *nitrate*.

Nitric acid and the nitrate ion are a perfect example of the nomenclature system that follows:

SUMMARY

For the total ionization of any oxyacid whose name ends in *-ic*

1) The formula of the anion is the formula of the acid without the hydrogen(s); with
2) A negative charge equal to the number of hydrogens in the acid; and
3) The name of the anion is the name of the central element of the acid changed to end in *-ate*.

To illustrate with a different acid, chloric acid, $HClO_3$, without the hydrogen is ClO_3. The difference is one hydrogen, so the charge on the ion is $1-$: ClO_3^-. To get the name, change chlor*ine* to chlor*ate*.

The names and formulas of five *-ic* acids and their corresponding *-ate* anions should be memorized. These acids, anions, and ionization equations are in Table 6.2. Hydrochloric acid is included to make the table a complete summary of the acid/anion combinations that illustrate the nomenclature system.

Bromine and iodine form oxyacids similar to the oxyacids of chlorine. In name and formula they may be substituted for chlorine in chloric acid and the anion derived from it.

EXAMPLE 6.5

Complete the name and formula blanks in the following table:

Acid Name	Acid Formula	Anion Formula	Anion Name
bromic acid	$HBrO_3$	BrO_3^-	Bromate
iodic acid	HIO_3	IO_3^-	iodate

---

*Oxoacid and oxoanion are alternate terms for oxyacid and oxyanion.

Acid Name	Acid Formula	Anion Formula	Anion Name
bromic acid	$HBrO_3$	BrO_3^-	bromate
iodic acid	HIO_3	IO_3^-	iodate

Thought process: Bromic acid corresponds with chloric acid, $HClO_3$. The acid and ion formulas come from substituting Br for Cl in the corresponding chlorine formulas. The anion name for an *-ic* acid is the name of the central element changed to end in *-ate*.

IO_3^- corresponds to ClO_3^-, the chlorate ion that comes from $HClO_3$, chloric acid. The acid and ion names and formulas are the same, except that iodine replaces chlorine. Hence, chlorate becomes iodate as the name of IO_3^-, chloric becomes iodic as the name of the acid, and $HClO_3$ becomes HIO_3 as the acid formula. The acid formula can also be derived directly from the anion formula. The ion has a 1− charge, so the acid must have one hydrogen: HIO_3.

SUMMARY

Table 6.2

Acids and Anions

Acid	Ionization Equation	Ion Name
Hydrochloric acid	$HCl \longrightarrow H^+ + Cl^-$	Chloride
Chloric acid	$HClO_3 \longrightarrow H^+ + ClO_3^-$	Chlorate
Nitric acid	$HNO_3 \longrightarrow H^+ + NO_3^-$	Nitrate
Sulfuric acid	$H_2SO_4 \longrightarrow 2\,H^+ + SO_4^{2-}$	Sulfate
Carbonic acid*	$H_2CO_3 \longrightarrow 2\,H^+ + CO_3^{2-}$	Carbonate
Phosphoric acid*	$H_3PO_4 \longrightarrow 3\,H^+ + PO_4^{3-}$	Phosphate

*The carbonic and phosphoric acid ionizations occur only slightly in water solutions. They are used here to illustrate the derivation of the formulas and names of the carbonate and phosphate ions, both of which are quite abundant from sources other than their parent acids.

EXAMPLE 6.6

Write the name and formula of the *-ic* acid of selenium and the name and formula of the anion formed by its total ionization, i.e., the loss of all hydrogens in the original acid.

Table 6.2 shows you the total ionization of sulfuric acid. Selenium (Z = 34) is just below sulfur in the periodic table, and we have already seen that the binary acid of selenium is like the binary acid of sulfur. It's true for the oxyacid also. The name and formula, please . . .

Selenate ion, SeO_4^{2-}

The name and formula correspond with sulfate ion and SO_4^{2-}.

Other Oxyacids and Their Oxyanions

Chlorine forms five acids that furnish a complete picture of the nomenclature of acids and the anions derived from their total ionization. Hydrochloric acid, HCl, and one oxyacid, chloric acid, $HClO_3$, have already been discussed. All five are assembled in Table 6.3. The table illustrates a *system* of nomenclature that begins with the *-ic* acid, whose name and formula you have memorized from Table 6.2. From there on, it is a system of prefixes (beginnings) and suffixes (endings) based on the number of oxygens in the *-ic* acid. Chlor*ic* acid and its chlor*ate* ion have three oxygens. Starting from there, and referring to Table 6.3:

Table 6.3
Prefixes and Suffixes in Acid and Anion Nomenclature (Acids and Anions of Chlorine Given as Examples)

Line	Oxygen Atoms Compared to *-ic* Acid and *-ate* Anion	Acid Prefix and/or Suffix (Example)	Anion Prefix and/or Suffix (Example)
1	One more $HClO_4$	*per-ic* (perchloric)	*per-ate* (perchlorate)
2	Same $HClO_3$	*-ic* (chloric)	*-ate* (chlorate)
3	One fewer $HClO_2$	*-ous* (chlorous)	*-ite* (chlorite)
4	Two fewer $HClO$	*hypo-ous* (hypochlorous)	*hypo-ite* (hypochlorite)
5	No oxygen HCl	*hydro-ic* (hydrochloric)	*-ide* (chloride)

SUMMARY

1) If the number of oxygens is one larger than the number in the *-ic* acid, the prefix *per-* is placed before both the acid and anion names: $HClO_4$ is *per*chloric acid and ClO_4^- is *per*chlorate ion.

2) If the number of oxygens is one smaller than the number in the *-ic* acid, the suffixes *-ic* and *-ate* are replaced with *-ous* and *-ite*. $HClO_2$ is chlor*ous* acid, and ClO_2^- is the chlor*ite* ion.

3) If the number of oxygens is one smaller than the number in the *-ous* acid (two smaller than the number in the *-ic* acid), the prefix *hypo-* is placed before both the acid and anion names, while keeping the *-ous* and *-ite* suffixes: HClO is *hypo*chlo*rous* acid and ClO⁻ is the *hypo*chlo*rite* ion.

As we have seen before, bromine and iodine can be substituted for chlorine in $HClO_3$ and ClO_3^-. These substitutions can also be made in the other acids of chlorine. Try your skill on the following:

E X A M P L E   6 . 7

Fill in the name and formula blanks in the following table. Try to do it without referring to Table 6.3, but use it if absolutely necessary.

Acid Name	Acid Formula	Anion Formula	Anion Name
periodic	*HIO₄*	*IO₄⁻*	*periodate*
hypobromous	HBrO	*BrO⁻*	*hypobromite*
iodous	*HIO₂*	IO_2^-	*iodite*

Acid Name	Acid Formula	Anion Formula	Anion Name
periodic	HIO_4	IO_4^-	periodate

Let's think about the top line only. The prefix *per-*, applied to the memorized formula of chloric acid, $HClO_3$, means one more oxygen atom. Therefore, perchloric acid is $HClO_4$. Substituting iodine for chlorine, we have periodic acid as HIO_4. Remove one hydrogen ion to get the anion formula, IO_4^-, with a negative charge equal to the number of hydrogens removed from the neutral molecule. The prefix *per-* is applied to the anion name as it is to the acid name. As perchloric acid → perchlorate ion, so periodic acid → periodate ion.

If you now want to reconsider your other entries, change them here on the other two lines,

Acid Name	Acid Formula	Anion Formula	Anion Name
	HBrO		
		IO_2^-	

Acid Name	Acid Formula	Anion Formula	Anion Name
periodic	HIO_4	IO_4^-	periodate
hypobromous	$HBrO$	BrO^-	hypobromite
iodous	HIO_2	IO_2^-	iodite

Reasoning processes are similar for all lines in the table. In the second line there are two fewer oxygen atoms than in chloric acid, $HClO_3$, and in the third line one fewer. Prefixes and suffixes match those for the corresponding chlorine substances in Table 6.3.

Fluorine forms only one oxyacid, hypofluorous acid. Its formula is written HOF. We will not be concerned with hypofluorous acid or the hypofluorite ion.

Nitric, sulfuric, and phosphoric acids have important variations with different numbers of oxygen atoms. We will limit ourselves to two of these. The acid and anion nomenclature system in Table 6.3 remains the same. See if you can apply it to this new situation.

EXAMPLE 6.8

Fill in the name and formula blanks in the following table:

Acid Name	Acid Formula	Anion Formula	Anion Name
Nitrous	HNO_2	NO_2^-	*Nitrite ion*
Sulfous / *Sulfurous*	H_2SO_3	SO_3^{2-}	sulfite

Acid Name	Acid Formula	Anion Formula	Anion Name
nitrous	HNO_2	NO_2^-	nitrite
sulfurous	H_2SO_3	SO_3^{2-}	sulfite

HNO_2 has one fewer oxygen atoms than nit*ric* acid, HNO_3, so its name must be nit*rous* acid. One hydrogen ion must be removed from the acid formula to produce the ion, NO_2^-. The anion from an *-ous* acid has an *-ite* suffix; nitrous acid → nitrite ion.

From the memorized sulfuric acid, H_2SO_4, you have the sulfate ion, SO_4^{2-}. The sulfite ion has one fewer oxygen, SO_3^{2-}. If the anion has a 2− charge, the acid must have two hydrogens: H_2SO_3. The name of the acid with one fewer oxygens than sulfuric acid is sulfurous acid.

The following example gives you the opportunity to practice what you have learned about acid and anion nomenclature. If you have memorized what is necessary, and know how to apply the rules that have been given, you will be able to write the required name and formulas with reference to nothing other than a periodic table. If you have really mastered the system, you will be able to extend it to the last substance in each column. These two substances have not been mentioned anywhere in this chapter.

EXAMPLE 6.9

For each of the following names, write the formula; for each formula, write the name.

phosphoric acid	CO_3^{2-}
sulfate ion	HF
bromous acid	NO_2^-
periodate ion	H_2SO_3
nitric acid	PO_4^{3-}
telluric acid (tellurium, Z = 52)	SeO_3^{2-} (Se, selenium, Z = 34)

———

phosphoric acid, H_3PO_4 (Table 6.2)	CO_3^{2-}, carbonate ion (Table 6.2)
sulfate ion, SO_4^{2-} (Table 6.2)	HF, hydrofluoric acid (Example 6.4)
bromous acid, $HBrO_2$ ($HClO_2$ and Table 6.3)	NO_2^-, nitrite ion (Example 6.8)
periodate ion, IO_4^- (Example 6.7)	H_2SO_3, sulfurous acid (Example 6.8)
nitric acid, HNO_3 (Table 6.2)	PO_4^{3-}, phosphate ion (Table 6.2)
telluric acid, H_2TeO_4	SeO_3^{2-}, selenite ion

References in parentheses tell where each name or formula may be found, or a starting point from which it may be figured out. The last item in each column includes elements from Group 6A (16), the same chemical family as sulfur. The formula of telluric acid matches that of sulfuric acid, H_2SO_4. From sulfuric acid the name of SO_4^{2-} is sulfate ion. One fewer oxygen atom makes it SO_3^{2-}, sulfite ion. Substitution of selenium for sulfur in name and formula gives SeO_3^{2-}, selenite ion.

SUMMARY

Table 6.4 summarizes this section. It is in the form of Groups 4A through 7A (14 through 17) of the periodic table. Each entry in the table shows the formula of an acid to the left of an arrow. To the right is the formula of the anion that results from the total ionization of the acid. The table includes all of the acids and anions whose names and formulas have been identified specifically, or can be figured out by the nomenclature system we have described. The key acids and anions—the ones that are the basis of this part of the nomenclature system—are highlighted in blue. If you have memorized the

Table 6.4

Acids and the Anions Derived from Their Total Ionization

	4A 14	5A 15	6A 16	7A 17
	$H_2CO_3 \longrightarrow CO_3^{2-}$	$HNO_3 \longrightarrow NO_3^-$ $HNO_2 \longrightarrow NO_2^-$		$HOF \longrightarrow OF^-$ $HF \longrightarrow F^-$
		$H_3PO_4 \longrightarrow PO_4^{3-}$	$H_2SO_4 \longrightarrow SO_4^{2-}$ $H_2SO_3 \longrightarrow SO_3^{2-}$ $H_2S \longrightarrow S^{2-}$	$HClO_4 \longrightarrow ClO_4^-$ $HClO_3 \longrightarrow ClO_3^-$ $HClO_2 \longrightarrow ClO_2^-$ $HClO \longrightarrow ClO^-$ $HCl \longrightarrow Cl^-$
		$H_3AsO_4 \longrightarrow AsO_4^{3-}$	$H_2SeO_4 \longrightarrow SeO_4^{2-}$ $H_2SeO_3 \longrightarrow SeO_3^{2-}$ $H_2Se \longrightarrow Se^{2-}$	$HBrO_4 \longrightarrow BrO_4^-$ $HBrO_3 \longrightarrow BrO_3^-$ $HBrO_2 \longrightarrow BrO_2^-$ $HBr \longrightarrow Br^-$
			$H_2TeO_4 \longrightarrow TeO_4^{2-}$ $H_2TeO_3 \longrightarrow TeO_3^{2-}$ $H_2Te \longrightarrow Te^{2-}$	$HIO_4 \longrightarrow IO_4^-$ $HIO_3 \longrightarrow IO_3^-$ $HIO_2 \longrightarrow IO_2^-$ $HIO \longrightarrow IO^-$ $HI \longrightarrow I^-$

key acids and anions and understand the system, you should be able to figure out any name or formula in this table, given the formula or name.

6.6 NAMES AND FORMULAS OF ACID ANIONS

PG 6F   Given the name (or formula) of an ion formed by the stepwise ionization of a polyprotic acid from a Group 4A, 5A, or 6A (14 through 16) element, write its formula (or name).

Polyprotic acids do not lose their hydrogens all at once, but rather, one at a time. The intermediate anions produced are stable chemical species that are the negative ions in many ionic compounds. The hydrogen-bearing **acid anion,** as it is called, releases a hydrogen ion when dissolved in water, just like any other acid.

Baking soda, commonly found in kitchen cabinets, contains the acid anion HCO_3^-. It can be regarded as the intermediate step in the ionization of carbonic acid:

$$H_2CO_3 \xrightarrow{-H^+} HCO_3^- \xrightarrow{-H^+} CO_3^{2-}$$

Table 6.5

Names and Formulas of Anions Derived from the Stepwise Ionization of Acids

Acid	Ion	Names of Ions	
		Preferred	Other
H_2CO_3	HCO_3^-	Hydrogen carbonate	Bicarbonate Acid carbonate
H_2S	HS^-	Hydrogen sulfide	Bisulfide Acid sulfide
H_2SO_4	HSO_4^-	Hydrogen sulfate	Bisulfate Acid sulfate
H_2SO_3	HSO_3^-	Hydrogen sulfite	Bisulfite Acid sulfite
H_3PO_4	$H_2PO_4^-$	Dihydrogen phosphate	Monobasic phosphate
$H_2PO_4^-$	HPO_4^{2-}	Hydrogen phosphate	Dibasic phosphate

HCO_3^- is the hydrogen carbonate ion—a logical name, since the ion is literally a hydrogen ion bonded to a carbonate ion. The ion is also called the bicarbonate ion.

Phosphoric acid, H_3PO_4, has three steps in its ionization process:

$$H_3PO_4 \xrightarrow{-H^+} H_2PO_4^- \xrightarrow{-H^+} HPO_4^{2-} \xrightarrow{-H^+} PO_4^{3-}$$

$H_2PO_4^-$ is the dihydrogen phosphate ion, signifying two hydrogen ions attached to a phosphate ion. HPO_4^{2-} is the monohydrogen phosphate ion, or simply the hydrogen phosphate ion. It is essential that the prefix *di-* be used in naming the $H_2PO_4^-$ ion to distinguish it from HPO_4^{2-}, but the prefix *mono-* is usually omitted in naming HPO_4^{2-}.

If you recognize the logic of this part of the nomenclature system, you will be able to extend it to intermediate ions from the stepwise ionization of hydrosulfuric, sulfuric, and sulfurous acids. All of these are shown in Table 6.5.

6.7 NAMES AND FORMULAS OF OTHER ACIDS AND IONS

PG 6G   Given the name (or formula) of the ammonium or hydroxide ions, write the corresponding formula (or name).

Ammonium Ion, NH_4^+; Hydroxide Ion, OH^-   The ammonium and hydroxide ions are two of the most common ions. Both are binary ions. The ammonium ion, with its 1+ charge, shares many chemical properties with the monatomic cations formed by Group 1A (1) elements. Even its name, with an *-ium* ending, is similar. The *-ide* ending for the hydroxide ion and its 1− charge suggest a monatomic anion belonging in Group 7A (17). Again, the chemical behavior of the hydroxide ion is in many ways like that of the monatomic anions formed by Group 7A (17) elements. You will encounter the ammonium and hydroxide ions often. Their names and formulas should be memorized.

Acetic Acid and the Acetate Ion   Acetic acid, $HC_2H_3O_2$, is the component of vinegar that is responsible for its odor and taste. Its ionization equation is

$$HC_2H_3O_2 \longrightarrow H^+ + C_2H_3O_2^-$$

Notice that only the hydrogen written first in the formula ionizes; the others do not. This is typical of organic acids, which usually produce ions containing carbon, hydrogen, and oxygen. The $C_2H_3O_2^-$ ion is the acetate ion, as might be expected from the -*ic* ending in the name of the acid, acet*ic*.

Hydrocyanic Acid, HCN   When hydrocyanic acid ionizes, it produces the cyanide ion, CN^-. Both the acid and the anion are exceptions to the nomenclature rules. The acid name suggests a binary acid, and the anion name suggests a monatomic anion.

$H^+ - CN^-$

Other Ions   The performance goals identify the important ions whose names and formulas you should be able to recognize and write. There are many others, too. Some of these, plus the ions already discussed, are listed in Tables 6.6 and 6.7. We recommend that you use these tables as references.

Table 6.6
Cations

Ionic Charge: 1+	Ionic Charge: 2+	Ionic Charge: 3+
Alkali Metals: **Group 1A (1)**	**Alkaline Earths:** **Group 2A (2)**	**Group 3A (3)**
Li^+ Lithium	Be^{2+} Beryllium	Al^{3+} Aluminum
Na^+ Sodium	Mg^{2+} Magnesium	Ga^{3+} Gallium
K^+ Potassium	Ca^{2+} Calcium	
Rb^+ Rubidium	Sr^{2+} Strontium	
Cs^+ Cesium	Ba^{2+} Barium	
Transition Elements	**Transition Elements**	**Transition Elements**
Cu^+ Copper(I)	Cr^{2+} Chromium(II)	Cr^{3+} Chromium(III)
Ag^+ Silver	Mn^{2+} Manganese(II)	Mn^{3+} Manganese(III)
Polyatomic Ions	Fe^{2+} Iron(II)	Fe^{3+} Iron(III)
NH_4^+ Ammonium	Co^{2+} Cobalt(II)	Co^{3+} Cobalt(III)
Others	Ni^{2+} Nickel	
H^+ Hydrogen	Cu^{2+} Copper(II)	
or	Zn^{2+} Zinc	
H_3O^+ Hydronium	Cd^{2+} Cadmium	
	Hg_2^{2+} Mercury(I)	
	Hg^{2+} Mercury(II)	
	Others	
	Sn^{2+} Tin(II)	
	Pb^{2+} Lead(II)	

Table 6.7
Anions

Ionic Charge: 1−				Ionic Charge: 2−	Ionic Charge: 3−
Halogens: Group 7A (17)		**Oxyanions**		**Group 6A (16)**	**Group 5A (15)**
F^- Fluoride		ClO_4^- Perchlorate		O^{2-} Oxide	N^{3-} Nitride
Cl^- Chloride		ClO_3^- Chlorate		S^{2-} Sulfide	P^{3-} Phosphide
Br^- Bromide		ClO_2^- Chlorite		**Oxyanions**	**Oxyanion**
I^- Iodide		ClO^- Hypochlorite		CO_3^{2-} Carbonate	PO_4^{3-} Phosphate
Acid Anions		BrO_3^- Bromate		SO_4^{2-} Sulfate	
HCO_3^- Hydrogen carbonate		BrO_2^- Bromite		SO_3^{2-} Sulfite	
HS^- Hydrogen sulfide		BrO^- Hypobromite		$C_2O_4^{2-}$ Oxalate	
HSO_4^- Hydrogen sulfate		IO_4^- Periodate		CrO_4^{2-} Chromate	
HSO_3^- Hydrogen sulfite		IO_3^- Iodate		$Cr_2O_7^{2-}$ Dichromate	
$H_2PO_4^-$ Dihydrogen phosphate		NO_3^- Nitrate		**Acid Anion**	
Other Anions		NO_2^- Nitrite		HPO_4^{2-} Hydrogen phosphate	
SCN^- Thiocyanate		OH^- Hydroxide			
CN^- Cyanide		$C_2H_3O_2^-$ Acetate		**Diatomic Elemental**	
H^- Hydride		MnO_4^- Permanganate		O_2^{2-} Perioxide	

6.8 FORMULAS OF IONIC COMPOUNDS

PG 6H Given the name of any ionic compound made up of ions included in Performance Goals 6E through 6G, or other ions whose formulas are given, write the formula of that compound.

Chemical compounds are electrically neutral. For ionic compounds this means that the formula unit must have an equal number of positive and negative charges. A net zero charge is achieved by combining cations and anions in such numbers that positive and negative charges are balanced. This is done in two steps:

SUMMARY

1) Write the formula of the cation, followed by the formula of the anion, omitting the charges.
2) Insert subscripts to show the number of each ion needed in the formula unit to make the sum of the charges equal to zero.
   a) If only one ion is needed, omit the subscript.
   b) If a polyatomic ion is needed more than once, enclose the formula of the ion in parentheses and place the subscript after the closing parenthesis.

EXAMPLE 6.10 _____

Write the formulas of potassium chloride and potassium hydroxide.

These two compounds involve three ions. Begin by writing the formulas of the ions, including charges, so we can see clearly the ions we must combine.

K^+ Cl^- OH^-

You must now decide how many cations and anions to combine for each compound so the sum of their charges is equal to zero. The potassium ion has a 1+ charge; the chloride ion has a 1− charge. How many potassium ions must combine with how many chloride ions so the sum of the charges is zero? Similarly, how many potassium ions and how many hydroxide ions?

1 potassium ion + 1 chloride ion; 1 potassium ion + 1 hydroxide ion

In both cases, $1 + (-1) = 0$.

Now write the formulas of the two compounds, following the two steps shown above.

Potassium chloride, KCl; potassium hydroxide, KOH

There are no subscripts in these formulas because each ion appears only once.

EXAMPLE 6.11 _____

Write the formulas of calcium chloride and calcium hydroxide.

Again begin by writing the formulas of the three ions in the two compounds.

Ca^{2+} Cl^- OH^-

Now decide how many Ca^{2+} ions must combine with how many Cl^- ions to produce a total charge of zero in the formula for calcium chloride. Do the same for the Ca^{2+} and OH^- ions.

1 Ca^{2+} ion + 2 Cl^- ions; 1 Ca^{2+} ion + 2 OH^- ions

This time $2+ + 2(1-) = 0$ for both compounds.

Now follow the two-step procedure for both compounds. Be careful on where and how you use parentheses.

Calcium chloride, $CaCl_2$; calcium hydroxide, $Ca(OH)_2$

If you recognize the two ions, you have the name of the compound.

There is one place where you are apt to be uncertain about the name of a "familiar" ion. For example, what is the name of $FeCl_3$? Iron chloride is not an adequate answer. It fails to distinguish between the two possible charges on the iron ion. Is $FeCl_3$ iron(II) chloride or iron(III) chloride? To decide, you must reason from the known charge on the chloride ion, $1-$, and the fact that the total charge on the compound is zero. The formula has three chloride ions, so the total negative charge is $3-$. This must be balanced by three positive charges from the iron ion, so it must be an iron(III) ion. The compound is iron(III) chloride. If the formula had been $FeCl_2$, the name iron(II) chloride would have been reached by recognizing that the $2-$ of two chloride ions is balanced by the $2+$ of a single iron(II) ion.

In writing or speaking the name of an ionic compound containing a metal that is capable of more than one ionic charge, *the compound name includes the charge of that metal.*

EXAMPLE 6.16 _____

Write the name of each compound below.

LiBr	$NaHSO_3$
$Mg(IO_4)_2$	K_2HPO_4
$AgNO_3$	$ZnCO_3$
$MnCl_3$	HgS
Hg_2Br_2	$(NH_4)_2SeO_4$
	(selenium, Z = 34)

_____   _____

LiBr, lithium bromide $NaHSO_3$, sodium hydrogen sulfite

$Mg(IO_4)_2$, magnesium periodate K_2HPO_4, potassium hydrogen phosphate

$AgNO_3$, silver nitrate $ZnCO_3$, zinc carbonate

$MnCl_3$, manganese(III) chloride   HgS, mercury(II) sulfide

Hg_2Br_2, mercury(I) bromide $(NH_4)_2SeO_4$, ammonium selenate

In $MnCl_3$, three $1-$ charges from three Cl^- ions require $3+$ from the manganese ion, so it is the manganese(III) ion. Similarly, one $2-$ charge from the sulfide ion in HgS must be balanced by $2+$ from a mercury(II) ion. In Hg_2Br_2, the two $1-$ charges from two Br^- ions are balanced by the $2+$ charge from the diatomic mercury(I) ion. In $(NH_4)_2SeO_4$, selenium substitutes for its family member, sulfur, in sulfate ion, SO_4^{2-}, so SeO_4^{2-} is the selenate ion.

6.10 HYDRATES

PG 6J Given the formula of a hydrate, state the number of water molecules associated with each formula unit of the anhydrous compound.

6K Given the name (or formula) of a hydrate, write its formula (or name). (This performance goal is limited to hydrates of ionic compounds discussed in this chapter.)

Some compounds, when crystallized from water solutions, form solids that include water molecules as part of the crystal structure. Such water is referred to as **water of crystallization** or **water of hydration.** The compound is said to be **hydrated** and is called a **hydrate.** Hydration water can usually be driven from a compound by heating, leaving the **anhydrous compound.**

Copper(II) sulfate is an example of a hydrate. The anhydrous compound, $CuSO_4$, is a nearly white powder. Each formula unit of $CuSO_4$ combines with five water molecules in the hydrate, which is a dark blue crystal. The formula of the hydrate, $CuSO_4 \cdot 5\,H_2O$, and its name, copper sulfate 5-hydrate, illustrate the nomenclature system for hydrates. The equation for the dehydration of this compound is

$$CuSO_4 \cdot 5\,H_2O \rightarrow CuSO_4 + 5\,H_2O$$

EXAMPLE 6.17 _____

a) How many water molecules are associated with each formula unit of anhydrous sodium carbonate in $Na_2CO_3 \cdot 10\,H_2O$? Name the hydrate.

b) Write the formula of nickel chloride 6-hydrate if the formula of the anhydrous compound is $NiCl_2$.

_____

a) Ten; sodium carbonate 10-hydrate   b) $NiCl_2 \cdot 6\,H_2O$

6.11 SUMMARY OF THE NOMENCLATURE SYSTEM

Throughout this chapter we have emphasized the importance of memorizing certain names and formulas and some prefixes and suffixes. They are the basis for the *system* of chemical nomenclature. From there on, it is a matter of applying the system to the different names and formulas you meet. Table 6.8 (turn to page 160) summarizes all the ideas that have been presented in this chapter. It should help you to learn the nomenclature system.

6.12 COMMON NAMES OF COMMON CHEMICALS

In a typical day we encounter and use a large number of chemicals. Many chemicals are listed in Table 6.9 (turn to page 161), along with their common names, their chemical names, and their formulas. How many are familiar to you? How many did you learn to name in this chapter?

SUMMARY

Table 6.8

Summary of Nomenclature System

Substance	Name	Formula
Element	Name of element	Symbol of element; exceptions: H_2, N_2, O_2, F_2, Cl_2, Br_2, I_2
Compounds made up of two nonmetals	First element in formula followed by second, changed to end in *-ide*, each element preceded by prefix to show the number of atoms in the molecule	Symbol of first element in name followed by symbol of second element, with subscript to show number of atoms in the molecule
Binary acid	Prefix *hydro-* followed by name of second element changed to end in *-ic*	H followed by symbol of second element with appropriate subscripts
Oxyacid	Most common: middle element, changed to end in *-ic* One more oxygen than *-ic* acid: add prefix *per-* to name of *-ic* acid One less oxygen than *-ic* acid: change ending of *-ic* acid to *-ous* Two less oxygens than *-ic* acid: add prefix *hypo-* to name of *-ous* acid	H followed by symbol of nonmetal followed by O, each with appropriate subscript. MEMORIZE THE FOLLOWING: Chloric acid—$HClO_3$ Nitric acid—HNO_3 Sulfuric acid—H_2SO_4 Carbonic acid—H_2CO_3 Phosphoric acid—H_3PO_4
Monatomic cation	Name of element followed by "ion"; if element forms more than one monatomic cation, elemental name is followed by ion charge in Roman numerals and in parentheses	Symbol of element followed by superscript to indicate charge
Monatomic anion	Name of element changed to end in *-ide*	Symbol of element followed by superscript to indicate charge
Polyatomic anion from total ionization of oxyacid	Replace *-ic* in acid name with *-ate*, or replace *-ous* in acid name with *-ite*, followed by "ion"	Acid formula without hydrogen plus superscript showing negative charge equal to number of hydrogens removed from acid formula
Polyatomic anion from stepwise ionization of oxyacid	"Hydrogen" followed by name of ion from total ionization of acid ("dihydrogen" in the case of $H_2PO_4^-$)	Acid formula minus one (or two for H_3PO_4) hydrogens, plus superscript showing negative charge equal to number of hydrogens removed from acid formula
Other polyatomic ions	Ammonium ion Hydroxide ion	NH_4^+ OH^-
Ionic compound	Name of cation followed by name of anion	Formula of cation followed by formula of anion, each taken as many times as necessary to yield a net charge of zero (polyatomic ion formulas enclosed in parentheses if taken more than once)
Hydrate	Name of anhydrous compound followed by "X-hydrate," where X is number of water molecules associated with one formula unit of anhydrous compound	Formula of anhydrous compound followed by "·X H_2O" where X is number of water molecules associated with one formula unit of anhydrous compound

Table 6.9
Common Names of Chemicals

Common Name	Chemical Name	Formula
Alumina	Aluminum oxide	Al_2O_3
Baking soda	Sodium hydrogen carbonate	$NaHCO_3$
Bleach (liquid)	Hydrogen peroxide or	H_2O_2
	Sodium hypochlorite	$NaClO$
Bleach (solid)	Sodium perborate	$NaBO_3$
Bluestone	Copper(II) sulfate 5-hydrate	$CuSO_4 \cdot 5\ H_2O$
Borax	Sodium tetraborate 10-hydrate	$Na_2B_4O_7 \cdot 10\ H_2O$
Brimstone	Sulfur	S
Carbon tetrachloride	Tetrachloromethane	CCl_4
Chile saltpeter	Sodium nitrate	$NaNO_3$
Chloroform	Trichloromethane	$CHCl_3$
Cream of tartar	Potassium hydrogen tartrate	$KHC_4H_4O_6$
Diamond	Carbon	C
Dolomite	Calcium magnesium carbonate	$CaCO_3 \cdot MgCO_3$
Epsom salts	Magnesium sulfate 7-hydrate	$MgSO_4 \cdot 7\ H_2O$
Freon (refrigerant)	Dichlorodifluoromethane	CCl_2F_2
Galena	Lead(II) sulfide	PbS
Grain alcohol	Ethyl alcohol; ethanol	C_2H_5OH
Graphite	Carbon	C
Gypsum	Calcium sulfate 2-hydrate	$CaSO_4 \cdot 2\ H_2O$
Hypo	Sodium thiosulfate	$Na_2S_2O_3$
Laughing gas	Dinitrogen oxide	N_2O
Lime	Calcium oxide	CaO
Limestone	Calcium carbonate	$CaCO_3$
Lye	Sodium hydroxide	$NaOH$
Marble	Calcium carbonate	$CaCO_3$
MEK	Methyl ethyl ketone	$CH_3COC_2H_5$
Milk of magnesia	Magnesium hydroxide	$Mg(OH)_2$
Muriatic acid	Hydrochloric acid	HCl
Oil of vitriol	Sulfuric acid (conc.)	H_2SO_4
Plaster of Paris	Calcium sulfate $\frac{1}{2}$-hydrate	$CaSO_4 \cdot \frac{1}{2}\ H_2O$
Potash	Potassium carbonate	K_2CO_3
Pyrites (fool's gold)	Iron disulfide	FeS_2
Quartz	Silicon dioxide	SiO_2
Quicksilver	Mercury	Hg
Rubbing alcohol	Isopropyl alcohol	$(CH_3)_2CHOH$
Sal ammoniac	Ammonium chloride	NH_4Cl
Salt	Sodium chloride	$NaCl$
Salt substitute	Potassium chloride	KCl
Saltpeter	Potassium nitrate	KNO_3
Slaked lime	Calcium hydroxide	$Ca(OH)_2$
Sugar	Sucrose	$C_{12}H_{22}O_{11}$
TSP (trisodium phosphate)	Sodium phosphate	Na_3PO_4
Washing soda	Sodium carbonate 10-hydrate	$Na_2CO_3 \cdot 10\ H_2O$
Wood alcohol	Methyl alcohol; methanol	CH_3OH

EVERYDAY CHEMISTRY

"It's Elementary, My Dear Friend . . . ."

When the fictional "Consulting Detective" Sherlock Holmes made his famous comment to Dr. Watson about some matter being elementary, Holmes meant it couldn't have been any simpler. To a chemist also, the elements are the simplest pieces of matter. But while the elements are the simplest stuff of which chemistry is made, the efforts needed to isolate new elements are often heroic. The battles fought to name newly discovered elements remind us that chemistry is done by human beings, not by disinterested spectators of nature.

Working in Paris in 1898, Marie and Pierre Curie isolated polonium (Z = 84) and radium (Z = 88) from two *truckloads* of pitchblende ore from which the uranium had already been removed. They obtained this "useless" ore from the Austrian government for a trifling amount of money but had to pay shipping. A ton of this ore contained about 30 milligrams of polonium and radium.

Working in Darmstadt, Germany, in 1984, a research group synthesized a *single atom* of meitnerium (Z = 109), named for Lise Meitner, an Austrian physicist who helped develop the idea of nuclear fission. They bombarded $^{209}_{83}Bi$ with a gentle beam of iron atoms at "just the right energy" so the nuclei "kissed" and fused into the $^{266}_{109}Mt$. It took a week of bombardment to produce the single atom of meitnerium. The existence of an atom with mass 266 amu was confirmed using several independent techniques.

Traditionally, the discoverer of a new element has named it, but not until the discovery is confirmed by another laboratory. When chemistry was done by individuals or very small groups using their own money, confirming discovery of a new element was fairly easy, and naming a new element was unlikely to cause controversy.

Today there are only three laboratories in the world capable of synthesizing heavy elements. They are the Heavy Ion Research Laboratory in Darmstadt, Germany; the Joint Institute for Nuclear Research in Dubna, Russia; and the Lawrence Berkeley Laboratory/Lawrence Livermore National Laboratory in Northern California. These three research laboratories are intensely competitive, and the 45-year Cold War added a contentious layer of politics.

Beginning in 1958, the American, Russian, and German groups have had a heated controversy over which group has the more convincing evidence of discovery of some elements from 102 to 109 and therefore has the right to name them. The German claims for discovering elements 107–109 are not in doubt, but due to enormous financial and experimental obstacles, no outside laboratory has yet verified their work. An unwelcome precedent may have been set. If the work is so expensive, and only a few atoms of a new element are made (and then quickly decay), how can the cost be justified?

The Livermore laboratory is one of the seven major laboratories in the United States run by the Department of Energy. (The other six labs are Argonne, Brookhaven, Lawrence Berkeley, Los Alamos, Oak Ridge, and Sandia.) Together, these seven labs spend about $6 billion annually on research and development. A total of about $3 billion was spent on atomic weapons development, research, and testing at the three major weapons laboratories (Livermore, Los Alamos, and Sandia.) Do we need to spend $3 billion dollars a year when we won the superpower nuclear arms race by default? What should we expect for our money from these labs?

These Department of Energy labs are a tempting political target in years of budget deficits. But closing these labs would deprive the nation of gold mines of talented people using state-of-the-art analytical instruments. In the past, the DOE labs contributed hardware and software design for super computers and massive parallel computers; development of safe nuclear reactors; diagnosis and treatment procedures for cancer; techniques used in biochemical research; development of new structural materials. If we wish to become independent of imported oil, the energy research now being done at these labs is essential.

As the century ends, the DOE national laboratories will have been given new missions. What will these missions be? They could include anything from predicting global climate patterns (a current strength at Livermore) to new tools for nuclear medicine (Brookhaven). These missions will not evolve, however, until we have a national discussion on the goals we want these labs to accomplish. The discussion may not be conclusive, but it must occur.

CHAPTER 6 IN REVIEW

6.1 Introduction to Nomenclature

6.2 Formulas of Elements

  6A Given the name or the formula of an element in Figure 5.8, write the other.

6.3 Compounds Made from Two Nonmetals

  6B Given the name or the formula of a binary molecular compound, write the other.

  6C Given the name or the formula of water, write the other; given the name or the formula of ammonia, write the other.

6.4 Names and Formulas of Ions Formed by One Element

  6D Given the name or the formula of an ion in Figure 6.2, write the other.

6.5 Names and Formulas of Acids and the Ions Derived from Acids

  6E Given the name (or formula) of an acid of a group 4A–7A (14–17) element, or of an ion derived from such an acid, write its formula (or name).

6.6 Names and Formulas of Acid Anions

  6F Given the name (or formula) of an ion formed by the stepwise ionization of a polyprotic acid from a Group 4A, 5A, or 6A (14 through 16) element, write its formula (or name).

6.7 Names and Formulas of Other Acids and Ions

  6G Given the name (or formula) of the ammonium or hydroxide ions, write the corresponding formula (or name).

6.8 Formulas of Ionic Compounds

  6H Given the name of any ionic compound made up of ions included in Performance Goals 6D through 6G, or other ions whose formulas are given, write the formula of that compound.

6.9 Names of Ionic Compounds

  6I Given the formula of an ionic compound made up of identifiable ions, write the name of that compound.

6.10 Hydrates

  6J Given the formula of a hydrate, state the number of water molecules associated with each formula unit of the anhydrous compound.

  6K Given the name (or formula) of a hydrate, write its formula (or name). (This Performance Goal is limited to hydrates of ionic compounds discussed in this chapter.)

6.11 Summary of the Nomenclature System

6.12 Common Names of Common Chemicals

CHAPTER 6 KEY WORDS AND MATCHING SETS

Set 1

_____ Atom that has gained or lost electrons.

_____ Ion with a positive charge.

_____ Monatomic anion suffix.

_____ Ion written last in the formula of an ionic compound.

_____ Charge on an atom that has gained one electron.

_____ Ions with 2+ charge come from this group.

_____ Element present in all acids in this chapter.

_____ Associated with anion from an -ic acid.

_____ Group that forms ions with a 2− charge.

_____ Charge on an atom that has lost one electron.

_____ Part of formula of manganese(II) sulfate 2-hydrate.

1. Hydrogen
2. Anion
3. $2 H_2O$
4. Cation
5. 1−
6. 2A (2)
7. -ate suffix
8. Monatomic ion
9. 6A (16)
10. -ide
11. 1+

Set 2

_____	One more oxygen than *-ate*.	1. SO_2
_____	Nonmetal bonded to nonmetal.	2. *hypo-* (nonmetal) *-ous*
_____	Ammonia.	3. S_2O
_____	Sulfur dioxide.	4. *Per-* (nonmetal) *-ate*
_____	Ammonium ion	5. OH^-
_____	One fewer oxygen than *-ate*.	6. *-ite*
_____	Acid with two fewer oxygens than *-ic*.	7. NH_3
_____	Water.	8. H_2O
_____	Hydroxide ion.	9. Five
_____	Disulfur monoxide.	10. NH_4^+
_____	*Penta-*	11. Binary molecular compound

STUDY HINTS AND PITFALLS TO AVOID

As noted in the introduction to this chapter, the most important thing you can do to learn nomenclature is to learn the *system*. The system is based on some rules, prefixes, and suffixes that must be memorized. These can then be applied in writing the names and formulas of hundreds of chemical substances. This is by far the easiest and quickest way to learn how to write chemical names and formulas.

Remember that the *elements* nitrogen, oxygen, hydrogen, fluorine, chlorine, bromine, and iodine exist as diatomic molecules. Note the limitation. It refers to the *elements,* not to compounds in which the elements may be present.

Here are a few memory aids that may help in learning the number prefixes in Table 6.1: A *mono*poly is when *one* company controls an economic product or service. A *two*-wheel cycle is a *bi*cycle, but a chemist might call it a *di*cycle. No problem with *three* wheels: it's a *tri*cycle. No help on *tetra-* for four, unless you happen to remember that a four-sided solid is called a tetrahedron. The *Penta*-gon is the *five*-sided building in Washington that serves as headquarters for U.S. military operations. *Six* and *hex-* are the only number/prefix combination that has the letter ''x.'' If you change an ''s'' to an ''h,'' in September, then *Hep*tember, *Oct*ober, *Nov*ember, and *Dec*ember list the beginnings of what were once the *seven*th, *eigh*th, *nin(e)*th, and *ten*th months of the year.

Notice that number prefixes are *almost never* used in naming ionic compounds. The *di*hydrogen phosphate ion is the only exception in this chapter, and the *di*chromate ion is in Table 6.7. Number prefixes were used in the past, but today people (particularly test graders!) will look at

you strangely if you talk or write about aluminum trichloride.

Be sure you use parentheses correctly in writing formulas of ionic compounds. They enclose *polyatomic* ions used more than once, never a monatomic ion. [$BaCl_2$, not $Ba(Cl)_2$. Also $Ba(OH)_2$, not $BaOH_2$; but $NaOH$, not $Na(OH)$].

A charge, written as a superscript, is included in the formula of *every* ion. Without a charge it is not an ion. However, do not include ionic charge in the formula of an ionic compound. (Na_2S, not $Na_2^+S^{2-}$.)

An oxyacid should be named as an *acid,* not as an ionic compound. For example, HNO_3 is nitric acid, not hydrogen nitrate.

Be sure to use the charge of a cation when naming an ionic compound if the element forms more than one monatomic ion. Do *not* use the charge if the element forms only one cation.

To learn the nomenclature system correctly is the first of two steps. The second is to apply it correctly. To develop this skill, you must practice, practice, and then practice some more until you write names and formulas almost automatically.

The end-of-chapter questions that follow give ample opportunity for practice. Take full advantage of them. In particular, perfect your skill in writing formulas of ionic compounds by completing Formula Writing Exercises 1 and 2. Your ultimate self-test lies in the last group of questions where different kinds of substances are mixed. You must first identify the kind of substance it is, select the proper rule to apply, and then apply it correctly.

CHAPTER 6 QUESTIONS AND PROBLEMS

General Instructions: *Most of the questions in this chapter ask that you write the name of any species if the formula is given, or the formula if the name is given. You will be reminded of this briefly at the beginning of each such block of questions. You should try to follow these instructions without reference to anything except a* clean *periodic table, one that has* nothing *written on it. Names and/or atomic numbers are given in questions involving elements not shown in Figure 6.1. An asterisk (*) marks a substance containing an ion you are not expected to recognize. If you cannot predict what it is from the periodic table, refer to Table 6.6 or 6.7.*

SECTION 6.2

1) The stable form of seven elements is the diatomic molecule. Write the names and the formulas of those elements.

2) The elements of Group 0(18) are stable as monatomic atoms. Write their formulas.

Questions 5 to 10: Given names, write formulas; given formulas, write names.

3) Barium, silver, oxygen, magnesium.

4) Chromium, chlorine, beryllium, iron.

5) S, F_2, Ni, B.

6) Cu, Mn, N_2, Kr.

7) Argon, iodine, zinc, carbon.

8) Lead, silicon, bromine, sodium.

SECTION 6.3

Questions 9 and 10: Given names, write formulas; given formulas, write names.

9) HBr, ammonia, P_2O_3, dichlorine oxide, H_2O.

10) Phosphorus tribromide, hydrogen chloride, N_2O, NH_3, SO_2.

SECTION 6.4

Questions 11 to 14: Given names, write formulas; given formulas, write names.

11) Cu^+, I^-, S^{2-}, Hg_2^{2+}, K^+.

12) Cr^{3+}, P^{3-}, Zn^{2+}, Br^-, Ca^{2+}.

13) Iron(II) ion, hydride ion, magnesium ion, aluminum ion, oxide ion.

14) Nitride ion, cobalt(III) ion, lithium ion, mercury(II) ion, sulfide ion.

SECTION 6.5

15) How do you recognize the formula of an acid?

16) What element is present in all acids discussed in this chapter?

17) What is the meaning of the suffix *-protic* when speaking of acids?

18) Distinguish between acids that are monoprotic, diprotic, triprotic, and polyprotic.

19) Table 6.10 has space for half of the names and formulas covered by Performance Goal 6E. One name or formula appears on each line. Fill in the remaining blanks. The first line is completed as an example.

Table 6.10

Acid Name	Acid Formula	Ion Name	Ion Formula
Hydrochloric	HCl	Chloride	Cl^-
	HNO_3		
		Phosphate	
			S^{2-}
Nitrous			
	HIO_3		
		Telluride	
			ClO^-
Iodous			
	H_2Se		
		Perchlorate	
			I^-
	$HClO_2$		
		Selenate	
			BrO_2^-

20) Table 6.11 (see page 166) has spaces for the names and formulas covered by Performance Goal 6F that are not in Question 21. Fill in the remaining blanks.

Table 6.11

Acid Name	Acid Formula	Ion Name	Ion Formula
Sulfuric			
	H_2CO_3		
		Chlorate	
			F^-
Bromic			
	H_2SO_3		
		Arsenate	
			IO_4^-
Selenous			
	H_2TeO_3		
		Hypoiodite	
			BrO^-
Telluric			
	$HBrO_4$		
		Bromide	

SECTION 6.6

21) Explain how an anion can behave as an acid. Is it possible for a cation to be an acid?

22) Distinguish between total ionization and stepwise ionization as the terms are used in this chapter.

Questions 23 to 26: Given names, write formulas; given formulas, write names.

23) Hydrogen sulfite ion, hydrogen carbonate ion.

24) Hydrogen selenide ion, dihydrogen phosphate ion.

_____

25) $HSeO_3^-$, HTe^-.

26) HSO_4^-, HPO_4^{2-}.

SECTION 6.7

Questions 27 to 30: Given names, write formulas; given formulas, write names. Refer to Table 6.6 or 6.7 only if necessary.

27) Hydroxide ion, cadmium ion (Z = 48).

28) Acetate ion, rubidium ion (Z = 37).

_____

29) NH_4^+, CN^-.

30) $HC_2H_3O_2$, Ga^{3+} (Z = 31).

SECTION 6.8

The formula writing exercise in Table 6.12 should be completed now. When you have developed your skill in writing formulas of ionic compounds, test it by writing the formulas of the compounds in Questions 31 to 36.

31) Calcium hydroxide, ammonium bromide, potassium sulfate.

32) Lithium carbonate, barium phosphate, aluminum nitrate.

_____

33) Magnesium oxide, aluminum phosphate, sodium sulfate, calcium sulfide.

34) Lithium chloride, ammonium nitrate, barium bromide, magnesium phosphate.

_____

35) Barium sulfite, chromium(III) oxide, potassium periodate, calcium hydrogen phosphate.

36) Barium hypoiodite, copper(II) nitrate, magnesium hydrogen sulfate.

SECTION 6.9

The formula writing and chemical name exercise in Table 6.13 (see page 168) should be completed now. Then test your skill by naming the compounds in Questions 37 to 42.

37) Li_3PO_4, $MgCO_3$, $Ba(NO_3)_2$.

38) NH_4Cl, KOH, Na_2SO_4.

_____

39) KF, $NaOH$, CaI_2, $Al_2(CO_3)_3$.

40) CaS, $BaCO_3$, K_3PO_4, $(NH_4)_2SO_4$.

_____

41) $CuSO_4$, $Cr(OH)_3$, Hg_2I_2

42) $MgSO_3$, $Al(BrO_3)_3$, $PbCO_3$.

SECTION 6.10

43) Among the following, identify all hydrates and anhydrous compounds: $NiSO_4 \cdot 6\,H_2O$, KCl, $Na_3PO_4 \cdot 10\,H_2O$.

44) Distinguish between a hydrate and an anhydrous compound.

_____

45) Epsom salts have the formula $MgSO_4 \cdot 7\,H_2O$. How many water molecules are associated with one formula unit of $MgSO_4$? Write the chemical name of Epsom salts.

46) What is the name of $CaCl_2 \cdot 2\,H_2O$?

_____

47) Write the formulas of ammonium phosphate 3-hydrate and potassium sulfide 5-hydrate.

48) One hydrate of barium hydroxide contains eight molecules of water per formula unit of the anhydrous compound. Write the formula and name of this hydrate.

Table 6.12
Formula Writing Exercise No. 1

Instructions: For each box, write the chemical formula of the compound formed by the cation at the head of the column and the anion at the left of the row. Refer only to the periodic table when completing this exercise. Correct formulas are listed in "Answers to Questions and Problems," Chapter 6.*

Ions	Potassium	Calcium	Chromium(III)	Zinc	Silver	Iron(II)	Aluminum	Mercury(I)
Nitrate								
Sulfate								
Hypochlorite								
Nitride								OMIT
Hydrogen sulfide								
Bromite								
Hydrogen phosphate								
Chloride								
Hydrogen carbonate								
Acetate†								
Selenite‡								

*Some compounds in the table are unknown.

†The acetate ion is derived from the ionization of acetic acid, $HC_2H_3O_2$. The ion formula is listed in Table 6.7.

‡The selenite ion contains selenium, $Z = 34$.

Table 6.13

Formula Writing Exercise No. 2

Instructions: For each box, write the chemical formula and name of the compound formed by the cation at the head of the column and the anion at the left of the row. Correct formulas and names are listed in ''Answers to Questions and Problems,'' Chapter 6.*

Ions	Na^+	Mg^{2+}	Pb^{2+}	Cu^{2+}	Fe^{3+}	NH_4^+	Hg^{2+}	$Ga^{3+†}$
OH^-								
BrO^-								
CO_3^{2-}								
ClO_3^-								
HSO_4^-								
Br^-								
PO_4^{3-}								
IO_4^-								
S^{2-}								
$MnO_4^{-‡}$								
$C_2O_4^{2-‡}$								

* Some compounds in the table are unknown.

† Ga is the symbol for gallium, Z = 31.

‡ These ions are listed in Table 6.7.

Questions 49 to 78: Items in the remaining questions are selected at random from any section of the chapter. Unless marked with an asterisk (*), all names and formulas are included in the performance goals and should be found with reference to no more than a periodic table. Ions in compounds marked with an asterisk are included in Tables 6.6 and 6.7, or, if the unfamiliar ion is monatomic, the atomic number of the element is given. In all questions, given a name, write the formula; given a formula, write the name.

49) N^{3-}, $Ca(ClO_3)_2$, iron(III) sulfate, phosphorus pentachloride.

50) Co_2O_3, Na_2SO_3, mercury(II) iodide, aluminum hydroxide.

_____

51) Selenium dioxide (selenium, Z = 34), magnesium nitrite, $FeBr_2$, Ag_2O.

52) Barium chromate*, calcium sulfite, CuCl, $AgNO_3$.

_____

53) HS^-, $BeBr_2$, aluminum nitrate, oxygen difluoride.

54) Hg_2Cl_2, HIO_4, cobalt(II) sulfate, lead(II) nitrate.

_____

55) Magnesium nitride, lithium bromite, $NaHSO_3$, KSCN*.

56) Uranium trifluoride (uranium, Z = 92), barium peroxide*, $MnCl_2$, $NaClO_2$.

_____

57) HNO_2, $Zn(HSO_4)_2$, potassium cyanide*, copper(I) fluoride.

58) N_2O_3, $LiMnO_4$*, indium selenide (indium, Z = 49)*, mercury(I) thiocyanate.

_____

59) Strontium iodate (strontium, Z = 34)*, sodium hypochlorite, Rb_2SO_4* (rubidium, Z = 37), P_2O_5.

60) Hydrogen sulfite ion, potassium nitrate, $MnSO_4$, SO_3.

_____

61) $Ni(HCO_3)_2$, CuS, chromium(III) iodide, potassium hydrogen phosphate.

62) BrO_3^-, $Ni(OH)_2$, silver chloride, silicon hexafluoride.

_____

63) Cobalt(III) sulfate, iron(III) iodide, $Cu_3(PO_4)_3$, $Mn(OH)_2$.

64) Hypochlorous acid, chromium(II) bromide, $KHCO_3$, $Na_2Cr_2O_7$*.

_____

65) Al_2Se_3 (selenium, Z = 34)*, $MgHPO_4$, potassium perchlorate, bromous acid.

66) K_2TeO_4 (tellurium, Z = 52)*, $ZnCO_3$, chromium(II) chloride, acetic acid*.

_____

67) Perchlorate ion, barium carbonate, NH_4I, PCl_3.

68) Tellurate ion (tellurium, Z = 52), manganese(III) phosphate, $NaC_2H_3O_2$*, H_2S (two names are possible).

_____

69) SnO, $(NH_4)_2Cr_2O_7$*, sodium hydride, oxalic acid*.

70) Na_2O_2*, $NiCO_3$, iron(II) oxide, hydrosulfuric acid.

_____

71) Magnesium sulfate, mercury(II) bromite, $Na_2C_2O_4$*, $Mn(OH)_3$.

72) Zinc phosphide, cesium nitrate (cesium, Z = 55), NH_4CN, S_2F_{10}.

_____

73) ICl, $AgC_2H_3O_2$*, lead(II) dihydrogen phosphate, gallium fluoride (gallium, Z = 31).

74) $CdCl_2$ (cadmium, Z = 48)*, $Ni(ClO_3)_2$, cobalt(III) phosphate, calcium periodate.

_____

75) Tin(II) fluoride, potassium chromate*, LiH, $FeCO_3$.

76) HPO_4^{2-}, CuO, sodium oxalate*, ammonia.

_____

77) Mercury(I) ion, cobalt(II) chloride, SiO_2, $LiNO_2$.

78) Calcium dihydrogen phosphate, potassium permanganate*, NH_4IO_3, H_2SeO_4 (selenium, Z = 34).

MATCHING SET ANSWERS

Set 1: 8–4–10–2–5–6–1–7–9–11–3

Set 2: 4–11–7–1–10–6–2–8–5–3–9

EXAMPLE 7.1 _____

How many atoms of each element are in a formula unit of magnesium chloride? Of barium iodate?

Before you can answer these questions you need the formulas of magnesium chloride and barium iodate. Use only a periodic table as a guide. The formulas are . . .

_____   _____

magnesium chloride, $MgCl_2$; barium iodate, $Ba(IO_3)_2$

$MgCl_2$ comes from Mg^{2+} giving a 2+ charge, and 2 Cl^- giving a total 2− charge. The formula of barium iodate is developed in the same way.

Now you have both formulas. How many atoms of each element in each formula?

_____   _____

magnesium chloride, $MgCl_2$: 1 magnesium atom, 2 chlorine atoms

barium iodate, $Ba(IO_3)_2$: 1 barium atom, 2 iodine atoms, 6 oxygen atoms

The 2 iodine atoms and 6 oxygen atoms in $Ba(IO_3)_2$ are just like the 2 nitrogen atoms and 6 oxygen atoms in $Ca(NO_3)_2$: 2×1 for iodine, and 2×3 for oxygen.

7.2 MOLECULAR MASS; FORMULA MASS

> **PG 7B** Distinguish among atomic mass, molecular mass, and formula mass.
> **7C** Calculate the formula (molecular) mass of any compound whose formula is known or given.

In Section 5.5 you learned that the atomic mass of an element is the average mass of its atoms, expressed in atomic mass units, amu. But what about compounds? Is there such a thing as a "compound mass"? The answer is yes. It is based on the chemical formula of the compound. It is called the **formula mass** of the compound, or, in the case of molecular compounds, **molecular mass.** These terms are defined exactly the same way that atomic mass is defined: **Molecular (or formula) mass is the average mass of molecules (or formula units) compared to the mass of an atom of carbon-12 at exactly 12 atomic mass units.**

The formula mass of a compound is equal to the sum of all of the atomic masses in the formula unit:

$$\text{formula mass} = \Sigma \text{ atomic masses in formula unit} \qquad (7.1)$$

Σ is the Greek letter sigma. When used as a symbol, it means "the sum of all values of" whatever follows.

A word about significant figures in the calculation of formula mass: Nearly all of the problems in this book can be solved with formula mass calculated to the first decimal place. We therefore make the first decimal place the general standard throughout the book, but with three exceptions:*

_____

*This is an arbitrary standard that is followed in this text. If your instructor prefers a different standard, by all means adopt it.

First, if the calculated formula mass has only two significant figures to the first decimal, that mass is expressed to the second decimal. Only two stable substances have such formula masses: hydrogen, H_2, 2.02 amu, and lithium hydride, LiH, 7.95 amu.

Second, if a formula contains more than four atoms of an element, we use that element's atomic mass to four significant figures. This sometimes affects the first decimal when the masses are added and rounded off.

Third, if measured amounts are in four significant figures, formula masses are calculated to four significant figures. This requires the second decimal when the formula mass of the compound is less than 100 amu. There are a few such problems much later in the book; you need not be concerned with them now.

Here we illustrate the calculation of the molecular mass of carbon dioxide, CO_2. There are one carbon atom and two oxygen atoms in the formula unit. The formula mass is the sum of the atomic masses of these three atoms:

Element	Atoms in Formula	Atomic Mass	Mass in Formula	
Carbon	1	12.0 amu	12.0 amu	= 12.0 amu
Oxygen	2	16.0 amu	2(16.0 amu)	= 32.0 amu
			Total molecular mass	44.0 amu

Table 7.1
Calculator Procedure

AOS LOGIC	
Press	Display
12	12
+	12
2	2
×	2
16	16
=	44

RPN LOGIC	
Press	Display
12	12.00
ENTER	12.00
2	2.00
ENTER	2.00
16	16.00
×	32.00
+	44.00

We are accustomed to setting up addition problems in vertical columns, as shown. However, a horizontal setup of the problem is convenient because when read from left to right it matches the typical calculator sequence:

$$12.0 \text{ amu C} + 2(16.0 \text{ amu}) \text{ O} = 44.0 \text{ amu } CO_2$$

The keyboard sequence is given in Table 7.1. From this point on, our formula mass calculation setups are written horizontally.

EXAMPLE 7.2

Calculate the formula mass of (a) magnesium sulfate and (b) aluminum sulfide.

First you need the formulas. They are . . .

_____

magnesium sulfate, $MgSO_4$ aluminum sulfide, Al_2S_3

Mg^{2+} and SO_4^{2-} combine on a 1:1 ratio. With aluminum sulfide, it takes three 2– charges from S^{2-} to balance two 3+ charges from Al^{3+}.

Let's work on the $MgSO_4$ first. Using the preceding formula and the periodic table for atomic masses, write a horizontal setup for the problem, but do not calculate the answer yet.

_____

Magnesium sulfate, commonly known as Epsom salts, is used in various different ways in the making of fabrics, including fireproofing them, weighting cotton and silk, and dyeing and printing calicos.

Aluminum sulfide is a potentially hazardous compound because when it is exposed to moisture it releases toxic hydrogen sulfide gas.

1 Mg atom + 1 S atom + 4 O atoms = 24.3 amu + 32.1 amu + 4(16.0 amu)

Now use your calculator to find the sum. Do it, if you can, without writing any other numbers; just the final answer.

———————   ———————

$MgSO_4$: 24.3 amu + 32.1 amu + 4(16.0 amu) = 120.4 amu

If you had any difficulty getting the correct answer on your calculator, take a few minutes to learn the technique. Learn it NOW!

Next write the horizontal setup for Al_2S_3 and find the formula mass on your calculator.

———————   ———————

Al_2S_3: 2(27.0 amu) + 3(32.1 amu) = 150.3 amu

EXAMPLE 7.3 ————————————————————————————————————

Ammonium nitrate is used in making such common products as anesthetics, matches, fireworks, and fertilizer.

Ammonium sulfate is a familiar fertilizer that is sold in most garden shops.

Calculate the formula mass of ammonium nitrate and ammonium sulfate.

First, we need the formulas. Careful. In writing the formula of an ionic compound, it is always the formula of the cation followed by the formula of the anion. All three ions in these compounds are polyatomic. Use parentheses as necessary.

———————   ———————

ammonium nitrate, NH_4NO_3 ammonium sulfate, $(NH_4)_2SO_4$

Both the ammonium ion, NH_4^+, and the nitrate ion, NO_3^-, contain nitrogen. They are not combined in writing the formula, however; each ion keeps its identity in the compound formula. In ammonium sulfate, it takes two ammonium ions, each with a 1+ charge, to balance the 2− charge of a single sulfate ion, SO_4^{2-}.

Now count up the atoms of each element in NH_4NO_3 and calculate its formula mass.

———————   ———————

Two N atoms, four H atoms, and three O atoms

NH_4NO_3: 2(14.0 amu) + 4(1.0 amu) + 3(16.0 amu) = 80.0 amu

In calculating formula mass it makes no difference which ions the nitrogen atoms come from.

Now the formula mass of $(NH_4)_2SO_4$:

———————   ———————

$(NH_4)_2SO_4$: 2(14.0 amu) + 8(1.008 amu) + 32.1 amu + 4(16.0 amu) = 132.2 amu

This is a formula that has more than four hydrogen atoms, so hydrogen's contribution to formula mass is calculated from 1.008 amu per atom.

EXAMPLE 7.4

The stable form of elemental chlorine is a diatomic (two-atom) molecule whose formula is Cl_2. Table sugar, $C_{12}H_{22}O_{11}$, is a molecular compound. Calculate the molecular masses of these substances.

Remember that *molecular* mass is the same as *formula* mass, except that it is used when speaking of molecular substances. Be sure you find the *formula* mass of Cl_2.

Elemental chlorine, Cl_2, is one of the most important industrial chemicals. It is used to bleach wood pulp in manufacturing paper, to bleach textiles, in the manufacture of plastics, and to purify drinking water.

Cl_2: $2(35.5 \text{ amu}) = 71.0 \text{ amu}$

$C_{12}H_{22}O_{11}$: $12(12.01 \text{ amu}) + 22(1.008 \text{ amu}) + 11(16.00 \text{ amu}) = 342.30 \text{ amu}$

Note that the *atomic* mass of chlorine, Cl, is 35.5 amu, but the *molecular* (*formula*) mass of molecular chlorine, Cl_2, is two times 35.5 amu. This illustrates an important point: *Always calculate formula (or molecular) mass exactly as the formula is written.* It follows that the formula must be written correctly! . . . In $C_{12}H_{22}O_{11}$, there are more than four atoms of all elements, so their atomic masses are used to four significant figures. By our "first decimal" standard, rounding off to 342.3 amu would be acceptable.

7.3 THE MOLE CONCEPT

PG 7D Define the *mole.* Identify the number that corresponds to one mole.
7E Given the number of moles (or units) in any sample, calculate the number of units (or moles) in the sample.

In the real world of humans who buy sugar in pounds or in kilograms, the formula mass of sugar at 342.3 amu is not very important. The amu is much, much too tiny to be useful to the average person. Even industrial and laboratory chemists work with chemical quantities that can be weighed. Their underlying purpose, though, is to combine reactants in the same *numbers* of formula units that appear in the chemical equation. The number of formula units in a sample is proportional to the mass of the sample—just as the mass of milk is proportional to the number of milk cartons in the shopping cart. This makes it possible to count atoms, molecules, and formula units by weighing.

To organize the "counting by weighing" idea, scientists have "invented" the SI unit for amount of substance. This unit is the **mole (mol): One mole is that amount of any substance that contains the same number of units as the number of atoms in exactly 12 grams of carbon-12.** The number is called **Avogadro's number, N,** in honor of the man whose interpretation of gases led to an early method for estimating atomic weights. Figure 7.1 shows 1 mole of several substances.

The value of N, the number of atoms in exactly 12 grams of carbon-12, has been determined by experiment. To three significant figures there are 6.02×10^{23} units per mole:

$$1 \text{ mole of any substance} = 6.02 \times 10^{23} \text{ units of that substance} \qquad (7.2)$$

This is one huge number! It staggers the imagination. To get some appreciation of the size of Avogadro's number, imagine a cubic box that is 29 cm—about 11½ inches—

LEARN IT NOW You have just been introduced to what is probably the most important concept in chemistry, the *mole.* It is the basis of, or somehow involved in, nearly every chemistry problem for the rest of this book. Take the time to learn about and understand the mole—*now!*

(a)

(b)

Figure 7.1

One mole of elements and one mole of compounds made up of the same elements (plus oxygen, in some cases). (a) Elements (clockwise from the top): carbon, zinc, nickel, calcium, iodine, sulfur, copper; (center) mercury; bromine. (b) Compounds (clockwise, beginning with red compound): mercury(I) iodide, calcium bromide, copper(II) bromide, copper(II) sulfide, zinc sulfide, nickel carbonate; (center) copper(II) sulfate.

long, high, and wide. At normal conditions, that box contains approximately one mole of molecules that make up the mixture we call air. If we were to close the box and connect it to the finest vacuum pump that has ever been made, we could remove nearly ''all'' of the molecules originally in the box—about 99.999999999% of them. The number of molecules still in the box—0.000000001% of the original number—would be about 6 *trillion:* 6,000,000,000,000! And what is 6 trillion? If you were to distribute the molecules equally among every man, woman, and child living on earth today, each person would receive about 1300 molecules!

To get a better idea of the mole and how it can be used, let us compare it to a more familiar counting unit, the dozen. Suppose a dozen were defined as the number of eggs in the standard carton in which they are sold in the local supermarket. By experiment (walking down to the store, opening a box of eggs, and counting them), you can determine that this number is 12. Now, when you buy eggs, you can specify the number in terms of dozens. When did you ever see a shopping list with 24 eggs on it? Would it not be 2 dozen eggs? Just as 2 dozen eggs means $2 \times 12 = 24$ eggs, 2 moles of carbon atoms means $2 \times 6.02 \times 10^{23}$ carbon atoms. If you ever have difficulty understanding ''mole'' in a sentence, substitute the word ''dozen'' and you will see what it means.

Avogadro's number, 6.02×10^{23} units per mole, is a ''per'' relationship, and therefore a dimensional analysis conversion factor between units and moles. The unit path is units $\leftrightarrow$ moles. There are 12 units per dozen of anything. That also is a conversion factor. Two parallel problems are (1) how many eggs are in 3 dozen eggs and (2) how many carbon atoms are in 3 moles of carbon? Both are one-step conversions:

FLASHBACK The use of per relationships to convert between quantities that are directly proportional to each other was introduced in Section 3.3.

$$3 \text{ doz eggs} \times \frac{12 \text{ eggs}}{1 \text{ doz eggs}} = 36 \text{ eggs};$$

$$3 \text{ mol C atoms} \times \frac{6.02 \times 10^{23} \text{ C atoms}}{1 \text{ mol C atoms}} = 1.81 \times 10^{24} \text{ C atoms}$$

EXAMPLE 7.5

How many moles of water are in 1.67×10^{22} water molecules?

How would you calculate the number of dozens of eggs in 48 eggs? This problem is solved in exactly the same way. "Plan" the problem and calculate the answer.

GIVEN: 1.67×10^{22} molecules water   WANTED: mol water
PATH: molecules → mol   FACTOR: 6.02×10^{23} molecules/mol

$$1.67 \times 10^{22} \text{ molecules water} \times \frac{1 \text{ mol water}}{6.02 \times 10^{23} \text{ molecules water}} = 0.0277 \text{ mol water}$$

✔ **QUICK CHECK 7.1**

What does one mole of carbon atoms have in common with one mole of oxygen atoms?

7.4 MOLAR MASS

PG 7F   Define *molar mass,* or interpret statements in which the term *molar mass* is used.
7G   Calculate the molar mass of any substance whose chemical formula is known.

The molar mass (MM) of a substance is the mass in grams of one mole of that substance. The units of molar mass follow from its definition: grams per mole (g/mol). Mathematically, the defining equation of molar mass is

$$MM \equiv \frac{mass}{mole} = \frac{g}{mol} \qquad (7.3)$$

FLASHBACK This per relationship yields a defining equation that is a conversion factor for dimensional analysis calculations. See Section 3.8.

The definitions of atomic mass, the mole, and molar mass are all directly or indirectly related to carbon-12. This leads to two important facts:

1) The mass of one atom of carbon-12—the atomic mass of carbon-12—is exactly 12 atomic mass units.
2) The mass of one mole of carbon-12 atoms is exactly 12 grams; its molar mass is exactly 12 grams per mole.

Notice that the atomic mass and the molar mass of carbon-12 are *numerically* equal. They differ only in units; atomic mass is measured in atomic mass units, and molar mass is measured in grams per mole. The same relationships exist between atomic and molar masses of all elements, between molecular masses and molar masses of molecular substances, and between formula masses and molar masses of ionic compounds. In other words:

The molar mass of any substance in grams per mole is numerically equal to the atomic or formula mass of that substance in atomic mass units.

For example, from sources in this chapter:

Source	Substance	Atomic/Formula Mass (amu)	Molar Mass (g/mol)
Section 7.2	O atoms	16.0	16.0
Example 7.2	$MgSO_4$	120.4	120.4
Example 7.4	Cl_2	71.0	71.0

If you can find the atomic or formula mass of a substance, change the units and you have its molar mass.

EXAMPLE 7.6

Bromine is extracted from seawater and used for bleaching fibers and silk and in the manufacture of medicinal bromine compounds.

Calcium fluoride occurs in nature and is known as *fluorspar*. It is used in making steel.

Calculate the molar mass of elemental bromine and of calcium fluoride.

Begin with the formulas of bromine and calcium fluoride. Set up the horizontal additions as if you were calculating formula mass, but use molar mass units.

Br_2: $2(79.9 \text{ g/mol Br}) = 159.8 \text{ g/mol Br}_2$

CaF_2: $40.1 \text{ g/mol Ca} + 2(19.0 \text{ g/mol F}) = 78.1 \text{ g/mol CaF}_2$

There is one mole of each substance in Figure 7.1. This means that each sample has the same number of atoms, molecules, or formula units. The mass of each sample is the molar mass of the substance.

✔ QUICK CHECK 7.2

a) What do 12.0 g of carbon atoms have in common with 16.0 g of oxygen atoms?
b) The molecular mass of the explosive, TNT, is 227 amu. What is the molar mass of TNT?

7.5 CONVERSION BETWEEN MASS, NUMBER OF MOLES, AND NUMBER OF UNITS

> **PG 7H** Given any one of the following for a substance whose formula is known, calculate the other two: mass, number of moles, number of formula units.

The "per" relationship in molar mass, grams per mole, means you can use dimensional analysis to convert from one to the other. Molar mass is the conversion factor. This one-step conversion is probably used more often than any other conversion in chemistry.

EXAMPLE 7.7 _____

You are carrying out a laboratory reaction that requires 0.0360 mole of barium chloride. How many grams of the compound do you weigh out?

Barium chloride is used in making paint pigment and in tanning leather.

Before you can use molar mass as a conversion factor, you must calculate its value. Do that for barium chloride.

_____   _____

$BaCl_2$: 137.3 g/mol Ba + 2(35.5 g/mol Cl) = 208.3 g/mol $BaCl_2$

It is now a one-step conversion from moles to grams, using molar mass as the conversion factor. "Plan" the problem and calculate the answer.

_____   _____

GIVEN: 0.0360 mol $BaCl_2$ WANTED: g $BaCl_2$
PATH: mol $BaCl_2 \rightarrow$ g $BaCl_2$ FACTOR: 208.3 g $BaCl_2$/mol $BaCl_2$

$$0.0360 \text{ mol } BaCl_2 \times \frac{208.3 \text{ g } BaCl_2}{1 \text{ mol } BaCl_2} = 7.50 \text{ g } BaCl_2$$

_____

Almost all problems involving quantities of chemicals require the formula and the molar mass of the chemical. Finding these were the starting steps in Example 7.7. From here on we refer to writing a formula and calculating its molar mass as the "starting steps." Then we are ready to use molar mass as a conversion factor.

The "starting steps" in a problem are writing the formula of the chemical concerned and calculating its molar mass.

EXAMPLE 7.8 _____

How many moles of aluminum sulfate are in 132 g of the compound?

The mass → mole conversion is by molar mass. It is a bit more challenging this time, but you can do it. Begin with the starting steps: Write the formula and calculate the molar mass of aluminum sulfate.

Among the many uses of aluminum sulfate are tanning leather, fireproofing and waterproofing cloth, treating sewage, and making antiperspirants.

_____   _____

$Al_2(SO_4)_3$: 2(27.0 g/mol Al) + 3(32.1 g/mol S) + 12(16.00 g/mol O) = 342.3 g/mol $Al_2(SO_4)_3$

The formula is a "two of the 3+'s" (two Al^{3+}) to balance "three of the 2−'s" (three SO_4^{2-}) at 6+ and 6−, like aluminum sulfide in Example 7.2.

Set up and solve the problem.

_____   _____

GIVEN: 132 g $Al_2(SO_4)_3$ WANTED: mol $Al_2(SO_4)_3$
PATH: g $Al_2(SO_4)_3 \rightarrow$ mol $Al_2(SO_4)_3$
FACTOR: 342.3 g $Al_2(SO_4)_3$/mol $Al_2(SO_4)_3$

$$132 \text{ g } Al_2(SO_4)_3 \times \frac{1 \text{ mol } Al_2(SO_4)_3}{342.3 \text{ g } Al_2(SO_4)_3} = 0.386 \text{ mol } Al_2(SO_4)_3$$

_____

Let's step back and look at some of the dimensional analysis changes we have made in the past few pages:

WHERE WE DID IT	CHANGES WE MADE	CONVERSION FACTORS WE USED
Section 7.3	*mol* ⟶ units	N: 6.02×10^{23} units/mol
Example 7.5	units ⟶ *mol*	N: 6.02×10^{23} units/mol
Example 7.7	*mol* ⟶ g	Molar mass
Example 7.8	g ⟶ *mol*	Molar mass

Notice that the mole is present in all four changes. In fact, the mole is the connecting link between grams and number of units. Using N for Avogadro's number, 6.02×10^{23}, and MM for molar mass, we have

$$\text{units} \longrightarrow \text{mol} \longrightarrow \text{g} \quad \text{or} \quad \text{g} \longrightarrow \text{mol} \longrightarrow \text{units}$$

Conversion factor: N   MM   MM   N

In other words, changing from units to mass or vice versa is a two-step dimensional analysis conversion; change the given quantity to moles, and then moles to the wanted quantity.

EXAMPLE 7.9

How many molecules are in 454 g (1 pound) of water?

We need the molar mass of water as one of the conversion factors. Complete the starting steps.

———

H_2O: $2(1.0$ g/mol H$) + 16.0$ g/mol O $= 18.0$ g/mol H_2O

Now, to see clearly the start, the finish, and the way to solve the problem, "plan" it: given, wanted, unit path, and factors.

———

GIVEN: 454 g H_2O WANTED: molecules H_2O
PATH: g H_2O → mol H_2O → molecules H_2O

FACTORS: 6.02×10^{23} molecules H_2O/mol H_2O; 18.0 g H_2O/mol H_2O

Set up and solve the problem.

———

$$454 \text{ g } H_2O \times \frac{1 \text{ mol } H_2O}{18.0 \text{ g } H_2O} \times \frac{6.02 \times 10^{23} \text{ molecules } H_2O}{1 \text{ mol } H_2O} = 1.52 \times 10^{25} \text{ molecules } H_2O$$

EXAMPLE 7.10 ————————————————————————

What is the mass of one billion billion ($1.00 \times 10^9 \times 10^9 = 1.00 \times 10^{18}$) molecules of ammonia?

You are on your own. Take it all the way.

Ammonia is an important industrial chemical that is used in refrigerants and in the manufacture of nitric acid, explosives, synthetic fibers, and fertilizers.

————— ————

GIVEN: 1.00×10^{18} molecules NH_3 WANTED: g NH_3
PATH: molecules $NH_3 \rightarrow$ mol $NH_3 \rightarrow$ g NH_3
FACTORS: 6.02×10^{23} molecules NH_3/mol NH_3; 17.0 g NH_3/mol NH_3

$$1.00 \times 10^{18} \text{ molecules } NH_3 \times \frac{1 \text{ mol } NH_3}{6.02 \times 10^{23} \text{ molecules } NH_3} \times \frac{17.0 \text{ g } NH_3}{1 \text{ mol } NH_3} = 2.82 \times 10^{-5} \text{ g } NH_3$$

This very small mass, about $\dfrac{6}{100,000,000}$ of a pound, suggests again the enormous number of molecules in a mole.

———

✔ **QUICK CHECK 7.3**

What are the "starting steps" in solving a chemical formula problem?

7.6 MASS RELATIONSHIPS BETWEEN ELEMENTS IN A COMPOUND: PERCENTAGE COMPOSITION

> **PG 7I** Calculate the percentage composition of any compound whose formula is known.
> **7J** For any compound whose formula is known, given the mass of a sample, calculate the mass of any element in the sample; or, given the mass of any element in the sample, calculate the mass of the sample or the mass of any other element in the sample.

The term *cent* refers to 100, as 100 cents in a dollar and 100 years in a century. **Percent therefore means "per 100."** Thus, **percent is the amount of one part of a mixture per 100 total parts in the mixture.** If the part whose percentage we wish to identify is A, then

$$\% \text{ A} \equiv \frac{\text{parts of A in mixture}}{100 \text{ total parts in mixture}} \tag{7.4}$$

Equation 7.4 is a defining equation for percentage. To calculate percentage, we use a more convenient form that is derived from Equation 7.4:

$$\% \text{ of A} = \frac{\text{parts of A}}{\text{total parts}} \times 100 \tag{7.5}$$

The ratio, (parts of A)/(total parts), is the fraction of the sample that is A. Multiplying that fraction by 100 gives percentage of A. To illustrate, in Example 7.6 you calculated the molar mass of calcium fluoride. The calculation setup was

$$CaF_2: 40.1 \text{ g/mol Ca} + 2(19.0 \text{ g/mol F}) = 78.1 \text{ g/mol } CaF_2$$

The part of a mole that is calcium is 40.1 g. The total mass of a mole is 78.1 g. The fraction of a mole that is calcium is 40.1/78.1. The percentage of calcium is therefore

$$\% \text{ Ca} = \frac{\text{parts of A}}{\text{total parts}} \times 100 = \frac{40.1 \text{ g Ca}}{78.1 \text{ g } CaF_2} \times 100 = 51.3\% \text{ Ca}$$

EXAMPLE 7.11

Calculate the percentage of fluorine in CaF_2.

This is an "equation" problem, using Equation 7.5.

_____

GIVEN: 2×19.0 g F; 78.1 g CaF_2 WANTED: % F

EQUATION: $\% \text{ F} = \dfrac{\text{g F}}{\text{g } CaF_2} \times 100 = \dfrac{2(19.0) \text{ g F}}{78.1 \text{ g } CaF_2} \times 100 = 48.7\% \text{ F}$

_____

The **percentage composition of a compound is the percentage by mass of each element in the compound.** The percentage composition of calcium fluoride is 51.3% calcium and 48.7% fluorine. As you have seen with CaF_2, percentage composition can be calculated from the same numbers that are used to find the molar mass of a compound.

If you calculate the percentage composition of a compound correctly, the sum of all percents must be 100%. This fact can be used to check your work. With calcium fluoride, 51.3% + 48.7% = 100.0%. When you apply this check, don't be concerned if you are high or low by 0.1%, or even 0.2% for a compound of high molar mass. This can result from legitimate roundoffs along the way.

EXAMPLE 7.12

Calculate the percentage composition of aluminum sulfate. Check your results.

You found the molar mass of aluminum sulfate in Example 7.8.

$$Al_2(SO_4)_3: 2(27.0 \text{ g/mol Al}) + 3(32.1 \text{ g/mol S}) + 12(16.0 \text{ g/mol O}) = 342.3 \text{ g/mol } Al_2(SO_4)_3$$

The numbers are bigger, but the procedure is just like Example 7.11.

_____

$$\frac{2(27.0) \text{ g Al}}{342.3 \text{ g } Al_2(SO_4)_3} \times 100 = 15.8\% \text{ Al} \qquad \frac{3(32.1) \text{ g S}}{342.3 \text{ g } Al_2(SO_4)_3} \times 100 = 28.1\% \text{ S}$$

$$\frac{12(16.0) \text{ g O}}{342.3 \text{ g } Al_2(SO_4)_3} \times 100 = 56.1\% \text{ O} \qquad 15.8\% + 28.1\% + 56.1\% = 100.0\%$$

_____

If you have the percentage composition of a compound, you can find the amount of any element in a known amount of the compound. One way to do this is to use percentage as a conversion factor, grams of element per 100 grams of compound. For example, if aluminum sulfate is 15.8% aluminum, the mass of aluminum in 88.9 g $Al_2(SO_4)_3$ is

$$88.9 \text{ g } Al_2(SO_4)_3 \times \frac{15.8 \text{ g Al}}{100 \text{ g } Al_2(SO_4)_3} = 14.0 \text{ g Al}$$

EXAMPLE 7.13

How many grams of fluorine are in 216 g of calcium fluoride?

In Example 7.11 you found that calcium fluoride is 48.7% fluorine. Solve the problem.

_____

GIVEN: 216 g CaF_2 WANTED: g F   PATH: g $CaF_2 \rightarrow$ g F
FACTOR: 48.7 g F/100 g CaF_2

$$216 \text{ g } CaF_2 \times \frac{48.7 \text{ g F}}{100 \text{ g } CaF_2} = 105 \text{ g F}$$

Are there three significant figures in the denominator, 100 g CaF_2? The 100 is a defined quantity, like 12 inches equal 1 foot. The total percent of anything is defined to be 100. The denominator therefore has an infinite number of significant figures.

EXAMPLE 7.14

An experiment requires that enough calcium fluoride be used to yield 1.91 g of calcium. How much calcium fluoride must be weighed out?

You know from Example 7.11 that 51.3% of a sample of CaF_2 is calcium. Complete the problem.

_____

GIVEN: 1.91 g Ca   WANTED: g CaF_2 PATH: g Ca $\rightarrow$ g CaF_2
FACTOR: 51.3 g Ca/100 g CaF_2

$$1.91 \text{ g Ca} \times \frac{100 \text{ g } CaF_2}{51.3 \text{ g Ca}} = 3.72 \text{ g } CaF_2$$

It is not necessary to know the percentage composition of a compound to change between mass of an element in a compound and mass of the compound. The masses of all elements in a compound and the mass of the compound itself are directly proportional to each other; they are related by "per" expressions. Once again, the molar mass figures for CaF_2 are

$$CaF_2: 40.1 \text{ g/mol Ca} + 2(19.0 \text{ g/mol F}) = 78.1 \text{ g/mol } CaF_2$$

From these numbers we conclude that:

g Ca $\propto$ g F   40.1 g Ca/38.0 g F

g Ca $\propto$ g CaF_2 40.1 g Ca/78.1 g CaF_2

g F $\propto$ g CaF_2 38.0 g F/78.1 g CaF_2

Any of these ratios, or their inverses, may be used as a conversion factor from the mass of one species to the mass of the other. For instance, to find the mass of CaF_2 that contains 1.91 g Ca (Example 7.14) from these numbers, we calculate

$$1.91 \; \cancel{\text{g Ca}} \times \frac{78.1 \text{ g CaF}_2}{40.1 \; \cancel{\text{g Ca}}} = 3.72 \text{ g CaF}_2$$

EXAMPLE 7.15 _____

Calculate the number of grams of fluorine in a sample of calcium fluoride that contains 2.01 g of calcium.

"Plan" the problem. Then write the calculation setup and find the answer.

_____   _____

GIVEN: 2.01 g Ca   WANTED: g F   PATH: g Ca $\rightarrow$ g F
FACTOR: 40.1 g Ca/38.0 g F

$$2.01 \; \cancel{\text{g Ca}} \times \frac{38.0 \text{ g F}}{40.1 \; \cancel{\text{g Ca}}} = 1.90 \text{ g F}$$

7.7 THE QUANTITATIVE MEANING OF A CHEMICAL FORMULA: A SUMMARY

A chemical formula conveys a large amount of information. You have learned to use this information to make conversions among masses, moles, and numbers of formula units, atoms, and ions. Not much has been said about ions; they will receive their share of attention later. At this point, note that the mass difference between ions and the atoms that make up the ions is so small that it may be disregarded. This is because electrons have so little mass.

Here we list all of the quantities that can be derived from one mole of $Ca(NO_3)_2$. This summarizes what can be learned from a formula, and also gives you the opportunity to test your formula calculation skill by confirming the given results.

$Ca(NO_3)_2$ is 24.4% Ca, 17.1% N, and 58.5% O. One mole of $Ca(NO_3)_2$ contains:

6.02×10^{23} $Ca(NO_3)_2$ formula units	1 mol $Ca(NO_3)_2$	164.1 g $Ca(NO_3)_2$
6.02×10^{23} Ca atoms	1 mol Ca atoms	40.1 g Ca atoms
1.20×10^{24} N atoms	2 mol N atoms	28.0 g N atoms
3.61×10^{24} O atoms	6 mol O atoms	96.0 g O atoms
6.02×10^{23} Ca^{2+} ions	1 mol Ca^{2+} ions	40.1 g Ca^{2+} ions
1.20×10^{24} NO_3^- ions	2 mol NO_3^- ions	124.0 g NO_3^- ions

7.8 SIMPLEST (EMPIRICAL) FORMULA OF A COMPOUND

PG 7K   Distinguish between a simplest (empirical) formula and a molecular formula.

Simplest Formulas and Molecular Formulas

Where do chemical formulas come from? They come from the same source as any fundamental chemical information, from experiments, usually performed in the laboratory. Among other things, chemical analysis can give us the percentage composition of a compound. Such data give us the **empirical formula,** a formula based on experiment, of the compound. This is a tentative formula that is also known as the **simplest formula.** We will use that term.

The percentage composition of ethylene is 85.7% carbon and 14.3% hydrogen. Its chemical formula is C_2H_4. The percentage composition of propylene is also 85.7% carbon and 14.3% hydrogen. Its formula is C_3H_6. These are, in fact, two of a whole series of compounds having the general formula C_nH_{2n}, where n is an integer. In ethylene and propylene, n = 2 and 3, respectively. All compounds in this series have the same percentage composition.

C_2H_4 and C_3H_6 are typical molecular formulas of real chemical substances. If, in the general formula, we let n = 1, the result is CH_2. This is the simplest formula for all compounds having the general formula C_nH_{2n}. The simplest formula shows the simplest ratio of atoms of the elements in the compound. All subscripts are reduced to their lowest terms; they have no common divisor.

Simplest formulas may or may not be molecular formulas of real chemical compounds. There happens to be no known stable compound with the formula CH_2—and there is good reason to believe that no such compound can exist. On the other hand, the molecular formula of dinitrogen tetroxide is N_2O_4. The subscripts have a common divisor, 2. Dividing by 2 gives the simplest formula, NO_2. This is also the molecular formula of a real chemical, nitrogen dioxide. In other words, NO_2 is both the simplest formula and the molecular formula of nitrogen dioxide, as well as the simplest formula of dinitrogen tetroxide.

C_2H_4 and C_3H_6, ethylene and propylene, are the building blocks from which the plastics polyethylene and polypropylene are made.

Nitrogen dioxide, NO_2, is responsible for one kind of chemical smog that produces a brown haze in the atmosphere.

EXAMPLE 7.16

Write SF after each formula that is a simplest formula. Write the simplest formula after each compound whose formula is not already a simplest formula:

C_4H_{10} C_2H_6O Hg_2Cl_2 $(CH)_6$ *distribute the six*

C_4H_{10}: C_2H_5 C_2H_6O: SF Hg_2Cl_2: HgCl $(CH)_6$: CH

Determination of a Simplest Formula

PG 7L   Given data from which the mass of each element in a sample of a compound can be determined, find the simplest (empirical) formula of the compound.

To find the simplest formula of a compound, we must find the whole number ratio of atoms of the elements in a sample of the compound. The numbers in this ratio are the subscripts in the simplest formula. The procedure by which this is done is as follows:

PROCEDURE

1) Find the masses of different elements in a sample of the compound.
2) Convert the masses into moles of atoms of the different elements.
3) Express the moles of atoms as the smallest possible ratio of integers.
4) Write the simplest formula, using the number for each atom in the integer ratio as the subscript in the formula.

It is usually helpful in a simplest formula problem to organize the calculations in a table with the following headings:

Element	Grams	Moles	Mole Ratio	Formula Ratio	Simplest Formula

We will use ethylene to show how to find the simplest formula of a compound from percentage composition. As noted, the compound is 85.7% carbon and 14.3% hydrogen. We need masses of elements in Step 1 of the procedure. Thinking of percent as the number of grams of one element per 100 g of compound, a 100-g sample must contain 85.7 g of carbon and 14.3 g of hydrogen. From this we see that *percentage composition figures represent the grams of each element in a 100-g sample of the compound.* These figures complete Step 1 of the procedure. They are entered into the first two columns of the table.

Element	Grams	Moles	Mole Ratio	Formula Ratio	Simplest Formula
C	85.7				
H	14.3				

We are now ready to find the number of moles of atoms of each element, Step 2 in the procedure. This is a one-step conversion from grams to moles, g $\rightarrow$ mol, as in Example 7.8.

Element	Grams	Moles	Mole Ratio	Formula Ratio	Simplest Formula
C	85.7	$\dfrac{85.7 \text{ g}}{12.0 \text{ g/mol}} = 7.14 \text{ mol}$			
H	14.3	$\dfrac{14.3 \text{ g}}{1.01 \text{ g/mol}} = 14.2 \text{ mol}$			

It is the ratio of these moles of atoms that must now be expressed in the smallest possible ratio of integers, Step 3 in the procedure. This is most easily done by *dividing each number of moles by the smallest number of moles.* In this problem the smallest number of moles is 7.14. Thus

Element	Grams	Moles	Mole Ratio	Formula Ratio	Simplest Formula
C	85.7	7.14	$\dfrac{7.14}{7.14} = 1.00$		
H	14.3	14.2 mol	$\dfrac{14.2}{7.14} = 1.99$		

The ratio of *atoms* of the elements in a compound is the same as the ratio of *moles* of atoms in the compound. To see this in a more familiar setting, the ratio of seats to wheels in a bicycle is 1/2. In four dozen bicycles there are four dozen seats and eight dozen wheels. The ratio of dozens is 4/8, which is the same as 1/2. Thus, the numbers in the Mole Ratio column are in the same ratio as the subscripts in the simplest formula.

When placed into a formula, the numbers must be integers. Accordingly, small roundoffs may be necessary to compensate for experimental errors. In this problem 1.00/1.99 becomes 1/2, and the empirical formula is CH_2.

Element	Grams	Moles	Mole Ratio	Formula Ratio	Simplest Formula
C	85.7	7.14	1.00	1	CH_2
H	14.3	14.2 mol	1.99	2	

If either quotient in the Mole Ratio column is not close to a whole number, the Formula Ratio may be found by multiplying both quotients by a small integer. You will be guided into this in the next example.

EXAMPLE 7.17

The mass of a piece of iron is 1.62 g. Exposed to oxygen under conditions in which oxygen combines with all of the iron to form a pure oxide of iron, the final mass increases to 2.31 g. Find the simplest formula of the compound.

As before, the masses of the elements in the compound are required. This time they must be obtained from the data. The number of grams of iron in the final compound is the same as the number of grams at the start. The rest is oxygen. How many grams of oxygen combined with 1.62 g of iron if the iron oxide produced has a mass of 2.31 g?

g oxygen = g iron oxide − g iron

2.31 g iron oxide − 1.62 g iron = 0.69 g oxygen

SUMMARY

To find the molecular formula of a compound:

1) Determine the simplest formula of the compound.
2) Calculate the molar mass of the simplest formula unit.
3) Determine the molar mass of the compound (which will be given at this time).
4) Divide the molar mass of the compound by the molar mass of the simplest formula unit to get n, the number of simplest formula units per molecule.
5) Write the molecular formula.

EXAMPLE 7.19

An unknown compound is found in the laboratory to be 91.8% silicon and 8.2% hydrogen. Another experiment indicates that the molar mass of the compound is 122 g/mol. Find the simplest and molecular formulas of the compound.

Start by finding the simplest formula.

Element	Grams	Moles	Mole Ratio	Formula Ratio	Simplest Formula
Si	91.8	3.27	1.00	2	
H	8.2	8.1	2.5	5	Si_2H_5

To use Equation 7.6, you must have the molar mass of the simplest formula unit. Find the molar mass of Si_2H_5.

61.2 g Si_2H_5/mol

Calculate the number of simplest formula units in the molecule and write the molecular formula.

$n = 122/61.2 = 2$ $(Si_2H_5)_2 = Si_4H_{10}$

EVERYDAY CHEMISTRY

How to Read a Food Label

There are about 600,000 food products available in the United States, made by a $360-billion a year industry. The 1990 Nutrition Labeling and Education Act requires food manufacturers to list the grams of fat, protein, cholesterol, sodium, and other materials in each serving of their products. These amounts are then to be placed in the context of a suggested 2000 calorie daily diet. There is to be an ingredient list, just as there is today, as this book is being written in June 1993. The ingredients are to appear in decreasing order by mass percent. As practiced today, this order is sometimes deceptive.

For example, an old-style label from a package of cookies sold through a college vending machine shows enriched wheat flour as the most abundant ingredient. Farther down the label you find sugar, honey, corn syrup, and apple juice, all of which are caloric sweeteners that are nutritionally equivalent to sugar. Thus the total sugar might be greater than the "main" ingredient, the wheat flour. There is no way to tell. The 1990 Act is supposed to eliminate this uncertainty. Readers of this book will have to determine for themselves if the goal has been met.

Among other things, the new food labels are to include total fat, saturated fat, cholesterol, total carbohydrate, sugars, dietary fiber, protein, sodium, and potassium. For each of these, the label is also to show the amount per serving, the percentage of recommended daily intake, and the recommended daily intake. In addition, the new rules include uniform definitions of "low fat" and "light." Another new item is percentage calories from fat. Let's look at that one more closely.

The Food and Drug Administration (FDA) and the American Heart Association recommend a daily diet in which no more than 30% of the calories come from fat. Fat is listed on a label in grams per serving. Each gram of fat accounts for 9 calories. These data can be used to calculate the percentage of fat from any food serving. For example, according to the label on a carton of whole milk, one serving accounts for a total of 160 calories, and it has 9 grams of fat. Thus,

$$\frac{9 \text{ g fat}}{\text{serving}} \times \frac{9 \text{ cal}}{\text{g fat}} = 81 \text{ fat cal/serving}$$

$$\% \text{ calories from fat} =$$
$$\frac{81 \text{ fat cal/serving}}{160 \text{ cal/serving}} \times 100 = 51\%$$

If you are conscious of the fat content in your diet, would you buy whole milk? That 51% of calories from fat seems excessive, doesn't it? But put it into perspective. If that serving of milk is used to dampen a serving of corn flakes (100 calories per serving), the 81 fat calories are now distributed among 260 total calories. The percentage of calories from fat drops to (81/260)100 = 31%, which is just over the recommended 30%. Add some fresh fruit to the breakfast and you add more total calories without increasing fat calories, and the percentage drops some more. Watch out, though, if you eat a couple of pieces of buttered toast with your meal.

But you want buttered toast. So you decide to reduce the fat from the milk by buying milk that is 2% fat instead of whole milk. You figure 2% is *much* lower than 51%. Wrong! It's lower, but not by the amount those numbers suggest. The 2% is the percentage *by mass* of the milk that is fat. Most food labels today describe fat content as percent by mass, not percent of calories from fat. A serving of 2% milk has 5 grams of fat and 140 calories. That calculates out to 32% calories from fat. That's the number that is to be compared to 51% for whole milk. Go another step to milk that is 1% fat by mass and you come out with 15% of the calories from fat. Skim milk is near zero percent calories from fat.

Figure 7.2 is a label on a "light" ice cream carton that complies with the regulations that take effect in May 1994. It clearly states the percentage of fat by mass and the percentage of fat from calories. The shopper can leave the calculator at home!

Isn't there an easier way to arrange a healthy, fat-conscious diet than dealing with all these percentages? Yes, there is. The FDA diet suggests a daily fat allowance of 65 grams. (65 grams × 9 calories per gram =

continued

EVERYDAY CHEMISTRY

continued

585 calories, which is 29% of the recommended 2000 calories per day.) Now your glass of whole milk simply accounts for 9 of those 65 grams, whether you drink the milk or put it on your corn flakes. Add the grams of fat, regardless of percentages, from everything else you eat. When you get to 65 grams of fat, become a vegetarian for the rest of the day and your diet will be acceptable— at least from the fat standpoint.

Perhaps the best advice is still the old suggestion, "Eat to live; don't live to eat." And read the labels.

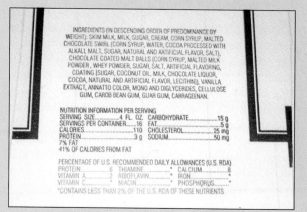

Figure 7.2
Label from a carton of "light" ice cream.

CHAPTER 7 IN REVIEW

7.1 The Number of Atoms in a Formula

7A Given the formula of a chemical compound, or a name from which the formula may be written, state the number of atoms of each element in the formula unit.

7.2 Molecular Mass; Formula Mass

7B Distinguish among atomic mass, molecular mass, and formula mass.

7C Calculate the formula (molecular) mass of any compound whose formula is known or given.

7.3 The Mole Concept

7D Define the *mole*. Identify the number that corresponds to one mole.

7E Given the number of moles (or units) in any sample, calculate the number of units (or moles) in the sample.

7.4 Molar Mass

7F Define *molar mass,* or interpret statements in which the term *molar mass* is used.

7G Calculate the molar mass of any substance whose chemical formula is known.

7.5 Conversion Between Mass, Number of Moles, and Number of Units

7H Given any one of the following for a substance whose formula is known, calculate the other two: mass, number of moles, number of formula units.

7.6 Mass Relationships Between Elements in a Compound: Percentage Composition

7I Calculate the percentage composition of any compound whose formula is known.

7J For any compound whose formula is known, given the mass of a sample, calculate the mass of any element in the sample; or, given the mass of any element in the sample, calculate the mass of the sample or the mass of any other element in the sample.

7.7 The Quantitative Meaning of a Chemical Formula: A Summary

7.8 Simplest (Empirical) Formula of a Compound

7K Distinguish between a simplest (empirical) formula and a molecular formula.

7L Given data from which the mass of each element in a sample of a compound can be determined, find the simplest (empirical) formula of the compound.

CHAPTER 7 KEY WORDS AND MATCHING SET

_7_	Amount of substance having same number of units as 12.0000 g ^{12}C.	1. "Per" expression
_2_	Mass, in atomic mass units, of one molecule of a substance.	2. Formula mass
_5_	Mass, in grams, of one mole of a substance.	3. Σ
_11_	Mass of formula unit compared to mass of carbon-12 atom.	4. Avogadro's number
_1_	Relationship between two quantities that can be used as a conversion factor in dimensional analysis.	5. Molar mass
_9_	Expression that shows the lowest possible ratio of atoms of different elements in a compound.	6. "Plan" the problem
_12_	Expression that shows the numbers of atoms in a molecule or formula unit of a compound.	7. Mole
_3_	The sum of all values of . . .	8. Percentage composition
_6_	Identify the given and wanted quantities in a problem, and the equation by which the problem may be solved by algebra, or the unit path and conversion factors by which the problem may be solved by dimensional analysis.	9. Simplest formula
_10_	Finding the formula and formula mass of substances involved in a chemical formula problem.	10. Starting steps
_4_	6.02×10^{23}	11. Molecular mass
_8_	Parts per hundred of each element in a compound.	12. Chemical formula

STUDY HINTS AND PITFALLS TO AVOID

It is helpful to recognize the place this chapter occupies in the order in which you are learning chemistry. In earlier chapters you learned an approach to solving chemistry problems and some beginning formula-writing skills. In this chapter you begin to apply them. You cannot solve a problem based on a chemical formula unless you have the correct formula. Nor can you write a chemical equation unless you can write the formulas of every species in the reaction. That is what you will do in the next chapter. In Chapter 9 you will add new material to the combined skills of Chapters 7 and 8 to find quantities in chemical changes. Later you will add to what you learn in Chapter 9. And so on . . .

The point of all this is that you recognize that you are building a foundation for what is to come. You will never "finish" with this chapter as long as you use or study chemistry.

Understanding the mole is absolutely essential to understanding chemistry. Take whatever time is necessary to get a clear understanding of Section 7.3. Know and understand how to use the mole. The conversion between mass and moles, in both directions, is probably the most important conversion in all of chemistry.

A strong suggestion: As you use dimensional analysis in solving problems, label each entry completely. Specifically, include the chemical formula of each substance in the calculation setup. Always calculate molar masses that correspond to the chemical formula. This is critical in the applications that lie ahead.

CHAPTER 7 QUESTIONS AND PROBLEMS

Many of the questions for this chapter require the formulas of chemical substances that are included in the performance goals in Chapter 6. Often these substances, elements, or compounds are identified by name only. This will give you practice in writing the formulas. If you are unable to do so, check back into Chapter 6 for the information you need. Formulas are given for all compounds not covered by Chapter 6 performance goals.

SECTION 7.1

1) How many atoms of each element are in a formula unit of each of the following: O_2; $Mg(NO_3)_2$; $(NH_4)_3PO_4$?

2) List the number of atoms of each element in a formula unit of each of the following substances: C_2H_5OH; $LiNO_3$; $Fe_3(SO_4)_2$.

SECTION 7.2

3) Distinguish between the terms *atomic mass, formula mass,* and *molecular mass.* Identify proper and improper uses of these terms.

4) Identify and define the units in which molecular mass and formula mass are expressed.

5) Which of the terms *formula mass, molecular mass,* or *atomic mass* is best suited for each of the following substances: CH_4, P_4, Na_3PO_4, Ne, Li_2SO_4, HBr?

6) Give the chemical formulas of three substances for which the term *atomic mass* would be appropriate. Then list three formulas for which molecular mass would be correct. Finally, write three formulas of substances for which *formula mass* would be best.

7) Calculate the formula mass of each substance in the list below. The formula mass of each substance is used in at least one problem that lies ahead.

Lithium	Oxygen
Ethanol, C_2H_5OH	Lithium chloride
Lithium nitrate	Aluminum oxide
Magnesium nitrate	Iron(III) sulfate
Lithium sulfide	Ammonium carbonate
Acetic acid, $HC_2H_3O_2$	Sodium acetate, $NaC_2H_3O_2$

8) Find the formula mass of each of the following substances:

Helium	Bromine
Diethyl ether, $(C_2H_5)_2O$	Ammonia

Potassium chlorate	Sodium carbonate
Zinc nitrate	Nickel phosphate
Cobalt(II) chloride	Copper(II) sulfate
Ammonium chromate, $(NH_4)_2CrO_4$	Barium bromate

SECTION 7.3

9) Formally, the mole is the SI unit for "amount of substance." If a mole is a unit, we must be able to count units. What, then, is the meaning of "two moles of water"?

10) It is perfectly logical to speak of 5 grams of silver and 5 grams of gold. Is it equally logical to speak of 5 moles of silver and 5 moles of gold? How are these expressions alike and/or different? Which would you rather have, 5 grams of gold or 5 moles of silver?

11) Does a mole represent a number? *Is* a mole a number? Explain your answers to these questions.

12) Most units we use to describe *amounts* of something can be used for all kinds of things. For example, we can talk about 15 grams of peas or 3 pounds of oranges. Would it be technically appropriate to speak of 15 moles of peas or 3 moles of oranges? Would it be practical to consider those amounts? Explain.

13) What is Avogadro's number, first by definition, and then by its numerical value? Did Avogadro assign the value to his number? If not, who did and by what right was the privilege transferred to this person?

14) What do 1.5 moles of table salt, NaCl, have in common with 1.5 moles of sugar, $C_{12}H_{22}O_{11}$? Do you suppose these amounts of these common chemicals might sit side-by-side on the same dinner table? (You can guess at the answer to this question now. It will be asked again in the General Questions, when you have more information on which to base an answer.)

15) How many moles are there in (a) 8.43×10^{23} atoms of calcium; (b) 3.20×10^{24} molecules of methane, CH_4; (c) 1.03×10^{23} formula units of iron(II) oxide?

16) Calculate the number of moles in (a) 7.67×10^{22} atoms of lead; (b) 6.05×10^{24} molecules of formaldehyde, HCHO; (c) 5.56×10^{24} formula units of mercury(I) chloride.

17) What is the number of atoms, molecules, or formula units in (a) 0.650 mole of manganese; (b) 0.949 mole of argon; (c) 38.2 moles of silver chloride?

18) Calculate how many atoms, molecules, or formula units are in (a) 2.49 moles of arsenic (Z = 33); (b) 0.0034 mole of sugar, $C_{12}H_{22}O_{11}$; (c) 30.1 moles of potassium permanganate, $KMnO_4$.

SECTION 7.4

19) How are the molar mass of molecules and molecular mass the same?

20) In what way are the molar mass of an ionic compound and the formula mass of the compound different?

---

21) Calculate the molar masses of all the substances in Question 7.

22) Find the molar masses of all substances listed in Question 8.

SECTION 7.5

Questions 23 to 26: Find the number of moles for each mass of substance given. (The formula masses for substances in Problem 23 were calculated in Problem 7.)

23) (a) 9.76 g O_2; (b) 5.09 g $Mg(NO_3)_2$; (c) 77.0 g Al_2O_3; (d) 956 g C_2H_5OH; (e) 0.493 g $(NH_4)_2CO_3$; (f) 43.0 g Li_2S.

24) (a) 83.5 g Be; (b) 187 g $C_3H_4Cl_4$; (c) 0.657 g $Ca(OH)_2$; (d) 4.93 g $CoBr_3$; (e) 9.90 g $(NH_4)_2Cr_2O_7$; (f) 38.2 g $Mg(ClO_4)_2$.

---

25) (a) 0.979 g KIO_3; (b) 86.8 g $BeCl_2$; (c) 203 g $Ni(NO_3)_2$.

26) (a) 91.6 g NaClO; (b) 188 g $Al(C_2H_3O_2)_3$; (c) 0.685 g Hg_2Cl_2.

Questions 27 to 30: Calculate the mass of each substance from the number of moles given. (The formula masses for substances in Problem 27 were calculated in Problem 7.)

27) (a) 0.967 mol LiCl; (b) 17.5 mol $HC_2H_3O_2$; (c) 8.60 mol Li; (d) 0.235 mol $Fe_2(SO_4)_3$; (e) 6.28 mol $NaC_2H_3O_2$.

28) (a) 0.245 mol $NaHCO_3$; (b) 0.0987 mol $AgNO_3$; (c) 1.69 mol Na_2HPO_4; (d) 0.309 mol $Ca(BrO_3)_2$; (e) 4.11 mol $(NH_4)_2SO_3$.

---

29) (a) 0.973 mol Li_2SO_4; (b) 2.84 mol $K_2C_2O_4$; (c) 0.231 mol $Pb(NO_3)_2$.

30) (a) 0.918 mol MnO_2; (b) 4.84 mol $Al(ClO_3)_3$; (c) 0.629 mol $CrCl_2$.

Questions 31 to 34: Calculate the number of atoms, molecules, or formula units that are in each given mass. (The

formula masses for substances in Problem 31 were calculated in Problem 7.)

31) (a) 69.2 g $LiNO_3$; (b) 0.515 g Li_2S; (c) 754 g $Fe_2(SO_4)_3$.

32) (a) 55.8 g $Be(NO_3)_2$; (b) 2.49 g Mn; (c) 0.0849 g C_3H_7OH.

---

33) (a) 0.0320 g I_2; (b) 411 g $C_2H_4(OH)_2$; (c) 1.89 g $Cr_2(SO_4)_3$.

34) (a) 8.80 g I; (b) 0.774 g C_9H_{20}; (c) 62.7 g $MnCO_3$.

---

35) Calculate the mass of (a) 3.40×10^{21} molecules of $C_{19}H_{37}COOH$; (b) 7.68×10^{24} atoms of fluorine, F; (c) 3.26×10^{23} formula units of nickel chloride.

36) Calculate the mass of (a) 8.52×10^{23} formula units of iron(II) oxide; (b) 7.68×10^{24} molecules of fluorine; (c) 6.37×10^{23} atoms of gold (Z = 79).

---

37) On a certain day, the Wall Street Journal quoted the price of gold in London as £226 (pounds) per troy ounce (1 troy ounce = 31.1 g). On the same day, the exchange rate between British pounds and U.S. dollars was $1.76 = £1.00. What was the price in U.S. cents of a single atom of gold on that day?

38) How many carbon atoms has a young man given to his bride-to-be if the engagement ring has a 0.634-carat diamond? There are 200 mg in a carat. (The price of diamonds doesn't seem so high when figured in dollars per atom.)

---

39) If one who sweetens her iced tea with the dietary sugar *fructose*, $C_6H_{12}O_6$, uses 0.65 grams, how many sugar molecules has she used?

40) The mass of a gallon of gasoline is about 2.7 kg. Assuming the gasoline is entirely octane, C_8H_{18}, calculate the number of molecules that can be held in a car with a 16.3 gallon fuel tank.

Chlorine and bromine are two of the elements that form stable diatomic molecules. As you answer Questions 41 and 42, keep in mind that the chemical formula of a molecule identifies precisely what is in the individual molecule.

41) (a) How many molecules are in 1.44 g Cl_2?
(b) How many atoms are in 1.44 g Cl_2?
(c) How many atoms are in 1.44 g Cl?
(d) What is the mass of 1.44×10^{23} atoms of Cl?
(e) What is the mass of 1.44×10^{23} molecules of Cl_2?

42) (a) What is the mass of 9.94×10^{23} atoms of Br?
(b) How many molecules are in 9.94 g Br_2?
(c) How many atoms are in 9.94 g Br?
(d) How many atoms are in 9.94 g Br_2?
(e) What is the mass of 9.94×10^{23} molecules of Br_2?

SECTION 7.6

43) Calculate the percentage composition of the following compounds, whose formula masses were calculated in Problem 7: (a) lithium chloride; (b) aluminum oxide; (c) acetic acid, $HC_2H_3O_2$; (d) magnesium nitrate; (e) iron(III) sulfate.

44) Calculate the percentage composition of: (a) aluminum bromide; (b) potassium cyanide, KCN; (c) zinc sulfate; (d) copper(II) hydroxide; (e) nickel acetate, $Ni(C_2H_3O)_2$.

---

45) Lithium fluoride is used as a flux when welding or soldering aluminum. How many grams of fluorine are in 688 grams of the compound?

46) Ammonium bromide is a raw material in the manufacture of photographic film. What mass of the compound contains 2.10 grams of bromine?

---

47) Potassium sulfate is a source of potassium in some fertilizers. How much potassium sulfate is required to furnish 311 grams of potassium?

48) Magnesium oxide is used in making bricks to line very-high-temperature furnaces. If a brick contains 2.92 kg of magnesium, what is the mass of the compound in the brick?

---

49) Zinc cyanide, $Zn(CN)_2$, is an important compound in zinc electroplating. If 391 g of the compound are introduced to a test solution in the laboratory, how many grams of zinc have been added to the bath?

50) Strontium nitrate, $Sr(NO_3)_2$, is responsible for the red color in fireworks displays. What mass of strontium is present in a rocket that contains 45.1 g $Sr(NO_3)_2$?

---

51) Molybdenum (Z = 42) is an important element in making steel alloys. It comes from an ore called wulfenite, $PbMoO_4$. What mass of pure wulfenite must be treated to obtain 874 kg Mo?

52) How much chlorine can be obtained from 45.1 grams of the insecticide calcium chlorate?

---

53) Methanol, CH_3OH, is used in the fuel for internal combustion engines. How many grams of carbon are in 70.6 grams of methanol?

54) Acetone, CH_3COCH_3, is a solvent that is widely used in manufacturing organic chemicals. How many grams of acetone contain 1.58 grams of hydrogen?

SECTION 7.8

55) Can C_5H_{10} be a molecular formula? Can it be a simplest formula? Explain why in each case.

56) Which of the following formulas can be simplest formulas: $H_2C_2O_4$; $C_{11}H_{24}$; HOOCCOOH; N_2O_4; P_4O_{10}? Write the simplest formulas for those compounds that are not already simplest formulas. Which of the formulas shown can be molecular formulas?

---

57) A certain compound is 52.2% carbon, 13.0% hydrogen, and 34.8% oxygen. Find the simplest formula of the compound.

58) Analysis of a compound shows it to be 49.4% potassium, 20.3% sulfur, and 30.3% oxygen. Find the simplest formula of the compound.

---

59) 13.51 grams of iron are exposed to a stream of oxygen until they react to produce 19.30 grams of a pure oxide of iron. What is the simplest formula of the product?

60) An 817-gram sample of a compound of lead and oxygen contains 741 grams of lead. What is the empirical formula of this compound?

---

61) A 10.73-gram sample of a white solid is analyzed. It is found to contain 0.54 gram of hydrogen, 3.75 grams of nitrogen, and the balance is oxygen. Find the simplest formula of the compound.

62) Nicotine, a harmful substance in cigarettes, is a compound of carbon, hydrogen, and nitrogen. A sample with a mass of 63.0 grams yields 46.7 grams of carbon and 10.9 grams of nitrogen in the laboratory; the remainder is hydrogen. Find the simplest formula of nicotine.

---

63) A compound is 23.1% carbon, 3.8% hydrogen, and 73.0% fluorine. Find the simplest formula of the compound.

64) The analysis of a compound gives the following results: 32.4% sodium, 0.7% hydrogen, 21.8% phosphorus, and the balance oxygen. What is the simplest formula of the compound?

---

65) A certain sugar is 40.0% carbon, 6.7% hydrogen, and 53.3% oxygen. The molar mass of the compound is 180 g/mol. Find both the simplest and molecular formulas of the compound.

66) A compound whose molar mass is 113 g/mol is 62.8% chlorine, 31.9% carbon, and 5.3% hydrogen. Determine the simplest and molecular formulas of the compound.

---

67) A 76.8-gram sample of a liquid widely used as a coolant in automobile engines is found to contain 30.7 grams of carbon, 5.1 grams of hydrogen, and the balance is oxygen. The molar mass of the compound is 60.0 g/mol. Find both the simplest and molecular formulas of the compound.

68) Analysis of 68.9 grams of a compound shows it to be mostly bromine—55.7 grams. Another 12.5 grams is carbon, and the remainder is hydrogen. Find the simplest formula of the compound. If its molar mass is 396 g/mol, what is the molecular formula of the compound?

GENERAL QUESTIONS
69) Distinguish precisely and in scientific terms the differences between items in each of the following groups:
(a) Atomic mass, molecular mass, formula mass, molar mass.
(b) Mole, molecule.
(c) Mole, Avogadro's number.
(d) Molecular formula, simplest formula.

70) Classify each of the following statements as true or false:
(a) The term *molecular mass* applies mostly to ionic compounds.
(b) Molar mass is measured in atomic mass units.
(c) In its practical application, a mole represents Avogadro's number.
(d) Grams are larger than atomic mass units; therefore molar mass is numerically larger than atomic mass.
(e) The molar mass of hydrogen is read directly from the periodic table, whether it be monatomic hydrogen, H, or hydrogen gas, H_2.
(f) A simplest formula is always a molecular formula, although a molecular formula may or may not be a simplest formula.

71) Assuming charcoal barbecue briquettes are 100% carbon, would you be able to barbecue steaks for your family and four guests with 1×10^{23} carbon atoms?

72) Calculate the number of S_8 molecules in 81.4 g S_8.

73) Let's try Question 14 again, now that you know about molar mass. Would it be reasonable to set a dinner table with one mole of salt, NaCl, in the salt shaker? How about one mole of sugar, $C_{12}H_{22}O_{11}$, in the sugar bowl?

74)* The quantitative significance of "take a deep breath" varies, of course, with the individual. When one person did so, she found that she inhaled 2.35×10^{22} molecules of the mixture of nitrogen and oxygen we call air. Assuming this mixture has an average molar mass of 29 g/mol, what is her apparent lung capacity in grams of air?

75)* Assuming gasoline to be pure octane, C_8H_{18}—actually it is a mixture of iso-octane and other hydrocarbons—an automobile getting 22.0 miles per gallon would consume 4.95×10^{23} molecules per mile. Calculate the mass of this amount of fuel.

76)* 9.050 grams of a certain compound containing only carbon and hydrogen are burned completely in oxygen. All the carbon is changed to 28.4 g CO_2 and all the hydrogen is changed to 11.6 g H_2O. What is the simplest formula of the original compound? (*Hint:* Find the grams of carbon and hydrogen in the original compound.)

MATCHING SET ANSWERS

7–11–5–2–1–9–12–3–6–10–4–8

QUICK CHECK ANSWERS

7.1 1 mol C atoms contains the same number of atoms as 1 mol O atoms.
7.2 a) 12.0 g C atoms contains the same number of atoms as 16.0 g O atoms. This is the same question as Quick Check 7.1, as 12.0 g C atoms and 16.0 g O atoms, the molar masses of both species, are both 1 mole of atoms.
  b) 227 g/mol

7.3 The starting steps in a chemical formula problem are writing the formula and calculating the molar mass of the substance.

8 Chemical Reactions and Equations

What is happening here? We could tell you quite accurately that powdered iron is burning as it is sprinkled into the flame of a burner. Rather than using words, a chemist might describe this reaction by writing a chemical equation: $4 \text{ Fe(s)} + 3 \text{ O}_2\text{(g)} \rightarrow 2 \text{ Fe}_2\text{O}_3\text{(s)}$. In this chapter you will learn that this is an example of a combination reaction. You will also learn about other kinds of reactions and how to write equations for them.

We are now ready to examine chemical reactions in detail. You will learn how to write equations for those reactions. Recall from Section 2.4 that an equation shows the formulas of the **reactants**—the starting substances that will be destroyed in the chemical change—written on the left side of an arrow, $\longrightarrow$. The formulas of the **products** of the reaction—the new substances formed in the chemical change—are written to the right of the arrow.

In this chapter you will learn to identify six different kinds of chemical changes. Writing a chemical equation is easier if you can classify the reaction as a certain type. You will be able to look at a particular combination of reactants, recognize what kind of reaction is possible, if any, and predict what products will be formed. The equation follows.

With these facts in mind, we suggest that you set your sights on the following goals as you begin this chapter:

1) Learn the mechanics of writing an equation.
2) Learn how to identify six different kinds of reactions.
3) Learn how to predict the products of each kind of reaction and write the formulas of those products.
4) Given potential reactants, write the equations for the probable reaction.

8.1 EVOLUTION OF A CHEMICAL EQUATION

If a piece of sodium is dropped into water, a vigorous reaction occurs (Fig. 8.1). A full description of the chemical change is "solid sodium plus liquid water yields hydrogen gas plus sodium hydroxide solution plus heat." That sentence is translated literally into a chemical equation:

$$Na(s) + H_2O(\ell) \longrightarrow H_2(g) + NaOH(aq) + heat \qquad (8.1)$$

The (s) after the symbol of sodium indicates it is a solid. Similarly, the (ℓ) after H_2O and the (g) after H_2 show that they are a liquid and a gas, respectively. When a substance is dissolved in water, the mixture is an "aqueous solution" and identified by (aq). ("**Aqueous**" comes from the Latin *aqua* for water.) These "**state symbols**" are sometimes omitted when writing equations, but they are included in most of the equations in this book. We suggest that you use or not use them according to the directions of your instructor.

Nearly all chemical reactions involve some energy transfer, usually in the form of heat. Generally, we omit energy terms from equations unless there is a specific reason for including them.

The Law of Conservation of Mass (Section 2.5) says that the total mass of the products of a reaction is the same as the total mass of the reactants. Atomic theory explains this by saying that atoms involved in a chemical change are neither created nor destroyed, but simply rearranged. Equation 8.1 does not satisfy this condition. There are two hydrogen atoms in H_2O on the left side of the equation, but three atoms of hydrogen on the right—two in H_2 and one in NaOH. The equation is not "**balanced.**"

An equation is balanced by placing a coefficient in front of one or more of the formulas, indicating it is used more than once. Hydrogen is short on the left side of Equation 8.1, so let's try two water molecules:

We examine the energy factor in a chemical reaction in Sections 9.5 and 9.6.

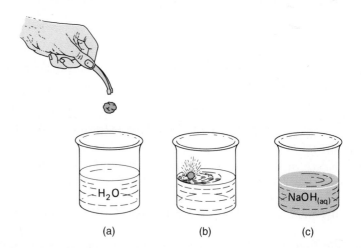

Figure 8.1

Sodium reacting with water. (a) A small piece of sodium is dropped into a beaker of water containing phenolphthalein, an indicator that turns pink in a solution of a metallic hydroxide. (b) Sodium forms a "ball" that dashes erratically over the water surface, releasing hydrogen as it reacts. Pink color near the sodium indicates that a sodium hydroxide (NaOH) solution is being formed in that local region. (c) Dissolved NaOH is now distributed uniformly through the solution, which is hot because of the heat released in the reaction. *Warning:* Do not "try" this experiment, as it is dangerous, potentially splattering hot alkali into eyes and onto skin and clothing.

$$\text{Na(s)} + 2\,\text{H}_2\text{O}(\ell) \longrightarrow \text{H}_2(\text{g}) + \text{NaOH(aq)} \qquad (8.2)$$

At first glance, this hasn't helped; indeed, it seems to have made matters worse. The hydrogen is still out of balance (four on the left, three on the right) and, furthermore, oxygen is now unbalanced (two on the left, one on the right). We are short one oxygen and one hydrogen atom on the right-hand side. But look closely. Oxygen and hydrogen are part of the same compound on the right, and there is one atom of each in that compound. If we take two NaOH units

$$\text{Na(s)} + 2\,\text{H}_2\text{O}(\ell) \longrightarrow \text{H}_2(\text{g}) + 2\,\text{NaOH(aq)} \qquad (8.3)$$

there are four hydrogens and two oxygens on both sides of the equation. These elements are now in balance. But, alas, the sodium has been *un*balanced. Correction of this condition, however, should be obvious:

$$2\,\text{Na(s)} + 2\,\text{H}_2\text{O}(\ell) \longrightarrow \text{H}_2(\text{g}) + 2\,\text{NaOH(aq)} \qquad (8.4)$$

The equation is now balanced. Note that, in the absence of a numerical coefficient, as with H_2, the coefficient is assumed to be 1.

Balancing an equation involves some important do's and don'ts that are apparent in this example:

DO: Balance the equation entirely by use of coefficients placed before the different chemical formulas.

DON'T: Change a correct chemical formula in order to make an element balance.

DON'T: Add some real or imaginary chemical species to either side of the equation just to make an element balance.

A moment's thought shows why the two "don'ts" are improper. The original equation expresses the *correct* formula of each species present. If you change a *correct* formula, it becomes *incorrect*. Usually the formula is for some nonexistent chemical. But even if the substance exists, or if you add something real to the equation, it is still wrong because the substance is not a species in the reaction.

✔ QUICK CHECK 8.1

Are the following true or false:

a) The equation $C_2H_4O + 3 O_2 \rightarrow 2 CO_2 + 2 H_2O$ is balanced.
b) The equation $H_2 + O_2 \rightarrow H_2O$ may be balanced by changing it to $H_2 + O_2 \rightarrow H_2O_2$.

8.2 BALANCING CHEMICAL EQUATIONS

PG 8A   Given an unbalanced chemical equation, balance it by inspection.

The balancing procedure in the preceding section is sometimes called "balancing by inspection." It is a trial-and-error method that succeeds in nearly all the reactions you are apt to meet in an introductory course. Most equations can be balanced without following a set procedure. However, if you prefer 1-2-3 . . . steps, the following work well, even with equations that are quite complicated.

PROCEDURE

1) Identify the "most complicated" formula, the formula having the largest number of atoms and/or the largest number of elements. Place a coefficient of "1" in front of that formula.
2) Balance the elements one at a time in the following order:
   a) Start with elements in the most complicated formula that are in only one other formula. Use fractional coefficients, if necessary.
   b) Save for last:
       (1) elements appearing in more than two formulas;
       (2) uncombined elements.
3) If fractions were used in Step 2, clear them by multiplying all coefficients by the lowest common denominator.
4) Remove any "1" coefficients that remain.
5) Check the entire equation.

Even if you use this procedure while learning how to balance equations, you will soon "see" how to get to a balanced equation more directly. Use these steps while they are helpful, but look for the shortcuts as soon as you can.

We now apply this procedure to the sodium-plus-water reaction in Section 8.1. All of the steps are listed first and then followed by comments or explanations for each step.

Steps	Reaction:	$Na(s) + H_2O(\ell) \longrightarrow H_2(g) + NaOH(aq)$
1	1 before most complicated	$Na(s) + H_2O(\ell) \longrightarrow H_2(g) + 1\,NaOH(aq)$
2a	Balance Na	$1\,Na(s) + H_2O(\ell) \longrightarrow H_2(g) + 1\,NaOH(aq)$
2a	Balance O	$1\,Na(s) + 1\,H_2O(\ell) \longrightarrow H_2(g) + 1\,NaOH(aq)$
2b	Balance H	$1\,Na(s) + 1\,H_2O(\ell) \longrightarrow \frac{1}{2}\,H_2(g) + 1\,NaOH(aq)$
3	Clear fractions	$2\,Na(s) + 2\,H_2O(\ell) \longrightarrow 1\,H_2(g) + 2\,NaOH(aq)$
4	Remove 1's	$2\,Na(s) + 2\,H_2O(\ell) \longrightarrow H_2(g) + 2\,NaOH(aq)$ (8.4)
5	Check equation	2 Na, 4 H, 2 O on each side of equation

1   NaOH is the most complicated formula because it has three atoms and three elements. It gets a coefficient of 1.

2a   Sodium and oxygen are balanced next because they appear in only one other formula. The 1 Na in NaOH is balanced by 1 Na on the left, and the 1 O in NaOH is balanced by 1 H_2O on the left.

Notice that the procedure in this section is not the same as the thought process by which we balanced the same equation in Section 8.1. There is no one "correct" way to balance an equation, but there are techniques that can shorten the process. Look for them in the examples ahead.

2b   Hydrogen is saved for last because it appears in three formulas and because it is uncombined as H_2. When starting to balance hydrogen, there are two H atoms in 1 H_2O on the left and one H atom in 1 NaOH on the right. The second H atom must come from H_2 on the right. To get one atom from H_2, we need ½ of an H_2 unit.

3   The fraction is cleared by multiplying the entire equation by 2.

4   Remove 1 coefficient.

5   Final check: 2 Na, 4 H, and 2 O on each side.

Is $Na(s) + H_2O(\ell) \rightarrow \frac{1}{2}\,H_2(g) + NaOH(aq)$ a legitimate equation? Yes and no. If you think of the equation as "1 Na atom reacts with 1 H_2O molecule to produce ½ an H_2 molecule and 1 NaOH unit," the equation is not legitimate. There is no such thing as "½ an H_2 molecule," any more than there is ½ an egg. But if you think about "1 mole of Na atoms reacts with one mole of H_2O molecules to produce ½ mole of hydrogen molecules and 1 mole of NaOH units," the fractional coefficient is reasonable. There can be ½ mole of hydrogen molecules just as there can be ½ dozen eggs.

There are times when it is necessary to use fractional coefficients, but none appear in this book. We therefore stay with the standard procedure of writing equations with whole-number coefficients. These coefficients should be written in the lowest terms possible; they should not have a common divisor. For example, although $4\,Na(s) + 4\,H_2O(\ell) \rightarrow 2\,H_2(g) + 4\,NaOH(aq)$ is a legitimate equation, it can and should be reduced to Equation 8.4 by dividing all coefficients by 2.

Notice that you can treat chemical equations exactly the same way you treat algebraic equations. Chemical formulas replace x, y, or other variables. You can multiply or divide an equation by some number by multiplying or dividing each term by that number. The order in which reactants or products is written may be changed; in an equation, $2\,H_2O + 2\,Na$ is the same as $2\,Na + 2\,H_2O$.

EXAMPLE 8.1 _____

Balance the equation

$$PCl_5(s) + H_2O(\ell) \longrightarrow H_3PO_4(aq) + HCl(aq)$$

Place a 1 in front of the most complicated formula—the formula with the greatest number of atoms and/or elements.

_____   _____

$$PCl_5(s) + H_2O(\ell) \longrightarrow 1\ H_3PO_4(aq) + HCl(aq)$$

Now begin to balance the elements in the most complicated formula that are in only one other formula.

_____   _____

$$1\ PCl_5(s) + 4\ H_2O(\ell) \longrightarrow 1\ H_3PO_4(aq) + HCl(aq)$$

Both phosphorus and oxygen are in one other formula. 1 P in H_3PO_4 requires 1 PCl_5, and 4 O in H_3PO_4 is satisfied by 4 H_2O. Hydrogen is in two other compounds, so we save that until last.

What other elements besides hydrogen can be balanced now? Balance them.

_____   _____

$$1\ PCl_5(s) + 4\ H_2O(\ell) \longrightarrow 1\ H_3PO_4(aq) + 5\ HCl(aq)$$

Chlorine is the only other element besides hydrogen that can be balanced at this point. 5 Cl in PCl_5 are balanced by 5 HCl.

Hydrogen remains.

_____   _____

Hydrogen is already balanced at 8 H on each side.

When no uncombined elements are present, the last element should already be balanced when you reach it. If not, look for an error on some earlier element.

Remove 1's and make a final check.

_____   _____

$$PCl_5(s) + 4\ H_2O(\ell) \longrightarrow H_3PO_4(aq) + 5\ HCl(aq)$$

1 P, 5 Cl, 8 H, and 4 O on each side.

EXAMPLE 8.2 _____

Balance the equation

$$BiOCl(aq) + H_2S(g) \longrightarrow Bi_2S_3(s) + H_2O(\ell) + HCl(aq)$$

Take it all the way without suggestions this time.

_____   _____

Reaction	$BiOCl(aq) + H_2S(g) \longrightarrow Bi_2S_3(s) + H_2O(\ell) + HCl(aq)$
1 before most complicated	$BiOCl(aq) + H_2S(g) \longrightarrow 1\ Bi_2S_3(s) + H_2O(\ell) + HCl(aq)$
Balance Bi	$2\ BiOCl(aq) + H_2S(g) \longrightarrow 1\ Bi_2S_3(s) + H_2O(\ell) + HCl(aq)$
Balance S	$2\ BiOCl(aq) + 3\ H_2S(g) \longrightarrow 1\ Bi_2S_3(s) + H_2O(\ell) + HCl(aq)$
Balance Cl	$2\ BiOCl(aq) + 3\ H_2S(g) \longrightarrow 1\ Bi_2S_3(s) + H_2O(\ell) + 2\ HCl(aq)$
Balance O	$2\ BiOCl(aq) + 3\ H_2S(g) \longrightarrow 1\ Bi_2S_3(s) + 2\ H_2O(\ell) + 2\ HCl(aq)$
Balance H	$6\ H \quad = \quad 4\ H \quad + 2\ H$
Remove 1's	$2\ BiOCl(aq) + 3\ H_2S(g) \longrightarrow Bi_2S_3(s) + 2\ H_2O(\ell) + 2\ HCl(aq)$
Check	2 Bi, 2 O, 2 Cl, 6 H, 3 S on both sides

Again, the last element is balanced by the time you get to it.

EXAMPLE 8.3

Balance the equation

$$C_2H_5COOH(aq) + O_2(g) \longrightarrow CO_2(g) + H_2O(\ell)$$

This equation is just a little bit tricky. See if you can avoid the traps.

1 before most complicated	$1\ C_2H_5COOH(aq) + O_2(g) \longrightarrow CO_2(g) + H_2O(\ell)$
Balance C	$1\ C_2H_5COOH(aq) + O_2(g) \longrightarrow 3\ CO_2(g) + H_2O(\ell)$
Balance H	$1\ C_2H_5COOH(aq) + O_2(g) \longrightarrow 3\ CO_2(g) + 3\ H_2O(\ell)$
Balance O	$1\ C_2H_5COOH(aq) + \frac{7}{2}\ O_2(g) \longrightarrow 3\ CO_2(g) + 3\ H_2O(\ell)$
Clear fractions	$2\ C_2H_5COOH(aq) + 7\ O_2(g) \longrightarrow 6\ CO_2(g) + 6\ H_2O(\ell)$

The traps are in the formula C_2H_5COOH. The symbols of all three elements appear twice. The formulas of organic compounds are often written this way in order to suggest the arrangement of atoms in the molecule, as you will see later. Students usually count carbon and hydrogen correctly, but they often overlook the oxygen in the original compound when selecting the coefficient for O_2.

Oxygen is often the last element balanced in an equation, and it is quite common for the remaining oxygen atoms to come from O_2 molecules. There are two atoms per molecule, so the number of oxygen *molecules* required is ½ the number of *atoms* required. If n is the number of atoms needed, the number of O_2 molecules is $n/2$. If n is an even number, the quotient is an integer and there is no fraction to clear. If n is odd, the fraction is usually left as an improper fraction, as $\frac{7}{2}$ in Example 8.3 rather than $3\frac{1}{2}$. When the equation is multiplied by 2, n becomes the integer coefficient for O_2.

8.3 WRITING CHEMICAL EQUATIONS

The general procedure for writing a chemical equation is:

PROCEDURE

1) Determine what kind of reaction it is.
2) Write the correct chemical formula for each reactant on the left and each product on the right.
3) Balance the equation.

Step 1 is completed for you in the title of the section that presents the six kinds of reactions we will study. The end-of-chapter questions include reactions that you will have to classify before writing the equation.

8.4 COMBINATION REACTIONS

> **PG 8B** Write the equation for the reaction in which a compound is formed by the combination of two or more simpler substances.

Your ability to classify a reaction as a certain type is a big help in predicting what products will form. You must know these before you can even begin to write the equation.

A reaction in which two or more substances combine to form a single product is a **combination reaction** or a **synthesis reaction.** The reactants are often elements, but sometimes compounds or both elements and compounds. The general equation for a combination reaction is

$$A + X \longrightarrow AX \qquad (8.5)$$

The reaction between sodium and chlorine to form sodium chloride is a combination reaction: $2\,Na(s) + Cl_2(g) \rightarrow 2\,NaCl(s)$.

EXAMPLE 8.4

Write the equation for the formation of water by direct combination of hydrogen and oxygen. (This is what happened when the dirigible *Hindenburg* burned in New Jersey in 1937. See Fig. 8.2.)

Figure 8.2

The end of the dirigible *Hindenburg* in May 1937. This event ended the use of explosive hydrogen in lighter-than-air craft. Today, the noble gas helium is used in the airships from which aerial views of major sporting events are displayed on television.

hydrogen + oxygen → water

This is a combination reaction in which two elements unite to form a compound. Write the formulas of the elements on the left side of an arrow and the formula of the product on the right.

$$H_2(g) + O_2(g) \longrightarrow H_2O(\ell)$$

You did remember to show hydrogen and oxygen as diatomic molecules, did you not?

Balance the equation in the manner described in the previous section.

$$H_2(g) + O_2(g) \longrightarrow 1\ H_2O(\ell)$$
$$1\ H_2(g) + O_2(g) \longrightarrow 1\ H_2O(\ell)$$
$$1\ H_2(g) + \tfrac{1}{2}\ O_2(g) \longrightarrow 1\ H_2O(\ell)$$
$$2\ H_2(g) + O_2(g) \longrightarrow 2\ H_2O(\ell)$$

We will no longer show the stepwise balancing of simple equations, but we will include it if the reaction is more complex.

EXAMPLE 8.5

Carbon dioxide is formed when charcoal (carbon) is burned in air, as in a backyard barbecue (Fig. 8.3). Write the equation for the reaction.

A word description of a reaction sometimes assumes that you already know something about it. This is an example. It assumes you know that when something burns in air, it is reacting chemically with the oxygen in the air. In other words, oxygen is an unidentified reactant, so it should appear on the left side of the equation along with carbon. With that hint, complete the equation.

$$C(s) + O_2(g) \longrightarrow CO_2(g)$$

Sometimes balancing an equation is easy, as when all coefficients are 1!

Compounds can react with each other in combination reactions too.

EXAMPLE 8.6

Solid magnesium hydroxide is formed when solid magnesium oxide combines with water. Write the equation.

$$MgO(s) + H_2O(\ell) \longrightarrow Mg(OH)_2(s)$$

Figure 8.3

Backyard barbecue. A simple combination reaction occurs in this device that is so widely used in summer.

SUMMARY

Equation-writing summary:

Reactants: Any combination of elements and/or compounds

Reaction type: Combination

Equation type: $A + X \longrightarrow AX$

Products: One compound

LEARN IT NOW The four points in this summary trace the thought process in writing an equation. After you examine the reactants, you will be able to decide what kind of a reaction it is and what kind of equation it will have. This also enables you to predict the products. This summary and those that follow it are gathered as a chapter summary in Section 8.10.

8.5 DECOMPOSITION REACTIONS

PG 8C Given a compound that is decomposed into simpler substances, either compounds or elements, write the equation for the reaction.

A **decomposition reaction** is the opposite of a combination reaction, in that a compound breaks down into simpler substances. The products may be any combination of elements and/or compounds. The general decomposition equation is

$$AX \longrightarrow A + X \qquad (8.6)$$

The discovery of oxygen by the heating of mercury(II) oxide is a typical decomposition reaction: $2\,HgO(s) \rightarrow 2\,Hg(\ell) + O_2(g)$.

EXAMPLE 8.7

Water decomposes into its elements when it is electrolyzed. Write the equation.

$$2\,H_2O(\ell) \longrightarrow 2\,H_2(g) + O_2(g)$$

This reaction is literally the reverse of the reaction in Example 8.4.

Many chemical changes can be made to go in either direction, as Examples 8.4 and 8.7 suggest. These are called **reversible reactions.** Reversibility is often indicated by a double arrow, $\rightleftarrows$. Thus

$$2\,H_2O(\ell) \rightleftharpoons 2\,H_2(g) + O_2(g) \qquad \text{and} \qquad 2\,H_2(g) + O_2(g) \rightleftharpoons 2\,H_2O(\ell)$$

are equivalent reversible equations.

EXAMPLE 8.8

A common laboratory procedure for producing oxygen is heating potassium chlorate. Solid potassium chloride is left behind.

Read the question carefully to be sure you identify the reactants and products correctly, and then write their formulas where they should be. Complete the equation.

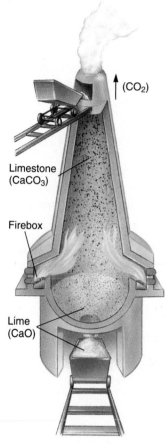

Limestone (CaCO₃)

Firebox

Lime (CaO)

Figure 8.4

Thermal decomposition of limestone (calcium carbonate). Calcium oxide, or "quicklime," is prepared by decomposing calcium carbonate in a large kiln at 800°C to 1000°C. Calcium oxide is among the most widely used chemicals in the United States, annual consumption being measured in millions of tons. Nearly one-half the CaO output is used in the steel industry, and much of the remainder is used to make "slaked lime," Ca(OH)₂, by reaction with water. Can you write the equation?

$$1 \ KClO_3(s) \longrightarrow KCl(s) + O_2(g)$$

$$1 \ KClO_3(s) \longrightarrow 1 \ KCl(s) + O_2(g)$$

$$1 \ KClO_3(g) \longrightarrow 1 \ KCl(s) + \tfrac{3}{2} O_2(g)$$

$$2 \ KClO_3(g) \longrightarrow 2 \ KCl(s) + 3 \ O_2(g)$$

Balancing is a little more difficult this time, so we have shown all the steps. In balancing oxygen there are 3 O's in one $KClO_3$, and they must come from O_2. This is the $n/2$ situation described earlier, where $n = 3$. The fractional coefficient disappears when the entire equation is multiplied by 2 in the last step.

E X A M P L E 8 . 9

Lime, CaO(s), and carbon dioxide gas are the products of the thermal decomposition of limestone, solid calcium carbonate (Fig. 8.4). ("Thermal" refers to heat. The reaction occurs at high temperature.)

Write the equation.

$$CaCO_3(s) \longrightarrow CaO(s) + CO_2(g)$$

S U M M A R Y

Equation-writing summary:

Reactants: One compound

Reaction type: Decomposition

Equation type: AX $\longrightarrow$ A + X

Products: Any combination of elements and/or compounds

8.6 COMPLETE OXIDATION OR BURNING OF ORGANIC COMPOUNDS

PG 8D Write the equation for the complete oxidation or burning of any compound containing only carbon and hydrogen, or carbon, hydrogen, and oxygen.

A large number of compounds, including petroleum products, alcohols, some acids, and carbohydrates, consist of two or three elements: carbon and hydrogen or carbon, hydrogen, and oxygen. When such compounds are burned in air they react with oxygen in the atmosphere. We say they are **oxidized,** a term that has other meanings in addition to "reacting with oxygen." The final products of a complete burning or oxidation are always the same: carbon dioxide, $CO_2(g)$, and water, $H_2O(g)$ or $H_2O(\ell)$, depending on the temperature at which the product is examined. The distinction is not important in this chapter, so we will use $H_2O(g)$ consistently.

In writing these equations you will be given only the identity of the compound that "burns" or is "oxidized." These words tell you the compound reacts with oxygen, so you must include it as a second reactant. The formulas of water and carbon dioxide appear on the right side of the equation. Thus the general equation for a complete oxidation (burning) reaction is always

$$C_xH_yO_z + O_2(g) \longrightarrow CO_2(g) + H_2O(g) \text{ [or } H_2O(\ell)] \qquad (8.7)$$

The burning of methane, the chief component of natural gas, is an example:

$$CH_4(g) + 2\,O_2(g) \longrightarrow CO_2(g) + 2\,H_2O(g)$$

As a rule, these equations are most easily balanced if you take the elements carbon, hydrogen, and oxygen in that order.

Both carbon dioxide, CO_2, and steam, $H_2O(g)$, are invisible. The white "smoke" commonly seen rising from chimneys and smokestacks is tiny drops of condensed H_2O. Black smoke comes from carbon that is not completely burned.

EXAMPLE 8.10

Write the equation for the complete burning of ethane, $C_2H_6(g)$.

$$1\,C_2H_6(g) + \quad O_2(g) \longrightarrow \quad CO_2(g) + \quad H_2O(g)$$
$$1\,C_2H_6(g) + \quad O_2(g) \longrightarrow 2\,CO_2(g) + \quad H_2O(g)$$
$$1\,C_2H_6(g) + \quad O_2(g) \longrightarrow 2\,CO_2(g) + 3\,H_2O(g)$$
$$1\,C_2H_6(g) + \tfrac{7}{2}\,O_2(g) \longrightarrow 2\,CO_2(g) + 3\,H_2O(g)$$
$$2\,C_2H_6(g) + 7\,O_2(g) \longrightarrow 4\,CO_2(g) + 6\,H_2O(g)$$

EXAMPLE 8.11

Write the equation for the complete burning of butanol, $C_4H_9OH(\ell)$, in air.

This equation is like the equation in Example 8.3—actually a little easier. Write the equation.

$$C_4H_9OH(\ell) + 6\,O_2(g) \longrightarrow 4\,CO_2(g) + 5\,H_2O(g)$$

If you did not get this equation, you probably counted only 9 hydrogen atoms in C_4H_9OH (there are 10), or overlooked the oxygen in C_4H_9OH in finding the coefficient of O_2. You also didn't check the final equation or you would have caught either error.

SUMMARY

Equation-writing summary:

    Reactants: Oxygen and a compound of carbon and hydrogen, or carbon, hydrogen, and oxygen

    Reaction type: Burning in air, or complete oxidation

Equation type: $C_xH_yO_z + O_2(g) \longrightarrow CO_2(g) + H_2O(g)$ [or $H_2O(\ell)$]

Products: $CO_2(g) + H_2O(g)$ [or $H_2O(\ell)$]

8.7 OXIDATION–REDUCTION (REDOX) REACTIONS— SINGLE REPLACEMENT TYPE

PG 8E   Given the reactants of a redox reaction ("single replacement" type only), write the equation for the reaction.

The "oxidation or burning" reactions of Section 8.6 are also oxidation–reduction reactions.

Many elements are capable of replacing ions of other elements from aqueous solutions. This is one kind of **oxidation–reduction reaction,** or **"redox" reaction.** The equation for such a reaction looks as if one element is replacing another in a compound. It is a **single replacement equation;** in fact, the reactions are sometimes called **single replacement reactions.** The general equation is

$$A + BX \longrightarrow AX + B \qquad (8.8)$$

Zinc is able to replace lead from a solution of a lead compound in a redox reaction:

$$Zn(s) + Pb(NO_3)_2(aq) \longrightarrow Pb(s) + Zn(NO_3)_2(aq)$$

Reactants in a single replacement equation are always an element and a compound. If the element is a metal, it replaces the metal or hydrogen in the compound. If the element is a nonmetal, it replaces the nonmetal in the compound. All three possibilities are in the next three examples.

EXAMPLE 8.12

Write the single replacement equation for the reaction between elemental calcium and hydrochloric acid (Fig. 8.5).

Begin by writing the formulas of the reactants to the left of the arrow.

—————   —————

$$Ca(s) + HCl(aq) \longrightarrow$$

Now decide which element in the compound, hydrogen or chlorine, will be replaced by the calcium. Reread the paragraph before this example if you need help. Then write the formulas of the products on the right side of the equation.

—————   —————

$$Ca(s) + HCl(aq) \longrightarrow H_2(g) + CaCl_2(aq)$$

A metal will replace a positive ion in a solution, which is the metal in the dissolved compound, or hydrogen if the dissolved compound is an acid. The displaced hydrogen is now an uncombined element and its formula is H_2.

Now balance the equation.

—————   —————

$$Ca(s) + 2\ HCl(aq) \longrightarrow H_2(g) + CaCl_2(aq)$$

Figure 8.5
The reaction of calcium metal and hydrochloric acid.

(a)

(b)

Figure 8.6
The reaction between copper and a
solution of silver nitrate, $AgNO_3$.

EXAMPLE 8.13 _____

Copper reacts with a solution of silver nitrate (Fig. 8.6). Write the equation for the reaction.

In this case an elemental metal is replacing a metal in a compound. The copper ion that forms is a copper(II) ion. Write the unbalanced equation.

_____   _____

$$Cu(s) + AgNO_3(aq) \longrightarrow Ag(s) + Cu(NO_3)_2(aq)$$

Now balance the equation.

_____   _____

$$Cu(s) + 2\,AgNO_3(aq) \longrightarrow 2\,Ag(s) + Cu(NO_3)_2(aq)$$

Example 8.13 gives us an opportunity to show you a balancing trick that can save you some time. Whenever a polyatomic ion is unchanged in a chemical reaction, the entire ion can be balanced as a unit. The nitrate ion is unchanged in

$$Cu(s) + AgNO_3(aq) \longrightarrow Ag(s) + Cu(NO_3)_2(aq)$$

The most complex formula is $Cu(NO_3)_2$. It has three "atoms": one copper and two nitrate ions. Copper is already balanced. $AgNO_3$ has one nitrate "atom" (ion), so it takes two $AgNO_3$ to balance the nitrates in one $Cu(NO_3)_2$. This, in turn, requires two Ag on the right. When you learn this technique, you will find it quicker and easier than balancing each element in a polyatomic ion separately. But remember the condition: The ion must be unchanged. All of it. An $NO_3{}^-$ compound on one side and an $NO_3{}^-$ *plus* an NO or some other nitrogen species on the other side will not work.

EXAMPLE 8.14 _____

Write the equation for the reaction between chlorine and a solution of sodium bromide.

This time the elemental reactant is a nonmetal, so it will replace the nonmetal in the compound. Write the balanced equation.

_____   _____

$$Cl_2(g) + 2\,NaBr(aq) \longrightarrow Br_2(aq) + 2\,NaCl(aq)$$

Just because an equation for a reaction can be written does not mean the reaction will occur. For example, if a piece of silver and a solution of copper(II) nitrate are given as reactants, you would produce the equation $2\,Ag(s) + Cu(NO_3)2(aq) \longrightarrow Cu(s) + 2\,AgNO_3(aq)$, just the reverse of the equation in Example 8.13. This reaction does not occur spontaneously, although it can be forced with help from a battery. To find which reaction "works," you must try the reactions in the laboratory. In Chapter 16 you will use the results of many experiments to predict which reactions will occur and which will not.

8.9   NEUTRALIZATION REACTIONS

PG 8G   Given the reactants in a neutralization reaction, write the equation.

We have seen that an acid is a compound that releases hydrogen ions, H^+. A substance that contains hydroxide ions, OH^-, is a **base.** When an acid is added to an equal amount of base, each hydrogen ion forms a chemical bond with a hydroxide ion to form a molecule of water. The acid and the base **neutralize** each other in a **neutralization reaction.**[*] An ionic compound called a **salt** is also formed; it usually remains in solution. The general equation is

$$\underset{\text{acid}}{HX(aq)} + \underset{\text{base}}{MOH(aq)} \longrightarrow \underset{\text{water}}{H_2O(\ell)} + \underset{\text{salt}}{MX(aq)} \qquad (8.10)$$

Neutralization reactions are described by double displacement equations, although it might not seem that way at first glance. The water molecule forms when the hydrogen ion from the acid combines with the hydroxide ion from the base. The double displacement character of the equation becomes clear if the formula of water is written HOH rather than H_2O. For the neutralization of hydrochloric acid by sodium hydroxide the equation is

$$HCl(aq) + NaOH(s) \longrightarrow HOH(\ell) + NaCl(aq) \qquad (8.11)$$

With the formula of water in conventional form, we have

$$HCl(aq) + NaOH(s) \longrightarrow H_2O(\ell) + NaCl(aq) \qquad (8.12)$$

Suggestion: When balancing a neutralization reaction equation, balance the ions in the salt first. If you do that correctly, you will have the same number of hydrogen and hydroxide ions. That is the number of water molecules that should be on the product side of the equation.

EXAMPLE 8.17

Write the equation for the neutralization of aqueous barium hydroxide with nitric acid.

The reactants are identified in the question. One product is water, and the other is the salt formed by the remaining ions. Begin with the unbalanced equation.

$$Ba(OH)_2(aq) + HNO_3(aq) \longrightarrow H_2O(\ell) + Ba(NO_3)_2(aq)$$

[*] There are other kinds of acids that do not contain hydrogen ions and bases that do not contain hydroxide ions. Reactions between them are also called neutralization reactions. The H^+ plus OH^- neutralization is the most common, and the only one we will consider here.

The salt is barium nitrate. It comes from the barium ion in $Ba(OH)_2$ and the nitrate ion from nitric acid.

Continue by balancing the barium and nitrate ions.

———

———

$$Ba(OH)_2(aq) + 2\,HNO_3(aq) \longrightarrow H_2O(\ell) + Ba(NO_3)_2(aq)$$

Barium is in balance in the unbalanced equation. It takes two HNO_3's to balance the two nitrates in $Ba(NO_3)_2$.

Now count the hydrogen and hydroxide ions on the left side of the equation. Are they equal? If so, that's how many water molecules are on the right. Complete the equation.

———

———

$$Ba(OH)_2(aq) + 2\,HNO_3(aq) \rightarrow 2\,H_2O(\ell) + Ba(NO_3)_2(aq)$$

Is everything balanced? In checking, you can still count the nitrates as a unit. In counting oxygen, you can skip the oxygens already balanced in the nitrates. If you prefer, you can count all the atoms separately. Take your choice.

———

———

1 Ba, 2 O, 4 H, and 2 NO_3^- on both sides; or 1 Ba, 8 O, 4 H, and 2 N on both sides.

Most hydroxides do not dissolve in water, but handbooks say they do dissolve in acids. What really happens is that they react with the acid in a neutralization reaction. The next equation shows one such reaction.

EXAMPLE 8.18 _____

Write the equation for the reaction between sulfuric acid and solid aluminum hydroxide.

Write the unbalanced equation.

———

———

$$H_2SO_4(aq) + Al(OH)_3(s) \longrightarrow H_2O(\ell) + Al_2(SO_4)_3(aq)$$

Balance the ions in the salt, aluminum and sulfate.

———

———

$$3\,H_2SO_4(aq) + 2\,Al(OH)_3(s) \longrightarrow H_2O(\ell) + Al_2(SO_4)_3(aq)$$

Two aluminum ions come from $Al_2(SO_4)_3$, and they are balanced by 2 $Al(OH)_3$. Similarly, three sulfate ions from the salt require 3 H_2SO_4.

$$3\ H_2SO_4(aq)\ +\ 2\ Al(OH)_3(s)\ \longrightarrow$$
$$H_2O(\ell)\ +\ Al_2(SO_4)_3(aq)$$

Count up the hydrogen and hydroxide ions. If they are equal, place the needed coefficient in front of H_2O.

$$3\ H_2SO_4(aq)\ +\ 2\ Al(OH)_3(s)\ \longrightarrow\ 6\ H_2O(\ell)\ +\ Al_2(SO_4)_3(aq)$$

Make a final check.

6 H, 3 $SO_4{}^{2-}$, 2 Al, and 6 O on each side; or, 6 H, 3 S, 18 O, and 2 Al on each side.

SUMMARY

Equation-writing summary:

    Reactants: Acid and a hydroxide base

    Reaction type: Neutralization

    Equation type: HX + MOH $\longrightarrow$ H_2O + MX (double displacement equation)

    Products: Water and a salt (ionic compound)

8.10 SUMMARY OF REACTIONS AND EQUATIONS

All of the equation-writing summaries at the ends of Sections 8.4 to 8.9 have been assembled into Table 8.1. The reactants (first column) are shown for each reaction type (second column). Each reaction type has a certain "equation type" (third column) that yields predictable products (fourth column).

Reading the column heads from left to right follows the "thinking order" by which reactants and products are identified. It will help you to "organize" your approach to writing equations. Given the reactants of a specific chemical change, you can fit them into one of the reactant boxes in the table, and thereby determine the reaction type. Once you know what kind of a reaction it is, you know the type of equation that describes it. You can then write the formulas of additional reactants, if any, on the left side of the arrow and the formulas of the products on the right. Balance the equation and you are finished.

We indicated earlier in the chapter that just because an equation can be written, it does not necessarily mean that the reaction will happen. But, by using the results of experiments that have been performed over many years, we can make reliable predictions. For example, we can predict with confidence that zinc will replace copper in a copper sulfate solution, but silver will not; or that pouring calcium nitrate solution into sodium fluoride solution will yield a precipitate, but pouring it into sodium bromide solution will not. We have deliberately refrained from making these predictions in this chapter; learning to write the equations is enough for now. Rest assured, however, that you will be able to think and predict like a chemist before you complete your study of this text.

SUMMARY

Table 8.1

Summary of Types of Reactions and Equations

Reactants	Reaction Type	Equation Type	Products
Any combination of elements and/or compounds that form one product	Combination	$A + X \rightarrow AX$ Combination	One compound
One compound	Decomposition	$AX \rightarrow A + X$ Decomposition	Any combination of elements and/or compounds
O_2* + compound of C and H or C, H, and O	Complete oxidation or burning	$C_xH_yO_z + O_2 \rightarrow CO_2 + H_2O$ Complete oxidation	CO_2* + H_2O*
Element + ionic compound or acid	Oxidation–reduction	$A + BX \rightarrow AX + B$ Single replacement	Element + ionic compound
Solution of ionic compound + acid or solution of second ionic compound	Precipitation	$AX + BY \rightarrow AY + BX$ Double replacement	Precipitate of ionic compound + acid or second ionic compound
Acid + hydroxide base	Neutralization	$HX + MOH \rightarrow HOH + MX$ Double replacement	Ionic compound (salt) + H_2O

*The reactant oxygen and the products carbon dioxide and water are usually not mentioned in the description of a reaction of this kind.

CHAPTER 8 IN REVIEW

8.1 Evolution of a Chemical Equation

8.2 Balancing Chemical Equations
8A Given an unbalanced chemical equation, balance it by inspection.

8.3 Writing Chemical Equations

8.4 Combination Reactions
8B Write the equation for the reaction in which a compound is formed by the combination of two or more simpler substances.

8.5 Decomposition Reactions
8C Given a compound that is decomposed into simpler substances, either compounds or elements, write the equation for the reaction.

8.6 Complete Oxidation or Burning of Organic Compounds

8D Write the equation for the complete oxidation or burning of any compound containing only carbon and hydrogen, or carbon, hydrogen, and oxygen.

8.7 Oxidation–Reduction (Redox) Reactions—Single Replacement Type
8E Given the reactants of a redox reaction ("single replacement" type only), write the equation for the reaction.

8.8 Precipitation Reactions
8F Given the reactants in a precipitation reaction, write the equation.

8.9 Neutralization Reactions
8G Given the reactants in a neutralization reaction, write the equation.

8.10 Summary of Reactions and Equations

CHAPTER 8 KEY WORDS AND MATCHING SETS

Set 1

_____ Used to identify a liquid in a chemical equation.

_____ Ionic compound that contains a hydroxide ion.

_____ Chemical change in which a substance reacts with oxygen to release CO_2 and H_2O.

_____ Used to identify an aqueous solution in a chemical equation.

_____ Acids and bases do this to each other.

_____ Nickname for oxidation–reduction.

_____ Chemical equation in which one element appears to replace another element in a compound.

_____ Chemical change described by an equation in which either side may be reactants and either side may be products.

_____ Chemical change in which two or more substances form a single product.

_____ Chemical change in which a solid substance forms when two solutions are combined.

_____ Identifies a species in a reaction as a solid, liquid, or gas, or a substance in aqueous solution.

1. (aq)
2. Combination reaction
3. Reversible reaction
4. State symbol
5. Single replacement
6. Precipitation reaction
7. Neutralize
8. Oxidation reaction
9. (ℓ)
10. Base
11. Redox

Set 2

_____ Describes a water solution.

_____ Used to identify a solid in a chemical equation.

_____ Chemical change in which an element appears to replace another element in a compound.

_____ Chemical equation in which ions in reactants appear to "change partners" to form products.

_____ Chemical change in which a substance reacts with oxygen in the air to release carbon dioxide, water, light, and heat.

_____ Reaction in which a single reactant forms at least two products.

_____ Solid substance that forms in a chemical reaction.

_____ Product other than water formed in a neutralization reaction.

_____ Used to identify a gas in a chemical equation.

_____ Number written before a chemical formula in an equation that tells how many formula units of the species are in the equation.

1. Precipitate
2. Oxidation–reduction reaction
3. Decomposition reaction
4. Coefficient
5. (g)
6. Burn
7. Double replacement equation
8. Salt
9. (s)
10. Aqueous

STUDY HINTS AND PITFALLS TO AVOID

Be careful about the "don'ts" in Section 8.1. It is very tempting to balance an equation quickly by changing a correct formula—or adding one that doesn't belong. Another device some creative students invent is slipping a coefficient into the middle of a correct formula, such as changing a correct $NaNO_3$ to $Na2NO_3$ to balance nitrates in $Ca(NO_3)_2$ on the other side. That doesn't work either. There is no such thing as $Na2NO_3$.

An equation-balancing exercise follows these study hints. We suggest you use it to practice this important skill.

Probably the best suggestion for studying this chapter is to focus on Table 8.1. Look for the "big picture." Knowing how things fit together helps in learning the de-

tails. Look at the *reactants*. They should tell you the *reaction type*. Each reaction type has a certain *equation type*, and that gives you the *products*. In order, these are the column heads, read left to right.

You might wish to practice this mentally by referring to the unclassified reactions beginning with Question 29. Based on the reactants described in each question, see if you can determine what kind of reaction it is and what its products are. Run through the whole list this way without writing equations until you feel sure of your ability. Then write and balance equations until you are completely confident that you can write any equation without hesitation.

EQUATION-BALANCING EXERCISE

Balance the following equations, for which correct chemical formulas are already written. Balanced equations are in the answer section.

1) $Li_2O + H_2O \longrightarrow LiOH$
2) $HgO \longrightarrow Hg + O_2$
3) $CaC_2 + H_2O \longrightarrow C_2H_2 + Ca(OH)_2$
4) $Zn(OH)_2 + H_2SO_4 \longrightarrow ZnSO_4 + H_2O$
5) $PbO_2 \longrightarrow PbO + O_2$
6) $Al + HCl \longrightarrow AlCl_3 + H_2$
7) $Fe_2(SO_4)_3 + Ba(OH)_2 \longrightarrow$
$\qquad BaSO_4 + Fe(OH)_3$
8) $Al + CuSO_4 \longrightarrow Al_2(SO_4)_3 + Cu$
9) $Mg + N_2 \longrightarrow Mg_3N_2$
10) $FeCl_2 + Na_3PO_4 \longrightarrow Fe_3(PO_4)_2 + NaCl$
11) $CaSO_4 \cdot 2H_2O \longrightarrow CaSO_4 + H_2O$

12) $C_3H_7CHO + O_2 \longrightarrow CO_2 + H_2O$
13) $NaHCO_3 + HCl \longrightarrow NaCl + CO_2 + H_2O$
14) $Bi(NO_3)_3 + NaOH \longrightarrow Bi(OH)_3 + NaNO_3$
15) $FeS + HBr \longrightarrow FeBr_2 + H_2S$
16) $P_4O_{10} + H_2O \longrightarrow H_3PO_4$
17) $CaCO_3 + H_3PO_4 \longrightarrow$
$\qquad Ca_3(PO_4)_2 + H_2O + CO_2$
18) $PCl_5 + H_2O \longrightarrow H_3PO_4 + HCl$
19) $CaI_2 + H_2SO_4 \longrightarrow CaSO_4 + HI$
20) $C_3H_7COOH + O_2 \longrightarrow CO_2 + H_2O$
21) $Mg(CN)_2 + HCl \longrightarrow HCN + MgCl_2$
22) $(NH_4)_2S + HBr \longrightarrow NH_4Br + H_2S$
23) $H_2SO_4 + NaC_2H_3O_2 \longrightarrow$
$\qquad Na_2SO_4 + HC_2H_3O_2$
24) $Fe + O_2 \longrightarrow Fe_2O_3$

CHAPTER 8 QUESTIONS AND PROBLEMS

For Questions 1 through 10, write the equation for each reaction described.

SECTION 8.4

1) Sodium peroxide, Na_2O_2, is formed from its elements.

2) Magnesium combines with oxygen to form magnesium oxide.

_____

3) Barium and bromine react to form barium bromide.

4) Rubidium (Z = 37) reacts with chlorine to form rubidium chloride.

_____

5) Calcium oxide and water form calcium hydroxide.

6) Copper(I) chloride is formed from copper and copper(II) chloride.

SECTION 8.5

7) Potassium chloride decomposes to its elements.

8) Hydrogen peroxide, H_2O_2, decomposes to oxygen and water.

_____

9) Carbonated beverages contain carbonic acid, an unstable compound that breaks down into carbon dioxide and water.

10) Iron(III) sulfide decomposes to its elements.

SECTION 8.6

For Questions 11 through 14, write the equation for the complete oxidation of the compound given.

11) C_4H_{10}

12) CH_3CHO _____

13) $C_7H_{15}OH$

14) $(C_2H_5)_2CO$

For Questions 15 through 28, write the equation for each reaction described.

SECTION 8.7

15) Potassium reacts vigorously with water to form hydrogen and a solution of potassium hydroxide.

16) Aluminum reacts with a solution of nickel nitrate. _____

17) Zinc is placed into hydrochloric acid.

18) Chlorine displaces iodine from potassium iodide solution.

SECTION 8.8

19) Silver sulfide precipitates when hydrogen sulfide gas is bubbled through silver nitrate solution.

20) A precipitate forms when solutions of magnesium sulfate and sodium hydroxide are combined. _____

21) When a solution of barium chloride is poured into a solution of ammonium carbonate, a precipitate results.

22) A bright canary-yellow precipitate forms when solutions of lead nitrate and potassium iodide are combined. _____

23)* Silver iodate precipitates when solutions of a sodium compound and a nitrate compound are combined.

24)* The reaction between the solution of a potassium compound and a chloride solution results in the precipitation of cadmium hydroxide, $Cd(OH)_2$.

SECTION 8.9

25) Lithium hydroxide neutralizes sulfuric acid.

26) Acetic acid, $HC_2H_3O_2$, reacts with barium hydroxide. _____

27) Potassium hydroxide solution reacts with hydrobromic acid.

28) Calcium hydroxide neutralizes sulfuric acid.

UNCLASSIFIED REACTIONS

Write the equation for the reaction described, or for the most likely reaction between given reactants. The kinds of reactions are arranged in a random order; that is, there are no matched pairs, as in the first 28 questions.

29) Glycerine, $C_3H_8O_3$, used in making soap, cosmetics, and explosives, is completely oxidized.

30) Sodium iodate solution is added to a solution of copper(II) sulfate.

31) Phosphorus tribromide is formed in the reaction between phosphorus and bromine.

32) Table sugar, $C_{12}H_{22}O_{11}$, is burned completely.

33) Chromium(III) nitrate is one of the products of the reaction between metallic chromium and a solution of tin(II) nitrate.

34) Calcium chlorite loses its oxygen to become calcium chloride.

35) Sulfurous acid decomposes spontaneously to sulfur dioxide and water.

36) Silicon tetrachloride is formed from its elements.

37) Powdered antimony ($Z = 51$) ignites when sprinkled into chlorine gas, producing antimony(III) chloride.

38) Acetaldehyde, CH_3CHO, a raw material used in the manufacture of vinegar, perfumes, dyes, plastics, and other organic materials, is oxidized.

39) The appearance of solid silver chromate, Ag_2CrO_4, indicates the end of a reaction when silver nitrate comes into contact with sodium chromate. (You should be able to figure out the formula of sodium chromate from the given formula of silver chromate.)

40) Fructose sugar, $C_6H_{12}O_6$, is completely oxidized.

41) Sodium phosphate and zinc chloride solutions react.

42) Silver nitrate solution is poured into potassium bromide solution.

43) A solution of potassium hydroxide reacts with a solution of zinc chloride.

44) Solutions of lead(II) nitrate and copper(II) sulfate are combined.

45) Only the first hydrogen comes off in the reaction between sulfamic acid, HNH_2SO_3, and potassium hydroxide.

46) Magnesium reacts with a solution of nickel chloride.

47) Pentane, C_5H_{12}, burns completely.

48) A solution of nickel chloride is poured into a solution of sodium carbonate.

49) When calcium hydroxide—sometimes called slaked lime—is heated, it forms calcium oxide—lime—and water vapor.

50) Fluorine reacts spontaneously with nearly all elements. Oxygen difluoride is the product when the second element is oxygen.

51) Cesium (Z = 55) reacts violently with water, releasing hydrogen, which usually explodes, and leaving a solution of cesium hydroxide. (*Hint:* Think of water as an acid whose formula is HOH.)

52) Lithium is added to a solution of manganese(II) chloride.

53) Sulfuric acid is produced when sulfur trioxide reacts with water.

54) Strontium hydroxide (Z = 38 for strontium) can be decomposed into strontium oxide and water.

55) Aluminum carbide, Al_4C_3, is the product of the reaction between its elements.

56) The concentration of sodium hydroxide solution can be found by reacting it with oxalic acid, $H_2C_2O_4$. (We're betting you can figure out the formula and the name of the ion that comes from oxalic acid.)

57) Ammonium sulfide solution is added to a solution of copper(II) nitrate.

58) Silver nitrate is added to a solution of sodium sulfide.

59) Silver nitrate solution is poured into a solution of potassium carbonate.

60) Bubbles form when metallic barium is placed in water. (See hint for Question 51.)

61) If manganese is placed in a solution of chromium(III) chloride, one of the reaction products is manganese(II) chloride.

62) Igniting a mixture of powdered iron(III) oxide and aluminum produces a vigorous reaction.

The remaining reactions are not readily placed into one of the six classifications used in this chapter. Nevertheless, enough information is given for you to write the equation.

63) Hydrogen and carbon dioxide are the products of the reaction between carbon monoxide and water.

64) Calcium sulfate is formed by the reaction between sulfur dioxide, calcium oxide, and oxygen.

65) Hydrogen is released and magnesium nitride is formed when magnesium reacts with ammonia.

66) Sulfur dioxide, water, and calcium sulfide react to produce calcium hydroxide and sulfur.

67)* Potassium permanganate, $KMnO_4$, reacts with potassium hydroxide to produce potassium manganate, K_2MnO_4, oxygen, and water.

68)* Chlorine can be produced in the laboratory by the reaction between potassium permanganate, $KMnO_4$, and hydrochloric acid. Manganese dioxide [also called manganese(IV) oxide], water, and potassium chloride are other products of the reaction.

69)* Chlorine is a by-product of the process by which magnesium is extracted from seawater. The chlorine is converted to hydrochloric acid by reacting it with natural gas, CH_4, and oxygen. Carbon monoxide is a second product of the reaction.

70)* Hydrogen is needed in the industrial process for making ammonia. It is obtained by the reaction between propane, C_3H_8, and steam. The carbon in the propane goes off as carbon dioxide.

MATCHING SET ANSWERS

Set 1: 9–10–8–1–7–11–5–3–2–6–4

Set 2: 10–9–2–7–6–3–1–8–5–4

QUICK CHECK ANSWERS

8.1 Both false. Equation a) has seven oxygen atoms on the left and six on the right. In b), *never* change a chemical formula to balance an equation. H_2O_2 is a real compound, but it has nothing to do with this reaction.

9

Quantity Relationships in Chemical Reactions

This is the precipitation of lead(II) chromate, PbCrO$_4$, that occurs when a solution of sodium chromate, Na$_2$CrO$_4$, is added to a solution of lead(II) nitrate, Pb(NO$_3$)$_2$. The equation for the reaction is Na$_2$CrO$_4$(aq) + Pb(NO$_3$)$_2$(aq) → PbCrO$_4$(s) + 2 NaNO$_3$(aq). The question that this chapter asks—and answers—is, "How many grams of lead chromate will precipitate if 123 g Na$_2$CrO$_4$ react?"

In Chapter 9 you will learn how to solve **stoichiometry** problems. A stoichiometry problem is a problem that asks, "How much or how many?" How many tons of sodium chloride must be electrolyzed to produce ten tons of sodium? How many kiloliters of chlorine at a certain temperature and pressure will be produced at the same time? How much energy is needed to do the job? And so forth . . .

A few introductory comments may help you to learn how to solve stoichiometry problems. The problem-solving strategy from Section 3.9 is used repeatedly. You will soon see that a series of "per" relationships link the given and wanted quantities in all problems in this chapter. Therefore the problems are solved by dimensional analysis.

As usual, our solutions to examples begin with the "plan"—identify the GIVEN and WANTED quantities, the unit PATH, and the conversion FACTORS. We know you will not write out, "Given: so many grams of X," and so forth, for every example. But you should *think* the given, wanted, and unit path. If you cannot think those steps clearly, then write them. Links that are missing in a thought process sometimes present themselves when the thoughts you do have are put on paper.

Stoichiometry problems can become long; most unit paths have three steps, and some four. But the problems are not difficult if you recognize the unit path, know the conversion factor for each step, and then apply the factors correctly. That's the skill you should develop in this chapter.

9.1 CONVERSION FACTORS FROM A CHEMICAL EQUATION

> **PG 9A** Given a chemical equation, or a reaction for which the equation is known, and the number of moles of one species in the reaction, calculate the number of moles of any other species.

A chemical equation may be interpreted quantitatively in two ways. The equation

$$PCl_5(s) + 4\,H_2O(\ell) \longrightarrow H_3PO_4(aq) + 5\,HCl(aq) \qquad (9.1)$$

may be read, "One PCl_5 molecule reacts with four H_2O molecules to produce one H_3PO_4 molecule and five HCl molecules." It also means, "One *mole* of PCl_5 molecules reacts with four *moles* of H_2O molecules to produce one *mole* of H_3PO_4 molecules and five *moles* of HCl molecules." To solve stoichiometry problems, we must think in terms of moles. It is only through moles that we can convert from one chemical in a reaction to another.

Unfortunately, we cannot measure moles—at least, not directly. Instead, we measure masses and volumes. These can be converted into moles. You already know how to do this with mass, using molar mass as a conversion factor (see Section 7.5). In this chapter we will use the mass–mole conversion, saving changes between volume and moles until Chapters 13 and 15.

The coefficients in a chemical equation give us the conversion factors to get from the moles of one substance to the moles of another substance in a reaction. For example, Equation 9.1 shows that four moles of H_2O are needed to react with each mole of PCl_5. In other words, the reaction uses 4 mol H_2O per mole PCl_5, or 4 mol H_2O/mol PCl_5. This "per" relationship is the conversion factor by which we may convert in either direction between moles of H_2O and moles of PCl_5. Similarly, it takes four moles of H_2O to produce one mole of H_3PO_4, 4 mol H_2O/mol H_3PO_4. Four moles of water also yield five moles of HCl, 4 mol H_2O/5 mol HCl. As always, the inverse of each conversion factor is also valid: 1 mol PCl_5/4 mol H_2O, 1 mol H_3PO_4/4 mol H_2O, and 5 mol HCl/4 mol H_2O.

If the number of moles of any species in a reaction, either reactant or product, is known, there is a one-step conversion to the moles of any other species. If 3.20 moles of PCl_5 react according to Equation 9.1, how many moles of HCl will be produced? The unit path is mol PCl_5 → mol HCl. From the several mole relationships that are available in the equation, the one that satisfies the unit path is 5 mol HCl/1 mol PCl_5:

$$3.20 \text{ mol } PCl_5 \times \frac{5 \text{ mol HCl}}{\text{mol } PCl_5} = 16.0 \text{ mol HCl}$$

Note that the 1 and 5 are exact numbers; they do not affect the significant figures in the final answer.

This method may be applied to any equation.

EXAMPLE 9.1

How many moles of oxygen are required to burn 2.40 moles of ethane, C_2H_6?

First you need an equation for this burning reaction.

_____

$$2 C_2H_6(g) + 7 O_2(g) \longrightarrow 4 CO_2(g) + 6 H_2O(g)$$

Now "plan" the problem, as you learned to do in Section 3.9. Identify the GIVEN and WANTED quantities and the unit PATH that connects them. Then figure out the needed conversion FACTOR from the equation.

_____

GIVEN: 2.40 mol C_2H_6 WANTED: mol O_2
PATH: mol C_2H_6 → mol O_2 FACTOR: 7 mol O_2/2 mol C_2H_6

The conversion factor comes from the "per" relationship between the given and wanted quantities, using their coefficients in the equation.

Complete the problem.

_____

$$2.40 \text{ mol } C_2H_6 \times \frac{7 \text{ mol } O_2}{2 \text{ mol } C_2H_6} = 8.40 \text{ mol } O_2$$

EXAMPLE 9.2

Ammonia is formed directly from its elements. How many moles of hydrogen are needed to produce 4.20 mol NH_3?

The procedure is exactly the same as in the previous example. Complete the problem.

_____

EQUATION: $N_2(g) + 3 H_2(g) \rightarrow 2 NH_3(g)$
GIVEN: 4.20 mol NH_3 WANTED: mol H_2
PATH: mol NH_3 → mol H_2 FACTOR: 3 mol H_2/2 mol NH_3

$$4.20 \text{ mol } NH_3 \times \frac{3 \text{ mol } H_2}{2 \text{ mol } NH_3} = 6.30 \text{ mol } H_2$$

FLASHBACK You wrote equations for burning reactions in Section 8.6. Recall the unnamed second reactant and the two products that are always formed.

9.2 MASS CALCULATIONS

PG 9B Given a chemical equation, or a reaction for which the equation can be written, and the number of grams or moles of one species in the reaction, find the number of grams or moles of any other species.

We are now ready to solve the problem that underlies the manufacture of chemicals and the design of laboratory experiments: How much product for so much raw material, or how much raw material for so much product? To solve this problem, you will tie together several skills:

You will write chemical formulas (Chapter 6).

You will calculate molar masses from the formulas (Section 7.4).

You will use molar masses to change mass to moles and moles to mass (Section 7.5).

You will use formulas to write chemical equations (Chapter 8).

You will use the equation to change from moles of one species to moles of another (Section 9.1).

And you will do all of these things in one problem!

The above list should impress upon you how much stoichiometry problems depend on other skills. If you have any doubt about these skills, you would find it helpful to review the sections listed.

Section 7.5 described writing the formula of a substance and calculating its molar mass as "starting steps" that had to be completed before you could convert from mass to moles or moles to mass. These steps are also performed for the given and wanted substances in stoichiometry problems. In addition, you must have the reaction equation to convert moles of given substances to moles of wanted substances. Therefore, we include writing the equation in the starting steps of a stoichiometry problem.

Once the starting steps are completed, the solution of these problems usually falls into a three-step "**stoichiometry pattern**." The unit path is

$$
\begin{array}{ccccccc}
\text{grams of} & & \text{moles of} & & \text{moles of} & & \text{grams of} \\
\text{given species} & \rightarrow & \text{given species} & \rightarrow & \text{wanted species} & \rightarrow & \text{wanted species} \\
& & \text{Step 1} & & \text{Step 2} & & \text{Step 3}
\end{array} \qquad (9.2)
$$

$$
\text{g G} \quad \times \quad \frac{\text{mol G}}{\text{g G}} \quad \times \quad \frac{\text{mol W}}{\text{mol G}} \quad \times \quad \frac{\text{g W}}{\text{mol W}} = \text{g W} \qquad (9.3)
$$

Source of conversion factor — molar mass of G — from equation — molar mass of W

In words, these three steps are:

PROCEDURE

1) Change grams of given species to moles, g G → mol G (Section 7.5).
2) Change moles of given species to moles wanted, mol G → mol W (Section 9.1).
3) Change moles of wanted species to grams, mol W → g W (Section 7.5).

LEARN IT NOW The stoichiometry pattern expressed in Equation 9.2 will be expanded to include gases in Chapter 13 and expanded again in Chapter 15 to include solutions. In both cases the conversion between grams and moles is replaced by a conversion between liters and moles. With gases you will use *molar volume*, expressed in liters per mole at a specified temperature and pressure. It is necessary to specify temperature and pressure because, as you saw in Chapter 4, the volume occupied by a given amount of gas varies greatly as these conditions change. In Chapter 15 you will use solution concentration expressed in *molarity*, whose units are moles per liter. The method by which those problems are solved is the same as the method for solving the problems in this chapter. LEARN IT NOW, and you will be ready to use it later.

Occasionally you may be given the mass of one substance and asked to find the number of moles of a second species. In this case the first two steps of the stoichiometry pattern complete the problem. Or, you may be given the moles of one substance and asked to find the grams of another. Steps 2 and 3 solve this problem.

The burning of ethane (Example 9.1) may be used to illustrate the method of stoichiometry.

EXAMPLE 9.3 _____

Natural gas is partly ethane. Therefore, this reaction occurs in your home every time you use a gas furnace, a gas range, or a gas water heater.

Calculate the number of grams of oxygen that are required to burn 155 g of ethane, C_2H_6, in the reaction $2\ C_2H_6(g) + 7\ O_2(g) \rightarrow 4\ CO_2(g) + 6\ H_2O(g)$.

SOLUTION

First, we complete the starting steps. The equation is given. The molar masses of the given and wanted substances must be calculated. They are 30.0 g C_2H_6/mol C_2H_6 and 32.0 g O_2/mol O_2. The "plan" for the problem is

GIVEN: 155 g C_2H_6 WANTED: g O_2
PATH: g $C_2H_6 \rightarrow$ mol $C_2H_6 \rightarrow$ mol $O_2 \rightarrow$ g O_2
FACTORS: 30.0 g C_2H_6/mol C_2H_6; 7 mol O_2/2 mol C_2H_6; 32.0 g O_2/mol O_2

In Step 1 we change the given quantity to moles. The setup begins

$$155\ \text{g}\ C_2H_6 \times \frac{1\ \text{mol}\ C_2H_6}{30.0\ \text{g}\ C_2H_6} \times \underline{\hspace{2cm}} \times \underline{\hspace{1.5cm}} =$$

If we calculated the answer to this point, we would have moles of C_2H_6. In Step 2 the setup is extended to convert moles of C_2H_6 to moles of O_2:

$$155\ \text{g}\ C_2H_6 \times \frac{1\ \text{mol}\ C_2H_6}{30.0\ \text{g}\ C_2H_6} \times \frac{7\ \text{mol}\ O_2}{2\ \text{mol}\ C_2H_6} \times \underline{\hspace{1.5cm}} =$$

If we calculated the answer to this point, we would have moles of O_2. In Step 3 the moles of oxygen are converted to grams:

$$155\ \text{g}\ C_2H_6 \times \frac{1\ \text{mol}\ C_2H_6}{30.0\ \text{g}\ C_2H_6} \times \frac{7\ \text{mol}\ O_2}{2\ \text{mol}\ C_2H_6} \times \frac{32.0\ \text{g}\ O_2}{1\ \text{mol}\ O_2} = 579\ \text{g}\ O_2$$

EXAMPLE 9.4

How many grams of oxygen are required to burn 3.50 moles of heptane, $C_7H_{16}(\ell)$?

Heptane is present in gasoline, so this reaction occurs in automobile engines.

Complete the starting steps. Be sure to read the problem carefully.

———

EQUATION: $C_7H_{16}(\ell) + 11\ O_2(g) \rightarrow 7\ CO_2(g) + 8\ H_2O(g)$
GIVEN: 3.50 mol C_7H_{16} WANTED: g O_2
PATH: mol $C_7H_{16} \rightarrow$ mol $O_2 \rightarrow$ g O_2
FACTORS: 11 mol O_2/mol C_7H_{16}; 32.0 g O_2/mol O_2

The given quantity is *moles* of heptane. In other words, the first step of the stoichiometry pattern is completed. The wanted quantity of O_2 is to be expressed in grams. Therefore, you need its molar mass for the mol $\rightarrow$ g conversion.

Set up the problem for changing 3.50 mol C_7H_{16} to mol O_2.

———

$$3.50\ \text{mol } C_7H_{16} \times \frac{11\ \text{mol } O_2}{1\ \text{mol } C_7H_{16}} \times \underline{\qquad} =$$

Now extend the setup to change moles of O_2 to grams.

———

$$3.50\ \text{mol } C_7H_{16} \times \frac{11\ \text{mol } O_2}{1\ \text{mol } C_7H_{16}} \times \frac{32.0\ \text{g } O_2}{1\ \text{mol } O_2} = 1.23 \times 10^3\ \text{g } O_2$$

EXAMPLE 9.5

How many moles of H_2O will be produced in the heptane-burning reaction that also yields 115 g CO_2?

Begin with the starting steps.

———

EQUATION: $C_7H_{16}(\ell) + 11\ O_2(g) \rightarrow 7\ CO_2(g) + 8\ H_2O(g)$
GIVEN: 115 g CO_2 WANTED: mol H_2O
PATH: g $CO_2 \rightarrow$ mol $CO_2 \rightarrow$ mol H_2O
FACTORS: 44.0 g CO_2/mol CO_2; 8 mol H_2O/7 mol CO_2

This time the wanted quantity is moles of CO_2, which is reached in the first two steps of the stoichiometry pattern.

Set up the first conversion, g $CO_2 \rightarrow$ mol CO_2, but do not calculate the answer.

———

of the given amount of reactant to product. Theoretical yield is always a calculated quantity, calculated by the principles of stoichiometry. In actual practice, factors such as impure reactants, incomplete reactions, and side reactions cause the **actual yield** (act) to be less than the calculated yield. The actual yield is a measured quantity, determined by experiment or experience.

If we know the actual yield found in the laboratory or in the production plant, and the theoretical yield calculated by stoichiometry, we can find the **percent yield**. This is the actual yield expressed as a percentage of the theoretical yield. As with all percentages, it is the part quantity (actual yield) over the whole quantity (theoretical yield) times 100:

$$\% \text{ yield} = \frac{\text{actual yield}}{\text{theoretical yield}} \times 100 \qquad (9.4)$$

If only 181 g CO_2 had been produced in Example 9.6 instead of the calculated 203 g, the percent yield would be

$$\% \text{ yield} = \frac{181 \text{ g}}{203 \text{ g}} \times 100 = 89.2\%$$

FLASHBACK   This is a specific form of general Equation 7.4 in Section 7.6:

$$\% \text{ of A} = \frac{\text{parts of A}}{\text{total parts}} \times 100$$

When a part quantity and a whole quantity are both given, percentage is calculated by substitution into this equation.

EXAMPLE 9.8

Barium sulfate is used in the manufacture of photographic papers and linoleum and as a color pigment in wallpaper.

A solution containing excess* sodium sulfate is added to a second solution containing 3.18 g of barium nitrate. Barium sulfate precipitates. (a) Calculate the theoretical yield of barium sulfate. (b) If the actual yield is 2.69 g, calculate the percent yield.

Calculating theoretical yield is a typical stoichiometry problem. Solve part (a).

EQUATION: $Ba(NO_3)_2(aq) + Na_2SO_4(aq) \rightarrow BaSO_4(s) + 2\,NaNO_3(aq)$
GIVEN: 3.18 g $Ba(NO_3)_2$ WANTED: g $BaSO_4$
PATH: g $Ba(NO_3)_2 \rightarrow$ mol $Ba(NO_3)_2 \rightarrow$ mol $BaSO_4 \rightarrow$ g $BaSO_4$
FACTORS: 261.3 g $Ba(NO_3)_2$/mol $Ba(NO_3)_2$; 1 mol $BaSO_4$/mol $Ba(NO_3)_2$;
233.4 g $BaSO_4$/mol $BaSO_4$

$$3.18 \text{ g } Ba(NO_3)_2 \times \frac{1 \text{ mol } Ba(NO_3)_2}{261.3 \text{ g } Ba(NO_3)_2} \times \frac{1 \text{ mol } BaSO_4}{1 \text{ mol } Ba(NO_3)_2} \times \frac{233.4 \text{ g } BaSO_4}{1 \text{ mol } BaSO_4} = 2.84 \text{ g } BaSO_4$$

Part (b) is solved by substitution into Equation 9.4.

GIVEN: 2.69 g (act); 2.84 g (theo)   WANTED: % yield

$$\text{EQUATION: } \% \text{ yield} = \frac{\text{actual yield}}{\text{theoretical yield}} \times 100 = \frac{2.69 \text{ g}}{2.84 \text{ g}} \times 100 = 94.7\%$$

---

*The word *excess* as used here means "more than enough." There is "more than enough" sodium sulfate to precipitate all of the barium in 3.18 grams of barium nitrate.

If *given the percent yield* and *either* the theoretical yield *or* the actual yield, the missing yield can be calculated from the given yield by dimensional analysis. The conversion factor is the percent yield in parts per hundred—or, specifically, grams actual yield/100 grams theoretical yield. For example, assume a manufacturer of magnesium hydroxide knows from experience that the percent yield is 81.3% from the production process. How many kilograms of the actual product should be expected if the theoretical yield is 697 kg? Working in kilograms rather than grams, we obtain

FLASHBACK You used percentage as a conversion factor this way in solving Examples 7.13 and 7.14 in Section 7.6. This is a specific example of a conversion factor from a defining equation and per relationship as introduced in Section 3.8.

GIVEN: 697 kg $Mg(OH)_2$ (theo)   WANTED: kg $Mg(OH)_2$ (act)
PATH: kg $Mg(OH)_2$ (theo) $\rightarrow$ kg $Mg(OH)_2$ (act)
FACTOR: 81.3 kg $Mg(OH)_2$ (act)/100 kg $Mg(OH)_2$ (theo)

$$697 \text{ kg Mg(OH)}_2 \text{ (theo)} \times \frac{81.3 \text{ kg Mg(OH)}_2 \text{ (act)}}{100 \text{ kg Mg(OH)}_2 \text{ (theo)}} = 567 \text{ kg Mg(OH)}_2 \text{ (act)}$$

A more likely problem for this maker of magnesium hydroxide is finding out how much raw material is required for a certain amount of product.

EXAMPLE 9.9

A manufacturer wants to prepare 800 kg $Mg(OH)_2$ (assume three significant figures) by the reaction $MgO(s) + H_2O(\ell) \rightarrow Mg(OH)_2(s)$. Previous production experience shows that the process has an 81.3% yield calculated from the initial MgO. How much MgO should be used?

A water mixture of magnesium hydroxide is commonly known as milk of magnesia.

SOLUTION

This is a two-part problem. We are given the actual yield. The stoichiometry problem for finding MgO must be based on theoretical yield. Therefore, we must first find the theoretical yield from the actual yield. Then we calculate the amount of reactant by stoichiometry.

"Planning" the first part of the problem gives us

GIVEN: 800 kg $Mg(OH)_2$ (act)   WANTED: kg $Mg(OH)_2$ (theo)
PATH: kg $Mg(OH)_2$ (act) $\rightarrow$ kg $Mg(OH)_2$ (theo)
FACTOR: 81.3 kg $Mg(OH)_2$ (act)/100 kg $Mg(OH)_2$ (theo)

$$800 \text{ kg Mg(OH)}_2 \text{ (act)} \times \frac{100 \text{ kg Mg(OH)}_2 \text{ (theo)}}{81.3 \text{ kg Mg(OH)}_2 \text{ (act)}} = 984 \text{ kg Mg(OH)}_2 \text{ (theo)}$$

Now we are ready to use the stoichiometry pattern to find the amount of MgO that is required to produce 984 kg $Mg(OH)_2$. The starting steps require the molar masses of $Mg(OH)_2$ and MgO as conversion factors. Once these are calculated, the "plan" and calculation follow:

GIVEN: 984 kg $Mg(OH)_2$ WANTED: kg MgO
PATH: kg $Mg(OH)_2$ $\rightarrow$ kmol $Mg(OH)_2$ $\rightarrow$ kmol MgO $\rightarrow$ kg MgO
FACTORS: 58.3 kg $Mg(OH)_2$/kmol $Mg(OH)_2$; 1 kmol MgO/kmol $Mg(OH)_2$;
      40.3 kg MgO/kmol MgO

$$984 \text{ kg Mg(OH)}_2 \times \frac{1 \text{ kmol Mg(OH)}_2}{58.3 \text{ kg Mg(OH)}_2} \times \frac{1 \text{ kmol MgO}}{1 \text{ kmol Mg(OH)}_2} \times \frac{40.3 \text{ kg MgO}}{1 \text{ kmol MgO}} = 6.80 \times 10^2 \text{ kg MgO}$$

This problem can be solved with a single calculation setup by adding the actual $\rightarrow$ theoretical conversion to the stoichiometry pattern. The 984 kg $Mg(OH)_2$ that starts the preceding setup is replaced by the calculation that produced the number—the setup that changed actual yield to theoretical yield. The two unit paths

$$kg\ Mg(OH)_2\ (act) \longrightarrow kg\ Mg(OH)_2\ (theo) \quad \text{and}$$
$$kg\ Mg(OH)_2\ (theo) \longrightarrow kmol\ Mg(OH)_2 \longrightarrow kmol\ MgO \longrightarrow kg\ MgO$$

are combined to give

$$kg\ Mg(OH)_2\ (act) \longrightarrow kg\ Mg(OH)_2\ (theo) \longrightarrow kmol\ Mg(OH)_2 \longrightarrow kmol\ MgO \longrightarrow kg\ MgO$$

The calculation setup becomes

$$800\ \cancel{kg\ Mg(OH)_2\ (act)} \times \frac{100\ \cancel{kg\ Mg(OH)_2\ (theo)}}{81.3\ \cancel{kg\ Mg(OH)_2\ (act)}} \times \frac{1\ \cancel{kmol\ Mg(OH)_2}}{58.3\ \cancel{kg\ Mg(OH)_2}} \times$$

$$\frac{1\ \cancel{kmol\ MgO}}{1\ \cancel{kmol\ Mg(OH)_2}} \times \frac{40.3\ kg\ MgO}{1\ \cancel{kmol\ MgO}} = 6.80 \times 10^2\ kg\ MgO$$

You may solve percent yield problems such as this as two separate problems, or as a single extended dimensional analysis setup. Either way, note that percent *yield* refers to the *product*, not to a reactant. Accordingly, your percent yield conversion should always be between actual and theoretical product quantities, not reactant quantities. Sometimes the conversion is at the beginning of the setup, as in Example 9.9, and sometimes at the end.

EXAMPLE 9.10 _____

Sodium sulfate is used in dyeing textiles and in the manufacture of glass and paper pulp.

A procedure for preparing sodium sulfate is summarized in the equation

$$2\ S(s) + 3\ O_2(g) + 4\ NaOH(aq) \longrightarrow 2\ Na_2SO_4(aq) + 2\ H_2O(\ell)$$

The percent yield in the process is 79.8%. Find the number of grams of sodium sulfate that will be recovered from the reaction of 36.9 g NaOH.

The starting steps will help you "plan" your strategy for solving this problem. Our "plan" will be for a single dimensional analysis setup from the given quantity to the wanted quantity.

_____   _____

GIVEN: 36.9 g NaOH   WANTED: g Na_2SO_4 (act)
PATH: g NaOH $\rightarrow$ mol NaOH $\rightarrow$ mol Na_2SO_4 $\rightarrow$ g Na_2SO_4 (theo) $\rightarrow$ g Na_2SO_4 (act)
FACTORS: 40.0 g NaOH/mol NaOH; 2 mol Na_2SO_4/4 mol NaOH;
         142.1 g Na_2SO_4/mol Na_2SO_4; 79.8 g Na_2SO_4 (act)/100 g Na_2SO_4 (theo)

Write the calculation setup to the theoretical yield. If you choose to solve the problem in two steps, calculate the theoretical yield. If you are going to write a single setup for the whole problem, do not calculate that value.

_____   _____

$$36.9 \text{ g NaOH} \times \frac{1 \text{ mol Na}_2\text{OH}}{40.0 \text{ g NaOH}} \times \frac{2 \text{ mol Na}_2\text{SO}_4}{4 \text{ mol NaOH}} \times \frac{142.1 \text{ g Na}_2\text{SO}_4 \text{ (theo)}}{1 \text{ mol Na}_2\text{SO}_4} \times \underline{\hspace{2cm}} =$$

If you are solving this problem in two steps, your answer to this point, the theoretical yield, is 65.5 g Na_2SO_4.

Now multiply by the percent yield conversion factor that changes theoretical yield to actual yield.

$$36.9 \text{ g NaOH} \times \frac{1 \text{ mol NaOH}}{40.0 \text{ g NaOH}} \times \frac{2 \text{ mol Na}_2\text{SO}_4}{4 \text{ mol NaOH}} \times \frac{142.1 \text{ g Na}_2\text{SO}_4 \text{ (theo)}}{1 \text{ mol Na}_2\text{SO}_4} \times \frac{79.8 \text{ g Na}_2\text{SO}_4 \text{ (act)}}{100 \text{ g Na}_2\text{SO}_4 \text{ (theo)}} = 52.3 \text{ g Na}_2\text{SO}_4$$

EXAMPLE 9.11

Sodium nitrate is produced from sodium nitrate by the reaction $2 \text{ NaNO}_3(s) \rightarrow 2 \text{ NaNO}_2(s) + \text{O}_2(g)$. How many grams of $NaNO_3$ must be used to produce an actual yield of 60.0 g $NaNO_2$ if the percent yield is 76.3%?

Take this one all the way.

GIVEN: 60.0 g $NaNO_2$ (act)   WANTED: g $NaNO_3$
PATH: g $NaNO_2$ (act) $\rightarrow$ g $NaNO_2$ (theo) $\rightarrow$ mol $NaNO_2$ $\rightarrow$ mol $NaNO_3$ $\rightarrow$ g $NaNO_3$
FACTORS: 76.3 g $NaNO_2$ (act)/100 g $NaNO_2$ (theo); 69.0 g $NaNO_2$/mol $NaNO_2$;
    2 mol $NaNO_3$/2 mol $NaNO_2$; 85.0 g $NaNO_3$/mol $NaNO_3$

$$60.0 \text{ g NaNO}_2 \text{ (act)} \times \frac{100 \text{ g NaNO}_2 \text{ (theo)}}{76.3 \text{ g NaNO}_2 \text{ (act)}} \times \frac{1 \text{ mol NaNO}_2}{69.0 \text{ g NaNO}_2} \times \frac{2 \text{ mol NaNO}_3}{2 \text{ mol NaNO}_2} \times \frac{85.0 \text{ g NaNO}_3}{1 \text{ mol NaNO}_3} = 96.9 \text{ g NaNO}_3$$

9.4 LIMITING REACTANT PROBLEMS

> **PG 9D** Given a chemical equation, or information from which it may be determined, and initial quantities of two or more reactants, (a) identify the limiting reactant; (b) calculate the theoretical yield of a specified product, assuming complete use of the limiting reactant; and (c) calculate the quantity of the reactant initially in excess that remains unreacted.

Zinc reacts with sulfur to form zinc sulfide: $\text{Zn}(s) + \text{S}(s) \rightarrow \text{ZnS}(s)$. Suppose you put three moles of zinc and two moles of sulfur into a reaction vessel and cause them to react until one is totally used up. How many moles of zinc sulfide will result? Also, how many moles of which element will remain unreacted?

This question is something like asking, ''How many pairs of gloves can you assemble out of 20 left gloves and 30 right gloves, and how many unmatched gloves, and for which hand, will be left over?'' The answer, of course, is 20 pairs of gloves. After you have assembled 20 pairs you run out of left gloves, even though you have 10 right gloves remaining.

It is very unusual for reactants in a chemical change to be present in the exact quantities that will react completely with each other. This condition is approached, however, in the process of titration, which you will consider in Section 15.12.

The same reasoning may be applied in the zinc sulfide question. The chemicals combine on a 1:1 mole ratio. If you start with three moles of zinc and two moles of sulfur, the reaction will stop when the two moles of sulfur are used up. Sulfur, **the reactant that is completely used up by the reaction, is called the limiting reactant.** One mole of zinc, the **excess reactant**, will remain unreacted.

The amount of product is limited by the moles of limiting reactant, and it must be calculated from that number of moles. According to the equation, each mole of sulfur produces one mole of zinc sulfide. If two moles of sulfur react, two moles of zinc sulfide are produced.

This entire analysis may be summarized as follows:

	Zn	+	S	→	ZnS
Moles at start	3		2		0
Moles used (−), produced (+)	−2		−2		+2
Moles at end	1		0		2

There are two widely used approaches to solving limiting reactant problems: One identifies the limiting reactant by comparing the number of moles of each reactant present with the number of moles of each reactant required by the chemical equation. The $Zn + S \rightarrow ZnS$ illustration is a simple example of the "comparison of moles" method. Unfortunately, calculations are more complicated when (1) the mole relationships from the equation are not in a simple 1:1 ratio and (2) quantities are expressed in grams.

The second approach calculates the amount of product that *can* be produced from each reactant quantity as if that reactant was the limiting reactant. Between the two results, the "smaller amount" is the correct answer. After that amount of product is formed, the reactant from which it was calculated—the limiting reactant—is all used up. The amount of excess reactant that remains is found by calculating the amount used by the limiting reactant, which is subtracted from the amount present initially.

We will describe both calculation methods. Your instructor may express a preference for one method, in which case that is the one you should study. Disregard the other. If the choice is left to you, you will probably find the "smaller amount" method easier right now because you already know what to do. The "comparison of moles" method, however, gives you an understanding of what happens to substances in a chemical change. This will be useful later in your study of chemistry. Answers in the back of the book are given in the comparison-of-moles method.

If your instructor does not tell you which of the two methods to use, and you must choose for yourself, we suggest that you look ahead to examples that will help you choose. Example 9.14 develops the comparison-of-moles method in detail, and Example 9.15 shows what the solved problem looks like when you do the problem without help. Examples 9.16 and 9.17 do the same for the smaller-amount method. Once you choose, study only the method you have chosen and ignore the other.

Comparison of Moles

The following example shows you what to do when the reactants do not react in a 1:1 mole ratio. Preparing a table, as in the $Zn + S \rightarrow ZnS$ example, usually helps.

EXAMPLE 9.12

When powdered antimony (Z = 51) is sprinkled into chlorine gas, antimony trichloride is produced: 2 Sb + 3 Cl$_2$ → 2 SbCl$_3$. If 0.167 mol Sb is introduced into a flask that holds 0.267 mol Cl$_2$,

When this reaction is performed in a test tube, little bursts of flame can be seen as the antimony falls through the chlorine.

a) How many moles of SbCl$_3$ will be produced if the limiting reagent reacts completely?

b) How many moles of which element will remain unreacted?

SOLUTION

	2 Sb	+	3 Cl$_2$	→	2 SbCl$_3$
Moles at start	0.167		0.267		0
Moles used (−), produced (+)					
Moles at end					

The limiting reactant is not easily recognized this time. One way to identify it is to select either reactant and ask, "If Reactant A is the limiting reactant, how many moles of Reactant B are needed to react with all of Reactant A?" This is just like Example 9.1. If the number of moles of Reactant B needed is more than the number available, Reactant B is the limiting reactant. If the number of moles of Reactant B needed is smaller than the number present, Reactant B is the excess reactant and Reactant A is the limiting reactant. The conclusion is the same regardless of which reactant is selected as A and which is selected as B. To illustrate:

If antimony is the limiting reactant, how many moles of chlorine are required to react with all of the antimony?

$$0.167 \text{ mol Sb} \times \frac{3 \text{ mol Cl}_2}{2 \text{ mol Sb}} = 0.251 \text{ mol Cl}_2$$

The table shows that there are 0.267 mol Cl$_2$ present, more than enough to react with all of the antimony. Antimony is the limiting reactant.

If chlorine is the limiting reactant, how many moles of antimony are required to react with all of the chlorine?

$$0.267 \text{ mol Cl}_2 \times \frac{2 \text{ mol Sb}}{3 \text{ mol Cl}_2} = 0.178 \text{ mol Sb}$$

The table shows that there are 0.167 mol Sb present, not enough to react with all of the chlorine. Antimony is therefore the limiting reactant.

Having established antimony as the limiting reactant by either of the above assumptions, the table can now be completed:

	2 Sb	+	3 Cl$_2$	→	2 SbCl$_3$
Moles at start	0.167		0.267		0
Moles used (−), produced (+)	−0.167		−0.251		+0.167
Moles at end	0		0.016		0.167

Do you recognize where the $+0.167$ mol $SbCl_3$ came from? It is like Example 9.1 again. If the reaction uses all of the limiting reactant, 0.167 mol Sb, how many moles of $SbCl_3$ will form? The moles of Sb and $SbCl_3$ are on a 1:1 ratio—the equation shows 2 mol Sb and 2 mol $SbCl_3$—so the moles of $SbCl_3$ produced are equal to the moles of Sb used, 0.167 mole.

EXAMPLE 9.13

Potassium nitrate is also used in making matches and explosives, including fireworks, gunpowder, and blasting powder.

How many moles of the fertilizer *saltpeter* (KNO_3) can be made from 7.94 mol KCl and 9.96 mol HNO_3 by the reaction

$$3\ KCl(s) + 4\ HNO_3(aq) \longrightarrow 3\ KNO_3(s) + Cl_2(g) + NOCl(g) + 2\ H_2O(\ell)$$

Also, how many moles of which reactant will be unused?

Because KNO_3 is the only product asked about, the other products need not appear in your tabulation. Setting up the table effectively ''plans'' this kind of problem. Do that, filling in only the ''Moles at start'' line.

	3 KCl	+	4 HNO$_3$	→	3 KNO$_3$
Moles at start	7.94		9.96		0
Moles used (−), produced (+)					
Moles at end					

Now identify the limiting reactant. ''Guess'' whether it is KCl or HNO_3. Test your guess by calculating the amount of the other reactant needed to react with all of what you think is the limiting reactant, as it was done in Example 9.12. When you are sure about the limiting reactant, insert the numbers on the second line.

	3 KCl	+	4 HNO$_3$	→	3 KNO$_3$
Moles at start	7.94		9.96		0
Moles used (−), produced (+)	−7.47		−9.96		+7.47
Moles at end					

This is how your calculations would have gone for either limiting reactant choice:

Limiting reactant: KCl

$$7.94\ \text{mol KCl} \times \frac{4\ \text{mol HNO}_3}{3\ \text{mol KCl}} = 10.6\ \text{mol HNO}_3$$

If KCl is the limiting reactant, 10.6 mol HNO_3 are required to react with all of the KCl. Only 9.96 mol HNO_3 are present. This is not enough HNO_3, so HNO_3 is the limiting reactant.

Limiting reactant: HNO$_3$

$$9.96\ \text{mol HNO}_3 \times \frac{3\ \text{mol KCl}}{4\ \text{mol HNO}_3} = 7.47\ \text{mol KCl}$$

If HNO_3 is the limiting reactant, 7.47 mol KCl are required to react with all of the HNO_3. 7.94 mol KCl are present. This is more than enough KCl, so HNO_3 is the limiting reactant.

The number of moles of each species at the end is simply the algebraic sum of the moles at the start and the moles used or produced. Complete the problem.

	3 KCl	+	4 HNO$_3$	→	3 KNO$_3$
Moles at start	7.94		9.96		0
Moles used (−), produced (+)	−7.47		−9.96		+7.47
Moles at end	0.47		0		7.47

7.47 mol KNO$_3$ are produced and 0.47 mol KCl remain unreacted.

Reactant and product quantities are not usually expressed in moles, but rather in grams. This adds one or more steps before and after the sequence in Example 9.13. We return to the antimony/chlorine reaction to illustrate the process.

EXAMPLE 9.14

Calculate the mass of SbCl$_3$ that can be produced by the reaction of 129 g Sb and 106 g Cl$_2$. Also find the number of grams of the element that will be left. (Z = 51 for Sb)

This time the table begins with the starting *masses* of all species, rather than moles. The masses must be converted to moles. It is convenient to add a line for molar mass too—the starting step idea. The first two lines are completed for you. Fill in the third.

	2 Sb	+	3 Cl$_2$	→	2 SbCl$_3$
Grams at start	129		106		0
Molar mass, g/mol	121.8		71.0		228.3
Moles at start					

	2 Sb	+	3 Cl$_2$	→	2 SbCl$_3$
Grams at start	129		106		0
Molar mass, g/mol	121.8		71.0		228.3
Moles at start	1.06		1.49		0

The conversion from mass of each reactant to moles is the usual g → mol division by molar mass.

The work thus far has brought us to what was the starting point of Example 9.13, the moles of each reactant before the reaction begins. Extend the table through the next two lines to find the moles of product and excess reactant after the limiting reactant is all used up. (See next page.)

	2 Sb	+	3 Cl$_2$	→	2 SbCl$_3$
Grams at start	129		106		0
Molar mass, g/mol	121.8		71.0		228.3
Moles at start	1.06		1.49		0
Moles used (−), produced (+)					
Moles at end					

	2 Sb	+	3 Cl$_2$	→	2 SbCl$_3$
Grams at start	129		106		0
Molar mass, g/mol	121.8		71.0		228.3
Moles at start	1.06		1.49		0
Moles used (−), produced (+)	−0.993		−1.49		+0.993
Moles at end	0.07		0		0.993
Grams at end					

The final step is to change moles to grams by means of molar mass.

	2 Sb	+	3 Cl$_2$	→	2 SbCl$_3$
Grams at start	129		106		0
Molar mass, g/mol	121.8		71.0		228.3
Moles at start	1.06		1.49		0
Moles used (−), produced (+)	−0.993		−1.49		+0.993
Moles at end	0.07		0		0.993
Grams at end	9		0		227

We are now ready to summarize the overall procedure for solving a limiting reactant problem:

PROCEDURE

1) Convert the grams of each reactant to moles.
2) Identify the limiting reactant.
3) Calculate the moles of each species that reacts or is produced.
4) Calculate the moles of each species that remains after the reaction.
5) Change the moles of each species to grams.

EXAMPLE 9.15 _____

A solution that contains 29.0 g of calcium nitrate is added to a solution that contains 33.0 g of sodium fluoride. Calculate the number of grams of calcium fluoride that will precipitate. How many grams of which reactant will remain unreacted?

One of the starting steps is the equation for the reaction. Write that first.

_____   _____

$$Ca(NO_3)_2(aq) + 2\,NaF(aq) \longrightarrow CaF_2(s) + 2\,NaNO_3(aq)$$

Now set up and solve the problem completely.

_____   _____

	$Ca(NO_3)_2$	+	2 NaF	→	CaF_2
Grams at start	29.0		33.0		0
Molar mass, g/mol	164.1		42.0		78.1
Moles at start	0.177		0.786		0
Moles used (−), produced (+)	−0.177		−0.354		+0.177
Moles at end	0		0.432		0.177
Grams at end	0		18.1		13.8

Smaller Amount Method

Consider once again the reaction $Zn + S \rightarrow ZnS$ when 3 mol Zn and 2 mol S are available to react. First find the number of moles of ZnS that may be formed by the complete reaction of all the zinc. Then calculate the number of moles of ZnS that may be formed by complete reaction of all the sulfur.

$$3\ \text{mol Zn} \times \frac{1\ \text{mol ZnS}}{1\ \text{mol Zn}} = 3\ \text{mol ZnS} \quad \Big| \quad 2\ \text{mol S} \times \frac{1\ \text{mol ZnS}}{1\ \text{mol S}} = 2\ \text{mol ZnS}$$

These setups show that there is enough zinc present to make 3 mol ZnS, but only enough sulfur for 2 mol ZnS. The reaction stops when sulfur, the limiting reactant, is used up. The limiting reactant is always the one that yields the smaller amount of product.

To find the amount of the excess reactant that remains, calculate the amount of excess reactant that will be used by the whole amount of the limiting reactant:

$$2\ \text{mol S} \times \frac{1\ \text{mol Zn}}{1\ \text{mol S}} = 2\ \text{mol Zn}$$

We started with 3 mol Zn. Reaction with the limiting reactant used up 2 mol Zn. The amount that remains is the starting quantity minus the quantity used: 3 mol Zn initially −2 mol Zn used = 1 mol Zn left.

The "smaller amount" method can be summarized as follows:

HNO_3, how many kilograms of NH_4NO_3 can be produced? Also, calculate the mass of unreacted reactant that is in excess.

---

45) $2\,NaIO_3 + 5\,NaHSO_3 \rightarrow 3\,NaHSO_4 + H_2O + I_2 + 2\,Na_2SO_4$ is the equation for one method of preparing iodine. If 6.00 g $NaIO_3$ react with 7.33 g $NaHSO_3$, how many grams of iodine can be produced? Which reactant will be left over? What mass will be left?

46) The well-known but now discredited insecticide, DDT, is made by the reaction $CCl_3CHO + 2\,C_6H_5Cl \rightarrow (ClC_6H_4)_2CHCCl_3$ (DDT) $+ H_2O$. In a laboratory test to determine percent yield, 2.19 g CCl_3CHO were reacted with 4.54 g C_6H_5Cl. Calculate the theoretical yield of DDT, identify the reactant that is in excess, and find the amount that is unreacted.

---

47) A mixture of tetraphosphorus trisulfide and powdered glass is in the white tip of strike-anywhere matches. The compound is made by the direct combination of the elements. If 133 g of phosphorus are mixed with the full contents of a 4-oz (126-g) bottle of sulfur, how many grams of compound can be formed? How much of which element will be left over?

48) Is 411 g Na_2CO_3 enough to neutralize 434 g HNO_3 in the reaction $Na_2CO_3 + 2\,HNO_3 \rightarrow 2\,NaNO_3 + CO_2 + H_2O$?

---

49)* In the recovery of silver from silver chloride waste (see Problem 30) a certain quantity of waste material is estimated to contain 216 g AgCl. The treatment tanks are charged with 139 g NaCN and 57 g Zn. Is there enough of the two reactants to recover all of the silver from the AgCl? If no, how many grams of silver chloride will remain? If yes, how many more grams of AgCl could have been treated by the available Zn and NaCN?

50)* The chemistry by which fluorides retard tooth decay is the hard, acid-resisting calcium fluoride layer that forms by the reaction $SnF_2 + Ca(OH)_2 \rightarrow CaF_2 + Sn(OH)_2$. If at the time of a treatment there are 192 mg $Ca(OH)_2$ on the teeth and the dentist uses a SnF_2 mixture that contains 415 mg SnF_2, has enough of the mixture been used to convert all of the $Ca(OH)_2$? If no, what minimum additional amount should have been used? If yes, by what number of milligrams was the amount in excess? (Sn is tin, Z = 50).

SECTION 9.5

51) Complete the following:
0.410 kJ = _____ kcal
25.4 cal = _____ J
802 kcal = _____ J

52) Complete the following:
65.9 J   = _____ cal
839 cal = _____ kJ
3.22 kcal = _____ kJ

---

53) 756 kJ of energy are released when 23 g of carbon are burned. Calculate the number of calories and kilocalories this represents.

54) 7.07 kJ are transferred in a reaction between sodium hydroxide and sulfuric acid. Express this in joules and calories.

---

55) 578 kcal of heat are removed from the body by evaporation each day. How many kilojoules are removed in a week?

56) 1085 kJ of heat are released when 65.8 g of sucrose are burned. How many calories are released for each gram burned?

SECTION 9.6

Questions 57–62: Thermochemical equations may be written in two ways, one with an energy term as a part of the equation, and alternately with ΔH set apart from the regular equation. In the questions that follow, write both forms of the equations for the reactions described. Recall that state designations are required for all substances in a thermochemical equation.

57) Energy is absorbed from sunlight in the photosynthesis reaction in which carbon dioxide and water vapor combine to produce solid sugar, $C_6H_{12}O_6$, and release oxygen. The amount of energy is 2820 kJ per mole of sugar formed.

58) When ethanol, C_2H_5OH, the alcohol in distilled spirits, burns to form gaseous carbon dioxide and liquid water, 1367 kJ of heat are released for each mole of ethanol burned.

---

59) The electrolysis of water is an endothermic reaction, absorbing 286 kJ for each mole of liquid water decomposed to its elements.

60) In "slaking" lime, CaO(s), by converting it to solid calcium hydroxide through reaction with water, 65.3 kJ of heat are released for each mole of calcium hydroxide formed.

---

61) The reaction in an oxyacetylene torch is highly exothermic, releasing 1310 kJ of heat for every mole of acetylene, $C_2H_2(g)$, burned. The end products are gaseous carbon dioxide and liquid water.

62) The extraction of elemental aluminum from aluminum oxide is a highly endothermic electrolytic process. The reaction may be summarized by an equation in which the oxide reacts with carbon to yield aluminum and carbon dioxide gas. The energy requirement is 540 kJ/mol of aluminum metal produced. (Caution on the value of ΔH.)

SECTION 9.7

63) Quicklime, the common name for calcium oxide, is made by heating limestone, $CaCO_3$, in a slowly rotating kiln about $2\frac{1}{2}$ meters in diameter and about 60 meters long. The reaction is $CaCO_3(s) + 178 \text{ kJ} \rightarrow CaO(s) + CO_2(g)$. How many kilograms of limestone can be decomposed by 4.72×10^5 kJ?

64) How much energy is required to decompose 76.7 mL H_2O into its elements? The thermochemical equation is $2 H_2O(\ell) + 572 \text{ kJ} \rightarrow 2 H_2(g) + O_2(g)$. (Recall that the density of water is 1.00 g/mL.)

65) The quicklime produced in Problem 63 is often converted to calcium hydroxide, or, as it is sometimes called, slaked lime, by an exothermic reaction with water: $CaO(s) + H_2O(\ell) \rightarrow Ca(OH)_2(s) + 66.5 \text{ kJ}$. How much energy is released in the slaking of 192 g CaO?

66) $\Delta H = 2.82 \times 10^3$ kJ for the photosynthesis reaction by which plants use energy from the sun to form carbohydrates from carbon dioxide and water. The equation is $6 CO_2(g) + 6 H_2O(\ell) \rightarrow C_6H_{12}O_6(s) + 6 O_2(g)$. How many grams of sugar can be formed by 959 kJ?

67) How much energy will be delivered to your automobile when you burn 25.0 kg of pure octane, C_8H_{18}, a principal component of gasoline? $\Delta H = -1.09 \times 10^4$ kJ for the reaction $2 C_8H_{18}(\ell) + 25 O_2(g) \rightarrow 16 CO_2(g) + 18 H_2O(\ell)$. (Watch your units.)

68) One of the fuels sold as "bottled gas" is butane, C_4H_{10}. How many kilograms of butane must be burned to release 9.92×10^4 kJ of energy if $\Delta H = -5.77 \times 10^3$ kJ for the reaction $2 C_4H_{10}(g) + 13 O_2(g) \rightarrow 8 CO_2(g) + 10 H_2O(\ell)$.

69)* Nitroglycerine is the explosive ingredient in industrial dynamite. Much of its destructive force comes from the sudden creation of large volumes of gaseous products. A great deal of energy is released too. $\Delta H = -6.17 \times 10^3$ kJ for the equation $4 C_3H_5(NO_3)_3(\ell) \rightarrow 12 CO_2(g) + 10 H_2O(g) + 6 N_2(g) + O_2(g)$. Calculate the energy that will be released in a blasting operation that uses 13.1 pounds of nitroglycerine.

70)* In 1866 a young chemistry student conceived the electrolytic method of obtaining aluminum from its oxide. This method is still used. $\Delta H = 1.97 \times 10^3$ kJ for the reaction $2 Al_2O_3(s) + 3 C(s) \rightarrow 4 Al(s) + 3 CO_2(g)$. The large amount of electrical energy required limits the process to areas of cheap power. How many kilograms of aluminum will be produced with 18.7 kw-hr (kilowatt-hours) if 1 kw-hr = 3.60×10^3 kJ?

GENERAL QUESTIONS

71) Distinguish precisely and in scientific terms the differences between items in each of the following groups:
a) Theoretical, actual, and percent yield
b) Limiting reactant, excess reactant
c) Molar mass, molar volume
d) Heat of reaction, enthalpy of reaction
e) H, ΔH
f) Chemical equation, thermochemical equation
g) Stoichiometry, thermochemical stoichiometry
h) Joule, calorie

72) Classify each of the following statements as true or false:
a) Coefficients in a chemical equation express the molar proportions among both reactants and products.
b) A stoichiometry problem can be solved with an unbalanced equation.
c) In solving a stoichiometry problem, the change from quantity of given substance to quantity of wanted substance is based on masses.
d) Percent yield is actual yield expressed as a percent of theoretical yield.
e) The quantity of product of any reaction can be calculated only through the moles of the limiting reactant.
f) ΔH is positive for an endothermic reaction, negative for an exothermic reaction.

73) One of the few ways of "fixing" nitrogen, meaning to make a nitrogen compound from the elemental nitrogen in the atmosphere, is by the reaction $Na_2CO_3 + 4 C + N_2 \rightarrow 2 NaCN + 3 CO$. Calculate the grams of Na_2CO_3 required to react with 187 g N_2.

74)* A student was given a 1.8220-g sample of a mixture of sodium nitrate and sodium chloride and asked to find the percentage of each compound in the mixture. She dissolved the sample and added a solution that contained an excess of silver nitrate, $AgNO_3$. The silver ion precipitated all of the chloride ion in the mixture as AgCl. It was filtered, dried, and weighed. Its mass was 2.323 g. What was the percentage of each compound in the mixture?

75)* 1.412 g of impure copper were dissolved in nitric acid to produce a solution of $Cu(NO_3)_2$. The solution went through a series of steps in which $Cu(NO_3)_2$ was changed to $Cu(OH)_2$, then to CuO, and then to a solution of $CuCl_2$. This was treated with an excess of a soluble phosphate, precipitating all the copper in the original sample as pure $Cu_3(PO_4)_2$. The precipitate was dried and weighed. Its mass was 2.714 g. Find the percent copper in the original sample.

76)* How many grams of magnesium nitrate, $Mg(NO_3)_2$, must be used to precipitate as magnesium hydroxide all of the hydroxide ion in 70.0 mL 17.0% NaOH, the density of which is 1.19 g/mL? The precipitation reaction is $2\,NaOH + Mg(NO_3)_2 \rightarrow Mg(OH)_2 + 2\,NaNO_3$.

77) How many grams of calcium phosphate will precipitate if excess calcium nitrate is added to a solution containing 1.89 g of sodium phosphate?

78) Emergency oxygen masks contain potassium superoxide, KO_2, pellets. When exhaled CO_2 passes through the KO_2, the following reaction occurs: $4\,KO_2(s) + 2\,CO_2(g) \rightarrow 2\,K_2CO_3(s) + 3\,O_2(g)$. The oxygen produced can then be inhaled, so no air from outside the mask is needed. If the mask contains 176 g KO_2, how many grams of oxygen can be produced?

79) Baking cakes and pastries involves the production of CO_2 to make the batter "rise." For example, citric acid, $H_3C_6H_5O_7$, in lemon or orange juice can react with baking soda, $NaHCO_3$, to produce the carbon dioxide gas: $H_3C_6H_5O_7(aq) + 3\,NaHCO_3(aq) \rightarrow Na_3C_6H_5O_7(aq) + 3\,CO_2(g) + 3\,H_2O(\ell)$.

a) If 8.00 g $H_3C_6H_5O_7$ react with 17.0 g $NaHCO_3$, how many grams of carbon dioxide will be produced?
b) How many grams of which reactant will remain unreacted?
c) Can you guess the name of $Na_3C_6H_5O_7$? Remember, it comes from cit*ric* acid.

80) A laboratory test of 13.9 g of aluminum ore yields 2.07 g of aluminum. If the aluminum compound in the ore is Al_2O_3, and it is converted to the metal by the reaction $2\,Al_2O_3(s) + 3\,C(s) \rightarrow 4\,Al(s) + 3\,CO_2(g)$, what is the percent Al_2O_3 in the ore?

81) How much energy is required to decompose 7.72 g $KClO_3$ according to the equation $2\,KClO_3(s) \rightarrow 2\,KCl(s) + 3\,O_2(g)$? $\Delta H = 89.5$ kJ for the reaction.

82)* If a solution of silver nitrate, $AgNO_3$, is added to a second solution containing a chloride, bromide, or iodide, the silver ion, Ag^+, from the first solution will precipitate the halide as silver chloride, silver bromide, or silver iodide. If excess $AgNO_3(aq)$ is added to a mixture of the above halides, it will precipitate them both, or all, as the case may be. A solution contains 0.197 g NaCl and 0.804 g NaBr. What is the smallest quantity of $AgNO_3$ that is required to precipitate both halides completely?

MATCHING SET ANSWERS

3–4–11–5–9–1–10–6–8–2–7

Atomic Theory and the Periodic Table: A Modern View

10

"Neon" signs are sealed glass tubes that contain different elements at low pressure. High-voltage electricity passing through a neon-filled tube gives the red color; a tube filled with argon and mercury gives the blue color. The same argon and mercury gases give the yellow-brown color when placed in a specially coated tube. The color we see is a combination of many distinct colors that can be separated by passing the light through a prism. Explaining why these colors were distinct led to quantum theory, our current model of atomic structure.

The four decades from 1890 to 1930 were a period of rapid progress in learning about the atom. We covered the first half of this period in Chapter 5; in this chapter we cover the second half.

In 1911, Rutherford's alpha scattering experiments were controversial. In Rutherford's atom, all the positive charge was crammed together in the dense, tiny nucleus. Like charges repel; the nucleus of the atom could not be stable, yet it was. The rules of classical physics that worked so well on large-scale systems did not work on atom-sized systems. What rules would replace classical physics?

Welcome to the world of the quantum.

10.1 THE BOHR MODEL OF THE HYDROGEN ATOM

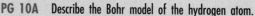

PG 10A	Describe the Bohr model of the hydrogen atom.
10B	Explain the meaning of quantized energy levels in an atom, and show how these levels are related to the discrete lines in the spectrum of that atom.
10C	Distinguish between ground state and excited state.

Niels Bohr

In 1913 Niels Bohr, a Danish physicist, suggested that an atom consists of an extremely dense nucleus that contains all of the atom's positive charge and nearly all of its mass. Negatively charged electrons of very small mass travel in orbits around the nucleus. The orbits are huge compared to the nucleus, which means that most of the atom is empty space. Bohr's model of the atom was based on many facts that were well known at the time. Among them were:

1) Light is part of the **electromagnetic spectrum** (Fig. 10.1). The spectrum has wave properties, such as velocity (v, or specifically for light, c), wavelength (λ), and frequency (ν) (Fig. 10.2).
2) White light produces a **continuous spectrum** when it passes through a prism (Fig. 10.1.) However, light from an element in a gas discharge tube produces a **line spectrum** consisting of separate, or **discrete,** lines of color (Fig. 10.3).
3) Light also has properties that suggest it is made up of individual bundles, or **quanta,** of energy. The energy, E, of each quantum is calculated by the equation $E = h\nu$, where h is a fixed number known as Planck's constant.

💡 FLASHBACK Two objects having opposite charges, one positive and one negative, attract each other. Two objects with the same charge repel each other. See Section 2.3.

4) Physics describes mathematical relationships among radii, speed, energy, and forces when one object moves in an orbit around another, as the planets move around the sun. Physics also describes forces of attraction between positively and negatively charged objects.

From these beginnings, Bohr boldly assumed that the energy possessed by the electron in a hydrogen atom and the radius of its orbit are **quantized.** A quantity that is quantized is limited to specific values; it may never be between two of those values. By contrast, an amount is **continuous** if it can have any value; between any two values there is an infinite number of other acceptable values. (See Figure 10.4.) A line spectrum is quantized, but the spectrum of white light is continuous. Bohr said that the electron in the hydrogen atom has **quantized energy levels.** This means that, at any instant, the electron may have one of several possible energies, but at no time may it have an energy between them.

Bohr calculated the values of his quantized energy levels by using an equation that contains an integer, n—1, 2, 3 . . . , and so forth. The results for the integers 1 to 4

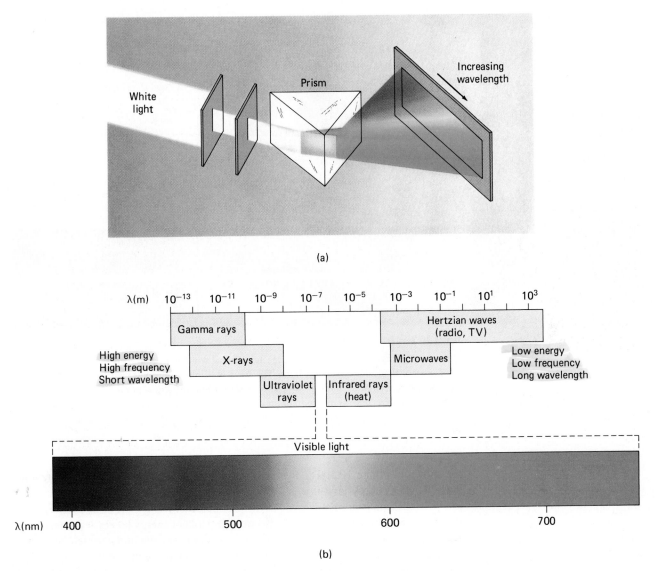

(a)

(b)

Figure 10.1

Dispersion of white light by a prism. (a) White light is passed through a slit and then through a prism. It is separated into a continuous spectrum of all wavelengths of visible light. (b) Visible light is only a small portion of the electromagnetic spectrum, covering a wavelength range of about 390 to 740 nm. The entire spectrum ranges from high-energy gamma rays with wavelengths as small as 10^{-13} m to low-energy radio waves at wavelengths as long as 10^3 m.

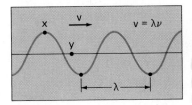

Figure 10.2

Wave properties. Waves may be represented mathematically by the curve shown. All waves have measurements such as wavelength, λ, the distance between corresponding points on consecutive waves; velocity, v, or, in the case of light, c, the linear speed of a point x on a wave; and frequency ν, the number of wave cycles that pass a point y each second. Reflection, refraction, and diffraction are other wave properties.

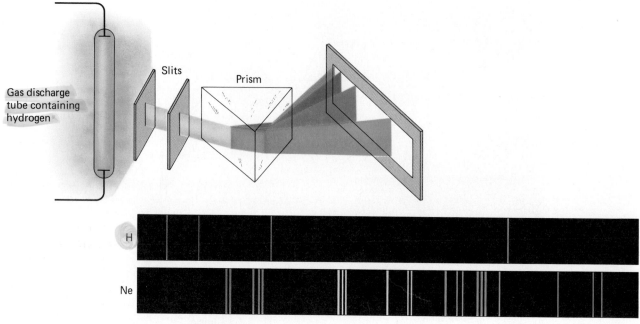

Figure 10.3

Dispersion of light from a gas discharge tube filled with hydrogen. This is like a "neon" light, except that neon gives red light. The magenta light from hydrogen is passed through a slit and then through a prism. It is separated into a line spectrum made up of four wavelengths of visible light. A photograph of these lines is shown beneath the diagram. Also shown is the neon spectrum. It is the combination of these colors that we see as the red light of a neon sign.

FLASHBACK   This is an example of minimization of energy, which is a driving force for physical and chemical change. See Section 2.4.

are shown in Figure 10.5. The lowest energy level, when $n = 1$, is called the **ground state.** Higher energy levels, when $n = 2$ or more, are called **excited states.**

The electron is normally found in the ground state. If the atom absorbs energy, the electron can be "excited," or raised to one of the higher energy levels. It cannot stay at that level, but falls back to the ground state, sometimes in one jump, sometimes in two or more. In doing this it releases light energy, $h\nu$, which is equal to the energy difference between the two levels. If this energy is in the visible part of the spectrum, it produces one of the lines in the spectrum of hydrogen.

Using different values of n in the equation, Bohr was able to calculate the energies of all known lines in the spectrum of the hydrogen atom. He also predicted additional

Figure 10.4

The quantum concept. Man on ramp can stop at any level above ground. His elevation above ground is not quantized. Man on stairs can stop only on a step. His elevation is quantized at h_1, h_2, h_3, or H.

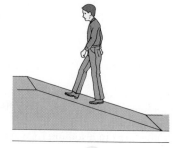
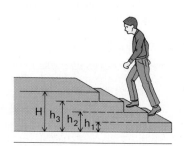

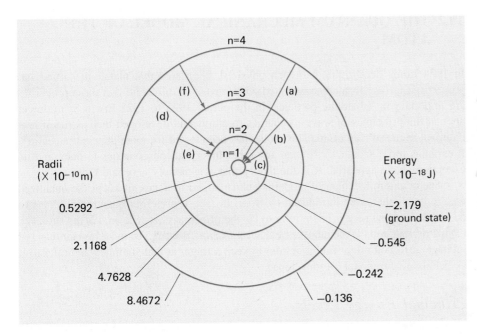

Figure 10.5

The Bohr model of the hydrogen atom. The electron is "allowed" to circle the nucleus only at certain radii and with certain energies, the first four of which are shown. An electron in the "ground state" $n = 1$ level can absorb the exact amount of energy to raise it to any other level, e.g., $n = 2, 3,$ or 4. An electron at such an "excited state" is unstable and drops back to the $n = 1$ level in one or more steps. Electromagnetic energy is radiated with each step. Jumps a to c are in the ultraviolet portion of the spectrum, d and e are in the visible range, and f is in the infrared region.

lines and their energies. When the lines were found, his predictions were proved to be correct. In fact, all of Bohr's calculations correspond with measured values to within one part per thousand. It certainly seemed that Bohr had found the answer to the structure of the atom.

There were problems, however. First, hydrogen is the *only* atom that fits the Bohr model. The model fails for any atom with more than one electron. Second, it is a fact that a charged body moving in a circle radiates energy. This means the electron itself should lose energy and promptly—in about 0.00000000001 second!—crash into the nucleus. This suggests that circular orbits violate the Law of Conservation of Energy (Section 2.5). So, for 13 years scientists did what they so often do when their best theory is contradicted by their experimental results. They accepted and used a theory they knew was only partly correct. They worked on the faulty parts, improved those parts, and finally replaced the old theory with more experimentally accurate ideas.

Niels Bohr made two huge contributions to the development of modern atomic theory. First, he suggested a reasonable explanation for atomic line spectra in terms of electron energies. Second, he introduced the idea of quantized electron energy levels in the atom. These levels appear in modern theory as **principal energy levels;** they are identified by the **principal quantum number, n.**

✔ QUICK CHECK 10.1

Identify the true statements, and rewrite the false statements to make them true.

a) Two of the following are quantized: (1) the speed of automobiles on a highway, (2) paper money in the United States, (3) canned soup on a grocery store shelf, (4) water coming from a faucet, (5) a person's weight.

b) Bohr described mathematically the orbits of electrons in a sodium (Z = 11) atom.

Erwin Schrödinger

10.2 THE QUANTUM MECHANICAL MODEL OF THE ATOM

In 1924 Louis de Broglie, a French physicist, suggested that matter in motion has properties that are normally associated with waves. He also said that these properties are important in subatomic particles. In the period 1925 to 1928 Erwin Schrödinger applied the principles of wave mechanics to atoms and developed the **quantum mechanical model of the atom.** This model has been tested for more than half a century. It explains more satisfactorily than any other theory all observations to date, and no exceptions have appeared. It is the theory that is generally accepted today.

The quantum mechanical model is mathematical in nature. It keeps the quantized energy levels that were introduced by Bohr. In fact, it uses four quantum numbers to describe electron energy. These refer to (1) the principal energy level, (2) the sublevel, (3) the orbital, and (4) the number of electrons in an orbital.* The model is summarized on page 261. You might find it helpful to keep a finger at that summary and refer to it as details of the model are developed.

Principal Energy Levels

> **PG 10D**   Identify the principal energy levels in an atom, and state the energy trend among them.

Following the Bohr model, principal energy levels are identified by the principal quantum number, n. The first principal energy level is $n = 1$, the second is $n = 2$, and so on. Mathematically, there is no end to the number of principal energy levels, but the seventh level is the highest occupied by ground state electrons in any element now known.

The energy possessed by an electron depends on the principal energy level it is in. In general, energies increase as the principal quantum number increases: $n = 1 < n = 2 < n = 3 \cdots n < 7$.

Sublevels

> **PG 10E**   For each principal energy level, state the number of sublevels, identify them, and state the energy trend among them.

For each principal energy level there are one or more **sublevels.** They are the *s, p, d,* and *f* **sublevels,** using initial letters that come from terms formerly used in spectroscopy.† A specific sublevel is identified by both the principal energy level and sublevel. Thus, the *p* sublevel in the third principal energy level is the 3*p* sublevel. An electron that is in the 3*p* sublevel may be referred to as a ''3*p* electron.''

The total number of sublevels within a given principal energy level is equal to n, the principal quantum number. For $n = 1$ there is one sublevel designated 1*s*. At $n = 2$

---

*The formal names of these numbers are principal, azimuthal, magnetic, and electron spin. We use the name and number of the principal quantum number, but not the other three. All, however, are described to the extent necessary to specify the distribution of electrons in an atom.

†The terms are ''sharp,'' ''principal,'' ''diffuse,'' and ''fundamental.'' They describe the appearance of the spectral lines in a gas discharge tube.

there are two sublevels, 2s and 2p. When n = 3 there are three sublevels, 3s, 3p, and 3d; and n = 4 has four sublevels, 4s, 4p, 4d, and 4f. Quantum theory describes sublevels beyond f when n = 5 or more, but these are not needed by elements known today.

For elements other than hydrogen, the energy of each principal energy level spreads over a range related to the sublevels. These energies increase in the order s, p, d, f. Thus, the energies for the

two sublevels at n = 2, the increasing order of energy is 2s < 2p;

three sublevels at n = 3, the increasing order of energy is 3s < 3p < 3d;

four sublevels at n = 4, the increasing order of energy is 4s < 4p < 4d < 4f.

Beginning with principal quantum numbers 3 and 4, the energy ranges overlap. This is shown in the margin. When "plotted" vertically, the highest n = 3 electrons (3d) are at higher energy than the lowest n = 4 electrons (4s). Note, however, that for the *same sublevel*, n = 3 electrons are always lower than n = 4 electrons: 3s < 4s; 3p < 4p; 3d < 4d.

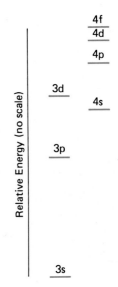

Electron Orbitals

PG 10F	Sketch the shapes of s and p orbitals.
10G	State the number of orbitals in each sublevel.

According to modern atomic theory, it is not possible to know at the same time both the position of an electron in an atom and its velocity. This means it is not possible to describe the "path" traveled by an electron. There are no clearly defined orbits, as in the Bohr atom. However, we can describe mathematically a region in space around a nucleus in which there is a "high probability" of finding an electron. These regions are called **orbitals.** Notice the uncertainty of the *orbital,* stated in terms of "probability," compared to a Bohr *orbit* that states exactly where the electron is, where it was, and where it is going. (See Fig. 10.6.)

Each sublevel has a certain number of orbitals. There is only one orbital for every s sublevel. All p sublevels have three orbitals, all d sublevels have five, and all f sublev-

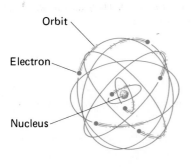

Bohr model

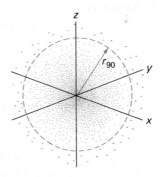

Quantum model

Figure 10.6

The Bohr model of the atom compared to the quantum model. Bohr described the electron as moving in circular orbits of fixed radii around the nucleus. The quantum model says nothing about the precise location of the electron nor the path in which it moves. Instead, each dot represents a possible location for the electron. The higher the density of dots is, the higher the probability that an electron is in that region. The r_{90} dimension is a "radius" such that 90% of the time the electron is inside the dashed line, and 10% of the time it is farther from the nucleus.

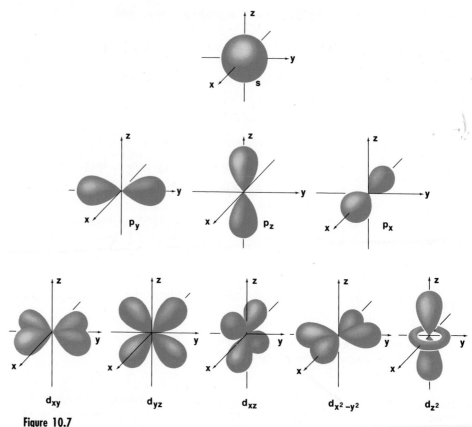

Figure 10.7

Shapes of electron orbitals according to the quantum mechanical model of the atom.

els have seven. This 1-3-5-7 sequence of odd numbers continues through higher sub-levels, although they are not needed to describe atoms of elements that are now known.

Figure 10.7 shows the shapes of the s, p, and d orbitals. The seven f orbitals have even more complex shapes. The x, y, and z axes around which these shapes are drawn are from the mathematics of the quantum theory. We will be concerned with the shapes of only the s and p orbitals.

All s orbitals have a spherical shape. As the principal quantum number increases, the size of the orbital increases. Thus, a $2s$ orbital is larger than a $1s$ orbital, $3s$ is larger than $2s$, and so forth. Similar increases in size through constant shapes are present with p, f, and d orbitals at higher principal energy levels.

The Pauli Exclusion Principle

PG 10H State the restrictions on the electron population of an orbital.

The last detail of the quantum mechanical model of the atom comes from the **Pauli exclusion principle.** Its effect is to limit the population of any orbital to two electrons. At any instant an orbital may be (1) unoccupied, (2) occupied by one electron, or (3) occupied by two electrons. No other occupancy is possible.

S U M M A R Y / Summary of the Quantum Model of the Atom

Principal Energy Levels Principal energy levels are identified by principal quantum number, n, a series of integers: $n = 1, 2, 3, \ldots 7$. *Generally,* energy increases with increasing n: $n = 1 < n = 2 < n = 3$.

Sublevels

Each principal energy level—each value of n—has n sublevels. These sublevels are identified by the principal quantum number followed by the letter $s, p, d,$ or f. Sublevels that are not needed by the elements that are known today are shown below in color.

Energy Trend	*Energy Level* n	Number of Sublevels	Identification of Sublevels
	1	1	1s
	2	2	2s, 2p
Increasing energy ↓	3	3	3s, 3p, 3d
	4	4	4s, 4p, 4d, 4f
	5	5	5s, 5p, 5d, 5f, 5g
	6	6	6s, 6p, 6d, 6f, 6g, 6h
	7	7	7s, 7p, 7d, 7f, 7g, 7h, 7i

Increasing energy →

For any given value of n, energy increases through the sublevels in the order of s, p, d, f: $2s < 2p$; $3s < 3p < 3d$; $4s < 4p < 4d < 4f$; etc.

Note: The *range* of energies in consecutive principal energy levels may overlap. Example: $4s < 3d < 4p$. However, for any given *sublevel*, energy and orbital size increase with increasing n: $1s < 2s < 3s \ldots$; $2p < 3p < 4p \ldots$; etc.

Orbitals and Orbital Occupancy

Each kind of sublevel contains a definite number of orbitals that begin with 1 and increase in order with odd numbers: $s, 1$; $p, 3$; $d, 5$; $f, 7$.

An orbital may be occupied by 0, 1, or 2 electrons, but never more than 2. Therefore, the maximum number of electrons in a sublevel is twice the number of orbitals in the sublevel.

Sublevel	Orbitals	Maximum Electrons per Sublevels
s	1	2
p	3	6
d	5	10
f	7	14

✔ QUICK CHECK 10.2

Identify the true statements, and rewrite the false statements to make them true. If possible, avoid looking at the summary of the quantum model.

a)   There is one s orbital when $n = 1$, two s orbitals when $n = 2$, three s orbitals when $n = 3$, and so on.

b)   All $n = 3$ orbitals are at lower energy than all $n = 4$ orbitals.

c)   There is no d sublevel when $n = 2$.

d)   There are five d orbitals at both the fourth and sixth principal energy levels.

10.3 ELECTRON CONFIGURATION

Figure 10.8 is an electron energy level diagram. Relative energies of sublevels are plotted vertically (not to scale), and principal quantum numbers are plotted horizontally. Each box represents one orbital, which may contain 0, 1, or 2 electrons. Different backgrounds separate the sublevels by periods in the periodic table.

Many chemical properties of an element depend on its **electron configuration,** the ground state distribution of electrons among the orbitals of a gaseous atom. Two rules guide the assignments of electrons to orbitals:

1)   At ground state the electrons fill the *lowest* energy orbitals available,

2)   No orbital can have more than two electrons.

Figure 10.8

Electron energy level diagram. Each column contains one box for each orbital in the $n = 1$, $n = 2$, $n = 3 \ldots n = 7$ principal energy levels. The boxes are grouped by sublevels s, p, d, and f, as far as required within each principal energy level. All sublevels are positioned vertically on the page according to a general energy level scale shown at the left. Any orbital that is higher on the page than a second orbital is therefore higher in energy than the second orbital. In filling orbitals from the lowest energy level, each orbital will generally hold two electrons before any orbital higher in energy accepts an electron. The scale at the right correlates energies of the sublevels with the periods of the periodic table.

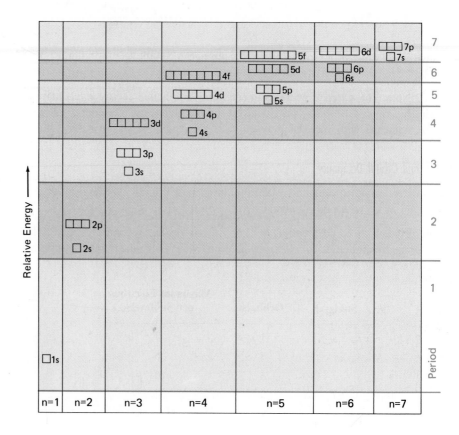

1 **H** $1s^1$																	2 **He** $1s^2$
3 **Li** $1s^2$ $2s^1$	4 **Be** $1s^2$ $2s^2$											5 **B** $1s^2$ $2s^22p^1$	6 **C** $1s^2$ $2s^22p^2$	7 **N** $1s^2$ $2s^22p^3$	8 **O** $1s^2$ $2s^22p^4$	9 **F** $1s^2$ $2s^22p^5$	10 **Ne** $1s^2$ $2s^22p^6$
11 **Na** [Ne] $3s^1$	12 **Mg** [Ne] $3s^2$											13 **Al** [Ne] $3s^23p^1$	14 **Si** [Ne] $3s^23p^2$	15 **P** [Ne] $3s^23p^3$	16 **S** [Ne] $3s^23p^4$	17 **Cl** [Ne] $3s^23p^5$	18 **Ar** [Ne] $3s^23p^6$
19 **K** [Ar] $4s^1$	20 **Ca** [Ar] $4s^2$	21 **Sc** [Ar]$4s^2$ $3d^1$	22 **Ti** [Ar]$4s^2$ $3d^2$	23 **V** [Ar]$4s^2$ $3d^3$	24 **Cr** [Ar]$4s^1$ $3d^5$	25 **Mn** [Ar]$4s^2$ $3d^5$	26 **Fe** [Ar]$4s^2$ $3d^6$	27 **Co** [Ar]$4s^2$ $3d^7$	28 **Ni** [Ar]$4s^2$ $3d^8$	29 **Cu** [Ar]$4s^1$ $3d^{10}$	30 **Zn** [Ar]$4s^2$ $3d^{10}$	31 **Ga** [Ar]$4s^2$ $3d^{10}4p^1$	32 **Ge** [Ar]$4s^2$ $3d^{10}4p^2$	33 **As** [Ar]$4s^2$ $3d^{10}4p^3$	34 **Se** [Ar]$4s^2$ $3d^{10}4p^4$	35 **Br** [Ar]$4s^2$ $3d^{10}4p^5$	36 **Kr** [Ar]$4s^2$ $3d^{10}4p^6$

Figure 10.9

Ground state electron configurations of neutral gaseous atoms.

$1s$ The one electron of a hydrogen atom (H, Z = 1) occupies the lowest energy orbital in any atom. Figure 10.8 shows this is the $1s$ orbital. The total number of electrons in any sublevel is shown by a superscript number. Therefore, the electron configuration of hydrogen is $1s^1$. Helium (He, Z = 2) has two electrons, and both fit into the $1s$ orbital. The helium configuration is $1s^2$.

These and other electron configurations to be developed appear in the first four periods of the periodic table in Figure 10.9.

$2s$ Lithium (Li, Z = 3) has three electrons. The first two fill the $1s$ orbital, as before. The third electron goes to the next orbital up the energy scale that has a vacancy. According to Figure 10.8, this is the $2s$ orbital. The electron configuration for lithium is therefore $1s^22s^1$. Similarly, beryllium (Be, Z = 4) divides its four electrons between the two lowest orbitals, filling both: $1s^22s^2$. These configurations are in Figure 10.9 too.

$2p$ The first four electrons of boron (B, Z = 5) fill $1s$ and $2s$ orbitals. The fifth electron goes to the next highest level, $2p$, according to Figure 10.8. The configuration for boron is $1s^22s^22p^1$. Similarly, carbon (C, Z = 6) has a $1s^22s^22p^2$ configuration.* The next four elements increase the number of electrons in the three $2p$ orbitals until they are filled with six electrons for neon (Ne, Z = 10). All of these configurations appear in Figure 10.9.

$3s$ and $3p$ The first ten electrons of sodium (Na, Z = 11) are distributed in the same way as the ten electrons in neon. The eleventh sodium electron is a $3s$ electron: $1s^22s^22p^63s^1$. The configurations for all elements whose atomic numbers are greater than 10 begin with the neon configuration, $1s^22s^22p^6$. This part of the configuration is often shortened to the **neon core,** represented by [Ne]. For sodium this becomes [Ne]$3s^1$; for magnesium (Mg, Z = 12), [Ne]$3s^2$; for aluminum (Al, Z = 13),

---

*Although we will not emphasize the point, the two $2p$ electrons occupy different $2p$ orbitals. In general, all orbitals in a sublevel are half-filled before any orbital is completely filled.

$[Ne]3s^23p^1$; and so on to argon (Ar, Z = 18), $[Ne]3s^23p^6$. The sequence is exactly as it was in Period 2. The neon core is used for Period 3 in Figure 10.9.

4s Potassium, (K, Z = 19) repeats at the 4s level the development of sodium at the 3s level. Its complete configuration is $1s^22s^22p^63s^23p^64s^1$. All configurations for atomic numbers greater than 18 distribute their first 18 electrons in the configuration of argon, $1s^22s^22p^63s^23p^6$. This may be shortened to the **argon core,** [Ar]. Accordingly, the configuration for potassium may be written $[Ar]4s^1$, and calcium (Ca, Z = 20) is $[Ar]4s^2$.

3d Figure 10.8 predicts that five 3d orbitals are next available for electron occupancy. The next three elements fill in order, as predicted, to vanadium (V, Z = 23): $[Ar]4s^23d^3$.* Chromium (Cr, Z = 24) is the first element to break the orderly sequence in which the lowest energy orbitals are filled. Its configuration is $[Ar]4s^13d^5$, rather than the expected $[Ar]4s^23d^4$. This is generally attributed to an extra stability found when all orbitals in a sublevel are half-filled or completely filled. Manganese (Mn, Z = 25) puts us back on the track, only to be derailed again at copper (Cu, Z = 29): $[Ar]4s^13d^{10}$. Zinc (Zn, Z = 30) has the expected configuration: $[Ar]4s^23d^{10}$. Examine the sequence for atomic numbers 21 to 30 in Figure 10.9 and note the two exceptions.

4p By now the pattern should be clear. Atomic numbers 31 to 36 fill in sequence in the next orbitals available, which are the 4p orbitals. This is shown in Figure 10.9.

Our consideration of electron configurations ends with atomic number 36, krypton. If we were to continue, we would find the higher s and p orbitals fill just as they do in Periods 2 to 4. The 4d, 4f, 5d, and 5f orbitals have several variations like those for chromium and copper, so their configurations must be looked up. But you should be able to reproduce the configurations for the first 36 elements—*not from memory nor from Figures 10.8 or 10.9, but by referring to the periodic table.*

Electron Configurations and the Periodic Table

PG 10I Use a periodic table to list electron sublevels in order of increasing energy.

FLASHBACK Recall from Section 5.6 that we are showing both the traditional A–B numbering scheme for groups in the periodic table and the 1 to 18 system that has been approved by the International Union of Pure and Applied Chemistry (IUPAC). Use—and "read"— the one selected by your isntructor.

Figure 10.9 shows that specific sublevels are filled in different regions of the periodic table. This is indicated by color in Figure 10.10. In Groups 1A (1) and 2A (2) the s sublevels are the highest occupied energy sublevels. The p orbitals are filled in order across Groups 3A (13) to 0 (18). The d electrons appear in the B groups and Group 8 (3 to 12). Finally, f electrons show up in the lanthanide and actinide series beneath the table.

Notice that when you read the periodic table from left to right across the periods in Figure 10.10, you get the order of increasing sublevel energy. The first period gives

*Some chemists prefer to write this configuration $[Ar]3d^34s^2$, putting the 3d before the 4s. This is equally acceptable. There is, perhaps, some advantage at this time in listing the sublevels in the order in which they fill. This same idea continues as the f sublevels are filled, but with frequent irregularities. We will not be concerned with f-block elements.

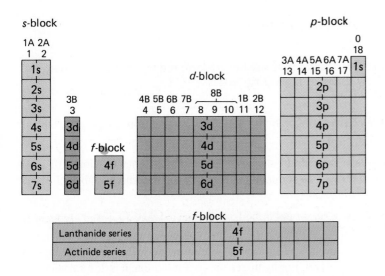

Figure 10.10

Arrangement of periodic table according to atomic sublevels. Highest energy sublevels occupied at ground state are s sublevels in Groups 1A and 2A (1 and 2). This region of the periodic table is the s-block. Similarly, p sublevels are the highest occupied sublevels in Groups 3A to 0 (13 to 18), the p-block. The d-block includes the B (3 to 12) groups whose highest occupied energy sublevels are d sublevels. Finally, the f-block is made up of the elements whose f sublevels hold the highest energy electrons.

only the 1s sublevel. Period 2 takes you through 2s and 2p. Similarly, the third period covers 3s and 3p. Period 4 starts with 4s, follows with 3d, and ends up with 4p; and so forth. Ignoring the minor variations in Periods 6 and 7, the complete list is

1s 2s 2p 3s 3p 4s 3d 4p 5s 4d 5p 6s 4f 5d 6p 7s 5f 6d 7p

Now compare this list with the order of increasing sublevel energy from Figure 10.8. They are exactly the same! *The periodic table therefore replaces the electron energy diagram as a guide to the order of increasing sublevel energy.*

Think, for a moment, how remarkable this is. Mendeleev and Meyer developed their periodic tables from the physical and chemical properties of the elements. They knew nothing of electrons, protons, nuclei, wave equations, or quantized energy levels. Yet, when these things were found some 60 years later, the match between the first periodic tables and the quantum mechanical model of the atom was nearly perfect.

If you are ever required to list the sublevels in order of increasing energy without reference to a periodic table, the following diagram taken from the summary of the quantum model may be helpful. Beginning at the upper left, the diagonal lines pass through the sublevels in the sequence required.

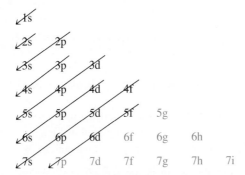

The sublevels shown in color are not needed for the elements known today, but the mathematics of the quantum mechanical model predict the order of increasing energy indefinitely.

Writing Electron Configurations

> **PG 10J** Referring only to a periodic table, write the ground state electron configuration of a gaseous atom of any element up to atomic number 36.

You can write the electron configuration of an atom with help from the periodic table if you can

1) List the sublevels in order of increasing energy (Performance Goal 10I) and
2) Establish the number of electrons in the highest occupied energy sublevel of the atom.

The number of electrons in the highest occupied energy sublevel of an atom is related to the position of the element in the periodic table. For atoms of all elements in Group 1A (1) that number is 1. This can be written as ns^1, where n is the highest occupied principal energy level. For hydrogen, n is 1; for lithium, n is 2; for sodium, 3; and so forth. In Group 2A (2) the highest occupied sublevel is ns^2. In Groups 3A (13) to 0 (18), the number of p electrons is found by counting from the left from 1 to 6. For the representative elements:

☀️ **FLASHBACK** Representative elements are those in the "A" groups of the periodic table, or Groups 1, 2, and 13–18 by the IUPAC system. The "B" groups (3–12) are transition elements (Section 5.6).

This same idea continues as the f sublevels are filled, but with frequent irregularities. We will not be concerned with f-block elements.

Group (A–B)	1A	2A			3A	4A	5A	6A	7A	0
Group (IUPAC)	1	2			13	14	15	16	17	18
s electrons	1	2	p electrons		1	2	3	4	5	6
Electron configuration:	ns^1	ns^2			np^1	np^2	np^3	np^4	np^5	np^6

A similar count-from-the-left order appears among the transition elements, in which the d sublevels are filled. There are interruptions, however. Among the $3d$ electrons the interruptions appear at chromium (Z = 24) and copper (Z = 29).

Group (A–B)	3B	4B	5B	6B	7B	←—8B—→			1B	2B
Group (IUPAC)	3	4	5	6	7	8	9	10	11	12
d electrons	1	2	3	5	5	6	7	8	10	10
Electron configuration	$3d^1$	$3d^2$	$3d^3$	$3d^5$	$3d^5$	$3d^6$	$3d^7$	$3d^8$	$3d^{10}$	$3d^{10}$

Fix these electron populations and their positions in the periodic table firmly in your thought now. Then cover both of these summaries and refer only to a full periodic table as you try the following example.

EXAMPLE 10.1

Write the electron configuration of the highest occupied energy sublevel for each of the following elements:

beryllium _____   phosphorus _____   manganese _____

_____   _____

beryllium, $2s^2$;   phosphorus, $3p^3$;   manganese, $3d^5$

Counting from the left, beryllium is the second box [Group 2A (2)] among the $2s$ sublevel elements, so its electron configuration is $2s^2$. Phosphorus is in the third box [Group 5A (15)] among the $3p$ sublevel elements, so its configuration is $3p^3$. Manganese is in the fifth box [Group 7B (7)] among the $3d$ sublevel elements, so its configuration is $3d^5$.

You are now ready to write electron configurations. The procedure follows.

PROCEDURE

1) Locate the element in the periodic table. From its position in the table, identify and write the electron configuration of its highest occupied energy sublevel. (Leave room for writing lower energy sublevels to its left.)
2) To the left of what has already been written, list all lower energy sublevels in order of increasing energy.
3) For each filled lower energy sublevel, write as a superscript the number of electrons that fill that sublevel. (There are two s electrons, ns^2; six p electrons, np^6; and ten d electrons, nd^{10}. Exceptions: For chromium and copper the $4s$ sublevel has only one electron, $4s^1$.)
4) Confirm that the total number of electrons is the same as the atomic number.

The last step checks the correctness of your final result. The atomic number is the number of protons in the nucleus of an atom, which is equal to the number of electrons. Therefore, the sum of the superscripts in an electron configuration, which is the total number of electrons in the atom, must be the same as the atomic number. For example, the electron configuration of oxygen (Z = 8) is $1s^2 2s^2 2p^4$. The sum of the superscripts is $2 + 2 + 4 = 8$, the same as the atomic number.

EXAMPLE 10.2

Write the complete electron configuration for chlorine (Cl, Z = 17).

SOLUTION

By steps from the previous procedure:

1) From its position in the periodic table [Group 7A (17), Period 3], the electron configuration of the highest occupied energy sublevel of chlorine is $3p^5$.
2) The sublevels having lower energies than $3p$ can be "read" across the periods from left to right in the periodic table, as in Figure 10.10: $1s\ 2s\ 2p\ 3s\ 3p^5$. If the neon core were to be used, these would be represented by $[Ne]3s\ 3p^5$.
3) A filled s sublevel has two electrons, and a filled p sublevel has six. Filling in these numbers yields $1s^2 2s^2 2p^6 3s^2 3p^5$, or $[Ne]3s^2 3p^5$.
4) $2 + 2 + 6 + 2 + 5 = 17 = Z$.

EXAMPLE 10.3

Write the complete electron configuration (no Group 0 core) for potassium (K, Z = 19).

First, what is the electron configuration of the highest occupied energy sublevel? (When you write the answer, leave space for the lower energy sublevels.)

$4s^1$ (Group 1A elements have one s electron.)

Now list to the left of $4s^1$ all lower energy sublevels in order of increasing energy.

_____   _____

$1s$ $2s$ $2p$ $3s$ $3p$ $4s^1$

Finally, add the superscripts that show how many electrons fill the lower energy sublevels. Check the final result.

_____   _____

$1s^22s^22p^63s^23p^64s^1$ $2 + 2 + 6 + 2 + 6 + 1 = 19 = Z$

Rewrite the configuration with a core from the closest Group 0 (18) element that has a smaller atomic number.

_____   _____

$[Ar]4s^1$

EXAMPLE 10.4 _____

Develop the electron configuration for cobalt, (Co, Z = 27).

This is your first example with d electrons. The procedure is the same. Write both a complete configuration and one with a Group 0 (18) core.

_____   _____

$1s^22s^22p^63s^23p^64s^23d^7$ or $[Ar]4s^23d^7$

By steps,

1) $3d^7$;

2) $1s$ $2s$ $2p$ $3s$ $3p$ $4s$ $3d^7$ or $[Ar]4s$ $3d^7$;

3) $1s^22s^22p^63s^23p^64s^23d^7$ or $[Ar]4s^23d^7$;

4) $2 + 2 + 6 + 2 + 6 + 2 + 7 = 27 = Z$.

10.4 VALENCE ELECTRONS

PG 10K	Using n for the highest occupied energy level, write the configuration of the valence electrons of any representative element.
10L	Write the Lewis (electron dot) symbol for an atom of any representative element.

FLASHBACK The vertical groups in the periodic table make up families of elements that have similar chemical properties (Section 5.6).

It is now known that many of the similar chemical properties of elements in the same column of the periodic table are related to the total number of s and p electrons in the highest occupied energy level. These are called **valence electrons.** In sodium, $1s^22s^22p^63s^1$, the highest occupied energy level is three. There is a single s electron in that sublevel, and there are no p electrons. Thus, sodium has one valence electron. With phosphorus, $1s^2 2s^2 2p^6 3s^2 3p^3$, the highest occupied energy level is again three. There are two s electrons and three p electrons, a total of five valence electrons.

Using n for any principal quantum number, we note that ns^1 is the configuration of the highest occupied principal energy level for all Group 1A (1) elements. All members of this family have one valence electron. Similarly, all elements in Group 5A (15) have the general configuration $ns^2 np^3$, and they have five valence electrons.

The highest occupied sublevels of all families of representative elements can be written in the form $ns^x np^y$. These are shown in the second row of Table 10.1. In all cases the number of valence electrons (third row) is the sum of the superscripts, x + y. Notice that for every group except Group 0 (18) the number of valence electrons is the same as the group number in the A–B system, or the same as the only or last digit in the IUPAC system.

EXAMPLE 10.5

Try to answer these questions without referring to anything; if you cannot, use only a full periodic table. (a) Write the electron configuration for the highest occupied energy level for Group 6A (16) elements. (b) Identify the group whose electron configuration for the highest occupied energy level is $ns^2 np^2$.

———   ———

(a) $ns^2 np^4$ (b) Group 4A (14)

(a) A Group 6A (16) element has six valence electrons, from the group number. The first two must be in the ns sublevel, and the remaining four must be in the np sublevel. (b) The total number of valence electrons is 2 + 2 = 4. The group is therefore 4A (14).

Another way to show valence electrons uses **Lewis symbols,** which are also called **electron dot symbols.** The symbol of the element is surrounded by that number of dots that matches the number of valence electrons. Dot symbols for the representative element in Period 3 are given in Table 10.1. Paired electrons, those that occupy the same

Table 10.1

Lewis Symbols of the Elements

Group (A–B) Group (IUPAC)	1A 1	2A 2	3A 13	4A 14	5A 15	6A 16	7A 17	0 18
Highest energy electron configuration	ns^1	ns^2	$ns^2 np^1$	$ns^2 np^2$	$ns^2 np^3$	$ns^2 np^4$	$ns^2 np^5$	$ns^2 np^6$
Number of valence electrons	1	2	3	4	5	6	7	8
Lewis symbol third period element	Na·	Mg:	Al:	·Si:	·P:	·S:	:Cl:	:Ar:

orbital, are usually placed on the same side of the symbol, and single occupants of an orbital are by themselves. This is not a hard and fast rule; exceptions are common if other positions better serve a particular purpose.

Group 0 (18) atoms have a full set of eight valence electrons, two in the *s* orbital and six in the *p* orbitals. This is sometimes called an **octet of electrons.** Elements in Group 0 (18) are particularly unreactive; only a few compounds of these elements are known. The filled octet is responsible for this chemical property.

EXAMPLE 10.6

Write electron dot symbols for the elements whose atomic numbers are 38 and 52.

Locate in the periodic table the elements whose atomic numbers are 38 and 52. Write their symbols. From the group each element is in, surround its symbol with the number of dots that matches the number of valence electrons.

————   ————

Sr :   · Te :

The periodic table gives Sr for the symbol of atomic number 38. The element is strontium, one isotope of which is a major problem in radioactive fallout. Strontium is in Group 2A (2), indicating two valence electrons. Te (tellurium) is the symbol for $Z = 52$. It is in Group 6A (16), so there are six electron dots.

10.5 TRENDS IN THE PERIODIC TABLE

Mendeleev and Meyer developed their periodic tables by trying to organize some of the recurring physical and chemical properties of the elements. Some of these properties are examined in this section.

Ionization Energy

A sodium atom (Na, $Z = 11$) has 11 protons and 11 electrons. One of the electrons is a valence electron. Mentally separate the valence electron from the other ten. This is pictured in the larger block of Figure 10.11. The valence electron, with its $1-$ charge, is still part of the neutral atom. The rest of the atom has 11 protons and 10 electrons (11 plus charges and 10 minus charges), giving it a net charge of $1+$. If we take the valence electron away from the atom, the particle that is left keeps that $1+$ charge. The particle is called a sodium ion. Its formula is Na^+.

It takes work to remove an electron from a neutral atom. Energy must be spent to overcome the attraction between the negatively charged electron and the positively charged ion that is left. **The energy required to remove one electron from a neutral gaseous atom of an element is the ionization energy of that element.**

Ionization energy is one of the more striking examples of a periodic property, particularly when graphed (Fig. 10.12). Notice the similarity of the shape of the graph between atomic numbers 3 and 10 (Period 2 in the periodic table) and atomic numbers 11 and 18 (Period 3). Notice also that the three peaks are elements in Group 0 (18) and the three low points are from Group 1A (1).

FLASHBACK This reviews Section 6.4 in which you learned that a monatomic (one-atom) ion is an atom that has gained or lost one, two, or three electrons. If the atom loses electrons, the ion has a positive charge and is a cation; if the atom gains electrons it becomes a negatively charged anion.

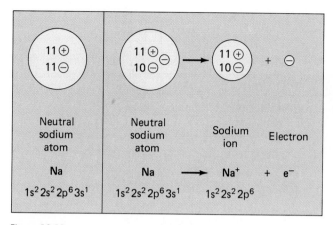

Figure 10.11

The formation of a sodium ion from a sodium atom.

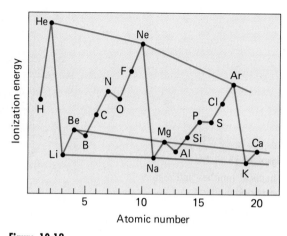

Figure 10.12

First ionization energy plotted as a function of atomic number, to show periodic properties of elements. The blue lines show that ionization energies of elements in the same family decrease as atomic number increases.

Observe the trends in Figure 10.12 too. As the atomic number increases within a period, the general trend in ionization energy is an increase. But there are interruptions at Groups 3A (3) and 6A (16). There are no exceptions in the trend between elements in the same group in the periodic table. The lines that connect elements in Groups 1A (1), 2A (2), and 0 (18) all slant down to the right. This indicates that ionization energies are lower as the atomic number increases within the group. If the graph is extended to the right, the same shapes and trends are found through the s and p sublevel portions of all periods.

The energy required to remove a second electron from an atom is its **second ionization energy,** and the third electron requires the **third ionization energy.** In all cases the ionization energy is very high when the valence electrons are gone and the next electron must be removed from a full octet of electrons. This adds to our belief that valence electrons are largely responsible for the chemical properties of the element.

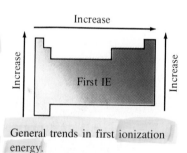

General trends in first ionization energy.

Chemical Families

Elements with similar chemical properties appear in the same group, or vertical column, in the periodic table. Several of these groups form **chemical families.** Family trends are most apparent among the representative elements. We limit our consideration to four families: the alkali metals in Group 1A (1), the alkaline earths in Group 2A (2), the halogens in Group 7A (17), and the noble gases in Group 0 (18). As these families are discussed, the symbol X is used to refer to any member of the family.

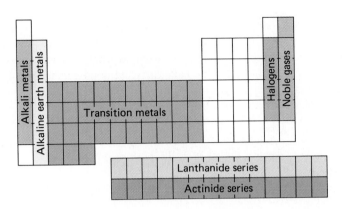

Figure 10.13

Chemical families and regions in the periodic table.

Noble Gases   Valence Electrons: $Z = 2$, ns^2 He; Others, ns^2np^6 $:\overset{..}{\underset{..}{X}}:$

The elements of Group 0 (18) are the **noble gases** (Fig. 10.13). In chemistry the word "noble" means unreactive. Only a few compounds of the noble gases are known, and none occur naturally.

The inactivity of the noble gases is believed to be the result of their filled valence electron sublevels. The two electrons of helium fill the $1s$ orbital, the only valence orbital that helium has. All other elements in the group have a full octet of electrons. This configuration apparently represents a "minimization of energy" arrangement of electrons that is very stable. The high ionization energies of a full octet have already been noted, so the noble gases resist forming positively charged ions. Nor do the atoms tend to gain electrons to form negatively charged ions.

The noble gases provide excellent examples of periodic trends in physical properties. Without exception, the density, melting point, and boiling point all increase as you move down the column in the periodic table.

Alkali Metals   Valence Electrons: ns^1 X·

With the exception of hydrogen and francium ($Z = 87$), Group 1A (1) elements are known as **alkali metals** (Fig. 10.13). (Francium is a radioactive element about which we know little.) The single valence electron is easily lost, forming an ion with a 1+ charge. All ions with a 1+ charge tend to combine with other elements in the same way. This is why the chemical properties of the elements in the family are similar.

Figure 10.12 shows that ionization energies of the alkali metals decrease as the atomic number increases. The higher the energy level is in which the ns^1 electron is located, the farther it is from the nucleus, and therefore the more easily it is removed. As a direct result of this, the **reactivity** of the element—that is, its tendency to react with other elements to form compounds—increases as you go down the column.

When an alkali metal atom loses its valence electron, the ion formed is **isoelectronic** with a noble gas atom. This means that it has the same electron configuration as a noble gas atom. (The prefix, *iso-*, means *same*, as in isotope.) For example, the electron configuration of sodium is $1s^2 2s^2 2p^6 3s^1$. If the $3s^1$ electron is removed, $1s^2 2s^2 2p^6$ is left. This is the configuration of the noble gas neon. In each case the alkali metal ion reaches the same configuration as the noble gas just before it in the periodic table. Its highest energy octet is complete—a highly stable electron distribution. The chemical properties of many elements can be explained in terms of their atoms becoming isoelectronic with a noble gas atom.

The "lighter-than-air" feature of an airship comes from the low-density helium that fills it. Helium is safe for this purpose because it is unreactive, unlike even lower-density hydrogen. See Figure 8.2, page 205.

Alkali metals do not normally look like metals. This is because they are so reactive that they combine with oxygen in the air to form an oxide coat, which hides the bright metallic lustre that can be seen in a freshly cut sample. These elements possess other common metallic properties too. For example, they are good conductors of heat and electricity and they are readily formed into wires and thin foils.

Distinct trends can be seen in the physical properties of alkali metals. Their densities increase as atomic number increases. Boiling and melting points generally decrease as you go down the periodic table. The single exception is cesium ($Z = 55$), which boils at a temperature slightly higher than the boiling point of rubidium ($Z = 37$).

Alkaline Earths   Valence Electrons: ns^2 X: Group 2A (2) elements are called **alkaline earths** or **alkaline earth metals.** Both the first and second ionization energies are relatively low, so the two valence electrons are given up readily to form ions with a 2+ charge. Again, the ions have the configuration of a noble gas. If magnesium, $[Ne]3s^2$, loses two electrons, only the $[Ne]$ core is left.

Trends like those noted with the alkali metals are also seen with the alkaline earths. Reactivity again increases as you go down the column in the periodic table. Physical property trends are less evident among the alkaline earths.

Halogens   Valence Electrons: ns^2np^5 :X: Four elements in Group 7A (17)—fluorine, chlorine, bromine, and iodine—make up the family known as the halogens, or "salt formers." (Astatine, $Z = 85$, is a radioactive element about which little is known.) The easiest way for a halogen to reach a full octet of electrons is to gain one. This gives it the configuration of a noble gas and forms an ion with a 1− charge. The tendency to gain the electron is greater the closer the added electron is to the nucleus. Consequently, reactivity is greatest for fluorine at the top of the group and least for iodine at the bottom.

Density, melting point, and boiling point all increase steadily with increasing atomic number among the halogens.

Hydrogen   Valence Electron: $1s^1$ H· You have probably wondered why hydrogen appears twice in our periodic table, at the top of Groups 1A (1) and 7A (17). Hydrogen is neither an alkali metal nor a halogen, although it shares some properties with both groups. Hydrogen combines with some elements in the same ratio as alkali metals, but the way the compounds are formed is different. Hydrogen atoms can also gain an electron to form an ion with a 1− charge, just like a halogen does. But other properties of hydrogen are not like those of the halogens. The way the periodic table is used makes it handy to have hydrogen in both positions, although it really stands alone as an element.

Atomic Size

PG 100 Predict how and explain why atomic size varies with position in the periodic table.

The upper part of Figure 10.14 shows the sizes of atoms of representative elements. Moving across the table from left to right, the atoms become smaller. Moving down the

Sodium (Na).

Magnesium (Mg).

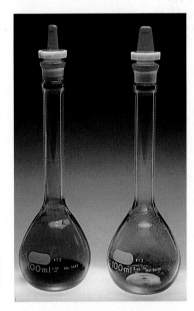
Bromine (Br$_2$; *left*) and iodine (I$_2$, *right*).

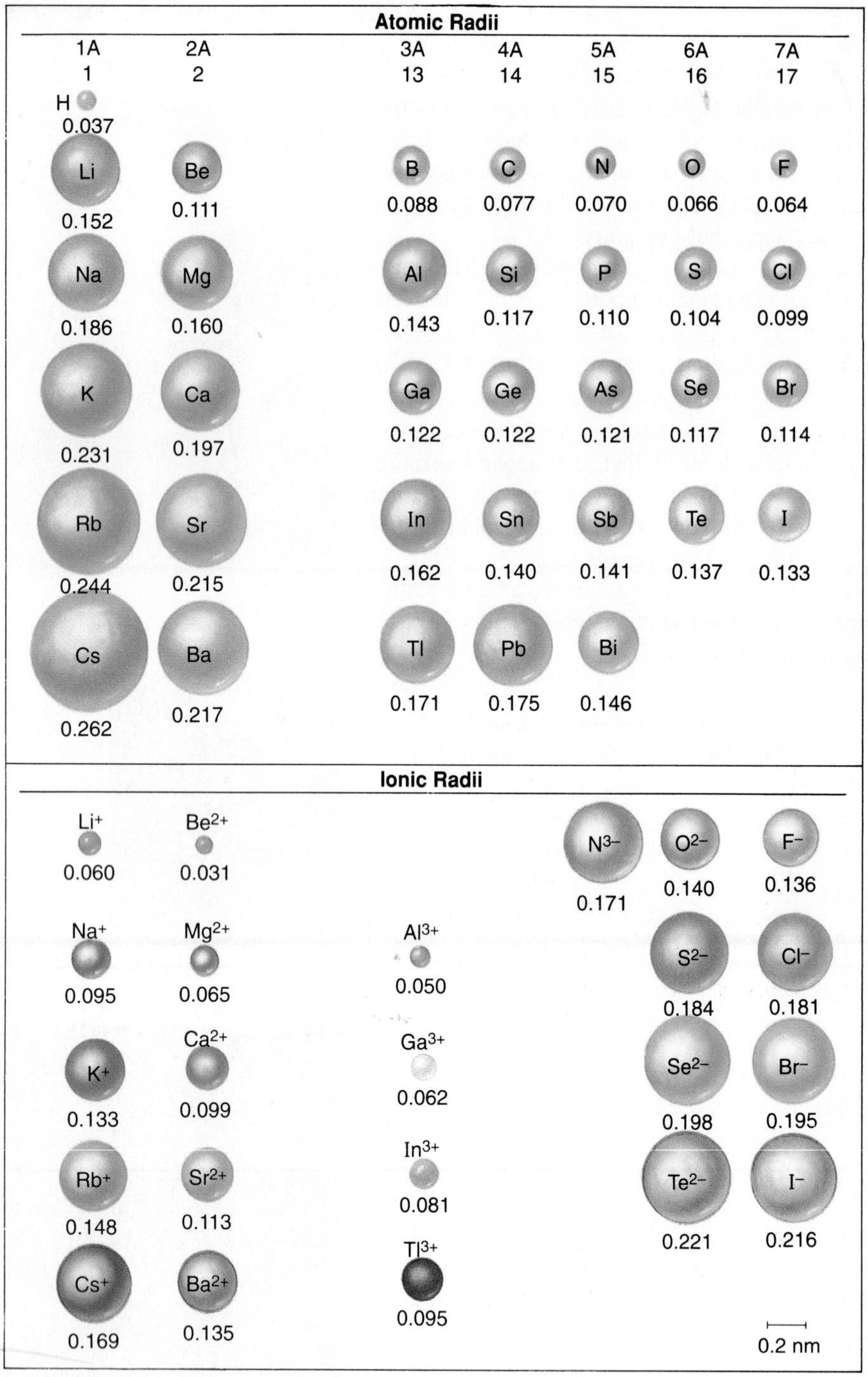

Figure 10.14

Sizes of atoms (upper section) and monatomic ions (lower section) of representative elements, expressed in nanometers. Ions of identical shading or coloring are isoelectronic. Trends to observe: (1) Sizes of atoms and ions increase with increasing atomic number in any group, i.e., going down any column. (2) Sizes of atoms decrease with increasing atomic number within a given period, i.e., going from left to right across any row. (3) Groups of isoelectronic ions, each shown in one color or shading, range from the right end of one period to the left end of the next period, and their sizes decrease with increasing atomic number. (4) Monatomic cations (positively charged ions) are smaller than the atom from which they come. (5) Monatomic anions (negatively charged ions) are larger than their parent atoms.

table in any group, atoms ordinarily increase in size. These observations are believed to be the result of three influences:

1) *Highest occupied principal energy level.* As valence electrons occupy higher and higher principal energy levels, they are generally farther from the nucleus and the atoms become larger. For example, the valence electron of a lithium atom is a $2s$ electron, whereas the valence electron of a sodium atom is a $3s$ electron. Sodium atoms are therefore larger.

2) *Nuclear charge.* Within any period, the valence electrons are all in the same principal energy level. As the number of protons in an atom increases, the positive charge in the nucleus increases. This pulls the valence electrons closer to the nucleus, so the atom becomes smaller. For example, the atomic number of sodium is 11 (11 protons in the nucleus), and the atomic number of magnesium is 12 (12 protons). The 12 protons in a magnesium atom attract the $3s$ valence electrons more strongly than the 11 protons of a sodium atom. The magnesium atom is therefore smaller.

3) *Shielding effect.* The attraction of the positively charged nucleus for the negatively charged outer electrons is partially canceled, or "shielded," by the repulsion of electrons in lower energy levels. Apparently this is less important than the number of occupied energy levels in determining atomic size. Even though a sodium atom has nearly four times as much nuclear charge, it is larger than a lithium atom.

SUMMARY

In summary, atomic size generally increases from right to left across any row of the periodic table and from top to bottom in any column. The smallest atoms are toward the upper right corner of the table, and the largest are toward the bottom left corner.

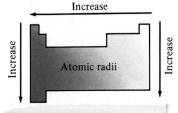

General trends in atomic radii of A Group elements with position in periodic table.

EXAMPLE 10.7

Referring only to a periodic table, list atomic numbers 15, 16, and 33 in order of increasing atomic size.

The preceding summary and marginal sketch should guide you into selecting the smallest and the largest of the three atoms.

———— ————

$16 - 15 - 33$

The smallest atom is toward the upper right (Z = 16) and the largest is toward the bottom left (Z = 33). Specifically, Z = 16 (sulfur) atoms are smaller than Z = 15 (phosphorus) atoms because sulfur atoms have a higher nuclear charge to attract the highest energy $3s$ and $3p$ electrons. The highest occupied energy level in a phosphorus atom is $n = 3$, but for a Z = 33 (arsenic) atom it is $n = 4$. Therefore, the phosphorus atom is smaller than the arsenic atom.

Sizes of Ions

> PG 10P Compare and explain the sizes of given isoelectronic monatomic ions.
> 10Q Compare and explain the sizes of given monatomic ions formed by elements in the same group in the periodic table.

The bottom part of Figure 10.14 shows the radii of monatomic ions of the representative elements. Like the sizes of atoms, the sizes of ions increase going down a column in the periodic table. Each step down represents an additional principal energy level that is occupied with electrons, yielding a larger radius. This matches the first of the three influences that determine the radius of an atom.

Unlike the sizes of atoms, the sizes of ions do not decrease across each period in the table. This is not an irregularity in periodic table trends, as it first appears, but a significant regularity in the sizes of isoelectronic species. Notice that for each isoelectronic group of ions, indicated by a color group in Figure 10.14, size decreases as atomic number increases—as the nuclear charge increases. This matches the second of the three influences listed earlier.

The third atomic size influence, the shielding effect of inner electrons, has no part in fixing ionic size among isoelectronic ions. They all have the same shielding because they have the same number of electrons.

The sizes of ions apparently are important in establishing some physical and chemical properties. For example, in some respects lithium is ''out of step'' with other alkali metals in its chemical properties. Beryllium tends to form bonds that are different from those formed by other alkaline earth metals. These and other features are partly the result of the smallness of lithium and beryllium ions.

Table 10.2

Some Physical and Chemical Properties of Metals and Nonmetals

Metals	Nonmetals
Lose electrons easily to form cations	Tend to gain electrons to form anions
1, 2, or 3 valence electrons	4 or more valence electrons
Low ionization energies	High ionization energies
Form compounds with nonmetals, but not with other metals	Form compound with metals and with other nonmetals
High electrical conductivity	Poor electrical conductivity (carbon in the form of graphite is an exception)
High thermal conductivity	Poor thermal conductivity; good insulator
Malleable (can be hammered into sheets)	Brittle
Ductile (can be drawn into wires)	Nonductile

Metals and Nonmetals

PG 10R Identify metals and nonmetals in the periodic table.

Both physically and chemically the alkali metals, alkaline earths, and transition elements are metals. Generally, an element is known as a metal if it can lose one or more electrons and become a positively charged ion. The larger the atom, the more easily the outermost electron is removed. Therefore, the metallic character of elements in a group increases as you go down a column in the periodic table.

It has also been noted that the size of an atom becomes smaller as the nuclear charge increases across a period in the table. The larger number of protons also holds the outermost electrons more strongly, making it more difficult for them to be lost. This makes the metallic character of elements *decrease* as you go from left to right across the period.

The properties of metals and nonmetals are compared in Table 10.2. Chemically, the distinction between metals and nonmetals, elements that lose electrons in chemical reactions and those that do not, is not a sharp one. It can be drawn roughly as a stair-step line beginning between atomic numbers 4 and 5 in Period 2 and ending between 84 and 85 in Period 6. (See Fig. 10.15.) Elements to the left of the line are metals, whereas those to the right are nonmetals.

Most of the elements next to the stair-step have some properties of both metals and nonmetals. They are often called **metalloids.** Included in the group are silicon and germanium, the semiconductors on which the electronics industry has been built. The Santa Clara Valley, south of San Francisco, used to be called "The Valley of Heart's Delight." However, the semiconductor silicon has become so important to the electronics industry in this area that it is now called "Silicon Valley."

General trends in metallic character of A Group elements with position in periodic table.

✔ QUICK CHECK 10.3

Take a look around you and count the number of things you take for granted that have transistors in them. To get you started, we'll list a few: Lamp dimmers, clocks that need batteries, battery chargers, radios, stereos and TVs that come on "instantly," telephones with push buttons, anything with a digital display. How many more can you find?

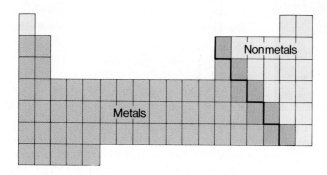

Figure 10.15

Metals and nonmetals. Green identifies elements that are metalloids. They have properties that are intermediate between those of metals and nonmetals.

SET 4

_____ Chemical family that consists of the most stable elements.

_____ Term that identifies energy required to remove an electron from a neutral gaseous atom.

_____ Chemical family that forms ions having a charge of 1+.

_____ Characteristic of an atom that tends to limit its size.

_____ Modified atom or group of chemically bonded atoms that has an electric charge.

_____ Size of Group 1A (1) atom compared to size of group 7A (17) atom in same period of the periodic table.

_____ Magnesium, calcium, strontium, and barium.

_____ Describes the way inner electrons modify the effect of nuclear charge on atomic size.

_____ Group of elements that have similar chemical properties.

_____ Elements in Group 7A (17) of the periodic table.

_____ Size of Group 1A (1) ion compared to size of group 7A (17) ion in same period of the periodic table.

1. Shielding effect
2. Smaller
3. Nuclear charge
4. Noble gases
5. Alkaline earths
6. Chemical family
7. Ion
8. Alkali metals
9. Halogens
10. Ionization
11. Larger

STUDY HINTS AND PITFALLS TO AVOID

It takes many words to describe the quantum model of the atom, even at this introductory level. The words are not easy to remember unless they are organized into some kind of pattern. The summary at the end of Section 10.2 gives you this organization.

Be sure you know what "quantized" means. Also understand the difference between a Bohr orbit (a fixed path the electron travels around the nucleus) and the quantum orbital (a mathematically defined region in space in which there is a high probability of finding the electron).

In writing electron configurations, we recommend that you use the periodic table to list the sublevels in increasing energy rather than the slanting-line memory device. The periodic table is the greatest organizer of chemical information there is, and every time you use it you strengthen your ability to use it in all other ways.

Understand well the three influences that determine atomic and ionic size. The same thinking appears with other properties later.

CHAPTER 10 QUESTIONS AND PROBLEMS

SECTION 10.1

1) Distinguish between a discrete line spectrum and a continuous spectrum. Into which classification would you place a rainbow?

2) Which end of the light spectrum, red or violet, has the highest energy? Which end has the longest wavelength? Which end travels faster? Which rays do you suppose cause sunburn?

3) Distinguish between something that is quantized and something that is not quantized, that is, continuous.

4) Is the light in the spectrum of white light quantized? How about the light from hydrogen in a gas discharge tube?

_____

5) Is milk from a cow quantized? How about when the milk is sold in the supermarket?

6) Nolan Ryan is one of the most famous fast-ball pitchers in the history of baseball—certainly one of the most enduring. Is the speed of his pitches quantized?

_____

7) Why do we believe that electron energies are quantized in an atom? Explain the lines in an atomic spectrum in terms of quantized energy levels.

8) In which direction are the energy levels of an electron quantized, as the electron passes from a low energy level to a high one, or when it drops from a high energy level to a lower one? Explain your answer.

---

9) What must be done to an atom, or what must happen to an atom, before it can emit light?

10) Why is it necessary for the change that is the answer to Question 9 to occur before an atom can emit light?

---

11) Distinguish between electrons in an excited state and those in a ground state. Which have the greater likelihood of giving off visible light?

12) Does an electron that drops from a high energy level to a lower level always emit light? If no, explain.

---

13) What major advances in our understanding of the atom came from Bohr's theory?

14) The Bohr model of the atom had shortcomings and today is not regarded as correct. What were these faults?

SECTION 10.2

15) What are "principal energy levels" in an atom? Are they related to the Bohr model of the atom? If so, how?

16) How are principal energy levels in an atom identified? How many levels are there, according to the quantum mechanical model of the atom?

---

17) Which principal energy level in an atom has the higher energy, $n = 1$ or $n = 2$?

18) List the principal energy levels 1 through 4 in order of decreasing energy.

---

19) Distinguish between principal energy levels and sublevels. How are sublevels identified?

20) How many sublevels are there when $n = 2$? When $n = 4$?

---

21) What sublevels are present when $n = 1$? When $n = 3$?

22) List the sublevels in order of increasing energy when $n = 3$.

---

23) How do s, p, d, and f sublevels compare in energy for any given value of n?

24) Describe the energy trend among sublevels as the value of n increases. How about d sublevels?

---

25) What is wrong with the statement, "The $3d$ sublevel has a higher energy than the $2d$ sublevel"?

26) What is wrong with the statement, "The electron travels around the atomic nucleus just as a planet travels around the sun"?

---

27) Describe the movement of an electron in an atom according to the quantum mechanical model.

28) Distinguish between an electron orbit and an electron orbital. Which term is used in modern atomic theory?

---

29) Describe by words and sketch the shapes of the s and p orbitals.

30) How many orbitals are in a $1s$ sublevel? In a $2s$ sublevel? In a $3d$ sublevel? In a $4d$ sublevel?

---

31) What is the significance of the Pauli exclusion principle?

32) List all possible numbers of electrons that can occupy any orbital.

---

33) List the maximum number of electrons that can occupy all orbitals in the $2s$ sublevel; the $3d$ sublevel; the $4d$ sublevel; the $5f$ sublevel.

34) What is the maximum number of electrons that can occupy all orbitals for which $n = 1$, that is, all orbitals in the first principal energy level? How many electrons can occupy all orbitals for which $n = 3$?

---

35) What general statement may be made about the energies of the principal energy levels? About the energies of the sublevels?

36) Show how and where energies of the principal energy levels overlap.

SECTION 10.3

37) What symbol indicates that a $2s$ orbital contains one electron? What symbol shows that a $2s$ orbital holds two electrons?

38) The question, "What symbol indicates that a $2s$ orbital contains three electrons?" is not legitimate—it has no correct answer. Why?

---

39) What element has the electron configuration $1s^2 2s^2 2p^4$? What period is it in? What group is it in?

40) $1s^2 2s^2 2p^6 3s^2 3p^2$ is the electron configuration of an element. Identify the element and state the group and period in which it appears in the periodic table.

11 Chemical Bonding: The Formation of Ionic Compounds and Molecules

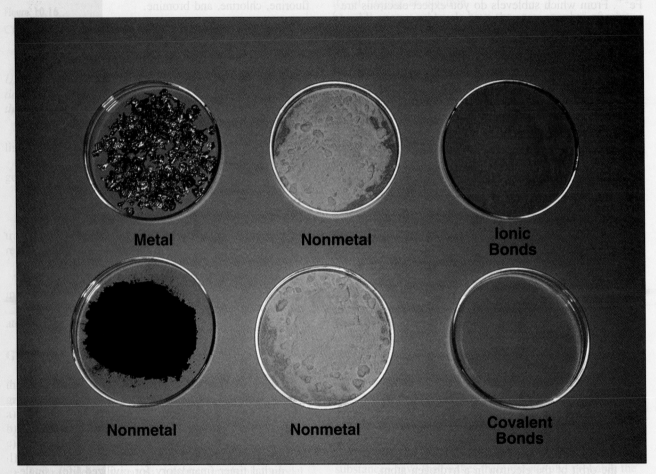

Metal   Nonmetal   Ionic Bonds

Nonmetal   Nonmetal   Covalent Bonds

Each container in this photo holds one mole—6.02×10^{23} formula units—of the substance shown. The top line indicates that when a metal combines with a nonmetal, the compound formed is held together by *ionic* bonds. In this case, gray cadmium combines with yellow sulfur to form orange cadmium sulfide. If two nonmetals combine, such as solid carbon and solid sulfur, a molecular compound held together by *covalent* bonds is produced. The compound is carbon disulfide, which is a liquid at room conditions.

In his atomic theory, John Dalton said that atoms of different elements combine to form compounds. He didn't say how they combined, or why. We now believe we understand *how* most chemical compounds are formed. *Why* atoms combine was touched upon in Section 2.4: "Minimization of energy is one of the driving forces that cause chemical reactions to occur." This "minimization" refers to the potential energy caused by attractions and repulsions between charged particles in the structure of the compound.

Chemical bond is a term that is used to identify any of several kinds of attractions among atomic, ionic, or molecular particles. In this chapter we consider the bonds between atoms in a compound. Later you will study other kinds of bonds.

Chemical bonds form and re-form when atoms, molecules, or ions collide. The first contact in a collision is between the outermost electrons of two particles. These are the valence electrons. Our study of chemical bonds therefore focuses on the role of the valence electrons in forming a bond between two atoms.

11.1 MONATOMIC IONS WITH NOBLE GAS ELECTRON CONFIGURATIONS

PG 11A Identify the monatomic ions that are isoelectronic with a given noble gas atom, and write the electron configuration of those ions.

In Section 10.5 you learned that elements in the same chemical family usually form monatomic ions having the same charge. They do this by gaining or losing valence electrons until they become isoelectronic with a noble gas atom and the octet of electrons is complete.

Table 11.1 shows how the electrons are gained and lost when nitrogen, oxygen, fluorine, sodium, magnesium, and aluminum form monatomic ions that are isoelectronic with a neon atom. The pattern built around neon is duplicated for other noble gases. Thus, phosphorus, sulfur, chlorine, potassium, calcium, and scandium (Z = 21) form ions that are isoelectronic with argon; and the three elements on either side of krypton form ions that duplicate the electron configuration of krypton.

The monatomic ions formed by hydrogen and lithium duplicate the electron configurations of helium with just two electrons: $1s^2$. Unlike other elements in Groups 2A (2) and 3A (13), beryllium and boron tend to form covalent bonds by sharing electrons, rather than forming ions. We will look at covalent bonds later in this chapter.

Figure 11.1 is a periodic table that summarizes the monatomic ions that are isoelectronic with noble gas atoms.

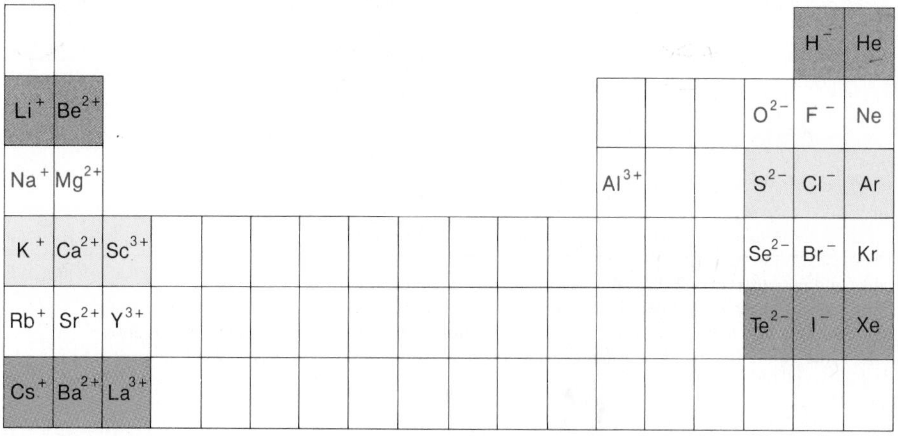

Figure 11.1

Monatomic ions that have noble gas electron configurations. Each color group includes one noble gas atom and the monatomic ions that are isoelectronic with that atom. The beryllium ion is included, although this element more commonly forms covalent bonds.

Table 11.1

Formation of Monatomic Ions that are Isoelectronic with Neon Atoms

Element	Atom	Electron(s)		Monatomic Ion	Atom/Ion Electron Count		
					Start	*Change*	*Final*
Nitrogen	$\begin{array}{c}7\,p^+\\7\,e^-\end{array}$	$+$	$e^- + e^- + e^-$ $\rightarrow$	$\begin{array}{c}7\,p^+\\10\,e^-\end{array}$	7	+3	10
Z = 7							
Group 5A (15)	$\cdot \ddot{\text{N}} \colon$	$+$	$\cdot(e^-) + \cdot(e^-) + \cdot(e^-)$ $\rightarrow$	$\left[\,\colon\!\ddot{\text{N}}\colon\right]^{3-}$ $\quad$ N^{3-}			
	$1s^2 2s^2 2p^3$	$+$	$3\,e^-$ $\rightarrow$	$1s^2 2s^2 2p^6$			
Oxygen	$\begin{array}{c}8\,p^+\\8\,e^-\end{array}$	$+$	$e^- + e^-$ $\rightarrow$	$\begin{array}{c}8\,p^+\\10\,e^-\end{array}$	8	+2	10
Z = 8							
Group 6A (16)	$\cdot \ddot{\text{O}} \colon$	$+$	$\cdot(e^-) + \cdot(e^-)$ $\rightarrow$	$\left[\,\colon\!\ddot{\text{O}}\colon\right]^{2-}$ $\quad$ O^{2-}			
	$1s^2 2s^2 2p^4$	$+$	$2\,e^-$ $\rightarrow$	$1s^2 2s^2 2p^6$			
Fluorine	$\begin{array}{c}9\,p^+\\9\,e^-\end{array}$	$+$	e^- $\rightarrow$	$\begin{array}{c}9\,p^+\\10\,e^-\end{array}$	9	+1	10
Z = 9							
Group 7A (17)	$\colon\!\ddot{\text{F}}\colon$	$+$	$\cdot(e^-)$ $\rightarrow$	$\left[\,\colon\!\ddot{\text{F}}\colon\right]^{-}$ $\quad$ F^{-}			
	$1s^2 2s^2 2p^5$	$+$	e^- $\rightarrow$	$1s^2 2s^2 2p^6$			
Neon	$\begin{array}{c}10\,p^+\\10\,e^-\end{array}$			$\begin{array}{c}10\,p^+\\10\,e^-\end{array}$	10		10
Z = 10							
Group 0 (18)	$\colon\!\text{Ne}\colon$			$\colon\!\text{Ne}\colon$			
	$1s^2 2s^2 2p^6$			$1s^2 2s^2 2p^6$			
Sodium	$\begin{array}{c}11\,p^+\\11\,e^-\end{array}$	$-$	e^- $\rightarrow$	$\begin{array}{c}11\,p^+\\10\,e^-\end{array}$	11	−1	10
Z = 11							
Group 1A (1)	Na $\cdot$	$-$	$\cdot(e^-)$ $\rightarrow$	Na$^+$			
	$1s^2 2s^2 2p^6 3s^1$	$-$	e^- $\rightarrow$	$1s^2 2s^2 2p^6$			
Magnesium	$\begin{array}{c}12\,p^+\\12\,e^-\end{array}$	$-$	$e^- + e^-$ $\rightarrow$	$\begin{array}{c}12\,p^+\\10\,e^-\end{array}$	12	−2	10
Z = 12							
Group 2A (2)	Mg $\cdot$	$-$	$\cdot(e^-) + \cdot(e^-)$ $\rightarrow$	Mg^{2+}			
	$1s^2 2s^2 2p^6 3s^2$	$-$	$2\,e^-$ $\rightarrow$	$1s^2 2s^2 2p^6$			
Aluminum	$\begin{array}{c}13\,p^+\\13\,e^-\end{array}$	$-$	$e^- + e^- + e^-$ $\rightarrow$	$\begin{array}{c}13\,p^+\\10\,e^-\end{array}$	13	−3	10
Z = 13							
Group 3A (13)	$\cdot$ Al $\cdot$	$-$	$\cdot(e^-) + \cdot(e^-) + \cdot(e^-)$ $\rightarrow$	Al^{3+}			
	$1s^2 2s^2 2p^6 3s^2 3p^1$	$-$	$3\,e^-$ $\rightarrow$	$1s^2 2s^2 2p^6$			

This table shows how monatomic anions (pink) and cations (yellow) that are isoelectronic with neon atoms (blue) are formed. A neon atom has ten electrons, including a full octet of valence electrons. Its electron configuration is $1s^2 2s^2 2p^6$. Nitrogen, oxygen, and fluorine atoms form anions by gaining enough electrons to reach the same configuration. Sodium, magnesium, and aluminum atoms form cations by losing valence electrons to reach the same configuration. Dots around each elemental symbol represent valence electrons.

You may wonder about the hydrogen ion, H^+. This ion does not normally exist by itself, but rather, it exists as a "hydrated" hydrogen ion, $H^+ \cdot H_2O$, commonly called the hydronium ion and written H_3O^+. The ion is not monatomic, and therefore it is not properly included in this section.

EXAMPLE 11.1 _____

Write the electron configurations for the calcium and chloride ions, Ca^{2+} and Cl^-. With what noble gases are these ions isoelectronic?

To begin, note the locations of calcium and chlorine in the periodic table. Write their Lewis symbols and from them state the number of electrons that must be gained or lost to achieve complete octets at the highest energy level.

_____   _____

Ca: must lose two electrons to achieve an octet, and :Cl: must gain one to achieve an octet.

You are now ready to answer the main questions: the electron configuration for Ca^{2+} and Cl^-, please, and the noble gases with which they are isoelectronic.

_____   _____

Both ions are isoelectronic with argon, $1s^2 2s^2 2p^6 3s^2 3p^6$

The calcium atom starts with the configuration $1s^2 2s^2 2p^6 3s^2 3p^6 4s^2$. In losing two electrons to yield Ca^{2+}, it reaches the electron configuration of argon. Chlorine, with configuration $1s^2 2s^2 2p^6 3s^2 3p^5$, must gain one electron to become Cl^-, which is isoelectronic with argon.

EXAMPLE 11.2 _____

Identify at least one more cation and one more anion that are isoelectronic with an atom of argon.

_____   _____

K^+, Sc^{3+}, S^{2-}, and P^{3-}

The formations of K^+, S^{2-}, and P^{3-} are identical to the formations of Na^+, O^{2-}, and N^{3-}, respectively, the ions immediately above them in the periodic table. Unless you referred to Figure 11.1, you probably did not include the scandium ion, Sc^{3+}, in your answer to the question. From the electron configuration of scandium ($Z = 21$), $1s^2 2s^2 2p^6 3s^2 3p^6 4s^2 3d^1$, you might have guessed the charge on the ion would be 3+ because removal of the three highest energy electrons leaves the configuration of the noble gas, argon. Your guess would have been correct.

✔ QUICK CHECK 11.1

Which ions among the following are isoelectronic with noble gas atoms?

$$Cu^{2+} \qquad S^{2-} \qquad Fe^{3+} \qquad Ag^+ \qquad Ba^{2+}$$

11.2 IONIC BONDS

We have been discussing the formation of monatomic ions as neutral atoms that "gain" or "lose" electrons. For most elements this is not a common event, but rather, an accomplished fact. The natural occurrence of many elements is in **ionic compounds**—compounds made up of ions—or solutions of ionic compounds. Nowhere, for example, are sodium or chlorine atoms to be found, but there are large natural deposits of sodium chloride (table salt) that are made up of sodium and chloride ions. The compound may also be obtained by evaporating seawater, which contains the ions in solution.

Sodium and chlorine are both highly reactive elements. If, after having been prepared from any natural source, they are brought together, they will react vigorously to form the compound sodium chloride. In that reaction a sodium atom literally loses an electron to become a sodium ion, Na^+, and a chloride atom gains an electron to become a chloride ion, Cl^-. Lewis diagrams show the electron transfer clearly:

$$Na \overset{\frown}{\cdot} + \cdot \overset{\cdot\cdot}{\underset{\cdot\cdot}{Cl}} : \longrightarrow Na^+ + \left[: \overset{\cdot\cdot}{\underset{\cdot\cdot}{Cl}} : \right]^- \longrightarrow NaCl \text{ crystal}$$

Ions are not always present in a $1:1$ ratio, as in sodium chloride. Calcium atoms have two valence electrons to lose to form a calcium ion, Ca^{2+}. But a chlorine atom receives only one electron when forming a chloride ion, Cl^-. It therefore takes two chlorine atoms to receive the two electrons from a single calcium atom:

$$Ca : + \quad \begin{matrix} \cdot \overset{\cdot\cdot}{Cl} : \\ \\ \cdot \overset{\cdot\cdot}{Cl} : \end{matrix} \longrightarrow Ca^{2+} + 2 \left[: \overset{\cdot\cdot}{\underset{\cdot\cdot}{Cl}} : \right]^- \longrightarrow CaCl_2 \text{ crystal}$$

FLASHBACK This section explains why the formulas of ionic compounds are what they are. These are the formulas you learned to write in Section 6.8. All of the electrons in all of the atoms in the formula unit of an ionic compound are accounted for.

The $1:2$ ratio of calcium ions to chloride ions is reflected in the formula of the compound formed, calcium chloride, $CaCl_2$. There are several combinations of charges that appear in ionic compounds, but they are always in such numbers that yield a compound that is electrically neutral.

Nearly all ionic compounds are solids at normal temperatures and pressures. The solid has a definite geometric structure called a **crystal.** Ions in a crystal are arranged so the potential energy resulting from the attractions and repulsions between them is at a minimum (Section 2.4). The precise form of the crystal depends on the kinds of ions in the compound, their sizes, and the ratio in which they appear. Figure 11.2 shows the structure of a sodium chloride crystal. The strong electrostatic forces that hold the ions in fixed position in the crystal are called **ionic bonds.**

Ionic crystals are not limited to monatomic ions; polyatomic ions—ions consisting of two or more atoms—also form crystal structures. Atoms in a polyatomic ion are held together by covalent bonds (Section 11.3). Figure 11.3 is a model of a calcium carbonate crystal. The formula of a carbonate ion, one of which is circled in Figure 11.3, is CO_3^{2-}. The carbon atom is surrounded by the three covalently bonded oxygen atoms. Not only is the carbonate ion a distinct unit in the structure of the crystal, it also behaves as a unit in many chemical changes.

FLASHBACK You used the unit-like behavior of polyatomic ions when balancing equations in Section 8.7.

The bonds in an ionic crystal are very strong. This is why nearly all ionic compounds are solids at room temperature. It takes a high temperature to break the many ionic bonds, free the ions from each other, and melt the crystal to become a liquid.

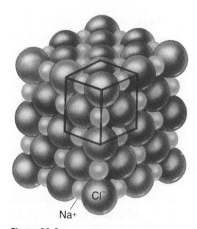

Figure 11.2

Arrangement of ions in sodium chloride. The red spheres represent sodium ions and the blue spheres are chloride ions. The number of positive charges is the same as the number of negative charges, making the crystal electrically neutral.

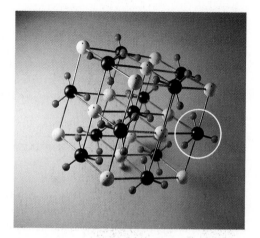

Figure 11.3

Model of a calcium carbonate crystal. The gray spheres represent calcium ions, Ca^{2+}. The black spheres with three red spheres attached (see circle) are carbonate ions, $CO_3{}^{2-}$. There are equal numbers of calcium and carbonate ions, yielding a compound that is electrically neutral.

Solid ionic compounds are poor conductors of electricity because the ions are locked in place in the crystal. When the substance is melted or dissolved, the crystal is destroyed. The ions are then free to move and able to carry electric current. Liquid ionic compounds and water solutions of ionic compounds are good conductors.

11.3 COVALENT BONDS

PG 11B Distinguish between ionic and covalent bonds.

When you eat, you probably have near you a salt shaker and a bowl of sugar. Let's study these two familiar white powders. Both dissolve easily in water, but the salt water conducts electricity, while the sugar water does not. Also, salt melts at 801°C to give a colorless liquid, while sugar doesn't melt at all. At only 160°C, sugar begins to char, emitting the aroma of caramel. (Indeed, that's how caramel is made.)

We know that ionic bonding explains the properties of many compounds, such as sodium chloride, quite well. However, other compounds such as sugar have properties so different from ionic compounds, we wonder what holds this "other kind" of compound together.

Hydrogen fluoride, HF, and methane, CH_4, are compounds of the other kind. Both are gases at room temperature and pressure. When condensed to liquids, they are nonconductors, like water. These are **molecular compounds,** whose ultimate structural unit is the individual **molecule.**

In 1916 G. N. Lewis proposed that two atoms in a molecule are held together by a **covalent bond** in which they share one or more pairs of electrons. The idea is that

Gilbert N. Lewis

when the bonding electrons can spend most of their time between two atoms, they attract *both* positively charged nuclei and "couple" the atoms to each other, much as two railroad cars are held together by the coupler between them. The result is a bond that is permanent until broken by a chemical change.

The simplest molecule and the simplest covalent bond appear in hydrogen, H_2. Using Lewis symbols, the formation of a molecule of H_2 can be represented as

$$H\cdot \; + \; \cdot H \longrightarrow H:H \quad \text{or} \quad H—H$$

FLASHBACK Lewis symbols use dots distributed around the symbol of an element to represent valence electrons. See Section 10.4.

The two dots or the straight line drawn between the two atoms represent the covalent bond that holds the atoms together. In modern terms we say that the *electron cloud* or *charge density* formed by the two electrons is concentrated in the region between the two nuclei. This is where there is the greatest probability of locating the bonding electrons. The atomic orbitals of the separated atoms are said to *overlap* (Fig. 11.4).

A similar approach shows the formation of the covalent bond between two fluorine atoms to form a molecule of F_2, and between one hydrogen atom and one fluorine atom to form an HF molecule:

$$:\!\overset{..}{F}\!\cdot \; + \; \cdot\!\overset{..}{F}\!: \; \longrightarrow \; :\!\overset{..}{F}\!:\!\overset{..}{F}\!: \quad \text{or} \quad :\!\overset{..}{F}\!—\!\overset{..}{F}\!: \quad \text{or} \quad F—F$$

$$H\cdot \; + \; \cdot\!\overset{..}{F}\!: \; \longrightarrow \; H:\!\overset{..}{F}\!: \quad \text{or} \quad H—\!\overset{..}{F}\!: \quad \text{or} \quad H—F$$

Fluorine has seven valence electrons. The $2s$ orbital and two of the $2p$ orbitals are filled, but the remaining $2p$ orbital has only one electron. The F_2 bond is formed by the overlap of the half-filled $2p$ orbitals of two fluorine atoms. In the HF molecule, the bond forms from the overlap of the half-filled $1s$ orbital of a hydrogen atom with the half-filled $2p$ orbital of a fluorine atom.

When used to show the bonding arrangement between atoms in a molecule, electron dot diagrams are commonly called **Lewis diagrams, Lewis formulas,** or **Lewis**

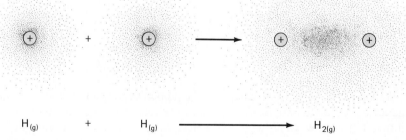

$$H_{(g)} \quad + \quad H_{(g)} \longrightarrow H_{2(g)}$$

Figure 11.4
The formation of a hydrogen molecule from two hydrogen atoms. Each dot represents the instantaneous position of an electron. The "charge clouds" of the $1s$ orbitals of two hydrogen atoms are said to overlap and form a covalent bond. The bonding electrons spend most of their time between the two nuclei, as suggested by the heavier density of electron position dots in that area.

structures. Notice that the unshared electron pairs of fluorine are shown for two of the three previous Lewis diagrams for F_2 and HF, but not for the third. Technically, they should always be shown, but they are frequently omitted when not absolutely needed. Unshared electron pairs are often called **lone pairs.**

When two bonding electrons are shared by two atoms, the electrons effectively "belong" to both atoms. They count as valence electrons for each bonded atom. Thus, each hydrogen atom in H_2 and the hydrogen atom in HF have two electrons, the same number as an atom of the noble gas helium. Each fluorine atom in F_2 and the fluorine atom in HF have eight valence electrons, matching neon and the other noble gas atoms.

These and many similar observations lead us to believe that the stability of a noble gas electron configuration contributes to the formation of covalent bonds. This generalization is known as the **octet rule,** or **rule of eight,** because each atom has "completed its octet." The tendency toward a complete octet of electrons in a bonded atom reflects the natural tendency for a system to move to the lowest energy state possible.

💡 **FLASHBACK** The correlation between minimization of energy, mentioned in Section 2.4 as a driving force for chemical change, and the full octet of valence electrons was noted in Section 10.5. Many laboratory measurements show that the energy of a system is reduced as bonds form in which the noble gas electron configuration is reached.

✔ **QUICK CHECK 11.2**

Identify the true statements and rewrite the false statements to make them true.

a) Atoms in molecular compounds are held together by covalent bonds.
b) A lone pair of electrons is not shared between two atoms.
c) Covalent bonds are common between atoms of two metals.
d) An octet of valence electrons usually represents a low energy state.

11.4 POLAR AND NONPOLAR COVALENT BONDS

PG 11C	Distinguish between polar and nonpolar covalent bonds.
11D	Given the electronegativities of all elements involved, rank bonds in order of increasing or decreasing polarity.
11E	Given the electronegativities of two elements, classify the bond between them as nonpolar covalent, polar covalent, or primarily ionic. If the bond is polar, state which end is positive and which end is negative.

As we might expect, the two electrons joining the atoms in the H_2 molecule are shared equally by the two nuclei. Another way of saying this is that the charge density is centered in the overlap region between the bonded atoms, as shown in Figure 11.4. **A bond in which the distribution of bonding electron charge is symmetrical, or centered, is said to be nonpolar.** A bond between identical atoms, as in H_2 or F_2, is always nonpolar.

In an HF molecule the charge density of the bonding electrons is shifted toward the fluorine atom and away from the hydrogen atom (Fig. 11.5). **A bond with an unsymmetrical distribution of bonding electron charge is a polar bond.** The fluorine atom in an HF molecule acts as a negative pole, and the hydrogen atom is a positive pole.

Bond polarity may be described in terms of the electronegativities of the bonded atoms. **The electronegativity of an element is a measure of the strength by which its atoms attract the electron pair it shares with another atom in a single covalent bond.** High electronegativity identifies an element with a strong attraction for bonding electrons.

Figure 11.5

A polar bond. Fluorine in a molecule of HF has a higher electronegativity than hydrogen. The bonding electron pair is therefore shifted toward fluorine. The nonsymmetrical distribution of charge yields a polar bond.

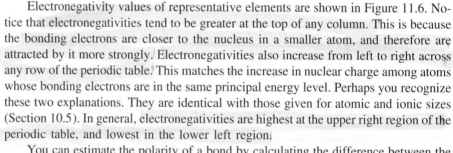

Figure 11.6

Electronegativities of the representative elements. Notice the trends in electronegativity values. They increase from left to right across any row of the table, and they increase from the bottom to the top of any column.

The concept of electronegativity was proposed by Linus Pauling, who is the winner of two Nobel prizes, one in Chemistry and one in Peace.

Electronegativity values of representative elements are shown in Figure 11.6. Notice that electronegativities tend to be greater at the top of any column. This is because the bonding electrons are closer to the nucleus in a smaller atom, and therefore are attracted by it more strongly. Electronegativities also increase from left to right across any row of the periodic table. This matches the increase in nuclear charge among atoms whose bonding electrons are in the same principal energy level. Perhaps you recognize these two explanations. They are identical with those given for atomic and ionic sizes (Section 10.5). In general, electronegativities are highest at the upper right region of the periodic table, and lowest in the lower left region.

You can estimate the polarity of a bond by calculating the difference between the electronegativity values for the two elements: The greater the difference the more polar the bond. In nonpolar H_2 and F_2 molecules, where two atoms of the same element are bonded, the electronegativity difference is zero. In the polar HF molecule the electronegativity difference is 1.9 (4.0 for fluorine minus 2.1 for hydrogen). A bond between carbon and chlorine, for example, with an electronegativity difference of $3.0 - 2.5 = 0.5$, is more polar than an H—H bond, but less polar than an H—F bond.

The more electronegative element toward which the bonding electrons are displaced acts as the "negative pole" in a polar bond. This is sometimes indicated by using an arrow rather than a simple dash, with the arrow pointing to the negative pole. In a bond between hydrogen and fluorine this is H $\leftrightarrow$ F. Another representation is $\delta-$ written in the region of the negative pole and $\delta+$ in the area with a positive charge. δ is a lowercase Greek delta. In this use it represents a "partial" negative or "partial" positive charge. Thus, for hydrogen fluoride, $^{\delta+}H—F^{\delta-}$.

EXAMPLE 11.3 _____

Using data from Figure 11.5, arrange the following bonds in order of increasing polarity, and circle the element that will act as the negative pole.

H—O	H—S
H—P	H—C

Locate the elements in the table, calculate the differences in electronegativity, and enter those differences.

———   ———

H—O: 3.5 − 2.1 = 1.4   H—S: 2.5 − 2.1 = 0.4

H—P: 2.1 − 2.1 = 0.0   H—C: 2.5 − 2.1 = 0.4

Now arrange the bonds in order from the least polar to the most polar.

———   ———

H—P, H—S, H—C, H—O   or   H—P, H—C, H—S, H—O

The H—P bond is nonpolar (electronegativity difference = 0). H—O is the most polar bond because it has the largest electronegativity difference (1.4). The other bonds, with equal electronegativity differences, have about the same polarity.

Now place an arrowhead on each bond that points toward the negative pole.

———   ———

H—P H ↔ S H ↔ C H ↔ O

Since sulfur, carbon, and oxygen are all more electronegative than hydrogen, the electron density in these three bonds is shifted away from hydrogen toward the other element. There is no electronegativity difference in the H—P bond, so neither atom will act as a negative pole.

Electronegativity numbers can also be used to predict whether a bond will be nonpolar covalent, polar covalent, or ionic. We must be cautious, however, not to give these predictions more credit than they deserve. In the first place, there is no sharp difference between the three bond classifications. The whole range of polarities passes *gradually* from a pure nonpolar covalent bond between identical atoms to the most ionic bond in cesium fluoride, CsF. Various authors use electronegativity differences anywhere from 1.7 to 1.9 as the crossover point between covalent and ionic bonds, which shows how arbitrary the classifications are. Only laboratory measurements can really determine the polarity of a bond. It is probably safe to say that:

SUMMARY

1) The only truly nonpolar bond is between identical atoms. If the electronegativity difference is less than 0.4, the bond is essentially nonpolar.
2) The bond between atoms with an electronegativity difference less than 1.7 is polar covalent.
3) An electronegativity difference greater than 1.9 identifies a bond that is primarily ionic.
4) If the electronegativity difference is from 1.7 to 1.9, the bond may be considered as very strongly polar covalent or slightly ionic.

EXAMPLE 11.4

Using Figure 11.5, classify each of the following bonds as nonpolar covalent, polar covalent, or ionic:

Li—F   C—I   P—Cl

First, determine each electronegativity difference.

_____   _____

Li—F, 3.0   C—I, 0.0   P—Cl, 0.9

Now the classification: nonpolar covalent, polar covalent, or ionic?

_____   _____

Li—F, ionic (electronegativity difference greater than 1.9)

C—I, nonpolar covalent (electronegativity difference, 0.0)

P—Cl, polar covalent (electronegativity difference less than 1.7)

Although the atoms are not identical, the electronegativity difference between carbon and iodine is negligible, so the bond may be considered nonpolar.

In Chapter 12 you will use bond polarities to predict the polarities of molecules. In Chapter 14 you will find that molecular polarity is largely responsible for the physical properties of many compounds.

11.5 MULTIPLE BONDS

So far our consideration of covalent bonds has been limited to the sharing of one pair of electrons by two bonded atoms. Such a bond is called a **single bond.** In many molecules, we find two atoms bonded by two pairs of electrons; this is a **double bond.** When two atoms are bonded by three pairs of electrons it is called a **triple bond.** All four electrons in a double bond and all six electrons in a triple bond are counted as valence electrons for each of the bonded atoms.

Probably the most abundant substance containing a triple bond is nitrogen, N_2. Its Lewis diagram may be thought of as the combination of two nitrogen atoms, each with three unpaired electrons:

$$:\overset{\cdot}{N}\cdot \; + \; \cdot\overset{\cdot}{N}: \; \longrightarrow \; :N::N: \quad or \quad :N{\equiv}N:$$

Counting the bonding electrons for both atoms, each nitrogen atom is satisfied with a full octet of electrons.

Experimental evidence supports the idea of **multiple bonds,** a general term that includes both double and triple bonds. A triple bond is stronger and the distance between bonded atoms is shorter than the same measurements for a double bond between the same atoms, and a double bond is shorter and stronger than a single bond. Bond strength is measured as the energy required to break a bond. The triple bond in N_2 is among the strongest bonds known. This is one of the reasons why elemental nitrogen is so stable and unreactive in the earth's atmosphere.

11.6 ATOMS THAT ARE BONDED TO TWO OR MORE OTHER ATOMS

Using hydrogen and fluorine as examples, we have seen that two atoms that have a single unpaired valence electron are able to form a covalent bond by sharing those electrons. What if an atom has two unpaired valence electrons? Can it form two bonds with two different atoms? The answer is yes. In fact, that is how a water molecule is formed. A hydrogen atom forms a bond with one of the two unpaired valence electrons in an oxygen atom, and a second hydrogen atom does the same with the second unpaired oxygen electron:

$$H \cdot + \cdot \ddot{O} \cdot + \cdot H \longrightarrow H : \ddot{O} : H \quad or \quad H - \ddot{O} - H$$

This is just like a hydrogen atom forming a bond with another hydrogen atom or with a fluorine atom.

A nitrogen atom has five valence electrons, three of which are unpaired. It therefore forms bonds with three hydrogen atoms to produce a molecule of ammonia, NH_3. Carbon has four valence electrons, only two of which are unpaired. This would lead us to expect that a carbon atom can form bonds with only two hydrogen atoms. In fact, all four electrons form bonds. The compound produced is methane, CH_4, the principal component of the natural gas burned as fuel in many homes. The Lewis diagrams for ammonia and methane are

$$H - \overset{..}{\underset{\underset{H}{|}}{N}} - H \quad and \quad H - \overset{\overset{H}{|}}{\underset{\underset{H}{|}}{C}} - H$$

ammonia   methane

All expectations should be checked experimentally. That carbon forms only two bonds is an example of a logical prediction that, when tested in the laboratory, is *not* confirmed. And we can be thankful for that! All organic life on earth has as its basis the ability of the carbon atom to form four bonds.

Multiple bonds appear in polyatomic molecules too. Ethylene, C_2H_4, the structural unit of the plastic polyethylene, has a double bond between two carbon atoms. There is a triple bond between carbon atoms in acetylene, C_2H_2, the fuel used in a welder's torch. The carbon atom in carbon dioxide, CO_2, is double bonded to two oxygen atoms. The Lewis diagrams for these compounds are

$$\underset{H}{\overset{H}{\diagdown}} C = C \underset{H}{\overset{H}{\diagup}} \qquad H - C \equiv C - H \qquad \ddot{O} = C = \ddot{O}$$

ethylene   acetylene   carbon dioxide

Notice that if both lone pairs and bonding electron pairs are counted, the octet rule is satisfied for all atoms except hydrogen in all Lewis diagrams in this section. Hydrogen, as usual, duplicates the two-electron count of the noble gas helium.

✔ QUICK CHECK 11.3

a) Is it possible, under the octet rule, for a single atom to be bonded by double bonds to each of three other atoms? Explain your answer.

b) What is the maximum number of atoms that can be bonded to the same atom and have that central atom conform to the octet rule? Explain.

c) Can a chlorine atom be bonded to two other atoms? Explain.

11.7 EXCEPTIONS TO THE OCTET RULE

Not all substances "obey" the octet rule. Two common oxides of nitrogen, NO and NO_2, have an odd number of electrons. It is therefore impossible to write Lewis diagrams for these compounds in which each atom is surrounded by eight electrons. Phosphorus pentafluoride, PF_5, places five electron-pair bonds around the phosphorus atom, and six pairs surround sulfur in SF_6.

Certain molecules whose Lewis diagrams obey the octet rule do not have the properties that would be predicted. Oxygen, O_2, was not used to introduce the double bond in Section 11.5 for that reason. On paper, O_2 appears to have an ideal double bond:

$$\ddot{\text{O}}{=}\ddot{\text{O}}$$

But liquid oxygen is *paramagnetic,* meaning that it is attracted by a magnetic field. (See Fig. 11.7. It shows liquid oxygen being held in place between the poles of a

Figure 11.7

A physical property of liquid oxygen. When liquid oxygen is poured into the field of a strong magnet, some of it is trapped and held between the poles of the magnet. All the oxygen you see is liquid. It looks just like water, and, outside the magnetic field, it behaves like water. But unlike water, the portion that is between the magnetic poles does not fall to the ground. This paramagnetism is due to the unpaired electrons in the oxygen molecule.

magnet.) This is characteristic of molecules that have unpaired electrons in their structures. This might suggest a Lewis diagram that has each oxygen surrounded by seven electrons:

$$:\ddot{O}—\ddot{O}:$$

But this conflicts with other evidence that the oxygen atoms are connected by something other than a single bond. In essence, it is impossible to write a single Lewis diagram that satisfactorily explains all of the properties of molecular oxygen.

Two other substances for which satisfactory octet rule diagrams can be drawn, but are contradicted experimentally, are the fluorides of beryllium and boron. We might even expect BeF_2 and BF_3 to be ionic compounds, but laboratory evidence strongly supports covalent structures having the Lewis diagrams:

$$:\ddot{F}—Be—\ddot{F}: \qquad \begin{array}{c} :\ddot{F} \quad \ddot{F}: \\ \diagdown \diagup \\ B \\ | \\ :\ddot{F}: \end{array}$$

In these compounds the beryllium and boron atoms are surrounded by two and three pairs of electrons, respectively, rather than four.

> Oxygen and the fluorides of beryllium and boron are additional examples showing that all predictions must be confirmed experimentally. A specific step in the progress of chemistry begins and ends in the laboratory. But each ending is a new beginning, leading to new predictions and new experiments to confirm them. Thus, the progress continues.

CHAPTER 11 IN REVIEW

11.1 Monatomic Ions with Noble Gas Electron Configurations

11A Identify the monatomic ions that are isoelectronic with a given noble gas atom, and write the electron configuration of those ions.

11.2 Ionic Bonds

11.3 Covalent Bonds

11B Distinguish between ionic and covalent bonds.

11.4 Polar and Nonpolar Covalent Bonds

11C Distinguish between polar and nonpolar covalent bonds.

11D Given the electronegativities of all elements involved, rank bonds in order of increasing or decreasing polarity.

11E Given the electronegativities of two elements, classify the bond between them as nonpolar covalent, polar covalent, or primarily ionic. If the bond is polar, state which end is positive and which end is negative.

11.5 Multiple Bonds

11.6 Atoms That Are Bonded to Two or More Other Atoms

11.7 Exceptions to the Octet Rule

CHAPTER 11 KEY WORDS AND MATCHING SETS

Set 1

_____ A covalent bond that has a symmetrical distribution of electrical charge.

_____ Represents an unpaired and unshared electron in a Lewis diagram.

_____ The kind of bond that is present between two atoms that have an electronegativity difference of 0.7.

_____ Rigid geometric arrangement of ions and/or molecules.

_____ Attractions between atoms, ions, or molecules.

_____ Represents a bonding electron pair in a Lewis diagram.

_____ Pure substance that is made up of tiny particles that are electrically neutral.

_____ Two valence electrons in an atom that do not bond the atom to another atom in the same molecule.

_____ A bond in which the electronegativity difference between bonded atoms is 2.3.

_____ One "explanation" of what happens to orbitals of two atoms when they form a covalent bond.

_____ A pure substance that is made up of ions that are held in a rigid crystal structure.

1. Lone pair
2. Crystal
3. Overlap
4. Chemical bonds
5. Nonpolar bond
6. Dash
7. Molecular compound
8. Dot
9. Ionic compound
10. Polar bond
11. Ionic bond

Set 2

_____ The tendency of atoms to form chemical bonds by surrounding themselves with eight valence electrons.

_____ Attractive force between atoms that share electrons in a molecule.

_____ A covalent bond in which two atoms share two pairs of electrons.

_____ A term for subatomic particles that appear to be concentrated in the region between two covalently bonded atoms.

_____ Smallest unit particle of a substance.

_____ A measure of the strength by which atoms of an element attract shared electron pairs in a single covalent bond.

_____ A covalent bond in which two atoms share six electrons.

_____ Sketch that shows bonding electrons and arrangement of atoms in a molecule.

_____ A term that describes double and triple bonds without distinguishing between them.

_____ A substance that is attracted to a magnetic field.

1. Molecule
2. Double bond
3. Triple bond
4. Paramagnetic
5. Electron cloud
6. Octet rule
7. Multiple bond
8. Covalent bond
9. Lewis diagram
10. Electronegativity

STUDY HINTS AND PITFALLS TO AVOID

In Chapter 6 you learned that binary compounds made up of two nonmetals are molecular and binary compounds made up of a metal and a nonmetal are ionic. In this chapter you learned why.

Metal atoms have one, two, or three valence electrons. Monatomic cations form when metal atoms lose these electrons, usually reaching an outer electron configuration that is isoelectronic with a noble gas. Nonmetal atoms in Groups 5A–7A (15–17) have five, six, or seven valence electrons. These atoms form monatomic anions by gaining enough electrons to reach an electron configuration that matches that of a noble gas. These oppositely charged ions form ionic bonds between themselves, yielding the crystalline structure that is typical of an ionic compound.

Nonmetal atoms have too many electrons to lose them all and become cations, so both an anion and a cation cannot come from two nonmetal atoms. By sharing electrons, however, they form covalent bonds in which each atom reaches a noble gas structure. The compounds formed in this way are molecular.

Note that hydrogen is a nonmetal that has only one valence electron rather than four, five, six, or seven. Nevertheless, a hydrogen atom reaches the noble gas structure of helium with one additional electron. Hydrogen therefore forms either ionic or covalent bonds in the same way as other nonmetals.

You are not expected to memorize the electronegativities of the elements in Figure 11.5. However, you should know that electronegativities are higher at the top of any column and at the right of any row in the periodic table. From this you can often predict which of two bonds is more polar.

CHAPTER 11 QUESTIONS AND PROBLEMS

SECTION 11.1

1) Referring only to a periodic table, identify those fourth period elements that form monatomic ions that are isoelectronic with a noble gas atom. Write the symbol for each such ion (example: Mg^{2+} in the second period).

2) Write the electron configuration for each fourth period monatomic ion identified in Question 1. Also identify the noble gas atoms having the same configuration.

___

3) Identify two monatomic cations that are isoelectronic with argon.

4) Identify by symbol two monatomic anions that are isoelectronic with neon.

___

5) Write the symbols of two ions that are isoelectronic with a calcium ion.

6) Write the symbols of two ions that are isoelectronic with a bromide ion.

___

7)* A monatomic ion with a 1+ charge has the electron configuration $1s^2 2s^2 2p^6 3s^2 3p^6$. (a) What noble gas atom has the same electron configuration? (b) What is the monatomic ion with a 1+ charge that has this configuration? (c) Write the symbol of an ion with a 2− charge that is isoelectronic with the two above species.

8)* If the monatomic ions in Question 7 (b) and (c) combine to form a compound, what is the formula of that compound?

SECTION 11.2

9) Using Lewis symbols, show how ionic bonds are formed by atoms of sulfur and potassium, leading to the correct formula of potassium sulfide.

10) Magnesium nitride is an ionic compound. Sketch the transfer of electrons from magnesium atoms to nitrogen atoms that accounts for the chemical formula of magnesium nitride.

___

11) When magnesium and chlorine react to form an ionic compound, why are there two chlorine atoms for each magnesium atom?

12) When magnesium and sulfur react to form an ionic compound, why is there only one sulfur atom for each magnesium atom instead of two?

SECTION 11.3

13) Distinguish between ionic bonds and covalent bonds.

14) Explain why covalent bonds are called electron-sharing bonds. How is this commonly related to the electron configuration of a noble gas?

___

15) Considering bonds between the following pairs of elements, which are most apt to be ionic and which are most apt to be covalent: sodium and sulfur; fluorine and chlorine; oxygen and sulfur? Explain your choice in each case.

16) Compare the bond between sodium and bromine in sodium bromide with the bond between two bromine atoms in elemental bromine. Which bond is ionic, and which bond is covalent? Describe how each bond is formed.

17) Show how a covalent bond forms between an atom of chlorine and fluorine, yielding a molecule of chlorine fluoride.

18) Sketch the formation of two covalent bonds by an atom of oxygen in making a molecule of water.

19)* The bond between two metal atoms is neither ionic nor covalent. Explain, according to the octet rule, why this is so.

20)* Explain why the bond between two nonmetal atoms is more apt to be covalent than ionic, and why the bond between a metal atom and a nonmetal atom is more apt to be ionic than covalent.

SECTION 11.4

21) Distinguish between *polar* covalent bonds and *nonpolar* covalent bonds.

22) There are six diatomic (two-atom) molecules that can form between the elements hydrogen, bromine, and chlorine: H_2, Br_2, Cl_2, HBr, HCl, and BrCl. Will the bonds in any of these molecules be completely nonpolar? If so, which? If so, how will the electron cloud(s) of the nonpolar bond(s) differ from the other electron clouds?

23) List the following bonds in order of increasing polarity: K—Br; S—O; N—Cl; Li—F; C—C. Based on Figure 11.5, classify each bond as (a) nonpolar, (b) polar covalent, or (c) primarily ionic.

24) Arrange the following bonds in order of decreasing polarity: F—I; I—Br; I—I; Cl—Br; F—F.

25) For each bond in Question 23, identify the atom that acts as the negative pole.

26) For each bond in Question 24, identify the positive pole, if any.

27) What is electronegativity? Why are the noble gases usually not included in the electronegativity table?

28) What electronegativity trend appears from left to right in a period in the periodic table? From top to bottom in any group?

29)* Referring only to a full periodic table (specifically *not* Figure 11.5), arrange the following bonds in order of decreasing polarity: Ca—O; Al—O; Na—O; S—O; K—O. If any two bonds cannot be positively placed relative to each other, explain why.

30)* You may look at a full periodic table in answering this question, but not at Figure 11.5. Which bond, F—P or O—S, is less polar? Explain your answer.

SECTION 11.5

31) Distinguish among single, double, triple, and multiple bonds.

32) Double and triple bonds conform to the octet rule. How about quadruple (4) and quintuple (5) bonds? Why or why not? If either a quadruple or quintuple bond could form, suggest two atoms between which it might be found. (No such bond is known, but theoretically one that obeys the octet rule is possible.)

SECTION 11.6

33) What is the maximum number of atoms a central atom can bond to and still conform to the octet rule? What is the minimum number?

34) Draw Lewis diagrams of central atoms showing all possible combinations of single, double, and triple bonds that can form around an atom that conforms to the octet rule. It is not necessary to show the atom to which the central atom is bonded—just the bonds.

35) An atom, X, is bonded to another atom by a double bond. What is the largest number of *additional atoms* to which X may be bonded and still conform to the octet rule? What is the minimum number? Justify your answer.

36)* Atom A is bonded to atom B by a triple bond. What is the largest number of *additional atoms* that can be bonded to A if A is to conform to the octet rule? What is the minimum number of *atoms* that can bond to B if B is to conform to the octet rule? (Caution: Italics indicate that this could be considered a ''trick'' question.) Draw the Lewis diagram of a molecule that illustrates both answers, substituting symbols of real elements for A and B. Include, if you can, the symbols of (a) real element(s) that would make your diagram one for a whole molecule.

SECTION 11.7

37) Given a chemical formula, there is a simple requirement that, if met, enables you to tell at a glance whether or not a Lewis diagram that satisfies the octet rule is possible for that substance. What is that requirement?

38) Consider the compounds N_2O, NO, N_2O_3, NO_2, N_2O_4, N_2O_5, NOCl, NO_2Cl, and NF_3. Without actually drawing diagrams, use the criterion from Question 37 to separate these compounds into two groups, one for which Lewis diagrams are possible, and one for which they are not.

39)* How is it possible for central atoms to be surrounded by five or six bonding electron pairs when there are only four valence electrons from the *s* and *p* orbitals of any atom?

40)* There are two iodides of arsenic (Z = 33), AsI_3 and AsI_5. A Lewis diagram that conforms to the octet rule can be drawn for one of these, but not the other. Draw the diagram that is possible, using two dots rather than dashes to represent any lone pairs. From that diagram, see if you can figure out how the second molecule might be formed by covalent bonds, even though it violates the octet rule.

---

41)* From Period 3 down in the periodic table, binary compounds formed between an element in Group 3A (13) and an element in Group 7A (17) are ionic. BF_3 behaves more as a molecular compound than as an ionic compound. Suggest a reason for this. Why is AlF_3 more ionic than BF_3?

42)* Considering your answers to Question 41, what ionic/molecular trend would you expect in BCl_3, BBr_3, and BI_3? Would the compounds be increasingly ionic or increasingly molecular? Explain.

GENERAL QUESTIONS

43) Distinguish precisely and in scientific terms the differences between items in each of the following groups:
a) Ionic compound, molecular compound.
b) Ionic bond, covalent bond.
c) Lone pair, bonding pair (of electrons).
d) Nonpolar bond, polar bond.
e) Single, double, triple, multiple bonds.

44) Classify each of the following statements as true or false:
a) A single bond between carbon and nitrogen is polar covalent.
b) A bond between phosphorus and sulfur will be less polar than a bond between phosphorus and chlorine.
c) The electronegativity of calcium is less than the electronegativity of aluminum.
d) Strontium (Z = 38) ions, Sr^{2+}, are isoelectronic with bromide ions, Br^-.
e) The monatomic ion formed by selenium (Z = 34) is expected to be isoelectronic with a noble gas atom.
f) Most elements in Groups 4A (14) do not normally form monatomic ions.

g) Multiple bonds can form only between atoms of the same element.
h) If an atom is triple bonded to another atom, it may still form a bond with one additional atom.
i) An atom that conforms to the octet rule can bond to no more than three other atoms if one bond is a double bond.
j) Only valence electrons can participate in forming bonds.
k) An atom can be surrounded by more than four pairs of electrons.

45). What is the electron configuration of the hydrogen ion, H^+? Explain your answer to this question.

46) Identify the pairs among the following that are not isoelectronic: (a) Ne and Na^+, (b) S^{2-} and Cl^-, (c) Mg^{2+} and Ar, (d) K^+ and S^{2-}; (e) Ba^{2+} and Te^{2-} (Z = 52).

47) Which bond formed between atoms of two elements whose atomic numbers are given would be expected to be the most ionic: (a) 8 and 16, (b) 11 and 35, (c) 17 and 20, (d) 3 and 53, (e) 9 and 55?

48) Which orbitals of each atom overlap in forming a bond between bromine and oxygen?

49) Do ionic bonds appear in molecular compounds? Do covalent bonds appear in ionic compounds?

50) Is there any such thing as a completely nonpolar bond? If yes, give an example. Is there any such thing as a completely ionic bond? If yes, give an example. Explain both answers.

51) If you did not have an electronegativity table, could you predict the relative electronegativities of elements whose positions are [diagram A, B] in the periodic table? What about elements whose positions are [diagram X, Y]? In both cases, explain why or why not.

MATCHING SET ANSWERS

QUICK CHECK ANSWERS

11.1 S^{2-} and Ba^{2+}. The other ions have d sublevel electrons not present in the noble gas atom with the next lower atomic number.

11.2 a, b, and d: True. c: Covalent bonds are common between atoms of two nonmetals.

11.3 a) No. Three double bonds would be 6 electron pairs, placing 12 electrons around the central atom.

b) Four atoms can be bonded by single bonds to the same central atom. At 2 electrons per bond, there would be 8 electrons around the atom, a full octet.

c) A chlorine atom has only one unpaired electron, so it would appear, from the information in this chapter, that it could form only one bond. You will learn in the next chapter that one atom can contribute *both* electrons in forming a bond. This makes it possible for chlorine to form single bonds with four other atoms.

The Structure and Shape of Molecules

12

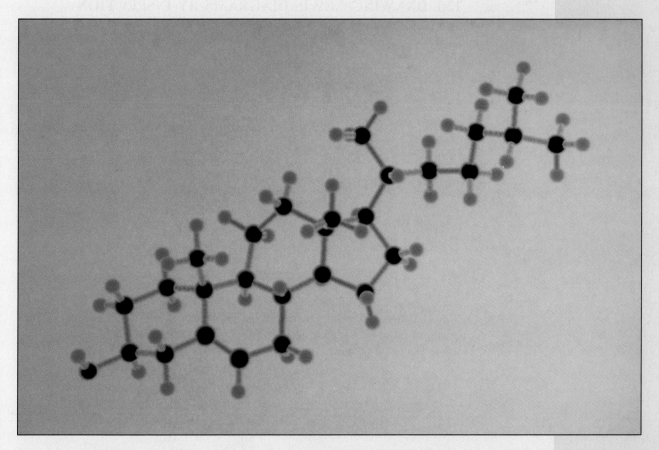

This picture of a cholesterol molecule was drawn on a computer. It suggests some of the ways atoms are arranged in complex organic molecules. Actually, compared to many biochemical molecules, cholesterol is not very complex.

You learned in Chapter 11 that atoms in molecular compounds and polyatomic ions are held together by covalent bonds. Lewis diagrams showed, in two dimensions, how the atoms are connected. However, Lewis diagrams do not show how the atoms are *arranged* in three dimensions—the actual shape of the molecule. In this chapter we study that arrangement, the structure and shape of molecules. It begins with the Lewis diagram. And we begin by learning how to draw Lewis diagrams.

12.1 DRAWING LEWIS DIAGRAMS BY INSPECTION

PG 12A Draw the Lewis diagram for any molecule or polyatomic ion made up of representative elements (Sections 12.1 and 12.2).

Lewis diagrams can often be drawn by inspection if all atoms have the electron configuration of a noble gas. Each atom except hydrogen has an octet of electrons; hydrogen has two electrons, matching helium. In Section 11.3 you saw how the single $1s$ electron of hydrogen forms a bond with the single unpaired $2p$ electron of fluorine to form an HF molecule:

$$H\cdot \ + \ \cdot \ddot{\underset{\cdot\cdot}{F}}: \ \longrightarrow \ H:\ddot{\underset{\cdot\cdot}{F}}: \quad or \quad H\!-\!\ddot{\underset{\cdot\cdot}{F}}:$$

In general, the unpaired valence electrons from two different atoms are capable of pairing to form covalent bonds. They do so until all atoms reach the electron configuration of a noble gas. The Lewis symbols for oxygen, nitrogen, and carbon are

$$\cdot \ddot{\underset{\cdot}{O}}: \qquad \cdot \ddot{N} \cdot \qquad \cdot \dot{C} \cdot$$

Oxygen has two unpaired electrons, so it bonds with two hydrogen atoms in forming a water molecule, H_2O. Similarly, nitrogen's three unpaired electrons can bond to three hydrogen atoms to form an ammonia molecule, NH_3. Carbon atoms have four valence electrons, all of which bond to hydrogen to form a methane molecule, CH_4.* The Lewis diagrams for these compounds are

$$
\begin{array}{ccc}
H\!-\!\ddot{O}: & H\!-\!\ddot{N}\!-\!H & H\!-\!\overset{\displaystyle H}{\underset{\displaystyle H}{\overset{|}{\underset{|}{C}}}}\!-\!H \\[2mm]
\,|\; & \,|\; & \\
H & H & \\[2mm]
H_2O & NH_3 & CH_4
\end{array}
$$

EXAMPLE 12.1 _____

Draw Lewis diagrams for carbon tetrachloride, CCl_4, and phosphorus tribromide, PBr_3.

A carbon atom has four valence electrons. It requires four more electrons to complete its octet and gets them by forming four bonds. Write the Lewis symbol for chlorine. From that symbol state the number of electrons each chlorine atom must gain to complete its octet.

_____   _____

*A carbon atom, with its $1s^2 2s^2 2p^2$ electron configuration, actually has only two unpaired electrons, the two $2p$ electrons. It is a fact, however, that carbon atoms form covalent bonds with four hydrogen atoms. One explanation for this involves "hybridized" orbitals, a topic beyond the scope of this text.

$$\cdot \ddot{\underset{\cdot}{Cl}} :$$ The atom has seven valence electrons and needs one to reach eight.

Assuming that new bonds will be formed when each atom contributes one electron to the bond, draw the Lewis diagram of CCl_4.

———   ———

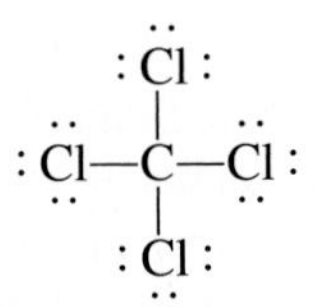

$$: \ddot{\underset{..}{Cl}} - \underset{\underset{: \ddot{\underset{..}{Cl}} :}{|}}{\overset{\overset{: \ddot{Cl} :}{|}}{C}} - \ddot{\underset{..}{Cl}} :$$ Each of four electrons from a carbon atom forms a bond with the unpaired electron from a separate chlorine atom. This is just like the Lewis diagram for methane, CH_4.

Now draw the Lewis symbols of a phosphorus atom and a bromine atom. For each, state the number of electrons required to complete the octet.

———   ———

$$\cdot \overset{..}{\underset{\cdot}{P}} \cdot$$ Three electrons required. $$\cdot \ddot{\underset{..}{Br}} :$$ One electron required.

Finally, the Lewis diagram for PBr_3.

———   ———

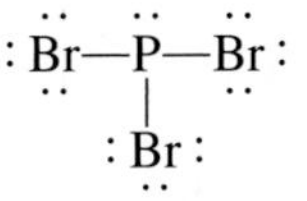

$$: \ddot{\underset{..}{Br}} - \underset{\underset{: \ddot{\underset{..}{Br}} :}{|}}{P} - \ddot{\underset{..}{Br}} :$$

Notice the similarity between this diagram and that for NH_3. The central elements are both in Group 5A (15) and have five valence electrons. The other element in each case requires one additional electron to reach a noble gas configuration. The inspection methods for drawing the diagrams are the same.

In all examples so far covalent bonds have been formed when each atom contributes one electron to the bonding pair. This is not always the case. Many bonds are formed where one atom contributes both electrons and the other atom offers only an empty orbital. This is called a **coordinate covalent bond.** To illustrate, an ammonium ion is produced when a hydrogen ion is bonded to the unshared electron pair of the nitrogen atom in an ammonia molecule:

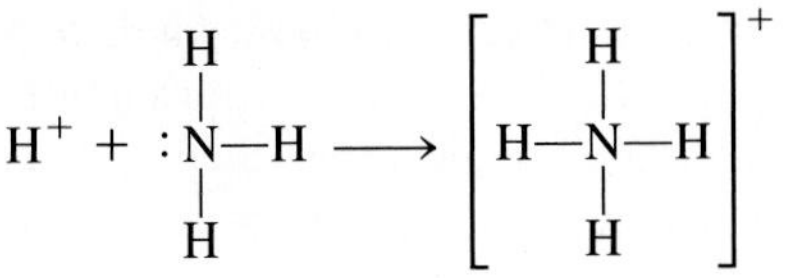

$$H^+ + : \underset{\underset{H}{|}}{\overset{\overset{H}{|}}{N}} - H \longrightarrow \left[\underset{\underset{H}{|}}{\overset{\overset{H}{|}}{H - N}} - H \right]^+$$

ammonia   ammonium ion

The four bonds in an ammonium ion are identical. This shows that a coordinate covalent bond is the same as a bond formed by one electron from each atom.

Notice that in drawing the Lewis diagram of an ion, the diagram is enclosed in brackets and the charge is shown as a superscript.

12.2 DRAWING COMPLEX LEWIS DIAGRAMS

Lewis diagrams are not readily drawn for some of the more complex molecules and polyatomic ions. The procedure that follows may be used to sketch the diagram for any species that obeys the octet rule. Each step is illustrated by drawing the Lewis diagram for the hydrogen carbonate ion, HCO_3^-.

LEARN IT NOW   Matching a representative element with its group number in the periodic table is a quick way to count the valence electrons in atoms of that element.

1) *Count the total number of valence electrons in the molecule or ion.*   Note that the number of valence electrons for a representative element is the same as its column number in the periodic table (or the final digit of the column number if you are using IUPAC* group numbers). If the species is an ion, the number of valence electrons must be adjusted to account for the charge on the ion. For each positive charge, subtract one electron; for each negative charge, add one electron.

    In HCO_3^- there is one valence electron from hydrogen, four from carbon, six from each oxygen, and one for the negative charge:

$$1 + 4 + 3(6) + 1 = 24$$

2) *Draw a tentative diagram for the molecule or ion, joining atoms by single bonds.* In some cases, only one arrangement of atoms is possible. In others, two or more structures are possible. Ultimately, chemical or physical evidence must be used to decide which structure is correct. A few general rules will help you to make diagrams that are most likely to be correct:
    a) A hydrogen atom always forms one bond. A carbon atom normally forms four bonds.
    b) When several carbon atoms appear in the same molecule, they are often bonded to each other. In some compounds they are arranged in a closed loop; however, we will avoid such cyclic compounds in this chapter.
    c) Make your diagram as symmetrical as possible. In particular, a compound or ion having two or more oxygen atoms and one atom of another nonmetal usually has the oxygen atoms arranged *around* the central nonmetal atom. If hydrogen is also present, it is usually bonded to an oxygen atom, which is then bonded to the nonmetal: H—O—X, where X is the nonmetal.

    HCO_3^- is described by Step 2c. The three oxygen atoms are placed around the carbon atom and the hydrogen atom is bonded to the oxygen:

$$H—O—C—O$$
$$\mid$$
$$O$$

3) *Determine the number of electrons available for lone pairs.*   Each bond in the tentative diagram accounts for two of the electrons in the final diagram. Therefore, subtract from the total number of electrons (Step 1) twice the number of bonds in the tentative diagram. The result is the number of electrons that are available for lone pairs.

    There are four bonds in the tentative HCO_3^- diagram. They account for $2 \times 4 = 8$ of the 24 valence electrons established in Step 1. $24 - 8 = 16$ electrons are left for lone pairs.

*International Union of Pure and Applied Chemistry.

4) *Place electron dots around each symbol except hydrogen, so the total number of electrons for each atom is eight. Stop when the supply of electrons determined in Step 3 has been used up.* Begin by filling in the valence orbitals of the outer atoms. If there are not enough electrons to fill all of the valence orbitals, leave some empty on the central nonmetal. If all orbitals are filled, the diagram is completed.

Lone pairs of electrons are added, first to the outer oxygen atoms, and then to the oxygen between H and C, and finally to the carbon atom, up to the total of the 16 electrons available (Step 3). The supply runs out at the third oxygen, leaving carbon with an incomplete octet—only three electron pairs.

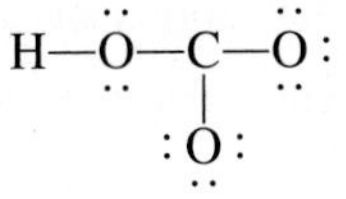

$$H-\overset{..}{\underset{..}{O}}-C-\overset{..}{\underset{..}{O}}:$$
$$\underset{\underset{..}{:\overset{}{O}:}}{|}$$

5) *If you did not have enough electrons to complete the octet for the central atom, move one or more lone pairs from an outer atom to form a double or triple bond with the central atom until all atoms have an octet.*

A lone pair may be moved from either or both of the outer oxygens to form a double bond with carbon. (See discussion below.) The Lewis diagram is now complete.

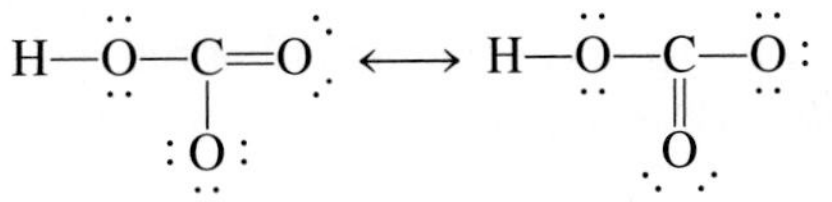

$$H-\overset{..}{\underset{..}{O}}-C=\overset{..}{\underset{..}{O}} \longleftrightarrow H-\overset{..}{\underset{..}{O}}-C-\overset{..}{\underset{..}{O}}:$$

The electron pair that was moved to form the double bond could have come from either outlying oxygen atom. The structure formed when a lone pair is moved from one or more identical outer atoms is a *resonance hybrid*. On paper, the bonds between the central carbon and the outlying oxygens look different. In fact, they are identical. Moreover, their strengths and lengths are between those found in true single and double bonds connecting the same two atoms.

It is customary to place a two-headed arrow between resonance diagrams, as shown above. However, further discussion of resonance is beyond the scope of an introductory text. Therefore, when we encounter a resonance structure, we will simply show one of the alternative diagrams, as given below.

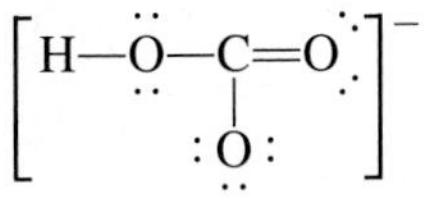

$$\left[H-\overset{..}{\underset{..}{O}}-C=\overset{..}{\underset{..}{O}}\right]^{-}$$

The square brackets enclose the diagram because HCO_3^- is an ion.

To summarize, the steps in drawing a complex Lewis diagram are as follows:

SUMMARY

1) Count the total number of valence electrons. Adjust for charge on ions.
2) Draw a tentative diagram. Join atoms by single bonds.
3) Subtract twice the number of single bonds (Step 2) from the total number of electrons available (Step 1) to get the electrons available for lone pairs.

4) Starting at the outer atoms, distribute the available lone pair electrons (Step 3) to complete the octet around each atom except hydrogen. If all octets are completed, the diagram is finished.

5) If there are not enough electrons to complete all octets, move one or more lone pairs from an outer atom to form a double or triple bond with the central atom until all atoms have an octet.

EXAMPLE 12.2

Write Lewis diagrams for the ClF molecule and the ClO^- ion.

Step 1 is to count the valence electrons for each species.

ClF: 7 (Cl) + 7 (F) = 14. ClO^-: 7 (Cl) + 6 (O) + 1 (charge) = 14

Chlorine and fluorine, both in Group 7A (17), have seven valence electrons. Oxygen in Group 6A (16) has six. In ClO^- there is one additional electron to account for the 1− charge on the ion.

Step 2 is to draw the tentative diagram. With two atoms, the only possible diagram has them bonded to each other:

Cl—F   and   Cl—O

Step 3 is to find out how many electrons are available for lone pairs. For each species the bond accounts for two of the total number of electrons. How many are left?

For each species 14 (total) − 2 × 1 (twice the number of bonds) = 12

Step 4 is to distribute electron pairs around each atom to complete its octet.

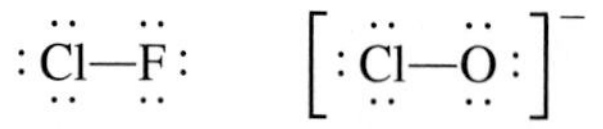

$: \overset{..}{\underset{..}{Cl}} - \overset{..}{\underset{..}{F}} :$ $\left[: \overset{..}{\underset{..}{Cl}} - \overset{..}{\underset{..}{O}} : \right]^-$

In both cases there is just the right number of electrons to complete the octet for both atoms. The diagrams are therefore complete. In the case of ClO^- the diagram is enclosed in brackets because it is an ion.

Did you notice how each step was the same for the two species in Example 12.2? This is because any two species that have (1) the same number of atoms and (2) the same number of valence electrons also have similar Lewis diagrams, whether they are molecules or polyatomic ions.

EXAMPLE 12.3

Draw the Lewis diagram for the sulfite ion, SO_3^{2-}.

Let's go for Steps 1 and 2: Get the total valence electron count and propose a tentative diagram.

$$6\ (S) + 3 \times 6\ (O) + 2\ (charge) = 26 \qquad O{-}S{-}O$$

The ion has a 2− charge, so two electrons must be added to the valence electrons of the atoms themselves. The diagram has the three oxygen atoms distributed around sulfur as the central atom, conforming to item 2c in the earlier detailed procedure.

How many electrons are available to distribute as lone pairs?

$$26\ (total) - 2 \times 3\ (twice\ the\ number\ of\ bonds) = 20$$

Distribute the 20 available electrons as ten electron pairs around the tentative diagram. If there are enough for all atoms, add the "finishing touch" to complete the diagram. If there are not enough, there will be an additional step.

There are enough electrons; the diagram is complete. The "finishing touch" is to surround the diagram with brackets and indicate the charge.

EXAMPLE 12.4

Draw the Lewis diagram for SO_2.

Complete the procedure through the first three steps, the tentative diagram and determination of the number of electrons to be distributed as electron pairs.

Total electrons: $6\ (S) + 2 \times 6\ (O) = 18$

Tentative diagram: $O{-}S{-}O$

Lone pair electrons: $18\ (total) - 2 \times 2\ (twice\ the\ number\ of\ bonds) = 14$

Now distribute the lone pair electrons to fill the octet of each atom. Complete the outer atoms first.

This time there are not enough electrons to complete the octet on the central atom.

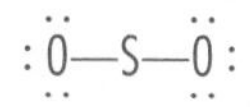

$$: \overset{\cdot\cdot}{\underset{\cdot\cdot}{O}} - S - \overset{\cdot\cdot}{\underset{\cdot\cdot}{O}} :$$

When there are not enough electrons to complete all octets, one or more of the intended lone pairs must be used to form a multiple bond. The central sulfur atom is short by two electrons, so moving one pair from either oxygen will satisfy sulfur while still being counted by oxygen. Complete the diagram (Step 5).

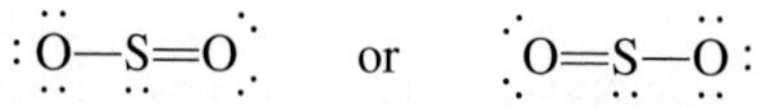

$$: \overset{\cdot\cdot}{\underset{\cdot\cdot}{O}} - S = \overset{\cdot\cdot}{O} : \quad \text{or} \quad \overset{\cdot\cdot}{O} = S - \overset{\cdot\cdot}{\underset{\cdot\cdot}{O}} :$$

The structure is a resonance hybrid. Either diagram is acceptable.

The rules we are following are readily applied to simple organic* molecules, which always contain carbon atoms, usually include hydrogen atoms, and may contain atoms of other elements, notably oxygen. If oxygen is present in an organic compound, it usually forms two bonds. If you remember that carbon forms four bonds and hydrogen forms one, and that two or more carbon atoms often bond to each other (Rules 2a and 2b at the beginning of this section), your tentative diagrams are most apt to be correct.

EXAMPLE 12.5 _____

Write the Lewis diagram for propane, C_3H_8.

Complete the example without guiding questions.

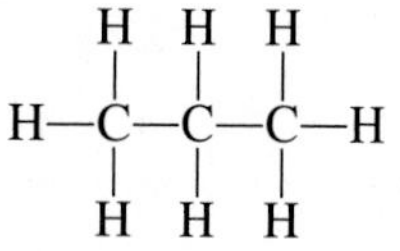

$$\begin{array}{ccc} H & H & H \\ | & | & | \\ H-C-C-C-H \\ | & | & | \\ H & H & H \end{array}$$

All 20 of the valence electrons—12 from three carbon atoms plus 8 from eight hydrogen atoms—are needed for the ten single bonds in the diagram. There are no lone pairs.

EXAMPLE 12.6 _____

Draw the Lewis diagram for acetylene, C_2H_2.

Take the first three steps this time—through determining the number of electrons available for forming lone pairs.

Total electrons: 2×4 (C) $+ 2 \times 1 = 10$

Tentative diagram: H—C—C—H

Lone pair electrons: 10 (total) $- 3 \times 2$ (twice the number of bonds) $= 4$

---

*Organic chemistry is the chemistry of compounds containing carbon, other than certain ''inorganic'' carbon compounds such as carbonates, CO and CO_2. See optional Section 12.7.

Now distribute the four lone pair electrons to complete the octets of both carbon atoms.

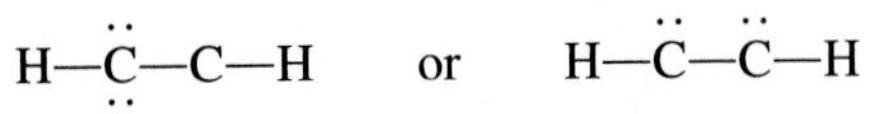

H—C̈—C—H   or   H—C̈—C̈—H

There is a double shortage of electrons this time. In the first diagram the second carbon is missing two electron pairs; in the second diagram both carbons are lacking one electron pair. The remedy is still the same, though. Move the available lone pairs into multiple bonding positions that will result in both carbons completing their octets.

H—C≡C—H

EXAMPLE 12.7

Draw a Lewis diagram for C_2H_6O.

Try to take this one all the way. Remember how many bonds carbon, hydrogen, and oxygen usually form. Also, bond the carbon atoms to each other.

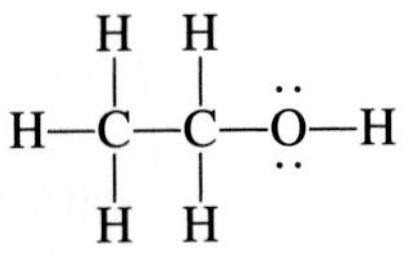

$$\begin{array}{cc} H & H \\ | & | \\ H—C—C—\ddot{O}—H \\ | & | \\ H & H \end{array}$$

If oxygen forms two bonds, it must form single bonds to two atoms or a double bond to one atom. If you attempt a double bond with this oxygen there will not be enough room for all the hydrogen atoms in the molecule.

By our insisting that the carbons be bonded to each other, the oxygen had to go between a carbon and a hydrogen. The total electron count is 20, and 16 of those make up the bonds in the molecule. The 4 that remain are lone pairs for oxygen.

The compound in Example 12.7 is ethyl alcohol. If we had not insisted that the carbon atoms be bonded to each other, the oxygen atom might have been placed between them:

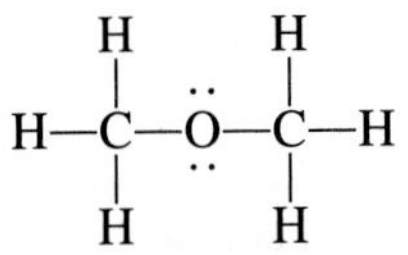

$$\begin{array}{ccc} H & & H \\ | & & | \\ H—C—\ddot{O}—C—H \\ | & & | \\ H & & H \end{array}$$

This is another well-known compound, dimethyl ether. Note that both compounds have the same molecular formula, C_2H_6O, but different structures. Compounds that have the same molecular formulas but different structures are **isomers** of each other. Isomers are distinctly different substances, each with its own unique set of properties. For example, ethyl alcohol boils at 78°C, while dimethyl ether boils at −25°C.

12.3 ELECTRON PAIR REPULSION: ELECTRON PAIR GEOMETRY

PG 12B Describe the electron pair geometry when a central atom is surrounded by two, three, or four electron pairs.

The shape of a molecule plays an important role in determining the physical and chemical properties of a substance. We will examine this role in later chapters in this book. In order to understand better and predict the shape/property relationship, we should know what is responsible for molecular shape. That is the purpose of this section and the next. *Discussion in these sections is limited to molecules having only single bonds.* Molecules with multiple bonds are considered in Section 12.5.

No single theory or model yet developed succeeds in explaining all the molecular shapes that have been observed in the laboratory. A theory that explains one group of molecules cannot explain another group. Each theory has its advantages—and limitations. Chemists therefore use them all within the areas to which they apply, fully recognizing that there is still much to learn about how atoms are assembled in molecules.

In this text we will explore one of the models used to explain **molecular geometry,** the more precise term used to describe the shape of a molecule. It is called the **valence shell electron pair repulsion theory, VSEPR.** VSEPR applies primarily to substances in which a second-period atom is bonded to two, three, and four other atoms. You may wonder why so much attention is focused on so few elements. The answer is that the second period includes carbon, nitrogen, and oxygen. Carbon alone is present in about 90% of all known compounds, and a large percentage of those include oxygen and/or nitrogen also. These elements warrant this kind of attention.

The basic idea of VSEPR is that the electron pairs we draw in Lewis diagrams repel each other in real molecules. Therefore they distribute themselves in positions

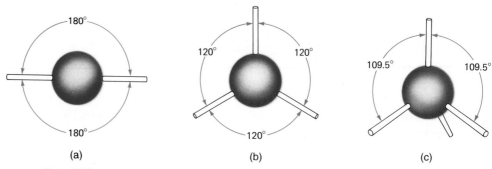

(a) (b) (c)

Figure 12.1

Electron pair geometry. "Tinker-toy" models show the arrangement of two, three, and four electron pairs (sticks) around a central atom (ball). (a) According to the electron pair repulsion principle, two sticks are as far from each other as possible when they are diametrically opposite each other. The geometry is linear, and the angle formed is 180°. (b) Three sticks are as far from each other as possible when equally spaced on a circumference of the ball. The sticks and the center of the ball are in the same plane and the angles are 120°. The geometry is trigonal (triangular) planar. (c) Four electron pairs are as far from each other as possible when arranged to form a tetrahedron. Each angle is 109.5°, which is sometimes called a "tetrahedral angle."

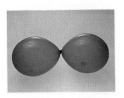

Figure 12.2
Structural pair geometries for two to four structural pairs. If one ties together several balloons of similar size and shape, they will naturally assume the geometries shown.

around the central atom that are as far away from each other as possible. These are the locations of lowest potential energy; they satisfy the "minimization of energy" tendency that we have noted is one driving force for natural change. This arrangement of electron pairs is called **electron pair geometry.** The electron pairs may be shared in a covalent bond, or they may be lone pairs; it makes no difference.

Earlier we drew Lewis diagrams in which carbon, nitrogen, or oxygen was the central atom. In all cases the central atom was surrounded by four pairs of electrons. In Section 11.7 you saw that beryllium and boron—also Period 2 elements—do not conform to the octet rule. The beryllium atom in BeF_2 is flanked by only two electron pairs; in BF_3, the boron atom has three electron pairs around it. Our question, then, is how do two, three, or four electron pairs distribute themselves around a central atom so they are as far apart as possible? This question is answered by identifying the "**electron pair angle,**" the angle formed by any two electron pairs and the central atom.

When electron pairs are as far apart as possible, all electron pair angles around the central atom are equal. The electron pair geometries that result from two, three, or four electron pairs are shown in Figure 12.1. The bond angles are derived by geometry.

Do we see these angles in our familiar, large-scale world? Indeed we do. We can duplicate these angles by simply tying together balloons of *similar size and shape.* See Figure 12.2. The balloons model identically sized and shaped electron orbitals. Once again, atom-sized properties are reproduced naturally on a larger scale.

The geometries are summarized below and in Table 12.1.

SUMMARY

1) If the central atom is surrounded by two electron pairs, the atom is on a line between the pairs. The electron pair geometry is **linear,** and the electron pair angle is 180°.
2) If the central atom is surrounded by three electron pairs, the atom is at the center of an equilateral triangle formed by the pairs. The geometry is **trigonal (triangular) planar,** and the electron pair angle is 120°.
3) If the central atom is surrounded by four electron pairs, the atom is at the center of a **tetrahedron** formed by the pairs. The geometry is **tetrahedral,** and the electron pair angle is 109.5°.

Table 12.1

Electron Pair and Molecular Geometries

Line	Electron Pairs	Bonded Atoms	Electron-Pair Geometry	Ball and Stick Model	Electron Pair and Bond Angle	Molecular Geometry	Lewis Diagram	Ball and Stick Model	Space Filling Model	Example	Actual Bond Angle
1	2	2	Linear		180°	Linear	A—B—A			BeF_2	180°
2	3	3	Trigonal (triangular) planar		120°	Trigonal (triangular) planar	A—B(—A)—A			BF_3	120°
3	4	4	Tetrahedral		109.5°	Tetrahedral	A—B—A			CH_4	109.5°
4	4	3	Tetrahedral		109.5°	Trigonal (triangular) pyramid or pyramidal	A—B̈—A			NH_3	107.5°
5	4	2	Tetrahedral		109.5°	Angular or bent	A—B̈—A or A—B̈—A			H_2O	104.5°

This summary and the caption to Figure 12.1 introduce the word *tetrahedron*. A tetrahedron is the simplest regular solid. A ''regular solid'' is a solid figure with identical faces. A cube is a regular solid that has six identical squares as its faces. A tetrahedron has four identical equilateral triangles for its faces, as shown in Figure 12.3. This geometric figure appears in all molecules in which carbon forms single bonds with four other atoms. Notice that a tetrahedron is a three-dimensional figure. It cannot be drawn exactly on the two-dimensional plane of a book page.

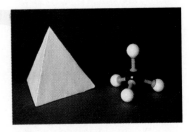

Figure 12.3

Tetrahedral models. The solid figure is a tetrahedron, the simplest regular solid. Its four faces are identical equilateral triangles. The model of methane, CH_4, has a tetrahedral structure. The carbon atom is in the middle of the tetrahedron, and a hydrogen atom is found at each of the four vertices.

12.4 MOLECULAR GEOMETRY

> **PG 12C** Given or having derived the Lewis diagram of a molecule or polyatomic ion in which a second-period central atom is surrounded by two, three, or four pairs of electrons, predict the molecular geometry around that atom.

Molecular geometry describes the shape of a molecule and the arrangement of *atoms* around a central atom. You might think of it as an ''atom geometry,'' in the same sense that the arrangement of electron pairs is the electron pair geometry. Thus, the **bond angle** is the angle between two bonds formed by the same central atom, as shown in Figure 12.4.

When all the electron pairs around a central atom are bonding pairs—when there are no lone pairs—the bond angles are the same as the electron pair angles. The molecular geometries are the same as the electron pair geometries described above. Also, the same terms are used to describe the shapes of the molecules. If the molecule contains one or two lone pairs, the bond angles are close to the electron pair angles predicted by the VSEPR theory, but slightly different. Different terms are needed to describe the shapes of these molecules.

We now describe the molecular geometries for all combinations of electron pairs and atoms that are connected to the central atom by single bonds. These descriptions are illustrated and summarized in Table 12.1. Line references are to line numbers in that table.

TWO ELECTRON PAIRS, TWO BONDED ATOMS

Two electron pairs, both bonding, yield the same electron pair and molecular geometries: linear (Line 1). A linear geometry has a 180° bond angle.

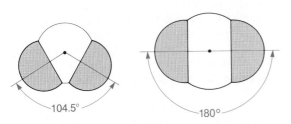

Figure 12.4

Bond angle. If an atom forms bonds with two other atoms, the angle between the bonds is the bond angle. In a water molecule the bonds form an angle of 104.5°. In a carbon dioxide molecule the bonds lie in a straight line. The bond angle is 180°.

THREE ELECTRON PAIRS, THREE BONDED ATOMS

Three electron pairs, all bonding, yield the same electron pair and molecular geometries: trigonal (triangular) planar (Line 2). Each bond angle is 120°.

FOUR ELECTRON PAIRS, FOUR BONDED ATOMS

Four electron pairs, all bonding, yield the same electron pair and molecular geometries: tetrahedral (Line 3). The tetrahedral methane molecule, CH_4, looks like a tall pyramid with a triangular base (Fig. 12.5a). Each bond angle is 109.5°—the tetrahedral angle.

FOUR ELECTRON PAIRS, THREE BONDED ATOMS

The four electron pairs retain their tetrahedral geometry, which is modified because only three of the electron pairs form bonds to other atoms (Line 4). The resulting shape is like a "squashed down" pyramid, called **trigonal (triangular) pyramidal** (Fig. 12.5b). The unshared electron pair is apparently drawn closer to

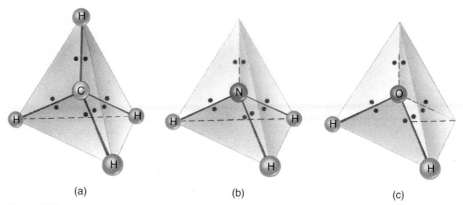

(a)   (b)   (c)

Figure 12.5

(a) Methane, CH_4, is a typical five-atom molecule. The four hydrogen atoms are at the corners of a tetrahedron, and the carbon atom is at its center. The molecule is three-dimensional. If the top hydrogen atom and the carbon atom are in the plane of the paper, the large hydrogen atom in the base is closer to you than the paper, and the other hydrogen atoms are behind the page. (b) Ammonia, NH_3, is a four-atom molecule having the shape of a pyramid with a triangular base. It is like the CH_4 molecule without the top hydrogen atom. The nitrogen atom is in the plane of the paper, the large hydrogen atom is in front of the paper, and the smaller hydrogen atoms are behind the page. Like the carbon atom in methane, the nitrogen atom in ammonia is surrounded by four electron pairs. (c) Water, H_2O, is a three-atom molecule with an angular shape. It is like the CH_4 molecule without the top and back right hydrogens, or like the ammonia molecule without the back right hydrogen, and the carbon or nitrogen atoms replaced with an oxygen atom. With only three atoms, the water molecule is two-dimensional. Like the carbon atom in methane and the nitrogen atom in ammonia, the oxygen atom in water is surrounded by four electron pairs. These pairs are *not* in the same plane as the three atoms; one pair is above that plane and the other is beneath it.

the central atom. It therefore exerts a stronger repulsion force on the three bonding pairs, pushing them closer together and reducing the bond angles slightly. In ammonia, NH_3, the bond angle is 107.5°.

FOUR ELECTRON PAIRS, TWO BONDED ATOMS

Again, the tetrahedral electron pair geometry is predicted, but it is further modified because only two of the electron pairs form bonds to other atoms (Line 5). The molecular geometry is **angular** or **bent** (Fig. 12.5c). The two lone pairs apparently exert a stronger repulsion than one pair, as the bond angle in water is 104.5°.

The five preceding paragraphs and their summary in Table 12.1 make you ready to predict some electron pair and molecular geometries around a central atom. We suggest the following procedure:

PROCEDURE

1) Draw the Lewis diagram.
2) Count the electron pairs around the central atom, both bonding and unshared.
3) Determine electron pair and molecular geometries. This is best done by reason rather than by memorization, reaching the following conclusions:
   a) Two electron pairs: electron pair and molecular geometries both linear. Bond angle, 180°.
   b) Three electron pairs: electron pair and molecular geometries both planar triangular. Bond angles, 120°.
   c) Four electron pairs: electron pair geometry tetrahedral. Bond angles, tetrahedral (109.5°) or approximately tetrahedral.
      (1) All electron pairs bonding: molecular geometry is tetrahedral.
      (2) Three electron pairs bonding, one lone pair: molecular geometry is trigonal pyramidal.
      (3) Two electron pairs bonding, two lone pairs: molecular geometry is angular (bent).

EXAMPLE 12.8

Predict the electron pair and molecular geometries of carbon tetrachloride, CCl_4.

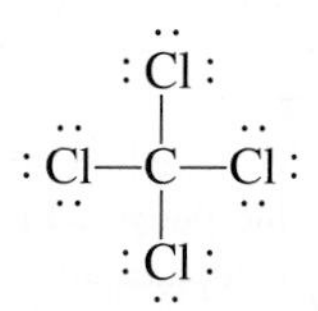

The Lewis diagram, drawn in Example 12.1, is shown at the right. From this you should establish the number of electron pairs around the central atom and the number of atoms bonded to the central atom. Both geometries follow.

With four electron pairs around carbon, all bonded to other atoms, both geometries are tetrahedral.

EXAMPLE 12.9 _____

Describe the shape of a molecule of boron trihydride, BH_3.

First draw the Lewis diagram. Remember that boron has only three valence electrons to contribute to covalent bonds. From the structure answer the question.

_____   _____

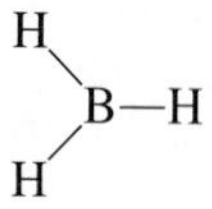

$$\begin{array}{c} H \\ \diagdown \\ B{-}H \\ \diagup \\ H \end{array}$$ trigonal planar

Three electron pairs yield both an electron pair geometry and a molecular geometry that are trigonal planar with 120° bond angles.

EXAMPLE 12.10 _____

Predict the electron pair geometry and shape of a molecule of dichlorine oxide, Cl_2O.

_____   _____

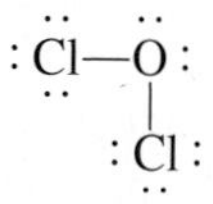

$$:\overset{..}{Cl}{-}\overset{..}{O}:\\ \underset{..}{\overset{|}{:Cl:}}$$ Electron pair geometry: tetrahedral; molecular geometry: bent

Oxygen has four electron pairs around it, yielding an electron pair geometry that is approximately tetrahedral. Only two of the electron pairs are bonded to other atoms, so the molecule is angular or bent. The structure is similar to that of water.

12.5 THE GEOMETRY OF MULTIPLE BONDS

Experimental evidence shows that the two or three electron pairs in a multiple bond behave as a single electron pair in establishing molecular geometry. This appears if we compare beryllium difluoride, carbon dioxide, and hydrogen cyanide, whose Lewis diagrams are

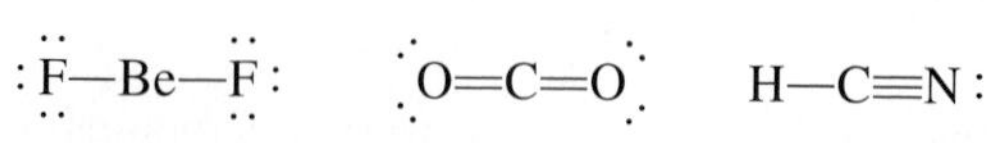

$$:\overset{..}{F}{-}Be{-}\overset{..}{F}:\qquad \overset{..}{O}{=}C{=}\overset{..}{O}\qquad H{-}C{\equiv}N:$$

All three molecules are linear; their bond angles are 180°. The two electron pairs in BeF_2 are as far from each other as possible. According to the VSEPR principle, this is responsible for the 180° bond angle in that compound. In carbon dioxide the carbon is flanked by two double bonds, and in hydrogen cyanide, one single bond and one triple. Evidently, the second and third electron pairs in double and triple bonds don't count when it comes to establishing molecular geometry.

Further evidence supporting this conclusion comes from comparing the bond angles in boron trifluoride and formaldehyde:

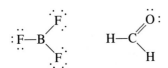

The shapes are both planar triangular with 120° bond angles. This is the angle predicted for three electron pairs under the VSEPR principle.

12.6 POLARITY OF MOLECULES

> **PG 12D** Given or having determined the Lewis diagram of a molecule, predict whether the molecule is polar or nonpolar.

We previously considered the polarity of covalent bonds. Now that we have some idea about how atoms are arranged in molecules, we are ready to discuss the polarity of molecules themselves. **A polar molecule is one in which there is an unsymmetrical distribution of charge,** resulting in + and − poles. A simple example is the HF molecule. The fact that the bonding electrons are closer to the fluorine atom gives the fluorine end of the molecule a partial negative charge, while the hydrogen end acts as a positive pole. (See Fig. 11.5, page 295.) In general, any diatomic molecule in which the two atoms differ from each other will be at least slightly polar. Other examples are HCl and BrCl. In both of these molecules the chlorine atom acts as a negative pole.

When a molecule has more than two atoms, we must know something about the bond angles in order to decide whether the molecule is polar or nonpolar. Consider, for example, the two triatomic molecules, BeF_2 and H_2O. Despite the presence of two strongly polar bonds, the linear BeF_2 *molecule* is nonpolar. Since the fluorine atoms are symmetrically arranged around Be, the two polar Be—F bonds cancel each other. This may be shown as

$$:\ddot{F} \leftrightarrow Be \leftrightarrow \ddot{F}:$$

in which the arrows point to the more electronegative atoms.

In contrast, the bent water molecule is polar; the two polar bonds do not cancel each other because the molecule is not symmetrical around a horizontal axis.

The negative pole is located at the more electronegative oxygen atom; the positive pole is midway between the two hydrogen atoms. In an electric field, water molecules tend to line up with the hydrogen atoms pointing toward the plate with the negative charge and the oxygen atoms toward the plate with the positive charge (Fig. 12.6).

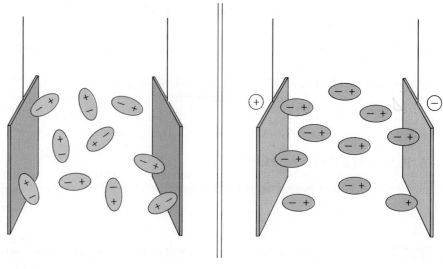

Field off   Field on

Figure 12.6

Orientation of polar molecules in an electric field. Two plates immersed in a liquid whose molecules are polar are connected through a switch to a source of an electric field. With the switch open, the orientation of the molecules is random (left). When the switch is closed, the molecules line up with the positive end toward the negative plate and the negative pole toward the positive plate.

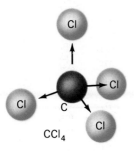

CCl_4

Nonpolar
(dipoles from polar bonds
cancel due to symmetry)

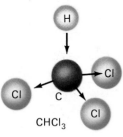

$CHCl_3$

Polar
(dipoles from polar bonds
do not cancel)

Figure 12.7

Polar and nonpolar molecules

Another molecule that is nonpolar despite the presence of polar bonds is CCl_4. The four C—Cl bonds are themselves polar, with the bonding electrons displaced toward the chlorine atoms. But because the four chlorines are symmetrically distributed about the central carbon atom (Fig. 12.7), the polar bonds cancel each other. If one of the chlorine atoms in CCl_4 is replaced by hydrogen, the symmetry of the molecule is destroyed. The chloroform molecule, $CHCl_3$, is polar.

From these observations we can state an easy way to decide whether a simple molecule is polar or nonpolar. If the central atom has no lone pairs and all atoms bonded to it are identical, the molecule is nonpolar. If these conditions are not met, the molecule is polar.

EXAMPLE 12.11

Is the BF_3 molecule polar? Is the NH_3 molecule polar?

The geometries of both of these molecules are described in Table 12.1. Consider BF_3 first. Sketch the Lewis diagram, with arrows pointing to the more electronegative element. Is the molecule polar?

BF$_3$ is nonpolar.

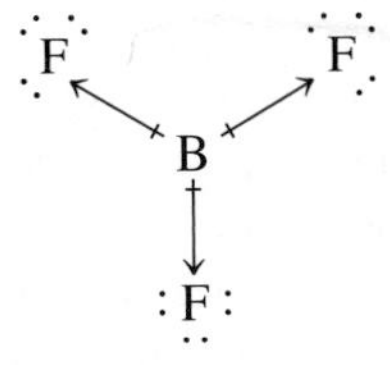

Even though fluorine is more electronegative than boron, the three fluorine atoms are arranged symmetrically around the boron atom. The polar bonds cancel.

Now sketch the trigonal pyramidal structure of NH$_3$, with arrows pointing to the more electronegative element. Is it polar or nonpolar?

———————   ———————

NH$_3$ is polar.

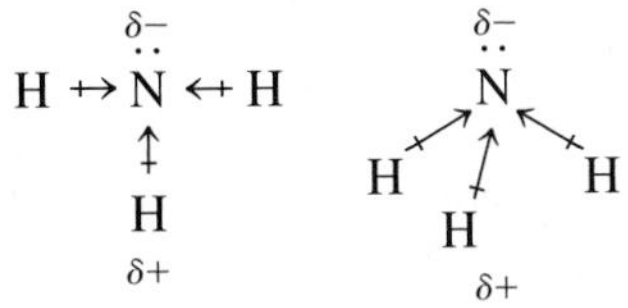

The bonding electrons in ammonia are displaced toward the more electronegative nitrogen atom. The bonds do not cancel in the unsymmetrical pyramidal shape, so the molecule is polar. The diagram at the right, which attempts to show the molecular shape, better suggests the charge displacement toward the nitrogen atom.

12.7 THE STRUCTURES OF SOME ORGANIC COMPOUNDS (OPTIONAL)

PG 12E   Distinguish between organic compounds and inorganic compounds.
  12F   Distinguish between hydrocarbons and other organic compounds.
  12G   On the basis of structure and the geometry of the identifying group, distinguish between alcohols, ethers, and carboxylic acids.

The Bonding Capabilities of the Carbon Atom

Organic chemistry is the chemistry of carbon compounds. Carbonates, cyanides, oxides of carbon, and a few other carbon-containing compounds are exceptions that are still classified as inorganic.

The property of carbon that qualifies it to define a whole branch of chemistry is the bonding capability of the carbon atom. Carbon atoms have four valence electrons that enable it to form four covalent bonds with other atoms. Most important is the fact that these bonds may be to other carbon atoms. As a result, many organic compounds have extremely long chains of carbon atoms. Atoms of no other element form long chains in this way.

The structure of molecules is the primary concern of organic research chemists. They know the physical and chemical characteristics of different structures. From that starting point, they "create," on paper or on a computer terminal, Lewis diagrams and molecular models of molecules that should have certain desirable properties. Then, beginning with the structures of chemicals that are known and available, they figure out chemical reactions that change the existing structures to those that are wanted. It is this kind of research that has led to many things that we use daily, including synthetic fabrics, plastics, and medical products.

In this section we will identify some common organic structures without attempting to study them in detail. We will begin with a Lewis diagram that you studied as an example in Section 12.2 and then turn to others you may have drawn in answering some of the end-of-chapter questions.

Hydrocarbons

The Lewis diagram of methane, CH_4, appears in Section 12.1. You drew the Lewis diagram of propane, C_3H_8, in Example 12.5. Methane and propane are **hydrocarbons, binary compounds of carbon and hydrogen.** They are members of a kind of hydrocarbon called **alkanes,** in which each carbon atom forms four single bonds. Alkanes differ from each other by the number of carbon atoms that form a carbon chain. Ethane, C_2H_6, lies between methane and propane. Compare the Lewis diagrams of the three compounds:

$$
\begin{array}{ccc}
\overset{\displaystyle H}{\underset{\displaystyle H}{H-\overset{|}{\underset{|}{C}}-H}} &
\overset{\displaystyle H\ \ H}{\underset{\displaystyle H\ \ H}{H-\overset{|}{\underset{|}{C}}-\overset{|}{\underset{|}{C}}-H}} &
\overset{\displaystyle H\ \ H\ \ H}{\underset{\displaystyle H\ \ H\ \ H}{H-\overset{|}{\underset{|}{C}}-\overset{|}{\underset{|}{C}}-\overset{|}{\underset{|}{C}}-H}} \\[2em]
CH_4 & C_2H_6 \ or \ CH_3CH_3 & C_3H_8 \ or \ CH_3CH_2CH_3 \\
\text{methane} & \text{ethane} & \text{propane}
\end{array}
$$

Two formulas are given for ethane. C_2H_6 is its molecular formula. CH_3CH_3 is called a **line formula.** It suggests the structure of the molecule as two $-CH_3$ groups bonded together. Divide the Lewis diagram between the carbon atoms and that's exactly what you have. The line formula for propane, $CH_3CH_2CH_3$, suggests correctly that the molecule is made up of two $-CH_3$ groups with a $-CH_2-$ group between them. Lewis diagrams for the next two alkanes, those with four and five carbon atoms, show that any number of $-CH_2-$ groups can be placed between two $-CH_3$ groups:

$$
\begin{array}{cc}
\overset{\displaystyle H\ \ H\ \ H\ \ H}{\underset{\displaystyle H\ \ H\ \ H\ \ H}{H-\overset{|}{\underset{|}{C}}-\overset{|}{\underset{|}{C}}-\overset{|}{\underset{|}{C}}-\overset{|}{\underset{|}{C}}-H}} &
\overset{\displaystyle H\ \ H\ \ H\ \ H\ \ H}{\underset{\displaystyle H\ \ H\ \ H\ \ H\ \ H}{H-\overset{|}{\underset{|}{C}}-\overset{|}{\underset{|}{C}}-\overset{|}{\underset{|}{C}}-\overset{|}{\underset{|}{C}}-\overset{|}{\underset{|}{C}}-H}} \\[2em]
C_4H_{10} \ or \ CH_3CH_2CH_2CH_3 & C_5H_{12} \ or \ CH_3CH_2CH_2CH_2CH_3 \\
or \ CH_3(CH_2)_2CH_3 & or \ CH_3(CH_2)_3CH_3 \\
\text{butane} & \text{pentane}
\end{array}
$$

Butane is the simplest alkane that exists as two distinct **isomers.** If we rearrange the atoms in butane, switching positions between the end —CH_3 group and the hydrogen on the second carbon, we get

H—C—C—C—C—H structure (n-butane)

C_4H_{10} *or* $CH_3CH_2CH_2CH_3$

n-butane

H—C—C—C—H structure (iso-butane)

C_4H_{10} *or* CH_3CHCH_3
CH_3

iso-butane

The two butanes are distinctly different chemical compounds, each having its own set of physical and chemical properties. Notice that the name *butane* is modified to distinguish between the isomers.

As the number of carbon atoms in an alkane increases, the number of isomers increases dramatically. There are three isomeric pentanes (five-carbon molecules), five isomeric hexanes (six-carbon molecules), nine heptanes (7 carbons), and 75 decanes (10 carbons). It is possible to draw over 300,000 isomeric structures for $C_{20}H_{42}$ and more than 100 million for $C_{30}H_{62}$. Obviously, not all of them have been prepared and identified! This does give us some idea, though, why there are more than ten times as many known organic compounds as there are inorganic compounds.

What do these molecules look like? What are their shapes and bond angles? First of all, they are three-dimensional and *they do not look like their two-dimensional Lewis diagrams.* All bond angles are tetrahedral. Thus what appears to be a nice straight line of carbon atoms in a Lewis diagram is actually a zig-zag chain of carbon atoms. This is seen in Figure 12.8, in which the model was deliberately arranged to show the zig-zag chain. Actually, if there are four or more carbon atoms in the molecule, it is most unlikely to find all the carbon atoms in the same plane. Instead the carbon chain twists and turns through a constantly changing three-dimensional pattern of its own, but always with tetrahedral bond angles.

The alkanes are one of five general groups of hydrocarbons. The cycloalkanes are like the alkanes in that each carbon atom is bonded to four other atoms by single bonds, but the carbon chain forms a closed ring. Alkenes differ from alkanes in having two carbon atoms that are connected by a double bond. Alkynes go one step further, with two carbon atoms linked with a triple bond. Finally there are the aromatic hydrocarbons that are based on the six-carbon planar ring structure of benzene. Questions at the end of the chapter include Lewis diagrams of some of these compounds.

Figure 12.8

Models of an isomer of butane, C_4H_{10}. The molecule shown, *n*-butane, has its four carbon atoms in a continuous chain.

Alcohols and Ethers

In Example 12.7 you drew the Lewis diagram for ethyl alcohol and were shown the diagram for its isomer, dimethyl ether. These diagrams, plus the Lewis diagram for water, are

H_2O *or* HOH
water

C_2H_6O *or* C_2H_5OH
ethyl alcohol *or* ethanol

C_2H_6O *or* CH_3OCH_3
dimethyl ether

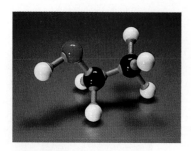

Figure 12.9

A model of ethanol (also called ethyl alcohol), C_2H_5OH.

All three molecules include an oxygen atom with two lone pairs of electrons that is single-bonded to two other atoms. This, as you learned in Section 12.4, yields a bent structure around the oxygen atom, as shown in Figure 12.9. In Section 12.6 you saw that a molecule with this structure is polar. This polarity is present in all three molecules, but to a diminishing extent from left to right.

The —OH part of an alcohol is called a **hydroxyl group.** Essentially, an **alcohol is an alkane in which a hydrogen atom is replaced by a hydroxyl group.** The hydroxyl group may be anyplace in the molecule, and there may be more than one. Other examples are

CH_3OH
methyl alcohol or methanol

C_3H_7OH
propyl alcohol or propanol

C_3H_7OH
2-propanol

$C_2H_4(OH)_2$
ethylene glycol

The chemical properties of an alcohol are the chemical properties of the hydroxyl group. Similar properties are present in all alcohols, though to different degrees. The physical properties of alcohols are also associated with the hydroxyl group, particularly among the smaller alcohol molecules. In large molecules, the properties of the hydrocarbon part of the molecule are usually more important.

Either or both of the —CH_3 groups in dimethyl ether may be replaced by longer carbon chains. Two additional examples are

C_4H_{10} *or* $(C_2H_5)_2O$
or C_2H_5—O—C_2H_5
diethyl ether

$C_4H_{10}O$ *or* $CH_3OC_3H_7$
methyl propyl ether

Carboxylic Acids

In end-of-chapter Questions 15 and 16 you are asked to draw Lewis diagrams for propionic acid, C_2H_5COOH, and formic acid, HCOOH. These formulas suggest the structure of the first and third members of a series of carboxylic acids, which contain the **carboxyl group,** —COOH. Adding the second member of the series, acetic acid, the Lewis diagrams are

$$
\begin{array}{ccc}
\mathrm{H-C}\Big\langle\!\!\begin{array}{c}\mathrm{O}\\ \mathrm{O-H}\end{array} &
\mathrm{H-\underset{\underset{\textstyle H}{|}}{\overset{\overset{\textstyle H}{|}}{C}}-C}\Big\langle\!\!\begin{array}{c}\mathrm{O}\\ \mathrm{O-H}\end{array} &
\mathrm{H-\underset{\underset{\textstyle H}{|}}{\overset{\overset{\textstyle H}{|}}{C}}-\underset{\underset{\textstyle H}{|}}{\overset{\overset{\textstyle H}{|}}{C}}-C}\Big\langle\!\!\begin{array}{c}\mathrm{O}\\ \mathrm{O-H}\end{array}
\end{array}
$$

HCOOH *or*	CH$_3$COOH *or*	C$_2$H$_5$COOH *or*
HCHO$_2$	HC$_2$H$_3$O$_2$	HC$_3$H$_5$O$_2$
formic acid	acetic acid	propanoic acid

As with hydrocarbons, alcohols, and ethers, the carbon chain may extend indefinitely.

Acetic acid is the best known of the carboxylic acids. You first met acetic acid in this book as one of the "other acids" in Section 6.7. After seeing the ionization equation,

$$HC_2H_3O_2 \longrightarrow H^+ + C_2H_3O_2^-$$

you read, "Notice that only the hydrogen written first in the formula ionizes; the others do not. This is typical of organic acids, which usually produce ions containing carbon, hydrogen, and oxygen." Now that you see the structure of acetic acid, you can see the difference between the ionizable hydrogen, which is bonded to an oxygen atom, and the other hydrogens, which are bonded to carbon atoms. The organic chemist's way of writing the same equation, along with the Lewis diagrams, shows this clearly:

$$\mathrm{CH_3COOH} \longrightarrow \mathrm{H^+} + \mathrm{CH_3COO^-}$$

$$
\mathrm{H-\underset{\underset{\textstyle H}{|}}{\overset{\overset{\textstyle H}{|}}{C}}-C}\Big\langle\!\!\begin{array}{c}\mathrm{O}\\ \mathrm{O-H}\end{array} \longrightarrow \mathrm{H^+} + \left[\mathrm{H-\underset{\underset{\textstyle H}{|}}{\overset{\overset{\textstyle H}{|}}{C}}-C}\Big\langle\!\!\begin{array}{c}\mathrm{O}\\ \mathrm{O-}\end{array}\right]^-
$$

acetic acid	$\longrightarrow$ hydrogen ion +	acetate ion

Carboxylic acids are weak acids, which means that they ionize only slightly in water. To the extent that they ionize, however, this is how it happens. When they react with bases in neutralization reactions, it is always the carboxylic hydrogen that reacts.

The geometry of the carboxylic group is shown in the model of acetic acid in Figure 12.10. The molecule is bent around the oxygen atom, forming an angle some-

FLASHBACK In Section 8.9 you learned how to write the equation for the reaction between an acid and a hydroxide base. This is the most common form of a neutralization reaction.

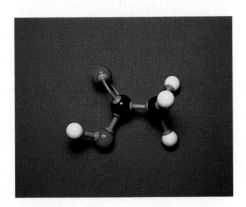

Figure 12.10
A model of acetic acid, CH$_3$COOH.

what smaller than the tetrahedral angle. In determining the geometry around the carbon atom, the double bond counts as if it were one of three single bonds. The geometry is trigonal planar with 120° bond angles. Bond angles around carbon atoms in the alkane part of the molecule are tetrahedral.

CHAPTER 12 IN REVIEW

12.1 Drawing Lewis Diagrams by Inspection
12A Draw the Lewis diagram for any molecule or polyatomic ion made up of representative elements (Sections 12.1 and 12.2).

12.2 Drawing Complex Lewis Diagrams

12.3 Electron Pair Repulsion: Electron Pair Geometry
12B Describe the electron pair geometry when a central atom is surrounded by two, three, or four electron pairs.

12.4 Molecular Geometry
12C Given or having derived the Lewis diagram of a molecule or polyatomic ion in which a second-period central atom is surrounded by two, three, or four pairs of electrons, predict the molecular geometry around that atom.

12.5 The Geometry of Multiple Bonds

12.6 Polarity of Molecules
12D Given or having determined the Lewis diagram of a molecule, predict whether the molecule is polar or nonpolar.

12.7 The Structures of Some Organic Compounds (Optional)
12E Distinguish between organic compounds and inorganic compounds.
12F Distinguish between hydrocarbons and other organic compounds.
12G On the basis of structure and the geometry of the identifying group, distinguish between alcohols, ethers, and carboxylic acids.

CHAPTER 12 KEY WORDS AND MATCHING SETS

Set 1

_____ Geometric term to describe the arrangement between a central atom and two other atoms bonded to it.

_____ Describes a molecule in which charge distribution is symmetrical.

_____ Electron pair geometry when two electron pairs are on opposite sides of an atom.

_____ Spatial arrangement of electrons around an atom.

_____ Electron pair geometry when an atom lies in the plane formed by three pairs of electrons.

_____ Describes a molecule having an unsymmetrical distribution of charge.

_____ Electron pair geometry when an atom is surrounded by four pairs of electrons.

1. Trigonal planar
2. Bond angle
3. Polar
4. Tetrahedral
5. Electron pair geometry
6. Nonpolar
7. Linear

Set 2

_____ Bond angle when a central atom is surrounded by four identical atoms.

_____ Spatial arrangement of atoms in a molecule.

_____ Molecular geometry when the bond angle is 180°.

_____ One explanation for the arrangement of electron pairs around a central atom.

_____ Molecular geometry when a central atom is bonded to three other atoms but surrounded by four electron pairs.

_____ Formed between an atom and two of its electron pairs.

_____ Molecular geometry associated with 120° bond angle.

_____ Molecular geometry around a central atom that has two lone pairs.

1. Trigonal planar
2. Electron pair angle
3. Tetrahedral
4. VSEPR
5. Linear
6. Molecular geometry
7. Bent
8. Trigonal pyramid

Sets 3 and 4 are from optional Section 12.7.

Set 3

_____ Hydrocarbons in which carbon atoms are in a closed loop and all carbon atoms form four single bonds.

_____ Example of a molecular formula.

_____ Hydrocarbons that contain a double bond between carbon atoms.

_____ Branch of chemistry based on carbon compounds.

_____ Hydrocarbons in which all carbon atoms form four single bonds.

_____ Example of a line formula.

_____ Hydrocarbons in which two carbon atoms are linked by a triple bond.

1. Alkenes
2. Alkanes
3. Cycloalkanes
4. Organic
5. Alkynes
6. C_6H_{14}
7. $CH_3(CH_2)_4CH_3$

Set 4

_____ Compounds whose properties are associated with the carboxyl group.

_____ Compounds containing only carbon and hydrogen.

_____ Molecular geometry around the oxygen atom in a hydroxyl group.

_____ Class of compounds in which an oxygen atom links two groups containing carbon and hydrogen atoms.

_____ Class of compounds that contains a hydroxyl group.

_____ Molecular geometry around the carbon atom in a hydroxyl group.

1. Bent
2. Acids
3. Trigonal planar
4. Ethers
5. Hydrocarbons
6. Alcohols

STUDY HINTS AND PITFALLS TO AVOID

There are many molecules for which you can draw two or more Lewis diagrams that satisfy the octet rule. Your diagram is most apt to be correct if you remember that (1) hydrogen always forms one bond and carbon almost always forms four; (2) two or more oxygen atoms are distributed around a central atom; (3) an oxygen atom is between a hydrogen atom and another nonmetal atom; and (4) your diagram is as symmetrical as possible.

The most common errors in Lewis diagrams are bonding oxygen atoms to each other and surrounding a central atom by three or five electron pairs. There are some compounds in which oxygen atoms are bonded to each other, but the number is not large. One of these appears in one of the questions that follow. The three- or five-electron pair errors most often occur when double bonds are present. Always check your final diagram to be sure all atoms conform to the octet rule.

Geometry places limits on the shapes of some molecules. If there are only two atoms, the geometry is linear; two points determine a line. If there are three atoms, they are either in a straight line (linear) or they are not (angular

or bent). Four atoms take you into the *possibility* of a three-dimensional molecule. That is why you must distinguish between trigonal planar and trigonal pyramidal. The adjective "trigonal" is necessary because some elements in the third and later periods form square planar and square pyramidal structures.

To distinguish between polar and nonpolar molecules, test the molecule for the two conditions that are required for nonpolarity. First, all atoms bonded to the central atom must be the same element. Second, there can be no lone pairs. If the molecule passes both tests, it is nonpolar; if it fails either test, it is polar.

CHAPTER 12 QUESTIONS AND PROBLEMS

SECTIONS 12.1 AND 12.2

Write Lewis diagrams for each of the following sets of molecules.

1) BrF, SF_2, PF_3.
2) HBr, H_2S, PH_3.

3) OH^-, ClO_3^-, NO_3^-.
4) OF_2, CO, CO_3^{2-}.

5) IO_2^-, H_3PO_4, HSO_4^-.
6) IO^-, BrO_4^-, H_2SO_4.

7) CH_3Cl, CHF_3, CBr_3I.
8) $CHBr_2I$, CH_2F_2, CH_2ClF.

There are two or more acceptable diagrams for most species in Questions 9 to 18.

9) $C_2H_4Cl_2$, C_2H_4BrCl, $C_3H_5F_2I$.
10) $C_2H_2Br_4$, $C_2H_2Br_2I_2$, $C_3H_4Cl_2I_2$.

11) C_4H_{10}, C_4H_8, C_2H_4O.
12) C_4H_6, CH_4O, $C_2H_6O_2$.

13) C_6H_{14}, C_3H_8O, $C_2H_2Cl_2$.
14) C_5H_{12}, C_5H_{10}, C_3H_6O.

15) Propionic acid, C_2H_5COOH.
16) Formic acid, HCOOH.

17) Hydrogen peroxide, H_2O_2; peroxide ion, O_2^{2-}.
18) HS^-, H_2S, H_2S_2, H_2S_3, H_2S_4.

SECTIONS 12.3 TO 12.5

Questions 19 to 30: For each molecule, or for the atom specified in a molecule, describe (a) the electron pair geometry and (b) the molecular geometry predicted by the electron pair repulsion theory.

19) BH_3, NF_3, HF.
20) BeH_2, CF_4, OF_2.

21) ClO^-, IO_3^-, NO_3^-.
22) IO_4^-, ClO_2^-, CO_3^{2-}.

23) Oxygen atom in C_2H_5OH.
24) Each carbon atom in C_2H_5OH.

25) Carbon atom in CH_3NH_2.
26) Nitrogen atom in CH_3NH_2.

27) Each carbon atom in C_2H_2.
28) Each carbon atom in C_2H_4.

29) Carbon atom in SCN^-.
30) Carbon atom in HCHO.

31) Draw a central atom with whatever combination of bonding and lone pair electrons that is necessary to yield a trigonal pyramid structure around that atom.

32) The Lewis diagram of a certain compound has the element E as its central atom. The bonding and lone pair electrons around E are shown. What is the molecular geometry around E?

SECTION 12.6

33) Identify two possible molecular geometries if an atom of X forms single covalent bonds with three atoms of Z. Can either molecule be polar? Explain your answer.

34) Predict the shapes of oxygen difluoride and beryllium hydride molecules. Which molecule, if either, might be nonpolar.

35) The nitrogen–fluorine bond in NF_3 has an electronegativity difference of 1.0. This is less than the 1.5 electronegativity difference between carbon and fluorine in CF_4. Yet NF_3 molecules are polar and CF_4 molecules are nonpolar. How can this be?

36) Is the carbon tetrafluoride molecule, CF_4, which contains four polar bonds (electronegativity difference 1.0) polar or nonpolar? Explain.

37) Compare the shapes and polarities of the following molecules: ClF, Cl_2, BrCl, ICl. In each case, identify the positive end of the molecule.

38) Describe the shapes and compare the polarities of HCl and HI molecules. In each case, identify the end of the molecule that is more positive.

39) As noted in the text, there are two possible Lewis diagrams for C_2H_6O. Both are real compounds: C_2H_5OH is ethanol, or ethyl alcohol, and CH_3OCH_3 is dimethyl ether. Sketch these molecules with arrows to indicate the direction of bond polarity around the oxygen atom. Predict the relative polarities of the molecule. What would you expect of the polarity of $C_5H_{11}OH$?

40) Sketch the water molecule, paying particular attention to the bond angle and using arrows to indicate the polarity of each bond. Then sketch the methanol molecule, $HOCH_3$, again using arrows to show bond polarity. Predict the approximate shape of both molecules around the oxygen atom. Also predict the relative polarities of the two molecules and explain your prediction.

The answers for Questions 41–44 are to be selected from the group of Lewis diagrams that follows. Each question may have more than one answer.

a) $\left[\begin{array}{c} H \\ | \\ H-N-H \\ | \\ H \end{array} \right]^+$ b) $: \ddot{Cl} - \ddot{O} - \ddot{Cl} :$

c) $\left[\begin{array}{c} H - \ddot{O} - H \\ | \\ H \end{array} \right]^+$ d) $: \ddot{F} - \overset{\displaystyle : \ddot{I} :}{\underset{\displaystyle : \ddot{I} :}{C}} - \ddot{I} :$

e) $H - Be - H$ f) $H - \overset{\displaystyle |}{\underset{\displaystyle H}{B}} - H$

41) Which species are linear?

42) Which species have tetrahedral shapes?

43) Identify all species that have trigonal planar geometries and all whose shapes are trigonal pyramids.

44) Which neutral molecules are polar?

SECTION 12.7

45) What distinguishes an organic compound from an inorganic compound?

46) What features of the carbon atom account for its ability to form so many chemical compounds?

47) Identify the hydrocarbons among the following: CH_3OH, $CH_3(CH_2)_6CH_3$, C_6H_6, $CH_2(NH_2)_2$, $C_{18}H_8$.

48) *Hydrocarbon* and *carbohydrate* are similar terms. They sound almost as if their syllables were reversed, and both are made up by combining two words or parts of words. Can you guess what those words or parts of words are for each term? Do you suppose a carbohydrate can be an example of a hydrocarbon, or vice versa?

49) Distinguish between the *structure* of an organic molecule and the *shape* of the molecule.

50) How are molecular structure and molecular shape shown in Lewis diagrams?

51)* Describe the shapes of C_2H_6 and C_2H_4. In doing so, explain why one molecule is planar (all atoms lie in the same plane) and the other molecule cannot be planar.

52)* Benzene, C_6H_6, is a planar compound in which the carbon atoms are arranged in a hexagon (six-sided regular polygon). A Lewis diagram of this molecule has a geometry that accounts for the observed *shape* of the molecule, including bond angles. Draw this diagram. (Experimental evidence shows that the diagram does not describe the *structure* of benzene accurately.)

53) Identify the structural features that distinguish between alkanes, alkenes, and alkynes.

54) How does an alkane differ from a cycloalkane?

55) How are alcohols and ethers similar to water in structure and shape? What distinguishes between alcohols and ethers?

56) Name and describe the group that is present in all alcohols. What does the presence of this group contribute to an alcohol molecule and its properties?

57) What are carboxylic acids? How do they differ from other acids, such as hydrochloric, sulfuric, and carbonic acid?

58) Identify the group that is present in carboxylic acids. Write a Lewis diagram for the group and describe the geometry around all central atoms within the group.

59) An inorganic chemist is most apt to write the formula of butyric acid as $HC_4H_7O_2$. How is an organic chemist more likely to write the formula? What information does one formula convey that the other does not?

60)* $NaCHO_2$ is how an inorganic chemist is most apt to write the formula of sodium formate. Write the formula of the formate ion as the inorganic chemist is likely to write it. Now draw a Lewis diagram of the formate ion, CHO_2^-. Compare the diagram with the Lewis diagram of formic acid, $HCHO_2$ (see Question 16). Is your formate ion diagram what you would expect from the ionization of formic acid?

GENERAL QUESTIONS

61) Distinguish precisely and in scientific terms the differences between items in each of the following groups:
a) Covalent bond, coordinate covalent bond
b) Molecular geometry, electron pair geometry
c) Angular geometry, trigonal planar geometry
d) Trigonal planar geometry, trigonal pyramidal geometry
e) Trigonal pyramidal geometry, tetrahedral geometry
f) Polar molecule, nonpolar molecule
g) Alkanes, alkenes, alkynes
h) Structural formula, condensed (line) formula, molecular formula
i) Alcohol, ether, carboxylic acid
j) Hydroxyl group, carboxyl group

62) Classify each of the following statements as true or false:
a) Molecular geometry around an atom may or may not be the same as electron pair geometry around the atom.
b) Electron pair geometry is the direct effect of molecular geometry.

c) If the geometry of a molecule is linear, the molecule must have at least one double bond.
d) A molecule with a double bond cannot have a trigonal pyramidal geometry.
e) A CO_2 molecule is linear, but an SO_2 molecule is angular.
f) A molecule is polar if it contains polar bonds.
g) A molecule having a central atom that has a lone pair of electrons is always polar.
h) A molecule having a central atom that has two lone pairs of electrons is always polar.
i) Carbon atoms normally form four bonds in organic compounds.
j) Hydrogen atoms never form double bonds.
k) Isomers have the same structural formulas but different molecular formulas.
l) Molecules cannot be made up of a ring of carbon atoms.

63)* Draw two different Lewis diagrams of C_3H_4.

64)* $H_2C_2O_4$ is the formula of oxalic acid. The two carbon atoms are bonded to each other, and the molecule is symmetrical. Draw the Lewis diagram.

65) One kind of C_5H_{10} molecule has its carbon atoms in a ring. Draw the Lewis diagram.

66) Draw Lewis diagrams for these five acids of chlorine: HCl, HClO, $HClO_2$, $HClO_3$, $HClO_4$.

67)* Compare Lewis diagrams for CCl_4, SO_4^{2-}, ClO_4^-, and PO_4^{3-}. Identify two things that are alike about these diagrams and how they are drawn. From these generalizations, can you predict the Lewis diagram of $S_eO_4^{2-}$ and CI_4?

68) Criticize the statement: "Carbon atoms in a normal alkane lie in a straight line."

69)* What are the shapes of the following: (a) *cis*-dibromoethene, $C_2H_2Br_2$; (b) acetylene, C_2H_2; (c) *n*-butane, $CH_3CH_2CH_2CH_3$.

MATCHING SET ANSWERS

Set 1: 2–6–7–5–1–3–4

Set 2: 3–6–5–4–8–2–1–7

Set 3: 3–6–1–4–2–7–5

Set 4: 2–5–1–4–6–3

The Ideal Gas Law and Gas Stoichiometry

13

The two balloons are filled with about the same mass of gas at the same temperature and pressure. The blue balloon is filled with helium (d = 0.16 g/L) while the red balloon is filled with argon (d = 1.6 g/L). The bear at the end of the rope is more dense than the atmosphere. (Charles D. Winters)

LEARN IT NOW Few general chemistry texts, if any, use a question-and-answer format for numerical examples like those in this book. Instead, examples are simply solved for you, with comments to explain how the problem is worked out. Consistent with our plans to make the later chapters of this book more like a general chemistry text, we begin by using worked-out examples in Chapter 13, but only for the simpler problems. Over several chapters we will increase the use of worked-out examples, eventually using them exclusively. However, we will continue the question-and-answer format for examples that do not involve calculations.

Our completely worked-out examples are almost always followed by quick checks that are problems. It is important in learning from solved examples that you test your understanding immediately after studying the example. We strongly suggest that you take the time to solve each quick check problem as soon as you come to it. That way you will LEARN IT—NOW!

13.1 THE COMBINED GAS LAWS REVISITED

In Chapter 4 we studied three gas laws for a fixed amount of any gas:

Gay-Lussac's Law: Pressure is directly proportional to absolute temperature, volume constant.

Charles' Law: Volume is directly proportional to absolute temperature, pressure constant.

Boyle's Law: Volume is indirectly proportional to pressure, temperature constant.

In Section 4.8, these laws were brought together as the "combined gas laws." The mathematical basis for the combination is that any quantity (absolute temperature, for

Figure 13.1

Law of Combining Volumes. Experiments show that when gases at the same temperature and pressure react, the reacting volumes are in a ratio of small, whole numbers. In the three reactions shown, the volumes of the individual gases are in the same ratio as the numbers of moles in the balanced equations. Compare the molecular and volume equations.

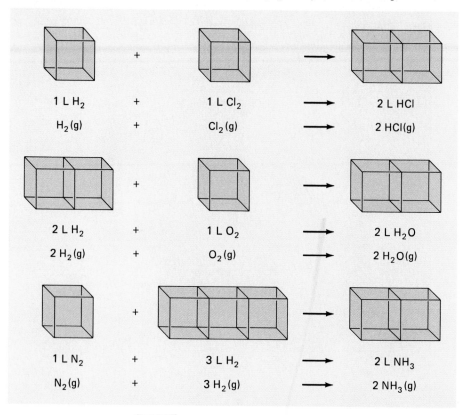

1 L H_2	+ 1 L Cl_2	→ 2 L HCl
$H_2(g)$	+ $Cl_2(g)$	→ 2 HCl(g)
2 L H_2	+ 1 L O_2	→ 2 L H_2O
2 $H_2(g)$	+ $O_2(g)$	→ 2 $H_2O(g)$
1 L N_2	+ 3 L H_2	→ 2 L NH_3
$N_2(g)$	+ 3 $H_2(g)$	→ 2 $NH_3(g)$

example) that is proportional to two or more other quantities (temperature and pressure in this case) is proportional to their product. Thus,

$$T \propto P \qquad T \propto V \qquad T \propto PV$$

where T is absolute temperature, P is pressure, and V is volume. Introducing a proportionality constant, k_c, and rearranging gives the equations

$$k_c T = PV \qquad \text{and} \qquad \frac{PV}{T} = k_c \qquad (4.13)$$

If subscripts 1 and 2 refer to two separate sets of pressure, temperature, and volume,

$$\frac{P_1 V_1}{T_1} = \frac{P_2 V_2}{T_2} \qquad (4.14)$$

Section 4.3 identified four measurable properties of gases, adding quantity to pressure, temperature, and volume. Quantity was constant in Chapter 4. This chapter opens by examining the effect of quantity as a fourth variable.

13.2 AVOGADRO'S LAW: VOLUME AND AMOUNT VARIABLE, PRESSURE AND TEMPERATURE CONSTANT

Early in the nineteenth century Gay-Lussac noticed that when gases react with each other, the reacting volumes are always in the ratio of small whole numbers *if the volumes are measured at the same temperature and pressure.* This observation is known as the **Law of Combining Volumes.** It extends to gaseous products, too. Several examples appear in Figure 13.1.

Shortly after Gay-Lussac's observations became known, Avogadro reasoned that they could be explained if **equal volumes of all gases** *at the same temperature and pressure* **contain the same number of molecules** (see Fig. 13.2). If the reacting molecules—or moles of molecules—react in a 1 : 1 ratio, and the reacting volumes also have

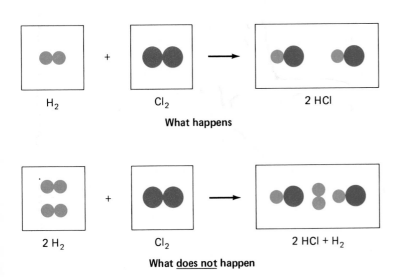

H₂   Cl₂   2 HCl

What happens

2 H₂   Cl₂   2 HCl + H₂

What <u>does not</u> happen

Figure 13.2

Avogadro's Law. The logic behind Avogadro's reasoning appears in these equations. In the first equation the same number of molecules in the equal reacting volumes uses all the molecules of both gases. In the second equation the number of molecules in the equal reacting volumes is not the same. This means that one reactant will be the limiting reagent (Section 9.4) and some of the other reactant will be left over. But that does *not* happen; it is contrary to experimental evidence. Therefore, equal volumes–equal molecules must be correct.

Count Amedeo Avogadro (1776–1856), depicted on a 1956 Italian postage stamp.

a 1:1 ratio, then the equal volumes of the *different gases* must have the same number of molecules. If both ratios are 1:2, then the larger volume must have twice as many molecules as the smaller volume of the other gas. It follows that **V ∝ n at constant temperature and pressure.** These statements are known as **Avogadro's Law.**

13.3 THE IDEAL GAS EQUATION

Having accumulated several different proportionalities between two or three of the four measurable properties of a gas, we now look for a single equation that ties them all together. We have seen from Avogadro's Law that V ∝ n; from Boyle's Law that V ∝ 1/P; and from Charles' Law that V ∝ T. If volume is proportional to three different quantities, it is also proportional to their product:

$$V \propto \frac{nT}{P} \tag{13.1}$$

Inserting a proportionality constant, R, yields an equation:

$$V = R\frac{nT}{P} \tag{13.2}$$

Rearranging gives the **ideal gas equation** in its most common form:

$$PV = nRT \tag{13.3}$$

This equation is also called the **ideal gas law.** It is one of the most important equations in this book. If you are a science or engineering major, you will probably meet Equation 13.3 in at least half a dozen other courses. The equation is exceptionally useful, and it should be memorized.

R is the **universal gas constant.** Its value is the same for any gas or mixture of gases that behaves like an ideal gas. It is significant to note that both Equation 13.3 and the value of R can be derived by applying the laws of physics to the ideal gas model. The fact that the same conclusion can be reached by theoretical calculations and by experiments makes us quite confident that the ideal gas model is correct.

To find the value of R experimentally, we need only to measure all four variables for any gas sample. It is a fact that one mole of any ideal gas occupies 22.4 liters at standard temperature and pressure. Solving Equation 13.3 for R and substituting values gives

$$R = \frac{PV}{nT} = \frac{1.00 \text{ atm} \times 22.4 \text{ L}}{1.00 \text{ mol} \times 273 \text{ K}} = 0.0821 \frac{\text{L} \cdot \text{atm}}{\text{mol} \cdot \text{K}} \tag{13.4}$$

$$R = \frac{PV}{nT} = \frac{760 \text{ torr} \times 22.4 \text{ L}}{1.00 \text{ mol} \times 273 \text{ K}} = 62.4 \frac{\text{L} \cdot \text{torr}}{\text{mol} \cdot \text{K}} \tag{13.5}$$

The value of R in Equation 13.4 is used for any problem in which pressure is given in atmospheres, and the value in Equation 13.5 is used if pressure is given in torr. The values of R should be memorized because they are used so frequently.*

FLASHBACK Recall from Section 4.8 that gases are often compared at "standard temperature and pressure (STP)," which is defined as 0°C and 1 atmosphere.

*Some instructors do not approve of Equation 13.5. If pressure is given in torr, they prefer that it be converted to atmospheres and then used in Equation 13.4. As usual, our recommendation is that you follow the instructions of your professor.

A useful variation of the ideal gas equation replaces n, the number of moles, by the mass of a sample, m, divided by molar mass, MM:

$$n = \frac{mass}{molar\ mass} = \frac{m}{MM} = \cancel{g} \times \frac{mol}{\cancel{g}} = mol$$

Thus

$$PV = nRT = \frac{m}{MM}RT = \frac{mRT}{MM} \qquad (13.6)$$

The ideal gas equation may be used to solve for any single variable when given values for all others. It may also be used to determine properties that are combinations of units, such as density and molar volume. We will examine several examples.

13.4 APPLICATIONS OF THE IDEAL GAS EQUATION

Determination of a Single Variable

> **PG 13A**   Given values for all except one of the variables in the ideal gas equation, calculate the value of that remaining variable.

If you know values for all variables except one in either form of the ideal gas equation, you can use algebra to find the value of the last variable. As with all problems to be solved algebraically, solve the equation for the wanted quantity. Then substitute the quantities that are given, including units, and calculate the answer.

Remember the importance of including units in solving problems by algebra. You will be performing many algebraic manipulations in solving ideal gas law problems. You will wind up with a calculation that involves four or five terms, and those terms will include five or six units; one term alone, R, includes four units. There is much opportunity for routine errors in algebra. However, if you include units in your setups, you will catch these errors before you even pick up your calculator.

EXAMPLE 13.1

What volume will be occupied by 0.393 mole of nitrogen at 0.971 atm and 24°C?

SOLUTION

To get a clear picture of the problem, we should "plan" it by listing what is given, what is wanted, and the mathematical connection between the two. The given measurements are clearly identified, but another given, R, calls for a decision. Which value of R should be used? The decision is based on the units of pressure. Pressure is specified in atmospheres, so we use 0.082 L · atm/mol · K, the value of R that includes atmospheres.

The "mathematical connection" between the given and the wanted is the ideal gas equation, solved for the wanted quantity. But which equation? That depends on whether the quantity is expressed in grams or moles. If in moles, use PV = nRT (Equation 13.3); if in grams, use PV = mRT/MM (Equation 13.6). In this case n is given, so PV = nRT, solved for the wanted quantity, V, is the mathematical connection.

The "plan" of the problem is therefore

GIVEN: 0.393 mol N_2; 0.971 atm; 24°C (297 K); 0.0821 L · atm/mol · K

WANTED: L N_2 EQUATION: $V = \dfrac{nRT}{P}$

Notice that Celsius degrees were changed immediately into kelvins so the change would not be forgotten.

Now the known values may be substituted and the answer calculated:

$$V = \frac{nRT}{P} = \frac{0.393 \text{ mol} \times \dfrac{0.0821 \text{ L} \cdot \text{atm}}{\text{mol} \cdot \text{K}} \times 297 \text{ K}}{0.971 \text{ atm}} = \frac{0.393 \text{ mol}}{0.971 \text{ atm}} \times \frac{0.0821 \text{ L} \cdot \text{atm}}{\text{mol} \cdot \text{K}} \times 297 \text{ K} = 9.87 \text{ L}$$

LEARN IT NOW Quick Check 13.1 is referred to later. Be sure to complete it at this time. Learn it—NOW!

✔ QUICK CHECK 13.1

How many moles of neon are in a 5.00-L gas cylinder at 18°C if they exert a pressure of 8.65 atm?

One of the most useful applications of the ideal gas equation is determining the molar mass of an unknown substance in the vapor state. The following example illustrates the method.

EXAMPLE 13.2 _____

1.67 grams of an unknown liquid are vaporized at a temperature of 125°C. Its volume is measured as 0.421 liter at 749 torr. Calculate the molar mass.

SOLUTION

This time pressure is given in torr, so we use 62.4 L · torr/mol · K for R. Molar mass is the wanted quantity, and it appears in the PV = mRT/MM form of the ideal gas equation. The "plan" and complete solution of the problem are:

GIVEN: 1.67 g; 125°C (398 K); 0.421 L; 749 torr; R = 62.4 L · torr/mol · K   WANTED: molar mass (g/mol)

$$\text{EQUATION: MM} = \frac{mRT}{PV} = \frac{1.67 \text{ g} \times \dfrac{62.4 \text{ L} \cdot \text{torr}}{\text{mol} \cdot \text{K}} \times 398 \text{ K}}{749 \text{ torr} \times 0.421 \text{ L}} = \frac{1.67 \text{ g}}{749 \text{ torr}} \times \frac{398 \text{ K}}{0.421 \text{ L}} \times \frac{62.4 \text{ L} \cdot \text{torr}}{\text{mol} \cdot \text{K}} = 132 \text{ g/mol}$$

✔ QUICK CHECK 13.2

Calculate the mass of 2.97 liters of methane, CH_4, measured at 71°C and 428 torr. (*Hint:* You can calculate the molar mass of methane from its formula.)

Gas Density

The customary unit for the density of a liquid or solid is grams per cubic centimeter, g/cm^3. The number of grams per cubic centimeter of a gas is so small, however, that gas densities are given in grams per liter, g/L. The quantities represented by these units, mass (m) and volume (V), both appear in Equation 13.6. The equation can therefore be solved for the density of a gas, m/V, in terms of pressure, temperature, molar mass, and R:

$$\frac{m}{V} = \frac{(MM)P}{RT} \tag{13.7}$$

In solving problems by dimensional analysis, molar mass is often used as a conversion factor. Calculating the molar mass of a known substance is one of the "starting steps" in such a problem. It is also a starting step when a problem is solved by algebra and molar mass is a variable in the equation. We will therefore include molar mass as a given value in the "plan" of a problem that is solved by algebra. This is done in the following example.

FLASHBACK Density is defined as mass per unit volume. Any mass unit over any volume units is therefore an acceptable unit for density. See Section 3.8.

EXAMPLE 13.3

Hydrogen is the least dense of all substances at a given temperature and pressure. Calculate its density at typical room conditions, 22°C and 751 torr.

SOLUTION

GIVEN: 22°C (295 K); 751 torr; 2.02 g/mol H_2; 62.4 L · torr/mol · K
WANTED: Density, m/V in g/L.

$$\frac{m}{V} = \frac{(MM)P}{RT} = \frac{2.02 \text{ g}}{1 \text{ mol}} \times \frac{751 \text{ torr}}{295 \text{ K}} \times \frac{1 \text{ mol} \cdot \text{K}}{62.4 \text{ L} \cdot \text{torr}} = 0.0824 \text{ g/L}$$

Notice that to divide by R, 62.4 L · torr/mol · K, you multiply by its inverse, 1 mol · K/62.4 L · torr.

✔ QUICK CHECK 13.3

Calculate the density of nitrogen at STP.

EXAMPLE 13.4

Find the molar mass of an unknown gas if its density is 1.45 g/L at 44°C and 0.331 atm.

SOLUTION

To solve this problem, you must interpret density in terms of its units. A density of 1.45 g/L means that 1 liter (V) of the gas has a mass (m) of 1.45 grams. Then the problem is just like Example 13.2, with different numbers.

GIVEN: 1.45 g; 44°C (317 K); 1.00 L; 0.331 atm; R = 0.0821 L · atm/mol · K
WANTED: molar mass (g/mol)

EQUATION: $MM = \dfrac{mRT}{PV} = \dfrac{1.45 \text{ g}}{1.00 \text{ \L}} \times \dfrac{317 \text{ \K}}{0.331 \text{ atm}} \times \dfrac{0.0821 \text{ \L} \cdot \text{atm}}{\text{mol} \cdot \text{\K}} = 114 \text{ g/mol}$

✔ QUICK CHECK 13.4

Calculate the molar mass of a gas if its STP density is 1.96 g/L.

Molar Volume

> **PG 13D** Calculate the molar volume of any gas at any specified temperature and pressure.
> **13E** Given the molar volume of a gas at a certain temperature and pressure and either number of moles or volume, calculate the other.

The ideal gas equation, PV = nRT, can be solved for the ratio of volume to moles, V/n. This ratio is called the **molar volume of a gas, the volume in liters of one mole:**

$$MV \equiv \frac{V}{n} = \frac{RT}{P} \qquad (13.8)$$

🔆 FLASHBACK Defining equations and their corresponding per relationships are summarized in Section 3.8.

Equation 13.8 is the defining equation for molar volume. From the equation, the units of molar volume are liters per mole, or L/mol.

Molar volume is very similar to molar mass, ". . . the mass in grams of one mole of [a] substance." There are also two important differences. They are:

The molar mass of a substance is constant, independent of temperature and pressure. Each substance has its own unique molar mass.

The molar volume of a gas is variable, depending on temperature and pressure. All gases have the same molar volume at the same temperature and pressure.

The dependence of molar volume on temperature and pressure is shown in Equation 13.8, where the two variables appear on the right side of the equation. Molar volume at a given temperature and pressure is found by direct substitution into the equation.

EXAMPLE 13.5

Calculate the molar volume of a gas at STP.

SOLUTION

GIVEN: 0°C (273 K); 1 mol (exactly); 0.0821 L · atm/mol · K
WANTED: molar volume, L/mol

EQUATION: $MV \equiv \dfrac{V}{n} = \dfrac{RT}{P} = \dfrac{273 \text{ \K}}{1 \text{ atm}} \times \dfrac{0.0821 \text{ L} \cdot \text{atm}}{\text{mol} \cdot \text{\K}} = 22.4 \text{ L/mol}$

Figure 13.3

Molar volume at standard temperature and pressure. The volume of the box in front of the student is 22.4 liters, the volume occupied by one mole of gas at 1 atmosphere and 0°C.

You have seen the number 22.4 before. It was given as the experimentally determined volume of one mole of any gas at STP that was used to calculate R (Equations 13.4 and 13.5). It is a number worth remembering: One mole of any gas *at STP* occupies 22.4 liters (Fig. 13.3).

✔ QUICK CHECK 13.5

Find the molar volume of a gas at 18°C and 8.65 atm.

If the molar volume of a gas is known, it may be used as a conversion factor to change moles to liters or liters to moles, L ↔ mol. This is just like using molar mass to change between grams and moles, g ↔ mol.

EXAMPLE 13.6

What volume is occupied by 4.21 moles of ethane at STP?

SOLUTION

GIVEN: 4.21 moles   WANTED: liters
PATH: mol ⟶ L
FACTOR: 22.4 L/mol at STP.

$$4.21 \; \cancel{mol} \times \frac{22.4 \text{ L}}{1 \; \cancel{mol}} = 94.3 \text{ L}$$

Notice that the identity of the gas, ethane, did not enter into the problem. Molar volume at a given temperature and pressure is the same for all gases.

There is another way to solve Example 13.8. We call it the "second method" to distinguish it from the molar volume method. This procedure also divides the problem into two separate problems. First, Step 1 of the stoichiometry pattern (L $H_2 \rightarrow$ mol H_2) is completed by solving the ideal gas law for n, substituting the given values, and calculating the moles of H_2:

GIVEN: 3.45 L H_2; 15.0 atm; 85°C (358 K); 0.0821 L · atm/mol · K.
WANTED: mol H_2

$$\text{EQUATION: } n = \frac{PV}{RT} = 3.45 \, \cancel{L} \times \frac{15.0 \, \cancel{atm}}{358 \, \cancel{K}} \times \frac{\text{mol} \cdot \cancel{K}}{0.0821 \, \cancel{L} \cdot \cancel{atm}} = 1.76 \text{ mol } H_2$$

The 1.76 mol H_2 now becomes a given in the second problem, "How many grams of NH_3 can be produced by 1.76 mol H_2?" This is calculated by applying dimensional analysis to the second and third steps of the stoichiometry pattern.

GIVEN: 1.76 mol NH_3 WANTED: g NH_3
PATH: mol $H_2 \rightarrow$ mol $NH_3 \rightarrow$ g NH_3
FACTORS: 3 mol H_2/2 mol NH_3; 17.0 g NH_3/mol NH_3

$$1.76 \, \cancel{\text{mol } H_2} \times \frac{2 \, \cancel{\text{mol } NH_3}}{3 \, \cancel{\text{mol } H_2}} \times \frac{17.0 \text{ g } NH_3}{1 \, \cancel{\text{mol } NH_3}} = 19.9 \text{ g } NH_3$$

The two steps in the "second method" are reversed when the wanted quantity is gas volume at specified temperature and pressure. You first complete Steps 1 and 2 of the stoichiometry pattern to get moles of wanted substance. Then the moles of wanted gas are converted to volume by the ideal gas equation. The volume given and volume wanted procedures are compared here:

PROCEDURE

Second Method: Volume Given	**Second Method: Volume Wanted**
1) Use ideal gas equation to change given volume to moles: n = PV/RT.	1) Calculate moles of wanted substance by Steps 1 and 2 of the stoichiometry pattern.
2) Use above result to calculate wanted quantity by Steps 2 and 3 of stoichiometry pattern.	2) Use ideal gas equation to change moles calculated above to volume: V = nRT/P.

As usual, if your instructor tells you which of the methods to use, use that one. If you are free to choose, select the one that is easiest for you. Whichever method you use, use it exclusively. The next two examples are solved by both methods. Follow only the one you will use, and disregard the other.

EXAMPLE 13.9

How many liters of CO_2, measured at 744 torr and 131°C, will be produced by the complete burning of 16.2 grams of butane, C_4H_{10}? The equation is

$$2 \, C_4H_{10}(g) + 13 \, O_2(g) \longrightarrow 8 \, CO_2(g) + 10 \, H_2O(g)$$

MOLAR VOLUME METHOD

Complete the first step: calculate the molar volume at 744 torr and 131°C.

_____   _____

GIVEN: 744 torr, 131°C (404 K), 62.4 L · torr/mol · K
WANTED: Molar volume, L/mol

EQUATION: $MV \equiv \dfrac{V}{n} = \dfrac{RT}{P} = \dfrac{404\,\cancel{K}}{744\,\cancel{torr}} \times \dfrac{62.4\,L \cdot \cancel{torr}}{mol \cdot \cancel{K}} = 33.9$ L/mol

Now you can use the 33.9 L/mol you have calculated as a conversion factor in the three-step stoichiometry pattern. Complete the problem.

_____   _____

GIVEN: 16.2 g C_4H_{10} WANTED: L CO_2
PATH: g $C_4H_{10} \rightarrow$ mol $C_4H_{10} \rightarrow$ mol $CO_2 \rightarrow$ g CO_2
FACTORS: 58.1 g C_4H_{10}/mol C_4H_{10}; 2 mol C_4H_{10}/8 mol CO_2; 33.9 L CO_2/mol CO_2

$16.2\,\cancel{g\,C_4H_{10}} \times \dfrac{1\,\cancel{mol\,C_4H_{10}}}{58.1\,\cancel{g\,C_4H_{10}}} \times \dfrac{8\,\cancel{mol\,CO_2}}{2\,\cancel{mol\,C_4H_{10}}} \times \dfrac{33.9\,L\,CO_2}{1\,\cancel{mol\,CO_2}} = 37.8$ L CO_2

SECOND METHOD

The wanted quantity is volume of gas. Therefore, complete the first step of the volume wanted procedure: Use stoichiometry to find the moles of CO_2 formed.

_____   _____

GIVEN: 16.2 g C_4H_{10} WANTED: mol CO_2
PATH: g $C_4H_{10} \rightarrow$ mol $C_4H_{10} \rightarrow$ mol CO_2
FACTORS: 58.1 g C_4H_{10}/mol C_4H_{10}; 2 mol C_4H_{10}/8 mol CO_2

$16.2\,\cancel{g\,C_4H_{10}} \times \dfrac{1\,\cancel{mol\,C_4H_{10}}}{58.1\,\cancel{g\,C_4H_{10}}} \times \dfrac{8\,mol\,CO_2}{2\,\cancel{mol\,C_4H_{10}}} = 1.12$ mol CO_2

Now complete the problem by calculating the volume that 1.12 mol CO_2 will occupy at 744 torr and 131°C.

_____   _____

GIVEN: 1.12 mol CO_2; 744 torr; 131°C (404 K); 62.4 L · torr/mol · K
WANTED: L CO_2.

EQUATION: $V = \dfrac{nRT}{P} = 1.12\,\cancel{mol}\,CO_2 \times \dfrac{62.4\,L \cdot \cancel{torr}}{\cancel{mol} \cdot \cancel{K}} \times \dfrac{404\,\cancel{K}}{744\,\cancel{torr}} = 37.9$ L CO_2

13.7 SUMMARY OF IMPORTANT INFORMATION ABOUT GASES

Our two chapters on gases are apt to leave you with a jumble of symbols and names and numbers that are difficult to organize. This section is added to help you organize that material, and also to indicate those concepts that are most important.

The Gas Laws   First, there are the proportionality laws. Four are important:

Name	Proportionality	Constant
Boyle	P and V, inversely proportional	T and n
Gay-Lussac	P and T, directly proportional	V and n
Charles	V and T, directly proportional	P and n
Avogadro	V and n, directly proportional	P and V

Notice that the P–V relationship is the only important *inverse* proportionality. All other combinations are directly proportional.

An important corollary of Avogadro's Law is that, at the same temperature and pressure, equal volumes of all gases contain the same number of molecules.

Dalton's Law of Partial Pressures is an easily remembered application of the principle that "the whole is equal to the sum of its parts." Expressed as an equation:

$$P = p_1 + p_2 + p_3 + \cdots \tag{13.10}$$

Other Equations   Of more than two dozen numbered equations in the two gas chapters, only a few need to be learned. One is Equation 13.10 for Dalton's Law. By far the most important equation is the ideal gas equation:

$$PV = nRT = \frac{mRT}{MM} \tag{13.3 and 13.6}$$

The molar volume of a gas can be found from an equation derived from the ideal gas equation:

$$MV \equiv \frac{V}{n} = \frac{RT}{P} \tag{13.8}$$

The combined gas laws for a fixed quantity of gas can also be derived from the ideal gas equation:

$$\frac{P_1V_1}{T_1} = \frac{P_2V_2}{T_2} \tag{4.14}$$

The equation to convert Celsius degrees to kelvins is used in almost all problems involving temperature:

$$K = {}^{\circ}C + 273 \tag{3.4}$$

Numbers   Four numbers should be memorized: the 273 that changes °C to K, two values for R, and the molar volume of a gas at STP:

$$R = 0.0821 \text{ L} \cdot \text{atm/mol} \cdot \text{K} = 62.4 \text{ L} \cdot \text{torr/mol} \cdot \text{K}$$

$$MV = 22.4 \text{ L/mol at STP}$$

CHAPTER 13 IN REVIEW

13.1 The Combined Gas Laws Revisited

13.2 Avogadro's Law: Volume and Amount Variable, Pressure and Temperature Constant

13.3 The Ideal Gas Equation

13.4 Applications of The Ideal Gas Equation:

13A Given values for all except one of the variables in the ideal gas equation, calculate the value of that remaining variable.

13B Calculate the density of a known gas at any specified temperature and pressure.

13C Given the density of an unknown gas at specified temperature and pressure, calculate the molar mass of that gas.

13D Calculate the molar volume of any gas at any specified temperature and pressure.

13E Given the molar volume of a gas at a certain temperature and pressure and either number of moles or volume, calculate the other.

13.5 Gas Stoichiometry

13F Given a chemical equation, or a reaction for which the equation can be written, and the mass of one species, or the volume of any gaseous species at specified temperature and pressure, find the mass of any other species, or volume of any other gaseous species at specified temperature and pressure.

13.6 Dalton's Law of Partial Pressures

13G Given the partial pressure of each component in a mixture of gases, find the total pressure.

13H Given the total pressure of a gaseous mixture and the partial pressures of all components except one, or information from which those partial pressures can be obtained, find the partial pressure of the remaining component.

13.7 Summary of Important Information About Gases

CHAPTER 13 KEY WORDS AND MATCHING SET

_____ Equal volumes of different gases at the same temperature and pressure contain the same number of molecules.

_____ If measured at the same temperature and pressure, reacting volumes of gases are in a ratio of small, whole numbers.

_____ Ideal gas equation solved for molar volume.

_____ Universal gas constant.

_____ $P = p_1 + p_2 + p_3 + \cdots$

_____ Ideal gas equation solved for molar mass.

_____ Represents molar volume.

_____ Ideal gas equation solved for gas density.

_____ Ideal gas equation expressed in mass of gas.

_____ Represents gas density.

_____ Ideal gas equation expressed in moles of gas.

1. R

2. PV = mRT/MM

3. Dalton's Law of Partial Pressures

4. Avogadro's Law

5. PV = nRT

6. (MM)P/RT

7. Gay-Lussac's Law

8. V/n

9. m/V

10. RT/P

11. mRT/PV

STUDY HINTS AND PITFALLS TO AVOID

The main study hints for this chapter have already appeared in the last section. If you have completed the examples and quick checks and mastered the concepts in the summary, you are well prepared to figure out anything you might need in regard to gases.

CHAPTER 13 QUESTIONS AND PROBLEMS

SECTION 13.2

1) Distinguish between Gay-Lussac's combining volume law and Avogadro's Law. Are the same or similar laws valid for solids or liquids? Why or why not?

2) Compare the volumes of 1.01 g H_2, 4.00 g He, 19.00 g F_2, and 20.2 g Ne.

SECTION 13.4

3) Find the pressure in atmospheres produced by 8.33 g CO_2 in a 5.00-L vessel at 36°C.

4) A 44.0-L cylinder contains 2.26 mol N_2 at 19°C. What pressure is exerted by the gas? Answer in torr.

5) The pressure caused by 4.02 mol NO at 21°C is 18.5 atm. What is the volume of the gas in liters?

6) A pressure of 776 torr is exerted by 11.9 g N_2O at 35°C. Calculate the volume of the container.

7) A 844-mL lecture bottle of methane, CH_4, is left with the valve slightly open. Assuming no air has mixed with the methane, how many moles of methane are left in the bottle after the pressure has become equal to atmospheric pressure of 749 torr at 21°C?

8) How many moles of Cl_2 are in a 1.85-L cylinder if the pressure is 1.37 atm at 23°C?

9) At what temperature will 0.119 mol of neon in a 1.97-L vessel exert a pressure of 805 torr?

10) At what Celsius temperature will oxygen have a density of 7.72 g/L and a pressure of 5.87 atmospheres?

11) How many grams of carbon dioxide must be placed in a 35.0-L tank to develop a pressure of 859 torr at 31°C?

12) Calculate the mass of ammonia in a 6.89-L cylinder if the pressure is 5.89 atm at 24°C.

13) If 0.681 g of an unidentified gas is placed in a 442-mL vessel at 49°C, it produces a pressure of 0.629 atm. Find the molar mass of the gas.

14) Calculate the molar mass of a gas if 7.46 g exert a pressure of 704 torr in a 3.74-L container at 31°C.

15) Calculate the volume of 3.37 g of oxygen at STP.

16) What volume will be occupied by 24.8 g of ethane, C_2H_6, at STP?

17) Calculate the mass of 115 L of fluorine at STP.

18) How many grams of hydrogen will be in a 1.75-L flask at STP?

19)* An organic compound has the following percentage composition: 55.8% carbon, 7.0% hydrogen, and 37.2% oxygen. If 3.26 g of the compound occupy 1.47 L at 160°C and 0.914 atm, find the molecular formula of the compound.

20)* A pure, unknown liquid is found to be 85.7% carbon and 14.3% hydrogen by weight. If 29.3 g of this liquid are vaporized in a 4.03-L flask at 260°C, the pressure is 2.84 atm. Determine the molecular formula of the compound.

21)* Air is a mixture of about 21% O_2 and 79% N_2, but it behaves as a pure gas with respect to pressure, volume, temperature, and amount. Estimate the effective "molar mass" of air from these figures. (*Hint:* Recall how you calculated atomic mass from percentages of isotopes in Chapter 5.)

22) The density of air is 1.18 g/L at 25°C and 1.00 atm. Calculate the effective "molar mass" of air from these data and compare it with the answer to Question 21.

23) Calculate the density of NO_2 at STP.

24) Find the STP density of argon.

25) The density of an unknown gas at STP is 2.09 g/L. What is its molar mass?

26) The STP density of an unidentified gas is 1.34 g/L. Calculate the molar mass of the gas.

27) A vessel whose volume is 1.68 L has a mass that is 1.94 g more when filled with a gas than its mass when empty. If the gas is at STP, what is its molar mass?

28)* A student introduces an unknown gas into a gas-weighing bottle. After adjusting the gas to STP, she finds that its mass is 247.647 g. She evacuates the bottle and weighs it again, this time getting 247.292 g. She then fills the bottle with water and finds that its mass is now 462.762 g. Find the molar mass of the gas.

29) What is the density of neon at 47°C and 743 torr?

30) The density of an unknown gas at 25°C and 0.923 atm is 1.43 g/L. Estimate the molar mass of the gas.

31) Temperature in a laboratory is held at a constant 20°C. What is the density of air in the room when atmospheric pressure is 0.994 atm? Assume the effective "molar mass" of air is 29 g/mol (see Questions 21 and 22).

32)* Use the density of air from Problem 31 to calculate the mass of air in a bedroom 11 ft × 13 ft × 8.0 ft. Answer first in grams, then in pounds. Use Table 3.3 in Section 3.6 for the needed conversion factors.

33)* Phosphorus vapor apparently consists of polyatomic molecules, the number of atoms in the molecule depending on the temperature. Measured at 790 torr, the vapor density is 2.74 g/L at 300°C and 0.617 g/L at 10000°C. Determine the molecular formulas at the two temperatures.

34) Just above its boiling point at 445°C, sulfur appears to be a mixture of polyatomic molecules. Above 100°C, however, there is but one structure. Find the formula of molecular sulfur if its vapor density is 0.70 g/L at 1150°C and 965 torr.

35) What is *molar volume*? We can say that the molar volumes of all gases are the same *if* certain conditions are met. What are these conditions and why are they important?

36) What is it about the molar volume of a gas that makes it different from the molar volumes of solids and liquids? How can this difference be accounted for?

37) Calculate the molar volume of a gas at (a) 381 torr and 49°C; (b) 0.774 atm and 5°C.

38) Find the molar volume of a gas at (a) 222°C and 1.43 atm; (b) 732 torr and −11°C.

39) Calculate the moles of gas in a 16.2-L sample if the molar volume is 45.5 L/mol.

40) The molar volume of a gas is 31.1 L/mol. How many moles are in 17.7 L?

41) What volume is occupied by 6.43 mol O_2 if its molar volume is 27.3 L/mol?

42) How many liters are occupied by 0.183 mol Cl_2 when the molar volume is 20.1 L/mol?

43)* What mathematical relationship is there between the densities and molar masses of two gases at a given temperature and pressure? Justify your answer. (*Hint:* Either molar volume or the ideal gas law may be used to predict and explain the relationship.)

44)* One of two gas cylinders is known to contain oxygen and the other nitrogen, but the tags that identify the contents of the cylinders have been removed. The densities of gas samples from the two tanks are measured at the same temperature and pressure. Gas A has the higher density, which is 1.29 g/L. Is gas A oxygen or nitrogen? How do you know? What is the density of the other gas?

SECTION 13.5

45) One source of sulfur dioxide used in making sulfuric acid comes from sulfide ores by the reaction $4\ FeS_2(s) + 11\ O_2(g) \rightarrow 2\ Fe_2O_3(s) + 8\ SO_2(g)$. Calculate the grams of FeS_2 that must react to produce 423 L of SO_2, measured at 1.48 atm and 384°C.

46) Considering natural gas in a laboratory burner to be pure methane, CH_4, calculate the number of grams of carbon dioxide that would result from the complete burning of 19.2 L of methane, measured at 0.813 atm and 26°C.

47) How many liters of hydrogen, measured at 0.940 atm and 32°C, will result from the electrolytic decomposition of 12.6 g of water?

48) Dolomite is used in the manufacture of refractory brick for lining very-high-temperature furnaces. It is processed through a rotary kiln in which carbon dioxide is driven off: $CaCO_3 \cdot MgCO_3 \rightarrow CaO \cdot MgO + 2\ CO_2$. For each kilogram of dolomite processed, how many liters of carbon dioxide escape to the atmosphere at 264°C and 1.09 atm?

49) The reaction chamber in a modified Haber process for making ammonia by direct combination of its element is operated at 550°C and 250 atm. How many grams of ammonia will be produced by the reaction of 97.0 L of nitrogen if introduced at the temperature and pressure of the chamber?

50) Carbon dioxide is released when cream of tartar reacts with baking powder: $KHC_4H_4O_6(s) + NaHCO_3(s) \rightarrow NaKC_4H_4O_6(s) + H_2O(g) + CO_2(g)$. If 4.83 g of $NaHCO_3$ react, what volume of CO_2 will be released at 1.04 atm in an oven that is at 343°C?

51)* When properly detonated, ammonium nitrate explodes violently, releasing hot gases: $NH_4NO_3(s) \rightarrow N_2O(g) + 2\ H_2O(g)$. Calculate the total volume of gas released at 975°C and 1.22 atm by the explosion of 26.8 g NH_4NO_3.

52)* Solder is an alloy of lead and tin. 4.77 g of solder are treated with nitric acid, causing the tin to react: $Sn(s) + 4 HNO_3(aq) \rightarrow SnO_2(s) + 4 NO_2(g) + 2 H_2O(\ell)$. The NO_2 produced has a volume of 1.94 L at 0.856 atm and 19°C. Calculate (a) the grams of tin that reacted and (b) the percentage composition of the solder.

---

53) One of the methods for making sodium sulfate, used largely in the production of kraft paper for grocery bags, involves passing air and sulfur dioxide from a furnace over lumps of salt. The equation is $4 NaCl + 2 SO_2 + 2 H_2O + O_2 \rightarrow 2 Na_2SO_4 + 4 HCl$. (Note the hydrochloric acid, an important by-product of the process.) If 2215 cubic feet of oxygen at 400°C and 835 torr react in a given period of time, how many cubic feet of sulfur dioxide react if measured at the same conditions?

54) Nitrogen dioxide is used in the chamber process for manufacturing sulfuric acid. It is made by direct reaction of oxygen with nitric oxide: $2 NO(g) + O_2(g) \rightarrow 2 NO_2(g)$. How many liters of nitrogen dioxide will be produced by the reaction of 207 L of oxygen, both gases being measured at atmospheric conditions, 18°C and 0.877 atm?

---

55) In the natural oxidation of hydrogen sulfide released by decaying organic matter, the following reaction occurs: $2 H_2S + 3 O_2 \rightarrow 2 SO_2 + 2 H_2O$. How many milliliters of oxygen at 3.52 atm and 81°C are required to react with 2.09 L of hydrogen sulfide measured at 31°C and 0.923 atm in a laboratory reproduction of the reaction?

56) Carbon monoxide is the gaseous reactant in a blast furnace that reduces iron ore to iron. It is produced by the reaction of coke with oxygen from preheated air. How many liters of atmospheric oxygen at an effective pressure of 0.240 atm and 29°C are required to produce 895 L of carbon monoxide at 440 torr and 1700°C? The equation is $2 C + O_2 \rightarrow 2 CO$.

SECTION 13.6
57) What is the *partial pressure* of one gas in a mixture? Use the ideal gas model to explain Dalton's Law of Partial Pressures.

58)* At 22°C, the partial pressure of methane, CH_4, is 373 torr in a mixture of methane and ethane, C_2H_6. If the total pressure is 746 torr, what is the partial pressure of ethane? Which gas, if either, is present in greatest mass? Explain.

---

59) Atmospheric pressure is the total pressure of the gaseous mixture called air. Atmospheric pressure is 754 torr on a day that the partial pressures of nitrogen, oxygen, and argon are 593 torr, 149 torr, and 7 torr, respectively. What is the partial pressure of all the miscellaneous gases in the air that day?

60) The partial pressures of different gases in a mixture are as follows: oxygen, 0.402 atm; nitrogen, 0.339 atm; helium, 0.106 atm; neon, 0.044 atm. What is the total pressure of the mixture?

---

61) The properties of oxygen are studied in the laboratory by collecting samples of the gas over water, as illustrated in Figure 13.5. In one experiment the total pressure of the oxygen and water vapor was 749 torr. The temperature was 26°C, at which the vapor pressure of water is 25 torr. What was the partial pressure of oxygen?

62)* The total volume of the gas collected in Problem 61 was 70.9 mL. If the moisture is removed from the mixture and the dry oxygen adjusted to STP, what will be the volume of the sample? Also, calculate the mass.

GENERAL QUESTIONS
63) Distinguish precisely and in scientific terms the differences between items in each of the following groups:
a) Avogadro's Law, law of combining volumes.
b) Ideal gas equation, ideal gas law.
c) Molar volume, molar mass.
d) Total pressure, partial pressure.

64) Determine whether each statement that follows is true or false:
a) The molar volume of a gas at 0.888 atm and 23°C is less than 22.4 L/mol.
b) To change moles of gas to liters, divide by RT/P.
c) The mass of 5.00 L of ammonia is more than the mass of 5.00 L of carbon monoxide if both volumes are measured at the same temperature and pressure.
d) The partial pressure of oxygen is less than the partial pressure of sulfur dioxide in a mixture of equal masses of the two gases.
e)* The partial pressures of oxygen and ammonia are about equal in a mixture of 43.5 g O_2 and 23.12 g NH_3.

65)* The compression ratio in an automobile engine is the ratio of gas pressure at the end of the compression stroke to the pressure at the beginning. Assume that the ideal gas law is obeyed and that compression occurs at constant temperature. The total volume of the cylinder in a compact automobile is 350 cm³, and the displacement (the reduction in volume during the compression stroke) is 309 cm³. What is the compression ratio in that engine?

66)* The volume of a bicycle tire is 1.5 L; assume it to be the same when "empty" or "full." The tire is to be pumped up to 60 psi gauge, using a bicycle tire pump that forces 0.39 L of air at 1.0 atm into the tire with each stroke. Assume that temperature remains constant at 22°C (295 K) during the process. Work in two significant figures, and use the rounded-off result from each step in the step that follows. Also, use 15 psi = 1 atm.

a) What is the gas pressure (atmospheres) in the tire when "empty"?
b) What will be the gas pressure (atmospheres) in the tire when "full"?
c) How many moles of "air" (gas mixture) are forced into the tire with each stroke of the pump?
d) How many moles of air are in the tire when "empty"?
e) How many moles of air are in the tire when "full"?
f) How many strokes of the pump are required to pump up the tire?

g) Why does each stroke of the pump become more difficult as you pump up a tire? (Assume you are tireless—and no pun is intended!)

67)* Suppose you were to use the bicycle pump described in Problem 54 to pump up a 41-L automobile tire to 30 psi gauge, with all other conditions being the same. How many strokes would be required?

MATCHING SET ANSWERS

4–7–10–1–3–11–8–6–2–9–5

QUICK CHECK ANSWERS

13.1 GIVEN: 5.00 L; 18°C (291 K); 8.65 atm; 0.0821 L · atm/mol · K   WANTED: moles (n)

$$\text{EQUATION: } n = \frac{PV}{RT} = \frac{8.65 \text{ atm} \times 5.00 \text{ L}}{\dfrac{0.0821 \text{ L} \cdot \text{atm}}{\text{mol} \cdot \text{K}} \times (273 + 18) \text{ K}} = 8.65 \text{ atm} \times \frac{\text{mol} \cdot K}{0.0821 \, L \cdot atm} \times \frac{5.00 \, L}{291 \, K} = 1.81 \text{ mol}$$

13.2 GIVEN: 2.97 L CH_4; 16.0 g/mol CH_4; 71°C (344 K); 428 torr; 62.4 L · torr/mol · K   WANTED: mass

$$\text{EQUATION: } m = \frac{(MM)PV}{RT} = \frac{\dfrac{16.0 \text{ g}}{1 \text{ mol}} \times 428 \text{ torr} \times 2.97 \text{ L}}{\dfrac{62.4 \text{ L} \cdot \text{torr}}{\text{mol} \cdot \text{K}} \times 344 \text{ K}}$$

$$= \frac{16.0 \text{ g}}{1 \, mol} \times \frac{428 \, torr}{344 \, K} \times \frac{mol \cdot K}{62.4 \, L \cdot torr} \times 2.97 \, L = 0.947 \text{ g } CH_4$$

13.3 GIVEN: 0°C (273 K); 760 torr; 28.0 g/mol N_2; 62.4 L · torr/mol · K   WANTED: Density, m/V, in g/L

$$\text{EQUATION: } \frac{m}{V} = \frac{(MM)P}{RT} = \frac{28.0 \text{ g}}{1 \, mol} \times \frac{760 \, torr}{273 \, K} \times \frac{mol \cdot K}{62.4 \text{ L} \cdot torr} = 1.25 \text{ g/L}$$

13.4 GIVEN: 1.96 g; 1.00 L; 273 K; 760 torr; 62.4 L · torr/mol · K   WANTED: Molar mass, g/mol

$$\text{EQUATION: } MM = \frac{mRT}{PV} = \frac{1.96 \text{ g}}{1.00 \, L} \times \frac{273 \, K}{760 \, torr} \times \frac{62.4 \, L \cdot torr}{\text{mol} \cdot K} = 43.9 \text{ g/mol}$$

13.5 GIVEN: 18°C (291 K); 8.65 atm; 0.0821 L · atm/mol · K   WANTED: L/mol.

EQUATION: $MV \equiv \dfrac{V}{n} = \dfrac{RT}{P} = \dfrac{291 \, \cancel{K}}{8.65 \, \cancel{atm}} \times \dfrac{0.0821 \, L \cdot \cancel{atm}}{mol \cdot \cancel{K}} = 2.76 \, L/mol$

13.6 GIVEN: 5.00 L   WANTED: mol   PATH: $L \rightarrow mol$
FACTOR: 2.76 L/mol (from Quick Check 13.5)

$$5.00 \, \cancel{L} \times \dfrac{1 \, mol}{2.76 \, \cancel{L}} = 1.81 \, mol$$

Does this answer look familiar? Well it might, because this is the same problem as Quick Check 13.1. This illustrates how molar volume shortens the change from liters to moles *when molar volume is known.* We will use this conversion in the next section.

13.7 GIVEN: 18.6 L SO_2 WANTED: g O_2 PATH: L $SO_2 \rightarrow$ mol $SO_2 \rightarrow$ mol $O_2 \rightarrow$ g O_2
FACTORS: 24.5 L SO_2/mol SO_2; 1 mol O_2/2 mol SO_2; 32.0 g O_2/mol O_2

$$18.6 \, \cancel{L \, SO_2} \times \dfrac{1 \, \cancel{mol \, SO_2}}{24.5 \, \cancel{L \, SO_2}} \times \dfrac{1 \, \cancel{mol \, O_2}}{2 \, \cancel{mol \, SO_2}} \times \dfrac{32.0 \, g \, O_2}{1 \, \cancel{mol \, O_2}} = 12.1 \, g \, O_2$$

13.8

	Volume	Temperature	Pressure	Amount
Initial value (1)	5.17 L	18°C; 291 K	49.0 atm	Constant
Final value (2)	V_2	35°C; 308 K	0.943 atm	Constant

$$5.17 \, \cancel{L \, Cl_2} \times \dfrac{49.0 \, \cancel{atm}}{0.943 \, \cancel{atm}} \times \dfrac{308 \, \cancel{K}}{291 \, \cancel{K}} \times \dfrac{2 \, L \, NOCl}{1 \, \cancel{L \, Cl_2}} = 569 \, L \, NOCl$$

13.9 $P_{O_2} = P - p_{H_2O} = 755 - 19.8 = 735 \, torr$

Liquids and Solids 14

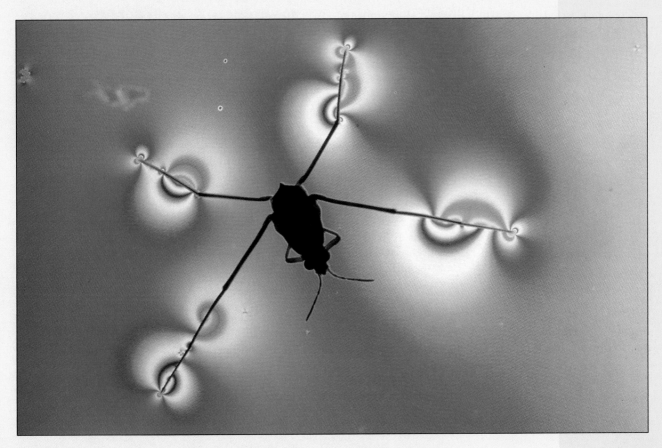

A waterstrider, *Gerris paludum,* walks on water. He is able to do this because of surface tension, a property of liquids that is described in this chapter. If a small quantity of a surfactant (**surf**ace **act**ive reac**tant**) were placed in the water, the surface tension would be reduced and the bug would sink. Household detergents are common surfactants.

14.1 PROPERTIES OF LIQUIDS

PG 14A Explain the differences between the physical behavior of liquids and gases in terms of the relative distances between molecules and the effect of those distances on intermolecular forces.

14B For two liquids, given comparative values of physical properties that depend on intermolecular attractions, predict the relative strengths of those attractions; or, given a comparison of the strengths of intermolecular attractions, predict the relative values of physical properties that are responsible for them.

The properties of liquids are easily observed and described—more so than the properties of gases. To understand liquid properties, however, it is helpful to compare the structure of a liquid with the structure of a gas. We saw in Chapter 4 that gas molecules are so far apart that attractive and repulsive forces between the particles are negligible. These forces are electrostatic in character. They are inversely related to the distance between the molecules; the closer the molecules, the stronger the forces. In a liquid, molecules are very close to each other. Consequently, the intermolecular attractions in a liquid are strong enough to affect its physical properties.

We can now compare the properties of liquids with four properties of gases that were listed in Section 4.1:

1) *Gases Can Be Compressed; Liquids Cannot.* Liquid molecules are "touchingly close" to each other. There is no space between them, so they cannot be pushed closer, as in the compression of a gas.

2) *Gases Expand to Fill Their Containers; Liquids Do Not.* The strong attractions between liquid molecules hold them together at the bottom of a container.

3) *Gases Have Low Densities; Liquids Have Relatively High Densities.* Density is mass per unit volume—mass divided by volume. If the molecules of a liquid are close together compared to the molecules of a gas, a given number of liquid molecules will occupy a much smaller volume than they occupy as a gas. The smaller denominator in the density ratio for a liquid means a higher value for the ratio.

4) *Gases Can Be Mixed in a Fixed Volume; Liquids Cannot.* When one gas is added to another, the molecules of the second gas occupy some of the space between the molecules of the first gas. There is no space between molecules of a liquid, so combining liquids must increase volume.

A liquid has several measurable properties whose values depend on intermolecular attractions, the tendency of the molecules to stick together. In fact, if you think in terms of "stick togetherness" of molecules with strong attractive forces, you can usually predict relative values of these properties for two liquids. We now examine some of these properties.

FLASHBACK This is the same vapor pressure that was mentioned in your study of Dalton's Law of Partial Pressure in Section 13.6.

Vapor Pressure Because of evaporation the open space above any liquid contains some molecules in the gaseous, or vapor, state. The partial pressure exerted by these gaseous molecules is called **vapor pressure.** If the gas space above the liquid is closed, the vapor pressure increases to a definite value called the **equilibrium vapor pressure.** In Section 14.3 you will study the mechanics by which that pressure is reached. Vapor pressure is inversely related to intermolecular attractions. If *stick togetherness* is high between liquid molecules, not much vapor escapes, so the vapor pressure is low.

S U M M A R Y

Liquids with relatively strong intermolecular attractions evaporate less readily, yielding lower vapor concentrations and therefore lower vapor pressures than liquids with weak intermolecular forces.

Molar Heat of Vaporization   It takes energy to overcome intermolecular attractions, separate liquid molecules from each other, and keep them apart. The energy required to vaporize one mole of a liquid at constant temperature and pressure is called **molar heat of vaporization.** The greater the *stick togetherness,* the greater the amount of energy that is needed.

S U M M A R Y

The molar heat of vaporization of a liquid with strong intermolecular attractions is higher than the molar heat of vaporization of a liquid with weak intermolecular attractions.

Boiling Point   Liquids can also be changed to gases by boiling. A liquid must be heated to make it boil. At the **boiling point,** the average kinetic energy of the liquid particles is high enough to overcome the forces of attraction that hold molecules in the liquid state. When *stick togetherness* is high, it takes more agitation (high temperature) to separate the molecules within the liquid, where boiling occurs.

S U M M A R Y

Liquids with strong intermolecular attractions require higher temperatures for boiling than liquids with weak intermolecular attractions.

The trends in vapor pressure, boiling point, and molar heat of vaporization are shown for several substances in Table 14.1.

Table 14.1
Physical Properties of Liquids

Substance	Vapor Pressure at 20°C (torr)	Normal Boiling Point (°C)	Heat of Vaporization (kJ/mol)	Intermolecular Forces
Mercury	0.0012	357	59	Strongest
Water	17.5	100	41	↑
Benzene	75	80	31	
Ether	442	35	26	↓
Ethane	27,000	−89	15	Weakest

FLASHBACK The ability of particles to change relative positions, or flow, is one of the properties that distinguishes a liquid from a gas or a solid. See Section 2.2.

Viscosity Molecules in a liquid are free to move about relative to each other; they "flow." Some liquids flow more easily than others. Water, for example, can be poured much more freely than syrup, and syrup more readily than honey. The ability of a liquid to flow is measured by its **viscosity.** Viscosity is an internal resistance to flow, and it is based on intermolecular attractions. More *stick togetherness* means higher viscosity, more "gooey-ness."

SUMMARY

Liquids with strong intermolecular attractions are generally more viscous than liquids with weak intermolecular attractions.

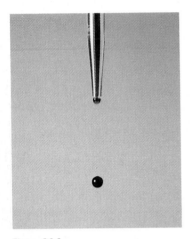

Figure 14.1
Spherical drops. A liquid drop is spherical in shape. For a given volume, a sphere has the smallest possible surface-to-volume ratio. The drop takes on the spherical shape because of surface tension.

Surface Tension When a liquid is broken into "small pieces" it forms spherical drops (Fig. 14.1). A sphere has the smallest surface area possible for a drop of any given volume. This tendency toward a minimum surface is the result of **surface tension.**

Within a liquid, each molecule is attracted in all directions by the molecules around it. At the surface, however, the attraction is nearly all downward, pulling the surface molecules into a sort of tight skin over a standing liquid or around a drop. This is surface tension. Its effect in water may be seen when a needle floats if placed gently on a still surface, or when small bugs run across the surface of a quiet pond (Fig. 14.2). High *stick togetherness* at the surface means more resistance to anything that would break through or stretch that surface.

SUMMARY

Liquids with strong intermolecular attractions have higher surface tension than liquids with weak intermolecular attractions.

Figure 14.2
Surface tension. Unbalanced downward attractive forces at the surface of a liquid pull molecules into a difficult-to-penetrate skin that will support small bugs or thin pieces of dense metals, such as a needle or razor blade. A bug literally runs *on* the water; it does not float *in* it. Molecules within the water are attracted in all directions, as shown.

✔ QUICK CHECK 14.1

a) What is the main difference between gases and liquids that accounts for the large differences in their properties?

b) Intermolecular attractions are stronger in A than in B. Which do you expect will have the higher surface tension, molar heat of vaporization, vapor pressure, boiling point, and viscosity?

c) X has a higher molar heat of vaporization than Y. Which do you expect will have a higher vapor pressure? Why?

14.2 TYPES OF INTERMOLECULAR FORCES

> **PG 14C** Identify and describe or explain dipole forces, dispersion forces, and hydrogen bonds.
> **14D** Given the structure of a molecule, or information from which it may be determined, identify the significant intermolecular forces present.
> **14E** Given the molecular structures of two substances, or information from which they may be obtained, compare or predict relative values of physical properties that are related to them.

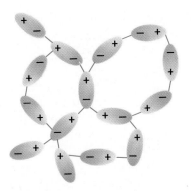

Figure 14.3
Dipole forces. Molecules tend to arrange themselves by bringing oppositely charged regions close to each other and forcing similarly charged regions far from each other.

It was stated in the last section that attractive forces between molecules are electrostatic in character; the attractions are between positive and negative charges. But molecules are electrically neutral. However, the *distribution* of electrical charge within the molecule is not always uniform or symmetrical. Some molecules are polar and some are nonpolar. In addition, some molecules are large and some are small. Molecular polarity and size both contribute to intermolecular attraction, and therefore to physical properties.

Three kinds of intermolecular forces can be traced to electrostatic attractions: dipole forces, dispersion forces, and hydrogen bonds.

1) *Dipole Forces.* A polar molecule is sometimes described as a **dipole.** The attraction between dipoles is between the positive pole of one molecule and the negative pole of another. Figure 14.3 shows the alignment of dipoles, one of several ways polar molecules attract each other.

Table 14.2 compares the boiling points of four pairs of substances that have about the same molecular size, indicated approximately by their molar masses. In each pair the boiling point of the substance with polar molecules is higher than the boiling point of the nonpolar substance. This is because polar molecules have stronger intermolecular attractions than nonpolar molecules.

💡 FLASHBACK Having drawn a Lewis diagram of a molecule (Sections 12.1 or 12.2), you can predict that it is polar if either of these conditions exist: A central atom has a lone pair of electrons, or a central atom is bonded to atoms of different elements. See Section 12.7.

Table 14.2
Boiling Points of Polar vs. Nonpolar Substances

Formulas	Polar or Nonpolar	Molecular Mass	Boiling Point (°C)	Formulas	Polar or Nonpolar	Molecular Mass	Boiling Point (°C)
N_2	Nonpolar	28	−196	GeH_4	Nonpolar	77	−90
CO	Polar	28	−192	AsH_3	Polar	78	−55
SiH_4	Nonpolar	32	−112	Br_2	Nonpolar	160	59
PH_3	Polar	34	−85	ICl	Polar	162	97

Figure 14.4

Dispersion forces. Electron clouds in molecules are constantly shifting. The temporary dipole in the left molecule in the top row ''induces'' the molecule next to it to become another temporary dipole (bottom row). The ''instantaneous dipoles'' are attracted to each other briefly. A small fraction of a second later the clouds shift again and continue to interact with each other or with other nearby molecules.

2) *Dispersion (London) Forces.* Attractions between substances with nonpolar molecules are called **dispersion forces,** or **London forces.** They are believed to be the result of shifting electron clouds within the molecules. If the electron movement in a molecule results in a temporary concentration of electrons at one side of the molecule, the molecule becomes a ''temporary dipole.'' This is shown in the left molecule in the top pair in Figure 14.4. The electrons repel the electrons in the molecule next to it, pushing them to the far side of that molecule. It becomes a second temporary dipole (bottom pair in Fig. 14.4). As long as these dipoles exist—a very small fraction of a second in each individual case—there is a weak attraction between them.

The strength of dispersion forces depends on the ease with which electron distributions can be distorted or ''polarized.'' Large molecules, with many electrons, or with electrons far removed from atomic nuclei, are more easily polarized than small molecules. Larger molecules are generally heavier. Consequently, intermolecular forces tend to increase with increasing molar mass among otherwise similar substances. Notice in Table 14.2 the increase in boiling points for both polar and nonpolar molecules as molar mass increases.

3) *Hydrogen Bonds.* Some polar molecules have intermolecular attractions that are much stronger than ordinary dipole forces. These molecules always have a hydrogen atom bonded to an atom that is small, is highly electronegative, and has at least one unshared pair of electrons. Nitrogen, oxygen, and fluorine are generally the only elements whose atoms satisfy these requirements (see Fig. 14.5).

The covalent bond formed between the hydrogen atom and the atom of nitrogen, oxygen, or fluorine is strongly polar. The electron pair is shifted away from the hydrogen atom toward the more electronegative atom. This leaves the hydrogen nucleus—nothing more than a proton—as a small, highly concentrated region of positive charge at the edge of a molecule. The negative pole of another molecule, which is the region near the nitrogen, oxygen, or fluorine atom, can get quite close to the hydrogen atom of the first molecule. This results in an extra strong attraction between the molecules. This kind of intermolecular attraction is a **hydrogen bond.**

FLASHBACK Electronegativity estimates the strength with which an atom attracts the pair of electrons that forms a bond between it and another atom. Covalent bonds are polar when there is a high electronegativity difference between the bonded atoms. See Section 11.4.

Electronegative Element	Lewis Diagram	Examples

Nitrogen — H—N̈— — H—N̈—H (Ammonia), H—N̈—C—H (Methylamine)

Oxygen — :Ö— / H — :Ö—H / H (Water), :Ö—C—H (Methanol)

Fluorine — H—F̈: — H—F̈: H—F̈: (Hydrogen fluoride)

Figure 14.5

Recognizing hydrogen bonding. Hydrogen bonds occur when a hydrogen atom is covalently bonded to a small atom that is highly electronegative and has one or more unshared electron pairs. Fluorine, oxygen, and nitrogen atoms fit this description. The hydrogen bond is between the atom of one of these elements in one molecule and the hydrogen atom of a nearby molecule.

Notice that a hydrogen bond is an *intermolecular* bond, a bond between different molecules. It is not a covalent bond between atoms in the *same* molecule. The dotted lines in Figure 14.6 represent hydrogen bonds between water molecules. While a hydrogen bond is much stronger than an ordinary dipole–dipole bond, it is roughly one tenth as strong as a covalent bond between atoms of the same two elements.

Of the three kinds of intermolecular attractions, hydrogen bonds are the strongest. When present between small molecules, hydrogen bonds are primarily responsible for the physical properties of a liquid. Dipole forces are next, and dispersion forces are the weakest of the three. Dispersion forces are present between all molecules. In small molecules dispersion forces are important only when the others are absent. But between large molecules—molecules that contain many atoms or even few atoms that have many electrons—dispersion forces are quite strong and often play the main role in determining physical properties.

Figure 14.7 summarizes the kinds of intermolecular forces and their effects on boiling points of similar compounds in three chemical families. We recommend that you study it carefully.

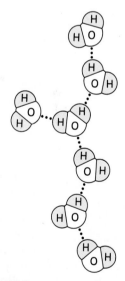

Figure 14.6

Hydrogen bonding in water. Intermolecular hydrogen bonds are present between the electronegative oxygen region of one molecule and the electropositive hydrogen region of a second molecule.

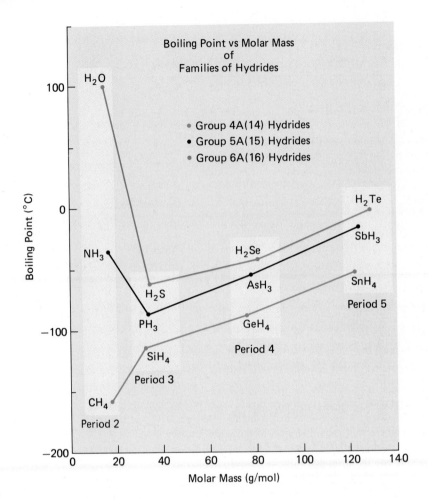

Figure 14.7

Intermolecular attractions illustrated by boiling points of hydrides. Liquids with strong intermolecular attractions usually boil at higher temperatures than liquids with weak intermolecular attractions. These attractions are caused by dipole forces, dispersion forces, and hydrogen bonding. Holding two of these variables essentially constant and changing the third, we can see how each variable affects the attractions by comparing boiling points.

Dipole forces The molecules in the middle small rectangle for the Period 4 hydrides, H_2Se, AsH_3, and GeH_4, are about the same size (nearly equal molar mass), and none of them has hydrogen bonding. They differ only in polarity. GeH_4 has tetrahedral molecules. They are nonpolar. The trigonal pyramidal molecules of AsH_3 are polar, but less so than angular H_2Se molecules. The least polar compound, GeH_4, has the lowest boiling point, and the most polar compound, H_2Se, has the highest boiling point. The same trend appears with the Period 3 and Period 5 hydrides. This indicates that, *other things being equal, intermolecular attractions increase as molecular polarity increases.*

Dispersion forces The Group 4A hydrides (blue line) all have tetrahedral structures. They are nonpolar, and they have no hydrogen bonding. The only intermolecular forces are dispersion forces. The molecules differ only in molecular size (mass), ranging from CH_4, the smallest, to SnH_4, the largest. The boiling points of the four compounds increase as their molecular sizes increase. Except for H_2O and NH_3, the same trend appears for Group 5A (black line) and Group 6A (red line) hydrides. This suggests that, *other things being equal, intermolecular attractions increase as molecular size increases.*

Hydrogen bonding The high boiling points of H_2O and NH_3 violate the trends in which small molecules boil at lower temperatures than large molecules that are otherwise similar. H_2O and NH_3 are the only two substances that have hydrogen bonding. This indicates that, *for small molecules in particular, hydrogen bonding causes exceptionally strong intermolecular attractions.*

✔ QUICK CHECK 14.2

Identify the true statements, and rewrite the false statements to make them true.

a) Dispersion forces are present only with nonpolar molecules.
b) All other things being equal, hydrogen bonds are stronger than dipole–dipole forces.
c) Dipoles have a net electrical charge.
d) Intermolecular forces are magnetic in character.
e) H_2O displays hydrogen bonding, but H_2S does not.

✔ QUICK CHECK 14.3

Determine the molecular geometry of each of the following, and from that, identify the major intermolecular force present:

a) CH_4;   b) CO_2;   c) OF_2;   d) HOCl

✔ QUICK CHECK 14.4

Identify the molecule in each pair below that you would expect to have the stronger intermolecular forces and state why.

a) CCl_4 or CBr_4;   b) NH_3 or PH_3

14.3 LIQUID–VAPOR EQUILIBRIUM

PG 14F Describe or explain the equilibrium between a liquid and its own vapor and the process by which it is reached.

In Section 4.3 you learned that temperature is a measure of the average kinetic energy of the particles in a sample. The range of kinetic energies at two different temperatures was shown graphically in Figure 4.5. This is reproduced here for a single temperature (Figure 14.8). Kinetic energy is plotted horizontally, and the fraction, or percentage, of the sample having a given kinetic energy is plotted vertically. Using the algebraic symbols on the graph, *y* molecules out of 100 have a kinetic energy represented by *x*. The area beneath the curve represents all of the particles in the sample, or 100% of the sample.

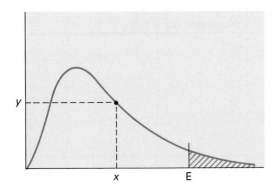

Figure 14.8
Kinetic energy distribution curve for a liquid at a given temperature. E is the escape energy, the minimum kinetic energy a molecule must have to break from the surface and "evaporate" into the gas phase. The total area between the curve and the horizontal axis represents the total number of molecules in the sample. The crosshatched area beneath the curve to the right of E represents the number of molecules having kinetic energy equal to or greater than E. At the temperature for which the graph is drawn, only a small fraction of all molecules has enough energy to evaporate.

To evaporate, or vaporize, a molecule must be at the surface of a liquid. It also must have enough kinetic energy to overcome the attractions of other molecules that would hold it in the liquid state. If E in Figure 14.8 represents this minimum amount of kinetic energy—we will call it the **escape energy**—only those surface molecules having that energy or more can get away. The fraction, or percentage, of all surface molecules having that much energy is given by the area beneath the curve to the right of E.

The rate at which a particular liquid evaporates depends on two things, temperature and surface area. If we think in terms of a unit area and hold temperature constant, the vaporization rate is also constant. These conditions will be assumed in the experiment about to be described.

If a liquid with weak intermolecular forces, such as benzene, is placed in an Erlenmeyer flask, which is then stoppered as in Figure 14.9, it begins to vaporize at constant rate. This rate is represented by the fixed length of the arrows pointing upward from the liquid in each view of the flask. It is also shown as the horizontal line in the graph of evaporation rate versus time.

At first (Time 0) the movement of molecules is entirely in one direction, from the liquid to the vapor. However, as the concentration of molecules in the vapor builds up, an occasional molecule hits the surface and reenters the liquid. The change of state from gas to a liquid is **condensation.** The rate of condensation per unit area at constant temperature depends on the concentration of molecules in the vapor state. At Time 1 there will be a small number of molecules in the vapor state, so the condensation rate will be more than zero, but much less than the evaporation rate. This is shown by the arrow lengths in the Time 1 flask of Figure 14.9.

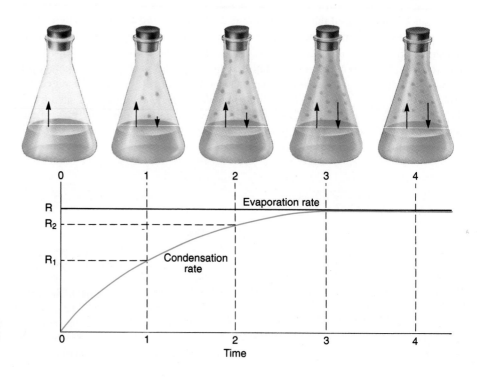

Figure 14.9

Development of a liquid–vapor equilibrium. The depth of shading in the vapor space indicates vapor concentration.

As long as the rate of evaporation is greater than the rate of condensation, the vapor concentration will rise over the next interval of time. Therefore, the rate of return from vapor to liquid rises with time (Time 2). Eventually, the rates of vaporization and condensation become equal (Time 3). The number of molecules moving from vapor to liquid in unit time just balances the number moving in the opposite direction. We describe this situation, **when opposing rates of change are equal,** as a condition of **dynamic equilibrium** between liquid and vapor. Once equilibrium is reached, the concentration of molecules in the vapor has a certain fixed value that does not change. The rates therefore remain equal (Time 4).

Changes that occur in either direction, such as the change from a liquid to a vapor and the opposite change from a vapor to a liquid, are called **reversible changes;** if the change is chemical, it is a **reversible reaction.** Chemists write equations describing reversible changes with a double arrow, one pointing in each direction. For example, the reversible change between liquid benzene, $C_6H_6(\ell)$, and benzene vapor, $C_6H_6(g)$, is represented by the equation

$$C_6H_6(\ell) \rightleftharpoons C_6H_6(g) \qquad (14.1)$$

If the ideal gas equation is solved for the partial pressure of a vapor

$$p = \frac{nRT}{V} = RT \times \frac{n}{V} \qquad (14.2)$$

we see that at constant temperature, the partial pressure is proportional to vapor concentration, n/V. When the vapor concentration becomes constant at equilibrium, the vapor pressure also becomes constant. **The partial pressure exerted by a vapor in equilibrium with its liquid phase at a given temperature is the equilibrium vapor pressure of the substance at that temperature.**

> "Humidity" is commonly blamed for discomfort on a hot, muggy day. This is because the *relative* humidity is high. Relative humidity is roughly equal to the ratio of the partial pressure of water vapor in the air to the equilibrium vapor pressure at the existing temperature: p_{H_2O}/p_{eq}. At equilibrium, this ratio is equal to 1. On a hot day, we perspire. As perspiration evaporates—an endothermic process—it cools us. If the partial pressure of water vapor in the air is close to the equilibrium vapor pressure, as it is on a humid day, comparable to Time 2 in Figure 14.9, the net rate of evaporation is low, so we lose the cooling effect.

The Effect of Temperature

> **PG 14G** Describe the relationship between vapor pressure and temperature for a liquid–vapor system in equilibrium; explain this relationship in terms of the kinetic molecular theory.

Equation 14.2 predicts that vapor pressure increases as temperature rises. This prediction is confirmed in the laboratory. Figure 14.10 shows how the vapor pressures of several substances change with temperature. Notice how a relatively small increase in temperature causes a large increase in vapor pressure. The vapor pressure of water, for example, is 18 torr at 20°C (293 K) and 55 torr at 40°C (313 K). The vapor pressure more than triples (increases by 200%) while absolute temperature increases by only about 7%.

Figure 14.11 explains the effect of temperature on vapor pressure. The curve labeled T_1 gives the kinetic energy distribution at one temperature, and T_2 is the curve for a higher temperature. As in Figure 14.8, E is the escape energy, the minimum energy a surface molecule must have to evaporate into the gas phase. The area beneath the curve to the right of E is the fraction of the sample that has enough kinetic energy to evaporate. As illustrated, the area for T_2 is about twice the area for T_1. We conclude that it is an *increase in the number of molecules with enough energy to evaporate* that is responsible for higher vapor pressure at higher temperature.

> A hot, humid day is often more uncomfortable than an even hotter day with the same absolute humidity, that is, the same water vapor pressure. As temperature rises, the equilibrium water vapor pressure rises. That increases the denominator of the relative humidity ratio while the numerator remains constant. At the lower relative humidity, comparable to Time 1 in Figure 14.9, the net evaporation rate is higher than at Time 2, and we once again enjoy its cooling effect.

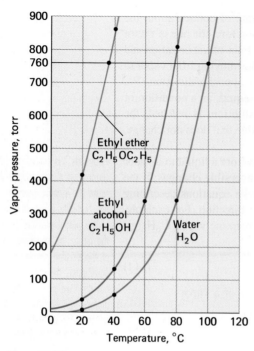

Figure 14.10

Vapor pressures of three liquids at different temperatures.

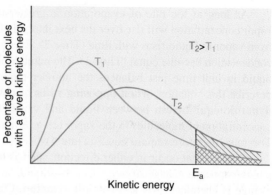

Figure 14.11

Kinetic energy distribution curves for the same liquid at two temperatures. Curve T_1 is for the lower temperature. As temperature is raised from T_1 to T_2, the average kinetic energy increases. The curve shifts to the right and is spread over a wider range. The area beneath the curve to the right of the escape energy, E, represents the fraction of the total number of surface molecules that have enough kinetic energy to evaporate. At T_1 only that fraction that is blue has the escape energy, but at higher temperature, T_2, the fraction is represented by red.

✔ QUICK CHECK 14.5

Identify the true statements, and rewrite the false statements to make them true.

a) A liquid–vapor equilibrium is reached when the amount of liquid is equal to the amount of vapor.
b) Rate of evaporation depends on temperature.
c) Equilibrium vapor pressure is higher at higher temperatures.

14.4 THE BOILING PROCESS

PG 14H Describe the process of boiling and the relationships among boiling point, vapor pressure, and surrounding pressure.

When a liquid is heated in an open container, bubbles form, usually at the base of the container where heat is being applied. The first bubbles are often air, driven out of solution by an increase in temperature. Eventually, when a certain temperature is reached, vapor bubbles form throughout the liquid, rise to the surface, and break. When this happens we say the liquid is boiling.

In order for a stable bubble to form in a boiling liquid, the vapor pressure within the bubble must be high enough to push back the surrounding liquid and the atmosphere above the liquid. The minimum temperature at which this can occur is called the

boiling point: The boiling point is that temperature at which the vapor pressure of the liquid is equal to the pressure above its surface. Actually, the vapor pressure within a bubble must be a tiny bit greater than the surrounding pressure, which suggests that bubbles probably form in local "hot spots" within the boiling liquid. The boiling temperature at one atmosphere—the temperature at which the vapor pressure is equal to one atmosphere—is called the **normal boiling point.** Figure 14.10 shows that the normal boiling point of water is 100°C; of ethyl alcohol, 79°C; and of ethyl ether, 35°C.

According to the definition, the boiling point of a liquid depends on the pressure above it. If that pressure is reduced, the temperature at which the vapor pressure equals the lower surrounding pressure comes down also, and the liquid will boil at that lower temperature. This is why liquids at higher altitudes boil at reduced temperatures. In mile-high Denver, where atmospheric pressure is typically about 634 torr, water boils at 95°C. It is possible to boil water at room temperature by creating a vacuum in the space above it. When pressure is reduced to 20 torr, water boils at 22°C, about 72°F. A method for purifying a compound that might decompose or oxidize at its normal boiling point is to boil it at reduced temperature in a vacuum, and then condense the vapor.

It is also possible to *raise* the boiling point of a liquid by *increasing* the pressure above it. The pressure cooker used in the kitchen takes advantage of this effect. By allowing the pressure to build up within the cooker, it is possible to reach temperatures as high as 110°C without boiling off the water. At this temperature food cooks in about half the time required at 100°C.

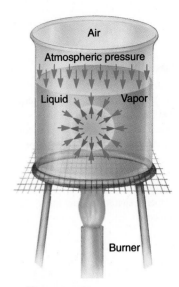

→ Vapor pressure
→ Atmospheric pressure

A liquid boils when its vapor pressure is slightly greater than the pressure above it. The temperature at which this occurs is the boiling point.

✔ QUICK CHECK 14.6

Are the following true or false?

a) Water can be made to boil at 60°F.
b) Bubbles can form anyplace in a boiling liquid.

14.5 WATER—AN "UNUSUAL" COMPOUND

Through much of this book you have seen trends and regularities among physical and chemical properties. Many of these have been related to the periodic table. Predictions have been based on these trends. A prediction is not reliable, though, until it is confirmed in the laboratory. Sometimes a substance does not behave as it is expected to, and we have to look further; but most substances fit into regular patterns.

Water does not fit.

Water is so common, so much a part of our daily lives, that it is hard to think of it as being unusual. But in terms of trends, unusual is exactly what water is. One example appears in Figure 14.7. Beginning at tellurium (Z = 52) in Group 6A (16), the boiling points of the hydrides drop as the molecules become smaller, as expected: −4°C for H_2Te, −42°C for H_2Se, and −62°C for H_2S. If the trend continued, the boiling point of H_2O should be about −72°C. Instead, it is +100°C. And that is only one example.

A close examination of the water molecule (Fig. 14.12) gives us some clues to explain this unique behavior. Aside from fluorine, oxygen is the most electronegative element there is. Therefore, the electrons forming each bond between hydrogen and oxygen are drawn strongly toward the oxygen atom, resulting in two very polar bonds—

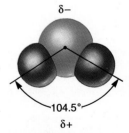

δ−

104.5°

δ+

Figure 14.12

The water molecule. The geometry of the water molecule and the polarity of its bonds make water molecules highly polar. In addition, water displays strong hydrogen bonding. These account for exceptionally strong intermolecular attractions that influence many properties of water.

Figure 14.13

Water molecules in solid form (ice) are held in a crystal pattern that has voids between them. When ice freezes, the crystal collapses, the molecules are closer together, and the liquid is more dense than the solid. This is why ice floats in water, a solid–liquid property shared by few other substances.

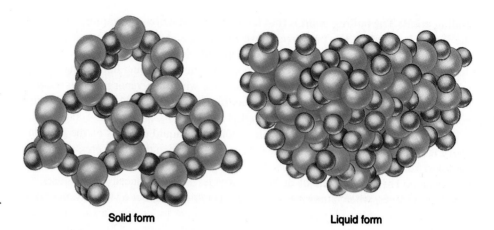

Solid form   Liquid form

Water is about the only substance we normally encounter in the solid, liquid, and gas phases. The gas phase is the water vapor in the air, the humidity mentioned earlier.

more polar than the bonds in other hydrides in the group. Furthermore, the 104.5° bond angle makes a strong dipole. Finally, add hydrogen bonding, which is probably the most important contributor to strong intermolecular attractions in water.

Among molecules of comparable size, water has several other unusual properties. Exceptionally high surface tension and heats of vaporization and fusion are among them. Its vapor pressure is particularly low, even compared to larger molecules whose vapor pressures you would expect to be low. Check the compounds in Figure 14.10, for example. We don't usually think of water as being viscous, but it is viscous when compared to substances with similar structures. Water dissolves a wider variety of gaseous, liquid, and solid substances than most solvents. You will see why this happens in the next chapter. Finally, the mere fact that water is a liquid at room conditions is unusual. It is one of a very small number of *inorganic* compounds (compounds without carbon) that exist as liquids at normal temperatures and pressures.

Water's most visible unusual property is that its solid form, ice, floats on its liquid form. Almost all substances expand—become less dense—when heated; and they contract—become more dense—when cooled. Water also becomes more dense as it is cooled—until it reaches 4°C. Below 4°C, it turns around and becomes less dense. When water freezes, there is about a 9% increase in volume as the molecules arrange themselves into an ''open'' crystal structure compared to the closer packing they have as a liquid. (See Fig. 14.13.) This expansion exerts enough force to break water pipes if the liquid is permitted to freeze in them.

Water is a most unusual compound. Much of life on earth could not exist but for its molecular structure and the unique properties it produces.

14.6 THE NATURE OF THE SOLID STATE

PG 14I   Distinguish between crystalline and amorphous solids.

Fortunately for these penguins, water is one of very few substances whose solid phase is *less* dense than the liquid.

A solid whose particles are arranged in a geometric pattern that repeats itself over and over in three dimensions is a **crystalline solid.** Each particle occupies a fixed position in the crystal. It can vibrate about that site but cannot move past its neighbors. The high degree of order often leads to large crystals that have a precise geometric shape. In ordinary table salt we can distinguish small cubic crystals of sodium chloride. Large,

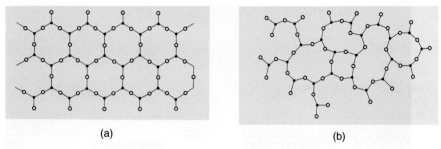

Figure 14.14
Crystalline (a) and amorphous (b) solids. Particles in a crystalline solid are arranged in a distinct geometric order. That order is absent in an amorphous solid.

beautifully formed crystals of such minerals as quartz (SiO_2) and fluorite (CaF_2) are found in nature.

In an **amorphous solid** such as glass, rubber, or plastic, there is no long-range order. Even though the arrangement around a particular site may resemble that in a crystal, the pattern does not repeat itself throughout the solid (Fig. 14.14). From a structural standpoint, we may regard an amorphous solid as intermediate between the crystalline and the liquid states. In many amorphous solids the particles have some freedom to move with respect to one another. The elasticity of rubber and the tendency of glass to flow when subjected to stress over a long period of time suggest that the particles in these materials are not rigidly fixed in position.

Crystalline solids have characteristic physical properties that can serve to identify them. Sodium chloride, for example, melts sharply at 801°C. This is in striking contrast to glass, which first softens and then slowly liquifies over a wide range of temperatures.

✔ QUICK CHECK 14.7

Identify the main structural difference between crystalline solids and amorphous solids.

14.7 TYPES OF CRYSTALLINE SOLIDS

PG 14J Distinguish among the following types of crystalline solids: ionic, molecular, network, and metallic.

Solids can be divided into four classes on the basis of their particle structure and the type of forces that hold these particles together in the crystal lattice.

IONIC CRYSTALS Examples are NaF, $CaCO_3$, AgCl, and NH_4Br. Oppositely charged ions are held together by strong electrostatic forces. As pointed out earlier, ionic crystals are typically high melting, frequently water-soluble, and have very low electrical conductivities. Their melts and water solutions, in which the ions can move around, conduct electricity readily. Figure 14.15 shows three ionic crystals.

FLASHBACK   Figures 11.2 and 11.3 in Section 11.2 show models of NaCl and $CaCO_3$, respectively. These are ionic crystals.

Recognize that combination specific heat and change of state problems are a *group* of problems in which each problem is related to one short stretch on the horizontal axis of the temperature-heat curve (Fig. 14.20). It helps a lot to sketch the graph, as you were guided to do in the last example. Individually, the problems are relatively easy. Whatever is under a sloped line is to be solved by the specific heat equation, $Q = m \times c \times \Delta T$. The calculation under a horizontal (change of state) line is $Q = m \times \Delta H_{vap}$ or $m \times \Delta H_{fus}$. When you get all the ΔH's calculated, add them for the final answer. It is in this step that the only pitfall in the chapter appears: Joules cannot be added to kilojoules! Be sure all energies are in the same units before they are added.

CHAPTER 14 QUESTIONS AND PROBLEMS

SECTION 14.1

1) Why will two gases mix with each other more quickly than two liquids?

2) Why is the liquid state of a substance more dense than the gaseous state?

—————

3) Why are intermolecular attractions weaker in the gaseous state than in the liquid state?

4) Explain why air is more easily compressed than water.

—————

5) How do intermolecular attractions influence the boiling point of a pure substance?

6) State and explain the relationship between intermolecular attractions and equilibrium vapor pressure.

—————

7) Why does molar heat of vaporization depend on the strength of intermolecular forces?

8) What relationship exists between viscosity and intermolecular forces?

—————

9) A tall glass cylinder is filled to a depth of 1 meter with water. Another cylinder is filled to a depth of 1 meter with motor oil. Identical ball bearings are dropped into each cylinder at the same instant. In which cylinder will the bearings reach the bottom first? Explain your prediction in terms of viscosity and intermolecular attractions.

10) Which liquid is more viscous, water or syrup? In which liquid do you suppose the intermolecular attractions are stronger? Explain.

—————

11) If water is spilled on a laboratory desktop, it usually spreads over the surface, wetting any papers or books that may be in its path. If mercury is spilled, it neither spreads nor makes paper wet, but forms little drops that are easily combined into pools by pushing them together. Suggest an explanation for these facts in terms of the apparent surface tension and intermolecular attractions in mercury and water.

12) A drop of honey and a drop of water of identical volumes are placed on a plate. The water drop forms a large shallow pool, but the honey drop forms a high circular blob with a much smaller diameter. Compare the surface tensions of the two liquids. Which liquid has stronger intermolecular attractions? Explain.

—————

13) The level at which a duck floats on water is determined more by the thin oil film that covers its feathers than by a body density that is lower than the density of water. The water does not "mix" with the oil, and therefore does not penetrate the feathers. If, however, a few drops of "wetting agent" are placed in the water near the duck, the poor duck will sink to its neck. State the effect of a wetting agent on surface tension and intermolecular attractions of water.

14) The cleansing ability of soap depends in large part on its ability to change the surface tension in water. How do you suppose soap affects the surface tension of water? Explain.

Questions 15 to 18: The table below gives the normal boiling and melting points for three nitrogen oxides.

	NO	N_2O	NO_2
Boiling point	−152°C	−88.5°C	+21.2°C
Melting point	−164°C	−90.8°C	−11.2°C

15)* Which of the three oxides would you expect to have the highest molar heat of vaporization? Explain how you reached your conclusion.

16) If samples of the three substances are side-by-side at a temperature at which all are solids and the temperature is raised gradually, list the compounds in the order at which they would begin to melt.

—————

17) Which of the three oxides would you expect to have a measurable vapor pressure at −90°C? Explain your answer.

18) Which of the three oxides would you expect to have the lowest viscosity at −90°C? Justify your conclusion.

SECTION 14.2

19) Under what circumstances are dispersion forces likely to produce stronger intermolecular attractions than dipole forces, and when are dispersion forces likely to be weaker?

20) Suggest a molecular structure in which hydrogen bonding may be present, but its contribution to intermolecular attractions is less than the contribution of dispersion forces. Justify your suggestion.

21) Identify the principal intermolecular forces in each of the following compounds: $NH(CH_3)_2$; CH_2F_2; C_3H_8; $BeCl_2$.

22) What are the principal intermolecular forces in each of the following compounds: HCl; C_3H_4; NCl_3; CH_3OH?

23) Compare dipole forces and hydrogen bonds. How are they different, and how are they similar?

24) Given an ionic compound and a polar molecular compound of about the same molar mass, which is more apt to have the higher melting point? Why? In explaining your answer, identify the interparticle forces present and the roles they play.

Questions 25 to 28: On the basis of molecular size, molecular polarity, and hydrogen bonding, predict for each pair of compounds the one that has the lower boiling point. State the reason for your choice. Assume molecular size is related to molar mass.

25) CH_4 and NH_3.
26) CF_4 and CI_4.

27) Ar and Kr.
28) CH_3OH and C_2H_5OH.

29) What feature of the hydrogen atom, when bonded to an appropriate second element, is largely responsible for the strength of hydrogen bonding between molecules?

30) Identify other elements to which hydrogen atoms must be bonded if hydrogen bonding is to be a significant intermolecular attraction. How are these elements different from other elements to which hydrogen might be bonded? Explain why the difference is important.

31) Of the three types of intermolecular forces, which one(s) (a) account for the abnormal properties of water; (b) increase with molecular size?

32) Of the three types of intermolecular forces, which one, when present, is most apt to be the strongest (a) in small molecules; (b) in large molecules?

33) Identify the intermolecular forces present in each of the following:

a) $H—C≡N$

b)

c)

d)

e)

34) Identify the intermolecular forces present in each of the following:

a) $O=C=O$ b)   c) $H—F$

d)   e)

35) Predict which compound, CO_2 or CS_2, has the lower melting and boiling points. Explain your prediction.

36) $C_6H_{13}OH$ has its carbon atoms in a continuous chain with the OH attached to an end carbon. Would you expect this compound to have a higher boiling point than C_2H_5OH? Explain your choice, including an identification of the principal intermolecular force in each compound.

SECTION 14.3

37) What is the meaning of *equilibrium*?
38) Why do we describe a liquid–vapor equilibrium as a *dynamic* equilibrium?

39) Explain why the rate of evaporation from a liquid depends on temperature.

40) Explain why the rate of condensation depends on concentration in the vapor state. Does temperature affect the rate of condensation? Explain either a yes or no answer.

---

41) A liquid in a beaker is placed in an air-tight cylinder. The liquid evaporates until equilibrium is established between the liquid and vapor states. Use the ideal gas equation to show that the partial pressure of the vapor depends on the temperature of the system.

42) A liquid in a beaker is placed in an air-tight cylinder. The liquid begins to evaporate. Use the ideal gas equation to show that at any instant *before equilibrium is reached* the partial pressure of the vapor depends on the vapor concentration at that instant. Assume temperature remains constant during the evaporation process.

---

43)* Assume the system described in Question 41 is at equilibrium. The piston in the cylinder is suddenly adjusted to reduce the volume of the cylinder. If there is no change in temperature, what changes will occur, if any, in the rates of evaporation and condensation?

44)* Assume the system described in Question 41 is at equilibrium. The piston in the cylinder is suddenly adjusted to increase the volume of the cylinder. If there is no change in temperature, what changes will occur, if any, in the *net* rates of evaporation and condensation?

---

45)* Three closed containers have different volumes: Container S is small, M is medium, and L is large. Beakers containing equal quantities of acetone, a volatile liquid, are placed in each box. All of the acetone evaporates in one container, but equilibrium is reached in the other two. (a) In which container does complete evaporation occur? (b) Compare the eventual vapor pressures in the three boxes. (c) Explain both answers.

46)* Three closed containers have identical volumes. A beaker containing a large quantity of ether, a highly volatile liquid, is placed in Container A. It evaporates until equilibrium is reached with a substantial amount of ether remaining. A beaker with a small amount of ether is placed in Container B. The ether all evaporates. A beaker with an intermediate amount of ether is placed in Container C. It evaporates until it reaches equilibrium with only a small amount of ether remaining. (a) Compare the final ether vapor pressures in the three containers. Explain your answer.

---

47)* The equilibrium vapor pressure of water at 24°C is 22.4 torr. A sealed flask contains air at 24°C and 757 torr and a glass vial filled with liquid water. The vial is broken, allowing some of the water to evaporate. What is the maximum pressure this system can reach?

48)* Suppose, after placing another water-filled vial in the same sealed flask described in Question 47, the air is evacuated from the flask. Again the vial is broken. Assuming temperature remains at 24°C, describe what will happen. In particular, what pressure will the system reach and where will the water initially in the vial be?

Questions 49 to 52 are based on the apparatus shown in Figure 14.21. Study the caption that explains how vapor pressure is measured, and then answer the questions.

49)* Why would the apparatus in Figure 14.21 be of little or no value in determining the equilibrium vapor pressure of water if used as described? Under what conditions might the apparatus give acceptable values for water vapor pressure?

50)* Assume you are about to use the apparatus in Figure 14.21 to determine the equilibrium vapor pressure of a volatile liquid. Would you expect the vapor pressure shown by the manometer to (a) increase rapidly at first, and then slowly as equilibrium is reached; (b) increase uniformly until equilibrium is reached; or (c) increase

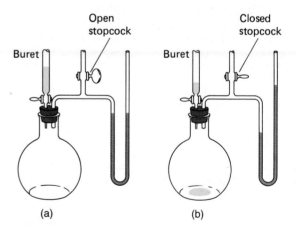

(a)   (b)

Figure 14.21

Measurement of vapor pressure. (a) The buret contains the liquid whose vapor pressure is to be measured. The flask, tubes, and manometer above the mercury in the left leg are all at atmospheric pressure through the open stopcock. The mercury in the open right leg of the manometer is also at atmospheric pressure, so the mercury levels are the same in the two legs. (b) To measure vapor pressure, stopcock is closed, trapping air in flask in space above the left mercury level. Liquid is introduced to flask from buret. Evaporation occurs until equilibrium is reached. Vapor causes increase in pressure which is measured directly by the difference in mercury levels.

slowly at first and rapidly as equilibrium is approached? Justify your answer.

_____

51)* After the system has reached equilibrium, as in Figure 14.21(b), an additional volume of liquid is introduced into the flask. Describe and explain what will happen to the pressure indicated by the manometer. Disregard any Boyle's Law effect; assume that the change in gas volume is negligible.

52)* Suppose that all of the liquid initially introduced to the flask evaporated. Why and how could this occur? Explain in terms of evaporation and condensation rates. How would the vapor pressure shown by the manometer compare with the equilibrium vapor pressure at the existing temperature? Is there further action that can be taken to complete the vapor pressure measurement, or is it necessary to start over? Justify your answer.

SECTION 14.4

53) Define boiling point. Draw a vapor pressure curve and locate the boiling point on it.

54) The vapor pressure of a certain compound is 609 torr at 20°C. Is the substance a gas or a liquid at 760 torr?

_____

55) Normally, a liquid is boiled by heating it. Suggest a second method and explain why it would work.

56)* An industrial reaction must be carried out in the liquid state at a temperature above the boiling point of the solution in which the reaction occurs. Suggest a way that this can be accomplished.

_____

57) Liquid feed water is delivered to modern boilers at a temperature well above the normal boiling point of water. Explain how this is possible.

58) The electric generators in most power plants are operated by steam-driven turbines. Spent steam—steam from which all the capability to drive turbines has been exhausted—is condensed to the liquid state, heated, and recycled as boiler feed water. Just before the spent steam is condensed its temperature may be as low as 32°C—about 90°F. How is this possible, seeing that the boiling point of water is 100°C, or 212°F?

_____

59) Explain why high-boiling liquids usually have high heats of vaporization.

60) Explain why low-boiling liquids usually have high vapor pressures.

_____

61) At 20°C the vapor pressure of substance M is 52 torr; of substance N, 634 torr. Which substance probably has the lower boiling point? the lower molar heat of vaporization?

62) The molar heat of vaporization of substance A is 26 kJ/mol; of substance B, 38 kJ/mol. Which substance would you expect to have the lower boiling point? the lower vapor pressure at 25°C?

SECTION 14.6

63) Is ice a crystalline solid or an amorphous solid? On what properties do you base your conclusion?

64) Compare amorphous and crystalline solids in terms of structure. How do crystalline and amorphous solids differ in physical properties? Explain the difference.

SECTION 14.7

Questions 65 and 66: For each solid whose physical properties are tabulated below, state whether it is most likely to be ionic, molecular, metallic, or a network solid.

	Solid	Melting Point	Water Solubility	Conductivity (Pure)	Type of Solid
65)	A	2300°C	Insoluble	Nonconductor	_____
	B	980°C	Soluble	Nonconductor	_____
66)	C	220°C	Insoluble	Nonconductor	_____
	D	1650°C	Insoluble	Excellent	_____

SECTION 14.8

(See Table 14.4 for heats of fusion and vaporization.)

67) A student is to find the heat of vaporization of isopropyl alcohol (rubbing alcohol). She vaporizes 61.2 g of the liquid at its boiling point and measures the energy required as 44.8 kJ. What heat of vaporization does she report?

68) A calorimetry experiment is performed in which it is found that 29.3 kJ are given off when 4.17 g of a substance condenses. What is the heat of vaporization of that substance?

_____

69) Calculate the energy released as 227 g of sodium vapor condense.

70) How much energy is needed to vaporize 21 g of copper at its normal boiling point?

_____

71) 79.4 kJ were released by the condensation of a sample of ethyl alcohol. If $\Delta H_{vap} = 0.880$ kJ/g, what was the mass of the sample?

72) What mass of hexane, a solvent used in rubber cement, can be boiled by 16.6 kJ if its heat of vaporization is 0.371 kJ/g?

_____

73) Acetone, C_3H_6O, is a highly volatile solvent sometimes used as a cleansing agent prior to vaccination. It evaporates quickly from the skin, making the skin feel

cold. How much heat is absorbed by 23.8 g of acetone as it evaporates if its molar heat of vaporization is 32.0 kJ/mol?

74) Dichlorodifluoromethane, CCl_2F_2, commonly known as Freon-12, is being phased out as a refrigerant for freezers, due to its role in ozone depletion. Calculate the amount of energy absorbed as 712 g of CCl_2F_2 vaporize. Its molar heat of vaporization is 35 kJ/mol.

75) Calculate the heat flow when 3.30 kg of lead freeze.

76) How much energy is required to melt 37.3 g of gold?

77) 36.9 g of an unknown metal release 2.51 kJ of energy in freezing. What is the heat of fusion of that metal?

78) 7.93 kJ are required to melt 52.5 g of naphthalene, which has been used in mothballs. What is the heat of fusion of naphthalene?

79) A piece of zinc releases 4.45 kJ while freezing. What is the mass of the sample?

80) Calculate the number of grams of silver that can be changed from a solid to a liquid by 9.72 kJ.

SECTION 14.9

(See Table 14.5 for specific heat values.)

81) Samples of two different metals, A and B, have the same mass. Both samples absorb the same amount of heat. The temperature of A increases by 10°C, and the sample of B increases by 12°C. Which metal has the higher specific heat?

82) If you are going to heat some water to boiling to prepare tea, will it take more time or less time if you start with hot water than if you start with cold, or will the times be the same? Explain.

83) Find the number of joules released as 467 grams of zinc cool from 68°C to 31°C.

84) How much heat is required to raise the temperature of 196 grams of lead from 22.8°C to 64.9°C?

85) How many kilojoules are required to cool 2.30 kilograms of gold from 88°C to 22°C?

86) The 2.55-kg blade of an iron sword has been forged in a fire and is being cooled from 416°C to 25°C. Calculate the heat flow from the blade.

87) The mass of some copper coins is 144 grams. The coins are at a temperature of 33°C. If they lose 1.47 kJ when they are tossed into a fountain and drop to the foun-

tain's water temperature, what is that temperature?

88) To what temperature will 545 grams of cobalt be raised if, beginning at 25.0°C, it absorbs 2.99 kJ of heat?

89)* A certain kind of rock is being checked for its ability to store heat in a solar heating system. A 3.62-kg piece is heated in an oven until it is at a uniform temperature of 92°C. It is then placed in a calorimeter that contains 9.96 kg of water at 17.1°C. The final temperature of the system is 28.0°C. Assuming no heat loss to the surroundings, find the specific heat of the rock.

90)* A calorimeter contains 72.0 grams of water at 17.2°C. A 141-gram piece of tin is heated to 87.0°C and dropped into the water. The entire system eventually reaches 23.5°C. Assuming all of the heat gained by the water comes from the cooling of the tin—no heat loss to the calorimeter or surroundings—calculate the specific heat of the tin.

SECTION 14.10

Figure 14.22 is a graph of temperature versus energy for a sample of a pure substance. Assume that letters J through P on the horizontal and vertical axes represent numbers, and that expressions such as R − S or X + Y + Z represent arithmetic operations to be performed with those numbers. The next ten questions are related to Figure 14.22.

91) Identify by letter the boiling and freezing points in Figure 14.22.

92) What values are plotted, both vertically and horizontally?

93) Identify all points on the curve in Figure 14.22 where the substance is entirely gas.

94) Identify in Figure 14.22 all points on the curve where the substance is entirely liquid.

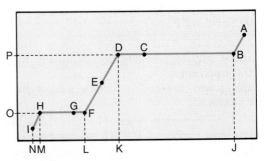

Figure 14.22

95) Identify in Figure 14.22 all points on the curve where the substance is partly solid and partly liquid.

96) Identify all points on the curve in Figure 14.22 where the substance is partly liquid and partly gaseous.

---

97) Describe the physical changes that occur as energy L − J is removed from the sample.

98) Describe what happens physically as the energy represented by J − K is added to the sample.

---

99) Using letters from the graph, show how you would calculate the energy required to boil the liquid at its boiling point.

100) Using letters from the graph, write the expression for the energy required to raise the temperature of the liquid from the freezing point to the boiling point.

---

101) A 127-gram piece of ice is removed from a refrigerator at −11°C. It is placed in a bowl where it melts and eventually warms to room temperature, 21°C. Calculate the amount of heat the sample has absorbed from the atmosphere.

102) How much energy is needed to raise the temperature of 632 g H_2O from 28°C to the boiling point, boil the water, and raise the steam temperature to 168°C?

---

103) A home melting pot is used for a metal casting hobby. At the end of a work period the pot contains 689 g Zn at 552°C. How much heat will be lost as the molten metal cools, solidifies, and cools further to room temperature, 21°C? Find the necessary data from the tables in the chapter.

104)* A 62.3-g aluminum ice tray in a home refrigerator holds 439 g of water. Calculate the energy that must be removed from the tray and its contents to reduce the temperature from 21°C to 0°C, freeze the water, and drop the temperature of the tray and the ice to −11°C. The average specific heat of aluminum is 0.88 J/g · °C over the temperature range involved.

---

105) A certain "white metal" alloy of lead, antimony, and bismuth melts at 264°C, and its heat of fusion is 29 J/g. Its average specific heat is 0.21 J/g · °C as a liquid and 0.27 J/g · °C as a solid. How much energy is required to heat the 941 kg of that alloy in a melting pot from a starting temperature of 26°C to its operating temperature, 339°C?

106) Find the heat flow when 34.3 kg of iron are drawn from a blast furnace at 1695°C and poured into a mold where it cools, freezes, and cools further to a shop temperature of 29°C.

GENERAL QUESTIONS

107) Distinguish precisely and in scientific terms the differences between items in each of the following groups.
a) Intermolecular forces, chemical bonds.
b) Vapor pressure, equilibrium vapor pressure.
c) Molar heat of vaporization, heat of vaporization.
d) Dipole forces, dispersion forces, London forces, hydrogen bonds.
e) Evaporation, vaporization, boiling.
f) Evaporation, condensation.
g) Fusion, solidification.
h) Boiling point, normal boiling point.
i) Amorphous solid, crystalline solid.
j) Ionic, molecular, network, and metallic crystals.
k) Heat of vaporization, heat of condensation.
l) Heat of fusion, heat of solidification.
m) Specific heat, heat of vaporization, heat of fusion.

---

108) Classify each of the following statements as true or false.
a) Intermolecular attractions are stronger in liquids than in gases.
b) Substances with weak intermolecular attractions generally have low vapor pressures.
c) Liquids with high molar heats of vaporization usually are more viscous than liquids with low molar heats of vaporization.
d) A substance with a relatively high surface tension usually has a very low boiling point.
e) All other things being equal, hydrogen bonds are weaker than dispersion or dipole forces.
f) Dispersion forces become very strong between large molecules.
g) Other things being equal, nonpolar molecules have stronger intermolecular attractions than polar molecules.
h) The essential feature of a dynamic equilibrium is that the rates of opposing changes are equal.
i) Equilibrium vapor pressure depends on the concentration of a vapor above its own liquid.
j) The heat of vaporization is equal to the heat of fusion, but with opposite sign.
k) The boiling point of a liquid is a fixed property of the liquid.
l) If you break (shatter) an amorphous solid, it will break in straight lines, but if you break a crystal, it will break in curved lines.
m) Ionic crystals are seldom soluble in water.
n) Macromolecular crystals are nearly always soluble in water.
o) The numerical value of molar heat of vaporization is always larger than the numerical value of heat of vaporization.

p) The units of heat of fusion are kJ/g · °C.

q) The temperature of water drops while it is freezing.

r) Specific heat is concerned with a change in temperature.

109)* The labels have come off the bottles of two white crystalline solids. You know one is sugar and the other is potassium sulfate. Suggest a safe test by which you could determine which is which.

110) Identify the intermolecular attractions in CH_3OH and CH_3F. Which of the two substances do you expect will have the higher boiling point and which will have the higher equilibrium vapor pressure? Justify your choices.

111)* The melting point of an amorphous solid is not always a definite value as it should be for a pure substance. Suggest a reason for this.

112) Under what circumstances might you find that a substance having only dispersion forces is more viscous than a substance that exhibits hydrogen bonding?

113)* Why does dew form overnight?

114)* It is a hot summer day and Chris wants a glass of lemonade. There is none in the refrigerator, so a new batch is prepared from freshly squeezed lemons. When finished, there are 175 grams of lemonade at 23°C. That is not a very refreshing temperature, so it must be cooled with ice. But Chris doesn't like ice in lemonade! Therefore, just enough ice is used to cool the lemonade to 5°C. Of course, the ice will melt and reach the same temperature. If the ice starts at −8°C, and if the specific heat of lemonade is the same as that of water, how many grams of ice does Chris use? Assume there is no heat transfer to or from the surroundings. Answer in two significant figures.

MATCHING SETS ANSWERS

Set 1: 8–3–6–9–2–7–1–4–5   Set 2: 8–9–7–3–5–6–2–4–1   Set 3: 7–9–4–1–3–8–5–2–6

QUICK CHECK ANSWERS

14.1   a) Gas molecules are widely separated compared to liquid molecules. b) A will have the higher surface tension, molar heat of vaporization, boiling point, and viscosity, all of which usually accompany strong intermolecular forces. B will have the higher vapor pressure, a property that is associated with weak intermolecular attractions. c) Y should have a higher vapor pressure. If X has a higher molar heat of vaporization than Y, it probably has stronger intermolecular attractions. That should cause X to have a lower vapor pressure.

14.2   b and e: True. a: Dispersion forces are present between all molecules. c: Dipoles have a nonsymmetrical distribution of electrical charge, but the net charge is zero. d: Intermolecular forces are electrical in character.

14.3   a) tetrahedral, dispersion; b) linear, dispersion; c) angular, dipole; d) angular, hydrogen bonding.

14.4   a) CBr_4 because a CBr_4 molecule is larger than a CCl_4 molecule; b) NH_3 because it has hydrogen bonding and PH_3 does not.

14.5   b and c: True. a: A liquid–vapor equilibrium is reached when the rate of evaporation is equal to the rate of condensation.

14.6   Both true.

14.7   Structural particles in a crystalline solid are arranged in a regular geometric order. In an amorphous solid the structural arrangement is irregular.

14.8   a: True. b: Macromolecules are usually poor conductors of electricity. c: A solid that melts at 152°C is probably a molecular crystal. d: A soluble molecular crystal is a nonconductor of electricity both as a solid and when dissolved.

14.9   GIVEN: 255 g H_2O; 2.26 kJ/g H_2O
WANTED: Q

EQUATION: $Q = m \times \Delta H_{fus}$

$$= 255 \text{ g } H_2O \times \frac{2.26 \text{ kJ}}{\text{g } H_2O} = 576 \text{ kJ}$$

14.10   GIVEN: 749 J; 23 J/g Pb   WANTED: g Pb
UNIT PATH: J → g

EQUATION: $m = \dfrac{Q}{\Delta H_{fus}} = 749 \text{ J} \times \dfrac{\text{g Pb}}{23 \text{ J}} = 33 \text{ g Pb}$

This problem can be solved either by equation, as shown, or by dimensional analysis, beginning the setup with 749 J.

Solutions 15

Solutions are found everywhere, both inside and outside the laboratory. The water solutions of many substances are beautifully colored. All natural waters are solutions. What we call "freshwater" has a very low concentration of dissolved substances, whereas the concentration is much higher with ocean water, or "saltwater." "Hard water" has calcium and magnesium salts dissolved in it. Rainwater is nearly pure, but even it is a solution of atmospheric gases in very low concentrations.

LEARN IT NOW In Chapter 15 we take another step in making this text more like the one you will use in your next chemistry course. Performance Goals no longer appear at the beginning of a section. For the rest of the course, you will not be able to focus your study of a section as you begin that study. You will have to decide for yourself what you must learn how to do. We don't take performance goals away from you altogether, though. They are still listed at the end of the chapter as a review.

What can you do to make up for this change? For one thing, join in its spirit by *not* looking at the Chapter in Review before you study each section. Instead, study the section and build an outline by whatever method you have been using. At the end of the section, *you* write the performance goals you think we might have written. Write them in action form, as ours are written. Describe precisely what you should be able to do after studying the section. When appropriate, state what information you must have in order to complete the task, such as the "givens" in a problem. At the end of the chapter, compare your performance goals with ours. See how well you have analyzed the text and identified what you are expected to have learned. In fact, the very act of writing down the performance goals will help you to learn. Learn it—NOW!

It has been said, and truly, that every instructor identifies at least some of the performance goals for every course. They appear to the student in the form of test questions. If you can figure out in advance what the teacher expects you to do, it is almost as good as having a copy of a test before you take it.

Another suggestion is appropriate at this time. The end of the school term approaches. If it is your custom to sell your used textbooks, don't sell this one if you are going to continue the study of chemistry. At least not until after the first term in your next course. In general chemistry you will be expected to know already the material in this text. Your new text will review that material, but very briefly. Most students who keep their prep chemistry book refer to it often. They report that it saves time and helps them to learn the more advanced material.

15.1 THE CHARACTERISTICS OF A SOLUTION

Solutions abound in nature. We are surrounded by the gaseous solution known as air. The oceans are aqueous (water) solutions of sodium chloride and other substances. Some of these substances are present in sufficient concentrations to make it commercially profitable to extract them. Magnesium is a notable example. Even what we call "fresh" water is a solution, although the concentrations are so low we tend to think of the water we drink as "pure." "Hard" water may be pure enough for human consumption, but there are enough calcium and magnesium salts present to form solid deposits in hot water pipes and boilers. Even rainwater is a solution; it contains dissolved gases. Oxygen is not very soluble in water, but what little there is in solution is mighty important to fish, who cannot survive without it.

A solution is a homogenous mixture. This implies uniform distribution of solution components, so that a sample taken from any part of the solution will have the same composition. Two solutions made up of the same substances, however, may have different compositions. A solution of ammonia in water, for example, may contain 1% ammonia by weight, or 2%, 5%, 20.3% . . . up to the 29% solution called "concentrated ammonia." This leads to variable physical properties, which are determined by the composition of a mixture.

A solution may exist in any of the three states, gas, liquid, or solid. Air is a gaseous solution, made up of nitrogen, oxygen, argon, and other gases in small amounts. In addition to oxygen in water, dissolved carbon dioxide in carbonated beverages is a familiar liquid solution of a gas. Alcohol in water is an example of the solution of two

The ocean is a liquid solution of many solids.

FLASHBACK Pure substances have definite, unchanging physical and chemical properties. The properties of a mixture, however, depend on how much of each component is in the mixture. As the composition changes, so do the properties. This is shown in Figure 2.2, Section 2.2.

liquids, and the oceans are liquid solutions of solids. Solid state solutions are common in the form of metal alloys.

Particle size distinguishes solutions from other mixtures. Dispersed particles in solutions, which may be atoms, ions, or molecules, are very small—generally less than 5×10^{-7} cm in diameter. Particles of this size do not settle on standing, and they are too small to be seen.

✔ QUICK CHECK 15.1

Which among the following are properties of a solution?

a) Definite percentage composition.
b) Variable physical properties.
c) Always made up of two pure substances.
d) Different parts can be detected visually.

15.2 SOLUTION TERMINOLOGY

In discussing solutions, we use a language of closely related and sometimes overlapping terms. We will now identify and define these terms.

Solute and Solvent When solids or gases are dissolved in liquids, the solid or gas is said to be the **solute** and the liquid the **solvent.** More generally, the solute is taken to be the substance present in a relatively small amount. The medium in which the solute is dissolved is the solvent. The distinction is not precise, however. Water is capable of dissolving more than its own weight of some solids, but the water continues to be called the solvent. In alcohol–water solutions, either liquid may be the more abundant and, in a given context, either might be called the solute or solvent.

Concentrated and Dilute A **concentrated** solution has a *relatively* large quantity of a specific solute per unit amount of solution, and a **dilute** solution has a *relatively* small quantity of the same solute per unit amount of solution. The terms compare concentrations of two solutions of the *same solute and solvent.* They carry no other quantitative meaning.

Solubility, Saturated, and Unsaturated **Solubility** is a measure of how much solute will dissolve in a given amount of solvent at a given temperature. It is sometimes expressed by giving the number of grams of solute that will dissolve in 100 grams (g) of solvent. A solution that can exist in equilibrium with undissolved solute is a **saturated** solution: A solution whose concentration corresponds to the solubility limit is therefore saturated. If the concentration of a solute is less than the solubility limit, it is **unsaturated.**

Supersaturated Solutions Under carefully controlled conditions, a solution can be produced in which the concentration of solute is greater than the normal solubility limit. Such a solution is said to be **supersaturated.** A supersaturated solution of sodium acetate, for example, may be prepared by dissolving 80 g of the salt in 100 g of water at about 50°C. If the solution is then cooled to 20°C without stirring, shaking, or other disturbance, all 80 g of solute will remain in solution even though the solubility at

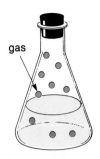

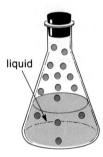

Figure 15.3

Effect of partial pressure of a gas on its solubility in a liquid. Both flasks represent saturated solutions. The solute gas concentration, and therefore its partial pressure, is lower in the top flask than in the second. Consequently, the solute concentration in solution is also lower in the top flask.

Stirring or agitating the solution prevents this buildup and maximizes the *net* dissolving rate.

3) At higher temperatures particle movement is more rapid, thereby speeding up all physical processes.

✔ QUICK CHECK 15.3

Assume that temperature remains constant while a solute dissolves until the solution becomes saturated. For a unit area

a) Is the rate of dissolving when the solution is one-third saturated more than, equal to, or less than the rate of dissolving when the solution is two-thirds saturated?

b) Is the rate of crystallization when the solution is one-third saturated more than, equal to, or less than the rate of crystallization when the solution is two-thirds saturated?

c) Is the *net* rate of dissolving when the solution is one-third saturated more than, equal to, or less than the net rate of dissolving when the solution is two-thirds saturated?

15.4 FACTORS THAT DETERMINE SOLUBILITY

The extent to which a particular solute dissolves in a given solvent depends on three things:

1) Strength of intermolecular forces within the solute, within the solvent, and between the solute and solvent.

2) The partial pressure of a solute gas over a liquid solvent.

3) The temperature.

Intermolecular Forces   Solubility is among the physical properties that are associated with intermolecular forces caused by molecular geometry. Generally speaking, *if forces between A molecules are about the same as the forces between B molecules, A and B will probably dissolve in each other.* From the standpoint of these forces, the molecules appear to be able to replace each other. On the other hand, if the intermolecular forces between A molecules are quite different from the forces between B molecules, it is unlikely that they will dissolve in each other.

Consider, for example, hexane, C_6H_{14}, and decane, $C_{10}H_{22}$. Each substance has only dispersion forces. The forces are roughly the same for the two substances, which are soluble in each other. Neither, however, is soluble in water or methanol, CH_3OH, two liquids that exhibit strong hydrogen bonding. But water and methanol are soluble in each other, again supporting the correlation between solubility and similar intermolecular forces.

💡 FLASHBACK   The partial pressure of one gas in a mixture of gases is the pressure that one gas would exert if it alone occupied the same volume at the same temperature. See Section 13.6.

Partial Pressure of Solute Gas over Liquid Solvent   Changes in partial pressure of a solute gas over a liquid solution have a pronounced effect on the solubility of the gas (Fig. 15.3). This is sometimes startlingly apparent on opening a bottle of a carbonated beverage. Such beverages are bottled under a carbon dioxide partial pressure that is slightly greater than one atmosphere, which increases the solubility of the gas. This is what is meant by "carbonated." As the pressure is released on opening, solubility decreases, resulting in bubbles of carbon dioxide escaping from the solution.

In an ideal solution the solubility of a gaseous solute in a liquid is directly proportional to the partial pressure of the gas over the surface of the liquid. An equilibrium is reached that is similar to the vapor pressure equilibrium described in Section 14.3 and the solid-in-liquid equilibrium discussed in Section 15.3. Neither the partial pressure nor the total pressure caused by other gases affects the solubility of the solute gas. This is what would be expected for an ideal gas, where all molecules are widely separated and completely independent.

Pressure has little or no effect on the solubility of solids or liquids in a liquid solvent. None of the events described in Figure 15.2 are influenced by gas pressure above the liquid surface.

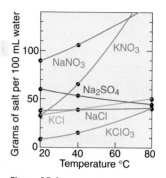

Figure 15.4

Temperature-solubility curves for various salts in water.

Temperature   Temperature exerts a major influence on most chemical equilibria, including solution equilibria. Consequently, solubility depends on temperature. Figure 15.4 indicates that the solubility of most solids increases with rising temperature, but there are notable exceptions. The solubilities of gases in liquids, on the other hand, are generally lower at higher temperatures. The explanation of the relationship between temperature and solubility involves energy changes in the solution process, as well as other factors.

✔ QUICK CHECK 15.4

If given the structural formulas of two substances, list the things you would look for to predict whether or not one would dissolve in the other. For each item listed, state the conditions under which solubility would be more probable.

15.5 SOLUTION CONCENTRATION: PERCENTAGE

The concentration of a solution tells how much solute is present per given amount of solution or a given amount of solvent. As a "per" expression, concentration has the form of a fraction, or a **concentration ratio.** Amount of solute appears in the numerator and may be in grams, moles, or equivalents (eq), a unit that is introduced later in this chapter. Quantity of solvent or solution is in the denominator and may be in mass or volume units. In general, concentration is

$$\frac{\text{quantity of solute (g or mol or eq)}}{\text{quantity of solution (g or L)}} \quad or \quad \frac{\text{quantity of solute (g or mol)}}{\text{quantity of solvent (kg)}}$$

We begin with percentage by mass. Percentage concentration is based on the concentration ratio g solute/g solution.

If a solution concentration is given in percent, you may assume it to be percent by mass unless specifically stated otherwise. By definition, percentage by mass is grams of solute per 100 grams of solution. The defining equation is

$$\% \text{ by mass} \equiv \frac{\text{g solute}}{100 \text{ g solution}} \tag{15.1}$$

A better way to calculate solution percentage is to multiply the concentration ratio g solute/g solution by 100:

$$\% \text{ by mass} = \frac{\text{g solute}}{\text{g solution}} \times 100 = \frac{\text{g solute}}{\text{g solute} + \text{g solvent}} \times 100 \qquad (15.2)$$

Be careful about the denominator. If a problem gives the mass of solute and mass of solvent, be sure to add them to get the mass of solution.

EXAMPLE 15.1

When 125 grams of solution were evaporated to dryness, 42.3 grams of solute were recovered. What was the percentage of solute?

SOLUTION

GIVEN: 42.3 g solute; 125 g solution   WANTED: %

EQUATION: $\% = \dfrac{\text{g solute}}{\text{g solution}} \times 100 = \dfrac{42.3 \text{ g}}{125 \text{ g}} \times 100 = 33.8\%$

EXAMPLE 15.2

3.50 g KNO_3 are dissolved in 25.0 g H_2O. Calculate the percentage concentration of KNO_3.

GIVEN: 3.50 g KNO_3; 25.0 g H_2O WANTED: % KNO_3

EQUATION: $\% = \dfrac{\text{g solute}}{\text{g solute} + \text{g solvent}} \times 100 = \dfrac{3.50 \text{ g}}{3.50 \text{ g} + 25.0 \text{ g}} \times 100 = 12.3\% \ KNO_3$

EXAMPLE 15.3

You are to prepare 2.50×10^2 g 7.00% Na_2CO_3. How many grams of sodium carbonate and how many milliliters of water do you use? (The density of water is 1.00 g/mL.)

This time the given percentage can be used as a dimensional analysis conversion factor (Equation 15.1) to find the grams of Na_2CO_3. Calculate that quantity first.

GIVEN: 2.50×10^2 g solution   WANTED: g Na_2CO_3
PATH: g solution → g Na_2CO_3 FACTOR: 7.00 g Na_2CO_3/100 g solution

$2.50 \times 10^2 \ \text{g solution} \times \dfrac{7.00 \text{ g } Na_2CO_3}{100 \text{ g solution}} = 17.5 \text{ g } Na_2CO_3$

The mass of the solution is 2.50×10^2 g. The mass of solute in the solution is 17.5 g. The rest is water. What is the mass of the water? What is the volume of that mass of water?

g H_2O = g solution − g solute = 2.50×10^2 g − 17.5 g Na_2CO_3 = 232 g H_2O

At 1.00 g/mL H_2O, 232 g H_2O = 232 mL H_2O

15.6 SOLUTION CONCENTRATION: MOLARITY

In working with liquids, volume is easier to measure than mass. Therefore, a solution concentration based on volume is usually more convenient than one based on mass. **Molarity, M, is the moles of solute per liter of solution.** The concentration ratio is the defining equation:

$$M \equiv \frac{\text{moles solute}}{\text{liter solution}} = \frac{\text{mol}}{L} \qquad (15.3)$$

If a solution contains 0.755 mole of sulfuric acid per liter, we identify it as 0.755 M H_2SO_4. In words, it is "point 755 molar sulfuric acid." In a calculation setup we would write "0.755 mol H_2SO_4/L."

Notice that molarity is a "per" relationship. All of the calculation methods you have used before with per relationships can be used with molarity. Notice also that the units in the denominator are liters, but volume is often given in milliliters. To convert mL to L, divide by 1000—move the decimal three places to the left.

FLASHBACK Problems based on Equation 15.3 are like all other "defining equation" problems. Using Q for the quantity defined, N for the numerator, and D for the denominator, $Q \equiv N/D$. If you are given N and D, find Q by the equation. If given Q and either N or D, find the other by dimensional analysis, using Q as the conversion factor. See Section 3.8.

FLASHBACK This is a unit↔milliunit conversion, made by moving the decimal point three places. Dimensional analysis or the larger/smaller rule tells you which way it goes. See Section 3.4.

EXAMPLE 15.4 _____

How many grams of silver nitrate must be dissolved to prepare 5.00×10^2 mL 0.150 M $AgNO_3$?

SOLUTION

To "plan" the problem, we follow our usual procedure: Given data are changed to required units immediately, whenever possible. The mL → L change is made routinely in the givens.

GIVEN: 5.00×10^2 mL (0.500 L)   WANTED: g $AgNO_3$
PATH: L → mol $AgNO_3$ → g $AgNO_3$
FACTORS: 0.150 mol $AgNO_3$/L; 169.9 g $AgNO_3$/mol $AgNO_3$

In this problem, molarity is the conversion factor in the first step of the unit path, changing the given volume to moles of silver nitrate. Moles are then changed to mass, using molar mass as the second conversion factor.

$$0.500\,L \times \frac{0.150\ \text{mol } AgNO_3}{L} \times \frac{169.9\ \text{g } AgNO_3}{1\ \text{mol } AgNO_3} = 12.7\ \text{g } AgNO_3$$

A clearer understanding of molarity can be gained by mentally "preparing" the solution in Example 15.4. First weigh out the 12.7 g $AgNO_3$ (Fig. 15.5a). Then transfer the crystals to a 500-mL volumetric flask containing *less than* 500 mL of water (Fig. 15.5b). After dissolving the solute, add water to the 500-mL mark on the neck of the

 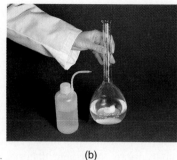 

(a)   (b)   (c)

Figure 15.5

Preparation of 500.0 mL 0.150 M $AgNO_3$. (a) Weigh out 12.7 g $AgNO_3$. (b) Dissolve the $AgNO_3$ in a partly filled 500-mL volumetric flask. (c) Dilute to 500.0 mL mark on the flask.

flask (Fig. 15.5c). Notice that molarity is based on the volume of *solution,* not the volume of *solvent.* This is why the solute is dissolved in less than 500 mL of water and then diluted to that volume.

EXAMPLE 15.5

15.8 g NaOH are dissolved in water and diluted to 1.00×10^2 mL. Calculate the molarity.

Begin by ''planning'' the problem.

GIVEN: 15.8 g NaOH; 1.00×10^2 mL (0.100 L)
WANTED: M   PATH: g NaOH $\rightarrow$ mol NaOH
FACTOR: 40.0 g NaOH/mol NaOH   EQUATION: M = mol/L

To find molarity from its definition (Equation 15.3), you need to know the volume of solution, which is given, and the moles of NaOH. Grams are given. The grams-to-moles conversion must be made before you can use the defining equation. How many moles of NaOH are in the solution?

$$15.8 \text{ g NaOH} \times \frac{1 \text{ mol NaOH}}{40.0 \text{ g NaOH}} = 0.395 \text{ mol NaOH}$$

Now you can plug into the defining equation for molarity.

$$\text{EQUATION: M} = \frac{\text{mol solute}}{\text{L}} = \frac{0.395 \text{ mol NaOH}}{0.100 \text{ L}} = 3.95 \text{ mol NaOH/L} = 3.95 \text{ M NaOH}$$

Example 15.5 can be solved in a single setup if we write the concentration ratio directly from the given data. One concentration ratio is the quantity of a solute/quantity of solution. In the previous example, 15.8 g NaOH is a quantity of solute, and 1.00×10^2 mL (0.100 L) is a quantity of solution. Thus, the concentration ratio is

$$\frac{15.8 \text{ g NaOH}}{0.100 \text{ L}}$$

To change this to molarity, grams in the numerator must be changed to moles. The unit path and calculation setup are

$$\frac{\text{g}}{\text{L}} \longrightarrow \frac{\text{mol}}{\text{L}} \quad or \quad \frac{(\text{g} \longrightarrow \text{mol})}{\text{L}}$$

$$\frac{15.8 \text{ g NaOH}}{0.100 \text{ L}} = \frac{15.8 \text{ g NaOH} \times \dfrac{1 \text{ mol NaOH}}{40.0 \text{ g NaOH}}}{0.100 \text{ L}} = \frac{15.8 \text{ g NaOH}}{0.100 \text{ L}} \times \frac{1 \text{ mol NaOH}}{40.0 \text{ g NaOH}} = 3.95 \text{ M NaOH}$$

We will write other concentration ratios directly from data later in the chapter.
Among the most important of calculations made with molarity is its use as a conversion factor in changing liters to moles or moles to liters. (Can you guess why this is an important conversion and how it is used?)

EXAMPLE 15.6 _____

Find the number of milliliters of 1.40 M solution that contain 0.287 mole of solute.

"Plan" and solve the problem.

_____   _____

GIVEN: 0.287 mol solute   WANTED: mL   PATH: mol → L → mL
FACTOR: 1.40 mol/L

$$0.287 \text{ mol} \times \frac{1 \text{ L}}{1.40 \text{ mol}} = 0.205 \text{ L} = 205 \text{ mL}$$

To change from units to milliunits, move the decimal three places to the right. Remember the larger/smaller idea. When in doubt, extend the setup one more step with Old Reliable, dimensional analysis.

EXAMPLE 15.7 _____

How many moles of solute are in 45.3 mL 0.550 M solution?

In essence, you have seen this example before. It is the first step in Example 15.4. "Plan" and solve the problem.

_____   _____

GIVEN: 45.3 mL (0.0453 L)   WANTED: mol   PATH: L → mol
FACTOR: 0.550 mol/L

$$0.0453 \text{ L} \times \frac{0.550 \text{ mol}}{\text{L}} = 0.0249 \text{ mol}$$

One equivalent of an acid is the quantity that yields one mole of hydrogen ions in a chemical reaction. One equivalent of a base is the quantity that reacts with one mole of hydrogen ions. Because hydrogen and hydroxide ions combine on a one-to-one ratio, one mole of hydroxide ions is one equivalent of base.

According to these statements, both one mole of HCl and one mole of NaOH are one equivalent. They yield, respectively, one mole of H^+ ions and one mole of OH^- ions. H_2SO_4, on the other hand, may have two equivalents per mole because it can release two moles of H^+ ions per mole of acid. Similarly, one mole of $Al(OH)_3$ may represent three equivalents because three moles of OH^- may react.

Notice that the number of equivalents in a mole of an acid depends on a specific reaction, not just the number of moles of H's in a mole of the compound. It is the number of H's that react that count. By controlling reaction conditions, phosphoric acid can have one, two, or, theoretically, three equivalents per mole:

$$NaOH(aq) + H_3PO_4(aq) \longrightarrow NaH_2PO_4(aq) + H_2O(\ell) \qquad \text{1 eq acid/mol} \quad (15.7)$$

$$2\,NaOH(aq) + H_3PO_4(aq) \longrightarrow Na_2HPO_4(aq) + 2\,H_2O(\ell) \qquad \text{2 eq acid/mol} \quad (15.8)$$

$$3\,NaOH(aq) + H_3PO_4(aq) \longrightarrow Na_3PO_4(aq) + 3\,H_2O(\ell) \qquad \text{3 eq acid/mol} \quad (15.9)$$

There is one equivalent per mole of base in each of the above reactions. Indeed, NaOH can have only one equivalent per mole, because there is only one mole of OH^- in one mole of NaOH. It is noteworthy that *the number of equivalents of acid and base in each reaction are the same*. In Equation 15.7, 1 mol H_3PO_4 gives up only 1 eq H^+, and it reacts with 1 mol NaOH, which is 1 eq NaOH. In Equation 15.8, 1 mol H_3PO_4 yields 2 eq H^+, and it reacts with 2 mol NaOH, which is 2 eq NaOH. In Equation 15.9 there are 3 eq H^+ and 3 mol NaOH, which is 3 eq NaOH.

EXAMPLE 15.10 _____

State the number of equivalents of acid and base per mole in each of the following reactions:

	eq acid/mol	eq base/mol

$$2\,HBr + Ba(OH)_2 \longrightarrow BaBr_2 + 2\,H_2O$$

$$H_3C_6H_5O_7 + 2\,KOH \longrightarrow K_2HC_6H_5O_7 + 2\,H_2O$$

Remember, you are interested only in the number of moles of H^+ or OH^- that *react*, not the number present, in one mole of acid or base. The formula of citric acid, $H_3C_6H_5O_7$, is written as an inorganic chemist is most apt to write it—with three ionizable hydrogens first:

_____   _____

	eq acid/mol	eq base/mol
$2\,HBr + Ba(OH)_2 \longrightarrow BaBr_2 + 2\,H_2O$	1	2
$H_3C_6H_5O_7 + 2\,KOH \longrightarrow K_2HC_6H_5O_7 + 2\,H_2O$	2	1

In the first equation each mole of $Ba(OH)_2$ yields two OH^- ions, so there are two eq/mol. Each mole of HBr produces one H^+, so there is one eq/mol. In the second equation there could be one, two, or three eq $H_3C_6H_5O_7$/mol, depending on how many ionizable hydrogens are released in the reaction. That number is two: $H_3C_6H_5O_7 \rightarrow$ $2\,H^+ + HC_6H_5O_7{}^{2-}$. The $2\,H^+$ ions released combine with the $2\,OH^-$ ions from 2 moles of KOH to form $2\,H_2O$ molecules. In KOH there is only $1\,OH^-$ in a formula unit, so there can be only one eq/mol.

Notice that, as with the three phosphoric acid neutralizations, *the number of equivalents of acid and base in each equation in Example 15.10 is the same*. In the first reaction 2 mol HBr is 2 eq, and 1 mol $Ba(OH)_2$ is 2 eq. In the second reaction 1 mol $H_3C_6H_5O_7$ is 2 eq and 2 mol KOH is 2 eq. The "same number of equivalents of all reactants" idea extends to the product species too. Once you find the number of equivalents of one species in a reaction, you have the number of equivalents of *all* species. It is this fact that makes normality such a useful tool in quantitative work.

The fact that the number of equivalents of all species in a chemical reaction is the same is *why* normality is a convenient concentration unit for analytical work. You will see how this becomes an advantage in Section 15.13.

It is sometimes convenient to find the **equivalent mass, g/eq,** of a substance, **the number of grams per equivalent.** Equivalent mass is similar to molar mass, g/mol. Equivalent mass is readily calculated by dividing molar mass by equivalents per mole:

$$\frac{g/mol}{eq/mol} = \frac{g}{mol} \times \frac{mol}{eq} = g/eq \qquad (15.10)$$

In ordinary acid–base reactions there are one, two, or three equivalents per mole. It follows that the equivalent mass of an acid or base is the same as, one half of, or one third of the molar mass. The molar mass of phosphoric acid is 98.0 g/mol. For the three reactions of phosphoric acid (Equations 15.7, 15.8, and 15.9) the equivalent masses are

Equation 15.7: $\dfrac{98.0\ g\ H_3PO_4/mol}{1\ eq\ H_3PO_4/mol} = \dfrac{98.0\ g\ H_3PO_4}{1\ eq\ H_3PO_4} = 98.0\ g\ H_3PO_4/eq\ H_3PO_4$

Equation 15.8: $\dfrac{98.0\ g\ H_3PO_4/mol}{2\ eq\ H_3PO_4/mol} = \dfrac{98.0\ g\ H_3PO_4}{2\ eq\ H_3PO_4} = 49.0\ g\ H_3PO_4/eq\ H_3PO_4$

Equation 15.9: $\dfrac{98.0\ g\ H_3PO_4/mol}{3\ eq\ H_3PO_4/mol} = \dfrac{98.0\ g\ H_3PO_4}{3\ eq\ H_3PO_4} = 32.7\ g\ H_3PO_4/eq\ H_3PO_4$

EXAMPLE 15.11

Calculate the equivalent masses of KOH (56.1 g/mol), $Ba(OH)_2$ (171.3 g/mol), and $H_3C_6H_5O_7$ (192.1 g/mol) for the reactions in Example 15.10.

KOH: $\dfrac{56.1\ g\ KOH}{1\ eq\ KOH} = 56.1\ g\ KOH/eq$

$Ba(OH)_2$: $\dfrac{171.3\ g\ Ba(OH)_2}{2\ eq\ Ba(OH)_2} = 85.7\ g\ Ba(OH)_2/eq$

$H_3C_6H_5O_7$: $\dfrac{192.1\ g\ H_3C_6H_5O_7}{2\ eq\ H_3C_6H_5O_7} = 96.1\ g\ H_3C_6H_5O_7/eq$

Just as molar mass makes it possible to convert in either direction between grams and moles, equivalent mass sets the path between grams and equivalents. In practice, it is often more convenient to use the fractional form for equivalent mass—the molar mass over the number of equivalents per mole. We use both setups in the next example, but only the fractional setup thereafter. If your instructor emphasizes equivalent mass as a quantity, you should, of course, follow those instructions.

EXAMPLE 15.12 _____

Calculate the number of equivalents in 68.5 g $Ba(OH)_2$.

The numbers you need are in Example 15.11. Complete the problem.

———   ———

Using equivalent mass, $68.5 \text{ g } Ba(OH)_2 \times \dfrac{1 \text{ eq } Ba(OH)_2}{85.7 \text{ g } Ba(OH)_2} = 0.799 \text{ eq } Ba(OH)_2$

The difference between 0.799 and 0.800 comes from the roundoff in the equivalent mass of $Ba(OH)_2$. Similar variations in problems solved by two methods appear in the examples ahead.

Using the fractional setup, $68.5 \text{ g } Ba(OH)_2 \times \dfrac{2 \text{ eq } Ba(OH)_2}{171.3 \text{ g } Ba(OH)_2} = 0.800 \text{ eq } Ba(OH)_2$

You are now ready to use the equivalent concept in normality problems.

EXAMPLE 15.13 _____

Calculate the normality of a solution that contains 2.50 g NaOH in 5.00×10^2 mL of solution.

SOLUTION

This is just like Example 15.5, except that moles in mol/L have been replaced by equivalents in eq/L. After converting 500 mL to liters, the concentration may be expressed in a concentration ratio of grams per liter, g/L. The ''plan'' for the problem then becomes

GIVEN: 2.50 g NaOH/0.500 L   WANTED: N
PATH: g/L → eq/L
FACTOR: 40.0 g NaOH/eq

$\dfrac{2.50 \text{ g NaOH}}{0.500 \text{ L}} \times \dfrac{1 \text{ eq NaOH}}{40.0 \text{ g NaOH}} = 0.125 \text{ N NaOH}$

In essence, grams per liter, g/L, has been divided by equivalent mass, g/eq.

EXAMPLE 15.14 _____

2.50×10^2 mL of a sulfuric acid solution contains 10.5 g H_2SO_4. Calculate its normality for the reaction $H_2SO_4 + 2\,NaOH \rightarrow Na_2SO_4 + 2\,H_2O$.

The procedure is the same. Express the concentration ratio in g/L and convert it to eq/L. In doing so, you must determine the number of equivalents in one mole of sulfuric acid.

———   ———

GIVEN: 10.5 g H_2SO_4/0.250 L   WANTED: N
PATH: g/L $\rightarrow$ eq/L   FACTOR: 98.1 g H_2SO_4/2 eq H_2SO_4

$$\frac{10.5 \ \cancel{\text{g } H_2SO_4}}{0.250 \text{ L}} \times \frac{2 \text{ eq } H_2SO_4}{98.1 \ \cancel{\text{g } H_2SO_4}} = 0.856 \text{ eq } H_2SO_4/L = 0.856 \text{ N } H_2SO_4$$

Both hydrogens are lost by H_2SO_4, so there are 2 eq/mol. The setup is the same as dividing the concentration in g/L by the equivalent mass of the acid:

$$\frac{10.5 \ \cancel{\text{g } H_2SO_4}}{0.250 \text{ L}} \times \frac{1 \text{ eq } H_2SO_4}{49.1 \ \cancel{\text{g } H_2SO_4}} = 0.855 \text{ eq } H_2SO_4/L = 0.855 \text{ N } H_2SO_4$$

Just as molarity provides a way to convert in either direction between moles of solute and volume of solution, normality offers a unit path between equivalents of solute and volume of solution.

EXAMPLE 15.15

How many equivalents are in 18.6 mL 0.856 N H_2SO_4?

"Plan" and solve the problem.

GIVEN: 18.6 ml (0.0186 L)   WANTED: eq
PATH: L $\rightarrow$ eq H_2SO_4 FACTOR: 0.856 eq H_2SO_4/L

$$0.0186 \ \cancel{\text{L}} \times \frac{0.856 \text{ eq } H_2SO_4}{\cancel{\text{L}}} = 0.0159 \text{ eq } H_2SO_4$$

Example 15.15 presents an important relationship involving normality—the product of volume (L) times normality (eq/L) is equivalents of solute:

$$V \times N = \cancel{\text{L}} \times \frac{\text{eq}}{\cancel{\text{L}}} = \text{eq} \qquad (15.11)$$

We will use this relationship in Section 15.13.
It would be nice to know how to prepare a solution of specified normality.

EXAMPLE 15.16

How many grams of phosphoric acid must be used to prepare 1.00×10^2 mL 0.350 N H_3PO_4 to be used in the reaction $H_3PO_4 + NaOH \rightarrow NaH_2PO_4 + H_2O$?

This is like Example 15.4, except that moles have been replaced by equivalents. Complete the problem.

GIVEN: 1.00×10^2 mL (0.100 L)   WANTED: g H_3PO_4
PATH: L $\rightarrow$ eq H_3PO_4 $\rightarrow$ g H_3PO_4
FACTORS: 0.350 eq H_3PO_4/L: 98.0 g H_3PO_4/mol H_3PO_4

$$0.100 \, \cancel{L} \times \frac{0.350 \, \cancel{\text{eq H}_3\text{PO}_4}}{\cancel{L}} \times \frac{98.0 \text{ g H}_3\text{PO}_4}{1 \, \cancel{\text{eq H}_3\text{PO}_4}} = 3.43 \text{ g H}_3\text{PO}_4$$

Only one of the three available hydrogens in H_3PO_4 reacts, so there is only 1 eq/mol.

15.9 SOLUTION CONCENTRATION: A SUMMARY

The most important guarantee for success in working with solution concentration is a clear understanding of the units in which it is expressed. These may be taken from the concentration ratio, which has the form

$$\frac{\text{quantity of solute (g or mol or eq)}}{\text{quantity of solution (g or L)}} \quad or \quad \frac{\text{quantity of solute (g or mol)}}{\text{quantity of solvent (kg)}}$$

Table 15.1 makes this form specific for percentage concentration, molarity, molality, and normality.

SUMMARY

Table 15.1
Summary of Solution Concentrations

Name (Symbol)	Mathematical Form
Percentage (%)	$\dfrac{\text{g solute}}{\text{g solute + g solvent}} \times 100$
Molarity (M)	$\dfrac{\text{mol solute}}{\text{L solution}}$
Molality (m)	$\dfrac{\text{mol solute}}{\text{kg solvent}}$
Normality (N)	$\dfrac{\text{eq solute}}{\text{L solution}}$

It is often convenient to write a concentration ratio directly from the data given in a problem. The data ratio g solute/L solution is readily converted to molarity or normality by dimensional analysis. Similarly, g solute/kg solvent can be changed to molality.

15.10 DILUTION PROBLEMS

Some common acids and bases are available in concentrated solutions that are diluted to a lower concentration for use. To dilute a solution you simply add more solvent. The number of moles of solute remains the same, but it is distributed over a larger volume.

Number of moles is V × M (Equation 15.4). Using subscript c for the concentrated solution and subscript d for the dilute solution, we obtain

$$V_c \times M_c = V_d \times M_d \qquad (15.12)$$

Equation 15.12 has two important applications. They are illustrated in the next two examples.

EXAMPLE 15.17 _____

How many milliliters of commercial hydrochloric acid, which is 11.6 molar, should be used to prepare 5.50 liters of 0.500 M HCl?

SOLUTION

GIVEN: $V_d = 5.50 \text{ L}_d$; $M_c = 11.6 \text{ mol/L}_c$; $M_d = 0.500 \text{ mol/L}_d$ WANTED: V_c

EQUATION: $V_c = \dfrac{V_d \times M_d}{M_c} = \dfrac{5.50 \, \cancel{L_d} \times 0.500 \, \cancel{\text{mol/L}_d}}{11.6 \, \cancel{\text{mol}}/L_c} = 0.237 \text{ L}_c = 237 \text{ mL}$

EXAMPLE 15.18 _____

50.0 mL H_2O are added to 25.0 mL 0.881 M NaOH. What is the concentration of the diluted solution?

There is a little trick to this question, but if you read the problem carefully, you will not be trapped. Complete the problem.

_____   _____

GIVEN: $V_c = 0.0250 \text{ L}_c$; $V_d = 0.0750 \text{ L}_d$; $M_c = 0.881 \text{ mol/L}_c$ WANTED: M_d

EQUATION: $M_d = \dfrac{V_c \times M_c}{V_d} = \dfrac{0.0250 \, \cancel{L_c} \times 0.881 \, \text{mol/}\cancel{L_c}}{0.0750 \text{ L}_d} = 0.294 \text{ M NaOH}$

The tricky part of this problem is the volume of the diluted solution. The problem states that 50.0 mL are added to 25.0 mL, giving the total volume of 75.0 mL, or 0.0750 L.

15.11 SOLUTION STOICHIOMETRY

The three steps for solving a stoichiometry problem are:

1) Convert the quantity of given species to moles.
2) Convert the moles of given species to moles of wanted species.
3) Convert the moles of wanted species to the quantity units required.

Using molarity as a conversion factor, we have another way to convert between a measurable quantity—volume of solution—and moles. (See Examples 15.6 and 15.7.) The combined unit paths are

FLASHBACK After writing the reaction equation, these three steps were used for solids in Section 9.2 and for gases in Section 13.5. The same procedure will now be used for solutions.

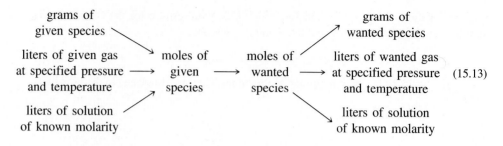

(15.13)

EXAMPLE 15.19

How many grams of lead(II) iodide will precipitate when excess potassium iodide solution is added to 50.0 mL of 0.811 M $Pb(NO_3)_2$?

"Plan" the problem, including the reaction equation.

$$Pb(NO_3)_2(aq) + 2\ KI(aq) \rightarrow PbI_2(s) + 2\ KNO_3(aq)$$

GIVEN: 50.0 mL (0.0500 L) $Pb(NO_3)_2$ WANTED: g PbI_2
PATH: L $Pb(NO_3)_2 \rightarrow$ mol $Pb(NO_3)_2 \rightarrow$ mol $PbI_2 \rightarrow$ g PbI_2
FACTORS: 0.811 mol $Pb(NO_3)_2$/L; 1 mol PbI_2/mol $Pb(NO_3)_2$; 461.0 g PbI_2/mol PbI_2

Begin the setup with the first conversion in the unit path. Do not calculate the answer.

$$0.0500\ \cancel{\text{L Pb(NO}_3)_2} \times \frac{0.811\ \text{mol Pb(NO}_3)_2}{1\ \cancel{\text{L Pb(NO}_3)_2}} \times \underline{\quad\quad} \times \underline{\quad\quad} =$$

Changing volume of solution into moles matches Example 15.7 and Equation 15.4.

The last two steps of the unit path complete the problem.

$$0.0500\ \cancel{L} \times \frac{0.811\ \cancel{\text{mol Pb(NO}_3)_2}}{1\ \cancel{L}} \times \frac{1\ \cancel{\text{mol PbI}_2}}{1\ \cancel{\text{mol Pb(NO}_3)_2}} \times \frac{461.0\ \text{g PbI}_2}{1\ \cancel{\text{mol PbI}_2}} = 18.7\ \text{g PbI}_2$$

EXAMPLE 15.20

Calculate the number of milliliters of 0.842 M NaOH that are required to precipitate as $Cu(OH)_2$ all of the copper ions in 30.0 mL 0.635 M $CuSO_4$.

The first two steps this time are the same as in the last example. Set up that far.

$$2\ NaOH(aq) + CuSO_4(aq) \rightarrow Cu(OH)_2(s) + Na_2SO_4(aq)$$

GIVEN: 30.0 mL (0.0300 L) $CuSO_4$ WANTED: mL NaOH
PATH: L $CuSO_4 \rightarrow$ mol $CuSO_4 \rightarrow$ mol NaOH $\rightarrow$ L NaOH
FACTORS: 0.635 mol $CuSO_4$/L; 2 mol NaOH/mol $CuSO_4$; 0.842 mol NaOH/L NaOH

$$0.0300\ \cancel{\text{L CuSO}_4} \times \frac{0.635\ \cancel{\text{mol CuSO}_4}}{1\ \cancel{\text{L CuSO}_4}} \times \frac{2\ \text{mol NaOH}}{1\ \cancel{\text{mol CuSO}_4}} \times \underline{\quad\quad} =$$

At this point you have the number of moles of NaOH. Its molarity may be used to change it to volume in milliliters, as in Example 15.6.

$$0.0300 \, \cancel{L} \times \frac{0.635 \, \cancel{\text{mol CuSO}_4}}{1 \, \cancel{L}} \times \frac{2 \, \cancel{\text{mol NaOH}}}{1 \, \cancel{\text{mol CuSO}_4}} \times \frac{1 \, \text{L}}{0.842 \, \cancel{\text{mol NaOH}}} = 0.0452 \, \text{L} = 45.2 \, \text{mL NaOH}$$

EXAMPLE 15.21 _____

How many liters of hydrogen, measured at STP, will be released by the complete reaction of 45.0 mL 0.486 M H_2SO_4 with excess granular zinc?

Do you recall the molar volume of a gas at STP? It is one of the numbers that were "worth remembering" from Section 13.7. Complete the problem.

_____

$$Zn(s) + H_2SO_4(aq) \rightarrow ZnSO_4(aq) + H_2(g)$$

GIVEN: 45.0 mL (0.0450 L) H_2SO_4 WANTED: L H_2
PATH: L $H_2SO_4 \rightarrow$ mol $H_2SO_4 \rightarrow$ mol $H_2 \rightarrow$ L H_2
FACTORS: 0.486 mol H_2SO_4/L; 1 mol H_2/mol H_2SO_4; 22.4 L H_2/mol H_2

$$0.0450 \, \cancel{L} \times \frac{0.486 \, \cancel{\text{mol H}_2\text{SO}_4}}{1 \, \cancel{L}} \times \frac{1 \, \cancel{\text{mol H}_2}}{1 \, \cancel{\text{mol H}_2\text{SO}_4}} \times \frac{22.4 \, \text{L H}_2}{1 \, \cancel{\text{mol H}_2}} = 0.490 \, \text{L H}_2$$

15.12 TITRATION USING MOLARITY

One of the more important laboratory operations in analytical chemistry is called **titration.** Titration is the very careful addition of one solution into another by means of a buret (Fig. 15.6). The buret accurately measures the volume of a solution required to react with a carefully measured amount of another dissolved substance. When that precise volume has been reached, an **indicator** changes color and the operator stops the flow from the buret. Phenolphthalein is a typical indicator for acid–base titrations. It is colorless in an acid solution and pink in a basic solution.

Titration can be used to **standardize** a solution, which means finding its concentration for use in later titrations. Sodium hydroxide cannot be weighed accurately because it absorbs moisture from the air and increases in weight during the weighing process. Therefore, it is not possible to prepare a sodium hydroxide solution whose molarity is known precisely. Instead, the solution is standardized against a weighed quantity of something that can be weighed accurately. Such a substance is called a **primary standard.** Oxalic acid 2-hydrate, $H_2C_2O_4 \cdot 2 \, H_2O$, is an example. When used to standardize sodium hydroxide, the equation is

$$H_2C_2O_4(aq) + 2 \, NaOH(aq) \longrightarrow Na_2C_2O_4(aq) + 2 \, H_2O(\ell)$$

Notice that the water in the hydrate is not a part of the equation. When one mole of $H_2C_2O_4 \cdot 2 \, H_2O$ dissolves, the hydrate water becomes a part of the solution and one mole of $H_2C_2O_4$ is available for reaction. The hydrate water must be taken into account in weighing the $H_2C_2O_4 \cdot 2 \, H_2O$, however.

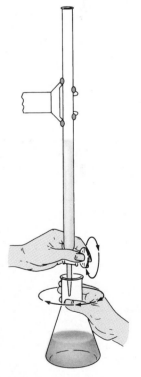

Figure 15.6

Titrating from a buret into a flask. By careful control of the valve, the chemist may deliver liquid from the buret to the flask in a steady stream, drop-by-drop, or in a single drop.

EXAMPLE 15.22 _____

1.18 g $H_2C_2O_4 \cdot 2 H_2O$ (126.0 g/mol) are dissolved in water and the solution is titrated with a solution of NaOH of unknown concentration. 28.3 mL NaOH(aq) are required to neutralize the acid. Calculate the molarity of the NaOH solution.

Begin your "plan" for this problem by listing the given and wanted quantities plus any conversion factors and equations you might use.

_____   _____

GIVEN: 1.18 g $H_2C_2O_4 \cdot 2 H_2O$; 28.3 mL (0.0283 L) NaOH   WANTED: M, mol NaOH/L
FACTORS: 126.0 g $H_2C_2O_4 \cdot 2 H_2O$/mol $H_2C_2O_4 \cdot 2 H_2O$; 2 mol NaOH/mol $H_2C_2O_4$
EQUATION: M = mol NaOH/L

Now let's think: What is wanted is mol NaOH/L. The volume of the solution is one of the givens. That takes care of the denominator, L. We still need the numerator, mol NaOH. That can be calculated by stoichiometry from selected parts of the information listed above. Let's try an intermediate "plan" that lists the given, path, and factors that will give moles of NaOH as the wanted. "Plan" only at this time.

_____   _____

GIVEN: 1.18 g $H_2C_2O_4 \cdot 2 H_2O$ WANTED: mol NaOH
FACTORS: 126.0 g $H_2C_2O_4 \cdot 2 H_2O$/mol $H_2C_2O_4 \cdot 2 H_2O$; 2 mol NaOH/mol $H_2C_2O_4$
PATH: g $H_2C_2O_4 \cdot 2 H_2O \rightarrow$ mol $H_2C_2O_4 \rightarrow$ mol NaOH

Now you are ready to use the stoichiometry pattern to the extent necessary to find mol NaOH.

_____   _____

$$1.18 \text{ g } H_2C_2O_4 \cdot 2 H_2O \times \frac{1 \text{ mol } H_2C_2O_4}{126.0 \text{ g } H_2C_2O_4 \cdot 2 H_2O} \times \frac{2 \text{ mol NaOH}}{\text{mol } H_2C_2O_4} = 0.0187 \text{ mol NaOH}$$

You now have both the numerator and denominator for the fraction in the equation that defines molarity. Complete the problem.

_____   _____

$$\text{EQUATION: M} = \frac{\text{mol}}{\text{L}} = \frac{0.0187 \text{ mol NaOH}}{0.0283 \text{ L}} = 0.661 \text{ M}$$

Ordinarily we would not calculate the intermediate answer, 0.187 mol NaOH. Instead, we would use only the setup, which represents moles of NaOH, and divide by liters of NaOH—or multiply by the inverse of volume, 1/0.0283 L:

$$\underbrace{1.18 \text{ g } H_2C_2O_4 \cdot 2 H_2O \times \frac{1 \text{ mol } H_2C_2O_4}{126.0 \text{ g } H_2C_2O_4 \cdot 2 H_2O} \times \frac{2 \text{ mol NaOH}}{\text{mol } H_2C_2O_4}}_{\text{mol}} \times \underbrace{\frac{1}{0.0283 \text{ L}}}_{1/\text{L}} = \underbrace{0.662 \text{ M NaOH}}_{\text{mol/L}}$$

The small difference in answers is caused by a roundoff in the intermediate answer, 0.0187 mol NaOH.

EXAMPLE 15.23

A potassium hydroxide solution is standardized by titrating against sulfamic acid, HSO_3NH_2 (97.1 g/mol). The equation is $HSO_3NH_2(aq) + KOH(aq) \rightarrow H_2O(\ell) + KSO_3NH_2(aq)$. 34.2 mL of the solution are used to neutralize 0.395 g HSO_3NH_2. Find the molarity of the KOH.

Set up and solve the entire problem.

GIVEN: 0.395 g HSO_3NH_2; 34.2 mL (0.0342 L) KOH   WANTED: M KOH
PATH: g $HSO_3NH_2 \rightarrow$ mol $HSO_3NH_2 \rightarrow$ mol KOH
FACTORS: 97.1 g HSO_3NH_2/mol HSO_3NH_2; 1 mol KOH/mol HSO_3NH_2
EQUATION: M = mol/L

$$0.395 \text{ g } \cancel{HSO_3NH_2} \times \frac{1 \text{ mol } \cancel{HSO_3NH_2}}{97.1 \text{ g } \cancel{HSO_3NH_2}} \times \frac{1 \text{ mol KOH}}{1 \text{ mol } \cancel{HSO_3NH_2}} \times \frac{1}{0.0342 \text{ L KOH}} = 0.119 \text{ M KOH}$$

Once a solution is standardized, it may be used to find the concentration of another solution. This is a widely used procedure in industrial laboratories.

EXAMPLE 15.24

A 25.0-mL sample of an electroplating solution is analyzed for its sulfuric acid concentration. It takes 46.8 mL of the 0.662 M NaOH from Example 15.22 to neutralize the sample. Find the molarity of the acid.

This time the first stoichiometry step begins with a volume of solution of known molarity rather than the mass of a solid. Complete the problem.

A concentrated NaOH solution is titrated into an acid that contains a large amount of phenolphthalein indicator.

$$2 \text{ NaOH}(aq) + H_2SO_4(aq) \rightarrow Na_2SO_4(aq) + 2 H_2O(\ell)$$

GIVEN: 46.8 mL (0.0468 L) NaOH   WANTED: M H_2SO_4
PATH: L NaOH $\rightarrow$ mol NaOH $\rightarrow$ mol H_2SO_4
FACTORS: 0.662 mol NaOH/L NaOH; 2 mol NaOH/mol H_2SO_4 EQUATION: M = mol/L

$$0.0468 \text{ L } \cancel{NaOH} \times \frac{0.662 \text{ mol } \cancel{NaOH}}{1 \text{ L } \cancel{NaOH}} \times \frac{1 \text{ mol } H_2SO_4}{2 \text{ mol } \cancel{NaOH}} \times \frac{1}{0.0250 \text{ L } H_2SO_4} = 0.620 \text{ M } H_2SO_4$$

15.13 TITRATION USING NORMALITY (OPTIONAL)

It was noted earlier that normality is a convenient concentration unit in analytical work. This is because of the fact pointed out in Section 15.8: *The number of equivalents of all species in a reaction is the same.* Hence, for an acid–base reaction

equivalents of acid = equivalents of base   (15.14)

There are two ways to calculate the number of equivalents (eq) in a sample of a substance. If you know the mass of the substance and its equivalent mass, use equivalent mass as a conversion factor to get equivalents, as in Example 15.12. If the sample is a solution and you know its volume and normality, multiply one by the other. $V \times N = eq$, according to Equation 15.11.

We illustrate normality calculations by repeating Examples 15.22 and 15.24, which were solved with molarity. In the first problem a primary standard is used to standardize a solution. In the second example the standardized solution is used to find the concentration of another solution.

EXAMPLE 15.25

1.18 g $H_2C_2O_4 \cdot 2 H_2O$ are dissolved in water and the solution is titrated with 28.3 mL of a solution of NaOH of an unknown concentration. Calculate the normality of the NaOH for the reaction $H_2C_2O_4(aq) + 2 NaOH(aq) \rightarrow Na_2C_2O_4(aq) + 2 H_2O(\ell)$.

Begin by "planning" the problem. Ordinarily, you include the molar mass of a substance in the "plan." When working in normality, however, we use equivalent mass, g/eq, which can be expressed as a ratio of molar mass to equivalents per mole (Equation 15.10). We suggest you write it that way in your "plan."

———   ———

GIVEN: 1.18 g $H_2C_2O_4 \cdot 2 H_2O$; 28.3 mL (0.0283 L) NaOH   WANTED: N, eq NaOH/L
FACTOR: 126.0 g $H_2C_2O_4 \cdot 2 H_2O$/2 eq   PATH: g $H_2C_2O_4 \cdot 2 H_2O \rightarrow eq$
EQUATION: N = eq/L

The equation shows that both ionizable hydrogens from $H_2C_2O_4$ are used, so there are two equivalents per mole of acid.

The equation is the defining equation for normality. To use it, we must know, or be able to find, both the numerator and the denominator quantities. The denominator is one of the givens. The numerator, equivalents of NaOH, is not. But the givens do include information with which to calculate the number of equivalents of $H_2C_2O_4$. And what is that equal to? The number of equivalents of NaOH. The reaction has the same number of equivalents of all species. If you can find one, you've found them all. Using only what you need from the above information, write the setup as far as the number of equivalents of NaOH, but do not calculate the intermediate answer.

———   ———

$$\text{1.18 g } H_2C_2O_4 \cdot 2 H_2O \times \frac{2 \text{ eq NaOH or } H_2C_2O_4}{126.0 \text{ g } H_2C_2O_4 \cdot 2 H_2O} \times \underline{} =$$

The setup represents equivalents of NaOH. Divide by volume to get normality. This can be done by multiplying by the inverse of volume, just as it was done in Example 15.22. Complete the problem.

———   ———

$$\text{1.18 g } H_2C_2O_4 \cdot 2 H_2O \times \frac{2 \text{ eq NaOH or } H_2C_2O_4}{126.0 \text{ g } H_2C_2O_4 \cdot 2 H_2O} \times \frac{1}{0.0283 \text{ L}} = 0.662 \text{ N NaOH}$$

The molarity in Example 15.22 and the normality in Example 15.25 are the same. This is always the case when there is one equivalent per mole. For an X molar solution

$$\frac{X \cancel{mol}}{L} \times \frac{1 \text{ eq}}{1 \cancel{mol}} = X \text{ eq/L}$$

Once you know the normality of one solution, you can use it to find the normality of another solution. Equation 15.11 (Section 15.8) indicates that the number of equivalents of a species in a reaction is the product of solution volume times normality. If the number of equivalents of all species in the reaction is the same, then

$$V_1 N_1 = V_2 N_2 \tag{15.15}$$

where subscripts 1 and 2 identify the reacting solutions. Solving for the second normality, we obtain

$$N_2 = \frac{V_1 N_1}{V_2} \tag{15.16}$$

That's all it takes to calculate normality in this follow-up to Example 15.24.

Equations 15.15 and 15.16 are the two key equations that make normality so useful in a laboratory in which the same titrations are run again and again.

EXAMPLE 15.26

A 25.0-mL sample of an electroplating solution is analyzed for its sulfuric acid concentration. It takes 46.8 mL of the 0.662 N NaOH from Example 15.25 to neutralize the sample. Find the normality of the acid.

GIVEN: V_1 = 46.8 mL (0.0468 L); N_1 = 0.662 eq/L; V_2 = 25.0 mL (0.0250 L)
WANTED: N_2

EQUATION: $N_2 = \dfrac{V_1 N_1}{V_2} = \dfrac{0.0468 \cancel{L_1} \times 0.662 \text{ eq}/\cancel{L_1}}{0.0250 \text{ L}_2} = 1.24 \text{ N H}_2\text{SO}_4$

In practice, a chemist is more apt to solve this problem in terms of milliequivalents:

$$\frac{46.8 \text{ mL} \times 0.662 \text{ meq/mL}}{25.0 \text{ mL}} = 1.24 \text{ N}$$

The normality in Example 15.26 is twice the molarity in Example 15.24. This is as it should be for 2 eq/mol. For an X molar solution

$$\frac{X \cancel{mol}}{L} \times \frac{2 \text{ eq}}{1 \cancel{mol}} = 2X \text{ eq/L}$$

15.14 COLLIGATIVE PROPERTIES OF SOLUTIONS (OPTIONAL)

A pure solvent has distinct physical properties, as does any pure substance. The introduction of a solute into the solvent affects these properties. The properties of the solution depend on the relative amounts of solvent and solute. It has been found experimentally that, in *dilute* solutions of certain solutes, the *change* in some of these properties is proportional to the molal concentration of the solute particles. **Solution properties that are determined only by the *number* of solute particles dissolved in a fixed quantity of solvent are called colligative properties.**

One of the reasons for putting salt on icy streets in winter is that some dissolves in whatever liquid is present. This lowers the freezing temperature and melts at least some of the ice or turns it into slush.

Freezing and boiling points of solutions are colligative properties. Perhaps the best-known example is the antifreeze used in the cooling systems of automobiles. The solute that is dissolved in the radiator water reduces the freezing temperature well below the normal freezing point of pure water. It also raises the boiling point above the normal boiling point.

The change in a freezing point is the **freezing point depression, ΔT_f,** and the change in a boiling point is the **boiling point elevation, ΔT_b.** The two proportionalities and their corresponding equations are

$$\Delta T_f \propto m, \qquad \Delta T_f = K_f m \qquad (15.17)$$

$$\Delta T_b \propto m, \qquad \Delta T_b = K_b m \qquad (15.18)$$

The proportionality constants, K_f and K_b, are, respectively, the **molal freezing point depression constant** and the **molal boiling point elevation constant.** Freezing and boiling point constants are properties of the solvent; they are the same, no matter what the solute may be. The freezing point constant for water is 1.86°C/m, and the boiling point constant is 0.52°C/m.

Some liberties in units and algebraic signs are taken when solving freezing and boiling point problems. Technically, °C·kg solvent/mol solute are the units for K_f or K_b. These units are usable, but they are awkward. The substitute, °C/m, is acceptable if we keep calculations based on Equations 15.17 and 15.18 separate from other calculations. We will follow this practice.

Again, technically, if the freezing point of the solvent is taken as the ''initial'' temperature and the always lower freezing point of the solution is the ''final'' temperature, ΔT_f and K_f must be negative quantities. In some texts they are regarded as such. Most chemists, however, use the word ''depression'' to identify clearly the direction the temperature is changing and treat both numbers as positives. We follow this practice too.

E X A M P L E   1 5 . 2 7 _____

Determine the freezing point of a solution of 12.0 g urea, $CO(NH_2)_2$, in 2.50×10^2 grams of water.

SOLUTION

GIVEN: 12.0 g $CO(NH_2)_2$; 2.50×10^2 g (0.250 kg) H_2O WANTED: T_f
FACTOR: 60.0 g $CO(NH_2)_2$/mol $CO(NH_2)_2$
EQUATIONS: m = mol solute/kg solvent; $\Delta T_f = K_f m$

To use Equation 15.17, we need to express the solution concentration in molality. Beginning with the concentration ratio, as in Example 15.8,

$$\frac{12.0 \text{ g } CO(NH_2)_2}{0.250 \text{ kg } H_2O} \times \frac{1 \text{ mol } CO(NH_2)_2}{60.0 \text{ g } CO(NH_2)_2} = 0.800 \text{ m } CO(NH_2)_2$$

EQUATION: $\Delta T_f = K_f m = \dfrac{1.86°C}{m} \times 0.800 \text{ m} = 1.49°C$

The freezing point depression is 1.49°C. The normal freezing point of water is 0°C. The freezing point of the solution is therefore 0°C − 1.49°C = −1.49°C.

Freezing point depression and/or boiling point elevation can be used to find the approximate molar mass of an unknown solute. The solution is prepared with measured masses of the solute and a solvent whose freezing or boiling point constant is known. The freezing point depression or boiling point elevation is found by experiment. The calculation procedure is:

PROCEDURE

1) Calculate molality from $m = \Delta T_f/K_f$ or $m = \Delta T_b/K_b$. Express as mol solute/kg solvent.
2) Express the concentration ratio from the given data in g solute/kg solvent.
3) Divide the second expression by the first (multiply the second by the inverse of the first):

$$\frac{\text{g solute/kg solvent}}{\text{mol solute/kg solvent}} = \frac{\text{g solute}}{\text{kg solvent}} \times \frac{\text{kg solvent}}{\text{mol solute}} = \frac{\text{g solute}}{\text{mol solute}}$$

EXAMPLE 15.28

The molal boiling point elevation constant of benzene is 2.5°C/m. A solution of 15.2 g of unknown solute in 91.1 g benzene boils at a temperature 2.1°C higher than the boiling point of pure benzene. Estimate the molar mass of the solute.

Calculate the molality of the solution (Step 1).

GIVEN: $\Delta T_b = 2.1°C$; $K_b = 2.5°C/m$; 91.1 g (0.0911 kg) benzene
WANTED: m (mol solute/kg benzene)

EQUATION: $m = \dfrac{\Delta T_b}{K_b} = \dfrac{2.1°C}{2.5°C/m} = 2.1°C \times \dfrac{m}{2.5°C} = 0.84\ m = 0.84$ mol solute/kg solvent

Now write the concentration ratio in g solute/kg solvent (Step 2) and divide it by the molality (Step 3).

$$\frac{15.2\text{ g solute}/0.0911\text{ kg benzene}}{0.84\text{ mol solute/kg benzene}} = \frac{15.2\text{ g solute}}{0.0911\text{ kg benzene}} \times \frac{\text{kg benzene}}{0.84\text{ mol solute}} = 2.0 \times 10^2\text{ g/mol}$$

CHAPTER 15 IN REVIEW

15.1   The Characteristics of a Solution

15.2   Solution Terminology

15A Distinguish among terms in the following groups:

   Solute and solvent
   Concentrated and dilute
   Solubility, saturated, unsaturated, and supersaturated
   Miscible and immiscible

15.3   The Formation of a Solution

15B Describe the formation of a saturated solution from the time excess solid solute is first placed into a liquid solvent.

15C Identify and explain the factors that determine the time required to dissolve a given amount of solute or to reach equilibrium.

15.4   Factors That Determine Solubility

15D Given the structural formulas of two molecular substances, or other information from which the strength of their intermolecular forces may be estimated, predict if they will dissolve appreciably in each other. State the criteria on which your prediction is based.

15E Predict how the solubility of a gas in a liquid will be affected by a change in the partial pressure of that gas over the liquid.

15.5   Solution Concentration: Percentage

15F Given grams of solute and grams of solvent or solution, calculate percentage concentration.

15G Given grams of solution and percentage concentration, calculate grams of solute and grams of solvent.

15.6   Solution Concentration: Molarity

15H Given two of the following, calculate the third: moles of solute (or data from which it may be found), volume of solution, molarity.

15.7   Solution Concentration: Molality (Optional)

15I Given two of the following, calculate the third: moles of solute (or data from which it may be found), mass of solvent, molality.

15.8   Solution Concentration: Normality (Optional)

15J Given an equation for a neutralization reaction, state the number of equivalents of acid or base per mole and calculate the equivalent mass of the acid or base.

15K Given two of the following, calculate the third: equivalents of acid or base (or data from which it may be found), volume of solution, normality.

15.9   Solution Concentration: A Summary

15.10 Dilution Problems

15L Given any three of the following, calculate the fourth: (a) volume of concentrated solution, (b) molarity of concentrated solution, (c) volume of dilute solution, (d) molarity of dilute solution.

15.11 Solution Stoichiometry

15M Given the quantity of any species participating in a chemical reaction for which the equation can be written, find the quantity of any other species, either quantity being measured in (a) grams, (b) volume of gas at specified temperature and pressure, or (c) volume of solution at specified molarity.

15.12 Titration Using Molarity

15N Given the volume of a solution that reacts with a known mass of a primary standard and the equation for the reaction, calculate the molarity of the solution.

15O Given the volumes of two solutions that react with each other in a titration, the molarity of one solution, and the equation for the reaction, calculate the molarity of the second solution.

15.13 Titration Using Normality (Optional)

15P Given the volume of a solution that reacts with a known mass of a primary standard and the equation for the reaction, calculate the normality of the solution.

15Q Given the volumes of two solutions that react with each other in a titration, the normality of one solution, and the equation for the reaction, calculate the normality of the second solution.

15.14 Colligative Properties of Solutions (Optional)

15R Given the freezing point depression or boiling point elevation and the molality of a solution, or data from which they may be found, calculate the molal freezing point constant or molal boiling point constant.

15S Given (a) the mass of solute and solvent in a solution; (b) the freezing point depression or boiling point elevation, or data from which they may be found; and (c) the molal freezing/boiling point constant of the solvent, find the approximate molar mass of the solute.

CHAPTER 15 KEY WORDS AND MATCHING SETS

Set 1 (Items in *italics* are from optional sections or marginal notes.)

_____ A term for liquids that do not mix with each other.

_____ Substance that gives a visual signal that a titration reaction has been completed.

_____ A solution in which there is a relatively high amount of solute per unit volume of solution.

_____ A solution that holds all the solute it is able to dissolve at a given temperature.

_____ A measure of the amount of solute that dissolves in a given amount of solvent.

_____ That part of a solution that is present in a relatively small amount.

_____ *Grams of solute per 100 mL of solution.*

_____ *Amount of acid that yields one mole of hydrogen ions.*

_____ *Freezing point of a solution compared to the freezing point of the pure solvent.*

_____ *Property of a solvent that depends on the amount of solute in the solution.*

1. Solute
2. Concentrated
3. Saturated
4. Percentage concentration by volume
5. Solubility
6. Freezing point depression
7. Immiscible
8. Colligative
9. Equivalent
10. Indicator

Set 2 (Items in *italics* are from optional sections.)

_____ A solution that is able to dissolve more solute.

_____ To determine accurately the concentration of a solution that will be used for titration reactions.

_____ Grams of solute per 100 grams of solution.

_____ Describes two liquids that are soluble in each other.

_____ A substance that can be weighed precisely for a reaction in which the concentration of a solution will be determined.

_____ A solution in which there is a relatively low amount of solute in a given volume of solution.

_____ *Moles of solute per kilogram of solvent.*

_____ *Number of grams in one equivalent of a substance.*

_____ *Difference between the boiling point of the solvent and the boiling point of the solution.*

_____ *Change in freezing point per mole of solute dissolved in 1 kilogram of solvent.*

1. Dilute
2. Equivalent mass
3. Standardize
4. Percentage concentration by mass
5. Boiling point elevation
6. Miscible
7. Unsaturated
8. Primary standard
9. Molality
10. Molal freezing point constant

Set 3 (Items in *italics* are from optional sections.)

_____ An unstable solution whose solute crystallizes if disturbed because of the excessive amount present.

_____ A condition in which solute particles are surrounded by water molecules.

_____ A homogeneous mixture.

_____ Medium, usually liquid, in which a substance is dissolved.

_____ Controlled introduction of a liquid into a vessel to measure the volume that reacts with a substance already in the vessel.

_____ Fraction made up of amount of solute over amount of solvent or solution.

_____ Moles of solute per liter of solution.

_____ *Amount of base that reacts with one mole of hydrogen ions.*

_____ *Equivalents of solute in a liter of solution.*

_____ *Molal boiling point constant.*

1. Normality

2. Hydrated

3. Titration

4. Concentration ratio

5. $\Delta T/m$

6. Solvent

7. Molarity

8. Supersaturated

9. Equivalent

10. Solution

STUDY HINTS AND PITFALLS TO AVOID

To solve solution problems easily, you must have a clear understanding of concentrations and the units in which they are expressed. Table 15.1 summarizes all of the concentrations used in this chapter. Study carefully the concentrations that have been assigned to you. Then practice with enough end-of-chapter problems until you have complete mastery of each performance goal.

Once in a while a student is tempted to change between moles and liters by using 22.4 L/mol or, even worse, 22.4 mol/L. The number 22.4 is so convenient, and the units look like just what is needed. But they are not; 22.4 applies to gases, not to solutions. And then, only to gases at STP.

CHAPTER 15 QUESTIONS AND PROBLEMS

SECTION 15.1

1) Every pure substance has a definite and fixed set of physical and chemical properties. A solution is prepared by dissolving one pure substance in another. Is it reasonable to expect that the solution will also have a definite and fixed set of properties that are different from the properties of either component? Explain your answer.

2) The text states that air is a solution. Does this solution include the gaseous H_2O that makes up the humidity in the air? Is your answer any different if the air is foggy? Explain.

_____

3) Can you ever see particles in a solution? Explain why or why not.

4) Identify the three kinds of particles in a solution.

SECTION 15.2

5) What is the difference between a solute and a solvent in a solution?

6) Distinguish between the solute and solvent in each of the following solutions: (a) vinegar, about 5% acetic acid in water; (b) a teaspoon of sugar in a cup of hot water; (c) 15% ammonia in water; (d) 65% alcohol in water. On what do you base your distinctions?

_____

7) Distinguish between a concentrated solution and a saturated solution. Can an unsaturated solution ever be a concentrated solution? Explain.

8) Distinguish between a dilute solution and an unsaturated solution. Explain.

_____

9) What is solubility? Can solubility be expressed in units other than grams of solute per 100 grams of solvent? If so, suggest another suitable expression.

10) It is a fact that the solubility of calcium iodide is 426 grams per 100 g H_2O. It is also a fact that the solubility of calcium iodide is 209 grams per 100 g H_2O. How can both of these statements be true?

---

11) When acetic acid, a clear, colorless liquid, and water are mixed, a clear, uniform, colorless liquid results. Is the acetic acid soluble in the water? Is acetic acid miscible in water? Explain your answers.

12) When sugar is stirred into water, a clear, uniform, colorless liquid results. Is sugar soluble in water? Is sugar miscible in water? Explain your answers. If your answers are different from those to Question 11, explain why they are different.

---

13)* In Chapter 2 you learned that physical properties must be employed to separate components of a mixture. Suggest a way to separate two immiscible liquids.

14)* Suggest a way to separate two miscible liquids. Do you know of any widespread industrial process in which this is done?

SECTION 15.3

15) What is a hydrated ion?

16) How is it that both cations and anions, positively charged ions and negatively charged ions, can be "hydrated" by the same substance, water?

---

17) The text describes forces that cause an ionic solute to dissolve in water. What are these forces?

18)* The text does *not* identify forces that are also present that must be overcome during the dissolving process. Can you imagine what they might be?

---

19) Describe the changes that occur between the time excess solute is placed into water and the time the solution becomes saturated.

20) The text states that a dynamic equilibrium is present when excess solute is in contact with a saturated solution. What does this mean? In particular, why is the equilibrium described as *dynamic?*

---

21) Compare the rate at which ions pass from solute to solution to the rate at which they pass from solution to solute when the solution is unsaturated.

22) At what time during the development of a saturated solution is the rate at which ions move from solvent to solute [solute(aq) → solute(s)] greater than the rate from solute to solvent [solute(s) → solute(aq)]?

---

23) Why can you not prepare a supersaturated solution by adding more solute and stirring until it dissolves?

24)* Silver acetate has a solubility of 2.52 g/100 g water at 80°C and 1.02 g/100 g water at 20°C. How do you prepare a supersaturated solution of silver acetate? Why does crystallization not occur as soon as the ion concentration is greater than the concentration of a saturated solution?

---

25) Identify three ways that you can reduce the amount of time required to dissolve a given amount of solute in a fixed quantity of solvent.

26) Explain how each of the acts in Question 25 speeds the dissolving process.

---

27) Why do people stir coffee after putting sugar into it? Would putting the coffee and sugar into a closed container and shaking it be as effective as stirring?

28)* Confectioner's sugar, used by bakers, is finely powdered. What advantage does it have over granular (crystalline) sugar in making bakery goods? Would there be any advantage or disadvantage in using it for sweetening coffee? Explain.

SECTION 15.4

Questions 29 to 32: Structural diagrams for several substances are given in Table 15.2.

29) Which of the following solutes do you expect to be more soluble in water than in cyclohexane: (a) formic acid, (b) benzene, (c) methylamine, (d) tetrafluoromethane? Explain your choice(s).

30) Which of the following solutes do you expect to be more soluble in cyclohexane than in water: (a) dimethyl ether, (b) hexane, (c) tetrachloroethene, (d) hydrogen fluoride? Explain your choice(s).

---

31) Which compound, glycerine or hexane (see Table 15.2), do you expect would be more miscible in water? Why?

32) Suppose you have a spot on some clothing and water will not take it out. If you have ethanol and cyclohexane (see Table 15.2, page 428) available, which would you choose as the more promising solvent to try? Why?

---

33) On opening a bottle of carbonated beverage, many bubbles are released and the sound of escaping gas is heard. This suggests that the beverage is bottled under high pressure. Yet, for safety reasons, the pressure cannot be much more than one atmosphere. What gas do you suppose is in the small space between the beverage and the cap of a bottle of carbonated beverage before the cap is removed?

34) What is responsible for the "hissing" sound and the bubbles when a bottle of carbonated beverage is opened (see Question 33)? How can so much "hiss" come from such a small volume of gas which cannot have an initial pressure much over one atmosphere?

Table 15.2

Lewis diagrams for Questions 29 to 32

H—F hydrogen fluoride	H O H water
$H-C$ O $O-H$ formic acid	Cl $C=C$ Cl Cl Cl tetrachloroethene
H $N-C-H$ H H H methylamine	F $F-C-F$ F tetrafluoromethane
H H H H H H H—C—C—C—C—C—C—H H H H H H H hexane	H H H—C—C—O H H H ethanol
H H H H—C—C—C—H O O O H H H glycerine	H C H—C C—H H—C C—H C H benzene*
H H C H H H—C C—H H—C C—H C H H cyclohexane	H O H C C H H H H dimethyl ether

*This diagram is not consistent with all of the properties of benzene, but it is adequate to predict benzene's ability to dissolve other substances or to be dissolved by other solvents.

SECTION 15.5

35) Find the percentage concentration of a solution prepared by dissolving 2.20 g barium chloride in 57.9 g of water.

36) If 837 grams of a solution contains 46.7 grams of solute, what is the percentage concentration?

37) How many grams of sodium sulfate are in 505 g 15.0% solution? How many grams of water?

38) If you are to prepare 75.0 g 7.25% NH_4NO_3 solution, how many grams of the salt do you weigh out, and in how many mL of water do you dissolve it?

SECTION 15.6

39) Potassium iodide is the additive in "iodized" table salt. Calculate the molarity of a solution prepared by dissolving 2.41 g of potassium iodide in water and diluting to 50.0 mL.

40) 7.50×10^2 mL of a solution contains 48.5 g of sodium sulfate, a chemical used in dyeing operations. What is the molarity of this solution?

41) 18.0 g of anhydrous nickel chloride are dissolved in water and diluted to 90.0 mL. 30.0 g of nickel chloride 6-hydrate are also dissolved in water and diluted to 90.0 mL. Identify the solution with the higher molar concentration and calculate its molarity.

42) The chemical name for the "hypo" used in photographic developing is sodium thiosulfate. It is sold as a 5-hydrate, $Na_2S_2O_3 \cdot 5\,H_2O$. What is the molarity of a solution prepared by dissolving 1.3×10^2 g of this compound in water and diluting the solution to 1350 mL?

43) Large quantities of silver nitrate are used in making photographic chemicals. Find the mass that must be used in preparing 2.50×10^2 mL 0.058 M $AgNO_3$.

44) Sodium carbonate is one of the most widely used sodium compounds. How many grams are required by an analytical chemist to prepare 2.50×10^2 mL 0.900 M Na_2CO_3?

45) Potassium hydroxide is used in making liquid soap, as well as many other things. How many grams would you use to prepare 2.50 L 1.40 M KOH?

46) How many grams of acetic acid, the odor and taste producer in vinegar, must be dissolved in water and diluted to 1.50×10^3 mL to yield 0.400 M $HC_2H_3O_2$?

47) What volume of concentrated sulfuric acid, which is 18 molar, is required to obtain 5.19 mol of the acid?

48) A laboratory reaction requires 0.0170 mol of HCl. How many milliliters of 0.746 M HCl would you use?

49) 0.132 M NaCl is to be the source of 8.33 g of dissolved solute. What volume of solution is needed?

50) Calculate the volume of concentrated ammonia solution, which is 15 molar, that contains 85.0 g of NH_3.

51) Calculate the moles of silver nitrate in 55.7 mL 0.204 M $AgNO_3$.

52) How many moles of solute are in 78.3 mL 1.26 M NaOH?

53) Despite its intense purple color, potassium permanganate is used in bleaching operations. How many moles are in 25.0 mL 0.0841 M $KMnO_4$?

54) 39.0 mL 0.548 M H_2SO_4 are used to titrate a base of unknown concentration. How many moles of sulfuric acid react?

---

55) The density of 3.30 M KSCN is 1.15 g/mL. What is its percentage concentration?

56)* The density of 12% H_2SO_4 is 1.08 g/mL. Calculate its molarity.

SECTION 15.7

57) Calculate the molal concentration of a solution of 44.9 g of naphthalene, $C_{10}H_8$, in 175 g of benzene, C_6H_6.

58) If 30.2 g of sugar, $C_{12}H_{22}O_{11}$, are dissolved in 2.50×10^2 mL of water, what is the molality of the solution?

---

59) Diethylamine, $(CH_3CH_2)_2NH$, is highly soluble in ethanol, C_2H_5OH. Calculate the number of grams of diethylamine that would be dissolved in 400 g of ethanol to produce 4.70 m $(CH_3CH_2)_2NH$.

60) If you are to prepare a 3.00-molal solution of urea in water, how many grams of urea, $CO(NH_2)_2$, would you dissolve in 75.0 mL of water?

---

61)* Methyl ethyl ketone, C_4H_8O, is a solvent popularly known as MEK that is used to cement plastics. How many grams of MEK must be dissolved in 1.00×10^2 mL of benzene, specific gravity 0.879, to yield a 0.254 molal solution?

62) Into what volume of water must 92.7 g of acetic acid, $HC_2H_3O_2$, be dissolved to produce 1.25 m $HC_2H_3O_2$?

SECTION 15.8

63) What is equivalent mass? Why can you state positively the equivalent mass of LiOH, but not H_2SO_4?

64) Explain why the number of equivalents in a mole of acid or base is not always the same.

---

65) State the number of equivalents in one mole of HNO_2; in one mole of H_2SeO_4 in $H_2SeO_4 \rightarrow H^+ + HSeO_4^-$. (Se is selenium, Z = 34.)

66) How many equivalents are in one mole of each of the following: HF; $H_2C_2O_4$ in the reaction $H_2C_2O_4 \rightarrow 2\,H^+ + C_2O_4^{2-}$?

---

67) State the maximum number of equivalents per mole of $Cu(OH)_2$, per mole of $Fe(OH)_3$.

68) What is the maximum number of equivalents in one mole of $Zn(OH)_2$ and RbOH? (Rb is rubidium, Z = 37.)

---

69) Calculate the equivalent masses of HNO_2 and H_2SeO_4 in Problem 65.

70) What are the equivalent masses of HF and $H_2C_2O_4$ in Problem 66?

---

71) What are the equivalent masses of $Cu(OH)_2$ and $Fe(OH)_3$ in Problem 67?

72) Find the equivalent masses of $Zn(OH)_2$ and RbOH in Problem 68.

---

73) What is the normality of the solution made when 2.25 g KOH are dissolved in water and diluted to 2.50×10^2 mL?

74) Calculate the normality of a solution prepared by dissolving 18.2 g of $HC_2H_3O_2$ in 2.50×10^2 mL of solution.

---

75) 6.69 g $H_2C_2O_4$ are dissolved in water, diluted to 2.00×10^2 mL, and used in a reaction in which it ionizes as follows: $H_2C_2O_4 \rightarrow H^+ + HC_2O_4^-$. What is the normality of the solution?

76) 10.7 g of $NaHCO_3$ are dissolved in 7.50×10^2 mL of solution. What is the normality in the reaction $NaHCO_3 + HCl \rightarrow NaCl + H_2O + CO_2$?

---

77) $NaHSO_4$ is used as an acid in the reaction $HSO_4^- \rightarrow H^+ + SO_4^-$. What mass of $NaHSO_4$ must be dissolved in 7.50×10^2 mL of solution to produce 0.200 N $NaHSO_4$?

78) How many grams of KOH must be used to prepare 5.00×10^2 mL 1.50 N KOH?

---

79)* $Na_2CO_3 \cdot 10\,H_2O$, is used as a base in the reaction $CO_3^{2-} + 2\,H^+ \rightarrow CO_2 + H_2O$. Calculate the mass of the hydrate needed to prepare 1.00×10^2 mL 0.500 N Na_2CO_3.

80)* Calculate the mass of $H_2C_2O_4 \cdot 2\,H_2O$, required for 5.00×10^2 mL 0.350 N $H_2C_2O_4$ for the reaction $H_2C_2O_4 + 2\,OH^- \rightarrow C_2O_4^{2-} + 2\,HOH$.

---

81) 73.1 mL 0.834 N NaOH has how many equivalents of solute?

82) How many equivalents are in 3.50 L 0.782 N H_2SO_4?

---

83) What volume of 0.492 N $KMnO_4$ contains 0.788 eq?

84) Calculate the volume of 0.326 N HCl that contains 0.346 eq.

SECTION 15.10

85) What is the molarity of the acetic acid solution if 45.0 mL 17 M $HC_2H_3O_2$ are diluted to 1.5 L?

86) Concentrated HCl is 12 molar. Find the molarity of the solution prepared by diluting 1.50×10^2 mL to 2.50 L.

_____

87) How many milliliters of concentrated nitric acid, 16 M HNO_3, will you use to prepare 7.50×10^2 mL 0.69 M HNO_3?

88) How many milliliters of concentrated ammonia, 15 M NH_3, would you dilute to 2.50×10^2 mL to produce 3.0 M NH_3?

_____

89) Calculate the volume of 18 M H_2SO_4 required to prepare 3.0 L 2.9 N H_2SO_4 for reactions in which the sulfuric acid is completely ionized.

90) How many milliliters of 12 M HCl must be diluted to 5.0 L to produce 0.75 N HCl?

SECTION 15.11

91) Calculate the grams of magnesium hydroxide that will precipitate from 25.0 mL 0.398 M $MgCl_2$ by the addition of excess NaOH solution.

92) How many grams of AgCl can be precipitated by adding excess NaCl to 65.0 mL 0.757 M $AgNO_3$?

_____

93)* The iron(III) ion content of a solution may be found by precipitating it as $Fe(OH)_3$, and then decomposing the hydroxide to Fe_2O_3 by heat. How many grams of iron(III) oxide can be collected from 35.0 mL 0.516 M $Fe(NO_3)_3$?

94) What mass of barium fluoride can be precipitated from 25.0 mL 0.465 M NaF by adding excess barium nitrate solution?

_____

95)* 25.0 mL of 0.269 M $NiCl_2$ are combined with 30.0 mL 0.260 M KOH. How many grams of nickel hydroxide will precipitate?

96)* 30.0 mL 0.350 M NaOH are added to 40.0 mL 0.190 M $CuSO_4$. How many grams of copper(II) hydroxide will precipitate?

_____

97) How many milliliters of 1.50 M NaOH must react with aluminum to yield 2.00 L of hydrogen, measured at 22°C and 789 torr, by the reaction $2 \, Al + 6 \, NaOH \rightarrow 2 \, Na_3AlO_3 + 3 \, H_2$? Assume complete conversion of reactants to products.

98) Calculate the volume of chlorine, measured at STP, that can be recovered from 65.0 mL 0.844 M HCl by the reaction $MnO_2 + 4 \, HCl \rightarrow MnCl_2 + 2 \, H_2O + Cl_2$, assuming complete conversion of reactants to products.

99) What volume of 0.842 M NaOH would react with 8.74 g of sulfamic acid, NH_2SO_3H, a solid acid with one replaceable hydrogen?

100) How many milliliters of 0.832 M HCl are needed to neutralize 1.46 g of sodium carbonate?

SECTION 15.12

101) $2 \, HC_7H_5O_2 + Na_2CO_3 \rightarrow 2 \, NaC_7H_5O_2 + H_2O + CO_2$ is the equation for a reaction by which a solution of sodium carbonate may be standardized. 5.038 g of $HC_7H_5O_2$ uses 51.89 mL of the solution in the titration. Find the molarity of the sodium carbonate.

102) Potassium hydrogen iodate is used as a primary standard in finding the concentration of a solution of potassium hydroxide by the reaction $KH(IO_3)_2 + KOH \rightarrow 2 \, KIO_3 + HOH$. What is the molarity of the base if 34.95 mL are required to titrate 1.587 g of the primary standard?

_____

103)* A student is to titrate solid maleic acid, $H_2C_4H_2O_4$ (two replaceable hydrogens), with 0.500 M NaOH. What is the maximum number of grams of maleic acid that can be used if the titration is not to exceed 50.0 mL?

104)* Oxalic acid 2-hydrate, $H_2C_2O_4 \cdot 2 \, H_2O$, is the primary standard used for finding the molarity of a sodium hydroxide solution. 3.804 g are dissolved in water and diluted to 500.0 mL. A 25.00-mL sample of that solution is titrated with the NaOH solution; 29.60 mL are required. Find (a) the molarity of the acid and (b) the molarity of the base. (Both replaceable hydrogens in the oxalic acid react.)

_____

105)* 17.02 g of $NaHCO_3$ are dissolved in water and diluted to 500.0 mL in a volumetric flask. 37.80 mL of that solution are required to titrate a 20.00 mL sample of sulfuric acid solution. What is the molarity of the acid? The reaction equation is $H_2SO_4 + 2 \, NaHCO_3 \rightarrow Na_2SO_4 + 2 \, H_2O + 2 \, CO_2$.

106)* What minimum number of grams of oxalic acid 2-hydrate would you specify for a student experiment involving a titration of no fewer than 18.0 mL 0.0900 M NaOH? The reaction is the same as in Question 104.

_____

107)* Calculate the hydroxide ion concentration in a 20.00-mL sample of an unknown if 14.75 mL 0.248 M H_2SO_4 are used in a neutralization titration.

108)* An analytical procedure for finding the chloride ion concentration in a solution involves the precipitation of silver chloride: $Ag^+ + Cl^- \rightarrow AgCl$. What is the molarity of the chloride ion if 16.15 mL 0.694 M $AgNO_3$ (the source of Ag^+) are needed to precipitate all of the chloride in a 25.00-mL sample of the unknown?

_____

109)* A student received a 599-mg sample of a mixture of Na_2HPO_4 and NaH_2PO_4. She is to find the percentage of each compound in the sample. After dissolving the mixture, she titrated it with 19.58 mL 0.201 M NaOH. If the only reaction is $NaH_2PO_4 + NaOH \rightarrow Na_2HPO_4 + HOH$, find the required percentages.

110)* A 855-mg sample of impure Na_2CO_3 was titrated with 47.06 mL 0.286 M HCl. Calculate the percent Na_2CO_3 in the sample.

SECTION 15.13

111) What is the normality of the sodium carbonate solution in Problem 101?

112) What is the normality of the base in Problem 102?

---

113) What is the normality of the acid in Problem 105?

114) Calculate the normalities of the acid and the base in Problem 104. Set up the problem completely from the data, not from the answers to Problem 104.

---

115) Calculate the normality of a solution of sodium carbonate if a 25.0-mL sample requires 39.8 mL 0.405 N H_2SO_4 in a titration.

116) What is the normality of an acid if 19.0 mL are required to titrate 15.0 mL 0.782 N NaOH?

---

117) 42.2 mL 0.402 N NaOH are required to titrate 50.0 mL of a solution of tartaric acid ($H_2C_4H_4O_6$) of unknown concentration. Find the normality of the acid.

118) 22.6 mL 0.406 N $AgNO_3$ are required to titrate the chloride ion in a 25.0-mL sample of nickel chloride solution. Find the normality of the nickel chloride.

---

119) Using a particular acid–base indicator, 25.0 mL of a phosphoric acid solution require 33.4 mL 0.196 N NaOH in a titration reaction. Find the normality of the acid.

120) Repeating the titration described in Problem 119, but with a different indicator, it is found that only 16.7 mL 0.196 N NaOH are required for 20.0 mL of the phosphoric acid solution. Calculate the normality of the acid and account for the difference in the answers in Problem 119 and this problem.

---

121) 1.21 g of an organic compound that functions as a base in reaction with sulfuric acid are dissolved in water and titrated with 0.170 N H_2SO_4. What is the equivalent mass of the base if 30.7 mL of acid are required in the titration?

122) 18.1 mL 0.483 N NaOH are required to titrate a solution prepared by dissolving 0.652 g of an unknown acid. What is the equivalent mass of the acid?

SECTION 15.14

123) Is the partial pressure exerted by one component of a gaseous mixture at a given temperature and volume a colligative property? Justify your answer, pointing out in the process what classifies a property as "colligative."

124) The specific gravity of a solution of KCl is greater than 1.00. The specific gravity of a solution of NH_3 is less than 1.00. Is specific gravity a colligative property? Why, or why not?

---

125) 27.2 g of analine, $C_6H_5NH_2$, are dissolved in 1.20×10^2 g of water. At what temperatures will the solution freeze and boil?

126) Calculate the boiling and freezing points of a solution of 40.7 g of glucose, $C_6H_{12}O_6$, in 85.0 g of water.

---

127) Calculate the freezing point of a solution of 2.12 g of naphthalene, $C_{10}H_8$, in 32.0 g of benzene, C_6H_6. Pure benzene freezes at 5.50°C, and its $K_f = 5.10$°C/m.

128) 4.40 g of paradichlorobenzene, $C_6H_4Cl_2$, are dissolved in 67.2 g of naphthalene, $C_{10}H_8$. Calculate the freezing point of the solution if pure naphthalene freezes at 80.2°C and K_f is 6.9°C/m.

---

129) What is the molality of a solution of an unknown solute in acetic acid if it freezes at 14.1°C? The normal freezing point of acetic acid is 16.6°C, and $K_f = 3.90$°C/m.

130) Calculate the molal concentration of an aqueous solution that boils at 100.75°C.

---

131) A solution of 16.1 g of an unknown solute in 6.00×10^2 g of water boils at 100.28°C. Find the molar mass of the solute.

132) If a solution prepared by dissolving 27.6 g of an unknown solute in 3.62×10^2 g of water freezes at −1.18°C, what is the approximate molar mass of the solute?

---

133) When 12.4 g of an unknown solute are dissolved in 90.0 g of phenol, the freezing point depression is 9.6°C. Calculate the molar mass of the solute if $K_f = 3.56$°C/m for phenol.

134) A solution of 11.2 g of an unknown solute dissolved in 78.3 g $C_{10}H_8$ freezes at 72.5°C. The normal freezing point of $C_{10}H_8$ is 80.2°C, and $K_f = 6.9$°C/m. Estimate the molar mass of the solute.

---

135) The normal freezing point of an unknown solvent is 28.7°C. A solution of 11.4 g of ethanol, C_2H_5OH, in 2.00×10^2 g of the solvent freezes at 22.5°C. What is the molal freezing point constant of the solvent?

136) The boiling point of a solution of 1.77 g of urea, NH_2CONH_2, in 18.4 g of a certain solvent is 3.83°C higher than the boiling point of the pure solvent. What is the molal boiling point constant of the solvent?

GENERAL QUESTIONS

137) Distinguish precisely and in scientific terms the differences among items in each of the following groups.
a)   Solute, solvent
b)   Concentrated, dilute
c)   Saturated, unsaturated, supersaturated
d)   Soluble, miscible
e)   (Optional) molality, molarity, normality
f)   (Optional) molar mass, equivalent mass
g)   (Optional) freezing point depression, boiling point elevation
h)   (Optional) molal freezing point constant, molal boiling point constant

138) Determine whether each statement that follows is true or false.
a)   The concentration is the same throughout a beaker of solution.
b)   A saturated solution of solute A is always more concentrated than an unsaturated solution of solute B.
c)   A solution can never have a concentration greater than its solubility at a given temperature.
d)   A finely divided solute dissolves faster because more surface area is exposed to the solvent.
e)   Stirring a solution increases the rate of crystallization.
f)   Crystallization ceases when equilibrium is reached.
g)   All solubilities increase at higher temperatures.
h)   Increasing air pressure over water increases the solubility of nitrogen in the water.

i)   An ionic solute is more apt to dissolve in a nonpolar solvent than in a polar solvent.
j)   (Optional) The molarity of a solution changes slightly with temperature, but the molality does not.
k)   (Optional) If an acid and a base react on a two-to-one mole ratio, there are twice as·many equivalents of acid as there are base in the reaction.
l)   The concentration of a primary standard is found by titration.
m)   Colligative properties of a solution are independent of the kinds of solute particles, but they are dependent on particle concentration.

139) When you heat water on a stove, small bubbles appear long before the water begins to boil. What are they? Explain why they appear.

140) Antifreeze is put into the water in an automobile to prevent it from freezing in winter. What does the antifreeze do to the boiling point of the water, if anything?

141)* 60.0 mL 0.322 M KI are combined with 20.0 mL 0.530 M $Pb(NO_3)_2$. (a) How many grams of PbI_2 will precipitate? (b) What is the final molarity of the K^+ ion? (c) What is the final molarity of the Pb^{2+} or I^- ion, whichever one is in excess?

142)* A solution has been defined as a homogeneous mixture. Pure air is a solution. Does it follow that the atmosphere is a solution? Explain.

143) Does percentage concentration of a solution depend on temperature?

144)* If you know either the percentage concentration of a solution or its molarity, what additional information must you have before you can convert to the other concentration?

MATCHING SET ANSWERS

Set 1: 7–10–2–3–5–1–4–9–6–8

Set 2: 7–3–4–6–8–1–9–2–5–10

Set 3: 8–2–10–6–3–4–7–9–1–5

QUICK CHECK ANSWERS

15.1 Only b. There may be more than two pure substances in a solution.

15.2 (a) The solutes are A and B; water is the solvent. (b) Degree of saturation cannot be estimated without knowing the solubility of the compound. (c) The question has no meaning because *concentrated* and *dilute* compare solutions of the same solute, not different solutes.

15.3 a) Equal. b) Less. The rate of crystallization increases as the concentration of the solution increases. c) More. The steadily increasing crystallization rate "subtracts from" the constant dissolving rate, reducing the net rate as time goes on. The net rate eventually reaches zero when equilibrium is reached.

15.4 Main criterion: Do substances have similar intermolecular attractions? If yes, they are probably soluble. Look for similarities in polarity, hydrogen bonding, and size, each of which contributes to similar intermolecular forces.

15.5 GIVEN: 13.5 g $C_6H_{12}O_6$ WANTED: mL H_2O PATH: g $C_6H_{12}O_6 \rightarrow$ mol $C_6H_{12}O_6 \rightarrow$ g $H_2O \rightarrow$ mL H_2O
FACTORS: 180.2 g/mol $C_6H_{12}O_6$; 0.255 mol KCl/kg H_2O; 1.00 g H_2O/mL H_2O

$$13.5 \text{ g } C_6H_{12}O_6 \times \frac{1 \text{ mol } C_6H_{12}O_6}{180.2 \text{ g } C_6H_{12}O_6} \times \frac{1000 \text{ g } H_2O}{0.255 \text{ mol } C_6H_{12}O_6} \times \frac{1.00 \text{ mL } H_2O}{1 \text{ g } H_2O} = 294 \text{ mL } H_2O$$

16 Reactions That Occur in Water Solutions: Net Ionic Equations

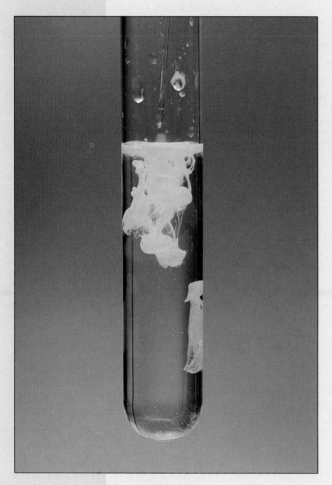

These photographs show the formation of precipitates when solutions are combined. If you were to write the double replacement equations for the two reactions, as you learned how to do in Chapter 8, you would get these equations:

$$2 \, KI(aq) + Pb(NO_3)_2(aq) \rightarrow PbI_2(s) + 2 \, KNO_3(aq) \qquad (NH_4)_2S(aq) + CuCl_2(aq) \rightarrow CuS(s) + 2 \, NH_4Cl(aq)$$

Are these accurate and real equations? Real, yes—and the only equations that can be used for stoichiometry problems. Accurate, well, not exactly. In Chapter 16 you will learn why these equations are not altogether correct and how to write the equations that tell exactly what happens in a chemical change.

In Chapter 8 you wrote chemical equations for some reactions that occur in water solutions. These equations, however, are not entirely accurate in describing the chemical changes that occur. Net ionic equations accomplish this by identifying precisely the particles in the solution that experience a change and the particles that are produced.

We begin by examining the electrical properties of a solution.

16.1 ELECTROLYTES AND SOLUTION CONDUCTIVITY

Suppose two metal strips, called **electrodes,** are placed in a liquid and wired to a battery and an electric light bulb (Fig. 16.1). One of the electrodes is given a positive charge by the battery, and the other electrode has a negative charge. If the liquid is pure water, nothing happens. If the liquid is a solution of a salt—an ionic compound—the bulb glows brightly. If the liquid is a sugar solution, the bulb does not glow.

The pure solvent, water, and the sugar solution are **nonconductors** because they do not conduct electricity. The salt solution, however, is an excellent **conductor** of electricity. What is it about the salt solution that gives it the ability to conduct an electric current?

An "electric current" is a movement of electric charge. In a metal, valence electrons are loosely held. When a surplus of electrons is introduced to one end of a wire, they repel the electrons in the wire, bumping them along from one end to the other. This is an electric current. It is the movement of negatively charged electrons.

There are no free-to-move electrons in a solution. What is present depends on the nature of the solute. A solid ionic solute, such as a salt, is made up of positively and negatively charged ions that are held in fixed positions relative to each other. These ions become free to move when the solute dissolves. The positively charged cations are attracted by the negatively charged electrode and repelled by the positively charged electrode. The cations therefore move toward the negatively charged electrode. In the same way, the negatively charged anions move toward the positively charged electrode. (See Fig. 16.2.)

FLASHBACK The "electron sea" model of a metallic crystal pictures a rigid structure of metal ions surrounded by electrons that can move freely among the ions. See Section 14.7 and Figure 14.19.

FLASHBACK Ions in an ionic solid are arranged in a crystal that has a distinct geometric pattern. They are held in place by ionic bonds resulting from electrostatic attractions and repulsions of nearby ions. Figures 11.2 and 11.3 (Section 11.2) show two such crystals.

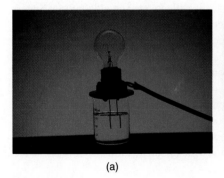
(a)

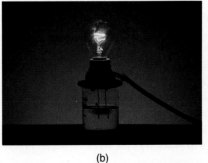
(b)

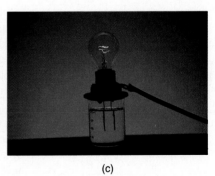
(c)

Figure 16.1
Nonelectrolytes, strong electrolytes, and weak electrolytes. The liquid in the beaker is used to "close" the electrical circuit. If the liquid is pure water or a solution of certain molecular compounds, the bulb does not light (a). Solutes whose solutions do not conduct electricity are nonelectrolytes. A soluble ionic salt is a strong electrolyte because its solution makes the bulb burn brightly (b). Some solutes are called weak electrolytes because their solutions conduct poorly and the bulb glows dimly (c).

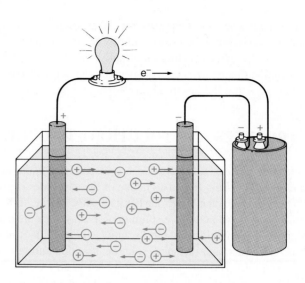

Figure 16.2

Conductivity in an ionic solution. If a solution conducts electricity, it is positive evidence that mobile ions are present. Positively charged cations are attracted to the negatively charged electrode, called a *cathode*. Similarly, negatively charged anions move to the positively charged electrode, called an *anode*.

It is the movement of ions that makes up an electric current in a solution. In fact, the ability of a solution to conduct electricity is regarded as positive evidence that ions are present in the solution.

The sugar solution is a nonconductor because no ions are present. The solute particles are neutral molecules. Being neutral, they do not move toward either electrode. Even if they did move, there would be no current because the molecules have no charge.

A solute whose solution is a good conductor is called a **strong electrolyte.** Sugar, whose solution is a nonconductor, is a **nonelectrolyte.** Some solutes are **weak electrolytes.** Their solutions conduct electricity, but poorly, permitting only a dim glow of the lamp in Figure 16.1. The term "electrolyte" is also applied generally to the solution through which current passes. The acid solution in an automobile battery is an electrolyte in this sense.

The equations that describe solution reactions include the ions and/or molecules that are the actual solute particles in the solution. These particles must be identified by their chemical formulas. A list of the particles in any solution is its **solution inventory.** You will learn how to write solution inventories for different kinds of solutes in the next two sections.

"Solution inventory" is not a widely used term, but it is a handy name by which to refer to dissolved solute particles that are present in greatest abundance.

✔ QUICK CHECK 16.1

Three compounds, A, B, and C, are dissolved in water, and the solutions are tested for electrical conductivity. Solution A is a poor conductor, B is a good conductor, and C does not conduct. Classify A, B, and C as electrolytes (strong, weak, non-) and state the significance of the conductivities of their solutions.

16.2 SOLUTION INVENTORIES OF IONIC COMPOUNDS

When an ionic compound dissolves, its solution inventory always consists of ions. These ions are identified simply by separating the compound into its ions. When sodium chloride dissolves, the solution inventory is sodium ions and chloride ions:

$$NaCl(s) \xrightarrow{H_2O} Na^+(aq) + Cl^-(aq) \qquad (16.1)$$

If the solute is barium chloride, the inventory is barium ions and chloride ions:

$$BaCl_2(s) \xrightarrow{H_2O} Ba^{2+}(aq) + 2\,Cl^-(aq) \qquad (16.2)$$

Notice that no matter where a chloride ion comes from, its formula is *always* Cl^-, never Cl_2^- or Cl_2^{2-}. The subscript after an ion in a formula, as the 2 in $BaCl_2$, tells us how many ions are present in the formula unit. This subscript is not part of the ion formula.

In this chapter you should include state symbols for all species in equations. It helps in writing net ionic equations. Also, remember to include the charge every time you write the formula of an ion.

EXAMPLE 16.1 _____

Write solution inventories for the following ionic compounds by writing their dissolving equations:

$$NaOH(s) \xrightarrow{H_2O}$$

$$K_2SO_4(s) \xrightarrow{H_2O}$$

$$(NH_4)_2CO_3(s) \xrightarrow{H_2O}$$

An H_2O above the arrow in an equation indicates that the reaction occurs in the presence of water.

_____   _____

$$NaOH(s) \xrightarrow{H_2O} Na^+(aq) + OH^-(aq)$$

$$K_2SO_4(s) \xrightarrow{H_2O} 2\,K^+(aq) + SO_4^{2-}(aq)$$

$$(NH_4)_2CO_3(s) \xrightarrow{H_2O} 2\,NH_4^+(aq) + CO_3^{2-}(aq)$$

Polyatomic and monatomic ions are handled in exactly the same way in writing solution inventories. In $(NH_4)_2CO_3$, notice that the subscript outside the parentheses tells us how many ammonium ions are present, while the subscript 4 inside the parentheses is part of the polyatomic ion formula.

EXAMPLE 16.2 _____

Write the solution inventories of the following ionic solutes *without* writing the dissolving equations: $MgSO_4$; $Ca(NO_3)_2$; $AlBr_3$; $Fe_2(SO_4)_3$.

This question is the same as Example 16.1. It asks simply for the products of the reaction without writing an equation. This is how you will write these inventories later in the chapter. Be sure to show the number of each kind of ion released by a formula unit—the coefficient if the equation were written. Also remember state symbols:

$MgSO_4$: $AlBr_3$:

$Ca(NO_3)_2$: $Fe_2(SO_4)_3$:

$MgSO_4$: $Mg^{2+}(aq) + SO_4^{2-}(aq)$ $AlBr_3$: $Al^{3+}(aq) + 3\,Br^-(aq)$

$Ca(NO_3)_2$: $Ca^{2+}(aq) + 2\,NO_3^-(aq)$ $Fe_2(SO_4)_3$: $2\,Fe^{3+}(aq) + 3\,SO_4^{2-}(aq)$

FLASHBACK The ionization of acetic acid was given in Section 12.7 as

Notice that only the hydrogen bonded to the oxygen leaves the molecule.

FLASHBACK The "ionization equation" for an acid has only the formula of the acid on the left, and the ions into which it separates on the right. If the acid is diprotic, the equation may show the release of just one or both of the ionizable hydrogens. A triprotic acid ionization equation may be written for one, two, or three hydrogens. See Sections 6.5 and 6.6.

FLASHBACK A reversible reaction is one in which the products, as an equation is written, change back to the reactants. Reversibility is indicated by a double arrow, one pointing in each direction. We have discussed reversible changes in studying the liquid–vapor equilibrium (Section 14.3) and in the formation of a saturated solution (Section 15.3).

✔ QUICK CHECK 16.2

Which ionic compounds have ions in their solutions and which do not?

16.3 STRONG ACIDS AND WEAK ACIDS

An acid, as we have used the term so far, is a hydrogen-bearing compound that releases hydrogen ions in water solution. Its formula is HX, H_2X, or H_3X, where X is any anion produced when the acid ionizes. X may be a monatomic ion, as when HCl ionizes and leaves Cl^-. X may contain oxygen, as the SO_4^{2-} from H_2SO_4. X may even contain hydrogen, as $H_2PO_4^-$ from the first ionization step from H_3PO_4. Organic acids usually contain hydrogen that is not ionizable. Acetic acid, $HC_2H_3O_2$, for example, ionizes to H^+ and the acetate ion, $C_2H_3O_2^-$. Do not be concerned if the anion is not familiar. It will behave just like the anion from any other acid.

Hydrochloric acid is the water solution of hydrogen chloride, a gaseous molecular compound. Hydrochloric acid is a **strong acid.** This means it is almost completely ionized in water and is an excellent conductor. The ionization equation is

$$HCl(g) \xrightarrow{H_2O} HCl(aq) \longrightarrow H^+(aq) + Cl^-(aq) \qquad (16.3)$$

Hydrofluoric acid is the water solution of hydrogen fluoride, another gaseous molecular compound. Hydrofluoric acid is a **weak acid.** It is only slightly ionized in water and is a poor conductor. The ionization process may be described by an equation similar to that for hydrochloric acid, but with an important difference:

$$HF(g) \xrightarrow{H_2O} HF(aq) \rightleftharpoons H^+(aq) + F^-(aq) \qquad (16.4)$$

The double arrows show that the ionization process is reversible. The longer arrow from right to left indicates that it is much more likely for hydrogen and fluoride ions to combine to form dissolved HF molecules than for the molecules to break into ions. In other words, the **major species** in the solution inventory are dissolved HF molecules. The **minor species,** present in much smaller numbers, are the H^+ and F^- ions. The low concentration of ions makes the solution a poor conductor.

In describing acids, we use the words "strong" and "weak" exactly the same way they are used in describing electrolytes. A strong acid is almost completely ionized in water, and a weak acid is ionized only slightly. *The solution inventory of a strong acid is the ions it forms.*

Technically, the solution inventory of a weak acid includes the acid molecule as the major species and the ions as minor species. However, in writing net ionic equations, we include only the major species. Therefore, *the solution inventory of a weak acid is the acid molecule.*

There are seven strong acids. Their names and formulas must be memorized. It helps to group them into three classifications. One acid, hydrochloric, fits into all three groups:

Three of the Best-Known Acids Are Strong	Three Hydrohalic Acids Are Strong	Three Chlorine Acids Are Strong
hydrochloric, HCl	hydrochloric, HCl	hydrochloric, HCl
nitric, HNO_3	hydrobromic, HBr	chloric, $HClO_3$
sulfuric, H_2SO_4	hydroiodic, HI	perchloric, $HClO_4$

To decide whether an acid is strong or weak, ask yourself, "Is it one of the strong acids?" If it is one of the "strong seven" acids listed here, then it is strong. If it is not one of the strong seven, it is weak.

The dividing line between strong and weak acids is arbitrary, and at least three acids are marginal in their classifications. Sulfuric acid is definitely strong in its first ionization step, but questionable in the second. In all reactions in this book the second ionization does occur, so we will regard it as a strong acid that releases both hydrogen ions. Both oxalic acid, $H_2C_2O_4$, and phosphoric acid, H_3PO_4, are marginal in their first ionizations, but definitely weak in the second and, for H_3PO_4, third. We regard them as weak, but we will avoid questions in which they might have to be classified.

EXAMPLE 16.3 ⎯⎯⎯⎯⎯⎯⎯⎯⎯⎯⎯⎯⎯⎯⎯⎯⎯⎯⎯

Write the solution inventories of the following acids: HNO_2, H_2SO_4, HI, $HC_3H_5O_2$.

You can, if necessary, write these inventories with equations, as in Example 16.1. It is better just to identify the ions or molecules, as in Example 16.2. If a single molecule produces more than one of a particular ion, as $Ca(NO_3)_2$ produced 2 NO_3^- ions in Example 16.2, show the number. Include state symbols.

⎯⎯⎯⎯⎯⎯⎯   ⎯⎯⎯⎯⎯⎯⎯

HNO_2: $HNO_2(aq)$ HI: $H^+(aq) + I^-(aq)$

H_2SO_4: $2 H^+(aq) + SO_4^{2-}(aq)$ $HC_3H_5O_2$: $HC_3H_5O_2(aq)$

HNO_2 and $HC_3H_5O_2$ are not among the seven strong acids, so their solution inventories are the molecules themselves. HI and H_2SO_4 are strong acids; they break up into ions. A common mistake with H_2SO_4 is to write $H_2^+(aq)$ or something like that for the hydrogen ion, $H^+(aq)$. Like the chloride ion in Equations 16.1 and 16.2, the hydrogen ion has the same formula no matter where it comes from.

⎯⎯⎯⎯⎯⎯⎯⎯⎯⎯⎯⎯⎯⎯⎯⎯⎯⎯⎯⎯⎯⎯⎯⎯⎯⎯⎯⎯⎯⎯

We can now summarize solution inventories as they have been described in this section and the last.

This summary of solution inventories identifies only the major species in the solution—the species that appear in net ionic equations.

SUMMARY

Ions make up the solution inventories of two kinds of substances:
    All soluble ionic compounds
    The ''strong seven'' acids
Neutral molecules are the solution inventory of everything else—primarily the three W's: weak acids, weak bases, and water.

☑ **QUICK CHECK 16.3**

What is the difference between a strong acid and a weak acid? How do you identify a strong acid?

16.4 NET IONIC EQUATIONS: WHAT THEY ARE AND HOW TO WRITE THEM

In Chapter 8 you learned how to write the equation for the precipitation reaction that occurs when a solution of lead nitrate is added to a solution of sodium chloride:

$$Pb(NO_3)_2(aq) + 2\,NaCl(aq) \longrightarrow PbCl_2(s) + 2\,NaNO_3(aq) \qquad (16.5)$$

In this chapter we call this kind of equation a **"conventional equation."**

    A conventional equation serves many useful purposes, including its essential role in solving stoichiometry problems. However, it falls short in describing a reaction that occurs in water solution. Usually, it does not describe the reactants and/or products correctly. Rarely does it describe accurately the chemical changes that occur. And it has no value in solving problems involving chemical equilibrium.

    To illustrate the shortcomings of a conventional equation, the solutions that react in Equation 16.5 contain no substances whose formulas are $Pb(NO_3)_2$ or $NaCl$. Actually present are the solution inventory ions, Pb^{2+} and NO_3^- in one solution and Na^+ and Cl^- in the other. The conventional equation doesn't tell you that. Nothing with the formula $NaNO_3$ is formed in the reaction. The Na^+ and NO_3^- ions are still there after the reaction. The conventional equation keeps that a ''secret'' too. The only substance in Equation 16.5 that is *really there* is solid lead chloride, $PbCl_2(s)$. If you perform the reaction you can see the precipitate.

    To write an equation that describes the reaction in Equation 16.5 more accurately, we replace the formulas of the dissolved substances with their solution inventories. This produces the **ionic equation:**

$$Pb^{2+}(aq) + 2\,NO_3^-(aq) + 2\,Na^+(aq) + 2\,Cl^-(aq) \longrightarrow$$
$$PbCl_2(s) + 2\,Na^+(aq) + 2\,NO_3^-(aq) \quad (16.6)$$

Notice that in order to keep the ionic equation balanced, the coefficients from the conventional equation are repeated. One formula unit of $Pb(NO_3)_2$ gives one Pb^{2+} ion and two NO_3^- ions, two formula units of $NaCl$ give two Na^+ ions and two Cl^- ions, and two formula units of $NaNO_3$ give two Na^+ ions and two NO_3^- ions.

    An ionic equation tells more than just what happens in a chemical change. It includes **spectator ions,** or simply **spectators.** A spectator is an ion that is present at the scene of a reaction but experiences no chemical change. It appears on both sides of the ionic equation. $Na^+(aq)$ and $NO_3^-(aq)$ are spectators in Equation 16.6. To change an ionic equation into a **net ionic equation,** you simply remove the spectators:

$$Pb^{2+}(aq) + 2\ Cl^-(aq) \longrightarrow PbCl_2(s) \qquad (16.7)$$

A net ionic equation indicates exactly what chemical change took place, and nothing else.

In Equations 16.5 to 16.7 you have the three steps to be followed in writing a net ionic equation. They are:

SUMMARY

1) Write the conventional equation, including designations of state—(g), (ℓ), (s), and (aq). Balance the equation.
2) Write the ionic equation by replacing each dissolved substance (aq) that is a strong acid or an ionic compound with its solution inventory. *Do not separate a weak acid into ions,* even though its state is (aq). Also, never change solids, (s), liquids, (ℓ), or gases, (g), into ions. Be sure the equation is balanced in both atoms and charge. (Charge balance is discussed in the next section.)
3) Write the net ionic equation by removing the spectators from the ionic equation. Reduce coefficients to lowest terms, if necessary. Be sure the equation is balanced in both atoms and charge.

✔ QUICK CHECK 16.4

After writing a conventional equation, including state designations (s), (ℓ), (g), or (aq), which species do you consider breaking into ions in writing the ionic equation? Of those, which *do* you break into ions, and which do you *not* break into ions?

16.5 REDOX REACTIONS THAT ARE DESCRIBED BY "SINGLE REPLACEMENT" EQUATIONS

A piece of zinc is dropped into sulfuric acid. Hydrogen gas bubbles out (Fig. 16.3). When the reaction ends, the vessel contains a solution of zinc sulfate. The question is, "What happened?" The answer lies in the net ionic equation.

The conventional equation (Step 1) is a single replacement equation:

$$Zn(s) + H_2SO_4(aq) \longrightarrow H_2(g) + ZnSO_4(aq)$$

H_2SO_4 is a strong acid and $ZnSO_4$ is a soluble ionic compound. Both have ions in their solution inventories. These inventories replace the compounds in the ionic equation (Step 2):

$$Zn(s) + 2\ H^+(aq) + SO_4^{2-}(aq) \longrightarrow H_2(g) + Zn^{2+}(aq) + SO_4^{2-}(aq) \quad (16.8)$$

The ionic equation remains balanced in both atoms and charge. The net charge is zero on both sides.

The final step (Step 3) in writing a net ionic equation is to rid the ionic equation of spectators. There is only one, the sulfate ion. Taking it away gives

$$Zn(s) + 2\ H^+(aq) \longrightarrow H_2(g) + Zn^{2+}(aq) \qquad (16.9)$$

Figure 16.3

The reaction between zinc and sulfuric acid.

💡 FLASHBACK A single replacement equation is one in which one uncombined element, A, appears to replace another element, B, in a compound: $A + BX \longrightarrow AX + B$. See Section 8.7 and Table 8.1.

This is the net ionic equation. This is what happened—and no more.

Notice that all equations are balanced. You have been balancing atoms since Chapter 8, but charge is something new. Neither protons nor electrons are created or destroyed in a chemical change, so the total charge among the reactants must be equal to the total charge among the products. In Equation 16.8, two plus charges in $2\,H^+(aq)$ added to two negative charges in $SO_4^{2-}(aq)$ give a net zero charge on the left. Two plus charges in $Zn^{2+}(aq)$ and two negative charges in $SO_4^{2-}(aq)$ on the right also total zero. The equation is balanced in charge.

Notice that the net charge does not have to be zero on each side for the equation to be balanced. The charges must be *equal*. In Equation 16.9 the net charge is $2+$ on each side.

EXAMPLE 16.4 _____

A reaction occurs when a piece of zinc is dipped into $Cu(NO_3)_2(aq)$ (Fig. 16.4). Write the conventional, ionic, and net ionic equations.

SOLUTION

We begin by writing the conventional equation (Step 1):

$$Zn(s) + Cu(NO_3)_2(aq) \longrightarrow Zn(NO_3)_2(aq) + Cu(s)$$

To get the ionic equation, we replace those species that are in solution, marked (aq), with their solution inventories. According to the summary in Section 16.3, only dissolved ionic compounds and strong acids are divided into ions. Copper(II) and zinc nitrates are ionic compounds. The ionic equation (Step 2) is therefore

$$Zn(s) + Cu^{2+}(aq) + 2\,NO_3^-(aq) \longrightarrow Zn^{2+}(aq) + 2\,NO_3^-(aq) + Cu(s)$$

The equation is balanced. There is one zinc on each side, an atom on one side and an ion on the other. Ditto for copper. Finally, there are two nitrate ions on each side. Net charge is zero on the two sides of the equation. The nitrates are the only spectators. Subtracting the spectators (Step 3) gives the net ionic equation:

$$Zn(s) + Cu^{2+}(aq) \longrightarrow Zn^{2+}(aq) + Cu(s)$$

Zinc and copper are balanced as before. This time net charge is $2+$ on each side of the equation. The equation is balanced in atoms and charge.

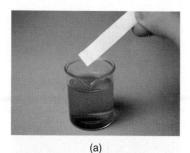

(a)

(b)

Figure 16.4

The reaction between zinc and a solution of copper(II) nitrate. (a) The zinc strip is shiny white before the reaction. (b) After dipping the zinc into the solution for about 2 seconds the strip is covered with tiny particles of copper that appear almost black when wet with the solution.

If you were to place a strip of copper into a solution of zinc nitrate and go through the identical thought process, the equations would be exactly the reverse of those in Example 16.4. Does the reaction occur in both directions? If not, in which way does it occur? And how can you tell?

Technically, both reactions occur, or can be made to occur. Practically, only the reaction in Example 16.4 takes place. The best way to find out which of two reversible reactions occurs is to try them and see. These experiments have been done, and the results are summarized in the **activity series** in Table 16.1. Under normal conditions, any element in the table will replace the dissolved ions of any element beneath it. Zinc is above copper in the table, so zinc will replace $Cu^{2+}(aq)$ ions in solution. Copper, being below zinc, will not replace $Zn^{2+}(aq)$ in solution.

Use Table 16.1 to predict whether or not a redox reaction will take place. If asked to write the equation for a reaction that does not occur, write NR for "no reaction" on the product side: $Cu(s) + Zn^{2+}(aq) \rightarrow NR$.

Table 16.1
Activity Series

Li K	Very reactive metals
Ba Ca Na	Will replace H_2 from H_2O
Mg Al Zn	Moderately reactive metals
Fe Ni Sn Pb	Will replace H_2 from acids
H_2	
Cu Ag Au	Weakly reactive metals

EXAMPLE 16.5

Write the conventional, ionic, and net ionic equations for the reaction that will occur, if any, between calcium and hydrochloric acid.

First, will a reaction occur? Check the activity series; then answer.

———

Yes. Calcium is above hydrogen in the series, so calcium will replace hydrogen ions from solution.

Write the conventional equation. Remember the state designation.

———

$Ca(s) + 2\,HCl(aq) \longrightarrow H_2(g) + CaCl_2(aq)$

Now write the ionic equation.

———

$Ca(s) + 2\,H^+(aq) + 2\,Cl^-(aq) \longrightarrow H_2(g) + Ca^{2+}(aq) + 2\,Cl^-(aq)$

Each $HCl(aq)$ yields one $H^+(aq)$ and one $Cl^-(aq)$. The conventional equation has two $HCl(aq)$, so there will be $2\,H^+(aq)$ and $2\,Cl^-(aq)$ in the ionic equation.

Now eliminate the spectators and write the net ionic equation.

———

$Ca(s) + 2\,H^+(aq) \longrightarrow H_2(g) + Ca^{2+}(aq)$

Chloride ion, $Cl^-(aq)$, is the only spectator.

Calcium reacts with hydrochloric acid.

EXAMPLE 16.6

Write the three equations for the reaction between nickel and a solution of $MgCl_2$, if any.

See what you can do this time without hints.

———

$Ni(s) + MgCl_2(aq) \longrightarrow NR$

Nickel is beneath magnesium in the activity series, so no redox reaction occurs.

EXAMPLE 16.7

Copper is placed into a solution of silver nitrate. Write the three equations.

———

$$Cu(s) + 2\,AgNO_3(aq) \longrightarrow 2\,Ag(s) + Cu(NO_3)_2(aq)$$

$$Cu(s) + 2\,Ag^+(aq) + 2\,NO_3{}^-(aq) \longrightarrow 2\,Ag(s) + Cu^{2+}(aq) + 2\,NO_3{}^-(aq)$$

$$Cu(s) + 2\,Ag^+(aq) \longrightarrow 2\,Ag(s) + Cu^{2+}(aq)$$

Figure 16.5 is a picture of this reaction.

EXAMPLE 16.8

If potassium is placed into water, hydrogen gas bubbles out. Write all three equations.

This reaction is more clearly seen if you write the formula of water as HOH. Treat it as a weak acid, with only the first hydrogen ionizable. Go for all three equations, but watch those state symbols.

$$2\,K(s) + 2\,HOH(\ell) \longrightarrow H_2(g) + 2\,KOH(aq)$$

$$2\,K(s) + 2\,HOH(\ell) \longrightarrow H_2(g) + 2\,K^+(aq) + 2\,OH^-(aq)$$

Water is a liquid molecular compound, HOH(ℓ). It does not break into ions. Never separate into ions a gas (g), liquid (ℓ), or solid (s). This particular ionic equation has no spectators, so it is also the net ionic equation. Figure 16.6 is a picture of this spectacular but dangerous reaction.

Figure 16.5

The oxidation of copper metal by silver ion. A clean piece of copper screen is placed in a solution of silver nitrate, $AgNO_3$. With time, the copper reduces Ag^+ to silver metal crystals and the copper is oxidized to Cu^{2+}. The blue color of the solution is due to the presence of aqueous copper(II) ion.

16.6 ION COMBINATIONS THAT FORM PRECIPITATES

An **ion-combination reaction** occurs when the cation from one reactant combines with the anion from another to form a particular kind of product compound. The conventional equation is a double displacement type in which the ions appear to "change

Figure 16.6

The reaction between potassium and water. The room was completely dark and all light for this photograph was produced by dropping a small piece of potassium into a beaker of water. (*Note*: THIS IS A POTENTIALLY EXPLOSIVE AND EXTREMELY DANGEROUS REACTION! DO NOT TRY IT.)

partners'': MY + NX → MX + NY. In this section the product is an insoluble ionic compound that settles to the bottom of the mixed solutions. A solid formed this way is called a **precipitate;** the reaction is a **precipitation reaction.**

FLASHBACK Double replacement equations were first used to describe precipitation reactions in Section 8.8. See also Table 8.1.

EXAMPLE 16.9

When hydrochloric acid and silver nitrate solutions are mixed, a white precipitate of silver chloride is produced. Develop the net ionic equation for the reaction.

The same three steps you used on redox reactions are applied to precipitation reactions. Write all three equations.

$$HCl(aq) + AgNO_3(aq) \longrightarrow HNO_3(aq) + AgCl(s)$$

$$H^+(aq) + Cl^-(aq) + Ag^+(aq) + NO_3^-(aq) \longrightarrow H^+(aq) + NO_3^-(aq) + AgCl(s)$$

$$Ag^+(aq) + Cl^-(aq) \longrightarrow AgCl(s)$$

$H^+(aq)$ and $NO_3^-(aq)$ are spectators in the ionic equation.

Let's try another example that has two interesting features at the end.

EXAMPLE 16.10

When solutions of silver chlorate, $AgClO_3(aq)$, and aluminum chloride, $AlCl_3(aq)$, are combined, silver chloride precipitates. Write the three equations.

Proceed as usual. When you get to the net ionic equation, which will have something new, see if you can figure out what to do about it. If necessary, read Step 3 of the procedure in Section 16.4.

$$3\,AgClO_3(aq) + AlCl_3(aq) \longrightarrow 3\,AgCl(s) + Al(ClO_3)_3(aq)$$

$$3\,Ag^+(aq) + 3\,ClO_3^-(aq) + Al^{3+}(aq) + 3\,Cl^-(aq) \longrightarrow 3\,AgCl(s) + Al^{3+}(aq) + 3\,ClO_3^-(aq)$$

$$3\,Ag^+(aq) + 3\,Cl^-(aq) \longrightarrow 3\,AgCl(s)$$

$$Ag^+(aq) + Cl^-(aq) \longrightarrow AgCl(s)$$

The net ionic equation this time has 3 for the coefficient of all species. The third step of the procedure says to reduce coefficients to lowest terms. The equation may be divided by 3, as shown. That leads to the second interesting feature.

Examples 16.9 and 16.10 produced the same net ionic equation even though the reacting solutions were completely different. Actually, only the spectators were different, but they are not part of the chemical change. Eliminating them shows that both reactions are exactly the same. It will be this way whenever a solution containing $Ag^+(aq)$ ion is added to a solution containing $Cl^-(aq)$ ion. Therefore, if asked to write

The $Ag^+ + Cl^- \to AgCl$ precipitation is pictured in Figure 8.7 (Section 8.8). The reactants for that photograph were NaCl and $AgNO_3$. The sodium and nitrate ions were spectators.

the net ionic equation for the reaction between such solutions, you can go directly to $Ag^+(aq) + Cl^-(aq) \to AgCl(s)$. (It is possible that a second reaction may occur between the other pair of ions, yielding a second net ionic equation.)

This simple and direct procedure may be used to write the net ionic equation for the precipitation of any insoluble ionic compound. The compound is the product, and the reactants are the ions in the compound. It is like writing a solution inventory equation (Equations 16.1 and 16.2, and Example 16.1) in reverse. Try it on the following example.

EXAMPLE 16.11

Write the net ionic equations for the precipitations of the following from aqueous solutions:

CuS:

$Mg(OH)_2$:

Li_3PO_4:

$$Cu^{2+}(aq) + S^{2-}(aq) \longrightarrow CuS(s)$$

$$Mg^{2+}(aq) + 2\,OH^-(aq) \longrightarrow Mg(OH)_2(s)$$

$$3\,Li^+(aq) + PO_4{}^{3-}(aq) \longrightarrow Li_3PO_4(s)$$

If we knew in advance those combinations of ions that yield insoluble compounds, we could predict precipitation reactions. These compounds have been identified in the laboratory. Table 16.2 shows the result of such experiments for a large number of ionic compounds. These solubilities have also been summarized in a set of "solubility rules" that your instructor may ask you to memorize. These rules are in Table 16.3.

In the next four examples use either table to predict precipitation reactions. We recommend that you use each table at least once.

EXAMPLE 16.12

Solutions of lead(II) nitrate and sodium bromide are combined. Write the net ionic equation for any precipitation reaction that may occur.

Start with the conventional equation.

$$Pb(NO_3)_2(aq) + 2\,NaBr(aq) \longrightarrow PbBr_2(\quad) + 2\,NaNO_3(\quad)$$

The spaces between parentheses after the products have been left blank. You must determine if the compound is soluble or insoluble. If soluble, the state symbol should be (aq); if insoluble, (s). Fill in the state symbols.

Table 16.2
Solubilities of Ionic Compounds* (S = Soluble; I = Insoluble)

Ions	Acetate	Bromide	Carbonate	Chlorate	Chloride	Fluoride	Hydrogen Carbonate	Hydroxide	Iodide	Nitrate	Nitrite	Phosphate	Sulfate	Sulfide	Sulfite
Aluminum	I	S		S	S	I		I	—	S		I	S	—	
Ammonium	S	S	S	S	S	S	S	—	S	S	S	S	S	S	S
Barium	S	S	I	S	S	I		S	S	S	S	I	I	—	I
Calcium	S	S	I	S	S	I		I	S	S	S	I	I	—	I
Cobalt(II)	S	S	I	S	S	—		I	S	S		I	S	I	I
Copper(II)	S	S	I	S	S	S		I		S		I	S	I	
Iron(II)	S	S	I		S	I		I	S	S		I	S	I	I
Iron(III)	—	S			S	I		I	S	S		I	S	—	
Lead(II)	S	I	I	S	I	I		I	I	S	S	I	I	I	I
Lithium	S	S	S	S	S	S	S	S	S	S	S	I	S	S	S
Magnesium	S	S	I	S	S	I		I	S	S	S	I	S	—	S
Nickel	S	S	I	S	S	S		I	S	S		I	S	I	I
Potassium	S	S	S	S	S	S	S	S	S	S	S	S	S	S	S
Silver	I	I	I	S	I	S		—	I	S	I	I	I	I	I
Sodium	S	S	S	S	S	S	S	S	S	S	S	S	S	S	S
Zinc	S	S	I	S	S	S		I	S	S		I	S	I	I

*Compounds having solubilities of 0.1 M or more at 20°C are listed as soluble (S); if the solubility is less than 0.1 M, the compound is listed as insoluble (I). A dash (—) identifies an unstable species in aqueous solution, and a blank space indicates lack of data. In writing equations for reactions that occur in water solution, insoluble substances, shown by I in this table, have the state symbol of a solid, (s). Dissolved substances, S in the table, are designated by (aq) in an equation.

Table 16.3
Solubility Rules for Ionic Compounds*

Most of the following compounds are Soluble	Exceptions	Most of the following compounds are Insoluble	Exceptions
Ammonium salts		Carbonates	NH_4^+ and alkali metals
Alkali metal salts (Column 1A)	Some Li^+	Phosphates	Na^+, K^+, and NH_4^+
Nitrates		Hydroxides	Ba^{2+} and alkali metals
Chlorides, bromides, iodides	Ag^+, Hg_2^{2+}, Pb^{2+}	Sulfides	NH_4^+ and salts of alkali metals
Acetates	Ag^+, Al^{3+}		
Chlorates			
Sulfates	Ba^{2+}, Sr^{2+}, Ca^{2+}, Pb^{2+}, Ag^+, Hg^{2+}, Hg_2^{2+}		

*For purposes of these rules, a compound is considered to be soluble if it dissolves to a concentration of 0.1 M or more at 20°C.

$$Pb(NO_3)_2(aq) + 2\,NaBr(aq) \longrightarrow PbBr_2(s) + 2\,NaNO_3(aq)$$

In Table 16.2 the intersection of the lead(II) ion line and the bromide column shows that lead(II) bromide is insoluble. Therefore, the designation (s) follows $PbBr_2$. The intersection of the sodium ion line and nitrate ion column shows that sodium nitrate is soluble in water. Hence, (aq) follows $NaNO_3$. Both conclusions are confirmed by the solubility rules: Lead(II) bromide is one of the three insoluble bromides, and all sodium and nitrate compounds are soluble.

Complete the example by writing the ionic and net ionic equations.

$$Pb^{2+}(aq) + 2\,NO_3{}^-(aq) + 2\,Na^+(aq) + 2\,Br^-(aq) \longrightarrow PbBr_2(s) + 2\,Na^+(aq) + 2\,NO_3{}^-(aq)$$
$$Pb^{2+}(aq) + 2\,Br^-(aq) \longrightarrow PbBr_2(s)$$

EXAMPLE 16.13

Write the net ionic equation for any reaction that will occur when solutions of nickel chloride and ammonium carbonate are combined.

This example does not ask for all three equations, but only the net ionic equation. You might wish to try for the net ionic equation without the other two, as in Example 16.11. If you are not ready for this step, write all three equations, as before. The three steps are shown in the answer.

$$NiCl_2(aq) + (NH_4)_2CO_3(aq) \longrightarrow NiCO_3(s) + 2\,NH_4Cl(aq)$$
$$Ni^{2+}(aq) + 2\,Cl^-(aq) + 2\,NH_4{}^+(aq) + CO_3{}^{2-}(aq) \longrightarrow NiCO_3(s) + 2\,NH_4{}^+(aq) + 2\,Cl^-(aq)$$
$$Ni^{2+}(aq) + CO_3{}^{2-}(aq) \longrightarrow NiCO_3(s)$$

To write the net ionic equation directly, you would have to decide from the reactants, $NiCl_2$ and $(NH_4)_2CO_3$, what the products would be in a double replacement equation. Exchanging the ions gives $NiCO_3$ and NH_4Cl. The solubility tables show that NH_4Cl is soluble and $NiCO_3$ is insoluble. $NiCO_3$ will precipitate. Therefore, write the formula of that compound on the right and the formulas of its ions on the left, and you have the net ionic equation.

EXAMPLE 16.14

Write the net ionic equation for any reaction that occurs between solutions of aluminum sulfate and calcium acetate, $Al_2(SO_4)_3(aq)$ and $Ca(C_2H_3O_2)_2(aq)$.

Be careful here!

$$Al_2(SO_4)_3(aq) + 3\,Ca(C_2H_3O_2)_2(aq) \longrightarrow 3\,CaSO_4(s) + 2\,Al(C_2H_3O_2)_3(s)$$
$$2\,Al^{3+}(aq) + 3\,SO_4{}^{2-}(aq) + 3\,Ca^{2+}(aq) + 6\,C_2H_3O_2{}^-(aq) \longrightarrow 3\,CaSO_4(s) + 2\,Al(C_2H_3O_2)_3(s)$$

This time *both* new combinations of ions precipitate. There are no spectators, so the ionic equation is the net ionic equation. Actually, there are two separate reactions taking place at the same time:

$$Al^{3+}(aq) + 3\ C_2H_3O_2^-(aq) \longrightarrow Al(C_2H_3O_2)_3(s)$$

$$Ca^{2+}(aq) + SO_4^{2-}(aq) \longrightarrow CaSO_4(s)$$

EXAMPLE 16.15

Write the net ionic equation for any reaction that occurs when solutions of ammonium sulfate and potassium nitrate are combined.

Again, be careful.

$$(NH_4)_2SO_4(aq) + KNO_3(aq) \longrightarrow NR$$

This time both new combinations of ions are soluble, so there is no precipitation reaction. If you complete the conventional equation, as you would have done in Chapter 8, and from it write the ionic equation, you will find that *all* ions are spectators.

16.7 ION COMBINATIONS THAT FORM MOLECULES

The reaction of an acid often leads to an ion combination that yields a molecular product instead of a precipitate. Except for the difference in the product, the equations are written in exactly the same way. Just as you had to recognize an insoluble product and not break it up in the ionic equations, you must now recognize a molecular product and not break it into ions. Water or a weak acid are the two kinds of molecular products you will find.

Neutralization reactions are the most common molecular product reactions.

EXAMPLE 16.16

Write the conventional, ionic, and net ionic equations for the reaction between hydrochloric acid and a solution of sodium hydroxide.

Proceed just as you did for precipitation reactions. Watch your state designations.

$$HCl(aq) + NaOH(aq) \longrightarrow H_2O(\ell) + NaCl(aq)$$

$$H^+(aq) + Cl^-(aq) + Na^+(aq) + OH^-(aq) \longrightarrow H_2O(\ell) + Na^+(aq) + Cl^-(aq)$$

$$H^+(aq) + OH^-(aq) \longrightarrow H_2O(\ell)$$

Water is the molecular product. It is not ionized and it is in the liquid state.

FLASHBACK The neutralization reactions in Section 8.9 were between acids and hydroxide bases. The products were a salt and water. The general equation is

$$HX + MOH \rightarrow MX + H_2O.$$

The acid in a neutralization may be a weak acid. You must then recall that the solution inventory of a weak acid is the acid molecule; it is not broken into ions.

EXAMPLE 16.17 _____

Write the three equations leading to the net ionic equation for the reaction between acetic acid, $HC_2H_3O_2(aq)$, and a solution of sodium hydroxide.

——————   ——————

$$HC_2H_3O_2(aq) + NaOH(aq) \longrightarrow H_2O(\ell) + NaC_2H_3O_2(aq)$$

$$HC_2H_3O_2(aq) + Na^+(aq) + OH^-(aq) \longrightarrow H_2O(\ell) + Na^+(aq) + C_2H_3O_2^-(aq)$$

$$HC_2H_3O_2(aq) + OH^-(aq) \longrightarrow H_2O(\ell) + C_2H_3O_2^-(aq)$$

The weak acid molecule appears in its molecular form in the net ionic equations.

_____

   When the reactants are a strong acid and the salt of a weak acid, that weak acid is formed as the molecular product. You must recognize it as a weak acid and leave it in molecular form in the ionic and net ionic equations.

EXAMPLE 16.18 _____

Develop the net ionic equation for the reaction between hydrochloric acid and a solution of sodium acetate, $NaC_2H_3O_2(aq)$.

——————   ——————

$$HCl(aq) + NaC_2H_3O_2(aq) \longrightarrow HC_2H_3O_2(aq) + NaCl(aq)$$

$$H^+(aq) + Cl^-(aq) + Na^+(aq) + C_2H_3O_2^-(aq) \longrightarrow HC_2H_3O_2(aq) + Na^+(aq) + Cl^-(aq)$$

$$H^+(aq) + C_2H_3O_2^-(aq) \longrightarrow HC_2H_3O_2(aq)$$

_____

   Compare the reactions in Examples 16.16 and 16.18. The only difference between them is that Example 16.16 has the hydroxide ion as a reactant and Example 16.18 has the acetate ion as a reactant. In the first case the molecular product is water, formed when the hydrogen ion bonds to the hydroxide ion. In the second case the molecular product is acetic acid, a weak acid, formed when the hydrogen ion bonds to the acetate ion.

   Just as you can write the net ionic equation for a precipitation reaction without the conventional and ionic equations, so you can write the net ionic equation for a molecule formation reaction. Again, you must recognize the product from the formulas of the reactants. The acid will contribute a hydrogen ion to the molecular product. It will form a molecule with the anion from the other reactant. If the molecule is water or a weak acid, you have the reactants and products of the net ionic equation.

EXAMPLE 16.19 _____

Without writing the conventional and ionic equations, write the net ionic equations for each of the following pairs of reactants:

H_2SO_4 and LiOH:

KNO_2 and HBr:

——————   ——————

$$H^+(aq) + OH^-(aq) \longrightarrow H_2O(\ell) \qquad H^+(aq) + NO_2^-(aq) \longrightarrow HNO_2(aq)$$

$Li^+(aq)$ and $SO_4^{2-}(aq)$ are spectators in the first equation for neutralization reaction. HNO_2 is recognized as a weak acid in the second reaction because it is not one of the seven strong acids. $K^+(aq)$ and $Br^-(aq)$ are spectators.

There are two points by which you identify a molecular product reaction: (1) one reactant is an acid, usually strong; (2) one product is water or a weak acid. One of the most common mistakes in writing net ionic equations is the failure to recognize a weak acid as a molecular product. If one reactant in a double replacement equation is a strong acid, you can be sure there will be a molecular product. If it isn't water, look for a weak acid.

16.8 ION COMBINATIONS THAT FORM UNSTABLE PRODUCTS

Three ion combinations yield molecular products that are not the products you would expect. Two of the expected products are carbonic and sulfurous acids. If hydrogen ions from one reactant reach carbonate ions from another, H_2CO_3, carbonic acid, should form:

$$2 H^+(aq) + CO_3^{2-}(aq) \longrightarrow H_2CO_3(aq)$$

But carbonic acid is unstable and decomposes to carbon dioxide gas and water. The correct net ionic equation is therefore

$$2 H^+(aq) + CO_3^{2-}(aq) \longrightarrow CO_2(g) + H_2O(\ell)$$

Sulfurous acid, H_2SO_3, decomposes in the same way to sulfur dioxide and water, but the sulfur dioxide remains in solution:

$$2 H^+(aq) + SO_3^{2-}(aq) \longrightarrow SO_2(aq) + H_2O(\ell)$$

The third ion combination that yields unexpected molecular products occurs when ammonium and hydroxide ions meet:

$$NH_4^+(aq) + OH^-(aq) \longrightarrow \text{``}NH_4OH\text{''}$$

In spite of printed labels, laboratory bottles with NH_4OH etched on them, and wide use of the name "ammonium hydroxide," no substance having the formula NH_4OH exists at ordinary temperatures. The actual product is a solution of ammonia molecules, $NH_3(aq)$. The proper net ionic equation is therefore

$$NH_4^+(aq) + OH^-(aq) \longrightarrow NH_3(aq) + H_2O(\ell)$$

The reaction is reversible, and actually reaches an equilibrium in which NH_3 is the major species and NH_4^+ and OH^- are minor species.

There is no system by which these three "different" molecular product reactions can be recognized. You simply must be alert to them and catch them when they appear. Once again, the predicted but unstable formulas are H_2CO_3, H_2SO_3, and NH_4OH.

Figure 16.7
The reaction between hydrochloric acid and a solution of sodium carbonate.

EXAMPLE 16.20

Write the conventional, ionic, and net ionic equations for the reaction between solutions of sodium carbonate and hydrochloric acid (Fig. 16.7).

$$2\ HCl(aq) + Na_2CO_3(aq) \longrightarrow 2\ NaCl(aq) + CO_2(g) + H_2O(\ell)$$

$$2\ H^+(aq) + 2\ Cl^-(aq) + 2\ Na^+(aq) + CO_3^{2-}(aq) \longrightarrow 2\ Na^+(aq) + 2\ Cl^-(aq) + CO_2(g) + H_2O(\ell)$$

$$2\ H^+(aq) + CO_3^{2-}(aq) \longrightarrow CO_2(g) + H_2O(\ell)$$

16.9 ION-COMBINATION REACTIONS WITH UNDISSOLVED SOLUTES

In every ion-combination reaction considered so far, it has been assumed that both reactants are in solution. This is not always the case. Sometimes the description of the reaction will indicate that a reactant is a solid, liquid, or gas, even though it may be soluble in water. In such a case write the correct state symbol after the formula in the conventional equation and carry the formula through all three equations unchanged.

A common example of this kind of reaction occurs with a compound that handbooks say is insoluble in water but soluble in acids. The net ionic equation shows why.

EXAMPLE 16.21

Write the net ionic equation to describe the reaction when solid aluminum hydroxide dissolves in hydrochloric acid, nitric acid, and sulfuric acid.

This is a neutralization reaction between a strong acid and a solid—and normally insoluble—hydroxide. Write the conventional equation for hydrochloric acid only.

$$3\ HCl(aq) + Al(OH)_3(s) \longrightarrow 3\ H_2O(\ell) + AlCl_3(aq)$$

Now write the ionic and net ionic equations. Remember that you replace only dissolved substances, designated (aq), with their solution inventories.

$$3 \, H^+(aq) + 3 \, Cl^-(aq) + Al(OH)_3(s) \longrightarrow 3 \, H_2O(\ell) + Al^{3+}(aq) + 3 \, Cl^-(aq)$$

$$3 \, H^+(aq) + Al(OH)_3(s) \longrightarrow 3 \, H_2O(\ell) + Al^{3+}(aq)$$

Now think a bit before writing the net ionic equation for the reaction between solid aluminum hydroxide and nitric acid. The question asks only for the net ionic equation. Can you write it directly, without the conventional and ionic equations? If not, write all three equations.

$$3 \, H^+(aq) + Al(OH)_3(s) \longrightarrow 3 \, H_2O(\ell) + Al^{3+}(aq)$$

This is the same as the equation for hydrochloric acid. Sulfuric acid will also produce the same equation. The chloride, nitrate, and sulfate ions are spectators in the three reactions.

16.10 SUMMARY OF NET IONIC EQUATIONS

Table 16.4 summarizes this entire chapter. The blue area is essentially the same as the last three rows of Table 8.1, Section 8.10. This is the table that summarized writing conventional equations.

SUMMARY

Table 16.4
Summary of Net Ionic Equations

Reactants (Conventional)	Reaction Type	Equation Type (Conventional)	Products (Conventional)	Reactants (Net Ionic)	Products (Net Ionic)
Element + salt *or* Element + strong acid	Oxidation-reduction	Single replacement	Element + salt	Element + ion	Element + ion
Two salts *or* Salt + strong acid *or* Salt + hydroxide base	Precipitation	Double replacement	Two salts	Two ions	Ionic precipitate
Strong acid + hydroxide base	Molecule formation, (H_2O), neutralization	Double replacement	Salt + H_2O	$H^+ + OH^-$	H_2O
Weak acid + hydroxide base	Molecule formation, (H_2O), neutralization	Double replacement	Salt + H_2O	Weak acid + OH^-	H_2O + anion from weak acid
Strong acid + salt of weak acid	Molecule formation, (weak acid)	Double replacement	Salt + weak acid	H^+ + anion of weak acid	Weak acid
Strong acid + carbonate *or* hydrogen carbonate	Unstable product + decomposition	Double replacement + decomposition	Salt + H_2O + CO_2 Salt + H_2O + CO_2	$H^+ + CO_3^{2-}$ $H^+ + HCO_3^-$	$H_2O + CO_2$ $H_2O + CO_2$
Strong acid + sulfite *or* hydrogen sulfite			Salt + H_2O + SO_2 Salt + H_2O + SO_2	$H^+ + SO_3^{2-}$ $H^+ + HSO_3^-$	$H_2O + SO_2$ $H_2O + SO_2$
Ammonium salt + hydroxide base	"NH_4OH" + decomposition	Double replacement + decomposition	Salt + NH_3 + H_2O	$NH_4^+ + OH^-$	$H_2O + NH_3$

CHAPTER 16 IN REVIEW

16.1   Electrolytes and Solution Conductivity
16A Distinguish among strong electrolytes, weak electrolytes, and nonelectrolytes.
16B Describe or explain electrical conductivity through a solution.

16.2   Solution Inventories of Ionic Compounds
16C Given the formula of an ionic compound, write the solution inventory when it is dissolved in water.

16.3   Strong Acids and Weak Acids
16D Explain why the solution of an acid may be a good conductor or a poor conductor of electricity.
16E Given the formula of a soluble acid, write the solution inventory when it is dissolved in water.

16.4   Net Ionic Equations: What They Are and How to Write Them

16.5   Redox Reactions That Are Described by "Single Replacement" Equations
16F Given two substances that may engage in a redox reaction and an activity series by which

the reaction may be predicted, write the conventional, ionic, and net ionic equations for the reaction that will occur, if any.

16.6   Ion Combinations That Form Precipitates
16G Predict whether or not a precipitate will form when known solutions are combined; if a precipitate forms, write the net ionic equation. (Reference to a solubility table may or may not be allowed.)
16H Given the product of a precipitation reaction, write the net ionic equation.

16.7   Ion Combinations That Form Molecules
16I Given reactants that yield a molecular product, write the net ionic equation.

16.8   Ion Combinations That Form Unstable Products
16J Given reactants that form H_2CO_3, H_2SO_3, or "NH_4OH" by ion combination, write the net ionic equation for the reaction.

16.9   Ion-Combination Reactions with Undissolved Solutes

16.10 Summary of Net Ionic Equations

CHAPTER 16 KEY WORDS AND MATCHING SETS

Set 1

_____ Present in solutions in relatively low concentrations.

_____ Equation showing full chemical formulas of all substances involved in a reaction.

_____ Equation that includes particles that are at the scene of a reaction, but undergo no change.

_____ Solute whose solution is a good conductor.

_____ List of particles in a solution.

_____ Reaction in which ions from different reactants join to form a new compound.

_____ The more abundant particles in a solution.

_____ Solute in a solution that is a poor conductor of electricity.

_____ Solid formed by ions of different solutes when their solutions are combined.

_____ Metal strip by which electric current enters or leaves a solution.

_____ Strong electrolyte that includes hydrogen ions in its solution inventory.

1. Strong acid
2. Major species
3. Solution inventory
4. Weak electrolyte
5. "Conventional equation"
6. Precipitate
7. Ion-combination reaction
8. Strong electrolyte
9. Minor species
10. Electrode
11. Ionic equation

Set 2

_____ HNO_2, H_2S, and $HC_2H_3O_2$ are examples.

_____ List of elements in which any element will replace from solution the dissolved ion of any second element that is beneath the first element.

_____ Solute which, when dissolved, yields a solution that does not conduct electricity.

_____ Particle that is at the scene of a reaction, but does not participate in the reaction.

_____ Movement of electric charge.

_____ Liquid or other substance through which an electric current does not flow.

_____ Solution inventory species that are responsible for a solution's ability to conduct an electric current.

_____ Chemical expression that includes only solute particles that change in a reaction.

_____ Reaction in which a solid is formed from a solution.

1. Spectator
2. Ions
3. Nonconductor
4. Net ionic equation
5. Precipitation reaction
6. Nonelectrolyte
7. Electric current
8. Weak acid
9. Activity series

STUDY HINTS AND PITFALLS TO AVOID

Table 16.4 summarizes this entire chapter. It should be a focal point of your study of net ionic equations. The upper left-hand corner of the table is taken from Table 8.1. It describes conventional equations you have been writing for some time. Be sure to see the connection between these equations and the expanded Table 16.4.

If a conventional equation, _including states,_ can be written, it can be converted into a net ionic equation by the three steps in Section 16.4. Check the activity series, solubility table or rules, or molecular products (water or a weak acid) to be sure there is a reaction. Look out for double reactions. Recognize unstable products (H_2CO_3, H_2SO_3, or ''NH_4OH'').

There are several pitfalls awaiting you in this chapter. Be careful of these:

1) Incorrect or missing states of reactants and products. Many incorrect solution inventories can be traced to not recognizing the state of some species.
2) Not recognizing weak acids as molecular products that _do not_ separate into ions. This is the winner for ''most common error in writing net ionic equations.''
3) Making diatomic ions because there are two atoms of an element in a compound. H_2^+ is the most common error.
4) Insufficient practice. Writing net ionic equations is a learning-by-doing skill. Making mistakes and learning from those mistakes is the usual procedure. Students who complete _both_ steps before the test are happier students.

CHAPTER 16 QUESTIONS AND PROBLEMS

SECTION 16.1

1) How does a weak electrolyte differ from a non-electrolyte?

2) Solute A is a weak electrolyte and solute B is a strong electrolyte. How do these electrolytes differ?

3) How can it be that all soluble ionic compounds are electrolytes but soluble molecular compounds may or may not be electrolytes?

4) Compare the passage of ''electricity'' through a wire and through a solution. What conclusion may be drawn if a liquid is able to carry an electric current?

Questions 5 through 12: Write the solution inventory for the water solution of each substance given:

SECTION 16.2

5) $MnCl_2$, $(NH_4)_2SO_4$, Na_2S.

6) $Mg(NO_3)_2$, $Ca(OH)_2$, LiF. _____

7) KNO_2, $NiSO_4$, K_3PO_4.
8) $FeCl_3$, NH_4NO_3, Li_2SO_3.

SECTION 16.3
9) HBr, $H_2C_4H_4O_4$, HNO_3.
10) HI, $HCHO_2$, $HC_4H_4O_6$. _____

11) HF, $HC_2H_3O_2$, $HClO_4$.
12) HCl, H_2SO_4, $H_3C_2H_5O_7$. _____

SECTION 16.5

Questions 13 through 18: For each pair of reactants, write the net ionic equation for any redox reaction that may be predicted by Table 16.1 (Section 16.5). If no redox reaction occurs, write NR.

13) $Cu(s) + Li_2SO_4(aq)$.
14) $Mg(s) + Al_2(SO_4)_3(aq)$. _____

15) $Ba(s) + HCl(aq)$.
16) $Zn(s) + AgNO_3(aq)$. _____

17) $Ni(s) + CaCl_2(aq)$.
18) $Pb(s) + Ca(NO_3)_2(aq)$.

Questions 19 through 26: For each pair of reactants given, write the net ionic equation for any precipitation reaction that may be predicted by Tables 16.2 and 16.3 (Section 16.6). If no precipitation occurs, write NR.

19) $Pb(NO_3)_2(aq) + KI(aq)$.
20) $NiCl_2(aq) + CuSO_4(aq)$. _____

21) $KClO_3(aq) + Mg(NO_2)_2(aq)$.
22) $FeCl_2(aq) + (NH_4)_2S(aq)$. _____

23) $AgNO_3(aq) + LiBr(aq)$.
24) $CoSO_4(aq) + NaOH(aq)$. _____

25) $ZnCl_2(aq) + Na_2SO_3(aq)$.
26) $BaCl_2(aq) + Na_2CO_3(aq)$. _____

27) Write the net ionic equations for the precipitation of each of the following insoluble ionic compounds from aqueous solutions: $PbCO_3$; $Ca(OH)_2$.
28) Write the net ionic equations for the precipitation of each of the following insoluble ionic compounds from aqueous solutions: MgF_2; $Zn_3(PO_4)_2$.

SECTION 16.7

Questions 29 through 34: For each pair of reactants

given, write the net ionic equation for the molecule formation reaction that will occur.

29) $NaNO_2(aq)$, $HI(aq)$.
30) $KC_4H_4O_6(aq)$, $HCl(aq)$. _____

31) $KC_3H_5O_3(aq)$, $HClO_4(aq)$.
32) $HBr(aq)$, $Ca(OH)_2(aq)$. _____

33) $H_2SO_4(aq)$, $RbOH(aq)$.
34) $HNO_3(aq)$, $NaC_6H_5O_7(aq)$. _____

SECTION 16.8

Questions 35 through 38: For each pair of reactants given, write the net ionic equation for the reaction that will occur.

35) $HBr(aq)$, $(NH_4)_2SO_3(aq)$.
36) $(NH_4)_2SO_4(aq)$, $LiOH(aq)$. _____

37) $H_3PO_4(aq)$, $Na_2SO_3(aq)$.
38) $HNO_3(aq)$, $K_2CO_3(aq)$. _____

The remaining questions include all types of reactions in this chapter. Use the activity series and solubility tables to predict whether or not redox or precipitation reactions will take place. If you find a "no reaction" combination, mark it NR.

39) Barium chloride and sodium sulfite solutions are combined in an oxygen-free atmosphere.
40) A piece of magnesium is dropped into a solution of zinc nitrate. _____

41) Copper(II) sulfate and sodium hydroxide solutions are combined.
42) A solution of potassium iodide is able to dissolve solid silver chloride and form a new precipitate. _____

43) Bubbles appear as hydrochloric acid is poured onto solid magnesium carbonate.
44) Aluminum nitrate solution is added to potassium phosphate solution. _____

45) Nitric acid is able to dissolve solid lead(II) hydroxide.
46) A lead strip is immersed in a solution of copper(II) sulfate. _____

47) Oxalic acid, $H_2C_2O_4(s)$, is neutralized by sodium hydroxide solution.
48) Silver nitrate solution is added to a solution of ammonium carbonate. _____

49) What happens if nickel is placed into hydrochloric acid?

50) When combined in an oxygen-free atmosphere, sodium sulfite solution and hydrochloric acid react.

---

51) Hydrochloric acid is poured into a solution of sodium hydrogen sulfite.

52) Ammonium chloride solution is added to a solution of potassium hydroxide.

---

53) Solutions of magnesium sulfate and ammonium bromide are combined.

54) A silver wire is suspended in magnesium nitrate solution.

---

55) Magnesium ribbon is placed in hydrochloric acid.

56) Sodium benzoate solution, $NaC_7H_5O_2$(aq), is treated with hydrochloric acid.

---

57) Solid nickel hydroxide is readily dissolved by hydrobromic acid.

58) Sulfuric acid neutralizes potassium hydroxide.

---

59) Sodium fluoride solution is poured into nitric acid.

60) Sulfuric acid is able to dissolve copper(II) hydroxide.

---

61) Silver wire is dropped into hydrochloric acid.

62) Dilute nitric acid is poured over solid barium hydroxide.

---

63) When metallic lithium is added to water, hydrogen is released.

64) Metallic zinc is dropped into a solution of silver nitrate.

---

65) Aluminum shavings are dropped into a solution of copper(II) nitrate.

66) A solution of sodium carbonate is poured into a solution of calcium nitrate.

---

67)* (a) Hydrochloric acid reacts with a solution of sodium hydrogen carbonate. (b) Hydrochloric acid reacts with a solution of sodium carbonate. (c) A limited amount of hydrochloric acid reacts with a solution of sodium carbonate, yielding $NaHCO_3$(aq) as one product.

68) Consider H_3PO_4 to be a weak acid and write the net ionic equations for Equations 15.7 and 15.8, page 410.

GENERAL QUESTIONS

69) Distinguish precisely and in scientific terms the differences among items in each of the following groups:
a) Strong electrolyte, weak electrolyte.
b) Electrolyte, nonelectrolyte.
c) Strong acid, weak acid.
d) Conventional, ionic, and net ionic equations.
e) Ion combination, precipitation, and molecule formation reactions.
f) Molecule formation and neutralization reactions.
g) Acid, base, salt.

70) Classify each of the following statements as true or false:
a) Electrons carry electricity through a solution.
b) Ions must be present if a solution conducts electricity.
c) Ions make up the solution inventory of a solution of an ionic compound.
d) There are no ions present in the solution of a weak acid.
e) Only seven important acids are weak.
f) Hydrofluoric acid, which is used to etch glass, is a strong acid.
g) Spectators are included in a net ionic equation.
h) A net ionic equation for a reaction between an element and an ion is the equation for a redox reaction.
i) A compound that is insoluble forms a precipitate when its ions are combined.
j) Precipitation and molecule formation reactions are both ion-combination reactions having double displacement conventional equations.
k) Neutralization is a special case of a molecule-formation reaction.
l) One product of a molecule-formation reaction is a strong acid.
m) Ammonium hydroxide is a possible product of a molecule-formation reaction.

MATCHING SET ANSWERS

Set 1: 9–5–11–8–3–7–2–4–6–10–1

Set 2: 8–9–6–1–7–3–2–4–5

QUICK CHECK ANSWERS

16.1 A is a weak electrolyte because its solution is a poor conductor. This indicates the presence of ions, but in relatively low concentration. B is a strong electrolyte because its solution is a good conductor. The solution contains ions in relatively high concentration. C is a nonelectrolyte because it does not conduct. Its solution contains no ions.

16.2 All soluble ionic compounds have ions in their solution inventories. Insoluble ionic compounds form no solutions, so they have no solution inventories.

16.3 Strong acids ionize nearly completely; weak acids only slightly. Only seven acids are strong: H_2SO_4, HNO_3, HCl, HBr, HI, $HClO_4$, and $HClO_3$.

16.4 Only species with (aq) after them can be broken into ions. Of those, ionic compounds and strong acids are separated; weak acids are not.

Acid–Base (Proton Transfer) Reactions

Did you know that your local grocery store is a major industrial outlet for acids and bases? It's a fact. They don't come in the kinds of bottles sold to college chemistry laboratories, but they are acids and bases nevertheless. If you doubt it, read some of the labels. The substances in the picture with the red background are all acidic, whereas those in front of the blue background are basic. The background color selection is deliberate. It corresponds to the colors acids and bases impart to litmus, one of the best known acid-base indicators.

LEARN IT NOW In Chapter 17 we take the next to the last step in removing the learning aids that you will not find in a general chemistry text. The Chapter in Review is gone. The text no longer offers performance goals to focus your study or to review it.

We hope that you have been writing your own performance goals for the last two chapters. If not, this is a good time to begin. Writing learning objectives as you study is an effective way to learn.

Here is a suggestion that can make writing your own performance goals even more effective: Find one or two other students who are willing to write learning objectives along with you. At the end of each assignment, exchange them. You will all benefit, not only from the act of writing, but also from giving and receiving ideas that you might overlook individually.

Whether you write your performance goals alone or with others, don't wait until the end of the chapter. Do it with each assignment, when the ideas are fresh in your thought. Learn it—NOW!

17.1 TRADITIONAL ACIDS AND BASES

Originally, the word **acid** described something that had a sour, biting taste. The tastes of vinegar (acetic acid) and lemon juice (citric acid) are typical examples. Substances with such tastes have other common properties too. They impart certain colors to organic substances, such as litmus, which is red in acid solutions; they react with carbonate ions and release carbon dioxide; they react with and neutralize a base; and they release hydrogen when they react with certain metals.

Traditionally, a **base** is something that tastes bitter. Bases impart a different color to organic substances (blue to litmus); they neutralize acids; and they form precipitates when added to solutions of most metal ions. Also, bases feel slippery, or "soapy."

To understand these two distinct groups, we must ask what features in their chemical structure and composition are responsible for their characteristic properties? This is the goal of the present chapter.

17.2 THE ARRHENIUS THEORY OF ACIDS AND BASES

In 1884 a brilliant young chemist, Svante Arrhenius, observed that all substances classified as acids contain hydrogen ions, H^+. An acid was thus identified as a **substance whose water solution contains a high concentration of hydrogen ions.** The hydroxide ion is present in all solutions then known as bases, so a base became **a solution that has a high concentration of hydroxide ions.** According to the **Arrhenius theory,** the properties of an acid are the properties of the hydrogen ion, and the properties of a base are the properties of the hydroxide ion.

Particularly noteworthy among the properties of acids and bases is their ability to neutralize each other. In Example 16.16 the net ionic equation for the reaction between a strong acid (HCl) and a strong base (NaOH) is a molecule-forming ion combination that yields water:

$$H^+(aq) + OH^-(aq) \longrightarrow H_2O(\ell)$$

Svante August Arrhenius

If the acid is weak, as in Example 16.17, the hydroxide ion essentially pulls a hydrogen ion right out of the un-ionized acid molecule:

$$HC_2H_3O_2(aq) + OH^-(aq) \longrightarrow H_2O(\ell) + C_2H_3O_2^-(aq)$$

The carbon dioxide from the reaction of an acid with a carbonate is the end result of the combination of hydrogen and carbonate ions. The expected product is carbonic acid, H_2CO_3. However, in Section 16.8 you learned that this is one of the "unstable products" of an ion combination. It decomposes into carbon dioxide and water (Example 16.20):

$$2\,H^+(aq) + CO_3^{2-}(aq) \longrightarrow CO_2(g) + H_2O(\ell)$$

Hydrogen ions of acids and hydrogen gas occupy a unique position in the activity series (Section 16.5). All other members of this series are metals and metal ions that engage in single-replacement-type redox reactions. The reaction of the hydrogen ion appears in the net ionic equation for the reaction of calcium with hydrochloric acid (Example 16.5):

$$Ca(s) + 2\,H^+(aq) \longrightarrow H_2(g) + Ca^{2+}(aq)$$

Table 16.2 shows that only the hydroxides of barium and the alkali metals are soluble. This is why these solutions are the only strong Arrhenius bases. It also explains the basic property that precipitates form when a strong base is combined with the salt of most metals. The net ionic equation for the precipitation of magnesium hydroxide is typical (Example 16.11):

$$Mg^{2+}(aq) + 2\,OH^-(aq) \longrightarrow Mg(OH)_2(s)$$

✔ QUICK CHECK 17.1

According to the Arrhenius theory of acids and bases, how do you recognize an acid and a base?

17.3 THE BRÖNSTED–LOWRY THEORY OF ACIDS AND BASES

In Section 6.5 you learned to recognize an acid as "a hydrogen-bearing compound that, when dissolved in water, loses one or more hydrogen ions, H^+." It was further pointed out that a hydrogen ion is a single proton. Using gaseous hydrogen chloride as an example, these dissolving and ionization processes can be represented as

$$HCl(g) \xrightarrow{\ H_2O\ } H^+(aq) + Cl^-(aq)$$

The symbol (aq) indicates that the ions are in a water solution. In Section 15.3 you learned that ions in a water solution are hydrated, or surrounded by polar water molecules. At this point we must give particular attention to the hydration of the hydrogen ion.

It is highly unlikely that a water solution can have individual protons swimming around in it. Those tiny high concentrations of positive charge are strongly attracted to the negative portion of polar water molecules. In fact, by using one of the unshared electron pairs in a water molecule to form a covalent bond, the proton—the hydrogen ion—can become isoelectric with the noble gas helium:

$$H^+ + \ddot{O}-H \longrightarrow \left[H-\ddot{O}-H \right]^+$$
$$\quad\quad\quad | \quad\quad\quad\quad\quad | \quad\quad$$
$$\quad\quad\quad H \quad\quad\quad\quad\quad H \quad\quad$$

$$H^+ + H_2O \longrightarrow H_3O^+$$

The species produced, H_3O^+, is known as the **hydronium ion.**

There is considerable evidence to support the belief that "hydronium ion, H_3O^+" better describes the species in an acid than "hydrogen ion, H^+." Nevertheless, most chemists and most chemistry textbooks continue to refer to the hydrogen ion, H^+, with the understanding that it is really a hydrated hydrogen ion. When the occasion demands, as it does in this section, they switch to the hydronium ion, H_3O^+. Indeed, there is a sizable minority of chemists and textbooks that refer only to the hydronium ion and its formula. In this book we have chosen to stand with the majority, but, as usual, we advise you to follow your instructor if he or she prefers the alternative, H_3O^+.

From the hydronium ion standpoint, the equation for the ionization of HCl(g) includes the water molecule as a reactant:

$$H_2O(\ell) + HCl(g) \longrightarrow H_3O^+(aq) + Cl^-(aq) \quad\quad (17.1)$$

$$H-\ddot{O}: + H-\ddot{C}l: \longrightarrow \left[H-\ddot{O}-H \right]^+ + \left[:\ddot{C}l: \right]^-$$
$$\quad | \quad\quad\quad\quad\quad\quad\quad\quad\quad\quad | \quad\quad$$
$$\quad H \quad\quad\quad\quad\quad\quad\quad\quad\quad\quad H \quad\quad$$

$$\begin{array}{cc} \text{base} & \text{acid} \\ \text{proton} & \text{proton} \\ \text{receiver} & \text{donor} \end{array}$$

The structural equation suggests that the bond between the hydrogen and chlorine in HCl is broken, releasing a hydrogen ion—a proton. The proton moves to the water molecule where it uses an unshared electron pair to form a new bond. In 1923 two men, Johannes N. Brönsted and Thomas M. Lowry, independently described an **acid–base reaction as a proton-transfer reaction in which a proton is transferred from the acid to the base. The acid is a proton donor, and the base is a proton acceptor.** This acid–base concept is known as the **Brönsted–Lowry theory.**

According to Brönsted and Lowry, anything that can receive a proton is a base. In Equation 17.1, the proton is transferred from the hydrogen chloride molecule to the water molecule. The HCl molecule is the acid, the proton donor. The acceptor of the proton is the water molecule; water is a base in Equation 17.1.

The transfer of a proton in the reaction between water and ammonia is more apparent if we use HOH for the formula of water:

$$NH_3(aq) + HOH(\ell) \rightleftharpoons NH_4^+(aq) + OH^-(aq) \qquad (17.2)$$

base proton receiver   acid proton donor

This time water is an acid. It donates a proton to ammonia, the base. A substance that can behave as an acid in one case and a base in another, as water does in Equations 17.1 and 17.2, is said to be **amphoteric.**

The double arrow in Equation 17.2 indicates that the reaction is reversible. It suggests correctly that the reaction reaches a chemical equilibrium. When a reversible reaction is read from left to right, the **forward reaction,** or the reaction in the **forward direction,** is described; from right to left, the change is the **reverse reaction,** or in the **reverse direction.** In 1 M $NH_3(aq)$, less than 1% of the ammonia is changed to ammonium ions. NH_3 is the major species in the solution inventory, and NH_4^+ and OH^- are the minor species. The reverse reaction is therefore said to be **favored.** The favored direction of an equilibrium points to the major species.

HCl is an acid, a proton donor, in Equation 17.1. NH_3 is a base, a proton acceptor, in Equation 17.2. Do you suppose that if we were to put HCl and NH_3 together, the HCl would give its proton to NH_3? The answer, yes, is readily evident if bottles of concentrated ammonia and hydrochloric acid are opened next to each other (see Fig. 17.1). The solid product is often found on the outside of reagent bottles in the laboratory. Conventional and Lewis diagram equations describe the reaction:

FLASHBACK Reversible changes and the double arrow were first discussed in Section 14.3 in connection with the liquid–vapor equilibrium. An equilibrium is dynamic; the changes continue to occur, but at equal rates in the opposite directions.

FLASHBACK The major species in a solution inventory are the molecules or ions from a single solute that are present in greater concentration. Particles with a low concentration are minor species. (Section 16.3)

Figure 17.1

Reaction between ammonia and hydrogen chloride. Ammonia gas, $NH_3(g)$, and hydrogen chloride gas, HCl(g), escape from concentrated solutions. When the gases come into contact with each other, they form solid NH_4Cl, which appears as white ''smoke.''

$$NH_3(g) + HCl(g) \longrightarrow NH_4Cl(s) \tag{17.3}$$

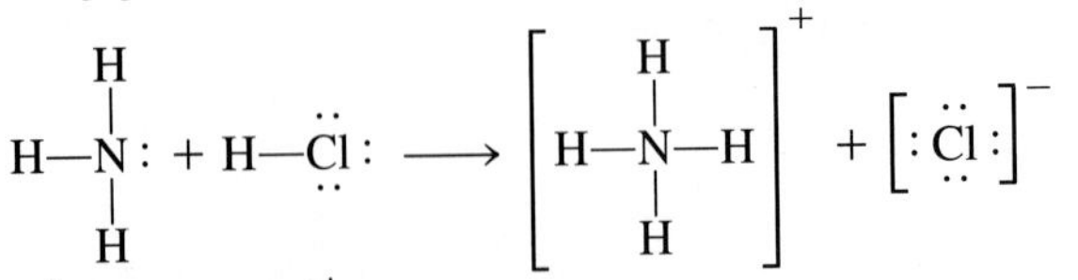

base
proton
receiver

acid
proton
donor

$NH_4Cl(s)$ is an assembly of NH_4^+ and Cl^- ions. Equation 17.3 is an acid–base reaction in the absence of both water and the hydroxide ion.

Reviewing Equations 17.1 to 17.3, we find that all fit into the general equation for a Brönsted–Lowry proton-transfer reaction:

$$\underset{\substack{\text{base}\\\text{proton}\\\text{receiver}}}{B} + \underset{\substack{\text{acid}\\\text{proton}\\\text{donor}}}{HA} \longrightarrow HB^+ + A^- \tag{17.4}$$

In this equation, note that the charges are not "absolute" charges. They indicate, rather, that the acid species, in losing a proton, leaves a species having a charge one less than the acid, and that the base, in gaining a proton, increases by 1 in charge.

☑ **QUICK CHECK 17.2**

What is the difference between a Brönsted–Lowry acid and an Arrhenius acid? between a Brönsted–Lowry base and an Arrhenius base? Are all Brönsted–Lowry bases also Arrhenius bases? Are all Arrhenius bases also Brönsted–Lowry bases? Explain.

17.4 THE LEWIS THEORY OF ACIDS AND BASES (OPTIONAL)

Look at the Lewis diagrams of the bases in Equations 17.1 to 17.3. In each case the structural feature of the base that permits it to receive the proton is an unshared pair of electrons. A hydrogen ion has no electron to contribute to the bond, but it is able to accept an electron pair to form the bond. According to the **Lewis theory** of acids and bases, **a Lewis base is an electron-pair donor,** and a **Lewis acid is an electron-pair acceptor.**

A Lewis acid is not limited to a hydrogen ion. A common example is boron trifluoride, which behaves as a Lewis acid by accepting an unshared pair of electrons from ammonia:

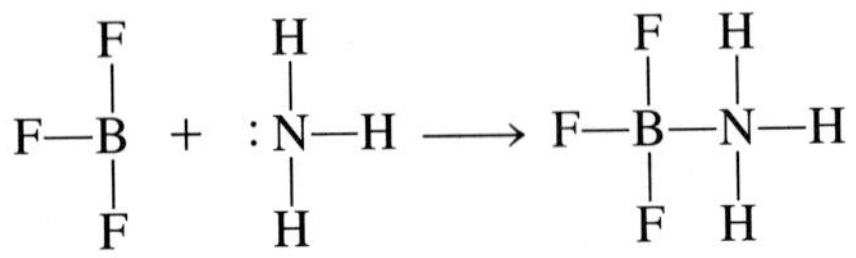

acid

base

Other examples of Lewis acid–base reactions appear in the formation of some complex ions and in organic reactions. You will study these in more advanced courses.

✔ QUICK CHECK 17.3

By the Brönsted–Lowry theory water can be either an acid or a base. Can water be a Lewis acid? a Lewis base? Explain.

SUMMARY / (Sections 17.2–17.4)

The identifying features of acids and bases according to the three acid–base theories are summarized below:

Theory	Acid	Base
Arrhenius	Hydrogen ion	Hydroxide ion
Brönsted–Lowry	Proton donor	Proton acceptor
Lewis	Electron-pair acceptor	Electron-pair donor

Our principal interest in acid–base chemistry is in aqueous solutions, where the Brönsted–Lowry theory prevails. The balance of the chapter is limited to the proton-transfer concept of acids and bases.

17.5 CONJUGATE ACID–BASE PAIRS

It was noted that Equation 17.2 reaches equilibrium. The fact is that most acid–base reactions reach equilibrium. Accordingly, Equation 17.5 is a rewrite of Equation 17.4, except that a double arrow is used to show the reversible character of the reaction. Look carefully at the reverse reaction:

$$B + HA \rightleftharpoons \underset{\substack{\text{acid} \\ \text{proton} \\ \text{donor}}}{HB^+} + \underset{\substack{\text{base} \\ \text{proton} \\ \text{receiver}}}{A^-} \tag{17.5}$$

Is not HB^+ donating a proton to A^- in the reverse reaction? In other words, HB^+ is an acid in the reverse reaction, and A^- is a base. From this we see that the products of any proton-transfer acid–base reaction are *another* acid and base for the reverse reaction.

Combinations such as acid HA and base A^- that result from an acid losing a proton or a base gaining one are called **conjugate acid–base pairs.** Any two substances that differ by one H^+ are a conjugate acid–base pair. In the forward reaction of Equation 17.2, $HOH(\ell)$ releases a hydrogen ion, leaving the hydroxide ion:

$$HOH(\ell) \longrightarrow H^+(aq) + OH^-(aq)$$

Water is an acid, a proton donor. What is left after the proton is gone, $OH^-(aq)$, is the **conjugate base** of water. In the reverse direction, $OH^-(aq)$ is a base because it gains a

Equation 17.2:

$$NH_3 + HOH \rightleftharpoons NH_4^+ + OH^-$$

We will continue to use HOH as the formula of water in this chapter because it suggests the structure of the molecule better than the usual H_2O. This is a common practice with acids when structure is emphasized. The formula of hypochlorous acid, for example, is usually written HClO; but it may be written HOCl to show that oxygen links the hydrogen to chlorine. The very slight ionizations of both compounds are the same:

$$H—O—H \rightleftharpoons H^+ + [O—H]^-$$

$$H—O—Cl \rightleftharpoons H^+ + [O—Cl]^-$$

It is important to recognize that, no matter how the formula is written, water, hypochlorous acid, and all other acids are molecular compounds, not ionic.

proton. The **conjugate acid** of $OH^-(aq)$ is $HOH(\ell)$, the species formed when the base gains a proton.

In the other part of the reaction

$$NH_3(aq) + H^+(aq) \rightleftharpoons NH_4^+(aq)$$

the base, $NH_3(aq)$, in the forward direction gains a proton to form its conjugate acid, $NH_4^+(aq)$; and $NH_3(aq)$ is the conjugate base of acid $NH_4^+(aq)$ in the reverse direction. For the whole reaction two conjugate acid–base pairs can be identified:

$$NH_3(aq) + HOH(\ell) \rightleftharpoons NH_4^+(aq) + OH^-(aq) \tag{17.2}$$

conjugate acid–base pair

conjugate acid–base pair

You can write the formula of the conjugate base of any acid simply by removing a proton. If HCO_3^- acts as an acid, its conjugate base is CO_3^{2-}. You can also write the formula of the conjugate acid of any base by adding a proton. If HCO_3^- is a base, its conjugate acid is H_2CO_3. Notice that HCO_3^- is amphoteric. Any amphoteric substance has both a conjugate base and a conjugate acid.

EXAMPLE 17.1

a) Write the formula of the conjugate base of H_3PO_4.
b) Write the formula of the conjugate acid of $C_7H_5O_2^-$.

In (b) don't let $C_7H_5O_2^-$ confuse you just because it is unfamiliar. Just do what must be done to find the formula of a conjugate acid.

a) $H_3PO_4 \longrightarrow H^+ + H_2PO_4^-$, the conjugate base of H_3PO_4
b) $C_7H_5O_2^- + H^+ \longrightarrow HC_7H_5O_2$, the conjugate acid of $C_7H_5O_2^-$

Remove a proton to get a conjugate base, and add one to get a conjugate acid.

EXAMPLE 17.2

Nitrous acid engages in a proton-transfer reaction with formate ion, CHO_2^-:

$$HNO_2(aq) + CHO_2^-(aq) \rightleftharpoons NO_2^-(aq) + HCHO_2(aq)$$

Answer the questions about this reaction in the steps that follow.

For the forward reaction identify the acid and the base.

HNO_2 is the acid; it donates a proton to CHO_2^-, the base, or proton receiver.

Identify the acid and base for the reverse reaction.

$HCHO_2$ is the acid; it donates a proton to NO_2^-, the base, or proton receiver.

Identify the conjugate of HNO_2. Is it a conjugate acid or a conjugate base?

———————   ———————

NO_2^- is the conjugate *base* of HNO_2. It is the base that remains after the proton has been donated by the acid.

Identify the other conjugate acid–base pair, and classify each species as the acid or the base.

———————   ———————

CHO_2^- and $HCHO_2$ is the other conjugate acid–base pair. CHO_2^- is the base, and $HCHO_2$ is the conjugate acid—the species produced when the base accepts a proton.

EXAMPLE 17.3 ————————————————————————————

Identify the conjugate acid–base pairs in

$$HC_4H_5O_3 + PO_4^{3-} \rightleftharpoons HPO_4^{2-} + C_4H_5O_3^-$$

———————   ———————

$HC_4H_5O_3$ and $C_4H_5O_3^-$ is one conjugate acid–base pair; PO_4^{3-} and HPO_4^{2-} is the second pair.

17.6 RELATIVE STRENGTHS OF ACIDS AND BASES

In Section 16.3 the distinction was made between the relatively few **strong acids** and the many **weak acids.** Strong acids are those that ionize almost completely, whereas weak acids ionize but slightly. Hydrochloric acid is a strong acid; 0.10 M HCl is almost 100% ionized. Acetic acid is a weak acid; only 1.3% of the molecules ionize in 0.10 M $HC_2H_3O_2$.

In a Brönsted–Lowry sense an acid behaves as an acid by losing protons. The more readily protons are lost, the stronger the acid. A base behaves as a base by gaining protons. The stronger the attraction for protons, the stronger the base.

Let's write the ionization equations for hydrochloric and acetic acids, one above the other, with the stronger acid first:

| Strong acid | $HCl \rightleftharpoons H^+ + Cl^-$ | ? base |
| Weak acid | $HC_2H_3O_2 \rightleftharpoons H^+ + C_2H_3O_2^-$ | ? base |

The conjugate bases of the acids are on the right-hand sides of the equations. What is the relative strength of the two bases? Which is stronger, Cl^- or $C_2H_3O_2^-$? If a chloride ion gained a proton, it would form HCl, a strong acid that would immediately lose that proton. The Cl^- ion has a weak attraction for protons. It is a **weak base.** If the acetate ion gained a proton, it would form $HC_2H_3O_2$, a weak acid that holds its proton tightly. The $C_2H_3O_2^-$ ion has a strong attraction for protons. It is a **strong base.**

We can therefore complete the comparison between these acids and their conjugate bases:

Strong acid	$HCl \rightleftharpoons H^+ + Cl^-$	Weak base
Weak acid	$HC_2H_3O_2 \rightleftharpoons H^+ + C_2H_3O_2^-$	Strong base

Table 17.1 lists many acids and bases in this way. Acid strength decreases from top to bottom, and the base strength increases. By referring to this table, we can compare the relative strengths of different acids and bases.

NOTICE:

The stronger the acid, the weaker its conjugate base

and

The stronger the base, the weaker its conjugate acid

Table 17.1
Relative Strengths of Acids and Bases

Acid Strength	Acid Name	Acid Formula		Base Formula	Base Strength
↑ Increasing STRENGTH	Perchloric	$HClO_4$	$\rightleftharpoons$	$H^+ + ClO_4^-$	↑ Decreasing STRENGTH
	Hydroiodic	HI	$\rightleftharpoons$	$H^+ + I^-$	
	Hydrobromic	HBr	$\rightleftharpoons$	$H^+ + Br^-$	
	Hydrochloric	HCl	$\rightleftharpoons$	$H^+ + Cl^-$	
	Nitric	HNO_3	$\rightleftharpoons$	$H^+ + NO_3^-$	
	Sulfuric	H_2SO_4	$\rightleftharpoons$	$H^+ + HSO_4^-$	
	Hydronium ion	H_3O^+	$\rightleftharpoons$	$H^+ + H_2O$	
	Oxalic	$H_2C_2O_4$	$\rightleftharpoons$	$H^+ + HC_2O_4^-$	
	Sulfurous	H_2SO_3	$\rightleftharpoons$	$H^+ + HSO_3^-$	
	Hydrogen sulfate ion	HSO_4^-	$\rightleftharpoons$	$H^+ + SO_4^{2-}$	
	Phosphoric	H_3PO_4	$\rightleftharpoons$	$H^+ + H_2PO_4^-$	
	Hydrofluoric	HF	$\rightleftharpoons$	$H^+ + F^-$	
	Nitrous	HNO_2	$\rightleftharpoons$	$H^+ + NO_2^-$	
	Formic (methanoic)	$HCHO_2$	$\rightleftharpoons$	$H^+ + CHO_2^-$	
	Benzoic	$HC_7H_5O_2$	$\rightleftharpoons$	$H^+ + C_7H_5O_2^-$	
	Hydrogen oxalate ion	$HC_2O_4^-$	$\rightleftharpoons$	$H^+ + C_2O_4^{2-}$	
	Acetic (ethanoic)	$HC_2H_3O_2$	$\rightleftharpoons$	$H^+ + C_2H_3O_2^-$	
	Propionic (propanoic)	$HC_3H_5O_2$	$\rightleftharpoons$	$H^+ + C_3H_5O_2^-$	
	Carbonic	H_2CO_3	$\rightleftharpoons$	$H^+ + HCO_3^-$	
	Hydrosulfuric	H_2S	$\rightleftharpoons$	$H^+ + HS^-$	
	Dihydrogen phosphate ion	$H_2PO_4^-$	$\rightleftharpoons$	$H^+ + HPO_4^{2-}$	
	Hydrogen sulfite ion	HSO_3^-	$\rightleftharpoons$	$H^+ + SO_3^{2-}$	
	Hypochlorous	$HClO$	$\rightleftharpoons$	$H^+ + ClO^-$	
	Boric	H_3BO_3	$\rightleftharpoons$	$H^+ + H_2BO_3^-$	
	Ammonium ion	NH_4^+	$\rightleftharpoons$	$H^+ + NH_3$	
	Hydrocyanic	HCN	$\rightleftharpoons$	$H^+ + CN^-$	
	Hydrogen carbonate ion	HCO_3^-	$\rightleftharpoons$	$H^+ + CO_3^{2-}$	
	Monohydrogen phosphate ion	HPO_4^{2-}	$\rightleftharpoons$	$H^+ + PO_4^{3-}$	
	Hydrogen sulfide ion	HS^-	$\rightleftharpoons$	$H^+ + S^{2-}$	
	Water	HOH	$\rightleftharpoons$	$H^+ + OH^-$	
Decreasing ↓	Hydroxide ion	OH^-	$\rightleftharpoons$	$H^+ + O^{2-}$	Increasing ↓

EXAMPLE 17.4 _____

Use Table 17.1 to list the following acids in order of decreasing strength (strongest first): $HC_2O_4^-$, NH_4^+, H_3PO_4.

Find the three acids among those shown in the table and list them from the strongest (first) to the weakest (last).

_____   _____

H_3PO_4, $HC_2O_4^-$, NH_4^+

EXAMPLE 17.5 _____

Using Table 17.1, list the following bases in order of decreasing strength (strongest first): $HC_2O_4^-$, SO_3^{2-}, F^-.

_____   _____

SO_3^{2-}, F^-, $HC_2O_4^-$

The ion $HC_2O_4^-$ appears in both Examples 17.4 and 17.5, first as an acid and second as a base. It is the intermediate ion in the two-step ionization of oxalic acid, $H_2C_2O_4$. The $HC_2O_4^-$ ion is amphoteric.

17.7 PREDICTING ACID–BASE REACTIONS

A chemist likes to know if an acid–base reaction will occur when certain reactants are brought together. Obviously, there must be a potential proton donor and acceptor— there can be no proton-transfer reaction without both. From there the decision is based on the relative strengths of the conjugate acid–base pairs. The stronger acid and base are the most reactive. They *do* what they must do to behave as an acid and a base. The weaker acid and base are more stable—less reactive. It follows that *the stronger acid will always transfer a proton to the stronger base, yielding the weaker acid and base as favored species at equilibrium.* Figure 17.2 summarizes the proton transfer from the

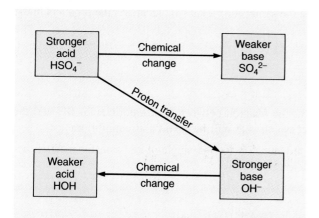

Figure 17.2

Predicting acid–base reactions from positions in Table 17.1. The spontaneous chemical change always transfers a proton from the stronger acid to the stronger base, both shown in pink. The products of the reaction, the weaker acid and the weaker base, are shown in blue. The direction, forward or reverse, that is favored is the one that has the weaker acid and base as products.

stronger acid to the stronger base from the standpoint of positions in Table 17.1.

Hydrogen sulfate ion, HSO_4^-, is a relatively strong acid that holds its proton weakly. Hydroxide ion, OH^-, is a strong base that attracts a proton strongly. If an HSO_4^- ion finds an OH^- ion, the proton will transfer from HSO_4^- to OH^-:

$$HSO_4^-(aq) + OH^-(aq) \rightleftharpoons SO_4^{2-}(aq) + HOH(\ell) \qquad (17.6)$$

Now identify the acid and the base for both the forward and reverse reactions. In the forward direction the acid is HSO_4^-. Its conjugate base for the reverse reaction is SO_4^{2-}. Similarly, OH^- is the base in the forward reaction, and HOH is the conjugate acid for the reverse reaction. In Equation 17.7 the acid–base roles for the different directions are shown with the letters A for acid and B for base:

$$HSO_4^-(aq) + OH^-(aq) \rightleftharpoons SO_4^{2-}(aq) + HOH(\ell) \qquad (17.7)$$
$$\quad\;\; A \qquad\qquad B \qquad\qquad\quad B \qquad\qquad A$$

Now compare the two acids in strength. HSO_4^- is near the top of the list, a much stronger acid than water. We therefore label HSO_4^- with SA for stronger acid and water with WA for weaker acid. Similarly, compare the bases: OH^- is a stronger base (SB) than SO_4^{2-} (WB):

$$HSO_4^-(aq) + OH^-(aq) \rightleftharpoons SO_4^{2-}(aq) + HOH(\ell) \qquad (17.8)$$
$$\quad\; SA \qquad\qquad SB \qquad\qquad\;\; WB \qquad\qquad WA$$

As you see, the weaker combination is on the right-hand side in this equation. This indicates that the reaction is favored in the forward direction. The proton transfers spontaneously from the strong proton donor to the strong proton receiver. The products that are in greater abundance are the weaker conjugate base and conjugate acid.

The following procedure is recommended in the prediction of acid–base reactions:

PROCEDURE

1) For a given pair of reactants write the equation for the transfer of *one* proton from one species to the other. (Do not transfer two protons.)
2) Label the acid and base on each side of the equation.
3) Determine which side of the equation has *both* the weaker acid and the weaker base (they must both be on the same side). That side identifies the products in the favored direction.

EXAMPLE 17.6

Write the net ionic equation for the reaction between hydrofluoric acid, HF, and the sulfite ion, SO_3^{2-}, and predict which side will be favored at equilibrium.

The first step is to write the equation for the single-proton-transfer reaction between HF and SO_3^{2-}. Complete this step.

$$HF(aq) + SO_3^{2-}(aq) \rightleftharpoons F^-(aq) + HSO_3^-(aq)$$

Next, we identify the acid and base on each side of the equation. Do so with letters A and B, as in the preceding discussion.

_____   _____

$$HF(aq) + SO_3{}^{2-}(aq) \rightleftharpoons F^-(aq) + HSO_3{}^-(aq)$$
$$\quad A \qquad\quad B \qquad\qquad\quad B \qquad\quad A$$

In each case the acid is the species with a proton to donate. It is transferred from acid HF to base $SO_3{}^{2-}$ in the forward reaction and from acid $HSO_3{}^-$ to base F^- in the reverse reaction.

Finally, determine which reaction, forward or reverse, is favored at equilibrium. It is the side with the weaker acid and base. Refer to Table 17.1.

_____   _____

The forward reaction is favored at equilibrium.

$HSO_3{}^-$ is a weaker acid than HF, and F^- is a weaker base than $SO_3{}^{2-}$. These species are the products in the favored direction.

EXAMPLE 17.7_____

Write the net ionic equation for the acid–base reaction between $HCO_3{}^-$ and ClO^- and predict which side will be favored at equilibrium.

Complete the example as before.

_____   _____

$$HCO_3{}^-(aq) + ClO^-(aq) \rightleftharpoons CO_3{}^{2-}(aq) + HClO(aq)$$

The reverse reaction will be favored. This conclusion is based on $HCO_3{}^-$ being a weaker acid than HClO and ClO^- being a weaker base than $CO_3{}^{2-}$.

Up to this point most attention has been given to the direction in which an equilibrium is favored. This does not mean that we can ignore the unfavored direction. Consider, for example, the reaction described by Equation 17.2: $NH_3(aq) + HOH(\ell) \rightleftharpoons NH_4{}^+(aq) + OH^-(aq)$. Although this reaction proceeds only slightly in the forward direction, many of the properties of household ammonia, its cleaning power in particular, depend on the presence of OH^- ions.

17.8 THE WATER EQUILIBRIUM

In the remaining sections of this chapter you will be multiplying and dividing exponentials, taking the square root of an exponential, and working with logarithms. We will furnish brief comments on these operations as we come to them. For more detailed instructions, see Appendix I, Parts B and C.

One of the most critical equilibria in all of chemistry is represented by the next-to-last line in Table 17.1, the ionization of water. Careful control of tiny traces of hydrogen and hydroxide ions marks the difference between success and failure in an untold

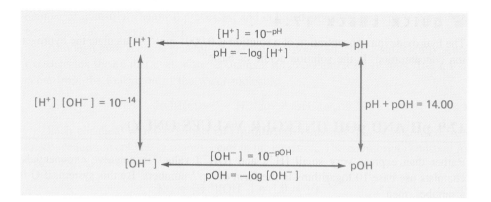

Figure 17.3

The "pH loop." Given the value for any corner of the pH loop, all other values may be calculated by progressing around the loop in either direction. Conversion equations are shown for each step.

EXAMPLE 17.10

Assuming complete ionization of 0.01 M NaOH, find its pH, pOH, $[OH^-]$, and $[H^+]$.

SOLUTION

a)   Starting with $[OH^-]$, if 0.01 mol of NaOH is dissolved in 1 L of solution, the concentration of the hydroxide ion is 0.01 molar: $[OH^-] = 0.01 = 10^{-2}$ M

b)   If $[OH^-] = 10^{-2}$, pOH $= -\log 10^{-2} = 2$.   (Equation 17.12)

c)   pH $= 14 - $ pOH $= 14 - 2 = 12$.   (Equation 17.16)

d)   If pH $= 12$, $[H^+] = 10^{-pH} = 10^{-12}$ M.   (Equation 17.13)

Two things are worth noting about Example 17.10. First, if we extend the problem by one more step, we complete the full pH loop. We began with $[OH^-]$ and went counterclockwise through pOH, pH, and $[H^+]$. The $[H^+]$ of 10^{-12} can be converted to $[OH^-]$ by Equation 17.10:

$$[OH^-] = \frac{K_w}{[H^+]} = \frac{10^{-14}}{10^{-12}} = 10^{-2} \text{ M}$$

This is the same as the starting $[OH^-]$. Completing the loop may therefore be used to check the correctness of the other steps in the process.

The second observation from Example 17.10 is that the loop may be circled in either direction. Starting with $[OH^-] = 10^{-2}$ and moving clockwise, we obtain

$$[H^+] = \frac{K_w}{[OH^-]} = \frac{10^{-14}}{10^{-2}} = 10^{-12} \text{ M}$$

It follows that pH $= 12$ and pOH $= 14 - 12 = 2$, the same results reached by circling the loop in the opposite direction.

You should now be able to make a complete trip around the pH loop.

EXAMPLE 17.11

The pH of a solution is 3. Calculate the pOH, $[H^+]$, and $[OH^-]$ in any order. Confirm your result by calculating the starting pH—by completing the loop.

You may go either way around the loop, but complete it whichever way you choose, making sure you return to the starting point.

Counterclockwise	**Clockwise**

From pH = 3, $[H^+] = 10^{-3}$ M   pOH = 14 − 3 = 11

$$[OH^-] = \frac{10^{-14}}{10^{-3}} = 10^{-11} \text{ M}$$ From pOH = 11, $[OH^-] = 10^{-11}$ M

From $[OH^-] = 10^{-11}$ M, pOH = 11 $$[H^+] = \frac{10^{-14}}{10^{-11}} = 10^{-3} \text{ M}$$

pH = 14 − 11 = 3   From $[H^+] = 10^{-3}$ M, pH = 3

Most of the solutions we work with in the laboratory and all of those involved in biochemical systems have pH values between 1 and 14. This corresponds to H^+ concentrations between 10^{-1} and 10^{-14} M, as shown in Table 17.2.

Let's pause for a moment to develop a "feeling" for pH—what it means. pH is a measure of acidity. It is an inverse sort of measurement; the higher the pH, the lower the acidity and vice versa. Table 17.3 brings out this relationship.

Table 17.2
pH Values of Common Liquids

Liquid	pH
Human gastric juices	1.0–3.0
Lemon juice	2.2–2.4
Vinegar	2.4–3.4
Carbonated drinks	2.0–4.0
Orange juice	3.0–4.0
Black coffee	3.7–4.1
Tomato juice	4.0–4.4
Cow's milk	6.3–6.6
Human blood	7.3–7.5
Seawater	7.8–8.3
Saturated $Mg(OH)_2$	10.5
Household ammonia	10.5–11.5
0.1 M Na_2CO_3	11.7
1 M NaOH	14.0

Table 17.3
pH and Hydrogen Ion Concentration

$[H^+]$	$[H^+]$	pH	Acidity or Basicity*
1.0	10^0	0	
0.1	10^{-1}	1	
0.01	10^{-2}	2	Strongly acid
0.001	10^{-3}	3	pH < 4
0.0001	10^{-4}	4	
0.0001	10^{-4}	4	
0.00001	10^{-5}	5	Weakly acid
0.000001	10^{-6}	6	$4 \le$ pH < 6
0.000001	10^{-6}	6	Neutral
0.0000001	10^{-7}	7	(or near neutral)
0.00000001	10^{-8}	8	$6 \le$ pH < 8
0.00000001	10^{-8}	8	
0.000000001	10^{-9}	9	Weakly basic
0.0000000001	10^{-10}	10	$8 \le$ pH < 11
0.0000000001	10^{-10}	10	
0.00000000001	10^{-11}	11	
0.000000000001	10^{-12}	12	Strongly basic
0.0000000000001	10^{-13}	13	$11 \le$ pH
0.00000000000001	10^{-14}	14	

*Ranges of acidity and basicity are arbitrary.

On examining Table 17.3, we see that each pH unit represents a factor of 10. Thus, a solution of pH 2 is 10 times as acidic as a solution with pH = 3, and 100 times as acidic as the solution of pH 4. In general, the relative acidity in terms of $[H^+]$ is 10^x, where x is the difference between the two pH measurements. From this we conclude that a 0.1 M solution of a strong acid, with pH = 1, is one million times as acidic as a neutral solution, with pH = 7. (One million is based on the pH difference, $7 - 1 = 6$. As an exponential, $10^6 = 1,000,000$.)

If you understand the idea behind pH, you should be able to make some comparisons.

EXAMPLE 17.12

Arrange the following solutions in order of decreasing acidity (i.e., highest $[H^+]$ first, lowest last): Solution A, pH = 8; Solution B, pOH = 4; Solution C, $[H^+] = 10^{-6}$; Solution D, $[OH^-] = 10^{-5}$.

SOLUTION

To make comparisons, all values should be converted to the same basis, pH, pOH, $[H^+]$, or $[OH^-]$. Because the question asks for a list based on acidity, we will find the $[H^+]$ for each solution.

$$[H^+] = 10^{-8} \text{ M for A, } 10^{-10} \text{ M for B, } 10^{-6} \text{ M for C, and } 10^{-9} \text{ M for D}$$

In arranging these $[H^+]$ values in decreasing order, remember that the exponents are negative:

Most	$10^{-6} > 10^{-8} > 10^{-9} > 10^{-10}$	Least
acidic	C A D B	acidic

✔ QUICK CHECK 17.5

In Solution W, pOH = 6; in Solution X, $[OH^-] = 10^{-12}$ M; in Solution Y, pH = 13; and in Solution Z, $[H^+] = 10^{-3}$ M. List these solutions in order of increasing pOH.

Various methods are used to measure pH in the laboratory (Fig. 17.4). Acid–base indicators have been mentioned already. Each indicator is effective over a specific pH

Figure 17.4

Measurement of pH. (a) A pH meter is a voltmeter calibrated to measure pH. (b) The color imparted to papers impregnated with certain dyes can be used for approximate measurements of pH.

(a)   (b)

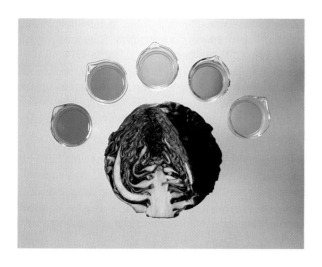

Purple cabbage boiled in water yields dyes that show a variety of colors over the pH range of 1 through 14. Shown (left to right) is purple cabbage juice in solutions of pH 1, 4, 7, 10, and 13.

range. Paper strips, impregnated with an indicator dye that functions over a range selected for the pH being measured, are widely used for rough pH readings. More accurate measurements are made with pH meters.

17.10 NONINTEGER pH–[H$^+$] AND pOH–[OH$^-$] CONVERSIONS

It is sad but true that real-world solutions do not come neatly packaged in concentrations that can be expressed as whole-number powers of 10. [H$^+$] is more apt to have a value such as 2.7×10^{-4} M, or the pH of a solution is more likely to be 6.24. The chemist must be able to convert from each of these to the other.

A pH number is a logarithm. Table 17.4 shows the logarithms of 3.45 multiplied by five different powers of 10: 0, 1, 2, 8, and 12. One column shows the value of the logarithm to seven decimals. Another column has the logarithms rounded off to the correct number of significant figures. Notice three things:

Table 17.4

Logarithms and Exponential Notation

Number		Logarithm	
Decimal Form	Exponential Notation	Value	Rounded Off
3.45	3.45×10^0	0.5378191	0.538
34.5	3.45×10^1	1.5378191	1.538
345	3.45×10^2	2.5378191	2.538
345,000,000	3.45×10^8	8.5378191	8.538
3,450,000,000,000	3.45×10^{12}	12.5378191	12.538

1) The mantissa of the logarithm—the number to the right of the decimal in the Value column—is always the same, 0.5378191. This is the logarithm of 3.45, the coefficient for each entry in the Exponential Notation column in Table 17.4.
2) The characteristic of the logarithm—the number to the left of the decimal in the Value column—is the same as the exponent in the Exponential Notation column.
3) All numbers in the first two columns are three-significant-figure numbers. This appears in both the decimal form of the number and in the coefficient when the number is written in exponential notation.

Item 2 shows that the digits to the left of the decimal in a logarithm—the characteristic—are related only to the exponent of the number when it is written in exponential notation. They have nothing to do with the coefficient of the number, which is where significant figures are expressed. Therefore, *in a logarithm the digits to the left of the decimal are not counted as significant figures. Counting significant figures in a logarithm begins at the decimal point.*

The significant figures in a number written in exponential notation are the significant figures in the coefficient. These show up in the *mantissa* of the logarithm. To be correct in significant figures, the coefficient and the mantissa must have the same number of digits. The correctly rounded off logarithms are in the right-hand column of Table 17.4. All numbers in that column are written in three significant figures. Only the digits after the decimal point are significant.

In working with pH, you will be finding logarithms of numbers smaller than 1. These logarithms are negative. The sign is changed to positive when the logarithm is written as a "p" value. Try one. Find $\log 3.45 \times 10^{-6}$. Enter the number into your calculator, and then press the "log" key. The display should read -5.462180905. If 3.45×10^{-6} represented $[H^+]$, the pH would be the opposite of -5.462180905, or 5.462 rounded off to three significant figures and with the sign changed.*

In the example that follows, the calculator sequence is given in detail. Although calculator keys may be marked differently, the procedure is essentially the same for AOS logic and RPN logic.

EXAMPLE 17.13

Calculate the pH of a solution if $[H^+] = 2.7 \times 10^{-4}$ M.

SOLUTION

$$pH = -\log (2.7 \times 10^{-4}) = 3.57$$

Press	Display
2.7	2.7
EE	2.7 00
4	2.7 04
+/−	2.7 -04
log	−3.568636236
+/−	3.568636236

The answer should be rounded off to two significant figures, 3.57.

---

*The mantissa appears to be different here than it was in Table 17.4, but really it is not when you trace its origin:

$$\log (3.45 \times 10^{-6}) = \log 3.45 + \log 10^{-6} = 0.538 + (-6) = -5.462$$

EXAMPLE 17.14

Find the pOH of a solution if its hydroxide ion concentration is 7.9×10^{-5} M.

pOH $= -\log (7.9 \times 10^{-5}) = 4.10$ (two significant figures)

There are two ways to change pH to [H$^+$]. The first is to raise 10 to the negative pH power. The second requires a 10^x key on the calculator. Simply enter the negative pH value and press the 10^x key.

EXAMPLE 17.15

The pOH of a solution is 6.24. Find [OH$^-$].

SOLUTION

[OH$^-$] = antilog $-6.24 = 10^{-6.24} = 5.8 \times 10^{-7}$ M

10^{-pOH} Sequence		10^x Sequence	
Press	**Display**	**Press**	**Display**
10	*10*	6.24	*6.24*
y^x	*10*	+/−	*−6.24*
6.24	*6.24*	10^x	*5.754399374 -07*
+/−	*−6.24*		
=	*5.754399374 -07*		

EXAMPLE 17.16

Find the hydrogen ion concentration of a solution if its pH is 11.62.

[H$^+$] = antilog $(-11.62) = 10^{-11.62} = 2.4 \times 10^{-12}$ M

EXAMPLE 17.17

[OH$^-$] $= 5.2 \times 10^{-9}$ M for a certain solution. Calculate in order pOH, pH, and [H$^+$], and then complete the pH loop by recalculating [OH$^-$] from [H$^+$].

$$pOH = -\log (5.2 \times 10^{-9}) = 8.28$$

$$pH = 14.00 - 8.28 = 5.72$$

$$[H^+] = \text{antilog } (-5.72) = 10^{-5.72} = 1.9 \times 10^{-6} \text{ M}$$

$$[OH^-] = \frac{1.0 \times 10^{-14}}{1.9 \times 10^{-6}} = 5.3 \times 10^{-9} \text{ M}$$

The variation between 5.2×10^{-9} M and 5.3×10^{-9} M comes from rounding off in expressing intermediate answers. If the calculator sequence is completed without rounding off, the loop returns to 5.2×10^{-9} M for $[OH^-]$.

CHAPTER 17 KEY WORDS AND MATCHING SETS

Set 1

Note: Two selections are from optional Section 17.4.

_____ Conjugate of a strong base.

_____ Reaction proceeds from left to right as equation is written.

_____ Tastes sour, turns litmus red, reacts with carbonates and some metals.

_____ Attracts protons strongly.

_____ Tastes bitter, turns litmus blue, forms precipitates with most metal ions.

_____ Electron-pair acceptor.

_____ $-\log [H^+]$.

_____ Reaction proceeds from right to left as equation is written.

_____ Attracts protons weakly.

_____ Electron-pair donor.

_____ Two species that differ by one proton.

1. Lewis acid

2. Weak base

3. Traditional base

4. Reverse reaction

5. Conjugate acid–base pair

6. Weak acid

7. Forward reaction

8. Traditional acid

9. Lewis base

10. pH

11. Strong base

Set 2

_____ Brönsted–Lowry acid–base reaction.

_____ At 25°C, equal to 10^{-14}.

_____ Value calculated from concentrations of substances in a system at equilibrium.

_____ Species left after acid has lost a proton.

_____ Solution with a high concentration of hydroxide ions.

_____ At 25°C, equal to $10^{-14}/[H^+]$.

_____ The direction in which a reversible reaction tends to proceed.

_____ Solution with a high concentration of hydrogen ions.

_____ At 25°C, equal to $14 - pH$.

_____ Species formed when a base accepts a proton.

_____ Hydronium ion.

1. Conjugate acid

2. pOH

3. Arrhenius base

4. Proton transfer

5. $[OH^-]$

6. H_3O^+

7. $[H^+][OH^-]$

8. Conjugate base

9. Arrhenius acid

10. Equilibrium constant

11. Favored

Set 3

_____ Proton acceptor.

_____ Loses protons easily.

_____ Solution in which $[OH^-] < [H^+]$.

_____ Symbol for the equilibrium constant of water.

_____ Conjugate of a weak acid.

_____ Behaves as an acid or a base.

_____ Solution in which $[H^+] < [OH^-]$.

_____ Holds protons strongly.

_____ Proton donor.

_____ Solution in which $[H^+] = [OH^-]$.

_____ Calculation in either direction of pH, pOH, $[OH^-]$, $[H^+]$, pH, in order, each from the value before it.

1. Basic

2. Amphoteric

3. Neutral

4. Strong base

5. Brönsted–Lowry base

6. Strong acid

7. "pH loop"

8. K_w

9. Acidic

10. Brönsted–Lowry acid

11. Weak acid

STUDY HINTS AND PITFALLS TO AVOID

You can always write the conjugate base of an acid by removing one H from the acid formula and reducing the acid charge by 1. Conversely, to write the formula of the conjugate acid of a base, add one H to the base formula and increase the charge by 1.

When writing the equation for a Brönsted–Lowry acid–base reaction, transfer only one proton to get the correct conjugate acid and base on the opposite side of the equation. Once the equation is written, each conjugate acid–base pair has the acid on one side and the base on the other side. The acid and base on the same side of the equation are *not* a conjugate pair.

A "p" number is the opposite (in sign) of the exponent of 10 when the concentration of an ion is given in exponential form. The exponents are always negative, so "p" numbers are positive. That gives an inverse character to the concentration of an ion and its "p" number. For H^+, the larger the $[H^+]$, the more acidic the solution and the smaller the pH.

Be careful of negative exponents. The more negative the exponent, the smaller the value. Even though 4 is larger than 3, -4 is smaller than -3. Therefore, 10^{-4} is smaller than 10^{-3}.

CHAPTER 17 QUESTIONS AND PROBLEMS

SECTIONS 17.1 TO 17.4

1) Identify at least two of the classical properties of acids and two of bases. For one base property and one acid property, show how it is related to the ion that is present in the base or acid.

2) What ion is present in solutions commonly regarded as acids, and what ion is present in solutions that are bases? Name three compounds whose water solutions are acids or bases because of these ions.

---

3) How does an Arrhenius base differ from a Brönsted–Lowry base? Are the same chemicals classified as bases in both systems? Explain or justify your answer.

4) Compare Brönsted–Lowry acids with Arrhenius acids. How are they different, and how are they alike? Give examples to justify your answer.

---

5) What do we mean when we say a Lewis acid is an electron-pair acceptor, and a Lewis base is an electron-pair donor? Give an example of each.

6) What is found in the structures of all Lewis acids? Lewis bases?

---

7) Diethyl ether and boron trifluoride react by forming a covalent bond between the molecules. Use the following "structural" equation to explain why this is a Lewis acid–base reaction.

8) The chemical properties of aluminum chloride, $AlCl_3$, are more typical of a molecular compound than an ionic compound. Draw Lewis diagrams of an $AlCl_3$ molecule and a chloride ion and show how they might combine to form an $[AlCl_4]^-$ ion. Explain how this is an example of a Lewis acid–base reaction.

SECTION 17.5

9) Write the formulas of the conjugate bases of HF and $H_2PO_4^-$. Write the formulas of the conjugate acids of NO_2^- and $H_2PO_4^-$.

10) What are the conjugate acids of OH^- and HCO_3^-? Write the formulas of the conjugate bases of H_3O^+ and HCO_3^-.

---

11) For the reaction

$$HSO_4^-(aq) + C_2O_4^{2-}(aq) \rightleftharpoons SO_4^{2-}(aq) + HC_2O_4^-(aq)$$

identify the acid and the base on each side of the equation—that is, the acid and base for the forward reaction and the acid and base for the reverse reaction.

12) Identify the acid and base for the forward reaction, and the acid and base for the reverse reaction, for

$$HNO_2{}^-(aq) + CN^-(aq) \rightleftharpoons NO_2{}^-(aq) + HCN(aq)$$

13) What are the conjugate acid–base pairs in Question 11?

14) What are the conjugate acid–base pairs in Question 12?

---

15) Identify both conjugate acid–base pairs in

$$HNO_2(aq) + C_3H_5O_2{}^-(aq) \rightleftharpoons NO_2{}^-(aq) + HC_3H_5O_2(aq)$$

16) Write the formulas for both conjugate acid–base pairs in

$$HSO_4{}^-(aq) + HC_2O_4{}^-(aq) \rightleftharpoons SO_4{}^{2-}(aq) + H_2C_2O_4(aq)$$

17) What are the conjugate acid–base pairs in

$$NH_4{}^+(aq) + HPO_4{}^{2-}(aq) \rightleftharpoons NH_3(aq) + H_2PO_4{}^-(aq)$$

18) Identify the conjugate acid–base pairs in

$$H_2PO_4{}^-(aq) + HCO_3{}^-(aq) \rightleftharpoons HPO_4{}^{2-}(aq) + H_2CO_3(aq)$$

SECTION 17.6

Refer to Table 17.1 when answering questions in this section.

19) How does a strong base differ from a weak base, according to the Brönsted–Lowry concept? Give two examples of a strong base and two examples of a weak base.

20) According to the Brönsted–Lowry concept, what is the difference between a strong acid and a weak acid? Give two examples that are strong acids and two examples that are weak acids.

---

21) Arrange the following acids in order of their increasing strength (weakest acid first): $HClO$, HOH, H_2SO_3, $HC_2O_4{}^-$.

22) Arrange the following bases in order of their decreasing strength (strongest base first): $CO_3{}^{2-}$, $H_2PO_4{}^-$, Br^-, $SO_4{}^{2-}$.

---

23) Arrange the following bases in order of their decreasing strength (strongest base first): Cl^-, ClO^-, $HSO_3{}^-$, H_2O, CN^-.

24) Arrange the following acids in order of their decreasing strength (strongest acid first): $NH_4{}^+$, HI, H_2O, $HSO_3{}^-$, $H_2C_2O_4$.

SECTION 17.7

For each acid and base given in this section, complete a proton-transfer equation for the transfer of one proton. Using Table 17.1, predict the direction in which the resulting equilibrium is favored.

25) $HC_3H_5O_2(aq) + PO_4{}^{3-}(aq) \rightleftharpoons$

26) $HC_7H_5O_2(aq) + SO_4{}^{2-}(aq) \rightleftharpoons$

---

27) $HSO_4{}^-(aq) + CO_3{}^{2-}(aq) \rightleftharpoons$

28) $HC_2H_2O_4(aq) + NH_3(aq) \rightleftharpoons$

---

29) $H_2CO_3(aq) + NO_3{}^-(aq) \rightleftharpoons$

30) $H_3PO_4(aq) + CN^-(aq) \rightleftharpoons$

---

31) $NO_2{}^-(aq) + H_3O^+(aq) \rightleftharpoons$

32) $H_2BO_3{}^-(aq) + NH_4{}^+(aq) \rightleftharpoons$

---

33) $HSO_4{}^-(aq) + HC_2O_4{}^-(aq) \rightleftharpoons$

34) $HPO_4{}^{2-}(aq) + HC_2H_3O_2(aq) \rightleftharpoons$

SECTION 17.8

35) What is the significance of the very small value of 10^{-14} for K_w, the equilibrium constant for the ionization of water?

36) If water ionizes, why does it not conduct electricity? Is water a nonconductor?

---

37) $[H^+] = 10^{-5}$ M and $[OH^-] = 10^{-9}$ M in a solution. Is the solution neutral, basic, or acidic? How do you know?

38) What is the meaning of the words, *acidic, basic,* and *neutral* when used in reference to solutions?

---

39) What is $[OH^-]$ in 0.01 M HCl? (*Hint:* Begin by finding $[H^+]$ in 0.01 M HCl.)

40) What is $[OH^-]$ in a solution in which $[H^+] = 10^{-12}$ M?

SECTION 17.9

41) Identify the ranges of the pH scale that are classified as strongly acidic, weakly acidic, strongly basic, weakly basic, and neutral, or close to neutral.

42) In which classification in Question 41 does each of the following solutions belong: (a) pH = 9; (b) pOH = 3; (c) pH = 7?

43) Select any integer from 1 to 14 and explain what is meant by saying that this number is the pH of a certain solution.

44) If the pH of a solution is 9.3, is the solution neutral, basic, or acidic? How did you reach your conclusion? List in order the pH values of a solution that is neutral, one that is basic, and one that is acidic.

Questions 45 to 52: The pH, pOH, [OH⁻], or [H⁺] is given for a solution. Find each of the other values. Also, classify each solution as strongly acidic, weakly acidic, neutral (or close to neutral), weakly basic, or strongly basic, as these terms are used in Table 17.3.

45) pH = 5
46) pOH = 6

47) $[OH^-] = 0.1$ M
48) $[H^+] = 10^{-3}$ M

49) pOH = 10
50) $[OH^-] = 10^{-12}$ M

51) $[H^+] = 10^{-7}$ M
52) pH = 9

SECTION 17.10

Questions 53 to 60: The pH, pOH, [OH⁻], or [H⁺] is given for a solution. Find each of the other values.

53) $[OH^-] = 2.5 \times 10^{-10}$ M
54) pH = 6.62

55) pH = 4.06
56) $[OH^-] = 1.1 \times 10^{-11}$ M

57) $[H^+] = 2.8 \times 10^{-1}$ M
58) pOH = 5.54

59) pOH = 7.40
60) $[H^+] = 7.2 \times 10^{-2}$ M

GENERAL QUESTIONS

61) Distinguish precisely and in scientific terms the differences between items in each of the following groups:
a)   Acid and base—by Arrhenius theory
b)   Acid and base—by Brönsted–Lowry theory
c)   Acid and base—by Lewis theory
d)   Forward reaction, reverse reaction
e)   Acid and conjugate base, base and conjugate acid
f)   Strong acid and weak acid
g)   Strong base and weak base
h) $[H^+]$ and $[OH^-]$
i)   pH and pOH

62) Classify each of the following statements as true or false:
a)   All Brönsted–Lowry acids are Arrhenius acids.
b)   All Arrhenius bases are Brönsted–Lowry bases, but not all Brönsted–Lowry bases are Arrhenius bases.
c) HCO_3^- is capable of being amphoteric.
d) HS^- is the conjugate base of S^{2-}.
e)   If the species on the right side of an ionization equilibrium are present in greater abundance than those on the left, the equilibrium is favored in the forward direction.
f) NH_4^+ cannot act as a Lewis base.
g)   Weak bases have a weak attraction for protons.
h)   The stronger acid and the stronger base are always on the same side of a proton-transfer reaction equation.
i)   A proton-transfer reaction is always favored in the direction that yields the stronger acid.
j)   A solution with pH = 9 is more acidic than one with pH = 4.
k)   A solution with pH = 3 is twice as acidic as one with pH = 6.
l)   A pOH of 4.65 expresses the hydroxide ion concentration of a solution in three significant figures.

63) Theoretically, can there be a Brönsted–Lowry acid–base reaction between OH^- and NH_3? If not, why not? If yes, write the equation.

64) Explain what amphoteric means. Give an example of an amphoteric substance, other than water, that does not contain carbon.

65) Very small concentrations of ions other than hydrogen and hydroxide are sometimes expressed with "p" numbers. Calculate pCl in a solution for which $[Cl^-] = 7.49 \times 10^{-8}$ M.

66)* Suggest a reason why the acid strength decreases with each step in the ionization of phosphoric acid: $H_3PO_4 \rightarrow H_2PO_4^- \rightarrow HPO_4^{2-}$.

67) Theoretically, can there be a Brönsted–Lowry acid–base reaction between SO_4^{2-} and F^-? If not, why not? If yes, write the equation.

68) Sodium carbonate is among the most important industrial bases. How can it be a base when it does not contain a hydroxide ion? Write the equation for a reaction that demonstrates its character as a base.

MATCHING SET ANSWERS

Set 1: 6–7–8–11–3–1–10–4–2–9–5

Set 2: 4–7–10–8–3–5–11–9–2–1–6

Set 3: 5–6–9–8–4–2–1–11–10–3–7

QUICK CHECK ANSWERS

17.1 An acid produces an H^+ ion and a base yields an OH^- ion.

17.2 Brönsted–Lowry (BL) and Arrhenius (AR) acids both yield protons; they are the same. AR bases all have hydroxide ions to receive protons; BL bases are anything that can receive protons. All AR bases are BL bases, but not all BL bases are AR bases.

17.3 Water can be a Lewis base because it has unshared electron pairs. It cannot be a Lewis acid because it has no vacant orbital to receive an electron pair from a Lewis base.

17.4 Given: $[OH^-] = 0.0001$ M (10^{-4} M).
Wanted: $[H^+]$

$$[H^+] = \frac{K_w}{[OH^-]} = \frac{10^{-14}}{10^{-4}} = 10^{-10} \text{ M}$$

The solution is basic.

17.5 Y (pOH = 1) < W (pOH = 6) < Z (pOH = 11) < X (pOH = 12)

18 Oxidation–Reduction (Electron-Transfer) Reactions

All common batteries convert chemical energy into electrical energy by means of an oxidation–reduction reaction. Another familiar battery is the storage battery found in automobiles.

18.1 ELECTROLYTIC AND VOLTAIC CELLS

In the opening section of Chapter 16 you learned that an electric current is passed through a fluid by the movement of charged particles, specifically, ions. The process is called **electrolysis,** the liquid through which the ions move is an **electrolyte,** and the container in which it all happens is an **electrolytic cell.** The "electricity" enters and leaves the cell through **electrodes,** which are usually, but not always, metal. The whole process is the result of chemical changes at the electrodes.

There are two kinds of cells in which electrolysis occurs. In the **voltaic cell,** also called a **galvanic cell,** the chemical changes are spontaneous. The voltaic cell is a "source" of electricity. In the electrolytic cell the changes happen only if the cell is connected to some outside source that "forces" the movement of ions by charging the electrodes. Figure 18.1 compares the two.

Both kinds of cells have many applications. Electrolytic cells are used to produce all of the Group 1A (1) and 2A (2) metals, with the exception of barium. Sodium chloride is electrolyzed commercially in an apparatus called the Downs cell to produce sodium and chlorine. Chlorine also comes from the electrolysis of sodium chloride solutions, after which the used electrolyte is evaporated to recover sodium hydroxide. Many common objects are made of metals that are electroplated with copper, nickel,

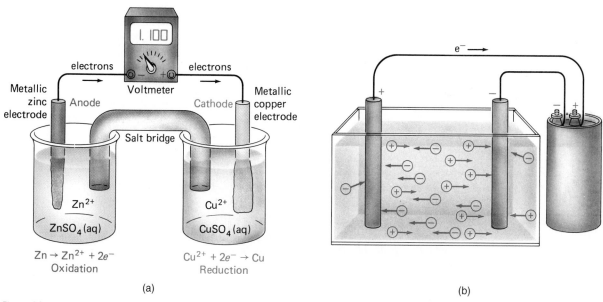

(a)   (b)

Figure 18.1

Voltaic and electrolytic cells. (a) A voltaic cell is one that causes an electric current that may light a flashlight, ring a bell, or start an automobile engine. The current is produced by spontaneous oxidation and reduction changes at the electrodes. The "salt bridge" is an ionic solution through which current flows without mixing the separate solutions. (b) The flow of electricity must be "forced" through an electrolytic cell by some outside source, such as a battery or generator. Chemical changes occur at the electrodes, but they are not spontaneous. In both cells the electrode at which oxidation occurs is the *anode,* and reduction occurs at the *cathode.*

chromium, zinc, tin, silver, gold, and other elements. The electrodeposits not only add beauty to the final product, but also protect the base metal from corrosion.

Voltaic cells were used to operate telegraph relays and doorbells back in the nineteenth century before electricity was generally available. The familiar "dry cell" for flashlights, toys, and other electrical devices is a voltaic cell, as are the longer lasting but more costly alkaline batteries.* When size is critical, as in calculators and watches, a silver cell may be used. All of these cells "run down" and must be replaced when the chemical reactions in them reach equilibrium. The "ni–cad" (nickel–cadmium) voltaic cell runs down too, but unlike the others, it can be recharged. A more familiar rechargeable battery is the lead storage battery used in automobiles.

18.2 ELECTRON-TRANSFER REACTIONS

In the cell in Figure 18.1(a), a strip of zinc is immersed in a solution of zinc ions, and a piece of copper is placed in a solution of copper ions. The solutions are connected by a "salt bridge," an electrolyte whose ions are not involved in the net chemical change. The two electrodes are connected by a wire. A voltmeter in the external circuit detects a flow of electrons from the zinc electrode to the copper electrode and also measures the "force" that moves the electrons through the circuit.

Where do the electrons entering the voltmeter come from, and where do they go on leaving? Four measurable observations answer that question. After the cell has operated for a period of time (1) the mass of zinc electrode decreases, (2) the Zn^{2+} concentration increases, (3) the mass of the copper electrode increases, and (4) the Cu^{2+} concentration decreases. The first two observations indicate that neutral zinc atoms lose two electrons to become zinc ions. Stated another way, zinc atoms are being divided into zinc ions and two electrons:

$$Zn(s) \longrightarrow Zn^{2+}(aq) + 2\,e^- \tag{18.1}$$

The electrons flow through the wire and the voltmeter to the copper electrode, where they join a copper ion to become a copper atom:

$$Cu^{2+}(aq) + 2\,e^- \longrightarrow Cu(s) \tag{18.2}$$

The chemical change that occurs at the zinc electrode is oxidation. **Oxidation is defined as the loss of electrons.** The reaction is described as a **half-reaction** because it cannot occur by itself. There must be a second half-reaction. The electrons lost by the substance **oxidized** must have some place to go. In this case they go to the copper ion, which is **reduced. Reduction is a gain of electrons.**

Equations 18.1 and 18.2 are **half-reaction equations.** If the half-reaction equations are combined—added algebraically—the result is the net ionic equation for the oxidation–reduction (redox) reaction:

$$Zn(s) \longrightarrow Zn^{2+}(aq) + 2e^- \tag{18.1}$$

$$\underline{Cu^{2+}(aq) + 2e^- \longrightarrow Cu(s)} \tag{18.2}$$

$$Zn(s) + Cu^{2+}(aq) \longrightarrow Cu(s) + Zn^{2+}(aq) \tag{18.3}$$

*Technically, a battery has two or more cells that are connected electrically. The term is also applied to a cell or combination of cells that furnishes electrical energy to any device.

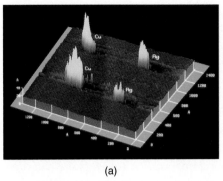
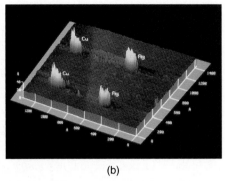

(a)   (b)

A scanning tunneling microscope of the world's smallest battery, about 70 nm on a side, about 0.01 the diameter of a red blood cell. In (a) the battery is not yet discharged; in (b), copper atoms have left the copper "pillars" and the silver terminals have been coated with a two-atom-thick plating of copper.

This chemical change is an **electron-transfer reaction.** Electrons have been transferred from zinc atoms to copper(II) ions. Notice that although no electrons appear in the final equation, the electron-transfer character of the reaction is quite clear in the half-reactions. Notice also that the number of electrons lost by one species is exactly equal to the number of electrons gained by the other.

If there is no need for the electrical energy that can be derived from this cell, the same reaction can be performed by simply dipping a strip of zinc into a solution of copper(II) ions. A coating of copper atoms quickly forms on the surface of the zinc. If the copper atoms are washed off the zinc and the zinc is weighed, its mass will be less than it was at the beginning. The concentration of copper ions in the solution goes down, and zinc ions appear. The half-reaction and net ionic equations are exactly as they are for the voltaic cell.

This same reaction was used in Example 16.4 as, "A reaction occurs when a piece of zinc is dipped into $Cu(NO_3)_2$(aq). Write the conventional, ionic, and net ionic equations." The conventional equation is a "single replacement" equation:

$$Zn(s) + Cu(NO_3)_2(aq) \longrightarrow Cu(s) + Zn(NO_3)_2(aq) \qquad (18.4)$$

Equation 18.3 is the net ionic equation produced in Example 16.4.

All of the single-replacement redox reactions encountered in Chapters 8 and 16 can be analyzed in terms of half-reactions. For example:

1) The evolution of hydrogen gas on adding zinc to sulfuric acid (Section 16.5):

Reduction: $2\,H^+(aq) + 2e^- \longrightarrow H_2(g)$
Oxidation: $\underline{\qquad\qquad Zn(s) \longrightarrow Zn^{2+}(aq) + 2e^-}$
Redox: $2\,H^+(aq) + Zn(s) \longrightarrow H_2(g) + Zn^{2+}(aq) \qquad (18.5)$

2) The preparation of bromine by bubbling chlorine gas through a solution of NaBr (Example 8.14):

Reduction: $Cl_2(g) + 2e^- \longrightarrow 2\,Cl^-$
Oxidation: $\underline{\qquad\quad 2\,Br^-(aq) \longrightarrow Br_2(\ell) + 2e^-}$
Redox: $Cl_2(g) + 2\,Br^-(aq) \longrightarrow 2\,Cl^-(aq) + Br_2(\ell) \qquad (18.6)$

These photographs of the zinc-copper(II) nitrate reaction appeared on page 442, next to Example 16.4. The upper photo shows the zinc before being dipped into the copper solution, and the bottom photo is after dipping. The black coating on the zinc is finely divided copper, which appears black when wet.

3) The formation of a "chemical pine tree" with needles of silver (Example 8.13 in Section 8.7 and Example 16.7 in Section 16.5) by placing a copper wire into a silver nitrate solution:

Reduction: $\quad\quad$ $2\,Ag^+(aq) + 2e^- \longrightarrow 2\,Ag(s)$

Oxidation: $\quad\quad\quad\quad\quad$ $Cu(s) \longrightarrow Cu^{2+}(aq) + 2e^-$

Redox: $\quad\quad$ $2\,Ag^+(aq) + Cu(s) \longrightarrow 2\,Ag(s) + Cu^{2+}(aq)$ $\quad\quad$ (18.7)

The development of Equation 18.7 needs special comment. The usual reduction equation for silver ion is $Ag^+(aq) + e^- \rightarrow Ag(s)$. Because two moles of electrons are lost in the oxidation reaction, *two moles of electrons must be gained in the reduction reaction*. As has already been noted, the number of electrons lost by one species must equal the number gained by the other species. It is therefore necessary to multiply the usual Ag^+ reduction equation by 2 to bring about this equality in electrons gained and lost. They then cancel when the half-reaction equations are added.

EXAMPLE 18.1

Combine the following half-reactions to produce a balanced redox reaction equation. Indicate which half-reaction is an oxidation reaction, and which is a reduction.

$$Co^{2+}(aq) + 2\,e^- \longrightarrow Co(s)$$
$$Sn(s) \longrightarrow Sn^{2+}(aq) + 2\,e^-$$

Reduction: $\quad\quad$ $Co^{2+}(aq) + 2e^- \longrightarrow Co(s)$

Oxidation: $\quad\quad\quad\quad$ $Sn(s) \longrightarrow Sn^{2+}(aq) + 2e^-$

Redox: $\quad\quad$ $Co^{2+}(aq) + Sn(s) \longrightarrow Co(s) + Sn^{2+}(aq)$

EXAMPLE 18.2

Combine the following half-reactions to produce a balanced redox equation. Identify the oxidation half-reaction and reduction half-reaction.

$$Fe^{2+}(aq) \longrightarrow Fe^{3+}(aq) + e^-$$
$$Al^{3+}(aq) + 3\,e^- \longrightarrow Al(s)$$

Oxidation: $\quad\quad$ $3\,Fe^{2+}(aq) \longrightarrow 3\,Fe^{3+}(aq) + 3e^-$

Reduction: $\quad\quad\quad$ $Al^{3+}(aq) + 3e^- \longrightarrow Al(s)$

Redox: $\quad\quad$ $3\,Fe^{2+}(aq) + Al^{3+}(aq) \longrightarrow 3\,Fe^{3+}(aq) + Al(s)$

In this example it is necessary to multiply the oxidation half-reaction equation by 3 in order to balance the electrons gained and lost.

Another reaction involving iron and aluminum introduces an additional technique.

EXAMPLE 18.3

Arrange and modify the following half-reactions as necessary, so they add up to produce a balanced redox equation. Identify the oxidation half-reaction and the reduction half-reaction.

$$Fe^{2+}(aq) + 2 e^- \longrightarrow Fe(s); \qquad Al(s) \longrightarrow Al^{3+}(aq) + 3 e^-$$

This will extend you a bit when it comes to balancing electrons. Two electrons are transferred for each atom of iron and three per atom of aluminum. In what ratio must the atoms be used to equate the electrons gained and lost? Multiply and add the half-reaction equations accordingly.

Reduction:	$3 Fe^{2+}(aq) + \cancel{6 e^-} \longrightarrow 3 Fe(s)$	
Oxidation:	$2 Al(s) \longrightarrow 2 Al^{3+}(aq) + \cancel{6 e^-}$	
Redox:	$3 Fe^{2+}(aq) + 2 Al(s) \longrightarrow 3 Fe(s) + 2 Al^{3+}(aq)$	

In this example electrons are transferred two at a time in the iron half-reaction and three at a time in the aluminum half-reaction. The simplest way to equate these is to take the iron half-reaction three times and the aluminum half-reaction twice. This gives six electrons for both half-reactions—just as two Al^{3+} and three O^{2-} balance the positive and negative charges in the ions making up the formula of Al_2O_3.

18.3 OXIDATION NUMBERS AND REDOX REACTIONS

The redox reactions that we have discussed up to this point have been rather simple ones involving only two reactants. With Equations 18.3 and 18.5–18.7 we can see at a glance which species has gained and which has lost electrons. Some oxidation–reduction reactions are not so readily analyzed. Consider, for example, a reaction that is sometimes used in the general chemistry laboratory to prepare chlorine gas from hydrochloric acid:

$$MnO_2(s) + 4 H^+(aq) + 2 Cl^-(aq) \longrightarrow Mn^{2+}(aq) + Cl_2(g) + 2 H_2O(\ell) \quad (18.8)$$

or the reaction, taking place in a lead storage battery, that produces the electrical spark to start an automobile:

$$Pb(s) + PbO_2(s) + 4 H^+(aq) + 2 SO_4^{2-}(aq) \longrightarrow 2 PbSO_4(s) + 2 H_2O(\ell) \quad (18.9)$$

Looking at these equations, it is by no means obvious which species are gaining and which are losing electrons.

"Electron bookkeeping" in redox reactions like Equations 18.8 and 18.9 is accomplished by using **oxidation states,** which are also called **oxidation numbers.** By following a set of rules, oxidation numbers may be assigned to each element in a molecule or ion. The rules are:

SUMMARY

1) The oxidation number of any elemental substance is 0 (zero).
2) The oxidation number of a monatomic ion is the same as the charge on the ion.

We can now state a broader definition of oxidation and reduction. **Oxidation is an increase in oxidation number; reduction is a reduction in oxidation number.** These definitions are more useful in identifying the elements oxidized and reduced when the electron transfer is not apparent. All you must do is find the elements that change their oxidation numbers and determine the direction of each change. One element must increase, and the other must decrease. (This corresponds with one species losing electrons while another gains.)

There are some techniques that enable you to spot quickly an element that changes oxidation number or to dismiss quickly some elements that do not change. These are:

1) An element that is in its elemental state must change. As an element on one side of the equation, its oxidation number is 0; as anything other than an element on the other side, it is *not* 0.
2) In other than elemental form, hydrogen is +1 and oxygen is −2. Unless they are elements on one side, they do not change. In more advanced courses you will have to be alert to the hydride, peroxide, and superoxide exceptions noted in the oxidation number rules.
3) A Group 1A (1) or 2A (2) element has only one oxidation state other than 0. If it does not appear as an element, it does not change. This observation is helpful when you must find the element oxidized or reduced in a conventional equation.

We will now use these ideas to find the elements oxidized and reduced in Equation 18.8,

$$MnO_2(s) + 4\,H^+(aq) + 2\,Cl^-(aq) \longrightarrow Mn^{2+}(aq) + Cl_2(g) + 2\,H_2O(\ell)$$

Chlorine is an element on the right, so it must be something else on the left. It is—the chloride ion, Cl^-. The oxidation number change is −1 to 0, an *increase*. Chlorine is *oxidized*.

Neither hydrogen nor oxygen appear as elements, so we conclude that they do not change oxidation state. That leaves manganese. Its oxidation state is +4 in MnO_2 on the left, and +2 as Mn^{2+} on the right. This is a *decrease*, from +4 to +2; manganese is *reduced*.

EXAMPLE 18.6

Determine the element oxidized and the element reduced in a lead storage battery, Equation 18.9:

$$Pb(s) + PbO_2(s) + 4\,H^+(aq) + 2\,SO_4^{2-}(aq) \longrightarrow 2\,PbSO_4(s) + 2\,H_2O(\ell)$$

Assign oxidation numbers to as many elements as necessary until you come up with the pair that changed. Then identify the oxidation and reduction changes. (Be careful. This one is a bit tricky.)

———   ———

Lead is both oxidized (0 in Pb to +2 in $PbSO_4$) and reduced (+4 in PbO_2 to +2 in $PbSO_4$).

The oxidation of lead can be spotted quickly, as it is an element on the left. You might have thought sulfur to be the element reduced, but its oxidation state is +6 in the sulfate ion whether the ion is by itself on the left, or part of a solid ionic compound on the right.

While oxidation number is a very useful device to keep track of what the electrons are up to in a redox reaction, we should emphasize that it has been "invented" to meet a need. It has no experimental basis. Unlike the charge of a monatomic ion, the oxidation number of an atom in a molecule or polyatomic ion cannot be measured in the laboratory. It is all very well to talk about "+4 manganese" in MnO_2 or "+6 sulfur" in the SO_4^{2-} ion, but take care not to fall into the trap of thinking that the elements in these species actually carry positive charges equal to their oxidation numbers.

S U M M A R Y / (Sections 18.2 and 18.3)

	Oxidation	Reduction
Change in electrons	Loss	Gain
Change in oxidation number	Increase	Decrease (Reduction)

18.4 OXIDIZING AGENTS (OXIDIZERS); REDUCING AGENTS (REDUCERS)

The two essential reactants in a redox reaction are given special names to indicate the roles they play. The species that accepts electrons is referred to as an **oxidizing agent, or oxidizer;** the species that donates the electrons so reduction can occur is called a **reducing agent,** or **reducer.** For example, in Equation 18.5

$$2\,H^+(aq) + Zn(s) \longrightarrow H_2(g) + Zn^{2+}(aq)$$

H^+ has accepted electrons from Zn—it has *oxidized* Zn to Zn^{2+}—and is therefore the oxidizing agent. Conversely, Zn has donated electrons to H^+—it has reduced H^+ to H_2—and is therefore the reducing agent. In Equation 18.8

$$MnO_2(s) + 4\,H^+(aq) + 2\,Cl^-(aq) \longrightarrow Mn^{2+}(aq) + Cl_2(g) + 2\,H_2O(\ell)$$

Cl^- is the reducer, reducing manganese from +4 to +2. The oxidizer is MnO_2—the whole compound, not just the Mn; it oxidizes chlorine from −1 to 0.

The following example summarizes the redox concepts.

E X A M P L E   1 8 . 7

Consider the redox equation

$$5\,NO_3^-(aq) + 3\,As(s) + 2\,H_2O(\ell) \rightarrow 5\,NO(g) + 3\,AsO_4^{3-}(aq) + 4\,H^+(aq)$$

a) Determine the oxidation number in each species: N: _____ in NO_3^-, and _____ in NO.

   As: _____ in As, and _____ in AsO_4^{3-}. H: _____ in H_2O, and _____ in H^+.

   O: _____ in NO_3^-, _____ in H_2O, _____ in NO, and _____ in AsO_4^{3-}.

b) Identify (1) the element oxidized _____   (2) the element reduced _____

   (3) the oxidizing agent _____   (4) the reducing agent _____

a) N: $+5$ in NO_3^-, and $+2$ in NO.
   As: 0 in As, and $+5$ in AsO_4^{3-}.
   H: $+1$ in both H_2O and H^+.
   O: -2 in all species.

b) (1) As is oxidized, increasing in oxidation number from 0 to $+5$.
   (2) N is reduced, decreasing in oxidation number from $+5$ to $+2$.
   (3) NO_3^- is the oxidizing agent, removing electrons from As.
   (4) As is the reducing agent, furnishing electrons to NO_3^-.

18.5 STRENGTHS OF OXIDIZING AGENTS AND REDUCING AGENTS

An oxidizing agent earns its title by its ability to take electrons from another substance. A **strong oxidizing agent** has a strong attraction for electrons. Conversely, a **weak oxidizing agent** attracts electrons only slightly. The strength of a reducing agent is measured by its ability to give up electrons. A **strong reducing agent** releases electrons readily, whereas a **weak reducing agent** holds on to its electrons.

One way to measure the "strength" of an element's ability to attract or release electrons is with a voltmeter, as shown in Figure 18.1. The voltage developed in a galvanic cell depends on how strongly the elements used as electrodes attract and release electrons. Table 18.2 has been assembled from many such measurements.

Table 18.2 is a list of oxidizing agents in order of decreasing strength on the left side of the equation and of reducing agents in order of increasing strength on the right side. The strongest oxidizing agent shown is fluorine, F_2, located at the top of the left

Table 18.2
Relative Strengths of Oxidizing and Reducing Agents

Oxidizing Agent		Reducing Agent
$F_2(g) + 2\,e^-$	$\rightleftharpoons$	$2\,F^-$
$Cl_2(g) + 2\,e^-$	$\rightleftharpoons$	$2\,Cl^-$
$\frac{1}{2}\,O_2(g) + 2\,H^+ + 2\,e^-$	$\rightleftharpoons$	H_2O
$Br_2(\ell) + 2\,e^-$	$\rightleftharpoons$	$2\,Br^-$
$NO_3^- + 4\,H^+ + 3\,e^-$	$\rightleftharpoons$	$NO(g) + 2\,H_2O$
$Ag^+ + e^-$	$\rightleftharpoons$	$Ag(s)$
$Fe^{3+} + e^-$	$\rightleftharpoons$	Fe^{2+}
$I_2(s) + 2\,e^-$	$\rightleftharpoons$	$2\,I^-$
$Cu^{2+} + 2\,e^-$	$\rightleftharpoons$	$Cu(s)$
$2\,H^+ + 2\,e^-$	$\rightleftharpoons$	$H_2(g)$
$Ni^{2+} + 2\,e^-$	$\rightleftharpoons$	$Ni(s)$
$Co^{2+} + 2\,e^-$	$\rightleftharpoons$	$Co(s)$
$Cd^{2+} + 2\,e^-$	$\rightleftharpoons$	$Cd(s)$
$Fe^{2+} + 2\,e^-$	$\rightleftharpoons$	$Fe(s)$
$Zn^{2+} + 2\,e^-$	$\rightleftharpoons$	$Zn(s)$
$Al^{3+} + 3\,e^-$	$\rightleftharpoons$	$Al(s)$
$Na^+ + e^-$	$\rightleftharpoons$	$Na(s)$
$Ca^{2+} + 2\,e^-$	$\rightleftharpoons$	$Ca(s)$
$Li^+ + e^-$	$\rightleftharpoons$	$Li(s)$

Oxidizing Agent — STRENGTH — Increasing / Decreasing

Reducing Agent — STRENGTH — Decreasing / Increasing

FLASHBACK Compare Table 18.2 to Table 17.1 in Section 17.6. In Table 17.1 the substances are listed according to their tendencies to release and receive protons; in Table 18.2 the substances are listed according to their tendencies to give or accept electrons. In both tables, the tendency listed is in decreasing order in the left column and in increasing order in the right column.

Chrome plating provides a pleasing, protective surface.

Corrosion, the result of many redox reactions, costs the United States about $30,000,000,000 each year.

column. Chlorine, Cl_2, listed just below fluorine, is used as a disinfectant in water supplies because of its ability to oxidize harmful organic matter. Notice that all equations in Table 18.2 are written as reduction half-reactions.

18.6 PREDICTING REDOX REACTIONS

As Table 17.1 enables us to write acid–base reaction equations and predict the direction that will be favored at equilibrium, Table 18.2 enables us to do the same for redox reactions. The redox table has a limitation, however. Acid–base reactions are all *single-proton-transfer* reactions and equations are automatically balanced if taken directly from the table. Redox half-reactions, on the other hand, frequently involve unequal numbers of electrons. They must be balanced as in Equation 18.7 and Example 18.2. The next five examples illustrate the process.

FLASHBACK Table 18.2 is the source of the activity series (Table 16.1 in Section 16.5). The activity series corresponds with the right-hand side of Table 18.2, from the bottom to the top. You used Table 16.1 to predict simple redox reaction when writing net ionic equations. Table 18.2 shows what happens in those reactions and includes more complex examples.

EXAMPLE 18.8

Write the net ionic equation for the redox reaction between the cobalt(II) ion, Co^{2+}, and metallic silver, Ag.

SOLUTION

First, as in a Brönsted–Lowry acid–base reaction there must be a proton giver and a proton taker, so in a redox reaction there must be an electron giver (reducer) and an electron taker (oxidizer). Consulting Table 18.2, we find Co^{2+} among the oxidizers and Ag among the reducers. The reduction half-reaction is taken directly from the table:

Reduction: $\qquad Co^{2+}(aq) + 2e^- \rightleftharpoons Co(s)$

To obtain the oxidation half-reaction for silver, we need to *reverse* the reduction half-reaction found in the table:

Oxidation: $\qquad Ag(s) \rightleftharpoons Ag^+(aq) + e^-$

Multiplying the oxidation equation by 2 to equalize electrons gained and lost, and adding to the reduction equation yields

Reduction: $Co^{2+}(aq) + 2e^- \rightleftharpoons Co(s)$

2 × Oxidation: $2 Ag(s) \rightleftharpoons 2 Ag^+(aq) + 2e^-$

Redox: $Co^{2+}(aq) + 2 Ag(s) \rightleftharpoons Co(s) + 2 Ag^+(aq)$

The principle underlying the prediction of the favored direction of a redox reaction is the same as for an acid–base reaction. The stronger oxidizing agent—strong in its attraction for electrons—will take the electrons from a strong reducing agent—strong in its tendency to donate electrons—to produce the weaker oxidizer and reducer. This is shown in Figure 18.2, which is strikingly similar to Figure 17.2, Section 17.7. In the reaction $2 H^+(aq) + Zn(s) \rightleftharpoons H_2(g) + Zn^{2+}(aq)$ (Equation 18.5), the positions in the table establish H^+ and Zn as the stronger oxidizer and reducer. The reaction is favored in the forward direction, yielding the weaker oxidizer and reducer, Zn^{2+} and H_2. (*Note:* We will use double arrows when necessary to indicate the equilibrium character of redox reactions.)

EXAMPLE 18.9

In which direction, forward (___) or reverse (___) will the redox reaction in Example 18.8 be favored?

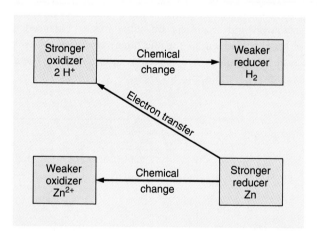

Figure 18.2

Predicting redox reactions from positions in Table 18.2. The spontaneous chemical change always transfers one or more electrons from the stronger reducing agent to the stronger oxidizing agent, both shown in pink. The products of the reaction, the weaker oxidizing and reducing agents, are shown in blue. The direction, forward or reverse, that is favored is the one that has the weaker oxidizing and reducing agents as products.

Reverse

Ag^+ is a stronger oxidizer than Co^{2+}, and is therefore able to take electrons from cobalt atoms. Also, cobalt atoms are a stronger reducer than silver atoms, and therefore readily release electrons to Ag^+. The weaker reducer and oxidizer, Ag and Co^{2+}, are favored.

EXAMPLE 18.10

Write the redox reaction equation between metallic copper and a strong acid, H^+, and indicate the direction that is favored.

Reduction:	$2 H^+(aq) + 2e^- \rightleftharpoons H_2(g)$
Oxidation:	$Cu(s) \rightleftharpoons Cu^{2+}(aq) + 2e^-$
Redox:	$2 H^+(aq) + Cu(s) \rightleftharpoons H_2(g) + Cu^{2+}(aq)$

The reverse reaction is favored.

EXAMPLE 18.11

Write the net ionic equation for the redox reaction between Al(s) and $Ni^{2+}(aq)$, and predict the favored direction, forward or reverse.

Reduction:	$Ni^{2+}(aq) + 2 e^- \rightleftharpoons Ni(s)$
Oxidation:	$Al(s) \rightleftharpoons Al^{3+}(aq) + 3 e^-$
$3 \times$ Reduction:	$3 Ni^{2+}(aq) + 6e^- \rightleftharpoons 3 Ni(s)$
$2 \times$ Oxidation:	$2 Al(s) \rightleftharpoons 2 Al^{3+}(aq) + 6e^-$
Redox:	$3 Ni^{2+}(aq) + 2 Al(s) \rightleftharpoons 3 Ni(s) + 2 Al^{3+}(aq)$

The forward reaction is favored.

One of the properties of acids listed in Section 17.1 is their ability to release hydrogen gas on reaction with certain metals. Judging from Example 18.10, copper is not among those metals. The metals that do release hydrogen are the reducers below hydrogen in Table 18.2. But there is more to the reactions between metals and acids than meets the eye.

EXAMPLE 18.12

Write the equation for the reaction between copper and nitric acid, and predict which direction is favored.

Copper is in Table 18.2, but you will search in vain for HNO_3. The solution inventory species of nitric acid (NO_3^- and H^+) are present, though. We will comment on the imbalance between hydrogen ions and nitrate ions shortly. This reaction summarizes our equation writing methods to this point. Take it all the way.

2 × Reduction: $2 NO_3^-(aq) + 8 H^+(aq) + 6e^- \rightleftharpoons 2 NO(g) + 4 H_2O(\ell)$

3 × Oxidation: $3 Cu(s) \rightleftharpoons 3 Cu^{2+}(aq) + 6e^-$

Redox: $2 NO_3^-(aq) + 8 H^+(aq) + 3 Cu(s) \rightleftharpoons 2 NO(g) + 4 H_2O(\ell) + 3 Cu^{2+}(aq)$

The forward reaction is favored.

Don't worry about those missing nitrate ions, the six unaccounted for from the eight moles of HNO_3 that furnished the $8 H^+$. They are in there as spectators, just enough to balance the $3 Cu^{2+}$.

18.7 REDOX REACTIONS AND ACID–BASE REACTIONS COMPARED

At this point it may be useful to pause briefly and point out how redox reactions resemble acid–base reactions.

1) Acid–base reactions involve a transfer of protons; redox reactions, a transfer of electrons.

2) In both cases the reactants are given special names to indicate their roles in the transfer process. An acid is a proton donor; a base is a proton acceptor. A reducing agent is an electron donor; an oxidizing agent is an electron acceptor.

3) Just as certain species (e.g., HCO_3^-, H_2O) can either donate or accept protons and thereby behave as an acid in one reaction and a base in another, certain species can either accept or donate electrons, acting as an oxidizing agent in one reaction and a reducing agent in another. An example is the Fe^{2+} ion, which can oxidize Zn atoms to Zn^{2+} in the reaction.

$$Fe^{2+}(aq) + Zn(s) \longrightarrow Fe(s) + Zn^{2+}(aq)$$

Fe^{2+} can also reduce Cl_2 molecules to Cl^- ions in another reaction:

$$Cl_2(g) + 2 Fe^{2+}(aq) \longrightarrow 2 Cl^-(aq) + 2 Fe^{3+}(aq)$$

4) Just as acids and bases may be classified as "strong" or "weak" depending on how readily they donate or accept protons, the strengths of oxidizing and reducing agents may be compared according to their tendencies to attract or release electrons.

5) Just as most acid–base reactions in solution reach a state of equilibrium, so most aqueous redox reactions reach equilibrium. Just as the favored side of an acid–base equilibrium can be predicted from acid–base strength, so the favored side of a redox equilibrium can be predicted from oxidizer–reducer strength.

18.8 WRITING REDOX EQUATIONS

Thus far we have considered only redox reactions for which the oxidation and reduction half-reactions are known. We are not always this fortunate. Sometimes we know only the reactants and products. Considering nitric acid, for example, suppose we know only that the product of the reduction of nitric acid is NO(g). How do we get from this information to the reduction half-reaction given in Table 18.2?

The steps for writing a half-reaction equation in an acidic solution are listed below. Each step is illustrated for the NO_3^- to NO change in Example 18.12.

SUMMARY

1) *After identifying the element oxidized or reduced, write a partial half-reaction equation with the element in its original form (element, monatomic ion, or part of a polyatomic ion or compound) on the left, and in its final form on the right:*

$$NO_3^-(aq) \longrightarrow NO(g)$$

2) *Balance the element oxidized or reduced.*
   Nitrogen is already balanced.

3) *Balance elements other than hydrogen or oxygen, if any.*
   There are none.

4) *Balance oxygen by adding water molecules where necessary.*
   There are three oxygens on the left and one on the right. Two water molecules are needed on the right:

$$NO_3^-(aq) \longrightarrow NO(g) + 2\,H_2O(\ell)$$

5) *Balance hydrogen by adding H^+ where necessary.*
   There are four hydrogens on the right and none on the left. Four hydrogen ions are needed on the left:

$$4\,H^+(aq) + NO_3^-(aq) \longrightarrow NO(g) + 2\,H_2O(\ell)$$

6) *Balance charge by adding electrons to the more positive side.*
   Total charge on the left is $+4 + (-1) = +3$; on the right, zero. Three electrons are needed on the left:

$$3\,e^- + 4\,H^+(aq) + NO_3^-(aq) \longrightarrow NO(g) + 2\,H_2O(\ell)$$

7) *Recheck the equation to be sure it is balanced in both atoms and charge.*

There are several good ways to write and balance complex half-reaction equations. If your instructor prefers a different way, by all means use it.

Notice that these instructions are for redox half-reactions in *acidic* solutions. The procedure is somewhat different with basic solutions, but we will omit that procedure in this introductory text.

When you have both half-reaction equations, proceed as in the earlier examples.

EXAMPLE 18.13

Write the net ionic equation for the redox reaction between iodide and sulfate ions in an acidic solution. The products are iodine and sulfur: $I^-(aq) + SO_4^{2-}(aq) \rightarrow I_2(s) + S(s)$.

First, using oxidation numbers, identify the element reduced and the element oxidized.

Sulfur is reduced (ox. no. change +6 to 0) and iodine is oxidized (−1 to 0).

Balance atoms first, and then charges, in the oxidation half-reaction:

$$I^-(aq) \longrightarrow I_2(s)$$

$$2\,I^-(aq) \longrightarrow I_2(s) + 2e^-. \qquad \text{(This one happens to be in Table 18.2.)}$$

Now for the reduction half-reaction:

$$SO_4^{2-}(aq) \longrightarrow S(s)$$

Sulfur is already in balance. The only other element is oxygen. According to Step 4, oxygen is balanced by adding the necessary water molecules. Complete that step.

$$SO_4^{2-}(aq) \longrightarrow S(s) + 4\,H_2O(\ell)$$

Four oxygen atoms in a sulfate ion require four water molecules.

Next comes the hydrogen balancing, using H^+ ions.

$$8\,H^+(aq) + SO_4^{2-}(aq) \longrightarrow S(s) + 4\,H_2O(\ell)$$

Finally, add to the positive side the electrons that will bring the charges into balance.

$$6\,e^- + 8\,H^+(aq) + SO_4^{2-}(aq) \longrightarrow S(s) + 4\,H_2O(\ell)$$

On the left there are 8 + charges from hydrogen ion and 2 − charges from sulfate ion, a net of 6 +. On the right the net charge is zero. Charge is balanced by adding 6 electrons to the left (positive) side.

Now that you have the two half-reaction equations, finish writing the net ionic equation as you did before.

$$2\,I^-(aq) \longrightarrow I_2(s) + 2e^-$$
$$8\,H^+ + SO_4^{2-}(aq) + 6e^- \longrightarrow S(s) + 4\,H_2O(\ell)$$

Reduction: $\qquad 8\,H^+(aq) + SO_4^{2-}(aq) + 6e^- \longrightarrow S(s) + 4\,H_2O(\ell)$

3 × Oxidation: $\qquad\qquad\qquad\qquad\qquad 6\,I^-(aq) \longrightarrow 3\,I_2(s) + 6e^-$

Redox: $\qquad 8\,H^+(aq) + SO_4^{2-}(aq) + 6\,I^-(aq) \longrightarrow 4\,H_2O(\ell) + 3\,I_2(s) + S(s)$

Let's check to make sure the equation is, indeed, balanced:

—the atoms balance (six I, one S, four O, and eight H atoms on each side);
—the charge balances ($+8 - 2 - 6 = 0 + 0 + 0$).

EXAMPLE 18.14

The permanganate ion, MnO_4^-, is a strong oxidizing agent that oxidizes chloride ion to chlorine in an acidic solution. Manganese ends up as a monatomic manganese(II) ion. Write the net ionic equation for the redox reaction.

This is a challenging example, but watch how it falls into place following the procedure that has been outlined. To be sure the question has been interpreted correctly, begin by writing an unbalanced skeleton equation. Put the identified reactants on the left side and the identified products on the right side.

$$MnO_4^-(aq) + Cl^-(aq) \longrightarrow Cl_2(g) + Mn^{2+}(aq)$$

The oxidation half-reaction is easiest. Write it next.

$$2\,Cl^-(aq) \longrightarrow Cl_2(g) + 2e^-$$

With a switch from iodine to chlorine, this is the same oxidation half-reaction as in the last example.

Now write the formulas of the starting and ending species for the reduction reaction on opposite sides of the arrow.

$$MnO_4^-(aq) \longrightarrow Mn^{2+}(aq)$$

Can you take the reduction to a complete half-reaction equation? First do oxygen, then hydrogen, and finally charge.

$$8\,H^+(aq) + MnO_4^-(aq) + 5e^- \longrightarrow Mn^{2+}(aq) + 4\,H_2O(\ell)$$

Oxygen: $MnO_4^-(aq) \longrightarrow Mn^{2+}(aq) + 4\,H_2O(\ell)$
       (four waters for four oxygen atoms).
Hydrogen: $8\,H^+(aq) + MnO_4^-(aq) \longrightarrow Mn^{2+}(aq) + 4\,H_2O(\ell)$
       (eight H^+ for four waters).
Charge: $+8 - 1$ on the left is $+7$; $+2 + 0$ on the right is $+2$. Charge is balanced by
       adding five negatives on the left, or five electrons, as in the final answer.

Now rewrite and combine the half-reaction equations for the net ionic equation.

2 × Reduction: $16\,H^+(aq) + 2\,MnO_4^-(aq) + 10e^- \longrightarrow 2\,Mn^{2+}(aq) + 8\,H_2O(\ell)$
5 × Oxidation: $10\,Cl^-(aq) \longrightarrow 5\,Cl_2(g) + 10e^-$

Redox: $16\,H^+(aq) + 2\,MnO_4^-(aq) + 10\,Cl^-(aq) \longrightarrow 2\,Mn^{2+}(aq) + 8\,H_2O(\ell) + 5\,Cl_2(g)$

Checking:
  —atoms balance (sixteen H, two Mn, eight O, and ten Cl);
  —charges balance ($+16 - 2 - 10 = +4$).
Can you imagine a trial-and-error approach to an equation such as this?

CHAPTER 18 KEY WORDS AND MATCHING SETS

Set 1

_____ Describes equation for either oxidation reaction or reduction reaction.

_____ Equal to oxidation number assigned to a monatomic ion.

_____ Liquid through which ions move in electrolysis.

_____ Substance that oxidizes another substance.

_____ Terminal in an electrolytic cell.

_____ Substance that gives up electrons easily.

_____ Oxidation number becomes smaller.

_____ Electrolyte that connects two parts of an electrolytic cell.

_____ Substance that holds electrons firmly.

_____ Chemical change involving loss of electrons.

_____ What must always be equal in an oxidation–reduction reaction.

_____ Oxidation number assigned to an uncombined element.

1. Salt bridge
2. Electric charge
3. Electrode
4. Weak reducing agent
5. Half-reaction
6. Electrons gained and lost
7. Oxidizing agent
8. Oxidation
9. Strong reducing agent
10. Electrolyte
11. Zero
12. Reduction

Set 2

_____ Substance that reduces the oxidation number of another substance.

_____ Device that produces electric current.

_____ Has weak attraction for electrons.

_____ Increase in oxidation number.

_____ Movement of charged particles through a fluid.

_____ Attracts electrons strongly.

_____ "Tool" used to keep track of electrons in oxidation–reduction reactions.

_____ Nickname for oxidation–reduction reaction.

_____ Container in which electrolysis occurs.

_____ Chemical change involving gain of electrons.

_____ What happens in an oxidation–reduction reaction.

1. Reduction
2. Reducing agent
3. Electron transfer
4. Strong oxidizing agent
5. Electrolytic cell
6. Weak oxidizing agent
7. Redox
8. Oxidation
9. Galvanic (voltaic) cell
10. Oxidation number
11. Electrolysis

STUDY HINTS AND PITFALLS TO AVOID

This is the last chapter in which Study Hints and Pitfalls to Avoid will appear.

Oxidation and reduction are so closely related and similarly defined that they are easily confused. "Oxidation" is not a common word outside a chemical sense, but "reduction" is—and we can take advantage of it:

If something is _reduced_, it gets _smaller_. If an element is _reduced_, its oxidation number becomes _smaller_. Oxidation is the opposite.

Watch out for negative numbers that get smaller. "Getting smaller" means becoming more negative, as from -1 to -3.

Students sometimes summarize the relationship between species oxidized/reduced with oxidizing/reducing agents by saying, "Whatever is oxidized is the reducing agent, and whatever is reduced is the oxidizing agent." *Caution:* This is true for *monatomic species only*. If the element being oxidized/reduced is a part of a polyatomic species, the entire compound or ion is the reducing/oxidizing agent.

There are several ways to balance complicated redox equations. We have shown you only one, and that is only for acidic solutions. If your instructor prefers another method, by all means use it. Whatever method you use, it takes practice to perfect it. You may question this while learning, but many students report that once they get the hang of it, balancing redox equations is fun!

CHAPTER 18 QUESTIONS AND PROBLEMS

SECTION 18.1

1) List as many of the things in your home that you can think of that are operated by voltaic cells.

2) What is electrolysis? Is it an electric current? Explain. What is the difference between electrolysis and the passage of electric current through a wire?

---

3) Distinguish among voltaic cells, electrolytic cells, and galvanic cells. Can any one of them operate either of the other two? Explain.

4) Examine Figure 2.4. Is the apparatus shown a cell? If so, which kind, voltaic, electrolytic, or galvanic? On what do you base your answers?

SECTION 18.2

5) Distinguish between oxidation and reduction. Why do we sometimes refer to oxidation–reduction reactions as *electron-transfer reactions*?

6) Of oxidation and reduction it has been said that you can't have one without the other. Why?

---

7) Classify each of the following half-reaction equations as oxidation or reduction half-reactions:
a) $Zn \rightarrow Zn^{2+} + 2\,e^-$
b) $2\,H^+ + 2\,e^- \rightarrow H_2$
c) $Fe^{2+} \rightarrow Fe^{3+} + e^-$
d) $NO + 2\,H_2O \rightarrow NO_3^- + 4\,H^+ + 3\,e^-$

8) Classify each of the following half-reaction equations as oxidation or reduction half-reactions:
a) $Sn^{4+} + 2\,e^- \rightarrow Sn^{2+}$
b) $Na \rightarrow Na^+ + e^-$
c) $Cl_2 + 2\,e^- \rightarrow 2\,Cl^-$
d) $O_2 + 4\,H^+ + 4\,e^- \rightarrow 2\,H_2O$

---

9) Identify each of the following half-reaction equations as a reduction half-reaction or an oxidation half-reaction:
a) Dissolving ozone:
$O_3 + H_2O + 2\,e^- \rightarrow O_2 + 2\,OH^-$

b) Dissolving gold (Z = 79):
$4\,Cl^- + Au \rightarrow AuCl_4^- + 3\,e^-$

10) Identify each of the following half-reaction equations as a reduction half-reaction or an oxidation half-reaction:
a) Automobile battery reaction:
$PbO_2 + SO_4^{2-} + 4\,H^+ + 2\,e^- \rightarrow PbSO_4 + 2\,H_2O$
b) Tarnishing silver: $2\,Ag + S^{2-} \rightarrow Ag_2S + 2\,e^-$

---

11) Combine the following half-reaction equations to produce a balanced redox equation:

$$Cr \longrightarrow Cr^{3+} + 3\,e^- \qquad Cl_2 + 2\,e^- \longrightarrow 2\,Cl^-$$

12) Combine the following half-reaction equations to produce a balanced redox equation:

$$Mg^{2+} + 2\,e^- \longrightarrow Mg \qquad Ni \longrightarrow Ni^{2+} + 2\,e^-$$

13) The half-reactions that occur at the electrodes of an alkaline cell, widely used in flashlights, calculators, and other common battery-operated devices, are

$$NiOOH + H_2O + e^- \longrightarrow Ni(OH)_2 + OH^-$$
$$Cd + 2\,OH^- \longrightarrow Cd(OH)_2 + 2\,e^-$$

Which equation is for the oxidation half-reaction? Write the overall equation for the cell.

14) $Pb + SO_4^{2-} \rightarrow PbSO_4 + 2\,e^-$ is the other half-reaction equation in an automobile battery. (The first equation is in Question 10.) Write the equation for the overall battery reaction.

SECTION 18.3

15) Give the oxidation number for the element whose symbol is underlined in each formula:
a) $\underline{Al}^{3+}$ $\underline{S}^{2-}$ $\underline{S}O_3^{2-}$ $Na_2\underline{S}O_4$
b) $\underline{N}_2O_3$ $\underline{N}O_3^-$ $\underline{Cr}O_4^{2-}$ $NaH_2\underline{P}O_4$

16) Give the oxidation number for the element whose symbol is underlined in each formula:
a) $\underline{Cl}O^-$ $\underline{Mg}^{2+}$ $KClO_3$ $\underline{Cl}^-$
b) $\underline{Mn}O_4^-$ $\underline{N}_2O_5$ $Na_2H\underline{P}O_4$ $\underline{N}H_4^+$

Questions 17 to 20: (1) Identify the element being oxidized or reduced; (2) state "oxidized" or "reduced"; and (3) show the change in oxidation number. Example: $2\,Cl^- \rightarrow Cl_2 + 2\,e^-$. Chlorine oxidized from -1 to 0.

17)
a) $Br_2 + 2\,e^- \rightarrow 2\,Br^-$
b) $Pb^{2+} + 2\,H_2O \rightarrow PbO_2 + 4\,H^+ + 2\,e^-$
c) $8\,H^+ + IO_4^- + 8\,e^- \rightarrow I^- + 4\,H_2O$
18)
a) $Co^{2+} \rightarrow Co^{3+} + e^-$
b) $2\,H^+ + SO_4^{2-} + 2\,e^- \rightarrow SO_3^{2-} + H_2O$
c) $Cu \rightarrow Cu^{2+} + 2\,e^-$

19)
a) $4\,H^+ + O_2 + 4\,e^- \rightarrow 2\,H_2O$
b) $NO_2 + H_2O \rightarrow NO_3^- + 2\,H^+ + e^-$
c) $2\,Cr^{3+} + 7\,H_2O \rightarrow Cr_2O_7^{2-} + 14\,H^+ + 6\,e^-$
20)
a) $F_2 + 2\,H^+ + 2\,e^- \rightarrow 2\,HF$
b) $MnO_2 + 4\,OH^- \rightarrow 2\,H_2O + MnO_4^{2-} + 2\,e^-$
e) $PH_3 \rightarrow 3\,H^+ + P + 3\,e^-$

SECTION 18.4

21) Identify the oxidizing and reducing agents in

$$Cl_2 + 2\,Br^- \longrightarrow 2\,Cl^- + Br_2$$

22) In the reaction between hydrogen and copper(II) oxide, identify the oxidizing and reducing agents:

$$CuO + H_2 \longrightarrow Cu + H_2O$$

23) $PbO_2 + Pb + 4\,H^+ \rightarrow 2\,PbSO_4 + 2\,H_2O$ is the equation for the reaction in a storage battery. Identify the reducing agent and the substance reduced. Also, what is the oxidizing agent, and what does it oxidize?

24) What are oxidized and reduced, and what are the oxidizing and reducing agents, in $3\,HNO_2 + BrO_3^- \rightarrow 3\,H^+ + 3\,NO_3^- + Br^-$?

SECTION 18.5

25) Which is the stronger reducing agent, Zn or Fe^{2+}? How do you know? What is the significance of one reducer being stronger than another?

26) Between H^+ and Ag^+, which is the weaker oxidizing agent? How do you reach your conclusion? Of what practical value is this information when it comes to chemical reactions?

27) Arrange the following oxidizers in order of *decreasing* strength—that is, strongest oxidizing agent first: Cu^{2+}, Fe^{2+}, Br_2, Na^+.

28) List the following reducing agents in order of *increasing* strength—weakest reducer first, strongest last: Fe^{2+}, Cl^-, Al, H_2.

SECTION 18.6

Questions 29 to 34: Write the redox equation for each pair of reactants given. Use Table 18.2 as a source of the needed half-reactions. Then predict which reaction, forward or reverse, is favored at equilibrium.

29) $I^- + Br_2 \rightleftharpoons$
30) $Co^{2+} + Fe^{2+} \rightleftharpoons$

31) $Br^- + H^+ \rightleftharpoons$
32) $Zn + Ni^{2+} \rightleftharpoons$

33) $Fe^{2+} + H_2O + NO \rightleftharpoons$
34) $H_2O + Ca^{2+} \rightleftharpoons$

SECTION 18.7

35) Explain how a strong acid is similar to a strong reducing agent. Also compare a strong oxidizer with a strong base.

36) Compare acid–base reactions with redox reactions. How are they similar, but what is their principal difference?

SECTION 18.8

Questions 37 to 46: Each "equation" shows an oxidizer and a reducer in a redox reaction, as well as their oxidized and reduced products. Write separate oxidation and reduction half-reaction equations, assuming that the reaction takes place in an acidic solution. Add them to produce a balanced redox equation.

37) $S_2O_3^{2-} + Cl_2 \rightleftharpoons SO_4^{2-} + Cl^-$
38) $Cr_2O_7^{2-} + Fe^{2+} \rightleftharpoons Fe^{3+} + Cr^{3+}$

39) $Sn + NO_3^- \rightleftharpoons H_2SnO_3 + NO_2$
40) $SO_2 + Ag^+ \rightleftharpoons SO_4^{2-} + Ag$

41) $C_2O_4^{2-} + MnO_4^- \rightleftharpoons CO_2 + Mn^{2+}$
42) $Br^- + BrO_3^- \rightleftharpoons Br_2$

43) $Cr_2O_7{}^{2-} + NH_4{}^+ \rightleftharpoons Cr_2O_3 + N_2$

44) $MnO_4{}^- + I^- \rightleftharpoons MnO_2 + I_2$

45) $As_2O_3 + NO_3{}^- \rightleftharpoons AsO_4{}^{3-} + NO$

46) $Zn + NO_3{}^- \rightleftharpoons Zn^{2+} + NH_4{}^+$

GENERAL QUESTIONS

47) Distinguish precisely and in scientific terms the difference between items in each of the following groups:

a) Oxidation, reduction (in terms of electrons)
b) Half-reaction equation, net ionic equation
c) Oxidation, reduction (in terms of oxidation number)
d) Oxidizing agent (oxidizer), reducing agent (reducer)
e) Electron-transfer reaction, proton-transfer reaction
f) Strong oxidizing agent, weak oxidizing agent
g) Strong reducing agent, weak reducing agent
h) Atom balance, charge balance (in equations)

48) Classify each of the following statements as true or false:

a) Oxidation and reduction occur at the electrodes in a voltaic cell.
b) The sum of the oxidation numbers in a molecular compound is zero, but in an ionic compound that sum may or may not be zero.

c) The oxidation number of oxygen is the same in all of the following: O^{2-}, $HClO_3{}^-$, $S_2O_3{}^{2-}$, NO_2.
d) The oxidation number of an alkali metal is always -1.
e) A substance that gains electrons and increases oxidation number is oxidized.
f) A strong reducing agent has a strong attraction for electrons.
g) The favored side of a redox equilibrium equation is the side with the weaker oxidizer and reducer.

49) One of the properties of acids listed in Section 17.1 is "the ability to react with certain metals and release hydrogen." Why is the property of acids limited to certain metals? Identify two metals that do release hydrogen from an acid and two that do not.

50)* When writing a half-reaction equation that takes place in acidic solution, why is it permissible to use hydrogen ions and water molecules without regard to their source?

51) It is sometimes said that in a redox reaction the oxidizing agent is reduced and the reducing agent is oxidized. Is this statement (a) always correct, (b) never correct, or (c) sometimes correct? If you select (b) or (c), give an example in which the statement is incorrect.

MATCHING SET ANSWERS

Set 1: 5–2–10–7–3–9–12–1–4–8–6–11

Set 2: 2–9–6–8–11–4–10–7–5–1–3

19 Chemical Equilibrium

In studying the liquid-vapor equilibrium in Chapter 14 and the solute-solution equilibrium in Chapter 15, you learned that equilibrium exists when forward and reverse reaction rates are the same in a reversible change. Similarly, chemical equilibrium depends on the rates of chemical reactions. This chapter opens with a brief study of the factors that determine reaction rate. One of them is concentration, as this photograph shows. In the left beaker, zinc is reacting relatively slowly with a dilute solution of sulfuric acid. In the right beaker, where the acid concentration is greater, the reaction is much faster and more vigorous.

19.1 THE CHARACTER OF AN EQUILIBRIUM

A careful review of the liquid–vapor equilibrium in Section 14.3 and the solution equilibrium in Section 15.3 reveals four conditions that are found in every equilibrium:

1) *The change is reversible and can be represented by an equation with a double arrow.* In a reversible change the substances on the left side of the equation produce the substances on the right, and those on the right are changed back into the substances on the left.
2) *The equilibrium system is "closed"—closed in the sense that no substance can enter or leave the immediate vicinity of the equilibrium.* All substances on either side of an equilibrium equation must remain to form the substances on the other side.
3) *The equilibrium is dynamic.* The reversible changes occur continuously, even though there is no appearance of change. By contrast, items in a static equilibrium are stationary, without motion, as an object hanging on a spring.
4) *The things that are* equal *in an equilibrium are the forward rate of change (from left to right in the equation) and the reverse rate of change (from right to left).* Specifically, amounts of substances present in an equilibrium are not necessarily equal.

✔ QUICK CHECK 19.1

Can a solution in a beaker that is open to the atmosphere contain a chemical equilibrium? Explain.

19.2 THE COLLISION THEORY OF CHEMICAL REACTIONS

If two molecules are to react chemically, it is reasonable to expect that they must come into contact with each other. What we see as a chemical reaction is the overall effect of a huge number of individual collisions between reacting particles. This view of chemical change is the **collision theory of chemical reactions.**

Figure 19.1 examines three kinds of molecular collisions for the imaginary reaction $A_2 + B_2 \rightarrow 2\ AB$. If the collision is to produce a reaction, the bond between A atoms in A_2 must be broken; similarly, the bond in the B_2 molecule must be broken. It takes energy to break these bonds. This energy comes from the kinetic energy of the molecules just before they collide. In other words, it must be a violent, bond-breaking collision. This is most apt to occur if the molecules are moving at high speed. Figure 19.1(a) pictures a reaction-producing collision.

Not all collisions result in a reaction; in fact, most of them do not. If the colliding molecules do not have enough kinetic energy to break the bonds, the original molecules simply bounce off each other with the same identity they had before the collision. Sometimes they may have enough kinetic energy, but only "sideswipe" each other and glance off unchanged (Fig. 19.1b). Other sufficiently energetic collisions may have an orientation that pushes atoms in the original molecules closer together rather than pulling them apart (Fig. 19.1c).

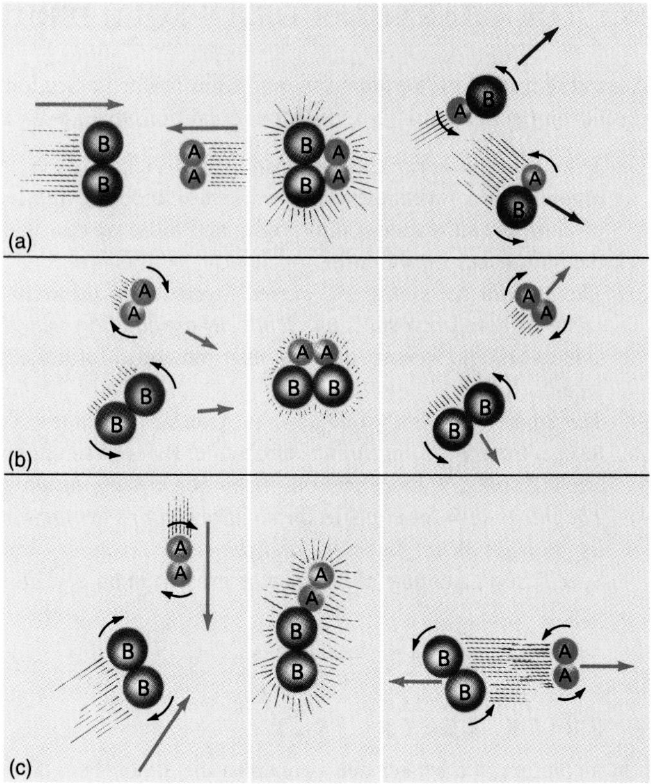

Figure 19.1
Molecular collisions and chemical reactions. (a) Reaction-producing collision between molecules A_2 and B_2, which have sufficient kinetic energy and proper orientation. (b) Glancing collision has proper orientation but not enough kinetic energy. There is no reaction. (c) Collision has enough kinetic energy, but poor orientation. There is no reaction.

To summarize: In order for an individual collision to result in a reaction, the particles must have (1) enough kinetic energy and (2) the proper orientation. The rate of a particular reaction depends on the frequency of effective collisions.

✔ QUICK CHECK 19.2

Describe three kinds of molecular collisions that will not result in a chemical change.

19.3 ENERGY CHANGES DURING A MOLECULAR COLLISION

When a rubber ball is dropped to the floor, it bounces back up. Just before it hits the floor, it has a certain amount of kinetic energy. It also has kinetic energy as it leaves the floor on the rebound. But during the time the ball is in contact with the floor, it slows down, even stops, and then builds velocity in an upward direction. During the period of reduced velocity the kinetic energy, $\frac{1}{2}mv^2$, is reduced, and even reaches zero at the turnaround instant. While the collision is in progress, the initial kinetic energy is changed to potential energy in the partially flattened ball. The potential energy is changed back into kinetic energy as the ball bounces upward.

It is believed that there is a similar conversion of kinetic energy to potential energy during a collision between molecules. This can be shown by a graph of energy versus a

💡 FLASHBACK The change from kinetic energy to potential energy and then back to kinetic energy for a bouncing ball is an example of the Law of Conservation of Energy (Section 2.5).

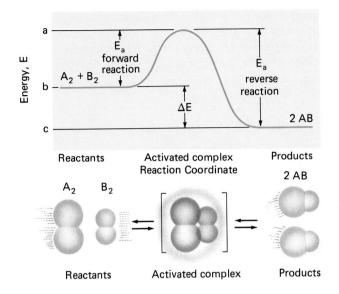

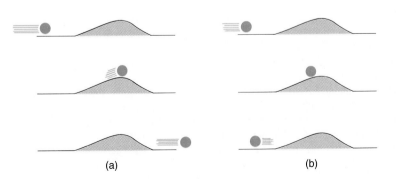

Figure 19.2

Energy-reaction graph for the reaction $A_2 + B_2 \rightarrow 2$ AB. E_a represents the activation energy for the forward and reverse reactions, as indicated. (Energy values a, b, and c are reference points in an end-of-chapter question.)

reaction coordinate that traces the energy of the system before, during, and after the collision (Fig. 19.2). The product energy minus the reactant energy is the ΔE for the reaction.

When two molecules are colliding, they form an **activated complex** that has a high potential energy in the hump of the curve. The increase in potential energy comes from the loss of kinetic energy during the collision. The activated complex is unstable. It quickly separates into two parts. If the collision is "effective" in producing a reaction, the parts will be product molecules; if the collision is ineffective, they will be the original reactant molecules.

The hump in Figure 19.2 is a potential energy barrier that must be surpassed before a collision can be effective. It is like rolling a ball over a hill. If the ball has enough kinetic energy to get to the top, it will roll down the other side (Fig. 19.3a). This corresponds to an effective collision in which the colliding molecules have enough kinetic energy to get over the potential energy barrier. However, if the ball does not have enough kinetic energy to reach the top of the hill, it rolls back to where it came from (Fig. 19.3b). If the colliding particles do not have enough kinetic energy to meet the potential energy requirement, the collision is ineffective.

Figure 19.3

Potential energy barrier. As the ball rolls toward the hill (potential energy barrier), its speed (kinetic energy) determines whether or not it will pass over the hill. Kinetic energy is changed to potential energy as the ball climbs the hill. In (a) it has more than enough kinetic energy to reach the top and roll down the other side. In (b) all the original kinetic energy is changed to potential energy before the ball reaches the top, so it rolls back down the same side.

⚡ FLASHBACK Activation energy is similar to the escape energy in the evaporation of a liquid, as described in Section 14.3. Only those molecules with more than a certain minimum kinetic energy are able to tear away from the bulk of the liquid and change to the vapor state.

The minimum kinetic energy needed to produce an effective collision is called **activation energy.** The difference between the energy at the peak of the Figure 19.2 curve and the reactant energies is the activation energy for the forward reaction. Similarly, the difference between the energy at the peak and the product energies is the activation energy for the reverse reaction.

✔ QUICK CHECK 19.3

In what sense is activation energy a ''barrier''?

19.4 CONDITIONS THAT AFFECT THE RATE OF A CHEMICAL REACTION

The Effect of Temperature on Reaction Rate

Chemical reactions are faster at higher temperatures. This can be seen in the kitchen in several ways. Food is refrigerated to slow down the chemical changes in spoiling. A pressure cooker reduces the time needed to cook some items in boiling water because water boils at a higher temperature under pressure. The opposite effect is seen in open cooking at high altitudes, where reduced atmospheric pressure allows water to boil at lower temperatures. Here cooking is slower, the result of reduced reaction rates.

Figure 19.4 explains the effect of temperature on reaction rates. The curve labeled T_1 gives the kinetic energy distribution among the particles in a sample at one temperature, and T_2 represents the distribution at a higher temperature. E_a is the activation energy, the minimum kinetic energy a particle must have to enter into a reaction-producing collision. It is the same at both temperatures. Only the fraction of the parti-

⚡ FLASHBACK The vapor pressure of a liquid rises as temperature increases. The boiling point of a liquid is the temperature at which its vapor pressure is equal to the pressure above the liquid. A higher pressure over the liquid requires a higher vapor pressure—and therefore a higher temperature—to boil the liquid. See Sections 14.3 and 14.4.

⚡ FLASHBACK In principle, this is the same curve as in Figure 14.11 (Section 14.3), which was used to explain the effect of temperature on the vapor pressure of a liquid. The total area beneath the curve represents the entire sample, so it is the same at all temperatures. At higher temperatures the curve flattens and shifts right. The *average* kinetic energy that corresponds with temperature is found along the horizontal axis.

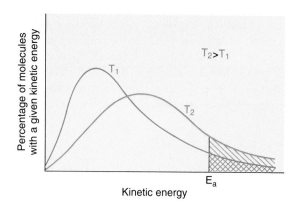

Figure 19.4

Kinetic energy distribution curves at two temperatures. E is the activation energy, the minimum kinetic energy required for a reaction-producing collision. Only the fraction of molecules represented by the total area beneath each curve to the right of the activation energy has enough kinetic energy to react. At lower temperature T_1 that area is crosshatched with red lines, and at higher temperature T_2 that area is crosshatched with blue lines. The larger fraction of molecules that is able to react is responsible for the higher reaction rate at higher temperatures.

cles in the sample represented by the area beneath the curve to the right of E_a is able to react. Compare the areas to the right of E_a. As illustrated, the fraction that is able to react is about $2\frac{1}{2}$ times as much for T_2 as for T_1. The rate of reaction is therefore greater at the higher temperatures.

The Effect of a Catalyst on Reaction Rate

Driving from one city to another over rural roads and through towns takes a certain amount of time. If a superhighway were to be built between the cities you would have available an alternative route that would be much faster. This is what a **catalyst** does. It provides an alternative "route" for reactants to change to products. The activation energy with the catalyst is lower than the activation energy without the catalyst. The result is that a larger fraction of the sample is able to enter into reaction-producing collisions, so the reaction rate increases. This is illustrated in Figure 19.5.

Catalysts exist in several different forms, and the precise function of many catalysts is not clearly understood. Some catalysts are mixed intimately in the reacting chemicals, while others do no more than provide a surface upon which the reaction may occur. In either case the catalyst is not permanently affected by the reaction. In other cases the catalyst participates in the reaction and undergoes a chemical change, but eventually it is regenerated in exactly the same amount as at the start.

The catalytic reaction that appears most often in beginning chemistry laboratories is the making of oxygen by decomposing potassium chlorate. Manganese dioxide is mixed in as a catalyst. A well-known industrial process is the catalytic cracking of

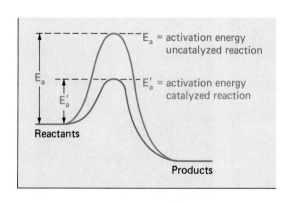

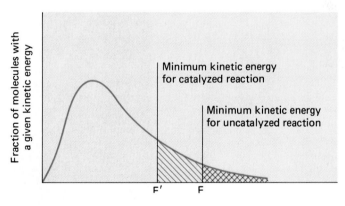

(a)

(b)

Figure 19.5

(a) The effect of a catalyst on activation energy and reaction rate. A catalyst provides a way for a reaction to occur that has a lower activation energy (blue curve) than the same reaction has without a catalyst (red curve). Molecules with a lower kinetic energy are therefore able to pass over the potential energy barrier. (b) The crosshatch area in each color represents the fraction of the total sample with enough energy to engage in reaction-producing collisions. The catalyzed area (blue) is much larger than the uncatalyzed area (red), so the catalyzed reaction rate is faster.

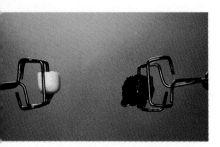

When heated, a sugar cube (sucrose, melting point 185°C) melts but does not burn. A sugar cube rubbed in cigarette ashes burns before it melts. Solid particles in cigarette ashes catalyze the combustion of sugar.

crude oil, in which large hydrocarbon molecules are broken down into simpler and more useful products in the presence of a catalyst. Biological reactions are controlled by catalysts called *enzymes.*

Some substances interfere with a normal reaction path from reactants to products, forcing the reaction to a higher activation energy route that is slower. Such substances are called **negative catalysts,** or **inhibitors.** Inhibitors are used to control the rates of certain industrial reactions. Sometimes negative catalysts can have disastrous results, as when mercury poisoning prevents the normal biological function of enzymes.

The Effect of Concentration on Reaction Rate

If a reaction rate depends on frequency of effective collisions, the influence of concentration is readily predictable. The more particles there are in a given space, the more frequently collisions will occur, and the more rapidly the reaction will take place.

The effect of concentration on reaction rate is easily seen in the rate at which objects burn in air compared to the rate of burning in an atmosphere of pure oxygen. If a burning splint is thrust into pure oxygen the burning is brighter, more vigorous, and much faster. In fact, the typical laboratory test for oxygen is to ignite a splint, blow it out, and then, while there is still a faint glow, place it in oxygen. It immediately bursts back into full flame and burns vigorously (see Fig. 19.6). Charcoal, phosphorus, and other substances behave similarly.

* * *

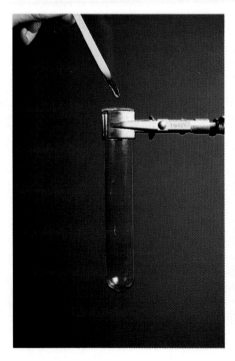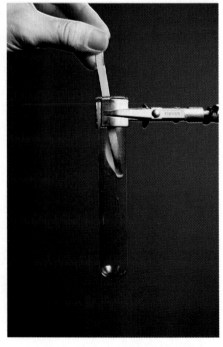

Figure 19.6
The glowing splint test for oxygen.

Three factors that influence the rate of a chemical reaction have been identified: temperature, catalyst, and concentration. In relating these factors to equilibrium considerations, we will consider only concentration and temperature. These variables affect forward and reverse reaction rates differently. A catalyst, on the other hand, has the same effect on both forward and reverse rates and, therefore, does not alter a chemical equilibrium. A catalyst does cause a system to reach equilibrium more quickly.

✔ QUICK CHECK 19.4

a) What happens to a reaction rate as temperature drops? Give two explanations for the change. State which one is more important and explain why.
b) How does a catalyst affect reaction rates?
c) Compare the reaction rates when a certain reactant is at high concentration and at low concentration. Explain the difference.

19.5 THE DEVELOPMENT OF A CHEMICAL EQUILIBRIUM

The role of concentration in chemical equilibrium may be illustrated by tracing the development of an equilibrium. The forward reaction in $A_2 + B_2 \rightleftharpoons 2\,AB$ is assumed to take place by the simple collision of A_2 and B_2 molecules, which separate as two AB molecules (Fig. 19.1). The reverse reaction is exactly the reverse process: two AB molecules collide and separate as one A_2 molecule and one B_2 molecule.

Figure 19.7 is a graph of forward and reverse reaction rates versus time. Initially, at Time 0, pure A_2 and B_2 are introduced to the reaction chamber. At the initial concentrations of A_2 and B_2 the forward reaction begins at a certain rate, F_0. Initially there are no AB molecules present, so the reverse reaction cannot occur. At Time 0 the reverse reaction rate, R_0, is zero. These points are plotted on the graph.

As soon as the reaction begins, A_2 and B_2 are consumed, thereby reducing their concentrations in the reaction vessel. As these reactant concentrations decrease, the forward reaction rate declines. Consequently, at Time 1, the forward reaction rate drops to F_1. During the same interval some AB molecules are produced by the forward reaction, and the concentration of AB becomes greater than zero. Therefore, the reverse reaction begins with the reverse rate rising to R_1 at Time 1.

At Time 1 the forward rate is greater than the reverse rate. Therefore, A_2 and B_2 are consumed by the forward reaction more rapidly than they are produced by the reverse reaction. The net change in the concentrations of A_2 and B_2 is therefore downward, causing a further reduction in the forward rate at Time 2. Conversely, the forward reaction produces AB more rapidly than the reverse reaction uses it. The net change in the concentration of AB is thus an increase. This, in turn, raises the reverse reaction rate at Time 2.

Similar changes occur over successive intervals until the forward and reverse rates eventually become equal. At this point a dynamic equilibrium is established. From this analysis we may state the following generalization:

For any reversible reaction in a closed system, whenever the opposing reactions are occurring at different rates, the faster reaction will gradually become slower, and the slower reaction will become faster. Finally, they become equal, and equilibrium is established.

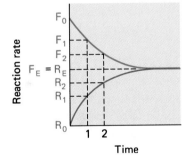

Figure 19.7

Changes in reaction rates during the development of a chemical equilibrium. The forward rate is shown in red and the reverse rate is blue.

💡 FLASHBACK The thought process behind this curve and analysis of a system reaching equilibrium is similar to the analysis of the development of the liquid–vapor equilibrium shown in Figure 14.9. See Section 14.3.

19.6 LE CHATELIER'S PRINCIPLE

In the last section you saw how concentrations and reaction rates jockey with each other until the rates become equal and equilibrium is reached. In this section we start with a system already at equilibrium and see what happens to it when the equilibrium is upset. To "upset" an equilibrium you must somehow make the forward and reverse reactions unequal, at least temporarily. One way to do this is to change the temperature. Another is to change the concentration of at least one substance in the system. A gaseous equilibrium can often be upset simply by changing the volume of the container.

How an equilibrium responds to a disturbance can be predicted from the concentration and temperature effects already considered. The predictions may be summarized in **Le Chatelier's Principle,** which says that **if an equilibrium system is subjected to a change, processes occur that tend to partially counteract the initial change, thereby bringing the system to a new position of equilibrium.**

We will now see how Le Chatelier's Principle explains three different equilibrium changes.

The Concentration Effect

The reaction of hydrogen and iodine to produce hydrogen iodide comes to equilibrium with hydrogen iodide as the favored species. The equal forward and reverse reaction rates are shown by the equal length arrows in the equation

$$H_2(g) + I_2(g) \rightleftharpoons 2\,HI(g)$$

> ☀ FLASHBACK Enclosing the formula of a substance in square brackets represents its concentration in moles per liter. This practice was introduced in Section 17.8 where you used brackets for hydrogen and hydroxide ion concentrations, $[H^+]$ and $[OH^-]$.

The sizes of the formulas represent the relative concentrations of the different species in the reaction. [HI] is greater than $[H_2]$ and $[I_2]$, which are equal.

If more HI is forced into the system, [HI] is increased. This raises the rate of the reverse reaction, in which HI is a reactant. This is indicated by the longer arrow from right to left:

$$H_2(g) + I_2(g) \xleftarrow{\hspace{1em}\rightharpoonup} 2\,HI(g)$$

The rates no longer being equal, the equilibrium is destroyed.

Now the changes described in italics at the end of the last section begin. The unequal reaction rates cause the system to "shift" in the direction of the faster rate—to the left, or in the reverse direction. As a result, H_2 and I_2 are made by the reverse reaction faster than they are used by the forward reaction. Their concentrations increase, so the forward rate increases. Simultaneously, HI is used faster than it is produced, reducing both [HI] and the reverse reaction rate. Eventually, the rates become equal at an intermediate value (note arrow lengths) and a new equilibrium is reached:

$$H_2(g) + I_2(g) \rightleftharpoons 2\,HI(g)$$

The sizes of the formulas indicate that all three concentrations are larger than they were originally, although [HI] has come back from the maximum it reached just after it was added.

The preceding example shows *why* an equilibrium shifts when it is disturbed in terms of concentrations and reaction rates. Le Chatelier's Principle makes it possible to predict the direction of a shift, forward or reverse, without such a detailed analysis. Just remember that the shift is always in the direction that tries to return the substance disturbed to its original condition.

EXAMPLE 19.1

The system $N_2(g) + 3 H_2(g) \rightleftharpoons 2 NH_3(g)$ is at equilibrium. Use Le Chatelier's Principle to predict the direction in which the equilibrium will shift if ammonia is withdrawn from the reaction chamber.

The equilibrium disturbance is clearly stated: Ammonia is withdrawn. In which direction, forward (__) or reverse (__), must the reaction shift to *counteract* the removal of ammonia—that is, to *produce* more ammonia to replace some of what has been taken away?

The shift will be in the *forward* direction.

Ammonia is the product of the forward reaction and therefore will be partially restored to its original concentration by a shift in the forward direction.

EXAMPLE 19.2

Predict the direction, forward or reverse, of a Le Chatelier shift in the equilibrium $CH_4(g) + 2 H_2S(g) \rightleftharpoons 4 H_2(g) + CS_2(g)$ caused by each of the following:

a) increase $[H_2S]$:   b) reduce $[CS_2]$:   c) increase $[H_2]$:

a) Forward, counteracting partially the increase in $[H_2S]$ by consuming some of it.
b) Forward, restoring some of the CS_2 removed.
c) Reverse, consuming some of the added H_2.

The Volume Effect

A change in the volume of an equilibrium system that includes one or more gases changes the concentrations of those gases. Usually—there is one exception, as you will see shortly—there is a Le Chatelier shift that partially offsets the initial change. Both the change and the adjustment involve the pressure exerted by the entire system.

From the ideal gas equation, $PV = nRT$, the pressure exerted by a gas at constant temperature is proportional to the concentration of the gas, n/V, measured in moles per liter:

$$P = RT \times \frac{n}{V} = \text{constant} \times \frac{n}{V}; \; P \propto \frac{n}{V} \qquad (19.1)$$

If volume is reduced, the denominator in n/V becomes smaller and the fraction becomes larger; thus pressure increases. Le Chatelier's Principle calls for a shift that will partially counteract the change—to reduce pressure. What change can reduce the pressure of the system at the new volume? The numerator in Equation 19.1 must be reduced; there must be fewer gaseous molecules in the system. In general:

If a gaseous equilibrium is compressed, the increased pressure will be partially relieved by a shift in the direction of fewer gaseous molecules; if the system is expanded, the reduced pressure will be partially restored by a shift in the direction of more gaseous molecules.

The coefficients of gases in an equation are in the same proportion as the number of gaseous molecules. We use this fact in predicting Le Chatelier shifts caused by volume changes. The following example shows how.

FLASHBACK The relationship between equation coefficients and the number of gaseous molecules comes from Avogadro's Law (Section 13.2). You used the relationship in stoichiometry problems in converting between volumes of different gases measured at the same temperature and pressure (Section 13.5).

EXAMPLE 19.3

Predict the direction of shift resulting from an expansion in the volume of the equilibrium $2 SO_2(g) + O_2(g) \rightleftharpoons 2 SO_3(g)$.

SOLUTION

First, will total pressure increase or decrease as a result of expansion? Boyle's Law indicates that pressure is inversely proportional to volume, so the total pressure will be less at the larger volume. The Le Chatelier shift must make up some of that lost pressure. How? By changing the number of gaseous molecules in the system. Will it take more molecules or fewer to raise pressure? If the pressure is proportional to the concentration, as indicated in Equation 19.1, the number of molecules, n, must increase to raise the pressure.

Now examine the reaction equation. Notice that a forward shift finds *three* reactant molecules, two SO_2 and one O_2, forming *two* SO_2 product molecules. The reverse shift has *two* reactant molecules yielding *three* product molecules. To increase the total number of molecules, then, the reaction must shift in the *reverse* direction: 2 molecules $\rightarrow$ 3 molecules.

EXAMPLE 19.4

The volume occupied by the equilibrium $SiF_4(g) + 2 H_2O(g) \rightleftharpoons SiO_2(s) + 4 HF(g)$ is reduced. Predict the direction of the shift in the position of equilibrium.

Will the shift be in the direction of more (___) gaseous molecules or fewer (___)?

———   ———

Fewer.

If volume is reduced, pressure increases. Increased pressure is counteracted by fewer molecules.

Now predict the direction of shift and justify your prediction by stating the numercial change in molecules from the equation.

———   ———

The shift is in the reverse direction. Molecules change from 4 on the right to 3 on the left. Note the molecule change is 4 to 3, not 5 to 3. Only gaseous molecules are involved in pressure adjustments, so the $SiO_2(s)$ doesn't count.

EXAMPLE 19.5

Returning to the familiar $H_2(g) + I_2(g) \rightleftharpoons 2\,HI(g)$, predict the direction of the shift that will occur because of a volume increase.

Take it all the way, but be careful.

_____   _____

There will be no shift because each side of the equation has two gaseous molecules.

When the number of gaseous molecules is the same, neither the number of molecules nor the pressure can be changed by a shift in the equilibrium. Increasing or reducing volume has no effect on the equilibrium.

The Temperature Effect

A change in temperature of an equilibrium will change both forward and reverse reaction rates, but the rate changes are not equal. The equilibrium is therefore destroyed temporarily. The events that follow are again predictable by Le Chatelier's Principle.

EXAMPLE 19.6

If the temperature of the equilibrium $PCl_5(g) \rightleftharpoons PCl_3(g) + Cl_2(g) + 92.5\,kJ$ is increased, predict the direction of the Le Chatelier shift.

In order to raise the temperature of something, it must be heated. We therefore interpret an increase in temperature as the "addition of heat," and a lowering of temperature as the "removal of heat." In applying Le Chatelier's Principle to a thermochemical equation, we may regard heat in much the same manner as any chemical species in the equation. Accordingly, if heat is *added* to the equilibrium system shown, in which direction must it shift to *use up*, or *consume*, some of the heat that was added? (If chlorine was *added*, in which direction would the equilibrium shift to use up some of the chlorine?) Forward (__); reverse (__).

_____   _____

The equilibrium must shift in the *reverse* direction to use up some of the added heat. An endothermic reaction consumes heat. As the equation is written, the reverse reaction is endothermic.

Including the energy term in the thermochemical equation, rather than showing ΔH separately (see Section 9.6), makes it easier to predict the Le Chatelier effect of a change in temperature. In the equation, energy can be thought of in the same way as a substance is considered with respect to its being "added," "removed," etc.

EXAMPLE 19.7

The thermal decomposition of limestone reaches the following equilibrium:

$$CaCO_3(s) + 176\,kJ \rightleftharpoons CaO(s) + CO_2(g)$$

Predict the direction this equilibrium will shift if the temperature is reduced: Forward (__); reverse (__).

_____   _____

The shift will be in the *reverse* direction. Reduction in temperature is interpreted as the removal of heat. The reaction will respond to replace some of the heat removed—as

the importance of associating any equilibrium constant expression with a specific chemical equation.

c) $K = \dfrac{[HCl]^4[O_2]}{[Cl_2]^2[H_2O]^2}$. The procedure for writing the equilibrium constant expression is the same no matter how complex the equation is.

So far all equilibrium constant expressions have been for equilibria in which all substances are gases. An equilibrium may also have solids, liquids, or dissolved substances as part of its equation. Solute concentrations are variable, and they appear in equilibrium constant expressions just like the concentrations of gases. If a liquid solvent or a solid is part of an equilibrium, however, its concentration is essentially constant. Its concentration is therefore omitted in the equilibrium constant expression. (We can, if you wish, say its constant value is "included" in the value of K.) Remember:

When writing an equilibrium constant expression, use only the concentrations of gases, (g), or dissolved substances, (aq). Do not include solids, (s), or liquids, (ℓ).

You have already seen one situation in which this rule has been applied. In Section 17.8 the equations for the ionization of water and its equilibrium constant were given as

$$H_2O(\ell) \rightleftharpoons H^+(aq) + OH^-(aq) \qquad K_w = [H^+][OH^-]$$

K_w has no denominator because the species on the left is a liquid. This example also shows the common practice of using a subscript to identify a constant for a particular kind of equilibrium. Other subscripts will appear shortly.

EXAMPLE 19.9

Write the equilibrium constant expression for each of the following:

a) $CaCO_3(s) \rightleftharpoons CaO(s) + CO_2(g)$
b) $Li_2CO_3(s) \rightleftharpoons 2\,Li^+(aq) + CO_3^{2-}(aq)$
c) $4\,H_2O(g) + 3\,Fe(s) \rightleftharpoons 4\,H_2(g) + Fe_3O_4(s)$
d) $HF(aq) \rightleftharpoons H^+(aq) + F^-(aq)$
e) $NH_3(aq) + H_2O(l) \rightleftharpoons NH_4^+(aq) + OH^-(aq)$

---

a) $K = [CO_2]$

b) $K = [Li^+]^2[CO_3^{2-}]$

c) $K = \dfrac{[H_2]^4}{[H_2O]^4}$

d) $K = \dfrac{[H^+][F^-]}{[HF]}$

e) $K = \dfrac{[NH_4^+][OH^-]}{[NH_3]}$

19.8 THE SIGNIFICANCE OF THE VALUE OF K

By definition, an equilibrium constant is a ratio—a fraction. The numerical value of an equilibrium constant may be very large, very small, or any place in between. While there is no defined intermediate range, equilibria with constants between 0.01 and 100 (10^{-2} to 10^2) will have appreciable quantities of all species present at equilibrium.

To see what is meant by "very large" or "very small" K values, consider an equilibrium similar to the hydrogen iodide system studied earlier. Substituting chlorine for iodine, the equilibrium equation is $H_2(g) + Cl_2(g) \rightleftharpoons 2\,HCl(g)$. At 25°C

$$K = \frac{[HCl]^2}{[H_2][Cl_2]} = 2.4 \times 10^{33}$$

This is a very large number—ten billion times larger than the number in a mole! The only way an equilibrium constant ratio can become so huge is for the concentration of one or more reacting species to be very close to zero. If the denominator of a ratio is nearly zero, the value of the ratio will be very large. A near zero denominator and large K means the equilibrium is favored overwhelmingly in the forward direction.

By contrast, if the equilibrium constant is very small, it means the concentration of one or more of the species on the right-hand side of the equation is nearly zero. This puts a near zero number in the numerator of K, and the equilibrium is strongly favored in the reverse direction.

SUMMARY

If an equilibrium constant is very large, the forward reaction is favored; if the constant is very small, the reverse reaction is favored. If the constant is neither large nor small, appreciable quantities of all species are present at equilibrium.

19.9 EQUILIBRIUM CALCULATIONS (OPTIONAL)

Equilibrium calculations cover a wide range of problem types. A thorough understanding of these is essential to understanding many chemical phenomena in the laboratory, in industry, and in living organisms. We will sample only a few of these in this section.

Two things should be written before attempting to solve any equilibrium problem. First is the equilibrium equation. Second is the equilibrium constant expression.

Solubility Equilibria

No ionic compounds are *completely* insoluble. It is appropriate, then, that we refer to "low-solubility compounds" rather than insoluble compounds.

The equilibrium equation for the dissolving of a low-solubility compound is very similar to the equation for the ionization of water. (See discussion in Section 19.7, just before Example 19.9.) For silver chloride, for example, the equilibrium and K equations are

$$AgCl(s) \rightleftharpoons Ag^+(aq) + Cl^-(aq) \qquad K_{sp} = [Ag^+][Cl^-] \qquad (19.4)$$

The equilibrium constant for a low-solubility compound is the **solubility product constant, K_{sp}.**

FLASHBACK The solubility table or rules you used in Section 16.6 to predict whether or not a precipitate will form carried a footnote indicating that a compound was considered to be "insoluble" if the molarity of the saturated solution is less than 0.1.

FLASHBACK The formation of a saturated solution of a low-solubility compound is exactly the same as the formation of a saturated solution of a soluble compound. It just happens more "quickly," that is, at lower ion concentrations. The process is described in Section 15.3.

EXAMPLE 19.10 _____

The chloride ion concentration of saturated silver chloride is 1.3×10^{-5} M. Calculate K_{sp} for silver chloride.

SOLUTION

Equation 19.4 is the equilibrium equation and K_{sp} expression. Stoichiometric reasoning plays a large part in solving equilibrium problems. Given information about the concentration of one species in a problem, the concentrations of other substances can be determined. In this case silver chloride has dissolved to the extent of releasing 1.3×10^{-5} moles of Cl^- per liter. According to the reaction equation, the moles of Ag^+ produced are the same as the moles of Cl^-. The concentrations at equilibrium must therefore both be 1.3×10^{-5}. Thus

$$K_{sp} = [Ag^+][Cl^-] = (1.3 \times 10^{-5})^2 = 1.7 \times 10^{-10}$$

EXAMPLE 19.11 _____

The solubility of magnesium fluoride is 73 mg/L. What is the indicated K_{sp}?

SOLUTION

The two equations come first:

$$MgF_2(s) \rightleftharpoons Mg^{2+}(aq) + 2\,F^-(aq) \qquad K_{sp} = [Mg^{2+}][F^-]^2$$

The K_{sp} equation requires concentrations in moles per liter. These can be calculated from the solubility. For the magnesium ion

$$\frac{0.073 \text{ g MgF}_2}{L} \times \frac{1 \text{ mol Mg}^{2+}}{62.3 \text{ g MgF}_2} = 1.17 \times 10^{-3} \text{ mol Mg}^{2+}/L = [Mg^{2+}]$$

(We have carried an extra significant figure at this point to approach more closely the digits that will normally be retained in a calculator.) The dissolving equation shows that twice as many fluoride ions as magnesium ions are released. Therefore

$$[F^-] = 2 \times [Mg^{2+}] = 2 \times 1.17 \times 10^{-3} = 2.34 \times 10^{-3} \text{ M}$$

The two concentrations are now substituted into the K_{sp} expression:

$$K_{sp} = [Mg^{2+}][F^-]^2 = (1.17 \times 10^{-3})(2.34 \times 10^{-3})^2 = 6.4 \times 10^{-9}$$

💡 **FLASHBACK** In Section 7.7 we listed 21 quantities that can be derived from the formula of $Ca(NO_3)_2$. Given any one, it is a one-step dimensional analysis conversion to any other. One unit path was g $Ca(NO_3)_2 \rightarrow$ mol Ca^{2+}. In exactly the same way, one mole of MgF_2 has a mass of 62.3 grams and contains one mole of Mg^{2+} ions, yielding a unit path of g $MgF_2 \rightarrow$ mol Mg^{2+}.

✔ **QUICK CHECK 19.5**

The solubility of cadmium hydroxide is 0.0016 g/L. Calculate the solubility product constant of $Cd(OH)_2$. (Z = 48 for cadmium.)

Solubility product constants have already been determined for most common salts, and their values may be found in handbooks. They are used in all sorts of problems, one of which is the reverse of the last two examples.

EXAMPLE 19.12 _____

Calculate the solubility of zinc carbonate, $ZnCO_3$, in (a) mol/L and (b) g/100 mL if $K_{sp} = 1.4 \times 10^{-11}$.

SOLUTION

$$ZnCO_3(s) \rightleftharpoons Zn^{2+}(aq) + CO_3^{2-}(aq) \qquad K_{sp} = [Zn^{2+}][CO_3^{2-}] = 1.4 \times 10^{-11}$$

If the $ZnCO_3$ is the only source of Zn^{2+} and CO_3^{2-} ions, stoichiometry dictates that they must be present in equal numbers and equal concentration. Furthermore, that concentration is the number of moles of solute that dissolve in 1 L of solution, or the solubility in moles per liter. Let the algebraic symbol s be solubility: s = solubility = $[Zn^{2+}] = [CO_3^{2-}]$. Use the value of K_{sp} to find s:

$$K_{sp} = [Zn^{2+}][CO_3^{2-}] = s^2 = 1.4 \times 10^{-11} \qquad s = 3.7 \times 10^{-6} \text{ mol/L}$$

The solubility can now be used to find the number of grams per 100 milliliters, or the number of grams in 0.100 L:

$$0.100 \text{ L} \times \frac{3.7 \times 10^{-6} \text{ mol}}{\text{L}} \times \frac{125.4 \text{ g } ZnCO_3}{1 \text{ mol } ZnCO_3} = 4.6 \times 10^{-5} \text{ g } ZnCO_3 \text{ per } 100 \text{ mL}$$

To solve $s^2 = 1.4 \times 10^{-11}$ for s, you must take the square root of both sides of the equation. Most calculators have a square root key. The calculator sequence is essentially

Key in 1.4
Press *EE*, *EXP*, or *EEX*
Key in *11*
Press $+/-$ or *CHS*
Press $\sqrt{\ }$

Suppose a soluble carbonate, such as Na_2CO_3, were to be dissolved in the saturated solution of zinc carbonate in Example 19.12. What would happen to the solubility of $ZnCO_3$? $[CO_3^{2-}]$ would increase. It would no longer be the same as $[Zn^{2+}]$ because the two ions would now be coming from different sources. According to Le Chatelier's Principle, the equilibrium should shift in the direction that would reduce $[CO_3^{2-}]$, which is in the reverse direction. In other words, less $ZnCO_3$ would dissolve; the solubility would be reduced.

The reduction in the solubility of a compound caused by one of its ions being present from a different source is an example of the **common ion effect.**

EXAMPLE 19.13

Calculate the solubility of $ZnCO_3$ in 0.010 M Na_2CO_3. Answer in moles per liter.

SOLUTION

The product of the concentrations of the two ions is the same, 1.4×10^{-11}, whether the concentrations are the same or different. Assuming the Na_2CO_3 is completely ionized, $[CO_3^{2-}]$ is 0.010 M. The zinc ion concentration gives the moles of $ZnCO_3$ dissolved per liter. Solving the K_{sp} expression for $[Zn^{2+}]$, we obtain

$$[Zn^{2+}] = \frac{K_{sp}}{[CO_3^{2-}]} = \frac{1.4 \times 10^{-11}}{0.010} = 1.4 \times 10^{-9} \text{ M}$$

As predicted by Le Chatelier's Principle, zinc carbonate is less soluble in a solution of sodium carbonate than in water, 3.7×10^{-6} M, from Example 19.12.

✔ **QUICK CHECK 19.6**

Find the solubility of AgBr (a) in water and (b) in 0.25 M NaBr. $K_{sp} = 5.0 \times 10^{-13}$ for AgBr. Answer in g/100 mL.

Ionization Equilibria

In Section 16.3 you learned that weak acids ionize only slightly when dissolved in water. If HA is the formula of a weak acid, its ionization equation and equilibrium constant expression are

$$HA(aq) \rightleftharpoons H^+(aq) + A^-(aq) \qquad K_a = \frac{[H^+][A^-]}{[HA]} \qquad (19.5)$$

The equilibrium constant is the **acid constant, K_a.** The undissociated molecule is the major species in the solution inventory, and the H^+ ion and the conjugate base of the acid (A^-) are the minor species.

The ionization of a weak acid is usually so small that it is negligible compared to the initial concentration of the acid. For example, if a 0.12 M acid is 3.0% ionized, the amount ionized is $0.030 \times 0.12 = 0.0036$ mol/L. When this is subtracted from the initial concentration and rounded off according to the rules of significant figures, the result is

$$0.12 - 0.0036 = 0.1164 = 0.12$$

In more advanced courses you will learn how to determine if the ionization has a negligible effect on the initial concentration. In this book we assume all ionizations are negligible *when subtracted from the initial concentration.* The ion concentrations by themselves, however, are not negligible.

If the ionization of HA is the only source of H^+ and A^- in the equilibrium of Equation 19.5, then $[H^+] = [A^-]$. (This is the same as $[Ag^+]$ being equal to $[Cl^-]$ in Example 19.1.) This makes it possible to calculate the percentage ionization and K_a from the pH of a weak acid whose molarity has been determined by titration.

EXAMPLE 19.14 _____

A 0.13 M solution of an unknown acid has a pH of 3.12. Calculate the percent ionization and K_a.

SOLUTION

FLASHBACK Equation 17.13:
$[H^+] = 10^{-pH}$

From Equation 17.13, $[H^+] = 10^{-3.12} = 7.6 \times 10^{-4}$ M. This is also $[A^-]$. This means that only 0.00076 of the 0.13 mol of acid in 1 L of solution is ionized. Expressed as percent

$$\frac{7.6 \times 10^{-4}}{0.13} \times 100 = 0.58\% \text{ ionized}$$

Substituting values of $[H^+]$ and $[A^-]$ into Equation 19.5 gives K_a:

$$K_a = \frac{[H^+][A^-]}{[HA]} = \frac{(7.6 \times 10^{-4})^2}{0.13} = 4.4 \times 10^{-6}$$

✔ QUICK CHECK 19.7

Find the acid constant of an unknown acid if the pH of a 0.22 M solution is 2.61.

As the K_{sp} values of low-solubility compounds are listed in handbooks, so are the K_a values of most weak acids. They can be used to determine $[H^+]$ and the pH of solutions of those acids. If the ionization of the acid is the only source of H^+ and A^-, multiplying both sides of Equation 19.5 by [HA] and substituting $[H^+]$ for its equal $[A^-]$ yields

$$[H^+][A^-] = [H^+]^2 = K_a[HA] \tag{19.6}$$

$$[H^+] = \sqrt{K_a[HA]} \tag{19.7}$$

EXAMPLE 19.15

What is the pH of a 0.20 molar solution of the acid in Example 19.14, for which $K_a = 4.4 \times 10^{-6}$?

SOLUTION

$[H^+]$ may be found by direct substitution into Equation 19.7. Its negative logarithm is the pH of the solution:

$$[H^+] = \sqrt{K_a[HA]} = \sqrt{(4.4 \times 10^{-6})(0.20)} = 9.4 \times 10^{-4} \qquad pH = 3.03$$

✔ QUICK CHECK 19.8

$K_a = 1.8 \times 10^{-5}$ for acetic acid, $HC_2H_3O_2$. Calculate the pH of 0.19 M $HC_2H_3O_2$.

Just as solubility equilibria can be forced in the reverse direction by the addition of a common ion, so can a weak acid equilibrium by adding a soluble salt of the acid. If A^- is added to the equilibrium in Equation 19.5, $[A^-]$ is no longer equal to $[H^+]$. To find the pH of such a solution, we solve Equation 19.5 for $[H^+]$:

$$[H^+] = K_a \times \frac{[HA]}{[A^-]} \tag{19.8}$$

Neither the [HA] nor the $[A^-]$ is changed significantly by the ionization, which is even smaller than its ionization in pure water. This is as predicted by the Le Chatelier effect of the common ion.

EXAMPLE 19.16

Find the pH of a 0.20 M solution of the acid in Examples 19.14 and 19.15 ($K_a = 4.4 \times 10^{-6}$) if the solution is also 0.15 M in A^-.

SOLUTION

Direct substitution into Equation 19.8 gives $[H^+]$. pH follows:

$$[H^+] = K_a \times \frac{[HA]}{[A^-]} = 4.4 \times 10^{-6} \times \frac{0.20}{0.15} = 5.9 \times 10^{-6} \, M$$

$$pH = -\log(5.9 \times 10^{-6}) = 5.23$$

The solution in Example 19.16 is a **buffer solution,** or, more simply, a **buffer.** A buffer is a solution that resists changes in pH because it contains relatively high concentrations of both a weak acid and a weak base. The acid is able to consume any OH^- that may be added, and the base can absorb H^+, both without significant change in either [HA] or [A$^-$]. For example, if 0.001 mol of HCl was dissolved in one liter of water, the pH would be 3. If the same amount of HCl was added to a liter of the buffer in Example 19.16, it would react with 0.001 mol of the A$^-$ present. The new concentration of A$^-$ would be 0.15 − 0.001 = 0.15, unchanged according to the rules of significant figures. An additional 0.001 mol of HA would be formed in the reaction, but that added to 0.20 is still 0.20. In other words, the [HA]/[A$^-$] ratio would be unchanged, so the [H$^+$] and pH would also be unchanged.

This all suggests that a buffer can be tailor-made for any pH simply by adjusting the [HA]/[A$^-$] ratio to the proper value. Solving Equations 19.5 for that ratio gives

$$\frac{[HA]}{[A^-]} = \frac{[H^+]}{K_a} \tag{19.9}$$

EXAMPLE 19.17

What [HA]/[A$^-$] ratio is necessary to produce a buffer with a pH of 5.00 if $K_a = 4.4 \times 10^{-6}$?

SOLUTION

If pH is 5.00, [H$^+$] = $10^{-5.00}$ M. Substituting into Equation 19.9, we obtain

$$\frac{[HA]}{[A^-]} = \frac{[H^+]}{K_a} = \frac{10^{-5.00}}{4.4 \times 10^{-6}} = 2.3$$

✔ QUICK CHECK 19.9

a) At what pH is a solution buffered if it is 0.37 molar in sodium hydrogen oxalate, NaHC$_2$O$_4$, and 0.28 molar in sodium oxalate, Na$_2$C$_2$O$_4$? (Do not be concerned about the formulas. The acid is the hydrogen oxalate ion, HC$_2$O$_4{}^-$, and its conjugate base is the oxalate ion, C$_2$O$_4{}^{2-}$.) $K_a = 6.4 \times 10^{-5}$ for HC$_2$O$_4{}^-$.

b) Calculate the ratio of molarities of benzoic acid to benzoate ion, [HC$_7$H$_5$O$_2$]/[C$_7$H$_5$O$_2{}^-$], that is needed to make a buffer whose pH is 4.95. $K_a = 6.5 \times 10^{-5}$ for benzoic acid.

Gaseous Equilibria

All the equilibria considered thus far in this section have been in aqueous solution. When an equilibrium involves only gases, the calculation principles and stoichiometric reasoning are the same as in solution equilibria. However, the changes in starting concentrations are not negligible. It often helps to trace these changes by assembling them into a table. The columns are headed by the species in the equilibrium just as they appear in the reaction equation. The three lines give the initial concentration of each substance, the change in concentration as the system reaches equilibrium, and the equilibrium concentration.

EXAMPLE 19.18

0.052 mole of NO and 0.054 mole of O$_2$ are placed in a 1.00-L vessel at a certain temperature. They react until equilibrium is reached according to the equation 2 NO(g) + O$_2$(g) ⇌ 2 NO$_2$(g). At equilibrium, [NO$_2$] = 0.028 M. Calculate K.

SOLUTION

From the equilibrium equation, $K = \dfrac{[NO_2]^2}{[NO]^2[O_2]}$

We begin by setting up the table and inserting all the given data.

	2 NO(g) +	O₂(g) ⇌	2 NO₂(g)
mol/L at start	0.052	0.054	0.000
mol/L change, + or −			
mol/L at equilibrium			0.028

To get any entry in the change (middle) line, find a species whose initial and final concentrations are known. In this case [NO₂] starts at 0.000 M and reaches 0.028 M. Its change is therefore +0.028 M. Inserting that value gives

FLASHBACK These three-line tables are like the three-line tables that were used for limiting reactant problems at the beginning of Section 9.4. In the earlier application, the quantity in the left column was moles. In its use here, the quantity is concentration in moles per liter.

	2 NO(g) +	O₂(g) ⇌	2 NO₂(g)
mol/L at start	0.052	0.054	0.000
mol/L change, + or −			+0.028
mol/L at equilibrium			0.028

We use stoichiometry to find the changes in the other two species. The coefficients in the equation show that the reaction that produces 0.028 mol NO₂ uses 0.028 mol NO (both coefficients are 2) and half as much O₂ (coefficient 1), or 0.014 mol O₂. Placing these in the table, we obtain

	2 NO(g) +	O₂(g) ⇌	2 NO₂(g)
mol/L at start	0.052	0.054	0.000
mol/L change, + or −	−0.028	−0.014	+0.028
mol/L at equilibrium			0.028

The final concentrations of the reactants are found by subtracting the amounts used from the starting concentrations–or adding them algebraically, as they appear in the table.

	2 NO(g) +	O₂(g) ⇌	2 NO₂(g)
mol/L at start	0.052	0.054	0.000
mol/L change, + or −	−0.028	−0.014	+0.028
mol/L at equilibrium	0.024	0.040	0.028

The equilibrium constant, K, may now be calculated by substituting the equilibrium concentrations into the equilibrium constant expression:

$$K = \frac{[NO_2]^2}{[NO]^2[O_2]} = \frac{0.028^2}{(0.024)^2(0.040)} = 34$$

✔ **QUICK CHECK 19.10**

HI is introduced to a reaction vessel at a concentration of 0.36 mol/L. It decomposes according to the equation $2\,HI(g) \rightleftharpoons H_2(g) + I_2(g)$ until equilibrium is reached when $[I_2] = 0.053$ mol/L. Calculate K.

CHAPTER 19 KEY WORDS AND MATCHING SETS

Set 1

_____ Term to indicate that no substance can enter or leave an equilibrium system.

_____ Term to describe the concept that reactions occur when particles come into contact with each other.

_____ What happens to kinetic energy when reacting particles are in contact with each other.

_____ What must be overcome by particle kinetic energies in order for an effective collision to occur.

_____ How an increase in temperature affects a reaction rate.

_____ Substance that increases a reaction rate without being permanently changed.

_____ Direction an equilibrium shifts when a reactant concentration is reduced.

_____ Direction an equilibrium shifts when volume is increased if the equation has three gaseous molecules and two solid molecules on the left and four gaseous molecules and one liquid molecule on the right.

1. Catalyst
2. Forward
3. Collision
4. Increases
5. Decreases
6. Reverse
7. Closed
8. Activation energy barrier

Set 2

_____ How a reduction in reactant concentration affects a reaction rate.

_____ Term to describe a particle collision that actually yields a reaction.

_____ Favored direction of an equilibrium that has a very small equilibrium constant.

_____ Term to indicate that reactions in an equilibrium are ongoing rather than completed.

_____ Substance that slows the rate of a reaction.

_____ Minimum particle kinetic energies required before a reaction can occur.

_____ Direction an equilibrium shifts when heated if the reaction is endothermic in the forward direction.

_____ What happens to potential energy between reacting particles during a collision.

1. Dynamic
2. Effective
3. Increases
4. Activation energy
5. Decreases
6. Inhibitor
7. Forward
8. Reverse

Set 3

NOTE: Four items in this set are from optional Section 19.9.

———— Symbol in a chemical equation that indicates that a reaction is reversible.

———— When a system at equilibrium is disturbed, it responds in such a way as to counteract partially the disturbance.

———— Describes what happens when more of a species already present in an aqueous equilibrium is added to the equilibrium.

———— Particle formed temporarily during a collision.

———— What are equal in a system at equilibrium.

———— Describes a solution that contains appreciable concentrations of both a weak acid and a weak base.

———— Name of an equilibrium constant for a very slightly soluble salt.

———— Ratio of product concentrations to reactant concentrations, each raised to an appropriate power, when a system is at equilibrium.

———— Class of solute whose equilibrium constant is K_a.

1. Double arrow

2. Common ion effect

3. Forward and reverse reaction rates

4. Equilibrium constant

5. Activated complex

6. Buffer

7. Solubility product constant

8. Le Chatelier's Principle

9. Weak acid

CHAPTER 19 QUESTIONS AND PROBLEMS

SECTION 19.1

1) In what way is a chemical equilibrium *dynamic?* Can you give an example—not necessarily chemical—of an equilibrium that is not dynamic? What word corresponding to *dynamic,* but having an opposite meaning, would be used to describe such an equilibrium?

2) What things are equal in an equilibrium? What does this mean in terms of the amount of each substance present in the equilibrium over a period of time?

3) Undissolved table salt is in contact with a saturated salt solution in (a) a sealed container and (b) an open beaker. Which system, if either, can reach equilibrium? Explain your answer.

4) What is meant by saying that an equilibrium is confined to a "closed system"? Can a closed system be open to the atmosphere? Explain.

5) A garden in a park has a fountain that discharges water into a pond. The pond overflows into a stream that cascades to the bottom of a small mound. The water is then pumped up into the fountain. Is the system a dynamic equilibrium? Explain your answer.

6) A river flows into a lake formed by a dam. Water flows through the dam's spillways as the river continues downstream. The water level of the lake is constant. Is this system a dynamic equilibrium? Explain your answer.

7) Explain why a molecular collision can be sufficiently energetic to cause a reaction, yet no reaction occurs as a result of that collision.

8) According to the collision theory of chemical reactions, what two conditions must be satisfied if a molecular collision is to result in a reaction?

SECTION 9.2

9) Is ΔE positive or negative for the forward reaction described in Figure 19.2? For a reaction in which heat is the only form of reaction energy, ΔE is equal to ΔH as ΔH is described in Section 9.6, Thermochemical Equations. Is the forward reaction of Figure 19.2 exothermic or endothermic? Use the letters a, b, and c on the vertical axis of Figure 19.2 to state algebraically the ΔE and activation energy of the reaction.

10) Sketch an energy-reaction graph for which the answers to the first two questions in Question 9 are the opposite of what they are in Question 9. Include points a, b, and c on the vertical axis and use them for the algebraic expressions of ΔE and the activation energy.

SECTION 19.3

11) Assume the reaction described by Figure 19.2 to be reversible. Compare the signs of the activation energies for the forward and reverse reactions. Which is positive and which is negative—or are they the same? If they are the same, are they positive or negative?

12) Assume the reaction described by Figure 19.2 is reversible. Compare the magnitude of the activation energies for the forward and reverse reactions. Which is greater, or are they equal?

13) What is an "activated complex"? Why can we not list the physical properties of a species that is an activated complex?

14) What is activation energy? How is activation energy related to the rate of a chemical reaction? Which has a higher reaction rate, a reaction with a large activation energy or a reaction with a small activation energy?

SECTION 19.4

15) How does a temperature change affect the rate of a reaction? Explain why the rate changes with both an increase and a decrease in temperature.

16) "At a given temperature, only a small fraction of the molecules in a sample has sufficient kinetic energy to engage in a chemical reaction." What is the meaning of that statement?

———

17) What is a catalyst? How does a catalyst affect reaction rates?

18) Compare the equilibria reached by two equilibrium systems that are identical in every way except that a catalyst is present in one and not in the other. State precisely any difference there may be in the equilibria and the processes by which they were reached.

———

19) For the hypothetical reaction A + B → C, what will happen to the rate of reaction if the concentration of A is decreased? What will happen if the concentration of B is increased? Explain why in both cases.

20) For the hypothetical reaction A + B → C, what will happen to the reaction rate if the concentration of A is decreased *and* the concentration of B is increased? Explain.

SECTION 19.5

If nitrogen and hydrogen are brought together at the proper temperature and pressure, they react until they reach equilibrium: $N_2(g) + 3 H_2(g) \rightleftharpoons 2 NH_3(g)$. Answer the next four questions with regard to the establishment of that equilibrium.

21) On a single set of coordinate axes, sketch graphs of the forward reaction rate versus time and the reverse reaction rate versus time from the moment the reactants are mixed to a point beyond the establishment of equilibrium.

22) What happens to the concentrations of the three species between the start of the reaction and the time equilibrium is reached?

———

23) When will the reverse reaction rate be at a maximum: at the start of the reaction, after equilibrium has been reached, or at some point in between?

24) When will the forward reaction rate be at a minimum: at the start of the reaction, after equilibrium has been reached, or at some point in between?

SECTION 19.6

25) If the system $2 SO_2(g) + O_2(g) \rightleftharpoons 2 SO_3(g)$ is at equilibrium and the concentration of O_2 is increased, predict the direction in which the equilibrium will shift. Justify or explain your prediction.

26) Predict the direction the equilibrium

$$SO_2(g) + NO_2(g) \rightleftharpoons SO_3(g) + NO(g)$$

will shift if the concentration of NO is reduced. Explain or justify your prediction.

———

27) If additional ammonia is pumped into the equilibrium system

$$4 NH_3(g) + 5 O_2(g) \rightleftharpoons 4 NO(g) + 6 H_2O(g)$$

in which direction will the reaction shift? Justify your answer.

28) Some $COCl_2$ is removed from the equilibrium $COCl_2(g) \rightleftharpoons CO(g) + Cl_2(g)$. In which direction will the equilibrium shift? Why?

———

29) Predict the direction of shift for the equilibrium $Cu(NH_3)_4^{2+}(aq) \rightleftharpoons Cu^{2+}(aq) + 4 NH_3(aq)$ if the concentration of Cu^{2+} is increased.

30) In which direction will the equilibrium

$$HC_2H_3O_2(aq) \rightleftharpoons H^+(aq) + C_2H_3O_2^-(aq)$$

shift if the concentration of H^+ is increased?

———

31) The volume of a container holding the equilibrium $4 H_2(g) + CS_2(g) \rightleftharpoons CH_4(g) + 2 H_2S(g)$ is reduced. Predict and explain the direction of the Le Chatelier shift.

32) Which direction will be favored if you enlarge the volume of the equilibrium

$$N_2O_3(g) \rightleftharpoons N_2O(g) + O_2(g)?$$

Explain.

———

33) In which direction will

$$CO(g) + H_2O(g) \rightleftharpoons CO_2(g) + H_2(g)$$

shift as a result of an increase in volume? Explain.

34) What Le Chatelier shift will occur to the equilibrium $C(s) + H_2O(g) \rightleftharpoons CO(g) + H_2(g)$ if its volume is reduced? Explain.

———

35) Which direction of the equilibrium

$$2 NO_2(g) \rightleftharpoons N_2O_4(g) + 59.0 kJ$$

will be favored if the system is cooled? Explain.

36) The equilibrium

$$CO_2(g) + H_2(g) + 41.4 \text{ kJ} \rightleftharpoons CO(g) + H_2O(g)$$

is heated. Predict the direction of the Le Chatelier shift. Explain.

---

37) If you wish to reduce the yield of SO_3 in the equilibrium

$$SO_2(g) + NO_2(g) \rightleftharpoons SO_3(g) + NO(g) + 41.8 \text{ kJ}$$

would you raise or lower the operating temperature? Explain.

38) If you wished to increase the relative amount of HI in the equilibrium $2 \text{ HI}(g) \rightleftharpoons H_2(g) + I_2(g) + 25.9 \text{ kJ}$, would you heat or cool the system? Explain.

---

39)* The solubility of calcium hydroxide is low; it reaches about 0.024 M at saturation. In acid solutions with many H^+ ions present, calcium hydroxide is quite soluble. Explain this fact in terms of Le Chatelier's Principle. (*Hint:* Recall what you know of reactions in which a molecular product is formed.)

40) Consider the equilibrium

$$4 \text{ NO}(g) + 6 \text{ H}_2O(g) + 905 \text{ kJ} \rightleftharpoons 4 \text{ NH}_3(g) + 5 \text{ O}_2(g)$$

Determine the direction of the Le Chatelier shift, forward or reverse, for each of the following actions: (a) add ammonia; (b) reduce temperature; (c) reduce volume; (d) add NO(g).

SECTION 19.7

Questions 41 to 50: For each equilibrium equation shown, write the equilibrium constant expression.

41) $CO(g) + H_2O(g) \rightleftharpoons CO_2(g) + H_2(g)$

42) $2 \text{ SO}_3(g) \rightleftharpoons 2 \text{ SO}_2(g) + O_2(g)$

---

43) $Zn_3(PO_4)_2(s) \rightleftharpoons 3 \text{ Zn}^{2+}(aq) + 2 \text{ PO}_4^{3-}(aq)$

44) $Cd(OH)_2(s) \rightleftharpoons Cd^{2+}(aq) + 2 \text{ OH}^-(aq)$

---

45) $Cu(NH_3)_4^{2+}(aq) \rightleftharpoons Cu^{2+}(aq) + 4 \text{ NH}_3(aq)$

46) $Ag(CN)_2^-(aq) \rightleftharpoons Ag^+(aq) + 2 \text{ CN}^-(aq)$

---

47) $CO(g) + H_2(g) \rightleftharpoons C(s) + H_2O(g)$

48) $2 \text{ H}_2S(g) + CH_4(g) \rightleftharpoons CS_2(g) + 4 \text{ H}_2(g)$

---

49) $HNO_2(aq) + H_2O(\ell) \rightleftharpoons H_3O^+(aq) + NO_2^-(aq)$

50) $HNO_2(aq) \rightleftharpoons H^+(aq) + NO_2^-(aq)$

---

51) "The equilibrium constant expression for a given reaction depends on how the equilibrium equation is written." Explain the meaning of that statement. You may, if you wish, use the equilibrium equation $2 \text{ NO}(g) + O_2(g) \rightleftharpoons 2 \text{ NO}_2(g)$ to illustrate your explanation.

52) Express the equilibrium in Question 48 in at least two different ways, that is, with two different equations. Write the equilibrium constant expressions for each equation. Are the constants numerically equal? Cite some evidence to support your answer.

SECTION 19.8

53) If silver nitrate solution is added to a sodium cyanide solution the following equilibrium will be reached: $Ag^+(aq) + 2 \text{ CN}^-(aq) \rightleftharpoons Ag(CN)_2^-(aq)$. For this equilibrium, $K = 5.6 \times 10^{18}$. In which direction is the equilibrium favored? Justify your answer.

54) $K = 2.3 \times 10^{-8}$ for $HCO_3^-(aq) + HOH(\ell) \rightleftharpoons H_2CO_3(aq) + OH^-(aq)$. In which direction is the reaction favored at equilibrium? Explain your answer.

---

55) If an equilibrium has a very small equilibrium constant, in which direction, forward or reverse, is it favored? Explain.

56) Hydrofluoric acid is a soluble weak acid. In water it ionizes partially and reaches equilibrium. Write the equilibrium equation for the ionization and the equilibrium constant expression. Will the value of the equilibrium constant be large or small? Justify your answer.

Questions 57 and 58: In Chapter 16 you learned how to write solution inventories and net ionic equations, based on the solubility of ionic compounds, the strength of acids, and the stability of certain ion combinations. Use these ideas to predict the favored direction of each equilibrium below. In each case, state whether you expect the equilibrium constant to be large or small.

57)* (a) $H_2SO_3(aq) \rightleftharpoons H_2O(\ell) + SO_2(aq)$

(b) $H^+(aq) + C_2H_3O_2^-(aq) \rightleftharpoons HC_2H_3O_2(aq)$

58) (a) $HCl(aq) \rightleftharpoons H^+(aq) + Cl^-(aq)$

(b) $BaSO_4(s) \rightleftharpoons Ba^{2+}(aq) + SO_4^{2-}(aq)$

SECTION 19.9

59) $Co(OH)_2$ dissolves in water to the extent of 3.7×10^{-6} mol/L. Find its K_{sp}.

60) The solubility of cadmium sulfide, CdS, is 8.8×10^{-14} mol/L. Calculate its K_{sp}.

---

61) 250 mL of water will dissolve 8.7 mg of silver carbonate. What is the K_{sp} of Ag_2CO_3?

62) 1.5×10^{-3} g CuBr dissolve in 150 mL of water to produce a saturated solution. Calculate the K_{sp} of CuBr.

---

63) Find the moles per liter and grams per 100 mL solubility of silver iodate, $AgIO_3$, if its K_{sp} is 2.0×10^{-8}.

64) $K_{sp} = 4.8 \times 10^{-9}$ for calcium carbonate. Calculate its solubility in (a) moles per liter and (b) grams per 100 mL.

---

65)* Find the solubility in mol/L of $Mn(OH)_2$ if its K_{sp} is 1.0×10^{-13}.

66)* For BaF_2, $K_{sp} = 1.7 \times 10^{-6}$. Calculate its solubility in mol/L.

---

67) What is the mol/L solubility of calcium oxalate in 0.22 M $Na_2C_2O_4$ if $K_{sp} = 2.4 \times 10^{-9}$ for CaC_2O_4?

68)* Calculate the number of grams of calcium oxalate that will dissolve in 750 mL 0.15 M $Ca(NO_3)_2$. $K_{sp} = 2.4 \times 10^{-9}$ for CaC_2O_4.

---

69) The pH of 0.22 M $HC_4H_5O_3$ (acetoacetic acid) is 2.12. Find its K_a and percent ionization.

70) pH = 1.93 for 1.0 M $HC_3H_5O_3$ (lactic acid). Calculate the K_a of lactic acid and percent ionization in a 1.0 molar solution.

---

71) Find the pH of 0.35 M $HC_2H_3O_2$. $K_a = 1.8 \times 10^{-5}$.

72) Calculate the pH of 0.12 M HNO_2. $K_a = 4.6 \times 10^{-4}$.

---

73)* 24.0 g of sodium acetate, $NaC_2H_3O_2$, are dissolved in 5.00×10^2 mL of 0.12 M $HC_2H_3O_2$ ($K_a = 1.8 \times 10^{-5}$). Calculate the pH of the solution.

74) What is the pH of 0.30 M $NaNO_2$ in 0.72 M HNO_2? $K_a = 4.6 \times 10^{-4}$ for HNO_2.

---

75) Find the ratio $[HC_2H_3O_2]/[C_2H_3O_2^-]$ that will yield a buffer in which pH = 4.18. $K_a = 1.8 \times 10^{-5}$.

76) What concentration ratio of benzoic acid to benzoate ion, $[HC_7H_5O_2]/[C_7H_5O_2^-]$, will produce a buffer having a pH of 4.65 if $K_a = 6.5 \times 10^{-5}$ for benzoic acid?

---

77)* 0.351 mol CO and 1.340 mol Cl_2 are introduced into a reaction chamber having a volume of 3.00 L. When equilibrium is reached according to the equation $CO(g) + Cl_2(g) \rightleftharpoons COCl_2(g)$, there are 1.050 mol Cl_2 in the chamber. Calculate K.

78) 0.10 mol PCl_3 and 0.087 mol Cl_2 are placed in a 1.5-L reaction vessel. They react to establish the equilib-

rium $PCl_3(g) + Cl_2(g) \rightleftharpoons PCl_5$. $[Cl_2] = 0.031$ M when equilibrium is reached. Calculate K at the temperature of the system.

GENERAL QUESTIONS

79) Distinguish precisely and in scientific terms the differences between items in each of the following groups.
a) Reaction, reversible reaction.
b) Open system, closed system.
c) Dynamic equilibrium, static equilibrium.
d) Activated complex, activation energy.
e) Catalyzed reaction, uncatalyzed reaction.
f) Catalyst, inhibitor.
g) Buffered solution, unbuffered solution.

80) Classify each of the following statements as true or false.
a) Some equilibria depend on a steady supply of a reactant in order to maintain the equilibrium.
b) Both forward and reverse reactions continue after equilibrium is reached.
c) Every time reaction molecules collide there is a reaction.
d) Potential energy during a collision is greater than potential energy before or after the collision.
e) The properties of an activated complex are between those of the reactants and the products.
f) Activation energy is positive for both the forward and reverse reactions.
g) Kinetic energy is changed to potential energy during a collision.
h) An increase in temperature speeds the forward reaction but slows the reverse reaction.
i) A catalyst changes the steps by which a reaction is completed.
j) An increase in concentration of a substance on the right side of an equation speeds the reverse reaction rate.
k) An increase in the concentration of a substance in an equilibrium increases the reaction rate in which the substance is a product.
l) Reducing the volume of a gaseous equilibrium shifts the equilibrium in the direction of fewer gaseous molecules.
m) Raising temperature results in a shift in the forward direction of an endothermic equilibrium.
n) The value of an equilibrium constant depends on temperature.
o) A large K indicates an equilibrium is favored in the reverse direction.

81) At Time 1 two molecules are about to collide. At Time 2 they are in the process of colliding, and their form is that of the activated complex. Compare the sum of their kinetic energies at Time 1 with the kinetic energy of the activated complex at Time 2. Explain your conclusions.

82) List three things you might do to increase the rate of the reverse reaction for which Figure 19.2 is the energy-reaction graph.

83)* The Haber process for making ammonia by direct combination of the elements is described by the equation $N_2(g) + 3 H_2(g) \rightarrow 2 NH_3(g) + 92$ kJ. If the purpose of the manufacturer is to make the greatest amount of ammonia in the least time, is he most apt to conduct the reaction at (a) high pressure or low pressure, (b) high temperature or low temperature? Explain your choice in each case.

84) Under proper conditions the reaction in Question 83 will reach equilibrium. Is the manufacturer apt to conduct the reaction under those conditions, that is, at equilibrium? Why or why not?

85)* The reaction in Question 83 has a yield of about 98% at 200°C and 1000 atm. Commercially, the reaction is performed at about 500°C and 350 atm, where the yield is only about 30%. Suggest reasons why operation at the lower yield is economically more favorable.

86)* The solubility of calcium hydroxide is low enough to be listed as "insoluble" in Table 16.2, but it is much more soluble than most of the other salts that are similarly classified. Its K_{sp} is 5.5×10^{-6}.

a) Write the equation for the equilibrium to which the K_{sp} is related.
b) If you had such an equilibrium, name at least two chemicals or general classes of chemicals that might be added to (1) reduce the solubility of $Ca(OH)_2$, (2) increase its solubility. Justify your choices.
c) Without adding a calcium or hydroxide ion, name a chemical or class of chemicals that would, if added, (1) increase $[OH^-]$, (2) reduce $[OH^-]$. Justify your choices.

87)* The table below lists several "disturbances" that may or may not produce a Le Chatelier shift in the equilibrium.

$$4 NH_3(g) + 7 O_2(g) \rightleftharpoons 6 H_2O(g) + 4 NO_2(g) + \text{energy}$$

If the disturbance is an immediate change in the concentration of any species in the equilibrium, place in the concentration column of that substance an "I" if the change is an increase, and a "D" if it is a decrease. If a shift will result, place F in the SHIFT column if the shift is in the forward direction, and R if it is in the reverse direction. Then determine what will happen to the concentrations of the other species because of the shift, and insert "I" or "D" for increase or decrease. If there is no Le Chatelier shift, write "None" in the SHIFT column, and leave the other columns blank.

Disturbance	Shift	$[NH_3]$	$[O_2]$	$[H_2O]$	$[NO_2]$
Add NO_2					
Reduce temperature					
Add N_2					
Remove NH_3					
Add a catalyst					

MATCHING SET ANSWERS

Set 1: 7–3–5–8–4–1–6–2   Set 2: 5–2–8–1–6–4–7–3   Set 3: 1–8–2–5–3–6–7–4–9

QUICK CHECK ANSWERS

19.1 An open beaker can contain a solution-type equilibrium if there is no significant evaporation of solvent.

19.2 "Glancing" collisions, collisions that have insufficient kinetic energy, and those that have improper orientation do not produce a chemical change.

19.3 A barrier is something that prevents or limits some event. Activation energy limits a reaction to that fraction of the intermolecular collisions that have enough kinetic energy and proper orientation.

19.4 a) As temperature drops, the fraction of collisions with enough kinetic energy to meet the activation energy requirement drops significantly, which reduces reaction rate. Collision frequency also drops, which reduces reaction rate slightly. b) A catalyst increases reaction rate by providing a reaction path with a lower activation energy. c) At higher concentrations collisions are more frequent, so reaction rate increases.

19.5 $Cd(OH)_2(s) \rightleftharpoons Cd^{2+}(aq) + 2\,OH^-(aq)$ $\qquad K_{sp} = [Cd^{2+}][OH^-]^2$

$$[Cd^{2+}] = \frac{0.0016\text{ g Cd(OH)}_2}{L} \times \frac{1\text{ mol Cd}^{2+}}{146.4\text{ g Cd(OH)}_2} = 1.1 \times 10^{-5}\text{ M}$$

$[OH^-] = 2 \times [Cd^{2+}] = 2 \times 1.1 \times 10^{-5} = 2.2 \times 10^{-5}\text{ M}$

$K_{sp} = [Cd^{2+}][OH^-]^2 = (1.1 \times 10^{-5})(2.2 \times 10^{-5})^2 = 5.3 \times 10^{-15}$

19.6 $AgBr(s) \rightleftharpoons Ag^+(aq) + Br^-(aq)$ $\qquad K_{sp} = [Ag^+][Br^-]$

a) Let $s = [Ag^+] = [Br^-]$ $\qquad s^2 = 5.0 \times 10^{-13}$ $\qquad s = 7.1 \times 10^{-7}$ mol AgBr/L

$$0.100\text{ L} \times \frac{7.1 \times 10^{-7}\text{ mol AgBr}}{1\text{ L}} \times \frac{187.8\text{ g AgBr}}{1\text{ mol AgBr}} = 1.3 \times 10^{-5}\text{ g AgBr/100 mL}$$

b) Solubility $= [Ag^+] = \dfrac{K_{sp}}{[Br^-]} = \dfrac{5.0 \times 10^{-13}}{0.25} = 2.0 \times 10^{-12}$ mol AgBr/L

$$0.100\text{ L} \times \frac{2.0 \times 10^{-12}\text{ mol AgBr}}{1\text{ L}} \times \frac{187.8\text{ g AgBr}}{1\text{ mol AgBr}} = 3.8 \times 10^{-11}\text{g AgBr/100 mL}$$

19.7 Let HA be the unknown acid. $HA(aq) \rightleftharpoons H^+(aq) + A^-(aq)$ $\qquad K_a = \dfrac{[H^+][A^-]}{[HA]}$

$[H^+] = 10^{-2.61} = 2.5 \times 10^{-3} = [A^-] =$ mol HA ionized per liter

$$K_a = \frac{(2.5 \times 10^{-3})(2.5 \times 10^{-3})}{0.22} = 2.7 \times 10^{-5}$$

19.8 $HC_2H_3O_2(aq) \rightleftharpoons H^+(aq) + C_2H_3O_2^-(aq)$ $\qquad K_a = \dfrac{[H^+][C_2H_3O_2^-]}{[HC_2H_3O_2]}$

$[H^+] = \sqrt{K_a[HA]} = \sqrt{(1.8 \times 10^{-5})(0.19)} = 1.85 \times 10^{-3}$ M (extra significant figure)

$pH = -\log[H^+] = -\log(1.85 \times 10^{-3}) = 2.73$ (two significant figures)

19.9 a) $[H^+] = K_a \times \dfrac{[HA]}{[A^-]} = 6.4 \times 10^{-5} \times \dfrac{[HC_2O_4^-]}{[C_2O_4^{2-}]} = 6.4 \times 10^{-5} \times \dfrac{0.37}{0.28} = 8.5 \times 10^{-5}$ M

$pH = -\log[H^+] = -\log(8.5 \times 10^{-5}) = 4.07$

b) $\dfrac{[HC_7H_5O_2]}{[C_7H_5O_2^-]} = \dfrac{[H^+]}{K_a} = \dfrac{10^{-4.95}}{6.5 \times 10^{-5}} = 0.17$

19.10 $2\,HI(g) \rightleftharpoons H_2(g) + I_2(g)$ $\qquad K = \dfrac{[H_2][I_2]}{[HI]^2}$

	2 HI(g) $\rightleftharpoons$	**H$_2$(g) +**	**I$_2$(g)**
mol/L at start	0.36	0.000	0.000
mol/L change, + or −	−0.106	+0.053	+0.053
mol/L at equilibrium	0.25	0.053	0.053

$$K = \frac{[H_2][I_2]}{[HI]^2} = \frac{(0.053)(0.053)}{0.25^2} = 0.045$$

Nuclear Chemistry

20

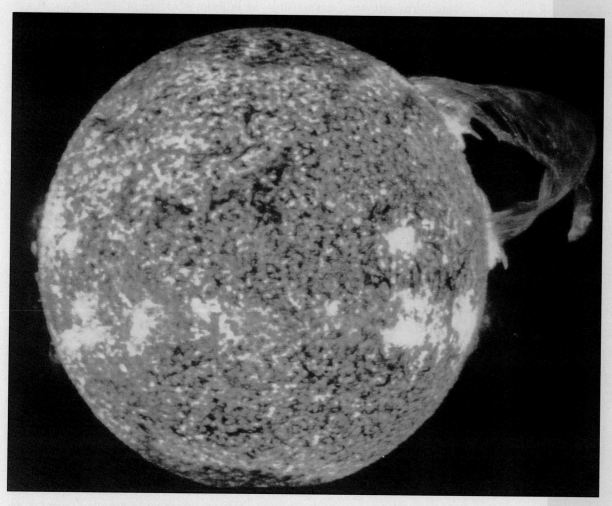

Our sun, like other stars, is a giant nuclear fusion reactor. It supplies energy to the earth from a distance of 93,000,000 miles. This photo was taken by NASA's Skylab 4.

20.1 THE DAWN OF NUCLEAR CHEMISTRY

Serendipity. This pleasant-sounding word refers to finding valuable things you are not looking for—an accidental discovery, in other words. Serendipity has been a part of many scientific discoveries. But what happened to Henri Becquerel in 1896 stands above them all. What he stumbled across now affects the lives of you and me and potentially every creature on this planet. Becquerel discovered nuclear chemistry.

Becquerel became interested in the penetrating power of X-rays soon after they were discovered—also accidentally—by Wilhelm Roentgen in 1895. Becquerel was also interested in phosphorescence, the phenomenon by which certain substances, called phosphors, glow after being exposed to light. He wondered if phosphorescent light can penetrate black paper as X-rays can. His plan was to put a uranium-containing phosphor on top of unexposed film wrapped in black paper and place them in sunlight. The paper would prevent the film from being exposed by the sunlight, so that if it were exposed at all, this would have to be from the phosphorescent rays that passed through the paper.

Alas, the day Becquerel chose for his experiment was cloudy. After waiting in vain for the sun to come out, he put his assembled material into a drawer to await the next sunny day. After several overcast days, Becquerel decided to develop the film to see if the initial cloudy-day experiment had caused even a trace of phosphorescent light to penetrate the paper. To his amazement, the film was highly exposed! The only explanation for this was that some sort of rays were leaving the uranium compound continuously, passing through the paper, and exposing the film. Sunlight and phosphorescence had nothing to do with the result.

This is how Becquerel discovered **radioactivity, the spontaneous emission of rays resulting from the decay, or breaking up, of an atomic nucleus.** Of course, Becquerel did not know about a "nucleus" in 1896. In fact, it was one of those rays coming from a radioactive source that Rutherford used in the experiment that led to the discovery of the nucleus (see caption of Fig. 5.3).

20.2 RADIOACTIVITY

Overall, there are 92 naturally occurring elements. (Elements with atomic numbers greater than 92 are not found in nature but have been made in laboratories and in some cases on a commercial production level.) The nuclei of atoms of all the elements are referred to generally as **nuclides.*** All the isotopes of the 92 elements in nature give us more than 300 different nuclides. Of these, over 260 are stable. They should last forever.

The remaining thirty-odd nuclides are radioactive. They are called **radionuclides.** Radionuclides are not stable. They are constantly changing, either into isotopes of the same element or into nuclides of different elements. Energy and a subatomic particle are also given off. The new nuclide may also be radioactive, leading to further decay, or it may be a stable isotope of some element.

Three products from radioactive decay have been identified. Initially they were called "rays." If a beam consisting of all three rays is aimed into an electric field, as in Figure 20.1a, the rays separate. One ray, called an **alpha ray,** or **α-ray,**† is attracted to the negatively charged plate, indicating that it has a positive charge. The α-ray has

---

*Some authors include the electrons outside the nucleus in the term *nuclide.*

†α, β, and γ are the Greek letters alpha, beta, and gamma.

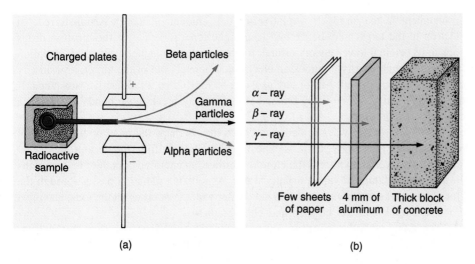

Figure 20.1

Alpha (α), beta (β) and gamma (γ) radioactive emissions. (a) Alpha and beta particles are deflected by an electric field, but gamma rays are not (b). The penetrating power of radioactive emissions increases in the order of alpha, beta, and gamma.

little penetrating power; it can be stopped by the outer layer of skin or a few sheets of paper (Fig. 20.1b). Alpha rays are now known to be nuclei of helium atoms, having the nuclear symbol $_2^4$He. They are commonly called **alpha particles,** or α-particles. The emission of an alpha particle is an **alpha decay reaction,** or simply an **alpha decay.**

The second kind of ray also turns out to be a beam of particles, but they are negatively charged and therefore attracted to the positively charged plate (Fig. 20.1a). Called **beta rays,** or β-rays, they have been identified as electrons. The nuclear symbol for a **beta particle,** or β-particle, is $_{-1}^0$e, indicating zero mass number and a -1 charge. β-particles have considerably more penetrating power than α-particles, but they can be stopped by a sheet of aluminum about 4 mm thick (Fig. 20.1b). The emission of a beta particle is a **beta decay reaction,** or **beta decay.**

The third kind of radiation is the **gamma ray,** or γ-ray. Gamma rays are not particles but very high-energy electromagnetic rays, similar to X-rays. Because of their high energy, gamma rays have high penetrating power. They can be stopped only by thick layers of lead or heavy concrete walls, as shown in Figure 20.1b. Not having an electric charge, gamma rays are not deflected by an electric field (Fig. 20.1a).

Some radioactive substances are harmful, but some have become valuable tools in industry, research, and medicine. In the following section we will discuss briefly the applications of radionuclides in medicine.

FLASHBACK If Sy is the symbol of an element, the nuclear symbol of an isotope of the element is

$$_{\text{atomic number}}^{\text{mass number}}\text{Sy}$$

The name of an isotope is the name of the element followed by the mass number, as $_{92}^{238}$U is uranium-238. (See Section 5.4.)

FLASHBACK The electromagnetic spectrum, which was described in Section 10.1 and Figure 10.1, includes X-rays, ultraviolet and infrared rays, visible light, microwaves, and radio and TV waves.

✔ QUICK CHECK 20.1

a) Write the nuclear symbol and electrical charge of an alpha particle and a beta particle.

b) List alpha, beta, and gamma rays in order of decreasing penetrating power.

20.3 THE DETECTION AND MEASUREMENT OF RADIOACTIVITY

When alpha, beta, or gamma radiation collides with an atom or molecule, some of the radiation energy is given to the target particle. The collision changes the electron

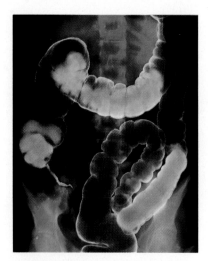

Film is still commonly used to detect radiation. Barium sulfate is given to patients needing gastrointestinal tract X-rays because it absorbs X-rays well and is not absorbed by the patient.

arrangement in the target. If a relatively small amount of energy is transferred, an electron in the target may be excited to a higher energy level. As the electron drops back to its original lower-energy orbital, it releases electromagnetic energy. If radiation transfers enough energy to knock an electron completely out of the target, a positively charged ion is produced. Air, or any gas, can be ionized by a radioactive substance. If the radiation strikes chemically bonded atoms, it may break those bonds and cause a chemical reaction.

There are several ways to detect radioactivity. Perhaps the most obvious, but not necessarily the most convenient, is exposing photographic film, the very property that led to its discovery. Another is the **cloud chamber,** an enclosed container that contains air and a supersaturated vapor, usually water. As ionizing radiation passes through the cloud chamber, some of the air is ionized. Water vapor condenses on the ions, leaving a cloud-like track that can be seen and photographed.

The **Geiger-Müller counter** (often shortened to **Geiger counter**) is the best-known instrument for measuring ionizing radiation. It consists of a tube filled with a gas, as shown in Figure 20.2. The gas is ionized by radiation passing through a thin glass window, permitting an electrical discharge between two electrodes. The current may be measured quantitatively on a meter. Some Geiger counters emit a "click" when radiation is found. Geiger counters are used to measure alpha and beta radiation.

Geiger counters do not measure gamma rays effectively because there are not enough gas particles to guarantee interaction with a neutral gamma ray. A **scintillation counter** uses a transparent solid, which has a higher particle density than a gas. This makes interaction with a gamma ray more likely. Particles in the solid absorb energy, and then release some of this energy in flashes of light, which can be counted. This light emission, which may continue for some time after exposure to gamma rays stops, is **phosphorescence.** This is what Becquerel was looking for when he discovered radioactivity.

A common example of a "scintillation counter" is a watch with luminous hands. Watch hands used to be painted with a mixture of radium salts, which are radioactive, and zinc sulfide. Zinc sulfide is a phosphor that glows as the radium atoms decay. This kind of watch emits both alpha and beta particles and will make a Geiger counter click. Newer luminous watches do not use radium salts, but compounds containing tritium,

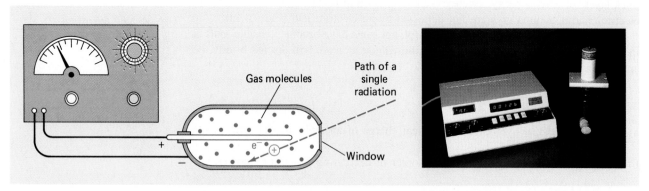

Figure 20.2

Geiger counter. The tube contains argon at low pressure. A high electrical potential is established between a positively charged wire in the center and the negatively charged case. Radiation enters the window and ionizes the argon. The ions move toward the electrodes, producing a measurable electrical pulse and an audible "click."

$^3_1$H, which are sealed in glass vials glued to the watch hands and face. Tritium is a low-energy beta emitter, which does not trigger a Geiger counter reading.

Two instruments used in medicine for detecting radioactivity are the gamma camera and the scanner. The gamma camera is placed over a target area and takes a snapshot of the area. The scanner moves while taking many pictures, each picture showing a ''slice'' of the area under study. These pictures may be combined to give a three-dimensional view of the interior of an organ. These processes are simple, usually without discomfort to the patient, and are often used instead of exploratory surgery.

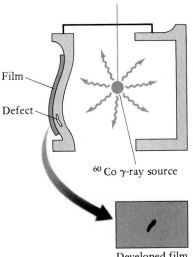

Film

Defect

^{60}Co γ-ray source

Developed film shows defect

A gamma-ray source to detect defects in cast metal parts. The developed photographic film is more strongly exposed where gamma rays passed through a defect.

✔ QUICK CHECK 20.2

How do Geiger counters and scintillation counters ''measure'' radioactivity? Precisely, what do they count? Do they count ''radiations'' or do they count something that is proportional to radiation? Suggest units for radioactivity as it would be measured by either of these counters.

20.4 THE EFFECTS OF RADIATION ON LIVING SYSTEMS

Shortly after radioactivity was discovered, it was thought that the radiations had certain curative powers. Radium compounds were made, and radium solutions were bottled and sold for drinking and bathing. That was before the harmful effects of radiation exposure were known—and some of the early users of these cures paid a dear price for acquiring that knowledge. Today's medical practitioners are much wiser, and they have devised sophisticated ways to use radionuclides to examine patients, diagnose their illnesses, and treat their disorders.

The harmful effects of radiation on living systems come from its ability to break chemical bonds and thereby destroy healthy tissue. Obviously, the wise course is to avoid exposure to destructive radiation! Destructive radiation also has its good side: It destroys *unhealthy* tissue too. By selectively radiating unhealthy tissue, it can be removed or reduced. This is the key to radiation therapy. The trick is to do it without damaging surrounding healthy tissue as well; or at least be sure that the destruction of the bad tissue is more helpful than the destruction of the good tissue is harmful. Modern practitioners have become pretty good at this, but there is still much room for improvement.

To describe quantitatively the effects of radiation on a living system you must measure: 1) how many disintegrations occur during the time of exposure; 2) how much energy is released with each disintegration; and 3) how much energy is absorbed by the living system.

The unit in which the intensity of radiation is measured is the **curie, Ci: One curie is defined as 3.7 × 10^{10} disintegrations per second.** This is the rate of radiation given off by one gram of pure radium-286. The curie is a very large amount of radioactivity, so millicuries (mCi) and microcuries (μCi) are more often used in laboratory measurements.

We use three units to measure the consequences of radiation. One is the **roentgen, R, the amount of ionizing radiation of any type that generates 2.09 × 10^9 ion pairs in one cm^3 of dry air.** This is a measure of radiation dosage, the total amount of radiation exposure. One roentgen is equal to an energy input of 0.00993 J/kg of air.

The **rad,** or **r**adiation **a**bsorbed **d**ose, is a measure of radiation absorbed by body tissue, not air. One rad corresponds to an energy input of 0.0100 J/kg, just slightly larger than a roentgen. Exposure to one roentgen typically produces an absorbed dosage of 0.92–0.97 rad.

The most important measure of radiation is the **rem,** or **r**oentgen **e**quivalent in **m**an. The rem corrects for the fact that different radiation sources cause different amounts of damage to a person. For example, one roentgen of radiation from an alpha source *outside the body* is not considered harmful because the alpha particle has low penetrating power and does not get through the skin.

So, how much radiation can we take? Well, more than about 600 rems in one dose is fatal. The radiation therapy used in fighting cancers gives a *total* dose *over time* in the 4000–7000 rem range. This radiation indeed damages healthy tissue surrounding a tumor, but healthy cells are more capable of repairing themselves than are malignant cells. Besides, it's better than the alternative.

Most of the radiation you encounter is measured not in rems, but in *milli*rems, mrem. For example, a chest X-ray is about 25 mrem, a complete diagnostic gastrointestinal X-ray series is about 2000 mrem, and a dental X-ray is about 0.5 mrem. The federal standard for occupational exposure is 5000 mrem/year.

Background radiation is everywhere. It's in the rocks, the soil, the water, and even in us. Although the atmosphere acts as a filter between us and the cosmos, we are constantly bombarded by cosmic rays. Indeed, the average exposure to background radiation is estimated at 360 mrem/year. Figure 20.3 shows the sources of this radiation. Note that exposure to the largest (by far!) source of background radiation is voluntary: tobacco smoke. At 9000 mrem per year, this is nearly twice the standard for occupational exposure. Unfortunately for smokers, $^{210}_{84}$Po is found in the tobacco plant, and $^{210}_{84}$Po is an alpha emitter *inside the body*, where it can be very harmful.

If you chose to live in a radiation-proof home to escape cosmic rays, you would still have to contend with the radiation that is naturally in your body. If you weigh 150

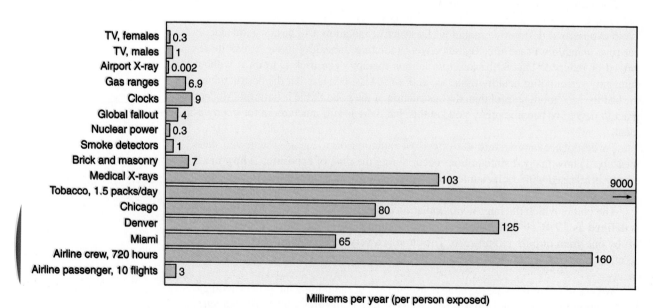

Millirems per year (per person exposed)

Figure 20.3

Typical radiation experienced by people in the United States each year.

Ionization smoke detector. Arrow points to chamber containing americium-241 inside detector.

pounds, you have about 225 grams of potassium ions in your body. Potassium ions are needed for nerve conduction and muscle (including the heart) contraction. Without potassium ions, you don't live. The natural abundance of $^{40}_{19}K$ is 0.0118% of all potassium atoms. A 150 pound man therefore has in his body about 4.1×10^{20} radioactive $^{40}_{19}K$ atoms. (You might like to confirm this statement by calculation. We'll show the calculation setup after the Quick Check answers.)

✔ QUICK CHECK 20.3

a) Recall your answer to Quick Check 20.2. Of the four units discussed in this section, the curie, the roentgen, the rad, and the rem, which do you think is most closely associated with the measurement of a Geiger or scintillation counter? Why?

b) In one sense, alpha radiation is the least harmful to humans; in another sense, it is very harmful. What makes the difference?

c) What is the easiest way you can protect yourself from the harmful effects of the various things that are included in ''background radiation''?

The roll on the right has been irradiated to preserve it. Both rolls were then left for two weeks. Half the grain on earth spoils in storage like the roll on the left.

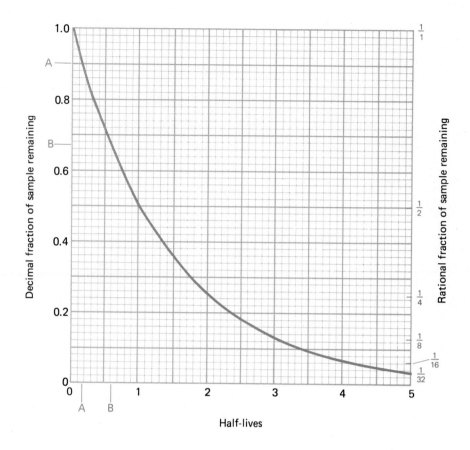

Figure 20.4

Half-life decay curve for a radioactive substance.

20.5 HALF-LIFE

The rate at which a radioactive substance decays is measured by its half-life, the time it takes for one-half of the radioactive atoms in a sample to decay. Each radionuclide has its own unique half-life, commonly written $t_{1/2}$. The units of $t_{1/2}$ are time units per half-life, or time/half-life. Time may be expressed in seconds, minutes, hours, days, or years.

Figure 20.4 is a graph of the fraction of an original sample that remains (vertical axis) after a number of half-lives (horizontal axis). The vertical axis at the left has a conventional scale, giving values in decimal fractions. The scale values on the axis on the right are the fractions that remain after each half-life period: $\frac{1}{2}$ after the first half-life; $\frac{1}{2}$ of $\frac{1}{2}$, or $\frac{1}{4}$ after the second; $\frac{1}{2}$ of $\frac{1}{4}$, or $\frac{1}{8}$ after the third; and so forth. Thus, the fraction of a sample that is still present after n half-lives is $(\frac{1}{2})^n$. If S is the starting quantity and R is the amount that remains after n half-lives, then

$$R = S \times (\tfrac{1}{2})^n \tag{20.1}$$

EXAMPLE 20.1

The half-life of $^{210}_{83}Bi$ is 5.0 days. If you begin with 16 grams of $^{210}_{83}Bi$ how many grams will you have 25 days (5 half-lives) later?

SOLUTION

This is a straightforward substitution into Equation 20.1. Recall, to raise a number to a power with a calculator, enter the number (0.5 for $\frac{1}{2}$), press y^x, then the power (5), and finally, =.

GIVEN: 16 g $^{210}_{83}$Bi; 5-half-lives   WANTED: g $^{210}_{83}$Bi remaining

EQUATION: $R = S \times (\frac{1}{2})^n = 16$ g $^{210}_{83}$Bi $\times (\frac{1}{2})^5 = 0.50$ g $^{210}_{83}$Bi

EXAMPLE 20.2

The half-life of $^{45}_{19}$K is 20 minutes. If you have a sample containing 2.1×10^3 micrograms (μg) of this isotope at noon, how many micrograms will remain at 3 o'clock in the afternoon?

SOLUTION

This is a two-step problem. The number of micrograms of $^{45}_{19}$K that remains is calculated by Equation 20.1. To use that equation, however, we need the number of half-lives, n. Because n is proportional to time, that number can be calculated by dimensional analysis. Thus:

a) GIVEN: 3 hours (noon to 3 PM)   WANTED: n (half-lives)
   PATH: hr → min → half-lives   FACTORS: 20 min/half-life; 60 min/hr

$$3 \text{ hr} \times \frac{60 \text{ min}}{\text{hr}} \times \frac{1 \text{ half-life}}{20 \text{ min}} = 9 \text{ half-lives} = n$$

b) GIVEN: $S = 2.1 \times 10^3$ μg $^{45}_{19}$K; n = 9 half-lives
   WANTED: μg $^{45}_{19}$K remaining (R)

EQUATION: $R = S \times (0.5)^n = 2.1 \times 10^3$ μg $^{45}_{19}$K $\times (0.5)^9 = 4.1$ μg $^{45}_{19}$K

To find the half-life of a radioactive isotope, you must determine starting and remaining quantities over a measured period of time. One way to interpret these data is to express the numbers as the fraction of the sample remaining, R/S. This is the vertical axis in Figure 20.4. Start with the fraction on the vertical axis, project horizontally to the curve, and then vertically to the number of half-lives on the horizontal axis. Divide time by half-lives to get the half-life of the substance: time/half-lives = $t_{1/2}$.

EXAMPLE 20.3

$^{125}_{52}$Sb has a longer half-life than many artificial radioisotopes. The mass of a sample is found to be 8.623 grams. The sample is set aside for 157 days, which is 0.43 year. At that time its mass is 7.762 g. Find $t_{1/2}$.

SOLUTION

GIVEN: $S = 8.623$ g $^{125}_{52}$Sb; $R = 7.762$ g $^{125}_{52}$Sb; time = 0.43 year
WANTED: $t_{1/2}$ (years/half-life)

Figure 20.5

The "Iceman," dubbed *Homo tyrolensis,* is thought to be the oldest human ever found, perhaps 2,000 years older than the mummy of King Tutankhamen.

To find the half-life, we need the time (0.43 year) and the number of half-lives. That number can be found from the quotient of R/S and Figure 20.4.

$$R/S = 7.762 \text{ g}/8.623 \text{ g} = 0.9002$$

This means that 90.02% of the sample remains after 0.43 year. Projecting from 0.9002 on the vertical axis of Figure 20.4 to the curve, and then down to the horizontal axis, we estimate the number of half-lives to be about 0.16. See points A on the curve.

Dividing, we obtain

$$\frac{0.43 \text{ year}}{0.16 \text{ half-life}} = 2.7 \text{ years per half-life} = t_{1/2}$$

This half-life rate of decay of radioactive substances is the basis of **radiocarbon dating,** by which the age of fossils is estimated. Carbon is the principal chemical element in all living organisms, both plant and animal. Most carbon atoms are carbon-12; but a small portion of the carbon in atmospheric carbon dioxide is carbon-14, a radioactive isotope with a half-life of 5.73×10^3 years. When a plant or animal is alive, it takes in this isotope from the atmosphere, while the same isotope in the organism is disappearing by nuclear disintegration. A "steady state" situation exists while the system lives, maintaining a constant ratio of carbon-14 to carbon-12. When the organism dies, the disintegration of carbon-14 continues, but its intake stops. This leads to a gradual reduction in the ratio of $^{14}_6C$ to $^{12}_6C$. By measuring the ratio and the amount of $^{14}_6C$ now present in a sample, we can calculate the $^{14}_6C$ present when the organism died. Figure 20.4 is then used to calculate the age of the sample.

EXAMPLE 20.4

The "Iceman" (Figure 20.5) is a Neolithic hunter whose frozen and intact remains were found in the Similaun Glacier on the Austrian/Italian border in the Tyrolean Alps. Analysis of the "Iceman" shows that for every 15.3 units of $^{14}_6C$ present at the time of his death, 8.1 remain today. How old is the "Iceman?"

SOLUTION

The quantities 15.3 and 8.1 refer to radioactive disintegrations detected by a Geiger counter or some other instrument. They express initial and final amounts, just as if they were expressed in grams. Thus

GIVEN: S = 15.3 units; R = 8.1 units; $t_{1/2} = 5.73 \times 10^3$ years/half-life
WANTED: years   PATH (after finding half-lives): half-lives → years

$R/S = 8.1 \text{ units}/15.3 \text{ units} = 0.53$ of the sample remains

From Figure 20.4, 0.53 of sample corresponds to 0.93 half-life (point B on the curve).

$$0.93 \; \cancel{\text{half-life}} \times \frac{5.73 \times 10^3 \text{ years}}{1 \; \cancel{\text{half-life}}} = 5.3 \times 10^3 \text{ years}$$

Carbon dating has produced evidence of modern human's presence on earth as long ago as 14,000 to 15,000 years. Similar dating techniques are also applied to mineral deposits. Analyses of geological deposits have identified rocks with an estimated age of 3.0 to 4.5 billion years, the latter figure being the scientist's estimate of the age of the earth. The oldest moon rocks analyzed to date indicate an age of about 3.5 billion years.

20.6 NUCLEAR EQUATIONS

When a radionuclide emits an alpha or beta particle, there is a **transmutation** of an element, a change from one element to another. This means that the remaining nuclide has a different atomic number, a different number of protons than the original nuclide. A nuclear change—actually a nuclear reaction—has occurred. The original substance (nuclide) has been destroyed and a new substance (nuclide) has been formed.

The emission of a gamma ray does not change the elemental identity of the nucleus, even though energy is released. In that sense, a gamma emission, by itself, is not a nuclear change. Therefore, for the remainder of this chapter we will consider only alpha and beta emissions in nuclear reactions.

As chemists write chemical equations to describe chemical changes, they write nuclear equations to describe nuclear changes. A nuclear equation shows the reactant nuclides and/or particles on the left and the product nuclides and/or particles on the right. The first step in the natural radioactive series observed by Becquerel is an alpha decay reaction. In it, a $^{238}_{92}U$ nucleus *disintegrates* or *decays* into a $^{4}_{2}He$ nucleus (alpha particle) and a $^{234}_{90}Th$ nucleus. The nuclear equation is

$$^{238}_{92}U \longrightarrow {}^{4}_{2}He + {}^{234}_{90}Th \tag{20.2}$$

Notice that this equation is "balanced" in both neutrons and protons. The total number of neutrons and protons is 238, the mass number of the uranium isotope. The total mass number of the two products is $234 + 4$, again 238. This accounts for all of the protons and neutrons in the reactant, $^{238}_{92}U$. In terms of protons only, the 92 in a uranium nucleus are accounted for by 90 in the thorium nucleus plus 2 in the helium nucleus. *A nuclear equation is balanced if the sums of the mass numbers on the two sides of the equation are equal, and if the sums of the atomic numbers are equal.*

The $^{234}_{90}Th$ nucleus resulting from the disintegration of uranium-238 is also radioactive. It is a beta decay reaction; it emits a beta particle, $_{-1}^{0}e$, and produces an isotope of protactinium, $^{234}_{91}Pa$:

$$^{234}_{90}Th \longrightarrow {}^{234}_{91}Pa + {}^{0}_{-1}e \tag{20.3}$$

In a beta particle emission the mass numbers of the reactant and product isotopes are the same, while the atomic number increases by 1. Although the actual process is more complex, it appears as if a neutron divides into a proton and an electron, and the electron is ejected.

The two disintegrations described in Equations 20.2 and 20.3 are only the first 2 of 14 steps that begin with $^{238}_{92}U$. There are eight α-particle emissions and six β-particle emissions, leading ultimately to a stable isotope of lead, $^{206}_{82}Pb$. This entire **natural radioactive decay series** is described in Figure 20.6. There are two other natural disintegration series. One begins with $^{232}_{90}Th$ and ends with $^{208}_{82}Pb$, and the other passes from $^{235}_{92}U$ to $^{207}_{82}Pb$.

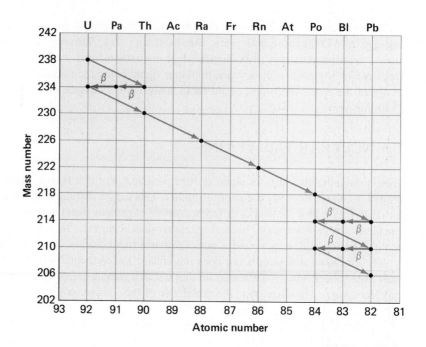

Figure 20.6

Radioactive decay series. This series begins with $^{238}_{92}U$, and after eight alpha emissions and six beta emissions, produces $^{206}_{82}Pb$ as a stable end product.

EXAMPLE 20.5

Write the nuclear equation for the changes that occur in the uranium-238 disintegration series when $^{226}_{88}Ra$ ejects an α-particle. Ra is the symbol for radium, one of the elements discovered by Pierre and Marie Curie in their study of radioactivity.

In writing a nuclear equation, one product will be the particle ejected. The mass number of the other product will be such that, when added to the mass number of the ejected particle, the total will be the mass number of the original isotope. What is the mass number of the second product of the emission of an alpha particle from a $^{226}_{88}Ra$ nucleus?

Marie Sklodowska Curie
Nobel Prize Physics, 1903 (with husband Pierre and Henri Becquerel) for discovery of radioactivity
Nobel Prize Chemistry 1911, for isolation of pure radium
Mother of Irene and Eve (Irene Curie-Joliot shared with her husband, Frederic, a Nobel prize in Chemistry, 1935)
First woman to teach at the Sorbonne, 1906
First woman to be appointed full professor at the Sorbonne, 1908

222.

The reactant isotope has a mass number of 226. It emits a particle having a mass number of 4. This leaves $226 - 4 = 222$ as the mass number of the remaining particle.

Now find the atomic number of the second product particle. The atomic number of the starting isotope is 88. It emitted a particle having two protons. How many protons are left in the nucleus of the other product?

———   ———

86.

If two protons are emitted from a nucleus having 88 protons, 86 remain.

You now know the mass number and the atomic number of the second product of an alpha particle emission from $^{226}_{88}Ra$. Using a periodic table, you can find the elemental symbol of this product and assemble all three symbols into the required nuclear equation.

———   ———

$$^{226}_{88}Ra \longrightarrow {}^{4}_{2}He + {}^{222}_{86}Rn$$

The second product in Example 20.5, $^{222}_{86}Rn$, is a radioactive isotope of the gas radon. Radon-222 further decays by emitting an alpha particle. In recent years there has been concern over radon-222 that enters building basements from the soil outside. Exposure to radon gas may increase your chances of developing lung cancer, particularly if you smoke.

EXAMPLE 20.6 _____

Write the nuclear equation for the emission of a β-particle from $^{210}_{83}Bi$.

The method is the same. Remember the beta particle, $_{-1}^{0}e$, has zero mass number, and an effective atomic number of -1. Both mass number and atomic number must be conserved in the equation.

———   ———

$$^{210}_{83}Bi \longrightarrow {}^{0}_{-1}e + {}^{210}_{84}Po$$

In the emission of a β-particle, which has effectively no mass, the mass number of the radioactive isotope and the product isotope are the same. The product isotope has an atomic number greater by one than the radioactive isotope, an increase of one proton. Po is the symbol that corresponds to atomic number 84. The element is polonium, the other element discovered by the Curies in their investigation of radioactivity. The name of the element was selected to honor Mme. Curie's native Poland.

✔ QUICK CHECK 20.4

Write the nuclear equation for the alpha decay of radon-222, $^{222}_{86}Rn$, the hazardous radioactive product in Example 20.5.

20.7 NUCLEAR REACTIONS AND ORDINARY CHEMICAL REACTIONS COMPARED

Now that you have seen the nature of a nuclear change and the type of equation by which it is described, we will pause to compare nuclear reactions with the others you have studied. There are four areas of comparison:

1) In ordinary chemical reactions the chemical properties of an element depend only on the electrons outside the nucleus, and the properties are essentially the same for all isotopes of the element. The nuclear properties of the various isotopes of an element are quite different, however. In the radioactive decay series beginning with uranium-238, $^{234}_{90}\text{Th}$ emits a β-particle, whereas a bit farther down the line $^{230}_{90}\text{Th}$ ejects an α-particle. Both $^{214}_{82}\text{Pb}$ and $^{210}_{82}\text{Pb}$ are β-particle emitters toward the end of the series, while the final product, $^{206}_{82}\text{Pb}$, has a stable nucleus, emitting neither alpha or beta particles, nor gamma rays.

2) Radioactivity is independent of the state of chemical combination of the radioactive isotope. The reaction of $^{210}_{83}\text{Bi}$ occurs for atoms of that particular isotope whether they are in pure elemental bismuth, combined in bismuth chloride, $BiCl_3$, bismuth sulfate, $Bi_2(SO_4)_3$, or any other bismuth compound, or if they happen to be present in the low-melting bismuth alloy used in sprinkler systems for fire protection in large buildings.

3) Nuclear reactions usually result in the formation of different elements because of changes in the number of protons in the nucleus of an atom. In ordinary chemical reactions the atoms keep their identity while changing from one compound as a reactant to another as a product.

4) Both nuclear and ordinary chemical changes involve energy, but the amount of energy for a given amount of reactant in a nuclear change is enormous—greater by several orders of magnitude, or multiples of ten—compared to the energies of ordinary chemical reactions.

20.8 NUCLEAR BOMBARDMENT AND INDUCED RADIOACTIVITY

In natural radioactive decay we find an example of the alchemist's get-rich-quick dream of converting one element to another. But the natural process for uranium does not yield the gold coveted by the alchemist; rather, it produces the element lead, with which the dreamer wanted to begin his transmutation. The question was still present after the discovery of radioactivity: Can we initiate the transmutation of one ordinarily stable element into another?

In 1919 Rutherford produced a "Yes" answer to that question. He found that he could "bombard" the nucleus of a nitrogen atom with a beam of alpha particles from a radioactive source, knocking a proton out of the nucleus and producing an atom of oxygen-17:

$$^{14}_{7}\text{N} + ^{4}_{2}\text{He} \longrightarrow ^{17}_{8}\text{O} + ^{1}_{1}\text{H}$$

The oxygen isotope produced is stable; the experiment did not yield any radioactive isotopes. Similar experiments were conducted with other elements, using high-speed alpha particles as atomic "bullets." It was found that most of the elements up to potassium can be changed to other elements by **nuclear bombardment.** None of the isotopes produced were radioactive.

One experiment during this period was first thought to yield a nuclear particle that emitted some sort of high-energy radiation, perhaps a gamma ray. In 1932 James Chadwick correctly interpreted the experiment and, in doing so, he became the first person to identify the neutron. The reaction comes from bombarding a beryllium atom with a high-energy α-particle:

$$\ce{^{9}_{4}Be + ^{4}_{2}He \longrightarrow ^{12}_{6}C + ^{1}_{0}n}$$

$^{1}_{0}n$ is the nuclear symbol for the neutron, with zero charge and a mass number of 1.

Two years later, in 1934, Irene Curie, the daughter of Pierre and Marie Curie, and her husband, Frederic Joliot, used high-energy α-particles to produce the first synthetic radionuclide. Their target was boron-10; the product was a radioactive nitrogen nucleus

$$\ce{^{10}_{5}B + ^{4}_{2}He \longrightarrow ^{13}_{7}N + ^{1}_{0}n}$$

Because this radionuclide is not found in nature, its decay is an example of **induced** or **artificial radioactivity.** When $^{13}_{7}N$ decays, it emits a particle having the mass of an electron and a charge equal to that of an electron, except that it is positive. This "positive electron" is called a **positron,** and it is represented by the symbol $^{0}_{1}e$. The decay equation is

$$\ce{^{13}_{7}N \longrightarrow ^{13}_{6}C + ^{0}_{1}e}$$

Today hundreds of radionuclides have been produced in laboratories all over the world. Many of these isotopes have been made in different kinds of **particle accelerators,** which use electric fields to increase the kinetic energy of the charged particles that bombard nuclei (Figure 20.7). Among the earliest and best-known accelerators is the

Figure 20.7

Particle accelerator at Argonne National Laboratory near Chicago. The outer ring, which is nearly four football fields in diameter, will house experiments for as many as 300 scientists at one time. (Photo courtesy of Argonne National Laboratory)

cyclotron, designed by E. O. Lawrence at the University of California, Berkeley. Other more powerful accelerators are approximately circular (over a mile in diameter).

One of the more exciting areas of research with bombardment reactions has been the production of elements that do not exist in nature. Except for trace quantities, no natural elements having atomic numbers greater than 92 have ever been discovered. In 1940, however, it was found that $^{238}_{92}\text{U}$ is capable of capturing a neutron:

$$^{238}_{92}\text{U} + ^{1}_{0}\text{n} \longrightarrow ^{239}_{92}\text{U} \tag{20.4}$$

The newly formed isotope is unstable, progressing through two successive β-particle emissions, yielding isotopes of the elements having atomic numbers 93 and 94:

$$^{239}_{92}\text{U} \longrightarrow ^{0}_{-1}\text{e} + ^{239}_{93}\text{Np (neptunium)} \tag{20.5}$$

$$^{239}_{93}\text{Np} \longrightarrow ^{0}_{-1}\text{e} + ^{239}_{94}\text{Pu (plutonium)} \tag{20.6}$$

Neptunium, plutonium, and all the other synthetic elements having atomic numbers greater than 92 are called the **transuranium elements.** All transuranium isotopes are radioactive, and a few have been isolated only in isotopes with very short half-lives and in extremely small quantities. Some of the bombardments yielding transuranium products use relatively heavy isotopes as bullets. For example, einsteinium-247 is produced by bombarding uranium-238 with ordinary nitrogen nuclei:

$$^{238}_{92}\text{U} + ^{14}_{7}\text{N} \longrightarrow ^{247}_{99}\text{Es} + 5\,^{1}_{0}\text{n}$$

FLASHBACK The Everyday Chemistry essay in Chapter 6 contains more recent information about this extremely specialized area of research. It describes some of the problems, such as cost, equipment, personnel, and time, required to produce tiny quantities of elements with atomic numbers up to 110—even a single atom of meitnerium ($Z = 109$).

✔ QUICK CHECK · 20.5

a) What is produced in a nuclear bombardment reaction?
b) What property must a particle have if it is to be used in a particle accelerator?

20.9 USES OF RADIONUCLIDES

Today there are hundreds, possibly thousands, of uses for synthetic radionuclides. The best known of these are in the field of medicine. People are not usually aware of others, although, at times, they may be close at hand. For example, do you have a smoke detector in your home? Battery-powered smoke detectors use a chip of americium-241, $^{241}_{95}\text{Am}$. The americium ionizes the air in the detector, which causes a small current to flow through the air. When smoke enters, it breaks the circuit and sets off the alarm, which is powered by a battery. With a half-life of 458 years, the americium doesn't need changing every year like the battery does.

Artificial radionuclides are used in food preservation. Worldwide, over 40 classes of foods are irradiated with gamma rays from cobalt-60, $^{60}_{27}\text{Co}$, or cesium-137, $^{137}_{55}\text{Cs}$. This process retards growth of bacteria, molds, and yeasts in foods, like heat pasteurization extends the shelf life of milk. Higher doses of gamma radiation sterilize foodstuffs, killing insects as well as bacteria, molds, and yeasts.

Industrial applications of radioisotopes include studies of piston wear and corrosion resistance. Petroleum companies use radioisotopes to monitor the progress of certain oils through pipelines. The thickness of thin sheets of metal, plastic, and paper is subject to continuous production control by using a Geiger counter to measure the amount of radiation that passes through the sheet; the thinner the sheet, the more radiation that will be detected by the counter. Quality control laboratories can detect small traces of radioactive elements in a metal part.

Scientific research is another major application of radioisotopes. Chemists use "tagged" atoms as *radioactive tracers* to study the mechanism, or series of individual steps, in complicated reactions. For example, by using water containing radioactive oxygen it has been determined that the oxygen in the glucose, $C_6H_{12}O_6$, formed in photosynthesis

$$6\,CO_2(g) + 6\,H_2O(\ell) \longrightarrow C_6H_{12}O_6(s) + 6\,O_2(g)$$

comes entirely from the carbon dioxide, and all oxygen from water is released as oxygen gas. Archaeologists use neutron bombardment to produce radioactive isotopes in an artifact, which makes it possible to analyze the item without destroying it. Biologists employ radioactive tracers in the water absorbed by the roots of plants to study the rate at which the water is distributed throughout the plant system. These are but a few of the many ingenious applications that have been devised for this useful tool of science.

20.10 NUCLEAR FISSION

In 1938, during the period when Nazi Germany was moving steadily toward war, dramatic and far-reaching events were taking place in her laboratories. A team made up of Otto Hahn, Fritz Strassman, and Lise Meitner was working with neutron bombardment of uranium. In the products of the reaction they were finding, surprisingly, atoms of barium and krypton, and other elements far removed in both atomic mass and atomic number from the uranium atoms and neutrons used to produce them. The only explanation was, at that time, unbelievable; the nucleus must be splitting into two nuclei of smaller mass. This kind of reaction is called **nuclear fission.**

In the fission of uranium-235 there are many products; it is not possible to write a single equation to show what happens. A representative equation is

$$^{235}_{92}U + ^{1}_{0}n \longrightarrow ^{94}_{38}Sr + ^{139}_{54}Xe + 3\,^{1}_{0}n$$

Notice that it takes a neutron to initiate the reaction. Notice also that the reaction produces *three* neutrons. If one or two of these collide with other fissionable uranium nuclei, there is the possibility of another fission or two. And the neutrons from those reactions can trigger others, repeatedly, as long as the supply of nuclei lasts. This is what is meant by a **chain reaction** (Fig. 20.8), in which a nuclear product of the reaction becomes a nuclear reactant in the next step, thereby continuing the process.

The number of neutrons produced in the fission of $^{235}_{92}U$ varies with each reaction. Some reactions yield two neutrons per uranium atom; others, like that above, yield three, and still others produce four or more. The average is about 2.5. If the quantity of uranium, or any other fissionable isotope, is large enough that most of the neutrons produced are captured within the sample, rather than escaping to the surroundings, the chain reaction will continue. The minimum quantity required for this purpose is called the **critical mass.**

Uranium-235 is capable of sustaining a chain reaction, but it makes up only 0.7% of all naturally occurring uranium. Therefore, it is not a very satisfactory source of

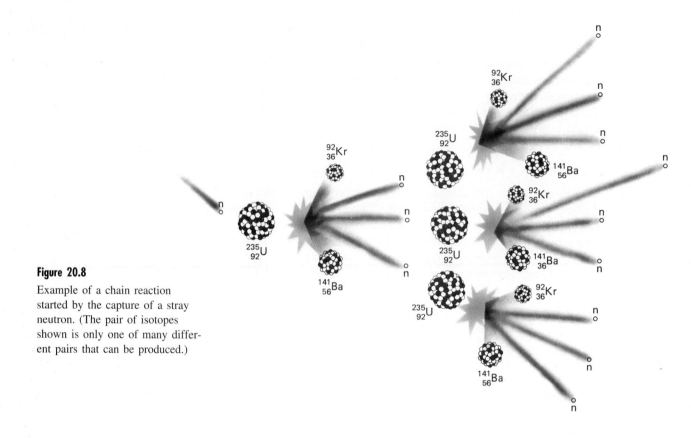

Figure 20.8

Example of a chain reaction started by the capture of a stray neutron. (The pair of isotopes shown is only one of many different pairs that can be produced.)

nuclear fuel. An alternative is the plutonium isotope, $^{239}_{94}Pu$, produced from $^{238}_{92}U$, the most abundant uranium isotope (Equations 20.4 to 20.6). $^{239}_{94}Pu$ has a long half-life (24,360 years) and is fissionable. It has been used in the production of atomic bombs and is also used in some nuclear power plants to generate electrical energy. It is made in a **breeder reactor,** the name given to a device whose purpose is to produce fissionable fuel from nonfissionable isotopes.

✔ QUICK CHECK 20.6

How do the products of a fission reaction compare with the reactants?

20.11 ELECTRICAL ENERGY FROM NUCLEAR FISSION

Aside from hydroelectric plants located on major rivers, most electrical energy comes from generators driven by steam. Traditionally, the steam comes from boilers fueled by oil, gas, or coal. The fast dwindling supplies of these fossil fuels, and the uncertainties surrounding the availability and cost of petroleum from the countries where it is so abundant, have turned attention to nuclear fission as an alternative energy source.

A diagram of a nuclear power plant is shown in Figure 20.9. The turbine, generator, and condenser are similar to those found in any fuel-burning power plant. The nuclear fission reactor has three main components: the fuel elements, control rods, and

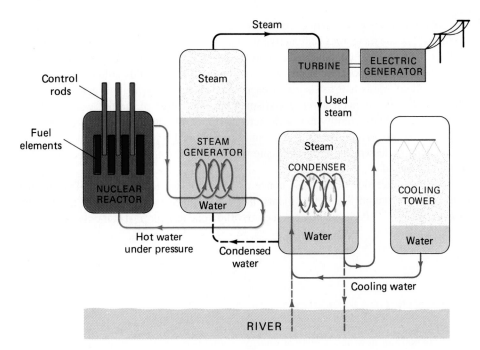

Figure 20.9

Schematic diagram of a nuclear power plant. Nuclear fission occurs in the reactor. Fission energy is used to heat water under pressure (red cycle), which changes turbine water to steam in the steam generator. High pressure steam (black cycle) drives the turbine, which in turn runs the electric generator that produces electric power. Spent steam from the turbine is changed to liquid water in the condenser and recycled back to the steam generator. Cooling water for the condenser (blue cycle) comes from a cooling tower, to which it is recycled. Make-up cooling water, and sometimes the cooling water itself, is drawn from a river, lake, or ocean.

moderator. The fuel elements are simply long trays that hold fissionable material in the reactor. As the fission reaction proceeds, fast-moving neutrons are released. These neutrons are slowed down by a moderator, which is water in the reactor illustrated. When the slower neutrons collide with more fissionable material the reaction is continued. The reaction rate is governed by cadmium control rods, which absorb excess neutrons. At times of peak power demand, the control rods are largely withdrawn from the reactor, permitting as many neutrons as necessary to find fissionable nuclei. When demand drops, the control rods are pushed in, absorbing neutrons and limiting the reaction.

The building, and even the continued use, of nuclear power plants faces stiff opposition from both individuals and public groups in the United States. The threat of an accident that might release large amounts of radiation over a densely populated area is the major concern. This fear became an actuality in Chernobyl, Republic of Ukraine, in 1986 when two water cooling systems failed. The chain of events that followed, including fire and a nonnuclear explosion, led to the release of radioactive gases that spread over much of Europe and into Asia. Cooling system problems were also behind the Three Mile Island accident in Pennsylvania in 1979. In this incident all radioactive substances were safely held within the reactor containment building, a safety feature in the design of U.S. power plants that is generally absent in the former USSR.

Whether getting energy from nuclear power plants is good or bad, ending the practice in the United States will not eliminate the danger. As Chernobyl demonstrated, the problem is global and cannot be solved by a single nation. In 1990, nuclear power plants produced 21% of the electricity in the United States. In 1990, nuclear power plants produced 75% of France's electricity, and 27% of Japan's electricity. Even if Americans chose to reduce their energy demands by 21%, other nations have chosen nuclear power as their way to become energy independent of imported oil. Nuclear power, for better or worse, is here to stay (Figure 20.10).

By 2010, 43% of Japan's electricity is expected to be generated by nuclear power plants.

Figure 20.10

Nuclear fuel. A uranium fuel pellet produces energy equal to the energy produced by about one ton of coal.

Even if a nuclear accident never occurs, there is still the problem of how and where to dispose of the dangerous radioactive wastes from nuclear reactors. One method is to collect them in large containers that may be buried in the earth. People who live and work near such disposal sites are seldom enthusiastic about this solution.

Finally, there is fear that some irresponsible government may use nuclear fuel for the manufacture of nuclear weapons, spreading the threat of atomic warfare. More frightening is the possibility that some terrorist group might steal the materials needed to build a bomb. While these threats cannot be removed from the earth today, perhaps they would be lessened if the large-scale production of nuclear fuel for electric power were eliminated.

On the other side of all these concerns is, of course, the question, "If we do not build and operate nuclear power plants, how else will the energy needs of the coming decades be met?" Perhaps the next section offers one answer, if it can only be reached in time.

20.12 NUCLEAR FUSION

There is nothing new about nuclear energy. Humans did not invent it. In fact, without knowing it, humanity has been enjoying its benefits since the beginning of recorded time, and before. In its common form, though, we do not call it nuclear energy. We call it solar energy. It is the energy that comes from the sun.

The energy the earth derives from the sun comes from another type of nuclear reaction called **nuclear fusion, in which two small nuclei combine to form a larger nucleus.** The smaller nuclei are "fused" together, you might say. The typical fusion reaction believed to be responsible for the heat energy radiated by the sun is represented by the equation

$$\,^2_1H + \,^3_1H \longrightarrow \,^4_2He + \,^1_0n$$

Fusion processes are, in general, more energetic than fission reactions. The fusion of one gram of hydrogen in the above reaction yields about four times as much energy as the fission of an equal mass of uranium-235. So far we have been able to produce only one kind of fusion reaction, and that has been the explosion of a hydrogen bomb.

Much research effort is being made to develop nuclear fusion as a source of useful energy. It has several advantages over a fission reactor. It presents more energy per given quantity of fuel. The isotopes required for fusion are far more abundant than those needed for fission. Best of all, fusion yields no radioactive waste, removing both the need for extensive disposal systems and the danger of accidental release of radiation to the atmosphere.

The main obstacle to be overcome before energy can be obtained from fusion is the extremely high temperature needed to start and sustain the reaction. The trigger for a hydrogen bomb is the heat generated by an atomic bomb. Furthermore, no substance known can hold fusion reactants at the needed temperature. Experiments are now in progress on "magnetic containment," in which the fuel is suspended in a magnetic field. Energy is then added by pulsing laser beams into the magnetic bottle. Temperatures as high as 40 million degrees Celsius have been reported, but could not be sustained.

Even if the technological obstacles to energy from fusion are overcome, time remains a serious problem. Only the most optimistic predictions foresee an operating plant in this century, and it will probably be well into the next before a significant portion of our energy needs can be supplied by fusion.

✔ QUICK CHECK 20.7

How do the products of a fusion reaction compare with the reactants?

Medicine and Radionuclides

Hospitals and larger medical clinics typically have a Department of Nuclear Medicine. This department is responsible for production, use, and disposal of radio-active materials used at the medical facility. Medical uses of radionuclides fall into two broad categories, diagnostic and therapeutic. A large hospital could use as many as 47 different radionuclides in as many as 194 diagnostic procedures and 29 therapeutic procedures.

Nearly all diagnostic radionuclides emit gamma rays, which are easily detected. A gamma ray is like a nuclear needle; it makes a quick, exceedingly narrow passage through the body, limiting damage to a small number of cells. These radionuclides must also have a short half-life to limit the time of the patient's exposure to radiation. The mechanism by which the body eliminates the radionuclide must be known. Finally, the chemical behavior of the radionuclide must not interfere with normal body functions.

Diagnostic Radionuclides

Nuclide		Half-Life	Emitted Particles	Uses
$^{51}_{24}Cr$	chromium-51	28 days	gamma	spleen imaging
$^{59}_{26}Fe$	iron-59	45 days	beta, gamma	bone marrow function
$^{99}_{43}Tc$	technetium-99	6 hours	gamma	bone, brain, liver, spleen imaging
$^{131}_{53}I$	iodine-131	8 days	beta, gamma	thyroid imaging
$^{201}_{81}Tl$	thallium-201	13 days	gamma	heart imaging

Therapeutic uses of radionuclides have different goals than diagnostic uses. Therapeutic radionuclides are used to destroy abnormal, usually cancerous, cells as selectively as possible. Cell poisons such as radiation destroy abnormal cells more rapidly than normal cells because abnormal cells divide more quickly than nor-mal cells. Therapeutic radionuclides are usually alpha or beta emitters. These decay particles cause heavy damage confined to a small area, due to their low penetrating power. In the body, an alpha or beta emitter is a nuclear bull in a cellular china shop.

Therapeutic Radionuclides

Nuclide		Half-Life	Emitted Particles	Uses
$^{32}_{15}P$	phosphorus-32	14 days	beta	some leukemias, widespread carcinomas
$^{60}_{27}Co$	cobalt-60	5.3 years	beta, gamma	external radiation source for cancer treatment
$^{90}_{39}Y$	yttrium-90	64 hours	beta, gamma	implanted in tumors
$^{131}_{53}I$	iodine-131	8 days	beta, gamma	thyroid cancer
$^{226}_{88}Ra$	radium-226	1620 years	alpha, gamma	implanted in tumors

CHAPTER 20 KEY WORDS AND MATCHING SETS

Set 1

_____ Unit of radiation intensity equal to 3.7×10^{10} disintegrations per second.

_____ Time in which 50% of a radionuclide disintegrates.

_____ Radiation unit that takes into account radiation damage to human tissue.

_____ Machine in which fissionable nuclear fuel is made from nuclides that are not fissionable.

_____ Radiation unit that generates 2.09×10^9 ion pairs in one cm^3 of dry air.

_____ Has atomic number greater than 92.

_____ Unit of radiation absorbed by body tissue.

_____ Sequence of steps in which a radionuclide found in nature eventually becomes a stable nuclide.

1. Rad
2. Curie
3. Transuranium element
4. Radioactive decay series
5. Breeder reactor
6. Half-life
7. Roentgen
8. Rem

Set 2

_____ Minimum amount of nuclear fuel to keep a chain reaction going.

_____ Measures radioactivity by counting flashes of light.

_____ Nuclear change in which two small nuclides combine to form a larger nuclide.

_____ Indicates a radioactive emission as a condensation track.

_____ Describes radioactivity from radionuclide not found in nature.

_____ Device in which ionized gas causes a "click."

_____ Nuclear change in which a large nuclide separates into two smaller nuclides.

_____ Estimating age of an object by measuring radioactive decay.

1. Induced or artificial
2. Scintillation counter
3. Radiocarbon dating
4. Critical mass
5. Cloud chamber
6. Geiger-Müller counter
7. Fusion
8. Fission

Set 3

_____ Symbol of a positron.

_____ Symbol of an α-particle.

_____ Symbol of a β-particle.

_____ Machine by which charged particles can be made to strike nuclides.

_____ Nuclear reaction in which one element changes into another.

_____ Reaction in which a nuclide is hit by an atomic nucleus or subatomic particle.

_____ Nuclear reaction that continues because a product of the reaction becomes a reactant in the next step.

1. Particle accelerator
2. Nuclear bombardment
3. $_{-1}^{0}e$
4. Transmutation
5. $_{2}^{4}He$
6. Chain reaction
7. $_{1}^{0}e$

CHAPTER 20 QUESTIONS AND PROBLEMS

SECTION 20.1 AND 20.2

1) What is a nuclide? How does a nuclide differ from an isotope?

2) Distinguish between a stable nucleus and a nucleus that is not stable.

---

3) *Decay* is a term used to describe what happens to a radioactive nucleus. What does *decay* mean in this sense?

4) What three kinds of emission are produced in radioactive decay? Describe the composition, or "structure," of each. Write the nuclear symbol, if any, for each kind of emission.

---

5) Compare the three principal forms of radioactive emission in terms of mass, electrical charge, and penetrating power.

6) One kind of radioactive emission is distinctly different from the other two. Identify the "different" one and tell how or why it is different.

SECTION 20.3

7) What happens, or might happen, when an emission from a radioactive substance collides with an atom or a molecule? Is this harmful?

8) Radiation is often described as "ionizing radiation." What is the meaning of this term? Is all radiation "ionizing" radiation? Justify your answer.

---

9) Identify some properties of radioactive emissions that are used in detecting and measuring them.

10) What is a scintillation counter? How does it work?

---

11) What is a Geiger-Müller counter (or simply Geiger counter)? How does it work?

12) How do Geiger and scintillation counters differ in their ways of telling an observer that an object is radioactive? Can either or both be used to measure radiation as well as to detect it? If so, precisely what is measured?

---

13) Distinguish between a gamma camera and a scanner.

14) What advantages do the gamma camera or scanner have over alternative medical procedures?

SECTION 20.4

15) Distinguish between the curie and the roentgen as radiation units.

16) Distinguish between the roentgen and the rad as radiation units.

---

17) Distinguish between the rad and the rem as radiation units.

18)* People who work with radioactive materials or in a radioactive atmosphere wear devices called dosimeters to monitor their exposure to radiation. Can you pick apart the word *dosimeter* and guess in which of the four units in Questions 15–17 its measurements are expressed? Is it the most useful of the four units for its purpose? If so, why; if not, why not? Also, if not, which unit would better serve the purpose, and why is it not used?

---

19) What is background radiation?

20) Is background radiation dangerous? What steps, if any, should we take to minimize our exposure to background radiation?

SECTION 20.5

21) What is the meaning of half-life as applied to radioactivity? What range of half-lives is found in natural radioactive decay series?

22)* If you were a supplier of radioactive materials to industrial or medical users, how would the half-life of different materials influence your decisions about how much, if any, you might keep in inventory for quick delivery to your customers? Explain.

---

23) What fraction of a radionuclide remains after the passage of five half-lives?

24) How many half-lives have passed when a radioactive substance has lost 7/8 of its radioactivity?

---

25)* Suppose you have a radionuclide A that goes through a two-step decay sequence, first to B, and then to C, which is stable. Suppose also that the half-life from A to B is six days, and the half-life from B to C is 1 day. Predict by listing in declining order, greatest to smallest, the amounts of A, B, and C that will be present (a) at the end of six days and (b) at the end of twelve days. Explain your prediction.

26)* Same question as Question 25, except that the half-life from A to B is one day, and from B to C is 6 days.

---

27) Calculate the amount of radionuclide that will be left after 33 minutes if the initial mass is 12.9 grams and its half-life is 11.0 minutes.

28) One way to measure rate of decay is by the rate of "clicks" from a Geiger counter. If, over a period of time, the click rate drops from 188 clicks per second to 47 clicks per second, how many half-lives will have passed? If the time period is 132 minutes, what is the half-life of the radionuclide?

_____

29) One of the more hazardous radioactive isotopes in the fallout of atomic bombs is strontium-90, for which the half-life is 28 years. If 654 g $^{90}_{38}$Sr fall on a family farm on the day a child is born in 1994, how many grams will still be on the land when the farmer's granddaughter is born in 2050? How about when the granddaughter marries on the same farm in 2070?

30) An artificial nuclide widely used in radiotherapy is $^{60}_{27}$Co, which has a half-life of 5.2 years.
a) What percentage of the stored amount of this isotope is lost annually due to radioactive decay?
b) What mass of a 105-gram sample of this nuclide will remain after 13 years?
c) Cobalt-62, with a half-life of 14 minutes, delivers even higher energy radiation than cobalt-60. Why isn't cobalt-62 used for radiotherapy instead of, or in addition to, cobalt-60?

_____

31) Uranium-235, the uranium isotope used in making the first atomic bomb, is the starting point of one of the natural radioactivity series. The next isotope in the series is thorium-231. At the beginning of a test period a sample contained 9.53 grams of the thorium isotope. After 83.2 hours only 1.05 grams of the original isotope remained. What is the half-life of thorium-231?

32) Iodine-131 is used for thyroid scans. The patient is scanned 24 hours after administration of the radionuclide, at which time the radioactivity of the iodine-131 is 92% of its initial value. What is the half-life of iodine-131 in hours?

_____

33) The half-life of $^{208}_{81}$Tl is 3.1 minutes. A 84.6-gram sample is studied in the laboratory.
a) How many grams of the isotope will remain after 12 minutes?
b) In how many minutes will the mass of $^{208}_{81}$Tl be 3.48 grams?

34) 16.3% of a sample of a radionuclide remain after the sample is first observed. If the half-life of the isotope is 17.2 minutes, how long has it been since that first observation?

_____

35) While excavating for the foundation of a new building, a contractor uncovered human skeletons in what turned out to be a burial ground for an ancient civilization. They were taken to a nearby university and submitted to radiocarbon dating analysis. There it was found that the bones emit radiation at a rate of 55% of the rate of a living organism. How many years ago did the specimen die? (Use 5730 years as the half-life of carbon-14.)

36)* A fragment of cloth found just outside of Jerusalem is believed to have been used by some person at about the beginning of the Christian era. Analysis shows that radiation from the fragment is 22.7 units, whereas radiation from a living specimen is 29.0 units when measured with the same instrument. Is it possible that the fragment might have come from the period believed? Justify your answer.

SECTION 20.6
37) What happens to the nucleus of an atom that experiences an alpha decay reaction? Compare the final nuclide with the original nuclide. Does the element undergo transmutation?

38) What happens to the nucleus of an atom that experiences a beta decay reaction? Compare the final nuclide with the original nuclide. Does the element undergo transmutation?

_____

39) Write nuclear equations for the beta emissions of $^{228}_{89}$Ac and $^{212}_{83}$Bi.

40) Write nuclear equations for the beta decay of $^{212}_{82}$Pb and $^{231}_{90}$Th.

_____

41) Write nuclear equations for the alpha decay of $^{216}_{84}$Po and $^{234}_{92}$U.

42) Write nuclear equations for the ejection of alpha particles from $^{238}_{90}$Th and from $^{212}_{84}$Po.

SECTION 20.7
43) Why is it possible to speak of the "chemical properties of lead," but not the "nuclear chemical properties of lead"?

44) How do the chemical properties of carbon-12 compare with the chemical properties of carbon-14? If there is a difference, explain why.

_____

45)* A sample of pure calcium chloride is prepared in a laboratory. A small but measurable amount of the calcium in the compound is made up of calcium-47 atoms, which are radioactive beta emitters with a half-life of 4.53 days. The compound is securely stored for a week in an inert atmosphere. When it is used at the end of that period, it is no longer pure. Why? With what element would you expect it to be contaminated?

46) A fundamental idea of Dalton's atomic theory is that atoms of an element can be neither created nor destroyed. We now know that this is not always true. Specifically, it is not true for uranium and lead atoms as they appear in nature. Are the numbers of these atoms increasing or decreasing? Explain.

_____

47) The radioactivity of a sample of dirt containing a uranium compound records 5000 counts per minute when measured with a Geiger counter. The sample is treated physically to isolate the uranium compound, which is then decomposed chemically into pure uranium. If you disregard any loss of radioactivity because of decay during the purification process, will the pure uranium still radiate at 5000 counts per minute, will it be more than 5000, or will it be less than 5000? Explain your answer.

48)* Two uranium sulfides have the formulas US and US_2. A laboratory prepares 50.0 grams of each compound. If the uranium in each compound has all of the isotopes of uranium in their normal distribution in nature, which, if either, will exhibit the greater amount of radioactivity? If the laboratory prepares 0.50 mole of each compound, which, if either, would be more radioactive? Explain both answers.

SECTION 20.8

49) Distinguish between nuclear reactions that begin by radioactivity and those that begin with nuclear bombardment. What is nuclear bombardment?

50) What distinguishes induced radioactivity from natural radioactivity? What can be said about the elements that are naturally radioactive compared to those that exhibit induced radioactivity?

---

51) What is a particle accelerator? What is its role in producing a nuclide that is radioactive?

52)* What does a particle accelerator do and how does it do it? What kinds of particles can be accelerated in a particle accelerator, and what kinds cannot? Explain.

---

53) How do transuranium elements differ from other elements? Where are the transuranium elements found on the periodic table?

54) Have all of the transuranium elements been discovered? If there is an element whose atomic number is 118, will it be a transuranium element? What sort of chemical properties would you expect it to have? Do you think it would be radioactive? Why?

Questions 55 and 56: Complete each nuclear bombardment equation by supplying the nuclear symbol for the missing species.

55)
$$^{44}_{20}Ca + {}^{1}_{1}H \longrightarrow ? + {}^{1}_{0}n$$
$$^{252}_{98}Cf + {}^{10}_{5}B \longrightarrow 5\,{}^{1}_{0}n + ?$$
$$^{106}_{46}Pd + {}^{4}_{2}He \longrightarrow {}^{109}_{47}Ag + ?$$

56)
$$^{98}_{42}Mo + {}^{2}_{1}H \longrightarrow ? + {}^{1}_{0}n$$
$$^{238}_{92}U + {}^{4}_{2}He \longrightarrow 3\,{}^{1}_{0}n + ?$$
$$? + {}^{2}_{1}H \longrightarrow {}^{60}_{27}Co + {}^{1}_{1}H$$

SECTIONS 20.10 TO 20.12

57) Distinguish between fission reactions and fusion reactions. Compare fission and fusion reactions with radioactive decay. How are they similar and how are they different?

58) Distinguish between an atomic bomb and a hydrogen bomb. Classify the explosions produced by each as a fission reaction or a fusion reaction.

---

59) What is a chain reaction? What essential feature must be present in a nuclear reaction before it can become a chain reaction?

60) Starting a chain reaction is one thing; keeping it going is another. What additional requirement is there if a chain reaction is to continue? By what term is this requirement identified?

---

61) What advantages do fusion reactions have over fission reactions as a source of nuclear power? If fusion reactions are more desirable than fission reactions, why don't we use them in power plants instead of fission reactions?

62) List some of the advantages and disadvantages of nuclear power plants compared to other sources of electrical energy. In your opinion, do the advantages outweigh the disadvantages?

GENERAL QUESTIONS

63) Distinguish precisely and in scientific terms the differences between items in each of the following groups.
a) Alpha, beta, and gamma radiation
b) X-rays, γ-rays
c) α-particle, β-particle
d) Curie (the unit), roentgen, rad, rem
e) Natural and induced radioactivity
f) Chemical reaction, nuclear reaction
g) Isotope, nuclide, radionuclide
h) Element, transuranium element
i) Nuclear fission, nuclear fusion
j) Atomic bomb, hydrogen bomb

64) Classify each of the following statements as true or false.
a) A radioactive atom decays in the same way whether or not the atom is chemically bonded in a compound.
b) The chemical properties of a radioactive atom of an element are different from the chemical properties of a nonradioactive atom of the same element.
c) α-rays have more penetrating power than β-rays.
d) α- and β-rays are particles, but a γ-ray is an "energy ray."
e) The curie measures radiation without respect to its effect.
f) For a given amount of radiation, 1 rad $>$ 1 roentgen.

g)   Radiation absorbed by living plant or animal tissue is measured in rems.

h)   Radioactivity is a nuclear change that has no effect on the electrons in nearby atoms.

i)   The number of protons in a nucleus changes when it emits a beta particle.

j)   The mass number of a nucleus changes in an alpha emission but not in a beta emission.

k)   Isotopes with higher atomic numbers generally have longer half-lives than isotopes with lower atomic numbers.

l)   The first transmutations were achieved by alchemists.

m)   Radioisotopes are made by bombarding a nonradioactive isotope with atomic nuclei or subatomic particles.

n)   The atomic numbers of products of a fission reaction are smaller than the atomic number of the original nucleus.

o)   Nuclear power plants are a safe source of electrical energy.

p)   The main obstacle to developing nuclear fusion as a source of electrical energy is a shortage of nuclei to serve as "fuel."

65)*   The examples and most of the problems having

to do with half-life compared initial and final quantities of a radioactive "substance" in terms of mass. This simplification is usually unrealistic. Can you suggest a reason why? If mass is not a suitable measure of an amount of radioactive matter, what is?

66)   A major form of fuel for nuclear reactors used to produce electrical energy is a fissionable isotope of plutonium. Plutonium is a transuranium element. Why is this element used instead of a fissionable isotope that occurs in nature?

67)   Why is half-life used for measuring rate of decay rather than the time required for the complete decay of a radioactive isotope?

68)   A ton of high grade coal has an energy output of about 2.5×10^7 kJ. The energy released in the fission of one mole of $^{235}_{92}$U is about 2.0×10^{10} kJ. How many tons of coal could be replaced by one pound of uranium-235, assuming the material and the technology were available?

MATCHING SET ANSWERS

Set 1: 2–6–8–5–7–3–1–4   Set 2: 4–2–7–5–1–6–8–3   Set 3: 7–5–3–1–4–2–6

QUICK CHECK ANSWERS

20.1 a) α: $^4_2$He, 2+; β: $^{\ 0}_{-1}$e, 1−.   b) Gamma, beta, alpha

20.2 Geiger and scintillation counters "count" individual radioactive emissions by measuring electric current (Geiger) or light intensity (scintillation) that is produced by the radiation. These are both proportional to the intensity of radiation and are interpreted in that way. This intensity can be expressed as so many counts per unit of time.

20.3 a) A radiation counting device records curies, a measure of the intensity of radiation. It is not concerned with a "dose" of radiation, as is the roentgen; how much is absorbed by body tissue, as is the rad; nor how much is absorbed in a human body, as is the

rem. b) Alpha radiation outside the body is unable to penetrate the skin, so it does no internal damage. Alpha radiation inside the body has nothing to stop it from damaging tissue. c) Don't smoke!

20.4 $^{222}_{86}$Rn → $^4_2$He + $^{218}_{84}$Po

20.5 a) Nuclear bombardment reactions produce isotopes that do not exist in nature. b) To be accelerated, a particle must have an electrical charge.

20.6 Isotopes produced in a fission reaction have smaller atomic numbers than the starting isotope.

20.7 An isotope produced by a fusion reaction has a larger atomic number than the starting isotopes.

NUMBER OF RADIOACTIVE $^{40}_{19}$K ATOMS IN A 150-POUND MAN

$$225 \ \cancel{g \ K} \times \frac{6.02 \times 10^{23} \ \cancel{K \ atoms}}{39.1 \ \cancel{g \ K}} \times \frac{0.0118 \ ^{40}_{19}K \ atoms}{100 \ \cancel{K \ atoms}} = 4.1 \times 10^{20} \ ^{40}_{19}K \ atoms$$

Organic Chemistry

Nearly everything in this picture of everyday products is made up of organic chemicals. Some of these products have a vegetable origin (apple, potatoes, paper), some are animal (leather glove, hamburger), some are both (petroleum), and some are synthetic (various plastic items). The only inorganic substances are the metal in the liquid plastic can and the iron on the cassette tape.

21.1 THE NATURE OF ORGANIC CHEMISTRY

Late in the eighteenth century, an emerging philosophic movement proposed the existence of an *elan vital,* or life force. (Mary Shelly, wife of the poet, wrote the novel *Frankenstein* in 1818 as an expression of this idea.) Every living organism contained this life force, which was responsible for production of the substances found only in living systems. These substances became known as **organic** compounds.

In 1828 Friedrich Wohler heated ammonium cyanate, which everyone agreed was not organic, and obtained urea, which is found in urine and which everyone agreed *is* organic. Organic chemistry then began to be known as the **chemistry of carbon compounds.** Carbonates, cyanides, oxides of carbon, and a few other compounds are exceptions that are still classified as inorganic.

Organic chemistry is not "different" from inorganic chemistry. It is only larger, even though all organic compounds contain the same element. All of the chemical principles we have studied for inorganic chemicals, such as bonding, reaction rates, equilibrium, and others, apply equally to organic chemicals.

21.2 THE MOLECULAR STRUCTURE OF ORGANIC COMPOUNDS

In 1965 the Chemical Abstracts Service* began assigning a registry number to each new substance reported. The total has passed 10,000,000, and about 95% of these new substances are organic. The distinguishing feature of organic chemistry is the carbon–carbon bond. Of all 109 elements, only carbon forms long chains of stable bonds with itself. This is why there are so many carbon compounds.

Table 21.1 summarizes the covalent bonding properties of carbon, hydrogen, oxygen, nitrogen, and the halogens, the elements most frequently found in organic compounds. All the bond geometries are indicated—the linear, planar, or three-dimensional shapes and, where constant, the actual bond angles. Of particular significance is the number of covalent bonds that atoms of the different elements can form. This is determined by the electron configuration of the atom. The bonding relationships of these elements are basic to your understanding of the structure of organic compounds.

When carbon forms four single bonds, they are arranged tetrahedrally around the carbon atom; the molecular geometry is tetrahedral (see Fig. 21.1). Recall from Chapter 12 that it is not possible to represent this three-dimensional shape accurately in a two-dimensional sketch. Thus the four bonds radiating from each carbon atom in

$$x-\underset{\underset{x}{|}}{\overset{\overset{x}{|}}{C}}-\underset{\underset{z}{|}}{\overset{\overset{z}{|}}{C}}-\underset{\underset{y}{|}}{\overset{\overset{y}{|}}{C}}-y$$

all form tetrahedral bond angles (109.5°). Furthermore, all three x positions are geometrically equal, three y positions are equal, and two z positions are equal. It follows that

Figure 21.1

Tetrahedral models. The solid figure is a tetrahedron, the simplest regular solid. Its four faces are identical equilateral triangles. The model of methane, CH_4, has a tetrahedral structure. The carbon atom is in the middle of the tetrahedron, and a hydrogen atom is found at each of the four vertices.

* *Chemical Abstracts* is a weekly summary of articles that have appeared in original research journals. It is published by the American Chemical Society.

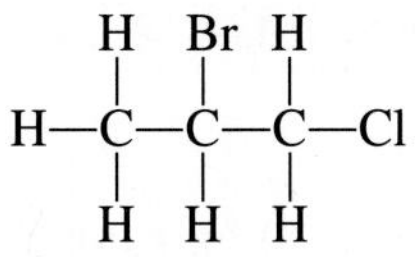

$$
\begin{array}{ccc}
& \text{H}\ \ \text{Br}\ \ \text{H} & \\
& |\ \ \ \ |\ \ \ \ | & \\
\text{H}-&\text{C}-\text{C}-\text{C}&-\text{Cl} \\
& |\ \ \ \ |\ \ \ \ | & \\
& \text{H}\ \ \text{H}\ \ \text{H} &
\end{array}
\qquad \text{and} \qquad
\begin{array}{ccc}
& \text{H}\ \ \text{H}\ \ \text{Cl} & \\
& |\ \ \ \ |\ \ \ \ | & \\
\text{H}-&\text{C}-\text{C}-\text{C}&-\text{H} \\
& |\ \ \ \ |\ \ \ \ | & \\
& \text{H}\ \ \text{Br}\ \ \text{H} &
\end{array}
$$

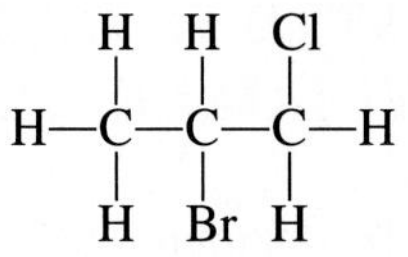

are the same compound.

Organic chemists do not usually show lone pair electrons in Lewis diagrams, unless there is specific reason for doing so. This is why they are not found around Br and Cl in the diagrams here.

HYDROCARBONS

The simplest organic compounds are the **hydrocarbons, binary compounds of carbon and hydrogen.** Two broad classifications of hydrocarbons are those called aliphatic hydrocarbons and aromatic hydrocarbons. The distinction between them will appear later. Aliphatic hydrocarbons are further subdivided into alkanes, cycloalkanes, alkenes, and alkynes. As you will see shortly, these classifications are all based on molecular structure.

Table 21.1 shows that a carbon atom is able to form four bonds to a maximum of four other atoms. A hydrocarbon in which each carbon atom is bonded to that maximum of four other atoms is **saturated** in the sense that there is no room for the carbon atom to form a bond to another atom. The table also shows that if a carbon atom is double- or triple-bonded, it is bonded to three or two other atoms. This is fewer than the maximum number of atoms the carbon atom is capable of bonding with. Such a hydrocarbon is **unsaturated** because it can form bonds to additional atoms.

Table 21.1
Bonding in Organic Compounds

Element	Number of Bonds*	Bond Geometry			
		Single Bond	**Double Bond**	**Double Bonds**	**Triple Bond**
Carbon	4	$\overset{|}{\underset{\diagdown}{\diagup}}$C Tetrahedral: 109.5° angles	$\diagdown$C= $\diagup$ Planar: 120° angles	=C= Linear: 180° angle	—C≡ Linear: 180° angle
Hydrogen	1	H—			
Halogens	1	:Ẍ—			
Oxygen	2	Ö Bent structure	Ö= Bent structure		
Nitrogen	3	N̈ Pyramidal structure	N̈ Bent structure		:N≡

*Number of bonds to which an atom of the element shown can contribute *one* electron.

FLASHBACK Table 21.1 is derived largely from Table 12.1, Sections 12.3 and 12.4. The earlier table includes pictures that show how molecular structure is related to electron pair geometry and the number of atoms bonded to the central atom.

21.3 SATURATED HYDROCARBONS: THE ALKANES AND CYCLOALKANES

In an **alkane, each carbon atom forms single bonds to four other atoms and there are no multiple bonds.** Most of the carbon atoms in an alkane are arranged in a continuous chain in which all bond angles are 109.5°, the tetrahedral angle. If *all* carbon atoms in the molecule are in the continuous chain, the compound is a **normal alkane.** In other alkanes some carbon atoms appear as "branches" off the main chain. They are isomers of the normal alkane having the same number of carbon atoms.

The first 10 alkanes are shown in Table 21.2. Careful examination of the formulas shows that they have the form C_nH_{2n+2}, where n is the number of carbon atoms. Notice also that the difference from one alkane to the next is one carbon and two hydrogen atoms, a $-CH_2-$ structural unit. A series of compounds in which each member differs from the members before it by a $-CH_2-$ unit is called a **homologous series.**

There are three ways to write the formula of an alkane. Using octane, the alkane with eight carbons, as an example, there is the molecular formula, C_8H_{18}. A molecular formula gives no information as to how atoms are arranged. A Lewis diagram, commonly referred to as a **structural formula** or **structural diagram,** shows that arrangement. A compromise between them is the **line formula,** which includes each CH_2 unit in the alkane chain. The line formula for octane is $CH_3CH_2CH_2CH_2CH_2CH_2CH_2CH_3$. A long line formula is usually shortened by grouping the CH_2 units: $CH_3(CH_2)_6CH_3$. This is called a **condensed formula.**

Table 21.2 also serves to introduce organic nomenclature. Notice that each alkane is named by combining a prefix and a suffix. The prefix indicates the number of carbon atoms in the compound, as shown in the two right columns. The suffix identifying an *alkane* is *-ane.* Thus, the name of methane comes from combining the prefix *meth-,* indicating one carbon, with the suffix *-ane,* indicating an alkane. The same prefixes are used to name other organic compounds and groups as well.

FLASHBACK Recall from Section 12.2 that isomers are two or more compounds that have the same molecular formula but different structures. They are distinctly different compounds, each with its own set of physical and chemical properties.

Table 21.2

The Alkane Series

Molecular Formula	Name	Number of Carbon Atoms	Prefix
CH_4	Methane	1	Meth-
C_2H_6	Ethane	2	Eth-
C_3H_8	Propane	3	Prop-
C_4H_{10}	Butane	4	But-
C_5H_{12}	Pentane	5	Pent-
C_6H_{14}	Hexane	6	Hex-
C_7H_{16}	Heptane	7	Hept-
C_8H_{18}	Octane	8	Oct-
C_9H_{20}	Nonane	9	Non-
$C_{10}H_{22}$	Decane	10	Dec-

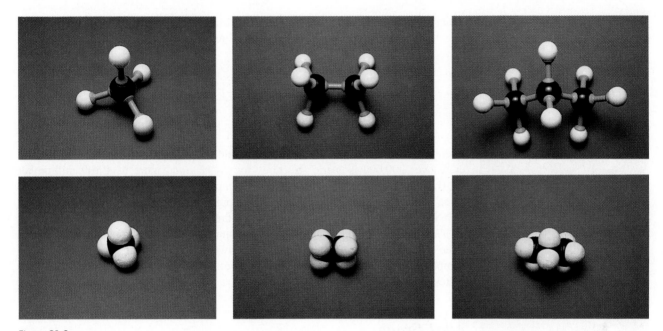

Figure 21.2

Space-filling models of the three simplest alkanes, methane, CH_4, ethane, C_2H_6, and propane, C_3H_8. There is a tetrahedral orientation of all four bonds around each carbon atom.

While there is only one possible structure for methane, ethane, and propane (Fig. 21.2), there are two possible isomers of butane, shown in Figure 21.3. Their Lewis diagrams are:

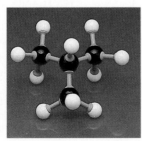

Figure 21.3

Models of two isomers of butane, C_4H_{10}. The molecule at the left, *n*-butane, has its four carbon atoms in a continuous chain. At the right, the molecule of *iso*butane has three carbon atoms in a row with the fourth carbon atom bonded to the middle atom of the three.

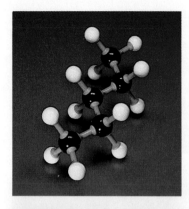

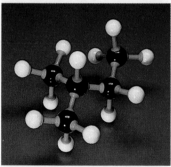

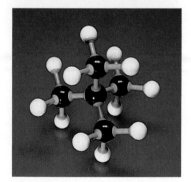

Figure 21.4

Models of three isomeric pentanes. Each molecule contains five carbon atoms and twelve hydrogen atoms: C_5H_{12}. The atoms are arranged in a different way in each of the three isomers.

Pentane has three isomers (Fig. 21.4). Their structures, showing only the carbon skeletons to make the diagrams less "cluttered," are

$$
\text{C—C—C—C—C} \qquad
\begin{matrix}
\text{C—C—C—C} \\
\mid \\
\text{C}
\end{matrix}
\qquad
\begin{matrix}
\text{C} \\
\mid \\
\text{C—C—C} \\
\mid \\
\text{C}
\end{matrix}
$$

n-pentane   isopentane   neopentane

There are 5 isomeric hexanes, 9 heptanes, and 75 possible decanes. It is possible to draw over 300,000 isomeric structures for $C_{20}H_{42}$ and more than 100 million for $C_{30}H_{62}$. Obviously, not all of them have been prepared and identified! This does give some idea, though, why there are so many organic compounds.

Notice beneath the diagrams above that the name of a normal alkane begins with *n-*. The *iso-* and *neo-* prefixes identify compounds in which a carbon atom is bonded to three or four other carbon atoms, respectively, but they have little use beyond these compounds. Instead, the International Union of Pure and Applied Chemistry (IUPAC) system is used. This system is based on *alkyl groups.*

Alkyl Groups

An **alkyl group is an alkane from which an end hydrogen atom has been removed.** If, on paper, we remove a hydrogen atom from methane, CH_4, we get $—CH_3$, where the dash indicates a bond that the alkyl is able to form with another atom. This $—CH_3$ group, appearing in the structural formula of a compound, is called a **methyl group.** The term is made up of the prefix *meth-* for one carbon (Table 21.2) and the suffix *-yl,* which is applied to alkyl groups. If we compare two compounds,

we see that the colored H in the first compound has been replaced by a $—CH_3$ group, or methyl group, in the second. If the replacement group has two carbon atoms

it is an ethyl group, $—C_2H_5$, one hydrogen short of ethane, C_2H_6. All the alkyl groups are similarly named.

Frequently, we wish to show a bonding situation in which *any* alkyl group may appear. The letter R is used for this purpose. Thus R—OH could be CH_3OH, C_2H_5OH, C_3H_7OH, or any other alkyl group attached to an —OH group.

Naming the Alkanes by the IUPAC System

We are now ready to describe the IUPAC system of naming isomers of the alkanes, as well as other compounds we will encounter shortly. The system follows a set of rules:

S U M M A R Y / IUPAC Names of Alkanes and Cycloalkanes

1) **Identify as the parent alkane the longest continuous chain.** For example, in the compound having the structure

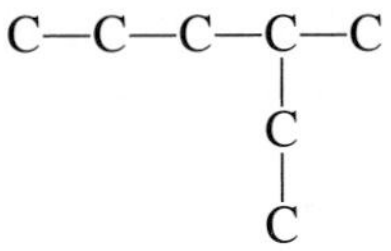

$$C-C-C-C-C$$
$$|$$
$$C$$
$$|$$
$$C$$

the longest chain is six carbons long, not five as you might first expect. This is readily apparent if we number the carbon atoms in the original representation of the structure and an equivalent layout:

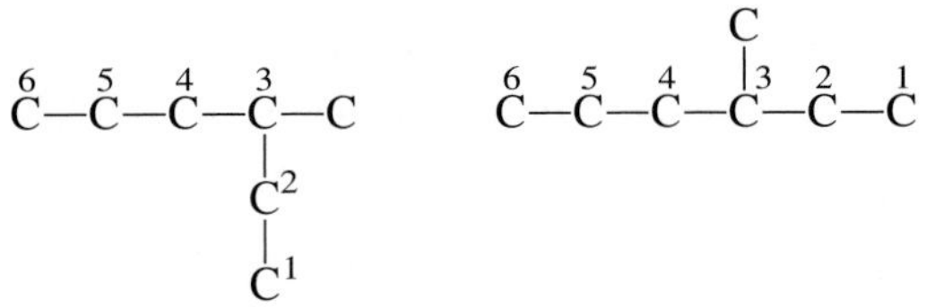

$$\overset{6}{C}-\overset{5}{C}-\overset{4}{C}-\overset{3}{C}-C \qquad \overset{6}{C}-\overset{5}{C}-\overset{4}{C}-\overset{3}{C}-\overset{2}{C}-\overset{1}{C}$$
$$\underset{C^2}{|}$$
$$|$$
$$C^1$$

with C over the 3 position in the second structure.

2) **Identify by number the carbon atom to which the alkyl group (or other species) is bonded to the chain.** In the example compound this is the *third* carbon, as shown. Notice that counting always begins at that end of the chain that places the branch on the *lowest* number carbon atom possible.

3) **Identify the branched group (or other species).** In this example the branch is a methyl group, —CH_3.

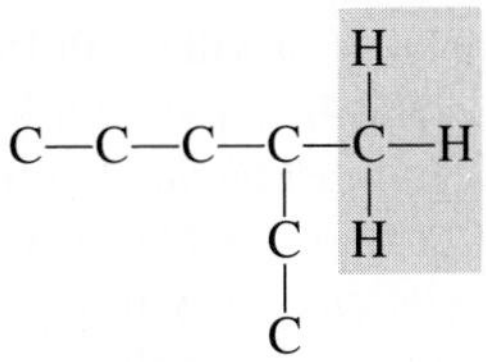

$$C-C-C-\overset{}{C}-\overset{\overset{H}{|}}{C}-H$$
$$\underset{C}{|} \; H$$
$$|$$
$$C$$

These three items of information are combined to produce the name of the compound, *3-methylhexane*. The 3 comes from the third carbon (Step 2); methyl comes from the branch group (Step 3); and hexane is the parent alkane (Step 1).

Sometimes the same branch appears more than once in a single compound. This situation is governed by the following rule:

4) **If the same alkyl group, or other species, appears more than once, indicate the**

number of appearances by di-, tri-, tetra-, etc., and show the location of each branch by number. For example

$$
\begin{array}{ccccc}
 & & \text{C} & \text{C} & \\
 & & | & | & \\
\text{C}-\text{C}-\text{C}-\text{C}-\text{C} & & & & \\
\end{array}
$$

is 2,3-dimethylpentane. In the other direction, to write the structural formula for 2,2,5-trimethylhexane, we would establish a six-carbon skeleton and attach methyl groups as required, two to the second carbon and one to the fifth:

$$
\begin{array}{ccccc}
 & \text{C} & & \text{C} & \\
 & | & & | & \\
\text{C}-\text{C}-\text{C}-\text{C}-\text{C}-\text{C} & & & & \\
 & | & & & \\
 & \text{C} & & & \\
\end{array}
$$

Twice above we have referred to "other species"—species other than an alkyl group that might be attached to a hydrocarbon. The most common species other than an alkyl is a halogen atom. If two chlorine atoms take the places of hydrogen atoms on the second and third carbon atoms of *n*-pentane, for example, or take the places of the methyl groups in 2,3-dimethylpentane above, the compound would be 2,3-dichloropentane. This leads to the next nomenclature rule:

5) **If two or more different alkyl groups, or other species, are attached to the parent chain, they are named in alphabetical order.** By this rule the compound

$$
\begin{array}{ccccc}
 & \text{Cl} & \text{Br} & & \\
 & | & | & & \\
\text{C}-\text{C}-\text{C}-\text{C}-\text{C} & & & & \\
\end{array}
$$

is 3-bromo-2-chloropentane. The structural diagram for 2,2-dibromo-4-chloroheptane is

$$
\begin{array}{ccccccc}
 & \text{Br} & & \text{Cl} & & & \\
 & | & & | & & & \\
\text{C}-\text{C}-\text{C}-\text{C}-\text{C}-\text{C}-\text{C} & & & & & & \\
 & | & & & & & \\
 & \text{Br} & & & & & \\
\end{array}
$$

FLASHBACK In Sections 14.1 and 14.2 we discussed the relationships between intermolecular forces and physical properties. Substances with strong intermolecular forces tend to have higher boiling points, heats of vaporization, viscosity, and surface tension, and lower vapor pressures. The forces, in increasing strength, were classified as dispersion (London) forces, dipole-dipole forces, and hydrogen bonds. Dispersion forces increase with molecular size, and with large molecules they may exert more influence on physical properties than dipole-dipole forces and hydrogen bonds.

Physical Properties of the Alkanes

Alkane molecules are all nonpolar. You will recall from Chapter 14 that intermolecular forces between nonpolar molecules increase with increasing molecular size. As a result, larger molecules have higher melting and boiling points. Among normal alkanes, those with continuous chains, compounds having fewer than five carbon atoms, have the weakest intermolecular attractions. They have low boiling points and are gases at room temperature. All are used as fuels; methane is the main constituent of "natural gas."

Intermolecular forces are stronger between larger alkanes from C_5H_{12} to $C_{17}H_{36}$. These higher-boiling compounds are liquids at room temperature. Several of the lower-molar-mass liquid alkanes containing five to ten carbon atoms are present in gasoline. Diesel fuel and lubricating oils are made up largely of higher-molar-mass liquid alkanes. Alkanes with molar masses greater than 300 are normally solids at room temperature.

Cycloalkanes

If a hydrogen atom is removed from each end carbon of a normal alkane, the two end carbons can bond to each other to form a ring or cycle of carbon atoms. This ring compound, in which all carbon atoms are saturated, is a **cycloalkane.** Cycloalkanes can form with a minimum of three carbon atoms; however, the resulting 60° bond angles are severely strained from the normal tetrahedral angle, so cyclopropane is unstable. The more common alkanes are those whose bond angles are close to or equal to the tetrahedral angle, such as cyclopentane and cyclohexane.

Three structural diagrams of cyclohexane are

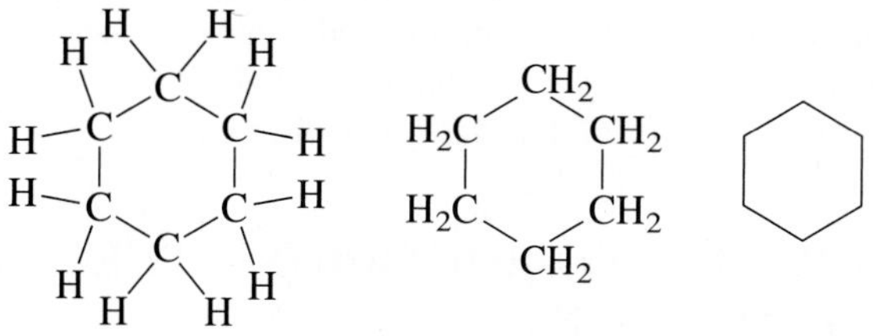

The left diagram is the most complete. The middle diagram illustrates the common practice of using —CH_2— when a carbon bonded to two hydrogen atoms is also bonded to two other atoms, as in a saturated chain. The diagram at the right is a skeleton diagram in which the vertex of each angle shows the relative location of a carbon atom. It is understood in such diagrams that each carbon atom forms as many additional bonds to hydrogen atoms as are necessary to bring its total number of bonds to four. Confusing? Only temporarily. Not nearly as confusing as it would be if this system were not used in the many highly complex structures in which cycloalkanes and other ring compounds appear. You will see some in the next chapter.

To name a cycloalkane, apply the prefix *cyclo-* to the name of the open-chain alkane with the same number of carbon atoms. The diagrams above have six carbon atoms in the ring. Hence the name, cyclohexane. If one or more alkyl groups or other species, such as a halogen, takes the place of one of the invisible hydrogen atoms in the ring, nomenclature rules 3 and 4 are applied. Thus

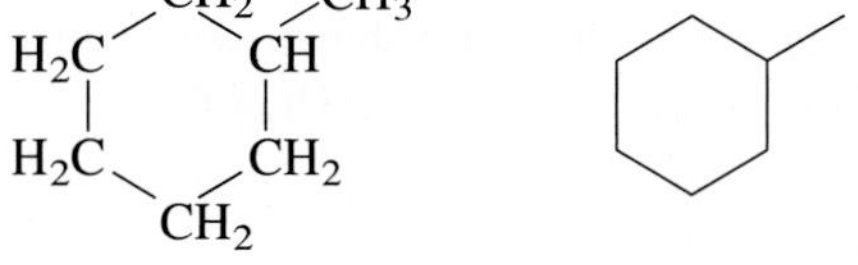

is methylcyclohexane. Notice that the methyl carbon in the skeleton diagram is not shown. Like the vertex in the polygon, a bond with no symbol at its end is understood to be to a carbon atom that forms additional bonds up to a total of four.

When two or more substitutions appear in a cycloalkane, their locations are determined by number. The ring carbon to which one substituent is bonded is number 1. Other numbers are assigned to give the lowest numbers possible to the other locations. Thus, the following compound is 1,3-dimethylcyclohexane, not 1,5-dimethylcyclohexane.

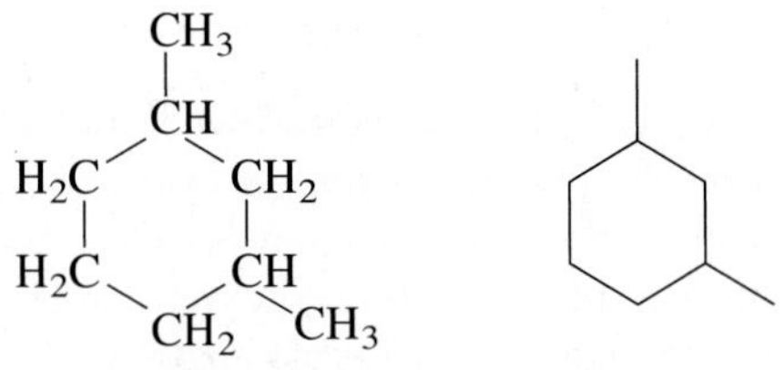

✔ QUICK CHECK 21.1

a) Identify the alkanes among C_7H_{16}, C_5H_{10}, $C_{11}H_{22}$, C_9H_{20}.
b) Write the formula of the alkyl group derived from pentane.
c) Write a structural diagram of 3,3-difluoro-4-iododecane.
d) Write a structural diagram of 1,1-diethylcyclohexane.

21.4 UNSATURATED HYDROCARBONS: THE ALKENES AND THE ALKYNES

Structure and Nomenclature

In a saturated hydrocarbon, each carbon atom is bonded to four other atoms. **Hydrocarbons in which two or more carbon atoms are (1) connected by a double or triple bond and (2) bonded to fewer than four other atoms are unsaturated.**

If one hydrogen atom, complete with its electron, is removed from each of two adjacent carbon atoms in an alkane (A below), each carbon is left with a single unpaired electron (B). These electrons may then form a second bond between the two carbon atoms (C):

Words or chemical symbols are sometimes placed above or above and below the arrow of an equation to indicate a substance whose presence is necessary for a reaction to proceed or to identify a reaction condition.

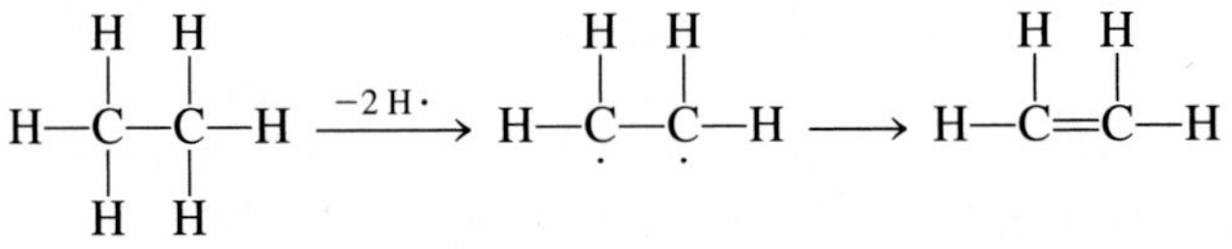

A B C

Each carbon atom is now bonded to three other atoms. **An aliphatic hydrocarbon in which two carbon atoms are bonded to three other atoms and double-bonded to each other is called an alkene.** Figure 21.5 shows two models of the simplest alkene.

Removal of another hydrogen atom from each of the double-bonded carbon atoms in an alkene yields a triple bond:

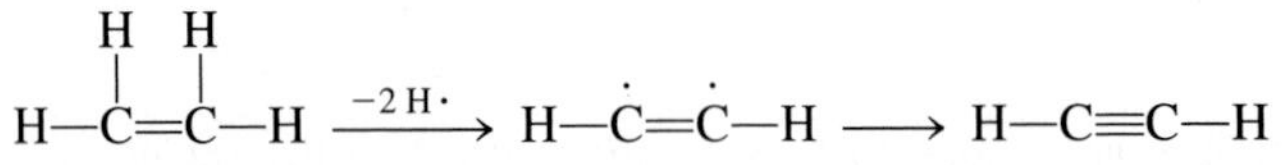

Each carbon atom is now bonded to two other atoms. **An aliphatic hydrocarbon in which two carbon atoms are triple bonded to each other is called an alkyne.** Models of acetylene, the most common alkyne, are shown in Figure 21.5.

Both the alkenes and the alkynes make up a new homologous series. Just as with the alkanes, each series may be extended by adding $—CH_2—$ units. Longer chains may have more than one multiple bond, but we won't consider such compounds in this text. The general formula for an alkene is C_nH_{2n}, and for an alkyne, C_nH_{2n-2}.

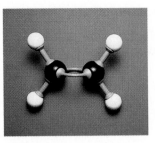   

Figure 21.5

Ball and stick and space-filling models of the first members of the alkene and alkyne hydro-carbon series, ethylene, C_2H_4 (left), and acetylene, C_2H_2 (right).

Table 21.3 gives the names and formulas of some of the simpler unsaturated hydrocarbons. The IUPAC nomenclature system for the alkenes matches that of the alkanes. The suffix designating the alkene hydrocarbon series is *-ene,* just as *-ane* identifies an alkane. For example, pentene is C_5H_{10}, hexene is C_6H_{12}, and octene is C_8H_{16}. The common names for the alkenes are produced similarly, except that the suffix is *-ylene.* These names for the smaller alkenes are firmly entrenched: C_2H_4 is almost always called ethylene, C_3H_6 is propylene, and C_4H_8 is butylene.

Acetylene, C_2H_2, is the first member of the alkyne series. Despite its *-ene* ending, acetylene is an alkyne, not an alkene. The IUPAC system is used for all alkynes except acetylene. So, for the alkynes with two, three, and four carbon atoms, the IUPAC names are ethyne, propyne, and butyne, respectively.

Table 21.3

Unsaturated Hydrocarbons

Hydrocarbon Series	n	Molecular	Structural	IUPAC	Common
			Formulas	**Names**	
Alkenes, C_nH_{2n}	2	C_2H_4	$\begin{array}{c}H\\ \diagdown \\ C{=}C \\ \diagup \diagdown\\ HH\end{array}$	Ethene	Ethylene
	3	C_3H_6	$\overset{H}{\underset{H}{>}}C{=}C{-}\overset{H}{\underset{H}{C}}{-}H$	Propene	Propylene
	4	C_4H_8	$\overset{H}{\underset{H}{>}}C{=}C{-}\overset{H}{\underset{H}{C}}{-}\overset{H}{\underset{H}{C}}{-}H$	Butene	Butylene
Alkynes, C_nH_{2n-2}	2	C_2H_2	$H{-}C{\equiv}C{-}H$	Ethyne	Acetylene
	3	C_3H_4	$H{-}C{\equiv}C{-}\overset{H}{\underset{H}{C}}{-}H$	Propyne	—

Isomerism Among the Unsaturated Hydrocarbons

There are two possible isomers of butyne, shown below. They differ by the position of the multiple bond. The melting and boiling points are given below their structures, to show that these are indeed different molecules.

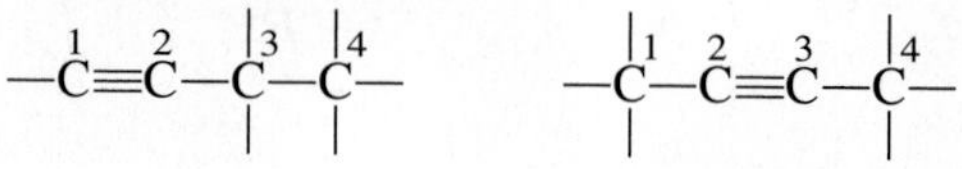

	1-butyne	2-butyne
melting point	−122.5°C	−32.3°C
boiling point	+8.1°C	+27°C

There are not two, but three distinct isomers of butene, C_4H_8. Depending on the location of the double bond, we have the two structures:

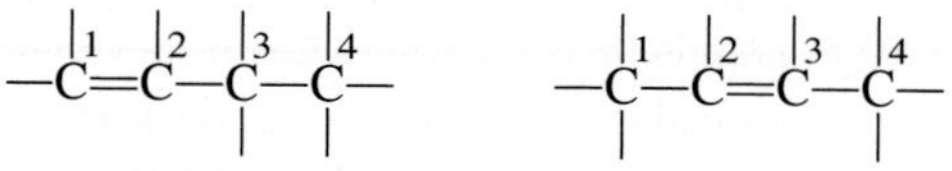

1-butene   2-butene

The third isomer is the result of *cis-trans* isomerism.

That part of a molecule that is on either side of a single bond may rotate freely around that bond as an axis. There is only limited rotation around a double bond. This leads to two possible arrangements around a double bond. The two methyl groups can be on the *same* side of the double bond, as in *cis*-2-butene, or on *opposite* sides, as in *trans*-2-butene. *Cis* and *trans* are prefixes meaning, respectively, *on this side* and *across*. Remember *trans* by associating it with a word like transcontinental, meaning across a continent. Another name for cis-trans isomers is **geometric isomers.**

The three C_4H_8 isomers are shown below, with their melting and boiling points, to show they are different substances.

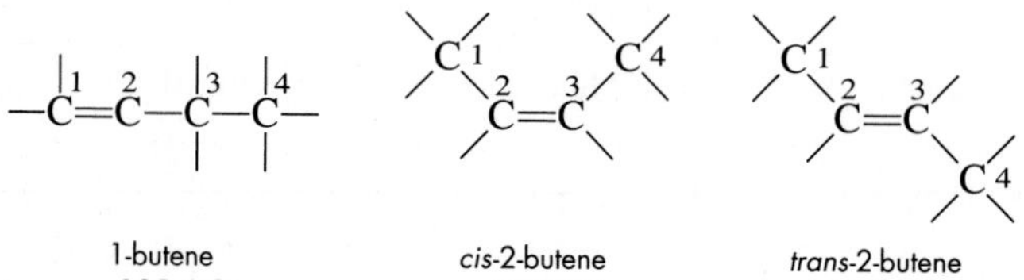

	1-butene	cis-2-butene	trans-2-butene
melting point	−185.4°C	−138.9°C	−105.6°C
boiling point	−6.3°C	+3.7°C	+0.9°C

The compound 1-butene shows that the carbon chain is numbered through the multiple bond so as to give the multiple bond the lowest possible number. The compound

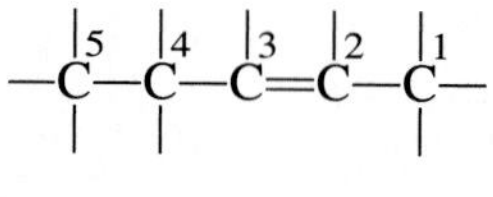

is 2-pentene, because the double bond is attached to the *second* carbon atom, counting from the right.

✔ QUICK CHECK 21.2

a) Identify the alkenes among the following: C_4H_6, C_2H_6, C_7H_{12}, C_8H_{16}.
b) Write a structural formula for *trans*-difluoroethylene, $C_2H_2F_2$.
c) Identify the alkynes among the following: C_4H_6, C_2H_6, C_7H_{12}, C_8H_{16}.
d) How many isomeric pentynes are there? Write a structural formula for 2-pentyne.

21.5 AROMATIC HYDROCARBONS

Initially, aliphatic compounds were associated with oils and fats, which contain long carbon chains. By contrast, the term **aromatic** was associated with a series of compounds found in such pleasant-smelling substances as oil of cloves, vanilla, wintergreen, cinnamon, and others. Ultimately, it was found that the key structure in aromatic hydrocarbons is the **benzene ring.** Now any hydrocarbon that does not contain a benzene ring—a hydrocarbon that is not an aromatic hydrocarbon—is an aliphatic hydrocarbon.

The simplest aromatic hydrocarbon is benzene, C_6H_6. Chemists have struggled for decades to find a structural diagram for benzene that is consistent with its physical and chemical properties, but without success. Two common forms are

The structure at the left satisfies the octet rule and predicts a planar molecule with 120° bond angles, which benzene has (see Fig. 21.6). However, the bonds in benzene are identical, rather than being alternating single and double bonds. The diagram at the right describes the bonding electrons as being "delocalized," belonging to the molecule as a whole rather than to specific bonds. It is this diagram that is most commonly used. As with cycloalkanes, each corner of the benzene molecule has a carbon atom, but it forms only one bond to an atom outside of the carbon ring. Unless indicated otherwise, this atom is assumed to be hydrogen.

Figure 21.6
Space-filling model of a benzene molecule, C_6H_6.

An alkyl group, halogen, or other species may replace a hydrogen in the benzene ring:

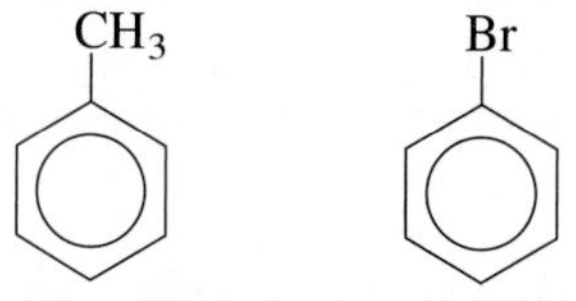

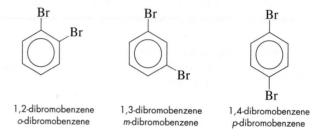

methylbenzene
toluene

bromobenzene

If two bromines substitute for hydrogens on the same ring, we must consider three possible isomers:

1,2-dibromobenzene
o-dibromobenzene

1,3-dibromobenzene
m-dibromobenzene

1,4-dibromobenzene
p-dibromobenzene

Two names are given for each isomer. The number system is the same as the system used to identify positions on a cyclohexane ring. It is more formal and serves any number of substituents. The other names are pronounced *ortho*-dibromobenzene, *meta*-dibromobenzene, and *para*-dibromobenzene. *Ortho-, meta-,* and *para-* are prefixes commonly used when two hydrogens have been replaced from the benzene ring. Relative to position X, the other positions are shown here:

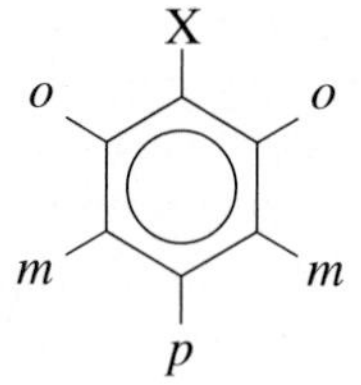

$$X$$

o o

m m

p

FLASHBACK Substances with intermolecular attractions that are roughly equal are most apt to be soluble in each other (Section 15.4). Similar molecular polarity contributes to similar intermolecular attractions.

The physical properties of benzene and its derivatives are quite similar to those of other hydrocarbons. The compounds are nonpolar, insoluble in polar solvents such as water, but generally soluble in nonpolar solvents. In fact, benzene is widely used as the solvent for many nonpolar organic compounds. Like other hydrocarbons of comparable molar mass, benzene is a liquid at room temperature.

✔ QUICK CHECK 21.3

a) Write the structural formula of 1,3,5-trifluorobenzene.
b) Many people buy mothballs that are made of *p*-dichlorobenzene. Write the structural formula for this substance.

21.6 SUMMARY OF THE HYDROCARBONS

Five types of hydrocarbons we have considered are summarized in Table 21.4.

SUMMARY

Table 21.4
Hydrocarbons

Type	Name	Formula	Saturation	Structure
Aliphatic open chain	Alkane	C_nH_{2n+2}	Saturated	$-\overset{\displaystyle \mid}{\underset{\displaystyle \mid}{C}}-$
	Alkene	C_nH_{2n}	Unsaturated	$\diagdown C = C \diagup$
	Alkyne	C_nH_{2n-2}	Unsaturated	$-C \equiv C-$
Aliphatic cyclic	Cycloalkane	C_nH_{2n}	Saturated	⬡
Aromatic	—	—	Unsaturated	⬡

21.7 SOURCES AND PREPARATION OF HYDROCARBONS

Almost all hydrocarbons are derived from that group of substances known as fossil fuels: coal, natural gas, and petroleum. These substances are natural products that have resulted from the decay of plants and animals that lived millions of years ago. We are familiar with natural gas as the most common fuel for home heating and cooking with gas stoves. This fuel is primarily methane, CH_4, plus a smaller but significant portion of ethane, C_2H_6. Coal is the main source of benzene; it is actually a by-product from the preparation of coke from coal. Other aromatic hydrocarbons are also recovered from the same process.

Petroleum is by far the largest source of the vast number of products broadly known as petrochemicals. Raw petroleum is a mixture of hydrocarbons containing up to 40 carbon atoms in the molecule. These large molecules are not useful in their natural form, but they are broken into smaller molecules in petroleum refineries. *Catalytic cracking* essentially "cracks" the long carbon chains into shorter molecules of five to ten carbon atoms. *Fractional distillation* separates hydrocarbons into "fractions" that boil at different temperatures. Lower alkanes, up to four or five carbon atoms per molecule, may be obtained in pure form by this method. The boiling points of larger alkanes are too close for their complete separation, so chemical methods must be used to obtain pure products.

FLASHBACK A "catalytic" process is one that occurs in the presence of a catalyst. In Section 19.4 a catalyst was identified as a substance that speeds the rate of a reaction without being permanently affected.

There are several industrial and laboratory methods by which alkanes may be prepared. One of the more important is the catalytic **hydrogenation** of an alkene. Hydrogenation is the reaction of a substance with hydrogen. The general reaction of the hydrogenation of an alkene is

$$C_nH_{2n} + H_2 \xrightarrow{\text{catalyst}} C_nH_{2n+2}$$

Unsaturated hydrocarbons are often prepared commercially from compounds derived originally from alkanes. Alkenes, for example, are produced from the *dehydration* of alcohols (Section 21.10) or the *dehydrohalogenation* of an alkyl halide. These two impressive terms describe very similar processes that are quite simple, at least in principle. Dehydration is the removal of a water molecule; dehydrohalogenation is the removal of a hydrogen atom and a halogen atom. For example, a water molecule may be separated from propyl alcohol, C_3H_7OH, to make propylene:

$$\underset{\text{propyl alcohol}}{H-\overset{\overset{\displaystyle H}{|}}{C}-\overset{\overset{\displaystyle H}{|}}{\underset{\underset{\displaystyle H}{|}}{C}}-\overset{\overset{\displaystyle H}{|}}{\underset{\underset{\displaystyle OH}{|}}{C}}-H} \xrightarrow{H_2SO_4} \underset{\text{propylene}}{H-\overset{\overset{\displaystyle H}{|}}{\underset{\underset{\displaystyle H}{|}}{C}}-\overset{\overset{\displaystyle H}{|}}{C}=\overset{\overset{\displaystyle H}{|}}{C}-H} + \underset{\text{water}}{HOH}$$

An alkyl halide is an alkane in which a halogen atom has been substituted for a hydrogen atom; or, viewed in another way, an alkyl halide is an alkyl group bonded to a halogen. The molecule is attacked with a base in the presence of an alcohol:

$$\underset{\text{propyl halide}}{H-\overset{\overset{\displaystyle H}{|}}{\underset{\underset{\displaystyle H}{|}}{C}}-\overset{\overset{\displaystyle H}{|}}{\underset{\underset{\displaystyle H}{|}}{C}}-\overset{\overset{\displaystyle H}{|}}{\underset{\underset{\displaystyle X}{|}}{C}}-H} + \underset{\text{base}}{KOH} \xrightarrow{\text{alcohol}} \underset{\text{propene}}{H-\overset{\overset{\displaystyle H}{|}}{\underset{\underset{\displaystyle H}{|}}{C}}-\overset{\overset{\displaystyle H}{|}}{C}=\overset{\overset{\displaystyle H}{|}}{C}-H} + \underset{\text{salt}}{KX} + \underset{\text{water}}{HOH}$$

One alkyne is of major importance—acetylene. It is produced commercially in a two-step process in which calcium oxide reacts with coke (carbon) at high temperatures to produce calcium carbide and carbon monoxide:

$$CaO(s) + 3\,C(s) \longrightarrow CaC_2(s) + CO(g)$$

Calcium carbide then reacts with water to produce acetylene:

$$CaC_2(s) + 2\,H_2O(\ell) \longrightarrow C_2H_2(g) + Ca(OH)_2(s)$$

21.8 CHEMICAL REACTIONS OF HYDROCARBONS

The combustibility—ability to burn in air—of the hydrocarbons is probably the chemical reaction most important to modern society. As components of liquid and gaseous fuels, hydrocarbons are among the most heavily processed and distributed chemical products in the world. When burned in an excess of air, the end products are water and carbon dioxide.

One major distinction separates the chemical properties of saturated hydrocarbons from those of the unsaturated hydrocarbons. By opening a multiple bond in an alkene or alkyne, the compound is capable of reacting by **addition,** simply by adding atoms of some element to the molecule. By contrast, an alkane molecule is literally saturated; there is no room for an atom to join the molecule without first removing a hydrogen atom. A reaction in which a hydrogen atom in an alkane is replaced by an atom of another element is called a **substitution** reaction.

Both alkanes and alkenes undergo *halogenation* reactions—reaction with a halogen. These reactions serve to show the difference between addition and substitution reactions:

Addition Reaction

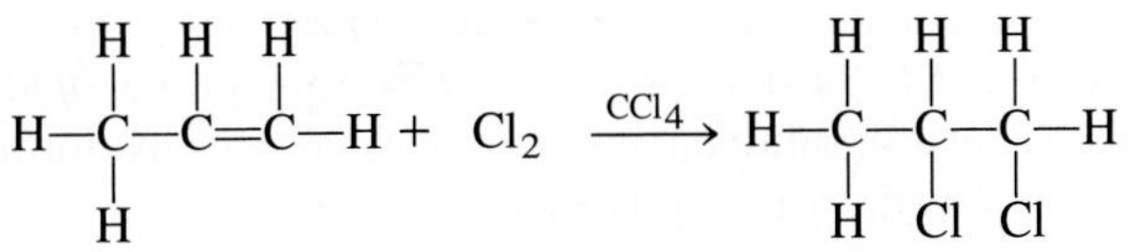

propene   chlorine   1,2-dichloropropane

The CCl$_4$ shown above the arrow is the solvent. It does not take part in the addition reaction.

Substitution Reaction

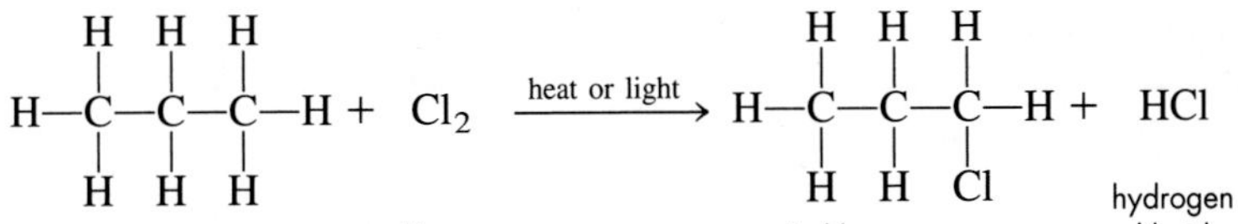

propane   chlorine   1-chloropropane   hydrogen chloride

The substituted chlorine atom may appear on either an end carbon atom or the middle carbon; the actual product is usually a mixture of 1-chloropropane and 2-chloropropane.

Normally, addition reactions are more readily accomplished than substitution reactions. This is hinted in the reaction conditions specified above. The addition of a halogen to an alkene will occur easily at room temperature, whereas the substitution of a halogen for a hydrogen in an alkane requires either high temperature or ultraviolet light. This shows that unsaturated hydrocarbons are more reactive than saturated hydrocarbons.

Hydrogenation is also an addition reaction. We have already indicated that the hydrogenation of an alkene may be used to produce an alkane. Hydrogenation of an alkyne is a stepwise process, which may often be controlled to give the intermediate alkene as a product:

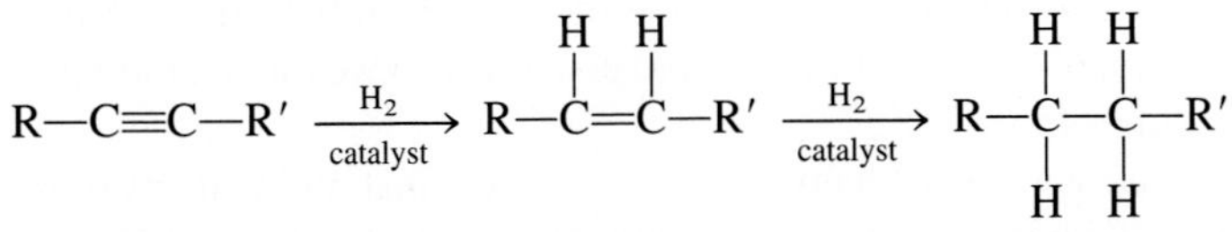

alkyne   alkene   alkane

Perhaps the most significant—and surprising—chemical property of benzene is that, despite its high degree of unsaturation, *it does not normally engage in addition reactions.* The classic nineteenth-century chemical tests for double bonds were reaction with bromine and with potassium permanganate to give addition products. Benzene gives neither reaction. The most important reaction of benzene itself is the substitution reaction in which one hydrogen is displaced from the benzene ring. Several substances may be used for substitution, including the halogens:

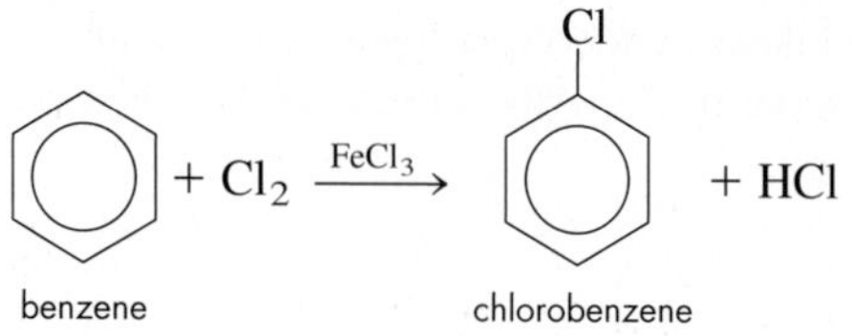

Substitutions with nitric and sulfuric acids yield, respectively, nitrobenzene and benzenesulfonic acid. Second substitutions on the same ring are possible although more difficult to bring about. Substitution reactions may also be performed on benzene derivatives, such as toluene, yielding isomers of nitrotoluene, for example. A triple nitro substitution produces 2,4,6-trinitrotoluene, better known simply as TNT.

✔ QUICK CHECK 21.4

Complete the following equations.

a) $C_4H_8 + Br_2 \rightarrow$

b) $C_6H_{14} + Cl_2 \rightarrow$

c) $C_6H_6 + Cl_2 \xrightarrow{FeCl_3} \quad + HCl$

21.9 USES OF HYDROCARBONS

It is impossible to overstate the importance of hydrocarbons in modern society. Alkanes move nations, both literally and figuratively—literally in transportation and figuratively in world politics. The oil crises of the 1970s and Operation Desert Storm in 1991 are recent events in which petroleum had a major impact on all industrialized nations. We rely heavily on petroleum products for the energy to heat homes, to operate anything electrical, to move people and goods, and to manufacture almost everything we use, including many things that are derived directly from alkanes.

How close to you right now are hydrocarbons or products derived from hydrocarbons? Well, unless you're sitting naked as you read this book, you're probably wearing some. If everything you have on is not made of wool, cotton, silk, rubber, hemp, bone, or leather, you are almost certainly wearing something synthetic that started as a hydrocarbon. And even if you are wearing clothes derived entirely from natural products, the processes by which the natural fibers were converted to wearable clothing used hydrocarbons in some form.

Look around you. How many things do you see that are made of or derived from hydrocarbons? To help you, look at Figure 21.7. The chart starts with alkanes, takes you through other hydrocarbons, and ends at the useful products in the bottom box in each column. Can you not see, touch, or smell at least one of those end products at this very instant?

Where on earth can you go to be totally separate from hydrocarbons? Almost nowhere. How about deep in a dense woods? No, the wonderful smell of a pine forest is from naturally occurring hydrocarbons called (for good reason) *pinenes*. On an all-

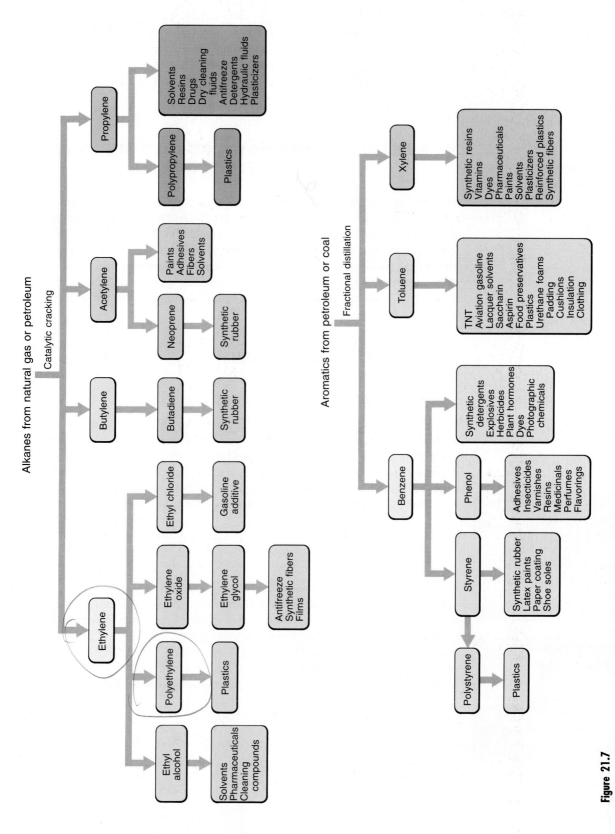

Figure 21.7

Some of the organic chemicals obtained from fossil fuels and their uses as raw materials.

wooden boat on an ocean or very large lake, out of sight of land? Perhaps, if the boat isn't painted with an oil-based paint commonly used on marine vessels. In the middle of a desert, on top of Mount Everest, or at the North or South Pole? Perhaps, but wherever it is, you'd better be dressed in clothing that was available in the 18th century!

ORGANIC COMPOUNDS CONTAINING OXYGEN AND NITROGEN

After carbon and hydrogen, the third element most commonly found in organic chemicals is oxygen. Capable of forming two bonds (Table 21.1), oxygen serves as a connecting link between two other elements or, double bonded, usually to carbon, as a terminal point in a **functional group.** A functional group is **an atom or group of atoms that establishes the identity of a class of compounds and determines its chemical properties.** In addition to several functional groups that contain oxygen, we will look briefly at two classes of nitrogen-bearing compounds that are so important in living organisms.

21.10 ALCOHOLS AND ETHERS

Structures and Names of Alcohols and Ethers

Beginning with a water molecule in the diagrams below, if we remove a hydrogen atom we have a **hydroxyl group.** This functional group identifies an **alcohol,** which is formed when the hydrogen atom is replaced with an alkyl group, R. Removal of both hydrogen atoms from a water molecule leaves the functional group of an **ether.** The ether molecule is completed when two alkyl groups bond to the oxygen. R′ in the diagram may be the same alkyl as R, or it may be a different alkyl.

Figures 21.8 and 21.9 show models of an alcohol and an ether, respectively. See if you can locate the functional group in each model. Alcohols are best known by their common names, which originate in the name of the alkyl group to which the hydroxyl group is bonded. This system names the alkyl group, followed by "alcohol." Thus, CH_3OH is *methyl alcohol* and C_2H_5OH is *ethyl alcohol.* Under IUPAC nomenclature rules for alcohols the *e* at the end of the corresponding alkane is replaced with the suffix *-ol* and the result is the name of the alcohol. Thus, methyl alcohol becomes *methanol,* and ethyl alcohol is formally *ethanol.*

Propyl alcohol has two isomers:

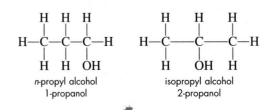

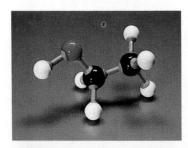

Figure 21.8

Models of ethanol (also called ethyl alcohol), C_2H_5OH.

These isomers are distinguished by stating the number of the carbon atom to which the hydroxyl group is bonded. Accordingly, *n*-propyl alcohol becomes 1-*propanol* and isopropyl alcohol is designated 2-*propanol*.

All ethers are called "ether," and identified specifically by naming first the two alkyl groups that are bonded to the functional group. If the groups are identical, the prefix *di-* may be used, as in diethyl ether.

Sources and Preparation of Alcohols and Ethers

Hydration of Alkenes   The major industrial source of several of our most important alcohols is the hydration of alkenes obtained from the cracking of petroleum. Beginning with ethylene, for example, the reaction may be summarized

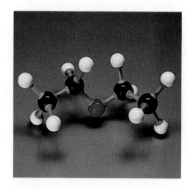

Figure 21.9
Model of diethyl ether, $C_2H_5OC_2H_5$.

$$H-\overset{\underset{|}{H}}{C}=\overset{\underset{|}{H}}{C}-H + HOH \longrightarrow H-\overset{\underset{|}{H}}{\underset{H}{C}}-\overset{\underset{|}{OH}}{\underset{|}{C}}-H$$

ethylene   ethyl alcohol

Fermentation of Carbohydrates   Making ethyl alcohol by the fermentation of sugars in the presence of yeast is probably the oldest synthetic chemical process known:

$$C_6H_{12}O_6 \xrightarrow{\text{yeast}} 2\,CO_2 + 2\,C_2H_5OH$$

glucose (sugar)   ethyl alcohol

A solution that is 95% ethyl alcohol (190 proof) may be obtained from the final mixture by fractional distillation. The mixture also yields two other products that are of commercial importance today: *n*-butyl alcohol and acetone (Section 21.11).

Under properly controlled conditions ethers can be prepared by dehydrating alcohols. At 140°C, and with constant alcohol addition to replace the ether as it distills from the mixture, ethyl ether is formed from two molecules of ethanol:

$$H-\overset{\underset{|}{H}}{\underset{H}{C}}-\overset{\underset{|}{H}}{\underset{H}{C}}-O-H + H-O-\overset{\underset{|}{H}}{\underset{H}{C}}-\overset{\underset{|}{H}}{\underset{H}{C}}-H \longrightarrow H-\overset{\underset{|}{H}}{\underset{H}{C}}-\overset{\underset{|}{H}}{\underset{H}{C}}-O-\overset{\underset{|}{H}}{\underset{H}{C}}-\overset{\underset{|}{H}}{\underset{H}{C}}-H + HOH$$

ethyl alcohol   ethyl alcohol   diethyl ether

Physical and Chemical Properties of Alcohols and Ethers

The structural similarity between water and alcohols suggests similar intermolecular forces and therefore similar physical properties. The lower alcohols (one to three carbon atoms) are liquids with boiling points ranging from 65°C to 97°C. These are comparable to water but well above the boiling points of alkanes of about the same molar mass. This is largely because of hydrogen bonding. Hydrogen bonding also accounts for the complete miscibility (solubility) between lower alcohols and water. As usual, boiling points rise with increasing molecular size. Solubility drops off sharply as the alkyl chain lengthens and the molecule assumes more of the character of the parent alkane.

FLASHBACK   Hydrogen bonding is the attraction between molecules in which a hydrogen atom is bonded to a nitrogen or oxygen atom that has at least one unshared pair of electrons (Section 14.3).

Figure 21.10

Some common alcohols. Carburetor cleaner contains methyl alcohol, or methanol, CH_3OH. The alcohol in alcoholic beverages is ethyl alcohol, or ethanol, C_2H_5OH. Rubbing alcohol is isopropyl alcohol, or 2-propanol, C_3H_7OH.

Ether molecules are less polar than alcohol molecules, and there is no opportunity for hydrogen bonding between them. Intermolecular attractions are therefore lower, as are the dependent boiling points. Up to three carbons, ethers are gases at room conditions. Diethyl ether, with four carbons, is a volatile liquid that boils at 35°C. The solubility of ether molecules in water is about the same as the solubility of isomeric alcohols, probably due to hydrogen bonding between the solute and the solvent.

The chemical properties of alcohols are essentially the chemical properties of the functional group, —OH. In some reactions the C—OH bond is broken, separating the entire hydroxyl group. The dehydration of alcohols to form alkenes (Section 21.7) is an example. In other reactions the O—H bond in the hydroxyl group is broken. These reactions will appear in the sections that follow.

Aside from combustion, ethers are relatively unreactive compounds, being quite resistant to attack by active metals, strong bases, and oxidizing agents. They are, however, highly flammable and must be handled cautiously in the laboratory.

Common Alcohols and Ethers

Methyl alcohol is an important industrial chemical with production measured in the billions of pounds annually. Figure 21.10 shows some common substances that are an alcohol or contain an alcohol. It is a raw material for the production of many chemicals, particularly formaldehyde, which is widely used in the plastics industry. It is also used in antifreezes, commercial solvents, and as a denaturant, or additive to ethyl alcohol to make it unfit for human consumption. Taken internally, methyl alcohol is a deadly poison, frequently causing blindness in less-than-lethal doses.

In addition to its uses in beverages, ethyl alcohol is used in organic solvents and in the preparation of various organic compounds such as chloroform and ether. Its production is also measured in the billions of pounds annually.

Other widely used alcohols include isopropyl alcohol, which is sold as rubbing alcohol, and *n*-butyl alcohol, used in lacquers in the automobile industry. Alcohols containing more than one hydroxyl group are also common. Permanent antifreeze in automobiles is ethylene glycol, which has two hydroxyl groups in the molecule (Fig. 21.11). Glycerine, or glycerol, a trihydroxyl alcohol, has many uses in the manufacture of drugs, cosmetics, explosives, and other chemicals.

The isolated word "ether" generally makes you think of the anesthetic that is so identified. This compound is diethyl ether, or simply ethyl ether; its line formula is C_2H_5—O—C_2H_5. Recently, its isomer, methyl propyl ether (Neothyl), CH_3—O—C_3H_7, has been gaining popularity as an anesthetic. It has fewer objectionable aftereffects than ethyl ether. Ethyl ether is used as a solvent for dissolving fats from foods and animal tissue in the laboratory. Because of its great combustibility, ethyl ether is also used as a cold-weather starting fluid for automobiles.

Figure 21.11

Engine coolant. The "permanent antifreeze" in nearly all engine coolants is ethylene glycol, a dihydroxyl alcohol: CH_2OHCH_2OH.

✔ **QUICK CHECK 21.5**

a) Write the name and structural formula of the functional group that identifies an alcohol.
b) Write the structural formula of the functional group that identifies an ether.
c) Draw Lewis diagrams for all compounds with the formula C_3H_8O. Identify these as alcohols or ethers.

21.11 ALDEHYDES AND KETONES

Aldehydes and **ketones** are characterized by the **carbonyl group,**

If at least one hydrogen atom is bonded to the carbonyl carbon, the compound is an aldehyde, RCHO; if two alkyl groups are attached, the compound is a ketone, R—CO—R′.

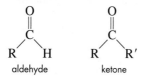

aldehyde   ketone

The simplest carbonyl compound is formaldehyde, HCHO, which has two hydrogen atoms bonded to the carbonyl carbon. If a methyl group replaces one of the hydrogens of formaldehyde, the result is acetaldehyde, CH_3CHO. Replacement of both formaldehyde hydrogens with methyl groups yields acetone:

formaldehyde   acetaldehyde   acetone

Figure 21.12 shows models of formaldehyde and acetone.

The lower aldehydes are best known by their common names. The IUPAC nomenclature system for aldehydes employs the name of the parent hydrocarbon, substituting the suffix *-al* for the final *e* to identify the compound as an aldehyde. Thus, the IUPAC name for formaldehyde is methanal, for acetaldehyde, ethanal, and so forth.

Ketones are named by one of two systems. The first duplicates the method of

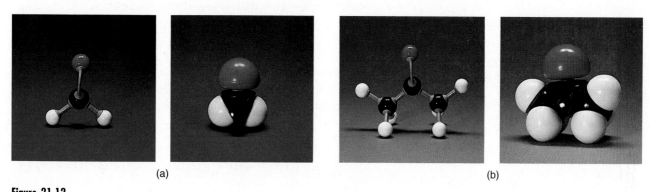

(a)   (b)

Figure 21.12

Models of (a) formaldehyde, $HCHO$, the simplest aldehyde and (b) acetone, CH_3OCH_3, the simplest ketone.

naming ethers: Identify each alkyl group attached to the carbonyl group, followed by the class name, ketone. Accordingly, methyl ethyl ketone has the structure

$$\underset{CH_3 \qquad C_2H_5}{\overset{\overset{\textstyle O}{\|}}{C}}$$

Under the IUPAC system the number of carbons in the longest chain carrying the carbonyl carbon establishes the hydrocarbon base, which is followed by *-one* to identify the ketone as the class of compound. Methyl ethyl ketone, having four carbons, would be called *butanone.* Two isomers of pentanone would be 2-*pentanone* and 3-*pentanone,* the number being used to designate the carbonyl carbon:

$$\underset{\underset{\text{2-pentanone}}{CH_3 \qquad CH_2CH_2CH_3}}{\overset{\overset{\textstyle O}{\|}}{C}} \qquad \underset{\underset{\text{3-pentanone}}{CH_3CH_2 \qquad CH_2CH_3}}{\overset{\overset{\textstyle O}{\|}}{C}}$$

Aldehydes and ketones may be prepared by oxidation of alcohols. If the product is to be a ketone, the alcohol must be a *secondary* alcohol, in which the hydroxyl group is bonded to an interior carbon:

$$\underset{\underset{\text{secondary alcohol}}{R'}}{\overset{H}{\underset{|}{R-\underset{|}{C}-OH}}} + \tfrac{1}{2}O_2 \longrightarrow \underset{\underset{\text{ketone}}{R'}}{R-\underset{|}{C}{=}O} + H_2O$$

Care must be taken not to overoxidize aldehyde preparations, since aldehydes are easily oxidized to carboxylic acids (see next section).

Aldehydes and ketones may also be produced by the hydration of alkynes. If the triple bond is on an end carbon, an aldehyde is produced; if between internal carbons, the result is a ketone. A typical reaction is the commercial preparation of acetaldehyde:

$$H-C{\equiv}C-H + HOH \longrightarrow \underset{\underset{\text{acetaldehyde}}{H}}{H-\underset{|}{\overset{|}{C}}-\underset{H}{\overset{O}{C}}}$$

The double bond of the carbonyl group can engage in addition reactions, just like the double bond in the alkenes. One such reaction is the catalytic hydrogenation of ketones to secondary alcohols, in which the hydroxyl group is bonded to a carbon atom *within* the chain:

$$\underset{\underset{\text{ketone}}{R'}}{R-\underset{|}{C}{=}O} + H_2 \xrightarrow{\text{catalyst}} \underset{\underset{\text{secondary alcohol}}{H}}{\overset{R'}{R-\underset{|}{\overset{|}{C}}-O-H}}$$

Oxidation reactions occur quite readily with aldehydes, but are resisted by ketones. When an aldehyde is oxidized, the product is a carboxylic acid:

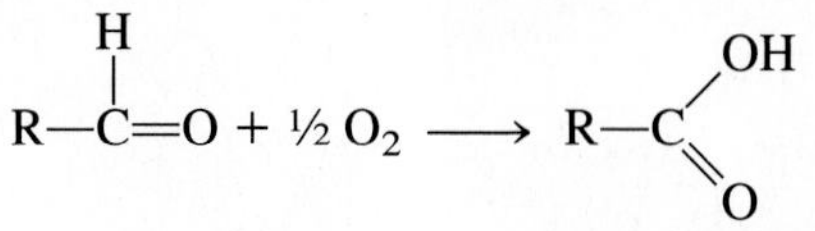

$$R-\overset{\overset{\displaystyle H}{|}}{C}=O + \tfrac{1}{2}O_2 \longrightarrow R-C\overset{\displaystyle OH}{\underset{\displaystyle O}{\diagdown}}$$

aldehyde   carboxylic acid

Formaldehyde is probably the best-known carbonyl compound. Large quantities (about 1 billion pounds/year) are made into polymers such as Bakelite, Formica, and Melmac. Cited as a likely carcinogen, use of formaldehyde to preserve biological specimens has virtually vanished. Acetaldehyde (ethanal) is used in the manufacture of organic compounds such as acetic acid and ethyl acetate. Other aldehydes you have probably encountered are benzaldehyde (almond flavor), cinnamaldehyde (cinnamon flavor), and vanillin (vanilla flavor).

Acetone is the most important ketone, with almost 2 million pounds produced yearly in the United States. It is a solvent used in the manufacture of other organic chemicals, drugs, and explosives. Acetone is also found in paint remover and nail polish remover. Methyl ethyl ketone (MEK) is used in the petroleum industry, as a lacquer solvent, and in nail polish remover. Most organic compounds whose names end in -one are ketones. You may be familiar with the anti-inflammatory cortisone, or the sex hormones progesterone and testosterone; all are ketones.

✔ QUICK CHECK 21.6

a) Write the name and structural formula of the functional group that identifies an aldehyde or a ketone.
b) Write Lewis diagrams that show the difference between an aldehyde and a ketone.
c) Write Lewis diagrams for all possible aldehyde or ketone isomers of $C_5H_{10}O$; identify the aldehydes and the ketones.

21.12 CARBOXYLIC ACIDS AND ESTERS

As we have seen, oxidation of an aldehyde produces a **carboxylic acid,** the general formula of which is frequently shown as RCOOH. The functional group, —COOH, is a combination of a carbonyl group and a hydroxyl group, rightly called the **carboxyl group.** You can probably pick out the carboxyl group in the acetic acid model in Figure 21.13a. In an **ester** the carboxyl carbon may be bonded to a hydrogen atom or an alkyl group, and the carboxyl hydrogen is replaced by another alkyl group, as shown here and in Figure 21.13b:

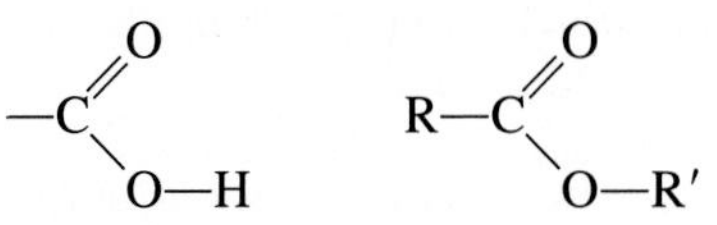

$$-C\overset{\displaystyle O}{\underset{\displaystyle O-H}{\diagdown}} \qquad\qquad R-C\overset{\displaystyle O}{\underset{\displaystyle O-R'}{\diagdown}}$$

carboxyl group   ester

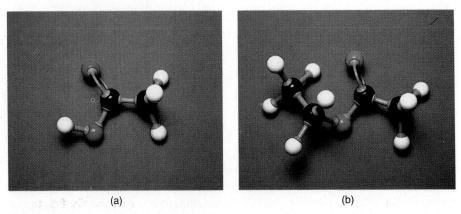

Figure 21.13
Models of (a) acetic acid, CH_3COOH, and (b) ethyl acetate, $CH_3COOC_2H_5$.

The geometry of the carboxyl group results in strong dipole attractions and hydrogen bonding between molecules. As a consequence, boiling points tend to be high compared to compounds of similar molecular mass. Formic acid, for example, boils at 100.5°C. Lower acids are completely miscible in water, but solubility drops off as the aliphatic chain lengthens and the molecule behaves more like a hydrocarbon.

Acetic acid, by far the most important of the carboxylic acids, is produced by the stepwise oxidation of ethanol, first to acetaldehyde and then to acetic acid:

Potassium permanganate, $KMnO_4$, is a strong oxidizing agent.

$$
\underset{\text{ethyl alcohol}}{H-\overset{\overset{\displaystyle H}{|}}{\underset{\underset{\displaystyle H}{|}}{C}}-\overset{\overset{\displaystyle H}{|}}{\underset{\underset{\displaystyle H}{|}}{C}}-OH}
\xrightarrow{KMnO_4}
\underset{\text{acetaldehyde}}{H-\overset{\overset{\displaystyle H}{|}}{\underset{\underset{\displaystyle H}{|}}{C}}-\overset{\overset{\displaystyle H}{|}}{C}=O}
\xrightarrow{KMnO_4}
\underset{\text{acetic acid}}{H-\overset{\overset{\displaystyle H}{|}}{\underset{\underset{\displaystyle H}{|}}{C}}-C\overset{\displaystyle OH}{\underset{\displaystyle O}{\Big\backslash\!\!\Big/}}}
$$

Carboxylic acids are weak acids that release a proton from the carboxyl group on ionization.* Acetic acid, for example, ionizes in water as follows:

$$CH_3COOH(aq) \rightleftharpoons CH_3COO^-(aq) + H^+(aq)$$

The ionization takes place but slightly; only about 1% of the acetic acid molecules ionize. The solution consists primarily of molecular CH_3COOH. This notwithstanding, acetic acid participates in typical acid reactions such as neutralization:

$$CH_3COOH(aq) + OH^-(aq) \longrightarrow HOH(\ell) + CH_3COO^-(aq)$$

and the release of hydrogen on reaction with a metal:

$$2\,CH_3COOH(aq) + Ca(s) \longrightarrow 2\,CH_3COO^-(aq) + Ca^{2+}(aq) + H_2(g)$$

Metal acetate salts may be obtained by evaporating the resulting solutions to dryness.

---

*In more advanced consideration of organic reactions, the term *acid* is also used in reference to Lewis acids (Section 17.4). This is why the adjective *carboxylic* is used to identify an organic acid containing the carboxyl group.

The reaction between an acid and an alcohol is called **esterification.** The products of the reaction are an ester and water. A typical esterification reaction is

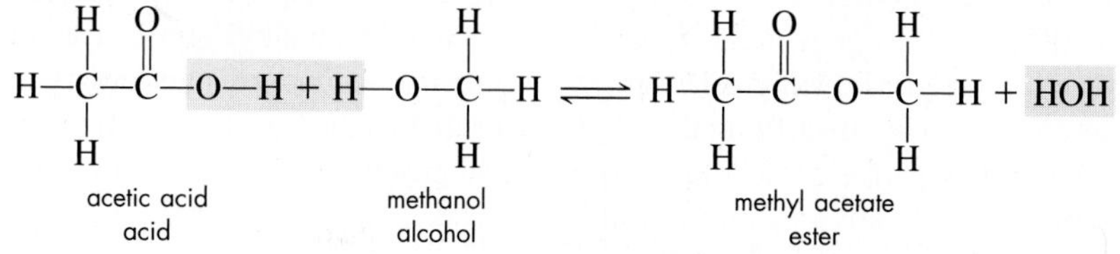

Notice how the water molecule is formed: *The acid contributes the entire hydroxyl group,* while the *alcohol furnishes only the hydrogen.*

The names of esters are derived from the parent alcohol and acid. The first term is the alkyl group associated with the alcohol; the second term is the name of the anion derived from the acid. In the preceding example, methyl alcohol (methanol) yields *methyl* as the first term, and acetic acid yields *acetate* as the second term.

Carboxylic acids engage in typical proton transfer acid–base type reactions with ammonia to produce salts. The ammonium salt so produced may then be heated, which causes it to lose a water molecule. The resulting product is called an *amide.* Compared to the original acid, an amide substitutes an $-NH_2$ group for the $-OH$ group of the acid (Section 21.13):

FLASHBACK Recall from Section 17.3 that a Brönsted-Lowry acid-base reaction involves the transfer of a proton from the acid to the base. An ammonia molecule, with an unshared electron pair on the nitrogen, is the proton receiver, the base, in this reaction.

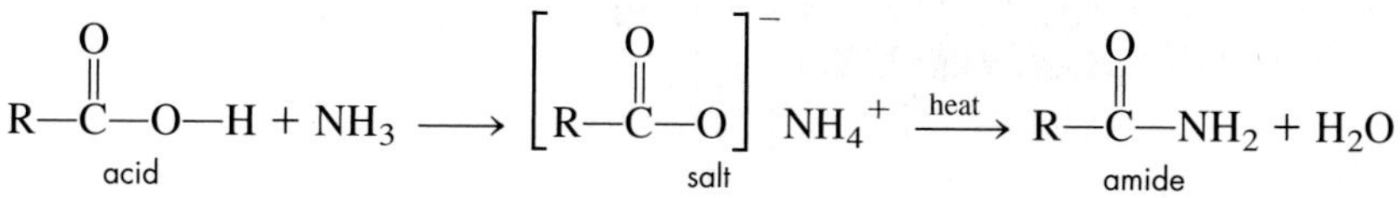

Formic acid and acetic acid are the two most important carboxylic acids. Formic acid is the source of irritation in the bite of ants and other insects, or the scratch of nettles. A liquid with a sharp, irritating odor, formic acid is used in manufacturing esters, salts, plastics, and other chemicals. Acetic acid is present to about 4% to 5% in vinegar and is responsible for its odor and taste. Acetic acid is among the least expensive organic acids, and is therefore a raw material in many commercial processes that require a carboxylic acid. Sodium acetate is one of several important salts of carboxylic acids. It is used to control the acidity of chemical processes and in the preparation of soaps and pharmaceutical agents.

Ethyl acetate and butyl acetate are two of the relatively few esters produced in large quantity. Both are used as solvents, particularly in the manufacture of lacquers. Other esters are involved in the plastics industry, and some find application in the medicinal fields. Esters are responsible for the odor of most fruit and flowers, leading to their use in the food and perfume industries.

✔ QUICK CHECK 21.7

a) Write the name and structural formula of the functional group that defines a carboxylic acid.
b) Describe in words the reactants and products of an esterification reaction.
c) Write Lewis diagrams for all possible carboxylic acid or ester isomers of $C_4H_8O_2$; identify the carboxylic acids and the esters.

21.13 AMINES AND AMIDES

Amines are organic derivatives of ammonia. An amine is formed by replacing one, two, or all three hydrogens in an ammonia molecule with an alkyl group. The number of hydrogens replaced distinguishes among a primary, secondary, and tertiary amine. Amines are named by identifying the alkyl groups that are bonded to the nitrogen atom, using appropriate prefixes if two or three identical groups are present, followed by the suffix *-amine*. Illustrative examples follow:

$$H—\overset{\displaystyle ..}{\underset{\displaystyle |}{N}}—H \qquad CH_3—\overset{\displaystyle ..}{\underset{\displaystyle |}{N}}—H \qquad CH_3—\overset{\displaystyle ..}{N}—C_2H_5 \qquad CH_3—\overset{\displaystyle ..}{\underset{\displaystyle |}{N}}—C_2H_5$$

ammonia   methylamine   ethylmethylamine   dimethylethylamine
                 primary amine   secondary amine   tertiary amine

Models of ammonia and the different methylamines are shown in Figure 21.14. An **amide** is a derivative of a carboxylic acid in which the hydroxyl part of the carboxyl group is replaced by an NH_2 group. For example

$$CH_3—\overset{\displaystyle O}{\overset{\displaystyle \|}{C}}\diagdown_{OH} \qquad \text{becomes} \qquad CH_3—\overset{\displaystyle O}{\overset{\displaystyle \|}{C}}\diagdown_{NH_2}$$

acetic acid   acetamide

by substitution of the $—NH_2$ in place of the $—OH$, as shown. An amide is named by replacing the *-ic acid* name of the acid with *amide*.

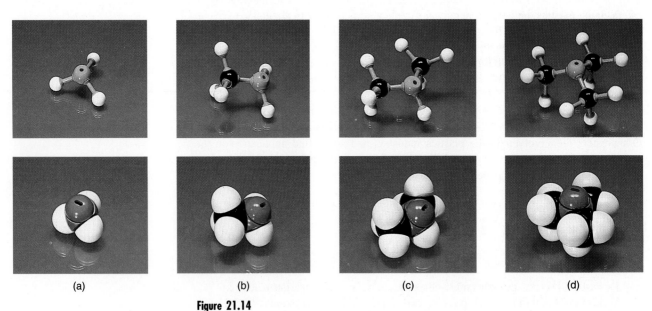

(a)   (b)   (c)   (d)

Figure 21.14

Models of (a) ammonia, NH_3, (b) methylamine, CH_3NH_2, (c) dimethylamine, $(CH_3)_2NH$, and (d) trimethylamine, $(CH)_3N$.

The amide structure appears in an important biochemical system, protein, as a connecting link between amino acids. The linkage is commonly called a *peptide linkage*. This linkage has the form

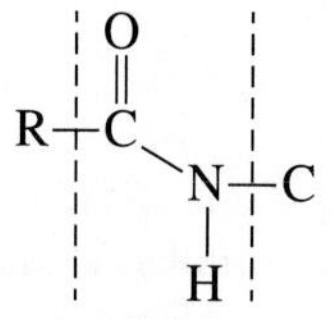

peptide linkage

An *amino acid* is an acid in which an amine group is substituted for a hydrogen atom in the molecule. The amino acids involved in the protein structure have the general formula

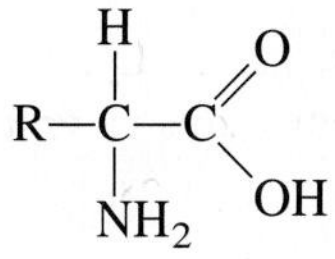

in which the amine and carboxyl groups are bonded to the same carbon atom. The peptide linkage is formed when the carboxyl group of one amino acid and the amine group of another combine by removing a water molecule:

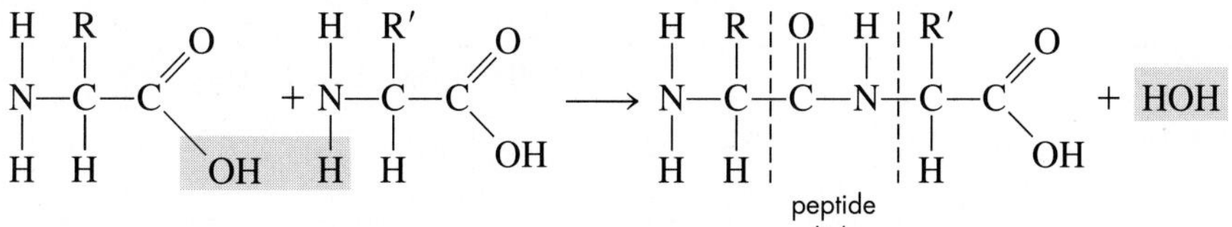

Proteins are chains of such links between as many as 18 different amino acids, producing huge molecules with molar masses ranging from about 34,500 to 50,000,000.

Dimethylamine and trimethylamine are used in making anion exchange resins. Many chemical products such as dyes, drugs, herbicides, fungicides, soaps, insecticides, and photographic developers are made from amines. The aromatic amine aniline (phenylamine) is used in dye making.

Every living thing has proteins, which are polymers of amino acids held together by amide bonds. These proteins perform many functions in a living system. We will discuss proteins in greater detail in Chapter 22.

✔ QUICK CHECK 21.8

a)   Write the structural formula of the functional group that defines an amine.
b)   Write the structural formula of the functional group that defines an amide.

21.14 SUMMARY OF THE ORGANIC COMPOUNDS OF CARBON, HYDROGEN, OXYGEN, AND NITROGEN

The eight types of organic compounds of carbon, hydrogen, oxygen, and nitrogen we have studied are summarized in Table 21.5.

SUMMARY

Table 21.5

Classes of Organic Compounds

Compound Class	General Formula	Functional Group	Names*
Alcohol	R—OH	—OH	Alkyl group + *alcohol;* methyl alcohol Alkane prefix + *-ol:* methanol
Ether	R—O—R′	 ![ether] O	Name both alkyl groups + *ether:* ethyl methyl ether Alkyl group + *-oxy-* + alkane: methoxyethane
Aldehyde	R—CHO	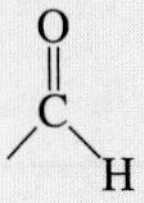 O‖ C—H	Common prefix + *-aldehyde:* formaldehyde Alkane prefix + *-al:* methanal
Ketone	R—CO—R′	 O‖ C	Name both alkyl groups + *ketone:* methyl ethyl ketone; methyl *n*-propyl ketone (Number carbonyl carbon) + alkane prefix + *-one:* butanone; 2-pentanone
Acid	R—COOH	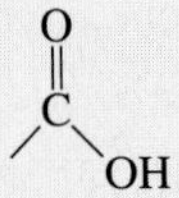 O‖ C—OH	Common name + acid: formic acid Alkane prefix + *-oic* + *acid:* methanoic acid
Ester	R—CO—OR′	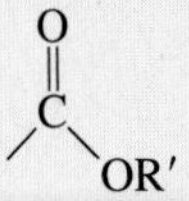 O‖ C—OR′	Alcohol alkyl group + acid anion: methyl acetate Alcohol alkyl group + acid alkane prefix + *-oate:* methyl ethanoate
Amine	RNH₂ R₂NH R₃N	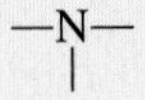 —N—	Name alkyl group(s) + *-amine:* methylamine *Amino-* + alkane: aminomethane
Amide	R—CONH₂	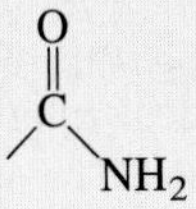 O‖ C—NH₂	Common acid prefix + *-amide:* formamide Alkane prefix + *-amide:* methanamide

*Common name followed by IUPAC name.

POLYMERS

21.15 ADDITION POLYMERS

The word *plastics* is commonly used to describe a large number of familiar substances, most of which were developed in the past 50 years. These materials are made up of huge molecules having molecular masses that sometimes run into the millions.

One way to make a large molecule is to connect many small chemical units, somewhat like links are joined to form a chain. In chemistry, each link is called a **monomer,** which, from the Greek, means having one part. The resulting chain is called a **polymer,** which means having many parts. The process by which polymers are formed is called **polymerization.**

Alkenes form **addition polymers,** which means that the monomers bond to each other without forming any other product. The polymerization of three molecules of an alkene occurs when one bond of the double bond between carbon atoms is broken in each molecule. Each carbon atom then has an unshared electron with which to form a bond with a neighboring molecule. The chain continues to build indefinitely in this way.

$$
\begin{array}{c}
\underset{\underset{H}{|}}{\overset{\overset{H}{|}}{C}}{=}\underset{\underset{R}{|}}{\overset{\overset{H}{|}}{C}} +
\underset{\underset{H}{|}}{\overset{\overset{H}{|}}{C}}{=}\underset{\underset{R}{|}}{\overset{\overset{H}{|}}{C}} +
\underset{\underset{H}{|}}{\overset{\overset{H}{|}}{C}}{=}\underset{\underset{R}{|}}{\overset{\overset{H}{|}}{C}}
\longrightarrow \cdot\underset{\underset{H}{|}}{\overset{\overset{H}{|}}{C}}{-}\underset{\underset{R}{|}}{\overset{\overset{H}{|}}{C}}\cdot +
\cdot\underset{\underset{H}{|}}{\overset{\overset{H}{|}}{C}}{-}\underset{\underset{R}{|}}{\overset{\overset{H}{|}}{C}}\cdot +
\cdot\underset{\underset{H}{|}}{\overset{\overset{H}{|}}{C}}{-}\underset{\underset{R}{|}}{\overset{\overset{H}{|}}{C}}\cdot
\longrightarrow {-}\underset{\underset{H}{|}}{\overset{\overset{H}{|}}{C}}{-}\underset{\underset{R}{|}}{\overset{\overset{H}{|}}{C}}{-}\underset{\underset{H}{|}}{\overset{\overset{H}{|}}{C}}{-}\underset{\underset{R}{|}}{\overset{\overset{H}{|}}{C}}{-}\underset{\underset{H}{|}}{\overset{\overset{H}{|}}{C}}{-}\underset{\underset{R}{|}}{\overset{\overset{H}{|}}{C}}{-}
\end{array}
$$

If R in the monomer is hydrogen, the monomer is ethylene, C_2H_4, and the polymer is polyethylene. Ethylene is the heart of the petrochemical industry. Over 40 billion pounds of ethylene were produced in the United States in 1992. From that mass about 15 billion pounds of polyethylene were produced. (That's about 60 pounds per person!) Polyethylene with a molar mass of about 15,000 is called low-density polyethylene. It is a supple material that folds and bends easily because the intermolecular (London) forces between the ''short'' carbon chains are weak. These properties make it ideal for sandwich bags and trash bags.

London forces increase as molecular size increases. Polymers with longer chains produce stronger interchain attractions and a corresponding increase in mechanical strength. Plastic milk bottles are made of polyethylene with a molar mass in the 250,000 range.

When covalent bonds form between carbon chains, a ''cross-linked'' polymer is formed and the physical properties change sharply. Cross-linked polyethylene is used for plastic screw caps on soda bottles. This plastic is rigid enough to mold as a solid and has enough mechanical strength to hold the screw thread needed to tighten the cap on the bottle.

If R in the above diagrams is a methyl group, $-CH_3$, the monomer is propylene, $CH_2{=}CHCH_3$, and the polymer is polypropylene. We expect the slightly larger R group to increase attractive forces between chains. Polypropylene is used to make plastic bottles, automobile battery cases, fabrics such as Herculon®, and indoor–outdoor carpeting.

Halogen-substituted alkenes can also polymerize. These molecules are polar, so the resulting polymer has dipole attractions in addition to London forces. A familiar example of such a polymer is Saran®, the food wrap that clings to itself. Saran is a **copolymer** because it is made from two monomers, $CH_2{=}CHCl$ and $CH_2{=}CCl_2$. The attractive forces between Saran polymer films come from the dipoles caused by the carbon–chlorine bonds. Table 21.6 lists more ethylene-based monomers that give addition polymers.

The attractive forces between noncross-linked polymer chains are weaker than covalent bonds. As a result, these polymers can be easily recycled, melted to break *only* the interchain attractive forces, and then molded into a new shape. These polymers are

Table 21.6

Common Polymers Formed from Ethene-like Monomers

Monomer Formula	Polymer Formula	Polymer Names	Uses
$CH_2=C-H$ / benzene ring / Styrene	$-CH_2-\overset{H}{\underset{}{C}}-CH_2-\overset{H}{\underset{}{C}}-$ with benzene rings	Polystyrene Styrofoam®	Insulation, packaging
$CH_2=C-H$ / $C\equiv N$ / Acrylonitrile	$-CH_2-\overset{H}{\underset{C\equiv N}{C}}-CH_2-\overset{H}{\underset{C\equiv N}{C}}-$	Polyacrylonitrile Orlon®, Acrilan®	Fabrics, rugs
$CH_2=C-H$ / $O-C-CH_3$ / O / Vinyl acetate	$-CH_2-\overset{H}{\underset{O-C-CH_3}{C}}-CH_2-\overset{H}{\underset{O-C-CH_3}{C}}-$ ($\overset{}{O}$)	Polyvinylacetate PVA	Chewing gum, paint, glues, safety glass
$CH_2=C-CH_3$ / $C-O-CH_3$ / O / Methyl methylacrylate	$-CH_2-\overset{CH_3}{\underset{C-O-CH_3}{C}}-CH_2-\overset{CH_3}{\underset{C-O-CH_3}{C}}-$ ($\overset{}{O}$)	Lucite® Plexiglass®	Contact lenses, molded transparent objects
$CH_2=C-H$ / Cl / Vinyl chloride	$-CH_2-\overset{H}{\underset{Cl}{C}}-CH_2-\overset{H}{\underset{Cl}{C}}-$	Polyvinylchloride PVC, Tedlar®	Floor tile, pipe
$CF_2=C-F$ / F / Tetrafluoroethylene	$-\overset{F}{\underset{F}{C}}-\overset{F}{\underset{F}{C}}-\overset{F}{\underset{F}{C}}-\overset{F}{\underset{F}{C}}-$	Teflon®	Coatings, gaskets, bearings
$CH_2=C-C\equiv N$ / $C-O-CH_3$ / O / Methyl α-cyanoacrylate	$-CH_2-\overset{C\equiv N}{\underset{C-O-CH_3}{C}}-CH_2-\overset{C\equiv N}{\underset{C-O-CH_3}{C}}-$ ($\overset{}{O}$)	Superglue Crazy Glue®	Adhesives, battlefield "stitches"

called **thermoplastics** and account for over 85% of the plastics sold. To be recycled and reused efficiently, the different thermoplastics must be separated by type. When done correctly, the resulting recycled plastic costs about half the price of new plastic. Thermoplastics do not degrade in land fills, but they shouldn't be put there in the first place. Polymers are too valuable to bury.

Polymers having cross-links between carbon chains are generally classed as **thermosets;** these cross-links are covalent bonds, like the bonds making up the polymer chains. As a result, these polymers cannot be easily melted and remolded; there is too much thermal decomposition. Work continues, however, to make a thermoset that can be easily recycled. The only addition polymer classified as a thermoset is "superglue."

✔ QUICK CHECK 21.9

The monomer of polyvinyl alcohol can be written as $CH_2\!=\!\underset{\underset{\displaystyle OH}{|}}{CH}$.

Draw a section of the expected addition polymer, which is used in glues and fibers. Show at least three monomer units. (When mixed with some borax, polyvinyl alcohol with molar mass about 10^5 turns into "Slime.")

21.16 CONDENSATION POLYMERS

In Section 21.12 you learned that carboxylic acids react with alcohols to form esters, and with ammonia and amines to form amides. In each reaction, a molecule of water is split out in a **condensation reaction.** These same reactions are used by polymer chemists to form polyesters and polyamides. However, to form the polymer chain by repeated condensation reactions, you must use a *di*carboxylic acid (two carboxyl groups) such as terephthalic acid and a *di*alcohol (two hydroxyl groups) such as ethylene glycol, as shown below.

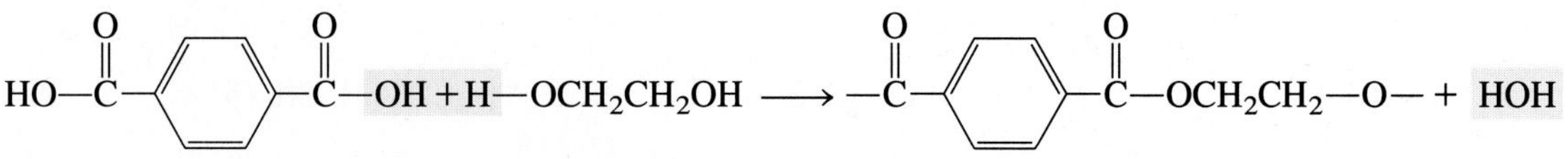

$$HO-\overset{\overset{\displaystyle O}{\|}}{C}-\!\!\!\!\langle\quad\rangle\!\!\!\!-\overset{\overset{\displaystyle O}{\|}}{C}-OH + H-OCH_2CH_2OH \longrightarrow -\overset{\overset{\displaystyle O}{\|}}{C}-\!\!\!\!\langle\quad\rangle\!\!\!\!-\overset{\overset{\displaystyle O}{\|}}{C}-OCH_2CH_2-O- + HOH$$

terephthalic acid   +   ethylene glycol $\longrightarrow$ "PET"

The linear polyester formed above is a polyethylene terephthalate, or PET. It has a molar mass of about 15,000. Because the ester group is polar, attractions between polymer chains are of a dipole type and are fairly strong. As a result PET polymers are used in fibers such as Dacron® or Fortrel®. Longer PET polymers are used for tire cords. Made into a Mylar® film and coated with magnetic particles, the PET becomes the base for audio and video recording tape. Finally, the two-liter soft drink bottles are a PET polymer. The United States produces about 4 billion pounds of PET each year.

The ethylene glycol dialcohol used above gives linear chains. If you add a third hydroxyl group, as in glycerol, the hydroxyl on the middle carbon can form ester bonds at an angle to the main chain, and so form a cross-linked polyester. Formation of cross-links increases molar mass dramatically, and is the most effective way of making a polymer with mechanical strength.

$$HOCH_2CHCH_2OH + HO-C \qquad C-OCH_2CHCH_2OH \longrightarrow$$

cross-linked polyester
(glyptal resin)

Cross-linked polyesters are called *glyptal resins* or *alkyd resins*. These resins have polar ester groups, and some free hydroxyl and carboxylic acid groups. As a result, they are water soluble. When you use a water-based paint, you are placing a small cross-linked polyester on the walls. As the paint dries, more cross-links form, and the film hardens, then cures. So when you paint a wall and the paint dries, stand back and gaze at the largest molecule you've ever seen. The cross-links form a tough film that is one giant molecule!

The best known synthetic polymer is nylon, a polyamide. Nylon was the brainchild of Dr. Wallace Carothers, who was hired away from Harvard University in 1928 by the Du Pont Company. Carothers was asked to develop a substitute for silk. Silk was known to be a protein, so Carothers' group studied ways of making amide bonds. In 1935, they prepared from adipic acid, a dicarboxylic acid, and hexamethylenediamine a product they called nylon 66. Note that each monomer has six carbons, hence the nylon 66 name.

adipic acid   hexamethylenediamine

nylon 66

The hydrogen bonding in an amide bond produces strong interchain attractive forces, and so nylon fibers have good tensile strength. Nylon with molar mass of about 10,000 can be made into a useful fiber. Nylon with molar mass above 100,000 has too much mechanical strength to be a fiber. This nylon, however, is mixed with glass fibers and used as valve covers in automobile engines. Current United States production of nylon is about 2.5 billion pounds annually. That's about 10 pounds per person.

✔ QUICK CHECK 21.10

A portion of the condensation polymer Kodel is given below. Give the structure of the starting materials from which Kodel can be made.

EVERYDAY CHEMISTRY

"In Which the Shape's the Thing . . ."

You've probably seen the terms *saturated, monounsaturated,* and *polyunsaturated* used on food labels to describe oils and fats in what we eat. Saturated fats (Figure 21.15) are cylindrical and tend to pack into masses that are solids at body temperature. Unsaturated fats and oils have a *cis* configuration around at least one double bond. These structures have a bend in them (Figure 21.16). These curved structures do not pack as easily as straight structures. As a result, the unsaturated fats and oils are liquids at body temperature.

Hydrogenation of vegetable oils (seen on food labels as "partially hydrogenated vegetable oils") changes some *cis* fatty acids into saturated fats and also isomerizes some *cis* double bonds into *trans* double bonds. These *trans* double bonds give a cylindrical molecule (Fig. 21.17), like saturated fats. Unfortunately, cylindrical fats, either saturated or *trans* unsaturated, are the fats that lead to blocked arteries. The bend in the fats makes all the difference in the blood.

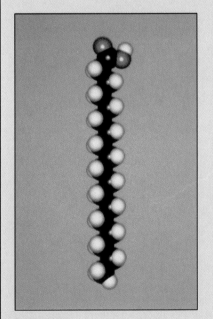

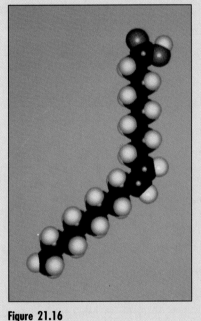

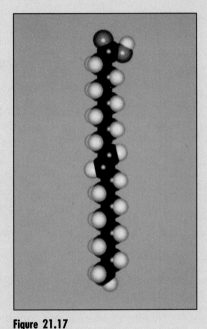

Figure 21.15
Space-filling model of stearic acid. Stearic acid, $CH_3(CH_2)_{16}COOH$. The COOH group is at the top of this model. In these models, carbon atoms are black, hydrogen atoms are white, and oxygen atoms are red.

Figure 21.16
Space-filling model of oleic acid Computer space-filling model of oleic acid, $CH_3(CH_2)_7C{=}C(CH_2)_7COOH$. The two hydrogen atoms attached to the *cis* double bond are yellow.

Figure 21.17
Space-filling model of *trans* unsaturated fatty acid. This is the *trans* isomer of oleic acid. The two hydrogens attached to the double bond are yellow.

CHAPTER 21 KEY WORDS AND MATCHING SETS

Set 1

_____ Describes compound with all bonding sites occupied, but can react by substitution.

_____ Structure of an ethyl group.

_____ Suffix for name of aliphatic hydrocarbon with a double bond.

_____ General symbol for an alkyl.

_____ Compounds with the same molecular formulas but different structures.

_____ Suffix in the IUPAC name of an alcohol.

_____ Describes a compound that can engage in addition reactions.

_____ Structure of a propyl group.

_____ Identifies C_6H_6 structure in an aromatic hydrocarbon.

_____ Suffix for the name of a saturated aliphatic hydrocarbon.

_____ Class of compounds in which a peptide linkage is found.

_____ Suffix for the name of an aliphatic hydrocarbon that has a triple bond.

1. _-ene_
2. Proteins
3. Isomers
4. Benzene ring
5. Saturated
6. _-yne_
7. R
8. $-C_3H_7$
9. _-ol_
10. $-C_2H_5$
11. Unsaturated
12. _-ane_

Set 2

_____ Hydrocarbon that includes a benzene ring.

_____ Two substituents are on opposite sides of a double bond.

_____ Each member differs from the next by the same structural unit.

_____ Method of separating low-boiling hydrocarbons based on their boiling points.

_____ Numbered position that corresponds with _meta-_ position on a benzene ring.

_____ Identifies an alkane without branches.

_____ Binary compound of hydrogen and carbon that has no benzene ring.

_____ Numbered position that corresponds with _para-_ site on a benzene ring.

_____ Both substituents are on the same side of a double bond.

_____ A group consisting of an alkane minus one hydrogen atom.

_____ Process for making short-chain hydrocarbons from long-chain hydrocarbons.

_____ Numbered position that corresponds with _ortho-_ site on a benzene ring.

1. _trans-_ isomer
2. 2
3. 3
4. 4
5. Fractional distillation
6. _cis-_ isomer
7. Aliphatic
8. Homologous series
9. Catalytic cracking
10. Normal
11. Aromatic
12. Alkyl

Set 3

_____ Prefix meaning three carbon atoms.

_____ Very long chain formed by combining small units without forming any other compound.

_____ Number of carbon atoms indicated by *hept-*.

_____ Describes a plastic that softens when heated and can be recycled.

_____ One part of a very long chain of similar parts.

_____ Prefix meaning two carbon atoms.

_____ Number of carbons indicated by *hex-*.

_____ Very long-chain product formed along with a second compound in polymerization.

_____ Prefix meaning one carbon atom.

_____ Describes a plastic that can neither be heated and re-formed nor recycled.

_____ Very long chain formed from two different units.

_____ Number of carbon atoms indicated by *pent-*.

1. Thermoplastic
2. Condensation polymer
3. Thermoset
4. Addition polymer
5. 5
6. 6
7. 7
8. *eth-*
9. Copolymer
10. *prop-*
11. Monomer
12. *meth-*

Set 4

_____ Has the general formula R—O—R′.

_____ Name of the functional group in aldehydes and ketones.

_____ Ring hydrocarbon in which each carbon is bonded to four atoms.

_____ Compound formed by reaction between an acid and an alcohol.

_____ An alkyl bonded to a hydroxyl group.

_____ Has the general formula C_nH_{2n+2}.

_____ Describes a group of atoms that identifies a class of organic compounds.

_____ Continuous chain hydrocarbon with twice as many hydrogen atoms as carbon atoms.

_____ A derivative of ammonia.

_____ General formula C_nH_{2n-2}

_____ Compound in which an —NH$_2$ group replaces the —OH in an acid.

_____ Name of the group identified as —COOH.

1. Functional group
2. Ester
3. Ether
4. Amide
5. Cycloalkane
6. Alkyne
7. Alcohol
8. Carboxyl
9. Amine
10. Carbonyl
11. Alkane
12. Alkene

CHAPTER 21 QUESTIONS AND PROBLEMS

SECTIONS 21.1 AND 21.2

1) Does today's definition of organic chemistry include the original definition of organic chemistry?

2) Would the cyanide ion or the carbonate ion be considered organic? What about the acetate ion?

3) What is the bond angle around a carbon atom with four single bonds? What word describes this molecular geometry?

4) What is the bond angle around a carbon atom with a double bond and two single bonds? What word describes this molecular geometry?

5) What is a hydrocarbon? Which, among the following, are hydrocarbons: CH_3OH; C_3H_4; C_8H_{10}; $CH_3CH_2CH_3$?

6) Outside of chemistry, what is the meaning of "saturated"? Use structural diagrams to show how the terms *saturated* and *unsaturated* are logically applied to hydrocarbons.

SECTION 21.3

7) Write the molecular formulas of the alkanes having 11 and 21 carbon atoms. How did you arrive at these formulas?

8) Explain why an alkane is an example of a homologous series.

---

9) Draw structural formulas for all the hydrocarbons with the formula C_4H_{10}.

10) Use your answer to Question 9 to explain what isomers are.

---

11) Write the molecular formula, line formula, and condensed formula of the normal alkane with seven carbon atoms. Also draw its structural formula.

12) Write the molecular formula, the line formula, and the condensed formula for the normal alkane having 12 carbon atoms.

---

13) What is an alkyl? What symbol is used to indicate that any alkyl may fit into an organic structure at the place it appears?

14) Write the molecular formula that represents the alkyls having 2 and 4 carbon atoms.

---

15) Write the molecular formula of butane. What is the name of $C_{10}H_{22}$?

16) What is the name of C_5H_{12}? Write the molecular formula of C_9H_{20}.

---

17) What is the IUPAC name of the molecule whose carbon skeleton is shown below?

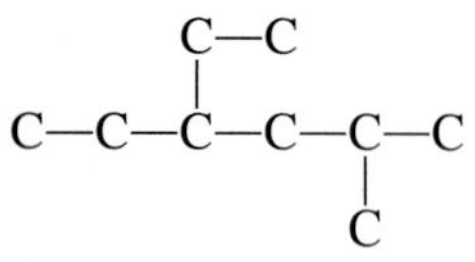

18) Draw the carbon skeleton of 3-ethyl-4-methylheptane.

---

19) Draw the carbon skeleton of 2,3-dimethylpentane.

20) Give the IUPAC name of the molecule shown below.

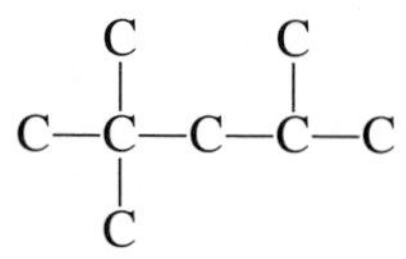

21) Both 1,1,1- and 1,1,2-trichloroethane are used industrially as fat and grease solvents. Draw the structural diagrams of these isomers.

22) Freon 114, an industrial refrigerant, has the IUPAC name 1,2-dichloro-1,1,2,2-tetrafluoroethane. Draw the structural diagram of this molecule.

---

23) How does a cycloalkane differ from a normal alkane?

24) Is the general formula of a cycloalkane the same as the general formula of an alkane, C_nH_{2n+2}? Draw any structural diagrams to illustrate your answer.

---

25) Draw the polygon that represents the structural diagram of cyclopentane.

26) Name the cycloalkane whose structural diagram is a square.

---

27) Draw a structural diagram of 1-chloro-2-iodocyclopentane.

28) Draw a structural diagram of 1-bromo-3-methylcyclohexane.

Questions 29–32: Write the names of the compounds whose carbon skeletons are given.

29)

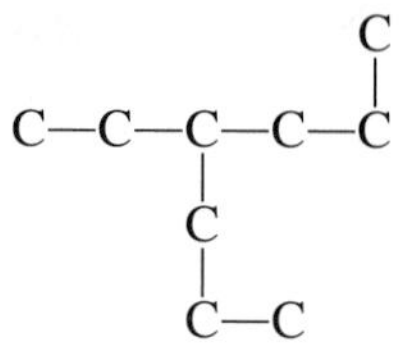

30)

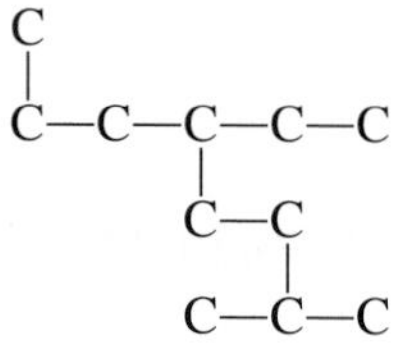

---

31)

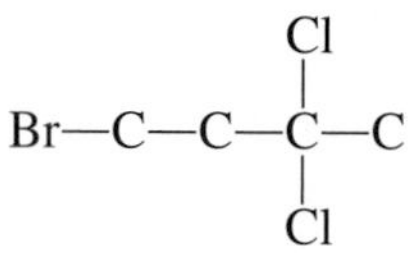

32)

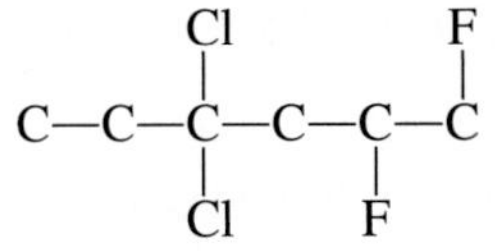

---

33) Draw the carbon skeleton of 1-chloro-2-iodocy-clopentane.

34) Draw the carbon skeleton of 1-bromo-3-methyl-cyclohexane.

---

35) Name the molecule whose carbon skeleton is drawn below.

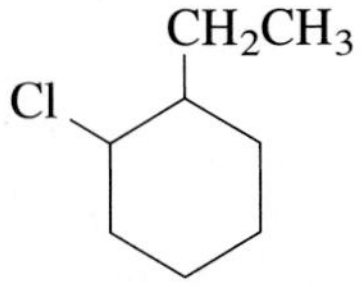

36) Give the IUPAC name of the molecule whose carbon skeleton is drawn below.

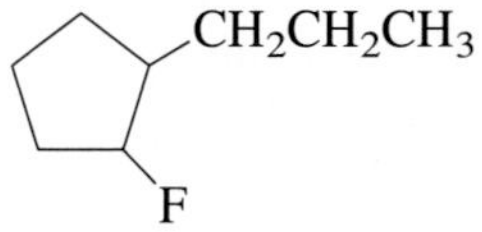

SECTION 21.4

37) What is the difference in bonding and in the general molecular formula between an alkene and an alkane with the same number of carbon atoms?

38) What is the difference in bonding and in general molecular formula between an alkyne and an alkane with the same number of carbon atoms?

---

39) Draw the structural formula of trichloroethene, a common dry cleaning solvent. Why isn't the IUPAC name for this substance 1,1,2-trichloroethene?

40) One starting material for Saran Wrap is 1,1-dichloroethene, also called vinylidene chloride. Draw the structural formula of this molecule.

---

41) Draw the structural formulas and explain in words the differences between *cis*-3-heptene and *trans*-3-heptene.

42) Draw the structural formula of 3-heptyne. Can a heptyne ever have *cis-trans* isomers?

---

43) The sex pheromone of the common house fly is *cis*-9-tricosene, where tricosene is the IUPAC name of a 23-carbon alkene. Draw the condensed formula of this molecule, marketed under the name Muscalure.

44) A molecule marketed as Disparlure, the sex pher-omone of the gypsy moth, is produced from the molecule shown below. Give the IUPAC name of this molecule. (An 18-carbon alkene is an octadecene.)

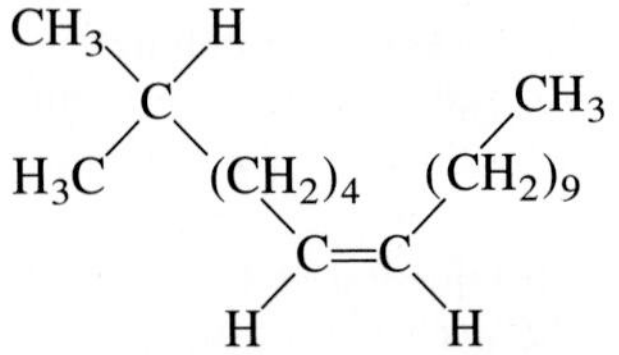

---

45) Give the IUPAC name of the following molecule:

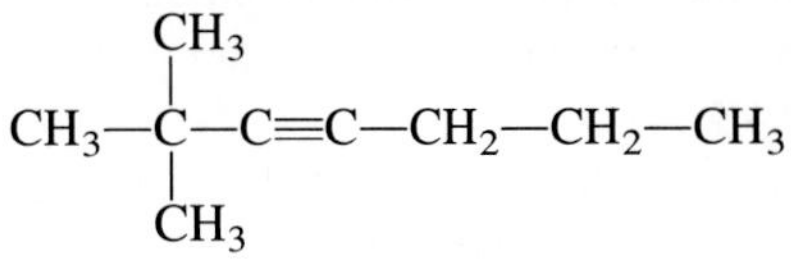

46) Give the IUPAC name of the following molecule:

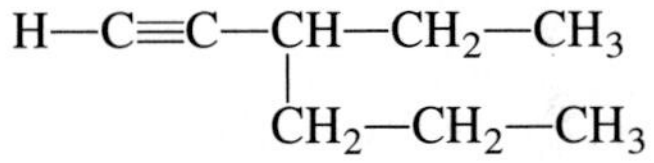

SECTION 21.5

47) Dimethylbenzenes have the common name *xylene*. Draw all possible xylene isomers and give their IUPAC names.

48) Trimethylbenzenes have the common name *mesitylene*. Draw all possible mesitylene isomers and give their IUPAC names.

---

49) Name the molecule given below.

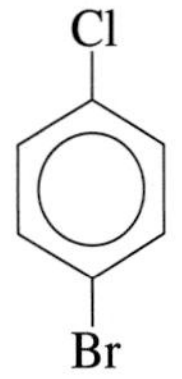

50) Name the molecule given below.

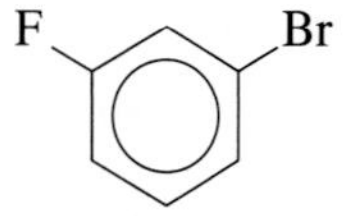

SECTION 21.8

51) Draw skeletal formulas for all the possible dichloro substitution products of propane and give their IUPAC names.

52) Draw skeletal formulas for all the possible dichloro substitution products of butane and give their IUPAC names.

___

53) Draw skeletal formulas for all the possible dichloro addition products of the butenes.

54) Draw skeletal formulas for all the possible dibromo addition products of the pentenes.

___

55) Write an equation for the hydrogenation of 2-butene. Does the *cis* or *trans* geometry of the butene starting material make a difference in the products obtained?

56) Would you get a different product from the hydrogenation of 1-butene than from the hydrogenation of *trans*-2-butene?

___

SECTION 21.10

57) Show how alcohols and ethers are structurally related to water.

58) Write the Lewis structures for all possible isomers with formula $C_4H_{10}O$. Identify them as alcohols or ethers.

___

59) Explain why ethers with formula $C_4H_{10}O$ have boiling points between 32 and 39°C, while alcohols with the same formula have boiling points between 82 and 118°C.

60) Explain why the ether with formula C_2H_6O is very slightly soluble in water while the alcohol with the same formula is infinitely soluble in water.

___

61) Write the structural formula for 2-hexanol.

62) Give the IUPAC name of

$$H-\overset{\overset{\displaystyle H}{|}}{\underset{\underset{\displaystyle H}{|}}{C}}-\overset{\overset{\displaystyle H}{|}}{\underset{\underset{\displaystyle H}{|}}{C}}-\overset{\overset{\displaystyle H}{|}}{\underset{\underset{\displaystyle H}{|}}{C}}-\overset{\overset{\displaystyle OH}{|}}{\underset{\underset{\displaystyle H}{|}}{C}}-\overset{\overset{\displaystyle H}{|}}{\underset{\underset{\displaystyle H}{|}}{C}}-H$$

___

63) Write the structural formula for butyl ethyl ether.

64) Write the structural formula for ethyl propyl ether.

___

65) Write a structural equation showing how propyl ether might be prepared from an alcohol.

66) How could isopropyl ether be prepared from an alcohol?

___

SECTION 21.11

67) Write structural formulas for propanal and propanone.

68) Why is there formaldehyde, but no "formanone"?

___

69) Use structural formulas to prepare acetone by oxidation of an alcohol.

70) Use structural formulas to prepare butanal from an alcohol.

SECTION 21.12

71) Write the structural formula for hexanoic acid, a wretched smelling molecule found in goat sweat.

72) Write the general formula for a carboxylic acid.

___

73) Write the equation for the reaction between propanoic acid and ethanol, and name the ester formed in this reaction.

74) Write the equation for the reaction between acetic acid and 1-propanol. Name the ester formed in this reaction. Do the same for the reaction between acetic acid and 2-propanol.

SECTION 21.13

75) Give structural formulas for all amines with the formula C_2H_7N and name them.

76) Give structural formulas for all amines with the formula C_3H_9N and name them.

___

77) Classify the amines from Problem 75 as primary, secondary, or tertiary.

78) Classify the amines from Problem 76 as primary, secondary, or tertiary.

___

79) Write the equation for the reaction between propanoic acid and ammonia. Name the product and give its functional group.

80) Write the equation for the reaction between propanoic acid and diethylamine.

SECTION 21.15

81) Draw three repeating units of the addition polymer made from the monomer

$$\overset{\overset{\displaystyle H}{|}}{\underset{\underset{\displaystyle H}{|}}{C}}=\overset{\overset{\displaystyle H}{|}}{\underset{\underset{\displaystyle Br}{|}}{C}}$$

82) Draw three repeating units of the addition polymer made from the monomer

83) The addition polymer shown below is used for ropes, fabrics, and indoor–outdoor carpeting. Give the structure of the monomer from which this polymer was made.

$$\left[CH_2-CH-CH_2-CH-CH_2-CH \right]_n$$
$$\underset{CH_3}{|}\underset{CH_3}{|}\underset{CH_3}{|}$$

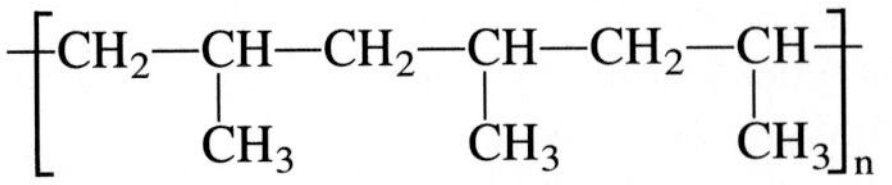

84) The addition polymer shown below is used to thicken motor oil. Give the structure of the monomer from which this polymer was made.

$$\left[CH_2-\overset{CH_3}{\underset{CH_3}{C}}-CH_2-\overset{CH_3}{\underset{CH_3}{C}}-CH_2-\overset{CH_3}{\underset{CH_3}{C}} \right]_n$$

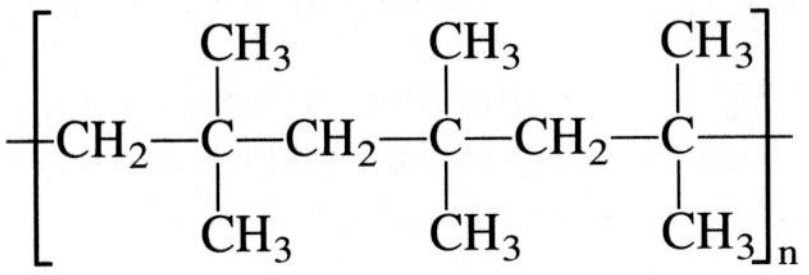

85) Draw three repeating units of the addition polymer made from chlorotrifluoroethene.

86) Plastic laboratory ware is usually an addition polymer made from 4-methyl-1-pentene. Draw three repeating units of this polymer.

SECTION 21.16

87) Lexan® is a polycarbonate ester condensation polymer that is transparent and nearly unbreakable. It is used in "bulletproof" windows (a 1-inch-thick Lexan plate will stop a .38 caliber bullet fired from 12 feet), football and motorcycle helmets, and the visors in astronauts' helmets. It is made from the two monomers shown below. Draw two repeating units of this polymer.

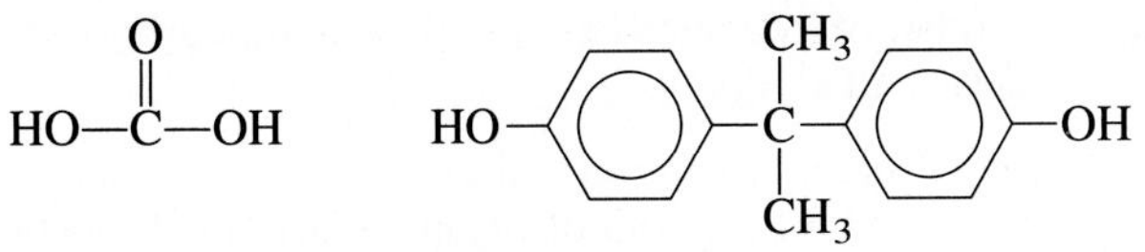

88) Draw two repeating units of the polyester formed from the two monomers given below.

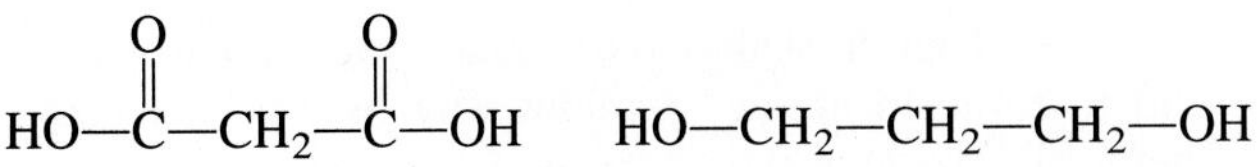

89) Kevlar® is a type of nylon called an *aramid* that contains aromatic rings. Because of its great mechanical strength, Kevlar is used in "bulletproof" clothing and in radial tires. The two monomers for Kevlar are below. Draw two repeating units of the Kevlar polymer.

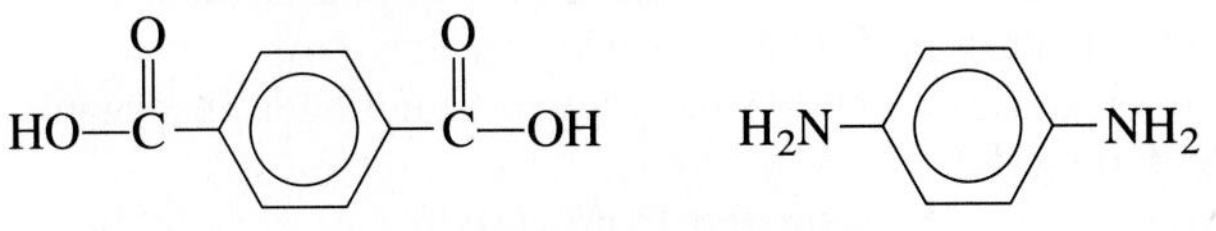

90) Nomex® is an aramid with great heat resistance; it is used in "fireproof" clothing worn by firefighters and racing drivers. The two monomers for Nomex are below. Draw two repeating units of the Nomex polymer.

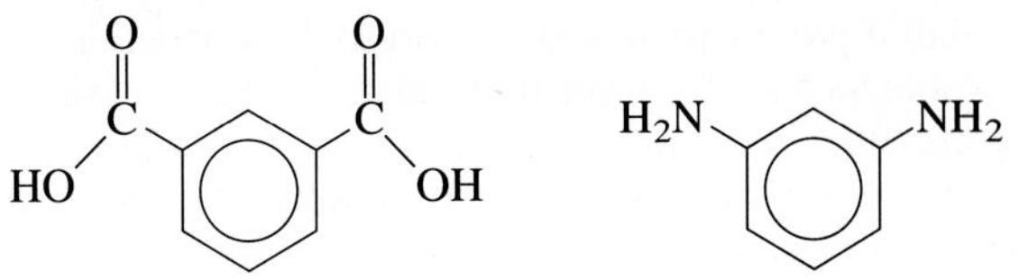

91) Give the monomers from which the following condensation polymer can be made.

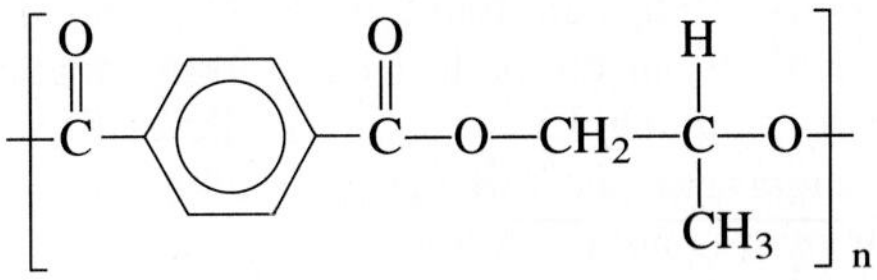

92) Give the monomers from which the following nylon polymer can be made.

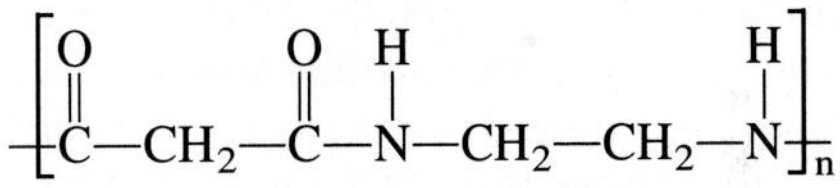

93) A leading nylon used in Europe is nylon 6, shown below. Nylon 6 is made by polymerization of a *single, difunctional* reactant. Draw the Lewis diagram of this reactant.

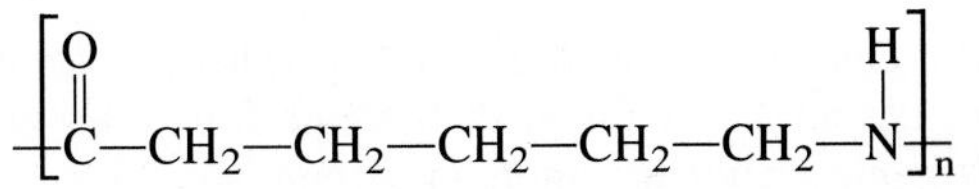

94) Stanyl® is nylon 46, widely used in Europe. Draw the Lewis diagrams of the Stanyl monomers.

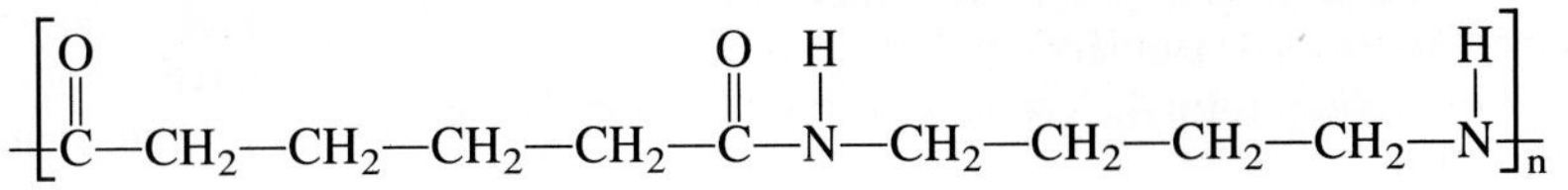

GENERAL QUESTIONS

95) Classify each of the following statements as true or false.

a) To be classified as organic, a compound must be or have been a part of a living organism.

b) Carbon atoms normally form four bonds in organic compounds.

c) Only an unsaturated hydrocarbon can engage in an addition reaction.

d) Members of a homologous series differ by a distinct structural unit.

e) Alkanes, alkenes, and alkynes are unsaturated hydrocarbons.

f) Alkyl groups are a class of organic compounds.

g) Isomers have the same molecular formulas but different structural formulas.

h) *Cis-, trans-* isomerism appears among alkenes but not alkynes.

i) Aliphatic hydrocarbons are unsaturated.

j) Aromatic hydrocarbons have a ring structure.

k) An alcohol has one alkyl group bonded to an oxygen atom, and an ether has two.

l) Carbonyl groups are found in alcohols and aldehydes.

m) An ester is an aromatic hydrocarbon, made by the reaction of an alcohol with a carboxylic acid.

n) An amine has one, two, or three alkyl groups substituted for hydrogens in an ammonia molecule.

o) An amide has the structure of a carboxylic acid, except that $-NH_2$ replaces $-OH$ in the carboxyl group.

p) A peptide link arises when a water molecule forms from a hydrogen from the $-NH_2$ group of one amino acid molecule and an $-OH$ from another amino acid molecule.

q) Forming cross-links in a polymer makes the polymer more likely to possess low mechanical strength.

r) Addition polymers are made by a reaction that gives off water as a second product.

s) Nylon and Dacron® are examples of condensation polymers.

96) Distinguish precisely and in scientific terms the differences among items in each of the following groups:

a) Organic chemistry, inorganic chemistry

b) Saturated and unsaturated hydrocarbons

c) Alkanes, alkenes, alkynes

d) Normal alkane, branched alkane

e) *Cis-, trans-,* isomers

f) Structural formula, condensed (line) formula, molecular formula

g) Addition reaction, substitution reaction

h) Alkane, alkyl group

i) Monomer, polymer

j) Aliphatic hydrocarbon, aromatic hydrocarbon

k) *Ortho-, meta-, para-*

l) Alcohol, aldehyde, carboxylic acid

m) Hydroxyl group, carbonyl group, carboxyl group

n) Primary, secondary, and tertiary alcohols

o) Alcohol, ether

p) Aldehyde, ketone

q) Carboxylic acid, ester

r) Carboxylic acid, amide

s) Amine, amide

t) Primary, secondary, tertiary amine

u) Monomer, polymer

v) Addition polymer, condensation polymer

97) What is the difference in bonding and in general molecular formula between an alkene and a cycloalkane with the same number of carbon atoms?

98) Draw all the isomers of C_4H_8.

99) Why is the delocalized structure (I) for benzene more appropriate than the cyclohexatriene structure (II)?

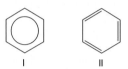

100) Show that the following statement is true. "Every alcohol with two or more carbons is an isomer of at least one ether."

101)* Explain why aldehydes are unstable with regard to oxidation, while ketones are stable. (HINT: Look at the number of hydrogen atoms bonded to the carbon that is bonded to the oxygen.)

102)* Chemists often use different isotopes such as oxygen-18 to chart the path of organic reactions. If acetic acid reacts with methanol that contains only oxygen-18, show where the oxygen-18 atom exists in the ester product.

103)* Can trimethylamine react with a carboxylic acid to form an amide? Explain why or why not using chemical equations.

104) Polymers are typically described as having a range of molecular masses, such as "polyvinyl alcohol, molecular mass 31,000–50,000." Why do polymers not have unique molecular masses, even though the monomer starting materials do have unique molecular masses?

105)* Write three repeating units of the addition polymer that can be made from acetylene. This material, called polyacetylene, conducts electricity because of the alternating single-, double-bond pattern in the main chain.

106)* Amides are often made by reaction of a carboxylic acid and an amine; this reaction is both acid-catalyzed and reversible. Use these facts to explain why nylon hosiery has a very short wear life in cities where "acid rain" is common.

107) A sample of polystyrene has an average molecular mass of 1,800,000. If a single styrene molecule has the formula C_8H_8, about how many styrene molecules are in a chain of this polystyrene?

MATCHING SET ANSWERS

Set 1: 5–10–1–7–3–9–11–8–4–12–2–6

Set 2: 11–1–8–5–3–10–7–4–6–12–9–2

Set 3: 10–4–7–1–11–8–6–2–12–3–9–5

Set 4: 3–10–5–2–7–11–1–12–9–6–4–8

QUICK CHECK ANSWERS

21.1 a) C_7H_{16} and C_9H_{20} are alkanes. b) $-C_5H_{11}$.

c)

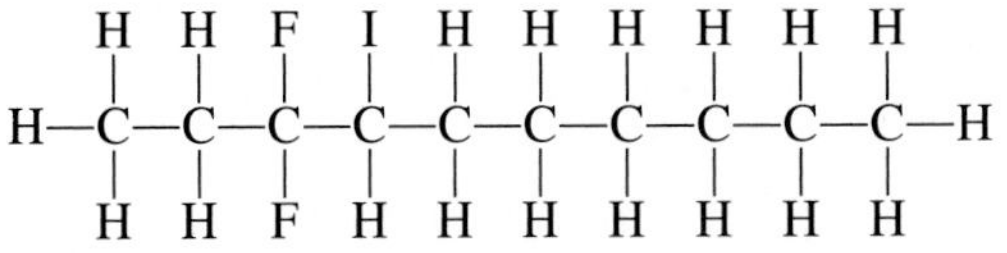

d) 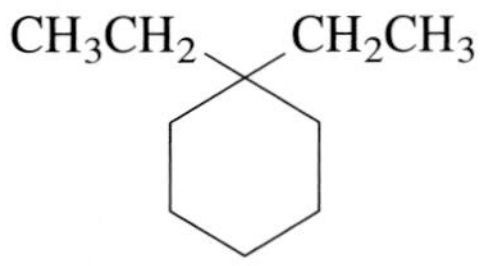 CH_3CH_2 CH_2CH_3

21.2 a) C_8H_{16} is the only alkene.

b)

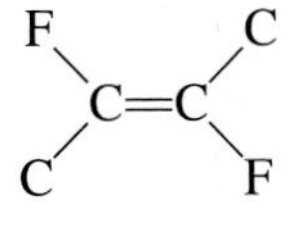

c) C_4H_6 and C_7H_{12} are the alkynes.

d) The isomeric pentynes are 1-pentyne and 2-pentyne. The structural formula for 2-pentyne is

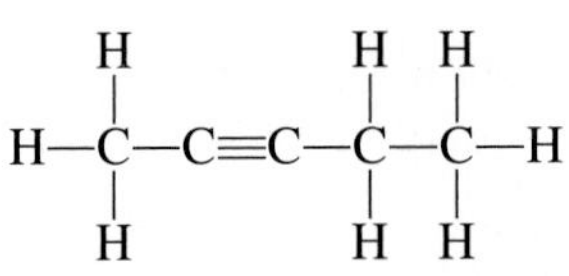

21.3

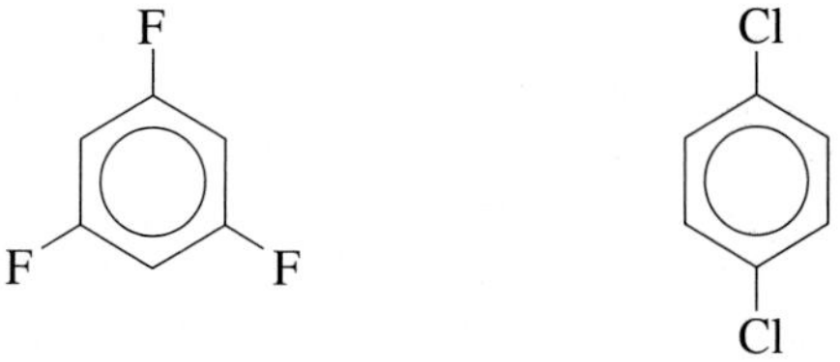

1,3,5-trifluorobenzene   *p*-dichlorobenzene

21.4 a) $C_4H_8 + Br_2 \rightarrow C_4H_8Br_2$

b) $C_6H_{14} + Cl_2 \rightarrow C_6H_{13}Cl + HCl$

c) $C_6H_6 + Cl_2 \xrightarrow{FeCl_3} C_6H_5Cl + HCl$

21.5 a) Hydroxyl group, $-OH$

b) $-O-$, where both bonds from oxygen are to carbon atoms.

c) Alcohols: $CH_3-CH_2-CH_2OH$; CH_3CHCH_3 with OH

Ether: $CH_3-O-CH_2CH_3$

21.6 a) Carbonyl group:

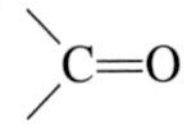

b) Aldehyde: 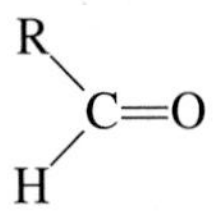
Ketone:

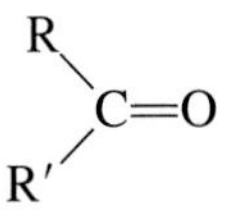

c) $CH_3CH_2CH_2CH_2-\overset{\overset{\displaystyle O}{\|}}{C}\diagdown_H$ aldehyde

$CH_3CH_2CH_2-\overset{\overset{\displaystyle O}{\|}}{C}-CH_3$ ketone

$CH_3CH_2-\overset{\overset{\displaystyle O}{\|}}{C}-CH_2-CH_3$ ketone

21.7 a) Carboxyl group: $-\overset{\diagup\overset{\displaystyle O}{\|}}{\underset{\diagdown O-H}{C}}$

b) Acid + alcohol → ester + water

c) $CH_3CH_2CH_2-\overset{\overset{\displaystyle O}{\|}}{C}-OH$ acid

$\overset{\overset{\displaystyle CH_3}{|}}{CH_3-CH}-\overset{\overset{\displaystyle O}{\|}}{C}-OH$ acid

$CH_3CH_2-\overset{\overset{\displaystyle O}{\|}}{C}-O-CH_2$ ester

$CH_3-\overset{\overset{\displaystyle O}{\|}}{C}-O-CH_2CH_3$ ester

$H-\overset{\overset{\displaystyle O}{\|}}{C}-O-CH_2CH_2CH_3$ ester

$H-\overset{\overset{\displaystyle O}{\|}}{C}-O-\overset{\overset{\displaystyle CH_3}{|}}{CH}-CH_3$ ester

21.8 a) $\overset{R_1-N-R_2}{\underset{\displaystyle R_3}{|}}$ One or two of the Rs may be H.

b) $-\overset{\diagup\overset{\displaystyle O}{\|}}{\underset{\diagdown NH_2}{C}}$

21.9 $-CH_2-\overset{|}{\underset{OH}{CH}}-CH_2-\overset{|}{\underset{OH}{CH}}-CH_2-\overset{|}{\underset{OH}{CH}}-CH_2-$

21.10 $HO-\overset{\overset{\displaystyle O}{\|}}{C}-$⬡$-\overset{\overset{\displaystyle O}{\|}}{C}-OH$; $HO-CH_2-$⬡$-CH_2-OH$

Biochemistry

22

This computer model depicts a small protein composed of 15 glycine units, in an α-helix conformation. In this chapter you will learn about molecules such as proteins, carbohydrates, fats, and steroids that you encounter every day.

Biochemistry is the study of life on a molecular level. In this chapter we take the view that biochemistry is 9/12ths chemistry. (Count the letters.) If you have studied Chapter 21, especially Sections 21.15 and 21.16, you're ready to tackle this chapter. If you haven't, read at least those last two sections. Return here with two ideas: 1) that large molecules can be made by joining together many smaller molecules, like links in a chain, and 2) that physical and chemical properties of large molecules can be studied and predicted like the properties of small molecules are studied and predicted.

Macromolecules are polymeric molecules with molar masses from about 5000 g/mole on up. In this chapter we introduce the major classes of biological macromolecules. We'll see that these classes all feature a modular assembly, made from surprisingly few different monomers.

22.1 AMINO ACIDS AND PROTEINS

The word **protein** comes from the Greek word *proteios*, meaning "of first importance." Among biological macromolecules, proteins are "first among equals" in terms of diverse functions in living systems. We'll begin at the bottom, starting with the 20 or so small molecules that join to form most proteins, then develop protein structure.

An **amino acid** is a small molecule characterized by a carboxyl group at one end of the carbon chain and an amine group on the carbon next to the carboxyl group.

FLASHBACK —NH_2 is the amine group in an amino acid, and —COOH is the carboxyl group. These structures were introduced in Sections 21.13 and 21.12, respectively.

$$H_2N-\underset{\underset{H}{|}}{\overset{\overset{R}{|}}{C}}-\overset{\overset{O}{||}}{C}-OH$$

There are 20 amino acids that are generally found in proteins. Amino acids differ by the identity of the R group and are divided into classes by the nonpolar or polar nature of the R group. Table 22.1 shows these acids. Note that each amino acid in the table is identified in three ways: the complete name, a three-letter abbreviation, and a single-letter abbreviation.

The symbol R stands for a group of atoms that complete a Lewis diagram.

Primary Protein Structure

Two amino acids can react with each other in two possible ways. For example, the reaction of glycine and alanine could give either glycylalanine, abbreviated Gly-Ala, or more succinctly, G-A

$$NH_2-\underset{glycine}{\overset{\overset{H}{|}}{CH}}-\overset{\overset{O}{||}}{C}-OH + NH_2-\underset{alanine}{\overset{\overset{CH_3}{|}}{CH}}-\overset{\overset{O}{||}}{C}-OH \longrightarrow NH_2-\overset{\overset{H}{|}}{CH}-\overset{\overset{O}{||}}{C}-\underset{glycylalanine}{NH}-\overset{\overset{CH_3}{|}}{CH}-\overset{\overset{O}{||}}{C}-OH + HOH$$

or alanylglycine, Ala-Gly, or A-G.

$$NH_2-\underset{alanine}{\overset{\overset{CH_3}{|}}{CH}}-\overset{\overset{O}{||}}{C}-OH + NH_2-\underset{glycine}{\overset{\overset{H}{|}}{CH}}-\overset{\overset{O}{||}}{C}-OH \longrightarrow NH_2-\overset{\overset{CH_3}{|}}{CH}-\overset{\overset{O}{||}}{C}-\underset{alanylglycine}{NH}-\overset{\overset{H}{|}}{CH}-\overset{\overset{O}{||}}{C}-OH + HOH$$

Table 22.1

The 20 Amino Acids Commonly Found in Proteins. (Both the three-letter and one-letter abbreviations are shown.)

Nonpolar R Groups							

Glycine	Gly	G	H—CH—COOH
			     NH_2
				Isoleucine	Ile	I	CH_3—CH_2—CH—CH—COOH
							     CH_3 NH_2

Alanine	Ala	A	CH_3—CH—COOH NH_2	Proline	Pro	P	—COOH

Valine	Val	V	CH_3—CH—CH—COOH
			    CH_3 NH_2
				Phenylalanine	Phe	F	CH_2—CH—COOH
							     NH_2

Leucine	Leu	L	CH_3—CH—CH_2—CH—COOH
			    CH_3 NH_2
				Methionine	Met	M	CH_3—S—CH_2CH_2—CH—COOH
							     NH_2

Polar but Neutral R Groups							

Serine	Ser	S	HO—CH_2—CH—COOH
			     NH_2
				Asparagine	Asn	N	NH_2—C—CH_2—CH—COOH
							   O NH_2

Threonine	Thr	T	CH_3—CH—CH—COOH
			    OH NH_2
				Glutamine	Gln	Q	NH_2—C—CH_2CH_2—CH—COOH
							   O NH_2

Cysteine	Cys	C	HS—CH_2—CH—COOH
			     NH_2
				Tryptophan	Trp	W	CH_2—CH—COOH
							     NH_2

Tyrosine	Tyr	Y	CH_2—CH—COOH NH_2				

Acidic R Groups				
Glutamic acid	Glu	E	HO—C—CH_2CH_2—CH—COOH O NH_2	
Aspartic acid	Asp	D	HO—C—CH_2—CH—COOH O NH_2	

Basic R Groups				
Lysine	Lys	K	NH_2—$CH_2CH_2CH_2CH_2$—CH—COOH NH_2	
Arginine	Arg	R	NH_2—C—NH—$CH_2CH_2CH_2$—CH—COOH NH NH_2	
Histidine	His	H	CH_2—CH—COOH NH_2	

FLASHBACK A peptide linkage is formed when the hydroxyl part of the carboxyl group of an amino acid molecule reacts with a hydrogen of the —NH$_2$ group of another amino acid molecule to form a molecule of water. The two amino acid molecules are then connected through the nitrogen atom. Lewis diagrams for this reaction appear in Section 21.13. Condensation polymers are formed when monomers combine in this way (see Section 21.16). The nitrogen atoms in the peptide linkages in the equations on page 608 and the diagrams to the right are circled.

In either case, the amino acid with the free carboxyl group is called the **C terminal** acid; the amino acid with the free amine group is called the **N terminal** acid. The bond between the acids is called a **peptide linkage.**

If we add a third amino acid, valine, to our dipeptides above, we could have glycylalanylvaline, Gly-Ala-Val, or G-A-V.

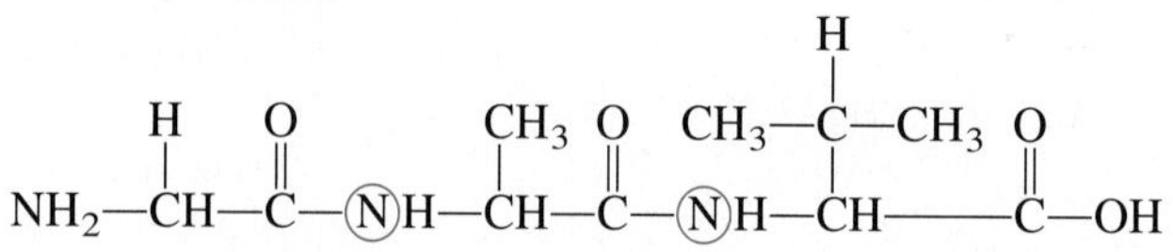

or valylglycylalanine, Val-Gly-Ala, or V-G-A

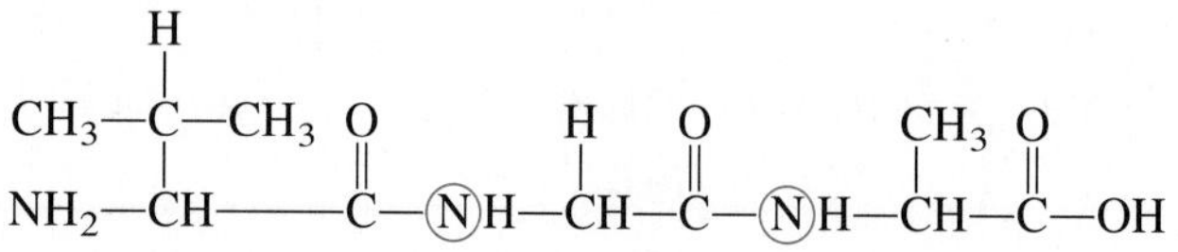

or valylalanylglycine, Val-Ala-Gly, or V-A-G

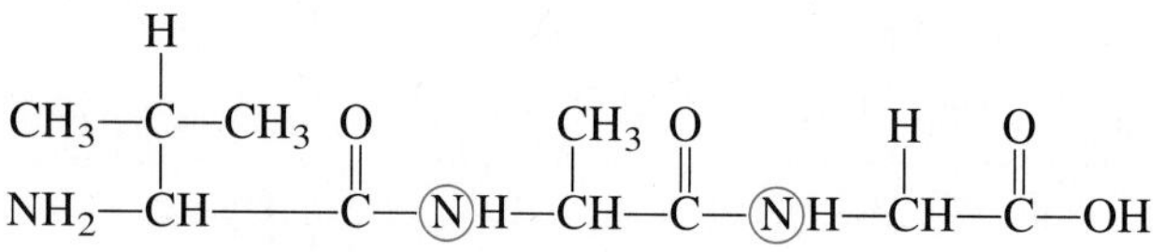

or alanylglycylvaline, Ala-Gly-Val, or A-G-V.

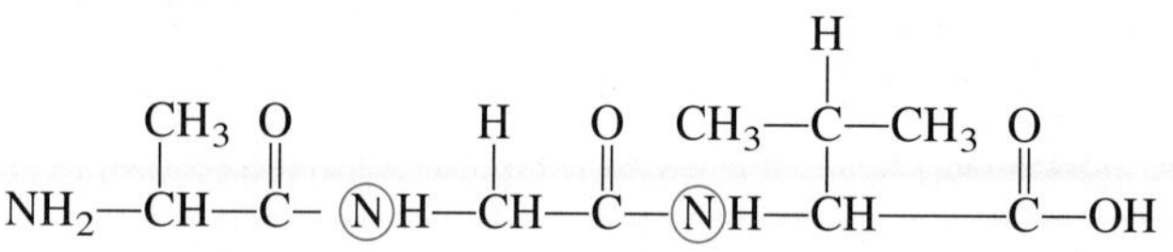

✔ QUICK CHECK 22.1

Using the single letter abbreviations A, V, and L, given in Table 22.1, list the tripeptides formed if alanine, valine, and leucine were all mixed together and peptide bonds allowed to form in all possible combinations.

You can see that the possible number of proteins, from only 20 amino acids, is vast. Molecules having fewer than about 40 amino acid residues are called **polypeptides;** a **protein** is a polypeptide chain with more than 40 amino acid residues. When we specify the order of amino acids in a protein, using Lewis structures or the abbreviations shown above, we have specified its **primary structure.**

No one is really sure where peptides end and proteins begin; they are overlapping terms.

Secondary Protein Structure

Metal springs can be made by taking wire and winding the wire into a spiral. In this operation, the primary structure ("electron sea" model, Section 14.7) of the metal remains unchanged. However, the **local conformation,** or arrangement in space, of the wire has changed. The spiral, or *helix,* is a regular local conformation that defines the **secondary structure** of the metal atoms in the spring.

Proteins have secondary structures also. The most important secondary structures are the α-helix and the β-pleated sheet. *Both structures reflect a maximum amount of hydrogen bonding;* these are the most stable conformations possible.

The α-helix is found in fibrous proteins like hair or wool or nails. These fibers are slightly elastic; stretching a hair, for example, will stretch the hydrogen bonds, but not break the amide bonds.

Look at the two figures on page 607. Both represent a small peptide with 15 glycine (the simplest amino acid, R = H) residues. The picture on the left emphasizes the main chain, with carbons shown in black and the amide nitrogen shown in blue. Note how the chain coils as if it were climbing a right-handed screw thread. No hydrogen atoms are shown in this figure.

The figure on the right includes the hydrogen atoms, which are shown in white. A hydrogen bond can be imagined as a nearly vertical line connecting an oxygen (red) atom, a hydrogen (white) atom, and a nitrogen (blue) atom. Study the figure to see that *every* oxygen and *every* nitrogen atom in the main chain is involved in hydrogen bonding, which governs the secondary structure of the protein. The α-helix allows the maximum amount of hydrogen bonding among the amino acids of a single protein chain. Not all amino acids polymerize to form an α-helix; the R groups of the amino acids must be small and nonpolar.

The tight spiral of the α-helix gives a rigid, rod-like protein. The α-keratins (hair, nails, skin) are examples of this structure. If fibers of an α-keratin are steamed and stretched, they almost double their length. This length increase occurs because the intrachain hydrogen bonds that stabilize the α-helix are broken, allowing the tightly curled helix to stretch into a more extended, zig-zag conformation called the β conformation. When adjacent protein chains are in this conformation, they share *inter*chain hydrogen bonds to form the **β-pleated sheet.**

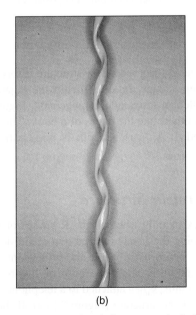

(a)　　　　　　　　　　　　　　(b)

(a) A curly telephone cord shows a helical structure when it is not stretched. (b) Stretch a telephone cord and the helix flattens to a pleat.

Glycogen, also called *animal starch* or *liver starch,* is the quick-acting carbohydrate reserve in mammals, including humans. Glycogen has a molar mass between 270,000 and 3,500,000 g/mole and resembles amylopectin. It is found in the liver of mammals and in rested muscle. When muscle activity occurs, glycogen is converted back to glucose 1-phosphate, which is then used for energy.

✔ QUICK CHECK 22.4

a) Using the structure for amylopectin, pick out an $\alpha(1 \rightarrow 4)$ bond. Point out which carbon in this bond is carbon-1 and which carbon is carbon-4.
b) Using the structure for amylopectin, pick out an $\alpha(1 \rightarrow 6)$ bond. Point out which carbon in this bond is carbon-1 and which carbon is carbon-6.

22.4 LIPIDS

Lipids are substances that are found in living organisms and are insoluble in water but soluble in nonpolar solvents. Lipids may be divided into three classes: 1) fats, oils, and phospholipids; 2) waxes; 3) lipids (usually) without ester groups, such as steroids. These three groups differ greatly in structure, but they share a common starting material, the unassuming acetic acid, CH_3COOH. We will see reminders of this origin several times in this section.

Fats and Oils

Fats and oils are mainly esters of long-chain (between 10 and 24 carbons) carboxylic acids; these acids are called **fatty acids.** The alcohol part of the ester comes from glycerol (also called glycerin), a trihydroxyl alcohol.

$$CH_3-CH_2-CH_2-CH_2-CH_2-CH_2-CH_2-CH_2-CH_2-CH_2-CH_2-\overset{\overset{\textstyle O}{\|}}{C}-O-CH_2$$

$$CH_3-CH_2-CH_2-CH_2-CH_2-CH_2-CH_2-CH_2-CH_2-\overset{\overset{\textstyle O}{\|}}{C}-O-CH$$

$$CH_3-CH_2-CH_2-CH_2-CH_2-CH_2-CH_2-CH_2-CH_2-CH_2-CH_2-\overset{\overset{\textstyle O}{\|}}{C}-O-CH_2$$

These esters are called triacylglycerols or triglycerides. Fats are triacylglycerols that are solids at room temperature; oils are triacylglycerols that are liquids at room temperature. The composition of the fatty acid parts of these esters varies with the organism that produced them. Plant oils are usually richer in unsaturated fatty acids than are animal fats.

Medical evidence correlates a diet high in saturated fats with hardening of the arteries (atherosclerosis) and possible heart attack. Table 22.2 lists the percentage of saturated fatty acids, monounsaturated fatty acids, and polyunsaturated fatty acids in some common dietary fats and oils. This table compares human ''depot'' (storage) fat with butter and margarines, then with several vegetable oils.

💡 FLASHBACK An unsaturated molecule has double or triple carbon–carbon bonds.

Table 22.2

Approximate Fatty Acid Composition of Common Fats and Oils

Fat or Oil	% Saturated (no double bonds)	% Monounsaturated (1 double bond)	% Polyunsaturated (> 1 double bond)
Human	35	55	10
Butter	66	31	4
Margarine, soft	18	37	45
Margarine, stick	21	46	33
Coconut	92	6	2
Palm kernel	81	18	1
Palm	47	43	10
Peanut	18	48	34
Olive	14	77	9
Corn	13	25	62
Canola	6	58	36

Human sex hormones

Oral contraceptives

Anabolics

Anti-inflammatories

Figure 22.4

In mammals, male and fema bolic steroids increase muscl matory properties. About 35(

The Canadian government is hoping that exports of canola oil (from a variety of rapeseed) to the United States will continue to increase.

There are two general trends hidden within this table. First, the higher the percentage of saturated fatty acids, the higher the melting point of the fat. Second, the oils are listed by climate, the oils from the warmer climates first, then those from the cooler climates. The warmer the climate, the higher the percentage of higher-melting saturated fatty acids.

Phospholipids

Like fats, **phospholipids** are esters of glycerol. However, a phospholipid has only two fatty acid residues; the third hydroxyl group is esterified to phosphoric acid, which with three hydrogens may form other ester bonds.

$$CH_3-(CH_2)_4-CH=CH-CH_2-CH=CH(CH_2)_7-\overset{\overset{\displaystyle O}{\|}}{C}-O-\overset{\displaystyle CH}{\underset{\displaystyle CH_2}{|}}-\overset{\overset{\displaystyle O}{\|}}{\underset{\displaystyle OH}{P}}-O-CH_2-CH_2-\overset{\overset{\displaystyle CH_3}{|}}{\underset{\displaystyle CH_3}{N}}-CH_3$$

$$CH_3-(CH_2)_{16}-\overset{\overset{\displaystyle O}{\|}}{C}-O-CH_2$$

Phospholipids have a polar head and long, nonpolar tails. Phospholipids are found in all animal and vegetable cells, where they are an important part of cell membranes. The phospholipid shown above is commonly called a **lecithin.** It is used widely in the cosmetic and food industries as an *emulsifier,* a substance that holds two immiscible liquids together as a suspension. You'll find lecithins in many food products such as chocolate, ice cream, margarine, and peanut butter.

see how these macromolecules differ. Which bonds can our enzymes break? Which bonds can our enzymes not break?

34) Why does glycogen contain $\alpha(1 \to 4)$ bonds rather than $\beta(1 \to 4)$ bonds?

SECTION 22.4

35) What are the three classes of lipids?

36) What physical property do the three classes of lipids share?

---

37) What physical property differentiates fats and oils?

38) Are oils usually obtained from animal sources or plant sources?

---

39) Use the letters A, B, and C to stand for three different fatty acids. Draw out all possible triacylglycerols you can make from these acids and one glycerol molecule.

40) Explain why the human body contains only fatty acids with an even number of carbon atoms.

---

41) Use the letters A and B to stand for two different fatty acids in a phospholipid. Draw out all possible phospholipids you can make from these two fatty acids.

42) Point out the polar part and the nonpolar part of lecithin.

SECTION 22.5

43) Name the two types of nucleic acid polymers present in cells.

44) In words, briefly explain the function of DNA and of RNA.

---

45) Draw the Lewis diagram for uracil.

46) Draw the Lewis diagram for thymine.

---

47) Draw the Lewis diagrams for adenine and thymine.

48) Draw the Lewis diagrams for guanine and thymine.

---

49) Draw the Lewis diagram for the sugar ribose.

50) Draw the Lewis diagram for the sugar deoxyribose. How does this sugar differ from the sugar in Question 49?

---

51) Draw the nucleoside adenosine, which is an adenine-ribose combination.

52) Draw the nucleoside deoxyadenosine, which is an adenine-deoxyribose combination.

---

53) Single-letter codes are often used to stand for the base in a nucleic acid. Draw the Lewis diagram for a DNA

fragment having the bases adenine-guanine-thymine, abbreviated A-G-T.

54) Draw the Lewis diagram for the DNA fragment that is *complementary* to the DNA fragment in Question 53.

---

55) Draw the Lewis diagram for an RNA fragment having the bases guanine-cytosine-uracil, abbreviated G-C-U.

56) Although RNA is single stranded, the strand sometimes folds back upon itself to give a complementary portion. What would be the complementary portion of the RNA fragment in Question 55?

---

57) Describe in words the role in protein synthesis of transfer RNA, tRNA.

58) Describe in words the role in protein synthesis of messenger RNA, mRNA.

GENERAL QUESTIONS

59) Distinguish precisely and in scientific terms the differences between items in each of the following groups.
a) Secondary and tertiary protein structures
b) α-helix and β-pleated sheet
c) Reversible and irreversible enzyme inhibitors
d) α-glucose and β-glucose
e) $\alpha(1 \to 4)$ and $\beta(1 \to 4)$ linkages
f) Fats and oils
g) Saturated fatty acids and unsaturated fatty acids
h) Triacylglycerols and phospholipids
i) Triacylglycerols and waxes
j) Waxes and steroids
k) Nucleoside and nucleotide
l) Thymine and uracil
m) α-helix and double helix

60) Biochemists use 120 g/mole as the "average molar mass" of a single amino acid in a protein. Hemoglobin is a small protein that transports oxygen from the lungs to the capillaries. The molar mass of hemoglobin is about 64,500 g/mole. Approximately how many amino acids make up hemoglobin?

61) The tobacco mosaic virus is a small virus that has been crystallized in a pure form. One complete virus has a molar mass about 40,000,000 g/mole, of which 38,000,000 g is protein. Use the data in Question 60 to determine the approximate number of amino acids in the proteins of a tobacco mosaic virus.

62) Which of the following macromolecules are polymers: cellulose, proteins, DNA, starch, RNA?

63) What are the monomer units for the polymers listed in Question 62, above?

This peanut butter, like al
ble products, contains no
terol; cholesterol is a pro
animal metabolism.

64) What element does protein supply us that carbohydrates or fats or oils do not supply us?

65) What element is found in DNA and RNA, but not in proteins?

66)* The structure of nylon 66 is given in Section 21.15. Use this structure to explain why nylon fabric is used in waterproof windbreakers while cotton, a cellulose material, is a fabric used where moisture must be absorbed.

67)* Rayon is an ester made by treating cotton with acetic acid. Use the cellulose structure given in Section

22.3 to draw a Lewis structure for a short rayon "molecule."

68)* The base content of a sample of pure mammalian DNA was analyzed to be 21% guanine. In this DNA sample, what is the percentage of cytosine? What is the percentage of adenine? What is the percentage of thymine?

69)* RNA, unlike DNA, is a single-stranded macromolecule. Would you expect the percentage guanine in RNA to equal the percentage cytosine? Why or why not?

MATCHING SET ANSWERS

Set 1: 8–3–12–1–6–14–10–2–4–11–9–7–5–13

Set 2: 10–4–13–11–1–12–3–6–2–7–9–8–5–14

QUICK CHECK ANSWERS

22.1 A-V-L, A-L-V, V-A-L, V-L-A, L-V-A, L-A-V
22.2 1) primary, local conformation, twists and turns
   2) α-helix
   3) β-pleated sheet
22.3 a and d true. b: The lock and key analogy explains why one enzyme helps a single reaction to occur faster. c: An enzyme substrate is the reactant in the enzyme-catalyzed reaction.
22.4 For parts a and b, first find in the amylopectin structure oxygen atoms having a bond drawn vertically, then continue as follows:
   a) The second bond from the oxygen atom is drawn up and to the right. The carbon atom directly above the oxygen is carbon-1; the carbon atom above right of this oxygen is carbon-4. (Remember that carbon-1 is the *only* carbon atom in glucose that is bonded to 2 oxygen atoms.)
   b) The second bond from the oxygen is drawn down and to the left. The carbon atom directly above the oxygen is carbon-1; the carbon atom below left (shown as CH_2—) is carbon-6.
22.5 a) The squeezable liquid margarine is the lowest of the three in saturated (higher-melting point) fatty acids and highest in unsaturated (lower-melting point) fatty acids. The soft solid has more saturated fatty acids than the liquid and less unsaturated fatty acids than the liquid. The stick margarine is highest in saturated fatty acids and lowest in unsaturated fatty acids. It most resembles butter, also sold in sticks.
   b) Alas, judging by its relatively high melting point, cocoa butter contains many, many saturated fatty acids.
   c) Steroids are characterized by four fused rings, as shown below:

22.6 a) The components of a nucleotide are: 1) a nitrogen-containing base, 2) a sugar, and 3) phosphate groups.
   b) RNA uses the sugar ribose and the bases adenine, cytosine, guanine, and uracil. DNA uses the sugar deoxyribose and the bases adenine, cytosine, guanine, and thymine.
   c) The only complementary DNA base pair is adenine-thymine (A-T).
   d) In replication, a DNA double helix unwinds. Each DNA strand serves as a template on which new complementary base pairs can form hydrogen bonds. After forming sugar–phosphate bonds, there are two double-stranded DNA molecules. Each new molecule is half old DNA strand, half newly constructed strand. This process is termed *semiconservative,* as half the strands of the new DNA molecules are conserved from the old DNA.

Appendix I
Chemical Calculations

A beginning student in chemistry is assumed to have developed calculation skills in earlier mathematics classes. Often chemistry is the first occasion for these skills to be put to the test of practical application. Experience shows that many students who may have learned these skills at one time, but have not used them regularly, can profit from a review of basic concepts. Others can benefit from a handy reference to the calculation techniques used in chemistry. This section of the Appendix is intended to meet these needs.

PART A
THE HAND CALCULATOR

Today every serious chemistry student uses a calculator to solve chemistry problems. A suitable calculator can (1) add, subtract, multiply, and divide; (2) perform these operations in exponential notation; (3) work with logarithms; and (4) raise any base to any power. Calculators that can perform these operations usually have other capabilities, too, such as squares and square roots, trigonometric functions, shortcuts for pi and percentage, enclosures, statistical features, and different levels of storage and recall.

Most calculators operate with one of two *logic* systems, each with its own order of operations. One is called the Algebraic Operating System (AOS), and the other is Reverse Polish Notation (RPN). In the examples that follow we will give *general* keyboard sequences for both systems as they are performed on calculators that are popularly used by students. Different brands may vary in details, particularly when some keys are used for more than one function. Some calculators offer an option on the number of digits to be displayed after the decimal point. With or without such an option, the number varies on different calculators. Accordingly, answers in this book may differ slightly from yours. Please consult the instruction book that accompanied your calculator for specific directions on these or other variations that may appear.

(You may wonder why you should not simply use your instruction book rather than the suggestions that follow. For complete mastery of your calculator you should do just that. If your present purpose is to learn how to use the calculator for chemistry, these instructions will be much easier. They also include practical suggestions that do not appear in a formal instruction book.)

One precautionary note before we begin: *A calculator should* never *be used as a substitute for thinking.* If a problem is simple and can be solved mentally, do it "in your head." You will make fewer mistakes. If you use your calculator, *think* your way through each problem and estimate the answer mentally. Suggestions on approximating answers are given in Part D of this Appendix. If the calculator answer appears reasonable, round it off properly and write it down. Then run through the calculation again to be sure you haven't made a keyboard error. Your calculator is an obedient and faithful servant that will do exactly what you tell it to do, but it is not responsible for the mistakes you make in your instructions.

The suggestion was made that you round off your answer properly before recording it. The reason for this is that calculator answers to many problems are limited in length only by the display. For example, $273 \div 45.6 = 5.9868421$ on one calculator. Some calculators can show more numbers, and therefore do. Usually, only the first three or four digits have meaning, and the others should be discarded. Procedures for deciding how many digits to write are given in Section 3.5, on significant figures.

Reciprocals, Square Roots, Squares, and Logarithms

Finding the reciprocal, square root, or square of a number is called a one-number function because only one number must be keyed into the calculator. In this case we are interested in finding $1/x$, $\sqrt{x}$, x^2, and log x. The procedures are the same for both operating systems:

1) Key in x.
2) Press function key ($1/x$, $\sqrt{x}$, x^2, or log x).

Example: Find $1/12.34$, $\sqrt{12.34}$, 12.34^2, and log 12.34.

Solution:

Problem	Press	Display
$1/12.34$	12.34	*12.34*
	$1/x$	*0.0810373*
$\sqrt{12.34}$	12.34	*12.34*
	$\sqrt{\ }$	*3.5128336*
12.34^2	12.34	*12.34*
	x^2	*152.2756*
log 12.34	12.34	*12.34*
	log	*1.0913152*

Antilogarithms, 10^x, and y^x

If x is the base 10 logarithm of a number, N, then N = antilog x = 10^x. Some calculators have a 10^x key that makes finding an antilogarithm a one-number function. Other calculators use an inverse function key, sometimes marked INV, to reverse the logarithm function. Again, finding an antilogarithm is a one-number function, but two function keys are used.

Example: Find the antilogarithm of 3.19.

Solution:

Press	Display		Press	Display
3.19	*3.19*		3.19	*3.19*
10^x	*1548.8166*		INV	*3.19*
			log	*1548.8166*

The y^x key can be used to raise any base, y, to any power, x. The procedure differs on the two operating systems, as the following example shows.

Example: Calculate $8.25^{0.413}$

AOS LOGIC		RPN LOGIC	
Press	**Display**	**Press**	**Display**
8.25	*8.25*	8.25	*8.25*
y^x	*8.25*	ENTER	*8.25*
.413	*0.413*	.413	*0.413*
=	*2.3905371*	y^x	*2.3905371*

Notice that, even though we always write a decimal fraction less than 1 with a zero before the decimal point, it is not necessary to enter such zeros into a calculator. The zeros are included in the display.

The y^x key can also be used to find an antilogarithm. In that case y = 10, and x is the given logarithm, the exponent to which 10 is to be raised.

Addition, Subtraction, Multiplication, and Division

For ordinary arithmetic operations the procedures for AOS and RPN logics are as follows:

AOS LOGIC	RPN LOGIC

The procedure for a common arithmetic operation is identical to the arithmetic equation for the same calculation. If you wish to add X to Y, the equation is X + Y = . The calculator procedure is:

1) Key in X.
2) Press the function key (+ for addition).
3) Key in Y.
4) Press =.

The display will show the calculated result.

Example: Solve 12.34 + 0.0567 = ?

Solution:

Press	Display
12.34	*12.34*
+	*12.34*
.0567	*0.0567*
=	*12.3967*

The numbers in the PRESS column are, in order, Steps 1, 2, 3, and 4 of the procedure. In Step 2, the function key is − for subtraction, × for multiplication, and ÷ for division.

The procedure for a common arithmetic operation is to key in *both* numbers and then tell the calculator what to do with them. If you wish to add X to Y, you enter X, key in Y, and instruct the calculator to add. The procedure is:

1) Key in X.
2) Press ENTER.
3) Key in Y.
4) Press the function key (+ for addition).

The display will show the calculated result.

Example: Solve 12.34 + 0.0567 = ?

Solution:

Press	Display
12.34	*12.34*
ENTER	*12.34*
.0567	*0.0567*
+	*12.3967*

The numbers in the PRESS column are, in order, Steps 1, 2, 3, and 4 of the procedure. In Step 4, the function key is − for subtraction, × for multiplication, and ÷ for division.

You may wish to confirm the following results on your calculator:

12.34 − 0.0567 = 12.2833   12.34 × 0.0567 = 0.699678   12.34 ÷ 0.0567 = 217.63668

Chain Calculations

A "chain calculation" is a series of two or more operations performed on three or more numbers. To the calculator the sequence is a series of two-number operations in which the first number is always the result of all calculations completed to that point. For example, in X + Y − Z the calculator first finds X + Y = A. The quantity A is already in and displayed by the calculator. All that needs to be done is to subtract Z from it. In the following example we deliberately begin with a negative number to illustrate the way such a number is introduced to the calculator. You simply press in the number, followed by the key that changes the sign, usually +/− or CHS. The example:
−2.45 + 18.7 + 0.309 − 24.6 = ?

AOS LOGIC		RPN LOGIC	
Press	**Display**	**Press**	**Display**
2.45	2.45	2.45	2.45
+/−	−2.45	CHS	−2.45
+	−2.45	ENTER	−2.45
18.7	18.7	18.7	18.7
+	16.25	+	16.25
.309	0.309	.309	0.309
−	16.559	+	16.559
24.6	24.6	24.6	24.6
=	−8.041	−	−8.04

Notice that it is not necessary to press the = key after each step *in a chain calculation involving only addition and/or subtraction.*

Combinations of multiplication and division are handled the same way. To solve 9.87 × 0.0654 ÷ 3.21:

AOS LOGIC		RPN LOGIC	
Press	**Display**	**Press**	**Display**
9.87	9.87	9.87	9.87
×	9.87	ENTER	9.87
0.0654	0.0654	.0654	0.0654
÷	0.645498	×	0.64549800
3.21	3.21	3.21	3.21
=	0.20108972	÷	0.20108972

Notice that it is not necessary to press the = key after each step *in a chain calculation involving only multiplication and/or division.*

Combination multiplication/division problems similar to the preceding one usually appear in the form of fractions in which all multipliers are in the numerator and all divisors are in the denominator. Thus, 9.87 × 0.0654 ÷ 3.21 is the same as $\dfrac{9.87 \times 0.0654}{3.21}$. In chemistry there are often several numerator factors and several denominator factors. A simple calculation that is easily completed "in your head" brings out some important facts about using calculators for chain calculations. Mentally, right now, calculate $\dfrac{9 \times 4}{2 \times 6} = ?$

There are several ways to get the answer. The most probable one, if you do it mentally, is to multiply 9 × 4 = 36 in the numerator, and then multiply 2 × 6 = 12 in the denominator. This changes the problem to 36/12. Dividing 36 by 12 gives 3 for the answer. This perfectly correct approach is often followed by the beginning calculator user when faced with numbers that cannot be multiplied and divided mentally. It is not the best method, however. It is longer and there is greater probability of error than is necessary.

In solving a problem such as (9 × 4)/(2 × 6), you can begin with any number and perform the required operations with other numbers in any order. Logically, you begin with one of the numerator factors. That gives you 12 different calculation sequences that yield the correct answer. They are

9 × 4 ÷ 2 ÷ 6	4 × 9 ÷ 2 ÷ 6
9 × 4 ÷ 6 ÷ 2	4 × 9 ÷ 6 ÷ 2
9 ÷ 2 ÷ 6 × 4	4 ÷ 2 ÷ 6 × 9
9 ÷ 2 × 4 ÷ 6	4 ÷ 2 × 9 ÷ 6
9 ÷ 6 × 4 ÷ 2	4 ÷ 6 × 9 ÷ 2
9 ÷ 6 ÷ 2 × 4	4 ÷ 6 ÷ 2 × 9

Practice a few of these sequences on your calculator to see how freely you may choose.

There is a common error in chain calculations that you should avoid. This is to interpret the above problem as 9 × 4 divided by 2 × 6, which is correct, but then punch it into the calculator as 9 × 4 ÷ 2 × 6, which is not correct. The calculator interprets these instructions as 9 × 4 = 36; 36 ÷ 2 = 18; 18 × 6 = 108. The last step should be 18 ÷ 6 = 3, as in the first setup in the previous list. *In chain calculations you must always* divide *by each factor in the denominator.*

There are over a hundred different sequences by which $\dfrac{7.83 \times 86.4 \times 291}{445 \times 807 \times 0.302}$ can be calculated. Practice some of them and see if you can duplicate the answer, 1.8152147.

You have seen that in multiplication and division you can take the factors in any order. This is possible in addition and subtraction too, provided that you keep each positive and negative sign with the number that follows it and treat the problem as an algebraic addition of signed numbers. When you mix addition/subtraction with multiplication/division, however, you must obey the rules that govern the order in which arithmetic operations are performed. Briefly, these rules are:

1) Simplify all expressions enclosed in parentheses.
2) Complete all multiplications and divisions.
3) Complete all additions and subtractions.

If your calculator is able to store and recall numbers, it can solve problems with a very complex order of operations. In this book you will find no such problems, but only those that require the simplest application of the first rule. Our comments will be limited to that application, and we will not use the storage capacity of your calculator, as the instruction book would probably recommend.

A typical calculation is

$$6.02 \times (22.1 - 48.6) \times 0.134$$

Recalling that factors in a multiplication problem may be taken in any order, you rearrange the numbers so the enclosed factor appears first:

$$(22.1 - 48.6) \times 6.02 \times 0.134$$

You may then perform the calculation in the order in which the numbers appear.

AOS LOGIC		RPN LOGIC	
Press	**Display**	**Press**	**Display**
22.1	*22.1*	22.1	*22.1*
−	*22.1*	ENTER	*22.1*
48.6	*48.6*	48.6	*48.6*
=	*−26.5*	−	*−26.5*
×	*−26.5*	6.02	*6.02*
6.02	*6.02*	×	*−159.53*
×	*−159.53*	.134	*0.134*
.134	*0.134*	×	*−21.37702*
=	*−21.37702*		

Notice that in a chain calculation involving *both* addition/subtraction *and* multiplication/division *it is necessary to press the = key after each addition or subtraction sequence in the AOS logic system before you proceed to a multiplication/division.*

Sometimes a factor in parentheses appears in the denominator of a fraction, where it is not easily taken as the first factor to be entered into the calculator. Again, such problems in this book are relatively simple. They may be solved by working the problem upside down and at the end using the 1/x key to turn it right side up. The process is demonstrated in calculating

$$\frac{13.3}{2.59(88.4 - 27.2)}$$

The procedure is to calculate $\dfrac{2.59(88.4 - 27.2)}{13.3}$ and find the reciprocal of the result.

AOS LOGIC		RPN LOGIC	
Press	**Display**	**Press**	**Display**
88.4	*88.4*	88.4	*88.4*
−	*88.4*	ENTER	*88.4*
27.2	*27.2*	27.2	*27.2*
=	*61.2*	−	*61.2*
×	*61.2*	2.59	*2.59*
2.59	*2.59*	×	*158.508*
÷	*158.508*	13.3	*13.3*
13.3	*13.3*	÷	*11.9179*
=	*11.917895*	1/x	*0.0839*
1/x	*.08390744*		

Exponential Notation

Modern calculators use exponential notation (Section 3.2) for very large or very small numbers. Ordinarily, if numbers are entered as decimal numbers, the answer appears as a decimal number. If the answer is too large or too small to be displayed, it "overflows" or "underflows" into exponential notation automatically. If you want the answer in exponential notation, even if the calculator can display it as a decimal number, you instruct the machine accordingly. The symbol on the key for this instruction varies with different calculators. EE and EEX are common.

A number shown in exponential notation has a space in front of the last two digits, which are at the right side of the display. These last two digits are the exponent. Thus, 4.68×10^{14} is displayed as 4.68 14; 2.39×10^{6} is 2.39 06. If the exponent is negative, a minus sign is present; 4.68×10^{-14} is 4.68 − 14.

To key 4.68×10^{14} and 4.68×10^{-14} into a calculator, proceed as follows.

AOS LOGIC		RPN LOGIC	
Press	**Display**	**Press**	**Display**
4.68	*4.68*	4.68	*4.68*
EE	*4.68 00*	EEX	*4.68 00*
14	*4.68 14*	14	*4.68 14*

The calculator display now shows 4.68×10^{14}. Use the next step only if you wish to change to a negative exponent, 4.68×10^{-14}.

+/−	$4.68-14$		CHS	$4.68-14$
Function key:	$4.68-14$		ENTER	$4.68-14$
+, −, ×, or ÷				

The calculator is now ready for the next number to be keyed in, as in earlier examples.

PART B
ARITHMETIC AND ALGEBRA

We present here a brief review of arithmetic and algebra to the point that it is used or assumed in this text. Formal mathematical statement and development are avoided. The only purpose of this section is to refresh your memory in those areas where it may be needed.

  1) ADDITION. a + b. Example:
2 + 3 = 5. The result of an addition is a **sum.**

  2) SUBTRACTION. a − b. Example:
5 − 3 = 2. Subtraction may be thought of as the addition of a negative number. In that sense, a − b = a + (−b). Example:

$$5 - 3 = 5 + (-3) = 2$$

The result of a subtraction is a **difference.**

  3) MULTIPLICATION. $a \times b = ab = a \cdot b = a(b) = (a)(b) = b \times a = ba = b \cdot a = b(a)$. The foregoing all mean that **factor** a is to be multiplied by factor b. Reversing the sequence of the factors, a × b = b × a, indicates that factors may be taken in any order when two or more·are multiplied together. Examples:

$$2 \times 3 = 2 \cdot 3 = 2(3) = (2)(3)$$
$$= 3 \times 2 = 3 \cdot 2 = 3(2) = 6$$

The result of a multiplication is a **product.**

Grouping of Factors (a)(b)(c) = (ab)(c) = (a)(bc). Factors may be grouped in any way in multiplication. Example:

$$(2)(3)(4) = (2 \times 3)(4) = (2)(3 \times 4) = 24$$

Multiplication by 1 n × 1 = n. If any number is multiplied by 1, the product is the original number. Examples:

$$6 \times 1 = 6; \quad 3.25 \times 1 = 3.25$$

Multiplication of Fractions

$$\frac{a}{b} \times \frac{c}{d} \times \frac{e}{f} = \frac{ace}{bdf}$$

If two or more fractions are to be multiplied, the product is equal to the product of the numerators divided by the product of the denominators. Example:

$$4 \times \frac{9}{2} \times \frac{1}{6} = \frac{4}{1} \times \frac{9}{2} \times \frac{1}{6}$$

$$= \frac{4 \times 9 \times 1}{1 \times 2 \times 6} = \frac{36}{12} = 3$$

  4) DIVISION. $a \div b = a/b = \dfrac{a}{b}$. The foregoing all mean that a is to be divided by b. Example:

$$12 \div 4 = 12/4 = \frac{12}{4} = 3$$

The result of a division is a **quotient.**

Special Case If any number is divided by the same or an equal number, the quotient is equal to 1. Examples:

$$\frac{4}{4} = 1; \qquad \frac{8-3}{4+1} = \frac{5}{5} = 1; \qquad \frac{n}{n} = 1$$

Division by 1 $\dfrac{n}{1} = n$. If any number is divided by 1, the quotient is the original number. Examples

$$\frac{6}{1} = 6; \qquad \frac{3.25}{1} = 3.25$$

From this it follows that any number may be expressed as a fraction having 1 as the denominator. Examples:

$$4 = \frac{4}{1}; \qquad 9.12 = \frac{9.12}{1}; \qquad m = \frac{m}{1}$$

  5) RECIPROCALS. If n is any number, the reciprocal of n is $\dfrac{1}{n}$; if $\dfrac{a}{b}$ is any fraction, the reciprocal of $\dfrac{a}{b}$ is $\dfrac{b}{a}$. The first part of the foregoing sentence is actually a special case of the second part: If n is any number, it is equal to $\dfrac{n}{1}$.

Its reciprocal is therefore $\dfrac{1}{n}$.

  A reciprocal is sometimes referred to as the **inverse** (more specifically, the multiplicative inverse) of a number.

This is because the product of any number multiplied by its reciprocal equals 1. Examples:

$$2 \times \frac{1}{2} = \frac{2}{2} = 1 \qquad n \times \frac{1}{n} = \frac{n}{n} = 1$$

$$\frac{4}{3} \times \frac{3}{4} = \frac{12}{12} = 1 \qquad \frac{m}{n} \times \frac{n}{m} = \frac{mn}{mn} = 1$$

Division may be regarded as multiplication by a reciprocal:

$$a \div b = \frac{a}{b} = a \times \frac{1}{b}$$

Example: $6 \div 2 = \frac{6}{2} = 6 \times \frac{1}{2} = 3$

$$a \div b/c = \frac{a}{b/c} = a \times \frac{c}{b}$$

Example: $6 \div \frac{2}{3} = \frac{6}{2/3} = 6 \times \frac{3}{2} = 9$

6)   SUBSTITUTION. If $d = b + c$, then $a(b + c) = ad$. Any number or expression may be substituted for its equal in any other expression. Example: $7 = 3 + 4$. Therefore, $2(3 + 4) = 2 \times 7$.

7)   "CANCELLATION." $\dfrac{ab}{ca} = \dfrac{\cancel{a}b}{c\cancel{a}} = \dfrac{b}{c}$. The process commonly called **cancellation** is actually a combination of grouping of factors (see 3), substitution (see 6) of 1 for a number divided by itself (see 4), and multiplication by 1 (see 3). Note the steps in the following examples:

$$\frac{xy}{yz} = \frac{yx}{yz} = \left(\frac{y}{y}\right)\left(\frac{x}{z}\right) = 1 \cdot \frac{x}{z} = \frac{x}{z}$$

$$\frac{24}{18} = \frac{6 \times 4}{6 \times 3} = \frac{6}{6} \times \frac{4}{3} = 1 \times \frac{4}{3} = \frac{4}{3}$$

Note that only *factors*, or *multipliers*, can be canceled. There is no cancellation in $\dfrac{a + b}{a + c}$.

8) ASSOCIATIVE PROPERTIES. An arithmetic operation is associative if the numbers can be grouped in any way.

Addition $a + (b - c) = (a + b) - c$. Addition, including subtraction, is associative. Example:

$$3 + 4 - 5 = (3 + 4) - 5 = 3 + (4 - 5) = 2$$

Multiplication $(a \times b)/c = a \times (b/c)$. Multiplication, including division (multiplication by an inverse), is associative. (See "Grouping of factors" under MULTIPLICATION above.) Example:

$$4 \times 6/2 = (4 \times 6)/2 = 4 \times (6/2) = 12$$

9) EXPONENTIALS. An exponential has the form B^p, where B is the **base** and p is the **power** or **exponent.** An exponential indicates the number of times the base is used as a factor in multiplication. For example, 10^3 means 10 is to be used as a factor 3 times:

$$10^3 = 10 \times 10 \times 10 = 1000$$

A negative exponent tells the number of times a base is used as a divisor. For example, 10^{-3} means 10 is used as a divisor 3 times:

$$10^{-3} = \frac{1}{10} \times \frac{1}{10} \times \frac{1}{10} = \frac{1}{10^3}$$

This also shows that an exponential may be moved in either direction between the numerator and denominator by changing the sign of the exponent:

$$\frac{1}{2^3} = 2^{-3} \qquad 3^4 = \frac{1}{3^{-4}}$$

Multiplication of Exponentials Having the Same Base $a^m \times a^n = a^{m+n}$. To multiply exponentials, add the exponents. Example:

$$10^3 \times 10^4 = 10^7$$

Division of Exponentials Having the Same Base $a^m \div a^n = \dfrac{a^m}{a^n} = a^{m-n}$. To divide exponentials, subtract the denominator exponent from the numerator exponent. Example:

$$10^7 \div 10^4 = \frac{10^7}{10^4} = (10^7)(10^{-4}) = 10^{7-4} = 10^3$$

Zero Power $a^0 = 1$. Any base raised to the zero power equals 1. The fraction $\dfrac{a^m}{a^m} = 1$ because the numerator is the same as the denominator. By division of exponentials, $\dfrac{a^m}{a^m} = a^{m-m} = a^0$.

Raising a Product to a Power $(ab)^n = a^n \times b^n$. When the product of two or more factors is raised to some power, each factor is raised to that power. Example:

$$(2 \times 5y)^3 = 2^3 \times 5^3 \times y^3$$
$$= 8 \times 125 \times y^3 = 1000y^3$$

Raising a Fraction to a Power $\left(\dfrac{a}{b}\right)^n = \dfrac{a^n}{b^n}$. When a fraction is raised to some power, the numerator and denominator are both raised to that power. Example:

$$\left(\frac{2x}{5}\right)^3 = \frac{2^3 x^3}{5^3} = \frac{8x^3}{125} = 0.064x^3$$

Square Root of Exponentials $\sqrt{a^{2n}} = a^n$. To find the square root of an exponential, divide the exponent by 2. Example: $\sqrt{10^6} = 10^3$. If the exponent is odd, see below.

Square Root of a Product $\sqrt{ab} = \sqrt{a} \times \sqrt{b}$. The square root of the product of two numbers equals the product of the square roots of the numbers. Example:

$$\sqrt{9 \times 10^{-6}} = \sqrt{9} \times \sqrt{10^{-6}} = 3 \times 10^{-3}$$

Using this principle, by adjusting a decimal point, you may take the square root of an exponential having an odd exponent. Example:

$$\sqrt{10^5} = \sqrt{10 \times 10^4} = \sqrt{10} \times \sqrt{10^4}$$
$$= 3.16 \times 10^2$$

The same technique may be used in taking the square root of a number expressed in exponential notation. Example:

$$\sqrt{1.8 \times 10^{-5}} = \sqrt{18 \times 10^{-6}}$$
$$= \sqrt{18} \times \sqrt{10^{-6}}$$
$$= 4.2 \times 10^{-3}$$

10) SOLVING AN EQUATION FOR AN UN-KNOWN QUANTITY: Most problems in this book can be solved by dimensional analysis methods. There are times, however, when algebra must be used, particularly in relation to the gas laws. Solving an equation for an unknown involves rearranging the equation so that the unknown is the only item on one side and only known quantities are on the other. "Rearranging" an equation may be done in several ways, but the important thing is that *whatever is done to one side of the equation must also be done to the other.* The resulting relationship remains an equality, a true equation. Among the operations that may be performed on both sides of an equation are addition, subtraction, multiplication, division, and raising to a power, which includes taking square root.

In the following examples a, b, and c represent known quantities, and x is the unknown. The object in each case is to solve the equation for x. The steps of the algebraic solution are shown, as well as the operation performed on both sides of the equation. Each example is accompanied by a practice problem that is solved by the same method. You should be able to solve the problem, even if you have not yet reached that point in the book where such a problem is likely to appear. Answers to these practice problems may be found at the end of Appendix I.

(1) $\qquad\qquad x + a = b$

$\qquad x + a - a = b - a \qquad$ Subtract a

$\qquad\qquad x = b - a \qquad$ Simplify

PRACTICE: *1) If $P = p_{O_2} + p_{H_2O}$, find p_{O_2}, when $P = 748$ torr and $p_{H_2O} = 24$ torr.*

(2) $\qquad ax = b$

$\qquad \dfrac{\cancel{a}x}{\cancel{a}} = \dfrac{b}{a} \qquad$ Divide by a, which is called the **coefficient** of x

$\qquad x = \dfrac{b}{a} \qquad$ Simplify

PRACTICE: *2) At a certain temperature, $PV = k$. If $P = 1.23$ atm and $k = 1.62$ L $\cdot$ atm, find V.*

(3) $\qquad \dfrac{x}{a} = b$

$\qquad \dfrac{\cancel{a}x}{\cancel{a}} = ba \qquad$ Multiply by a

$\qquad x = ba \qquad$ Simplify

PRACTICE: *3) In a fixed volume, $\dfrac{P}{T} = k$. Find P if $k = \dfrac{2.4 \text{ torr}}{K}$ and $T = 300$ K.*

PRACTICE: *4) For gases at constant volume, $\dfrac{P_1}{T_1} = \dfrac{P_2}{T_2}$. If $P_1 = 0.80$ atm at $T_1 = 320$ K, at what value of T_2 will $P_2 = 1.00$ atm?*

Note: Procedures (2) and (3) are examples of dividing both sides of the equation by the coefficient of x, which is the same as multiplying both sides by the inverse of the coefficient. This is best seen in a more complex example:

$$\frac{ax}{b} = \frac{c}{d}$$

$$\frac{a}{b}x = \frac{c}{d} \qquad \text{Isolate the coefficient of } x$$

$$\frac{b}{a} \times \frac{a}{b}x = \frac{c}{d} \times \frac{b}{a} \qquad \text{Multiply by the inverse of the coefficient of } x$$

$$x = \frac{cb}{da} \qquad \text{Simplify}$$

(4)
$$\frac{b}{ax} = \frac{d}{c}$$

$$\frac{ax}{b} = \frac{c}{d} \qquad \text{Invert both sides of the equation}$$

Proceed as in (3) above.

(5)
$$\frac{a}{b + x} = c$$

$$(b + x)\frac{a}{(b + x)} = c(b + x) \qquad \text{Multiply by } (b + x)$$

$$a = c(b + x) \qquad \text{Simplify}$$

$$\frac{a}{c} = \frac{\cancel{c}(b + x)}{\cancel{c}} \qquad \text{Divide by } c$$

$$\frac{a}{c} = b + x \qquad \text{Simplify}$$

$$\frac{a}{c} - b = x \qquad \text{Subtract } b$$

PRACTICE: *5) In how many grams of water must you dissolve 20.0 g of salt to make a 25% solution? The formula is*

$$\frac{g\ salt}{g\ salt + g\ water} \times 100 = \%;\ or$$

$$\frac{g\ salt}{g\ salt + g\ water} = \frac{\%}{100}$$

PART C
LOGARITHMS

Sections 17.8 and 17.9 are the only places in this text that use logarithms, and most of what you need to know about logarithms is explained at that point. Comments here are limited to basic information needed to support the text explanations.

The common logarithm of a number is the power, or exponent, to which 10 must be raised to be equal to the number. Expressed mathematically,

$$\text{If } N = 10^x, \text{ then } \log N = \log 10^x = x \qquad \text{(AP.1)}$$

The number 100 may be written as the base, 10, raised to the second power: $100 = 10^2$. According to the preceding equation, 2 is the logarithm of 10^2, or 100. Similarly, if $1000 = 10^3$, $\log 1000 = \log 10^3 = 3$. And if $0.0001 = 10^{-4}$, $\log 0.0001 = \log 10^{-4} = -4$.

Just as the powers to which 10 can be raised may be either positive or negative, so may logarithms be positive or negative. The changeover occurs at the value 1, which is 10^0. The logarithm of 1 is therefore 0. It follows that the logarithms of numbers greater than 1 are positive, and logarithms of numbers less than 1 are negative.

The powers to which 10 may be raised are not limited to integers. For example, 10 can be raised to the 2.45 power: $10^{2.45}$. The logarithm of $10^{2.45}$ is 2.45. Such a logarithm is made up of two parts. The digit or digits to the left of the decimal are the **characteristic.** The characteristic reflects the size of the number; it is related to the exponent of 10 when the number is expressed in exponential notation. The characteristic is 2 in 2.45. The digits to the right of the decimal make up the **mantissa,** which is the logarithm of the coefficient of the number when written in exponential notation. In 2.45 the mantissa is 0.45.

The number that corresponds to a given logarithm is its antilogarithm. In Equation AP.1, the antilogarithm of x is 10^x, or N. The antilogarithm of 2 is 10^2, or 100. The antilogarithm of 2.45 is $10^{2.45}$. The value of the antilogarithm of 2.45 can be found on a calculator, as described in Part A of Appendix I: antilog $2.45 = 10^{2.45} = 2.8 \times 10^2$. In this exponential form of the antilogarithm of 2.45 the characteristic, 2, is the exponent of 10, and the mantissa, 0.45, is the logarithm of the coefficient, 2.8. In terms of significant figures, the mantissa matches the coefficient. This is illustrated in Section 17.10.

Because logarithms are exponents, they are governed by the rules of exponents given in Section 9 of Part B of

Appendix I Chemical Calculations **A.9**

this Appendix. For example, the product of two exponentials to the same base is the base raised to a power equal to the sum of the exponents: $a^m \times a^n = a^{m+n}$. The exponents are added. Similarly, exponents to the base 10 (logarithms) are added to get the logarithm of the product of two numbers: $10^m \times 10^n = 10^{m+n}$. Thus

$$\log ab = \log a + \log b \qquad (AP.2)$$

In a similar fashion, the logarithm of a quotient is the logarithm of the dividend minus the logarithm of the divisor (or the logarithm of the numerator minus the logarithm of the denominator if the expression is written as a fraction):

$$\log a/b = \log a - \log b \qquad (AP.3)$$

Equation AP.1 is the basis for converting between pH and hydrogen ion concentration in Section 17.9—or between any "p" number and its corresponding value in exponential notation. This is the only application of logarithms in this text. In more advanced chemistry courses you will encounter applications of Equation AP.3 and others that are beyond the scope of this discussion.

Ten is not the only base for logarithms. Many natural phenomena, both chemical and otherwise, involve logarithms to the **base e,** which is 2.718. . . . Logarithms to base e are known as **natural logarithms.** Their value is 2.303 times greater than a base 10 logarithm. Physical chemistry relationships that appear in base e are often converted to base 10 by the 2.303 factor, although modern calculators make it just as easy to work in base e as in base 10. The "p" concept, however, uses base 10 by definition.

PART D
ESTIMATING CALCULATION RESULTS

A large percentage of student calculation errors would never appear on homework or test papers if the student would estimate the answer before accepting the number displayed on a calculator. *Challenge every answer.* Be sure it is reasonable before you write it down.

There is no single "right" way to estimate an answer. As your mathematical skills grow, you will develop techniques that are best for you. You will also find that one method works best on one kind of problem, and another method on another problem. The ideas that follow should help you get started in this important practice.

In general, estimating a calculated result involves rounding off the given numbers and calculating the answer mentally. For example, if you multiply 325 by 8.36 on your calculator, you will get 2717. To see if this answer is reasonable, you might round off 325 to 300, and 8.36 to 8. The problem then becomes 300×8, which you can calculate mentally to 2400. This is reasonably close to your calculator answer. Even so, you should run your calculator again to be sure you haven't made a small "typing" error.

If your calculator answer for the preceding problem had been 1254.5, or 38.975598, or 27,170, your estimated 2400 would signal that an error was made. These three numbers represent common calculation errors. The first comes from a mistake that may arise any time a number is transferred from one position (your paper) to another (your calculator). It is called transposition, and appears when two numerals are changed in position. In this case 325×3.86 (instead of 8.36) = 1254.5.

The answer 38.875598 comes from pressing the wrong function key: $325 \div 8.36 = 38.875598$. This answer is so unreasonable—$325 \times 8.36 =$ about 38!—that no mental arithmetic should be necessary to tell you it is wrong. But, like molten idols, calculators speak to some students with a mystic authority they would never dare to challenge. On one occasion neither student nor teacher could figure out why or how, on a test, the student used a calculator to divide 428 by 0.01, and then wrote down 7290 for the answer, when all he had to do was move the decimal two places!

The answer 27,170 is a decimal error, as might arise from transporting the decimal point and a number, or putting an incorrect number of zeros in numbers like 0.00123 or 123,000. Decimal errors are also apt to appear through an incorrect use of exponential notation, either with or without a calculator.

Exponential notation is a valuable aid in estimating results. For example, in calculating $41,300 \times 0.0524$, you can regard both numbers as falling between 1 and 10 for a quick calculation of the coefficient: $4 \times 5 = 20$. Then, thinking of the exponents, changing 41,300 to 4 moves the decimal four places left, so the exponential is 10^4. Changing 0.0524 to 5 has the decimal moving two places right, so the exponential is 10^{-2}. Adding exponents gives $4 + (-2) = 2$. The estimated answer is 20×10^2, or 2000. On the calculator it comes out to 2164.12.

Another "trick" that can be used is to move decimals in such a way that the moves cancel and at the same time the problem is simplified. In $41,300 \times 0.0524$, the decimal in the first factor can be moved two places left (divide by

100), and in the second factor, two places right (multiply by 100). Dividing and multiplying by 100 is the same as multiplying by $\dfrac{100}{100}$, which is equal to 1. The problem simplifies to 413×5.24, which is easily estimated as $400 \times 5 = 2000$. The same technique can be used to simplify fractions, too. $\dfrac{371,000}{6240}$ can be simplified to $\dfrac{371}{6.24}$ by moving the decimal three places left in both the numerator and denominator, which is the same as multiplying by 1 in the form $\dfrac{0.001}{0.001}$. An estimated $\dfrac{360}{6}$ gives 60 as the approximate answer. The calculator answer is 59.455128. Similarly, $\dfrac{0.000406}{0.000839}$ becomes about $\dfrac{4}{8}$, or 0.5; by calculator, the answer is 0.4839042.

ANSWERS TO PRACTICE PROBLEMS IN APPENDIX I

1) $p_{O_2} = P - p_{H_2O} = 748 - 24 = 724$ torr

2) $V = \dfrac{k}{P} = \dfrac{1.62\ \text{L atm}}{1.23\ \text{atm}} = 1.32$ L

3) $P = kT = \dfrac{2.4\ \text{torr}}{\cancel{K}} \times 300\ \cancel{K} = 720$ torr

4) $T_2 = \dfrac{T_1 P_2}{P_1} = \dfrac{(320\ \text{K})(1.00\ \cancel{\text{atm}})}{0.80\ \cancel{\text{atm}}} = 400$ K

5) Because of the complexity of this problem, it is easier to substitute the given values into the original equation and then solve for the unknown. The steps in the solution correspond to those on page A.8:

$$\frac{\text{g salt}}{\text{g salt} + \text{g water}} = \frac{\%}{100}$$

$$\frac{20.0}{20.0 + \text{g water}} = \frac{25}{100}$$

$$20.0 = 0.25(20.0 + \text{g water})$$

$$= 5.0 + 0.25(\text{g water})$$

$$20.0 - 5.0 = 0.25(\text{g water})$$

$$\text{g water} = \frac{15.0}{0.25} = 60\ \text{g water}$$

Appendix II
The SI System of Units

BASE UNITS

The International System of Units or *Système International* (SI), which represents an extension of the metric system, was adopted by the 11th General Conference of Weights and Measures in 1960. It is constructed from seven base units, each of which represents a particular physical quantity (Table AP-1).

Of the seven units listed in Table AP-1, the first five are particularly useful in general chemistry. They are defined as follows:

1) The *meter* was redefined in 1983 to be equal to the distance light travels in a vacuum in 1/299,792,458 second.
2) The *kilogram* represents the mass of a platinum-iridium block kept at the International Bureau of Weights and Measures at Sevres, France.
3) The *second* was redefined in 1967 as the duration of 9,192,631,770 periods of a certain line in the microwave spectrum of cesium-133.

Table AP-1
SI Base Units

Physical Quantity	Name of Unit	Symbol
1) Length	meter	m
2) Mass	kilogram	kg
3) Time	second	s
4) Temperature	kelvin	K
5) Amount of substance	mole	mol
6) Electric current	ampere	A
7) Luminous intensity	candela	cd

4) The *kelvin* is 1/273.16 of the temperature interval between the absolute zero and the triple point of water (0.01°C = 273.16 K).
5) The *mole* is the amount of substance that contains as many entities as there are atoms in exactly 0.012 kg of carbon-12.

PREFIXES USED WITH SI UNITS

Decimal fractions and multiples of SI units are designated by using the prefixes listed in Table 3.2. Those that are most commonly used in general chemistry are in boldface type.

DERIVED UNITS

In the International System of Units all physical quantities are expressed in the base units listed in Table AP-1 or in combinations of those units. The combinations are called **derived units.** For example, the density of a substance is found by dividing the mass of a sample in kilograms by its volume in cubic meters. The resulting units are kilograms per cubic meter, or kg/m^3. Some of the derived units used in chemistry are given in Table AP-2.

If you have not studied physics, the SI units of force, pressure, and energy are probably new to you. Force is related to acceleration, which has to do with changing the velocity of an object. One **newton** is the force that, when applied for one second, will change the straight-line speed of a 1-kilogram object by 1 meter per second.

A **pascal** is defined as a pressure of one newton acting on an area of one square meter. A pascal is a small unit, so pressures are commonly expressed in kilopascals (kPa), 1000 times larger than the pascal.

A **joule** (pronounced jo͞ol, as in pool) is defined as the work done when a force of one newton acts through a distance of one meter. *Work* and *energy* have the same

Table AP-2

SI Derived Units

Physical Quantity	Name of Unit	Symbol	Definition
Area	square meter	m^2	
Volume	cubic meter	m^3	
Density	kilograms per cubic meter	kg/m^3	
Force	newton	N	$kg \cdot m/s^2$
Pressure	pascal	Pa	N/m^2
Energy	joule	J	$N \cdot m$

units. Large amounts of energy are often expressed in kilojoules, 1000 times larger than the joule.

SOME CHOICES

It is difficult to predict the extent to which SI units will replace traditional metric units in the coming years. This makes it difficult to select and use the units that will be most helpful to the readers of this textbook. Add to that the authors' joy that, after nearly 200 years, the United States has finally begun to adopt the metric system, including some units that the SI system would eliminate, and their deep desire to encourage rather than complicate the use of metrics in their native land. With particular apologies to Canadian readers, who are more familiar with SI units than Americans are, we list the areas in which this book does not follow SI recommendations:

1) The SI unit of length is the *metre*, spelled in a way that corresponds to its French pronunciation. In America, and in this book, it is written *meter*, which matches the English pronunciation. "What's in a name?"

2) The SI volume unit, *cubic meter*, is huge for most everyday uses. The *cubic decimeter* is 1/1000 as large, and much more practical. When referring to liquids it is customary to replace this six-syllable name with the two-syllable *liter*—or *litre*, for the French spelling. (Which would you rather buy at the grocery store, two cubic decimeters of milk, or two liters?) In the laboratory the common units are again 1/1000 as large, the *cubic centimeter* for solids and the *milliliter* for liquids.

3) The *millimeter of mercury* has an advantage over the *pascal* or *kilopascal* as a pressure unit because the common laboratory instrument for "measuring" pressure literally measures millimeters of mercury. We lose some of the advantage of this "natural" pressure unit by using its other name, *torr*. Reducing eight syllables to one is worth the sacrifice. Again, "What's in a name?" For large pressures we continue to use the traditional *atmosphere*, which is 760 torr.

Answers to Questions and Problems

Chapter 2

1) Physical: a, c, d. Chemical: b, e.   3) b, d, e.

5) Chemical. The white and yolk of a fresh egg are changed to different substances in the boiling process.

7) Table tennis balls are lighter (less dense) and smoother than golf balls. Density and "smoothness" are both physical properties.

11) Gases are more easily compressed because of the large spaces between molecules.

13) Solids are most rigid because particles are in fixed positions.

15) A dense gas that is concentrated at the bottom of a container can be poured because its particles can move relative to each other. A single chunk of a solid cannot be poured because its particles are held in fixed position relative to each other. Chunks of solids, such as sugar crystals, can be poured.

17) Air in a confined space is uniformly distributed. It is homogeneous.

19) Faucet water may contain small solids that are filtered out in store-bought drinking water. Faucet water is apt to be heterogeneous, but drinking water should be homogeneous.

21) A fresh egg and linoleum are heterogeneous. Microscopically, brass is heterogeneous. Milk from a cow is heterogeneous; "homogenized" milk from the dairy is homogeneous.

23) Visibly, homogeneous; microscopically, heterogeneous.

25) Possible answers: "lead" in a lead pencil (graphite), diamond, charcoal, soot, and elemental carbon. Bubbles in a soft drink (carbon dioxide), limestone, all animal or vegetable food products are carbon compounds.

27) Sand at a beach is silicon dioxide.

29) Only compounds have more than one word in their names, so a, b, and c are compounds. All elements, including uranium and tin, and some compounds have one-word names.

31) Elements and compounds are pure substances, so all substances in Question 29 are pure substances.

33) Mixture of hydrogen and oxygen.

35) If sample can be separated into two or more pure substances by physical means, it is a mixture. If it can be separated into two or more pure substances by chemical means, it is a compound.

37) Water can be separated into two elements, so it must be a compound.

39) Boiling at a fixed temperature is a property of a pure substance, which may be either an element or a compound.

41) Taste the condensed liquid. (Taste is *not* a recommended analytical procedure. It is apt, in some cases, to produce surprising, unpleasant, harmful, and even fatal results.)

43) Pure substance, but heterogeneous.

45) Where would you rather be, in Hawaii or on a sugar beet farm? Buyers are influenced by mental images.

47) Mixture. Iron (black) can be separated from sulfur (yellow) by color, or by removing the iron with a magnet. Color and magnetism are physical properties.

49) Natural gas and petroleum products are compounds of carbon and hydrogen. Add oxygen and you get sugar and many other plant and animal food products. The properties of the compounds are independent of the properties of the elements.

51) Heating the substance yields a product of lower density, so it must be a new substance. The change is chemical, and the original substance is a compound. The formation of a yellow solid is a chemical change. The dissolving and recrystallization of the original substance are physical changes.

53) Elements: Ba, P, Cl, C, B. Compounds (elements in formula): NaF (2), NH_3 (2), $CuSO_4$ (3), $LiOH$ (3), $CaBr_2$ (2).

55)

	G, L, S	**P, M**	**HOM, HET**	**E, C**
Table salt	S	P	HOM	C
Mercury	L	P	HOM	E
Air in a closed jar	G	M	HOM	—
Sand on a beach	S	M	HET	—
Automobile exhaust	All, but mostly G	M	HET	—

57)

Contents of balloon	G	Either	HOM	E or —
Ice	S	P	Either	C
Baking soda	S	P	HOM	C
Ethyl alcohol	L	P	HOM	C
Wood	S	M	HET	—

59) Gravitational force is attraction only. Others may be attraction or repulsion. All three forces can exist simultaneously.

61) Reactants: Na_2SO_4 and $BaCl_2$; products: $BaSO_4$ and NaCl.

63) Ni is an elemental product, and $NiSO_4$ is a compound reactant.   65) Endothermic, a, d; exothermic, b, c, e.

67) Potential energy greatest at top, kinetic greatest just before hitting stream. Potential energy changes to kinetic energy, but total remains constant.

69) Positive moves toward negative. Potential energy decreases.   71) Equal mass by Law of Conservation of Mass.

73) Gaseous substance driven off by heat. (Substance is carbon dioxide.)

75) No violation. Energy converted to heat as ball flew through air, struck ground, and rolled.

80) True: e, f, i, j, l. False: a–d, g, h, k.

81) Nothing.   82) Those substances that are elements or compounds: mercury, water (including ice), and carbon.

83) Yes. See the illustration of the mixture of iron and sulfur on page 18 and the equation for the reaction between them to form the compound FeS on page 23.

84) Yes. Carbon monoxide and carbon dioxide are one example; water and hydrogen peroxide are another.

85) Rainwater is distilled water, a pure compound. Ocean water is a solution, a mixture of many substances in water, and therefore impure.

86) a)   The sample is neither an element nor a compound. If it were either, its composition would be the same throughout.
b,c) The contents are a heterogeneous mixture. Again, if it were homogeneous, its composition would be the same throughout.

87) Only certain sources of energy are usable by humans. It is these sources that must not be depleted.

88) The potential energy decreases in each case. From an energy standpoint, spontaneous change is always toward a lower potential energy.

89) Neither shape nor volume will be changed if you press the solid. Pressing the liquid balloon will change the shape, but not the volume. Pressing the gas-filled balloon will change the shape and also reduce the volume slightly for reasons that will become clear in Chapter 4.

90) Two substances cannot have identical physical and chemical properties. These are the properties that identify a pure substance and distinguish it from all other substances.

91) A nuclear reaction involves a conversion of mass to energy or vice versa, both of which contradict the conservation laws. However, the vast majority of reactions are not nuclear, and, within measurable limits, the conservation laws remain quite reliable.

92) The properties of iron, an element that is a pure substance, cannot be altered. Steel is an alloy, a mixture of iron and other substances. Its properties can be altered by varying the composition of the mixture.

93) Energy must be added to change liquid water into hydrogen and oxygen, and heat and/or light energy are given off when hydrogen and oxygen combine to form water. Both events suggest that the hydrogen and oxygen have more energy than liquid water.

94) At low temperatures, particles of carbon dioxide are not free to move relative to each other, which is a characteristic of a solid. At higher temperatures the particles move freely, which describes a gas. Liquid carbon dioxide at room temperature exists only under high pressure. Some fire extinguishers contain liquid carbon dioxide.

Chapter 3

NOTE: Cancellation marks have been omitted in the answer section so problem setups can be seen more clearly.

1) 8.26×10^{-2}; 2.63×10^6; 9.24×10^5 3) 0.000123; $751,000$; 0.00382

5) 4.66×10^{12}; 3.80×10^{-13}; 1.36×10^{-3}; 9.36×10^5 7) 498; 9.22×10^9; 5.68×10^{-12}; 1.17×10^{-2}

9) 4.27×10^4; 0.0188 11) 8.86×10^{-5}; 8.63×10^{14}

13) $4 \text{ miles} \times \dfrac{5280 \text{ ft}}{1 \text{ mile}} = 21{,}120 \text{ ft}$ 15) $0.375 \text{ mile} \times \dfrac{5280 \text{ ft}}{1 \text{ mile}} \times \dfrac{1 \text{ s}}{1.09 \times 10^3 \text{ ft}} = 1.82 \text{ s}$

17) $8.33 \text{ hr} \times \dfrac{9.31 \text{ miles}}{3 \text{ hr}} = 25.9 \text{ miles}$ 19) $\text{C\$29.95} \times \dfrac{\text{A\$1.00}}{\text{C\$1.21}} = \text{A\$24.75}$

21) $1 \text{ hc} = 2 \text{ sh} + 6 \text{ p} = 2 \text{ sh} \times \dfrac{12 \text{ p}}{\text{sh}} + 6 \text{ p} = 24 \text{ p} + 6 \text{ p} = 30 \text{ p}$ 27) $1 \text{ cL} \times \dfrac{1 \text{ L}}{100 \text{ cL}} = 0.01 \text{ L}$

29) $16 \text{ Mbytes} \times \dfrac{10^6 \text{ bytes}}{\text{Mbyte}} = 16 \times 10^6 \text{ bytes} = 1.6 \times 10^7 \text{ bytes}$ 31) Micrometer. Very short—about 1/250 inch.

33) $503 \text{ g} \times \dfrac{1 \text{ kg}}{1000 \text{ g}} = 0.503 \text{ kg}$; $197 \text{ g} \times \dfrac{1000 \text{ mg}}{\text{g}} = 1.97 \times 10^5 \text{ mg}$; $592 \text{ mg} \times \dfrac{1 \text{ g}}{1000 \text{ mg}} \times \dfrac{1 \text{ kg}}{1000 \text{ g}} = 5.92 \times 10^{-4} \text{ kg}$

35) $5.49 \text{ km} \times \dfrac{1000 \text{ m}}{1 \text{ km}} = 5.49 \times 10^3 \text{ m}$; $5.04 \text{ mm} \times \dfrac{1 \text{ m}}{1000 \text{ mm}} = 5.04 \times 10^{-3} \text{ m}$; $6.44 \text{ cm} \times \dfrac{1000 \text{ mm}}{100 \text{ cm}} = 64.4 \text{ mm}$

37) $10.3 \text{ cL} \times \dfrac{1 \text{ L}}{100 \text{ cL}} = 0.103 \text{ L}$; $3.41 \text{ L} \times \dfrac{1000 \text{ mL}}{\text{L}} = 3.41 \times 10^3 \text{ mL}$;

$5.60 \text{ mL} \times \dfrac{1 \text{ L}}{1000 \text{ mL}} = 0.00560 \text{ L}$; $9.68 \text{ mL} \times \dfrac{1 \text{ cm}^3}{1 \text{ mL}} = 9.68 \text{ cm}^3$

39) $2.96 \text{ Mm} \times \dfrac{10^6 \text{ m}}{\text{Mm}} = 2.96 \times 10^6 \text{ m}$; $2.72 \text{ ng} \times \dfrac{1 \text{ g}}{10^9 \text{ ng}} = 2.72 \times 10^{-9} \text{ g}$; $9.27 \text{ ML} \times \dfrac{10^6 \text{ L}}{\text{ML}} \times \dfrac{10^3 \text{ mL}}{\text{L}} = 9.27 \times 10^9 \text{ mL}$

41) 6–2–3–uncertain (3 or 4)–3–5–3–5   43) 80.4; 65.7; 11.8; 5.76×10^7; 5.59×10^{-3} or 0.00559

45) $4.98 \text{ g} + 8.0 \text{ g} + 0.939 \text{ g} + 144 \text{ g} = 157.919 \text{ g} = 158 \text{ g}$ 47) $102.189 \text{ g} - 91.27 \text{ g} = 10.919 \text{ g} = 10.92 \text{ g}$

49) $2 \text{ L} \times \dfrac{34.3 \text{ g}}{\text{L}} = 68.6 \text{ g}$; $2.74 \text{ L} \times \dfrac{34.3 \text{ g}}{\text{L}} = 94.0 \text{ g}$ 51) $1689 \text{ lb} \times \dfrac{1 \text{ kg}}{2.20 \text{ lb}} = 768 \text{ kg}$

53) $294 \text{ km} \times \dfrac{1 \text{ mile}}{1.61 \text{ km}} = 183 \text{ miles}$ 55) $325 \text{ mg} \times \dfrac{1 \text{ g}}{1000 \text{ mg}} \times \dfrac{\text{oz}}{28.3 \text{ g}} = 0.0115 \text{ oz}$

57) $6.24 \text{ carats} \times \dfrac{0.200 \text{ g}}{\text{carat}} \times \dfrac{\text{oz}}{28.3 \text{ g}} = 0.0441 \text{ oz}$ 59) $922 \text{ lb} \times \dfrac{\text{kg}}{2.20 \text{ lb}} = 419 \text{ kg}$
$= 1.25 \text{ g}$

61) $7 \text{ oz} \times \dfrac{1 \text{ lb}}{16 \text{ oz}} = 0.4 \text{ lb}$; $7.4 \text{ lb} \times \dfrac{1 \text{ kg}}{2.20 \text{ lb}} = 3.4 \text{ kg}$ 63) $132 \text{ cm} \times \dfrac{1 \text{ in.}}{2.54 \text{ cm}} = 52.0 \text{ in.} = 4 \text{ ft } 4 \text{ in.}$

65) $3.9 \times 10^2 \text{ miles} \times \dfrac{1.61 \text{ km}}{\text{mile}} = 6.3 \times 10^2 \text{ km}$

67) $8.5 \text{ in.} \times \dfrac{2.54 \text{ cm}}{\text{in.}} \times \dfrac{1000 \text{ mm}}{100 \text{ cm}} = 216 \text{ mm} > 210 \text{ mm}$; U.S. paper is wider.   69) $40.0 \text{ L} \times \dfrac{1 \text{ gal}}{3.785 \text{ L}} = 10.6 \text{ gal}$

71) $8.00 \text{ fl oz} \times \dfrac{1 \text{ qt}}{32 \text{ fl oz}} \times \dfrac{1 \text{ gal}}{4 \text{ qt}} \times \dfrac{\text{ft}^3}{7.48 \text{ gal}} \times \dfrac{64.4 \text{ lb}}{\text{ft}^3} \times \dfrac{454 \text{ g}}{\text{lb}} = 244 \text{ g}$

73)

Celsius	Fahrenheit	Kelvin
2	35	275
213	415	486
−19	−2	254
−131	−204	142
821	1510	548
−47	−52	226

75) 181°F   77) 20°C   79) 98.6°F   81) 1121 g/415 cm^3 = 2.70 g/cm^3

83) 11.2 kg = 11200 g. 11200 g/(7.45 cm × 19.7 cm × 9.7 cm) = 7.87 g/cm^3

85) 14.9 mL × $\dfrac{0.810\ g}{mL}$ = 12.1 g   87) 454 g × $\dfrac{mL}{0.915\ g}$ = 496 mL

90) True: a, c, d, f, g, j, k, m, n, p. False: b, e, h, i, l, o, q, r

93) 126 cans × $\dfrac{1\ lb}{21\ cans}$ × $\dfrac{454\ g}{lb}$ = 2.72 × 10^3 g   126 cans × $\dfrac{1\ lb}{21\ cans}$ × $\dfrac{454\ g}{lb}$ × $\dfrac{1\ cm^3}{2.7\ g}$ = 1.0 × 10^3 cm^3

94) 724.26 dol × $\dfrac{1\ hr}{6.25\ dol\ earned}$ × $\dfrac{100\ dol\ earned}{(100-23)\ dol\ take\ home}$ × $\dfrac{1\ shift}{4\ hr}$ × $\dfrac{1\ week}{5\ shifts}$ = 7.5 weeks = 8 weeks

Chapter 4

1) a) Small compared to space occupied. b) Particles not in contact. c) Particles can be pushed closer to each other.

3) Smoke particles being pushed around by moving gas particles, which are too small to be seen.

5) Same answer as 3.   7) Gas particles widely spaced so gas bubble less dense than liquid.

9) Particles move in all directions and collide with all walls of container.

11) New particles occupy space in between particles already present.   13) Volume, pressure.

19)

atm	5.52	4.15	0.715	0.626
psi	81.1	60.9	10.5	9.20
in. Hg	165	124	21.4	18.7
cm Hg	4.20 × 10^2	315	54.4	47.6
mm Hg	4.20 × 10^3	3.15 × 10^3	544	476
torr	4.20 × 10^3	3.15 × 10^3	544	476
Pa	5.59 × 10^5	4.20 × 10^5	7.25 × 10^4	6.34 × 10^4
kPa	559	4.20 × 10^2	72.5	63.4

21) 731 torr − 12 cm = 73.1 cm − 12 cm = 61 cm mercury = 6.1 × 10^2 torr

23) It might, but it would be hazardous. Vacuum flasks for this purpose are usually round, or at least convex over the entire outside surface. The unbalanced pressure exerted on the convex surface tends to push the glass into itself, a force the glass resists quite well. Glass is much weaker when the pressure is exerted on an inside concave surface. The glass is much more likely to explode than to implode (collapse).

25) Temperature is a measure of the average kinetic energy of the particles in a sample. The individual particles move at different speeds, which may change as the result of collisions with other particles. It is the *average* kinetic energy that is constant at a given temperature.

27) At 6°C you can swim, because water freezes at 0°C. You can skate at 260 K, which is −13°C.

29) Yes, unless it's raining. 295 K = 22°C = 72°F.   31) Pressure increases because temperature increases at constant volume.

33) At higher temperature particles move faster, hitting container walls more often and with greater force.

37) Gay-Lussac's Law states the direct proportionality between *absolute* temperature and pressure of a gas sample at constant volume. A graph of a direct proportionality passes through the origin; both values must be zero at the same time. Figure 4.7 shows this to be true for kelvins, not for degrees Celsius.

39) 912 torr × $\dfrac{300\ K}{341\ K}$ = 802 torr   41) 298 K × $\dfrac{3.00\ atm}{2.74\ atm}$ = 326 K = 53°C   43) 26.0 L × $\dfrac{287\ K}{312\ K}$ = 23.9 L

45) 16.0 L × $\dfrac{348\ K}{314\ K}$ = 17.7 L   47) 299 K × $\dfrac{42.1\ L}{38.5\ L}$ = 327 K = 54°C

49) In the car, the lower temperature reduced the gas pressure. Volume dropped to restore the pressure inside the balloon to equality with atmospheric pressure outside. In the house, the reverse processes occurred until the volume was equal to the store volume. The additional firmness inside the house compared to the store indicates higher house temperature, which raises pressure.

51) Squeezing bulb reduces gas volume, increasing pressure and forcing some air bubbles out of pipet. Releasing bulb increases volume and decreases pressure. Liquid pressure is then greater than gas pressure, so liquid enters pipet to reduce gas volume until liquid and gas pressures are equal.

53) $4.47 \text{ L} \times \dfrac{753 \text{ torr}}{977 \text{ torr}} = 3.45 \text{ L}$ 55) $39.2 \text{ L} \times \dfrac{804 \text{ cm Hg}}{664 \text{ cm Hg}} = 47.5 \text{ L}$

57) $9.50 \text{ atm} \times \dfrac{59.8 \text{ L}}{(59.8 + 3.16) \text{ L}} = 9.02 \text{ atm}$ 59) $712 \text{ torr} \times \dfrac{(2.40 + 0.26) \text{ L}}{2.40 \text{ L}} = 789 \text{ torr}$

63) Standard temperature and pressure are reference conditions by which amounts of gas can be compared.

65) $0.768 \text{ L} \times \dfrac{771 \text{ K}}{304 \text{ K}} \times \dfrac{1.01 \text{ atm}}{46.9 \text{ atm}} = 0.0419 \text{ L}$ 67) $4.01 \text{ atm} \times \dfrac{9.75 \text{ L}}{18.2 \text{ L}} \times \dfrac{307 \text{ K}}{349 \text{ K}} = 1.89 \text{ atm}$

69) $28.4 \text{ L} \times \dfrac{273 \text{ K}}{293 \text{ K}} \times \dfrac{755 \text{ torr}}{760 \text{ torr}} = 26.3 \text{ L}$ 71) $9.59 \text{ L} \times \dfrac{1 \text{ atm}}{0.231 \text{ atm}} \times \dfrac{314 \text{ K}}{273 \text{ K}} = 47.8 \text{ L}$

74) True: a–c, f, h, j; false: d, e, g, i   75) c, d, e, e.

76) An ideal gas is one that conforms to the gas laws described in this chapter. At low temperatures and high pressures particles would move more slowly and be closer together. Both of these conditions would lead to stronger intermolecular attractions, which are presumed to be negligible in an ideal gas, and cause the gas to depart from ideal behavior.

77) Assuming the meter reads volume of gas, you would reduce the volume of a given amount of gas by cooling it.

78) Particles of clay, a solid, remain in fixed position relative to each other. Pressing a balloon reduces the gas volume, increasing gas pressure above surrounding atmospheric pressure. On releasing the balloon, the above-atmospheric pressure pushes the balloon back to its original volume. The gas particles rearrange themselves and occupy the entire volume.

Chapter 5

3) The atoms lost by the log are present in the invisible gaseous products.

7) a) H_2O and KCl; b) potassium chloride; c) hydrogen atoms from HCl and oxygen and hydrogen atoms from KOH combined to form H_2O; potassium atoms from KCl and chlorine atoms combined to form KCl.

13) Alpha particles and atomic nuclei are positively charged. As an alpha particle approached a nucleus the repulsion between the positive charges deflected the alpha particle from its path.

23) Two isotopes of the same element cannot have the same mass number—the same total number of protons and neutrons—because all isotopes have the same number of protons. Two atoms of different elements can have the same mass numbers.

25)

Name of Element	Nuclear Symbol	Atomic Number	Mass Number	Number of		
				Protons	Neutrons	Electrons
Carbon	$^{13}_{6}\text{C}$	6	13	6	7	6
Silicon	$^{29}_{14}\text{Si}$	14	29	14	15	14
Calcium	$^{44}_{20}\text{Ca}$	20	44	20	24	20

29) Not all atoms of an element have the same mass.

31) Even though iodine atoms have one more proton than tellurium atoms, enough tellurium atoms have more neutrons than iodine atoms to make the average mass of tellurium atoms greater than the average mass of iodine atoms.

33) Percentage of second isotope is $100 - 60.4 = 39.6$.
$0.604 \times 68.9257 \text{ amu} + 0.396 \times 70.9249 \text{ amu} = 69.7 \text{ amu}$. Gallium, Ga

35) $0.6909 \times 62.9298 \text{ amu} + 0.3091 \times 64.9278 \text{ amu} = 63.55 \text{ amu}$. Copper.

37) $0.3707 \times 184.9530 \text{ amu} + 0.6293 \times 186.9560 \text{ amu} = 186.2 \text{ amu}$. Rhenium.

39) $0.6788 \times 57.9353 \text{ amu} + 0.2623 \times 59.9332 \text{ amu} + 0.0119 \times 60.9310 \text{ amu} +$
$\qquad\qquad 0.0366 \times 61.9283 \text{ amu} + 0.0108 \times 63.9280 \text{ amu} = 58.73 \text{ amu}$. Nickel.

41) $0.7577 \times 34.96885 \text{ amu} + 0.2423x \text{ amu} = 35.45 \text{ amu}$ $x = 36.95 \text{ amu}$ 43) 8. 4, 12, 20, 38, 56, 88.

45) a) Period 4, Group 3B (3). b) Period 2, Group 7A (17). c) Period 6, Group 1A (1).

47) 47.88; 102.9; 131.3.   49) 30.97; 20.18.

54) True: b (see comment), d, f, j, k, l, q, r, s. False: a, c, e, g, h, i, m, n, o, p. Regarding b, Dalton apparently made no specific comment about atomic diameters, but he did propose that all atoms of a given element are identical in every respect. This would include diameters.

55) What was left had to have a positive charge to account for the neutrality of the complete atom.

56) Variable e/m ratio suggests the positive portion of the atom consists of at least two particles, one with a positive charge and one neutral, present in varying number ratio and/or with different masses.

57) Planetary model of atom is similar to the solar system in that it obeys the same rules of classical physics, both conceptually and quantitatively. Major difference is that electron energies are quantized, whereas planetary orbits are not.

58) Chemical properties of isotopes of an element are identical.

59) 12.09899 amu. The difference in masses of the nuclear parts and the sum of the masses of protons and neutrons is what is responsible for nuclear energy in an energy-mass conversion.

60) Electrons, 0.0272%; protons, 49.95%; neutrons, 50.02%.

61) a) $\dfrac{12.01 \text{ amu}}{1 \text{ C atom}} \times \dfrac{1.66 \times 10^{-24} \text{ g}}{1 \text{ amu}} \times \dfrac{1 \text{ C atom}}{1.9 \times 10^{-24} \text{ cm}^3} = 1.0 \times 10^1 \text{ g/cm}^3$

   b) In packing carbon atoms into a crystal there are void spaces between the atoms. There are no voids in a single atom. (In fact, voids in diamond account for 66% of the total volume, and in graphite, 78%.)

   c) $1.9 \times 10^{-24} \text{ cm}^3/(1 \times 10^5)^3 = 2 \times 10^{-39} \text{ cm}^3$

   d) $\dfrac{12.01 \text{ amu}}{1 \text{ C nucleus}} \times \dfrac{1.66 \times 10^{-24} \text{ g}}{1 \text{ amu}} \times \dfrac{1 \text{ C nucleus}}{2 \times 10^{-39} \text{ cm}^3} = 1 \times 10^{16} \text{ g/cm}^3$

   e) $4 \times 10^{-5} \text{ cm}^3 \times \dfrac{1 \times 10^{16} \text{ g}}{1 \text{ cm}^3} \times \dfrac{1 \text{ lb}}{454 \text{ g}} \times \dfrac{1 \text{ ton}}{2000 \text{ lb}} = 4 \times 10^5 \text{ tons}$

Chapter 6

3) Ba, Ag, O_2, Mg.   5) Sulfur, fluorine, nickel, boron.

7) Ar, I_2, Zn, C.   9) Hydrogen bromide, NH_3, diphosphorus trioxide, Cl_2O, water.

11) Copper(I) ion, iodide ion, sulfide ion, mercury(I) ion, potassium ion.   13) Fe^{2+}, H^-, Mg^{2+}, Al^{3+}, O^{2-}.

15) The formula of an acid usually begins with H.   19) See completed Table 6.10, page A-20.

21) An anion or cation that contains hydrogen, such as HSO_4^- and NH_4^+, can lose the hydrogen and thus behave as an acid.

23) HSO_3^-, HCO_3^-.   25) Hydrogen selenite ion, hydrogen telluride ion.   27) OH^-, Cd^{2+}.

The following table contains the names and formulas of all compounds in Table 6.12. Some compounds in the table are not actually known.

	Potassium	Calcium	Chromium(III)	Zinc	Silver	Iron(II)	Aluminum	Mercury(I)
nitrate	KNO_3	$Ca(NO_3)_2$	$Cr(NO_3)_3$	$Zn(NO_3)_2$	$AgNO_3$	$Fe(NO_3)_2$	$Al(NO_3)_3$	$Hg_2(NO_3)_2$
sulfate	K_2SO_4	$CaSO_4$	$Cr_2(SO_4)_3$	$ZnSO_4$	Ag_2SO_4	$FeSO_4$	$Al_2(SO_4)_3$	Hg_2SO_4
hypochlorite	$KClO$	$Ca(ClO)_2$	$Cr(ClO)_3$	$Zn(ClO)_2$	$AgClO$	$Fe(ClO)_2$	$Al(ClO)_3$	$Hg_2(ClO)_2$
nitride	K_3N	Ca_3N_2	CrN	Zn_3N_2	Ag_3N	Fe_3N_2	AlN	—
hydrogen sulfide	KHS	$Ca(HS)_2$	$Cr(HS)_3$	$Zn(HS)_2$	$AgHS$	$Fe(HS)_2$	$Al(HS)_3$	$Hg_2(HS)_2$
bromite	$KBrO_2$	$Ca(BrO_2)_2$	$Cr(BrO_2)_3$	$Zn(BrO_2)_2$	$AgBrO_2$	$Fe(BrO_2)_2$	$Al(BrO_2)_3$	$Hg_2(BrO_2)_2$
hydrogen phosphate	K_2HPO_4	$CaHPO_4$	$Cr_2(HPO_4)_3$	$ZnHPO_4$	Ag_2HPO_4	$FeHPO_4$	$Al_2(HPO_4)_3$	Hg_2HPO_4
chloride	KCl	$CaCl_2$	$CrCl_3$	$ZnCl_2$	$AgCl$	$FeCl_2$	$AlCl_3$	Hg_2Cl_2
hydrogen carbonate	$KHCO_3$	$Ca(HCO_3)_2$	$Cr(HCO_3)_3$	$Zn(HCO_3)_2$	$AgHCO_3$	$Fe(HCO_3)_2$	$Al(HCO_3)_3$	$Hg_2(HCO_3)_2$
acetate	$KC_2H_3O_2$	$Ca(C_2H_3O_2)_2$	$Cr(C_2H_3O_2)_3$	$Zn(C_2H_3O_2)_2$	$AgC_2H_3O_2$	$Fe(C_2H_3O_2)_2$	$Al(C_2H_3O_2)_3$	$Hg_2(C_2H_3O_2)_2$
selenite	K_2SeO_3	$CaSeO_3$	$Cr_2(SeO_3)_3$	$ZnSeO_3$	Ag_2SeO_3	$FeSeO_3$	$Al_2(SeO_3)_3$	Hg_2SeO_3

29) Ammonium ion, cyanide ion. 31) $Ca(OH)_2$, NH_4Br, K_2SO_4. 33) MgO, $AlPO_4$, Na_2SO_4, CaS.

35) $BaSO_3$, Cr_2O_3, KIO_4, $CaHPO_4$. 37) Lithium phosphate, magnesium carbonate, barium nitrate.

The following table contains the formulas and names of all compounds in Table 6.13.

Na^+	Mg^{2+}
NaOH, sodium hydroxide	$Mg(OH)_2$, magnesium hydroxide
NaBrO, sodium hypobromite	$Mg(BrO)_2$, magnesium hypobromite
Na_2CO_3, sodium carbonate	$MgCO_3$, magnesium carbonate
$NaClO_3$, sodium chlorate	$Mg(ClO_3)_2$, magnesium chlorate
$NaHSO_4$, sodium hydrogen sulfate	$Mg(HSO_4)_2$, magnesium hydrogen sulfate
NaBr, sodium bromide	$MgBr_2$, magnesium bromide
Na_3PO_4, sodium phosphate	$Mg_3(PO_4)_2$, magnesium phosphate
$NaIO_4$, sodium periodate	$Mg(IO_4)_2$, magnesium periodate
Na_2S, sodium sulfide	MgS, magnesium sulfide
$NaMnO_4$, sodium permanganate	$Mg(MnO_4)_2$, magnesium permanganate
$Na_2C_2O_4$, sodium oxalate	MgC_2O_4, magnesium oxalate

Pb^{2+}	Cu^{2+}
$Pb(OH)_2$, lead(II) hydroxide	$Cu(OH)_2$, copper(II) hydroxide
$Pb(BrO)_2$, lead(II) hypobromite	$Cu(BrO)_2$, copper(II) hypobromite
$PbCO_3$, lead(II) carbonate	$CuCO_3$, copper(II) carbonate
$Pb(ClO_3)_2$, lead(II) chlorate	$Cu(ClO_3)_2$, copper(II) chlorate
$Pb(HSO_4)_2$, lead(II) hydrogen sulfate	$Cu(HSO_4)_2$, copper(II) hydrogen sulfate
$PbBr_2$, lead(II) bromide	$CuBr_2$, copper(II) bromide
$Pb_3(PO_4)_2$, lead(II) phosphate	$Cu_3(PO_4)_2$, copper(II) phosphate
$Pb(IO_4)_2$, lead(II) periodate	$Cu(IO_4)_2$, copper(II) periodate
PbS, lead(II) sulfide	CuS, copper(II) sulfide
$Pb(MnO_4)_2$, lead(II) permanganate	$Cu(MnO_4)_2$, copper(II) permanganate
PbC_2O_4, lead(II) oxalate	CuC_2O_4, copper(II) oxalate

Fe^{3+}	NH_4^+
$Fe(OH)_3$, iron(III) hydroxide	NH_4OH, ammonium hydroxide
$Fe(BrO)_3$, iron(III) hypobromite	NH_4BrO, ammonium hypobromite
$Fe_2(CO_3)_3$, iron(III) carbonate	$(NH_4)_2CO_3$, ammonium carbonate
$Fe(ClO_3)_3$, iron(III) chlorate	NH_4ClO_3, ammonium chlorate
$Fe(HSO_4)_3$, iron(III) hydrogen sulfate	NH_4HSO_4, ammonium hydrogen sulfate
$FeBr_3$, iron(III) bromide	NH_4Br, ammonium bromide
$FePO_4$, iron(III) phosphate	$(NH_4)_3PO_4$, ammonium phosphate
$Fe(IO_4)_3$, iron(III) periodate	NH_4IO_4, ammonium periodate
Fe_2S_3, iron(III) sulfide	$(NH_4)_2S$, ammonium sulfide
$Fe(MnO_4)_3$, iron(III) permanganate	NH_4MnO_4, ammonium permanganate
$Fe_2(C_2O_4)_3$, iron(III) oxalate	$(NH_4)_2C_2O_4$, ammonium oxalate

Hg^{2+}	Ga^{3+}
$Hg(OH)_2$, mercury(II) hydroxide	$Ga(OH)_3$, gallium hydroxide
$Hg(BrO)_2$, mercury(II) hypobromite	$Ga(BrO)_3$, gallium hypobromite
$HgCO_3$, mercury(II) carbonate	$Ga_2(CO_3)_3$, gallium carbonate
$Hg(ClO_3)_2$, mercury(II) chlorate	$Ga(ClO_3)_3$, gallium chlorate
$Hg(HSO_4)_2$, mercury(II) hydrogen sulfate	$Ga(HSO_4)_3$, gallium hydrogen sulfate
$HgBr_2$, mercury(II) bromide	$GaBr_3$, gallium bromide
$Hg_3(PO_4)_2$, mercury(II) phosphate	$GaPO_4$, gallium phosphate
$Hg(IO_4)_2$, mercury(II) periodate	$Ga(IO_4)_3$, gallium periodate
HgS, mercury(II) sulfide	Ga_2S_3, gallium sulfide
$Hg(MnO_4)_2$, mercury(II) permanganate	$Ga(MnO_4)_3$, gallium permanganate
HgC_2O_4, mercury(II) oxalate	$Ga_2(C_2O_4)_3$, gallium oxalate

TABLE 6.10 (completed)

Acid Name	Acid Formula	Ion Name	Ion Formula
Hydrochloric	HCl	Chloride	Cl^-
Nitric	HNO_3	Nitrate	NO_3^-
Phosphoric	H_3PO_4	Phosphate	PO_4^{3-}
Hydrosulfuric	H_2S	Sulfide	S^{2-}
Nitrous	HNO_2	Nitrite	NO_2^-
Iodic	HIO_3	Iodate	IO_3^-
Hydrotelluric	H_2Te	Telluride	Te^{2-}
Hypochlorous	HClO	Hypochlorite	ClO^-
Iodous	HIO_2	Iodite	IO_2^-
Hydroselenic	H_2Se	Selenide	Se^{2-}
Perchloric	$HClO_4$	Perchlorate	ClO_4^-
Hydroiodic	HI	Iodide	I^-
Chlorous	$HClO_2$	Chlorite	ClO_2^-
Selenic	H_2SeO_4	Selenate	SeO_4^{2-}
Bromous	$HBrO_2$	Bromite	BrO_2^-

39) Potassium fluoride, sodium hydroxide, calcium iodide, aluminum carbonate.
41) Copper(II) sulfate, chromium(III) hydroxide, mercury(I) iodide.
43) $NiSO_4 \cdot 6\,H_2O$ and $Na_3PO_4 \cdot 10\,H_2O$ are hydrates, and KCl is anhydrous.   45) 7; Magnesium sulfate 7-hydrate.
47) $(NH_4)_3PO_4 \cdot 3\,H_2O$; $K_2S \cdot 5\,H_2O$.   49) Nitride ion, calcium chlorate, $Fe_2(SO_4)_3$, PCl_5.
51) SeO_2, $Mg(NO_2)_2$, iron(II) bromide, silver oxide.   53) Hydrogen sulfide ion, beryllium bromide, $Al(NO_3)_3$, OF_2.
55) Mg_3N_2, $LiBrO_2$, sodium hydrogen sulfite, potassium thiocyanate.   57) Nitrous acid, zinc hydrogen sulfate, KCN, CuF.
59) $Sr(IO_3)_2$, $NaClO$, rubidium sulfate, diphosphorus pentoxide.
61) Nickel hydrogen carbonate, copper(II) sulfide, CrI_3, K_2HPO_4.
63) $Co_2(SO_4)_3$, FeI_3, copper(II) phosphate, manganese(II) hydroxide.
65) Aluminum selenide, magnesium hydrogen phosphate, $KClO_4$, $HBrO_2$.
67) ClO_4^-, $BaCO_3$, ammonium iodide, phosphorus trichloride.
69) Tin(II) oxide, ammonium dichromate, NaH, $H_2C_2O_4$.
71) $MgSO_4$, $Hg(BrO_2)_2$, sodium oxalate, manganese(III) hydroxide.   73) Iodine chloride, silver acetate, $Pb(H_2PO_4)_2$, GaF_3.
75) SnF_2, K_2CrO_4, lithium hydride, iron(II) carbonate.   77) Hg_2^{2+}, $CoCl_2$, silicon dioxide, lithium nitrite.

Chapter 7

1) Two oxygen atoms; one magnesium atom, two nitrogen atoms, and six oxygen atoms; three nitrogen atoms, twelve hydrogen atoms, one phosphorus atom, and four oxygen atoms.

5) Molecular mass, CH_4, P_4, and HBr because these are molecular substances; atomic mass, Ne; formula mass, all substances, but particularly Na_3PO_4 and Li_2SO_4 because neither of the other two terms technically fit these ionic compounds.

7) Units are amu for all substances: Li, 6.94; C_2H_5OH, $2(12.0) + 6(1.008) + 16.0 = 46.0$; $LiNO_3$, $6.9 + 14.0 + 3(16.0) = 68.9$; $Mg(NO_3)_2$, $24.3 + 2(14.0) + 6(16.00) = 148.3$; Li_2S, $2(6.9) + 32.1 = 45.9$; $HC_2H_3O_2$, $4(1.0) + 2(12.0) + 2(16.0) = 60.0$; O_2, $2(16.0) = 32.0$; $LiCl$, $6.9 + 35.5 = 42.4$; Al_2O_3, $2(27.0) + 3(16.0) = 102.0$; $Fe_2(SO_4)_3$, $2(55.9) + 3(32.1) + 12(16.00) = 400.1$; $(NH_4)_2CO_3$, $2(14.0) + 8(1.008) + 12.0 + 3(16.0) = 96.1$; $NaC_2H_3O_2$, $23.0 + 2(12.0) + 3(1.0) + 2(16.0) = 82.0$.

9) The amount of water that contains two times Avogadro's number of molecules, or 12.04×10^{23} molecules.

11) A mole is not a number by definition, but the definition effectively assigns to a mole the number 6.02×10^{23}.

13) A mole is the number of carbon-12 atoms in exactly 12 grams of carbon-12. The number is 6.02×10^{23}.

NOTE: Problems 15 and 17 have three parts that are solved in the same way; only the substances and the numbers are different. The setup for Part a) is given in detail. The setups for the remaining parts show numbers only, without units, except for the final answer. Your setups should be like the setup of Part a), complete with units. Answers to several identical-setup problems are printed this way, but in blue to alert you to the absence of units.

15) a) 8.43×10^{23} Ca atoms $\times \dfrac{1\ \text{mol Ca}}{6.02 \times 10^{23}\ \text{Ca atoms}} = 1.40$ mol Ca

   b) $3.20 \times 10^{24}/6.02 \times 10^{23} = 5.32$ mol CH_4 c) $1.03 \times 10^{23}/6.02 \times 10^{23} = 0.171$ mol FeO

17) a) 0.650 mol Mn $\times \dfrac{6.02 \times 10^{23}\ \text{Mn atoms}}{\text{mol Mn}} = 3.91 \times 10^{23}$ Mn atoms

   b) $0.949 \times 6.02 \times 10^{23} = 5.71 \times 10^{23}$ Ar atoms   c) $38.2 \times 6.02 \times 10^{23} = 2.30 \times 10^{25}$ formula units of AgCl

19) Molar mass and molecular mass are numerically equal.   21) They are numerically equal, but the units are grams per mole.

23)* a) 9.76 g $O_2 \times \dfrac{1\ \text{mol } O_2}{32.0\ \text{g } O_2} = 0.305$ mol O_2 b) $5.09/148.3 = 0.0343$ mol $Mg(NO_3)_2$

   c) $77.0/102.0 = 0.755$ mol Al_2O_3 d) $956/46.0 = 20.8$ mol C_2H_5OH

   e) $0.493/96.1 = 5.13 \times 10^{-3}$ mol $(NH_4)_2CO_3$ f) $43.0/45.9 = 0.937$ mol Li_2S

25) a) KIO_3: $39.1 + 126.9 + 3(16.0) = 214$ g/mol.   As in Problem 23a, $0.979/214 = 4.57 \times 10^{-3}$ mol KIO_3

   b) $BeCl_2$: $9.0 + 2(35.5) = 80.0$ g/mol $86.8/80.0 = 1.09$ mol $BeCl_2$

   c) $Ni(NO_3)_2$: $58.7 + 2(14.0) + 6(16.00) = 182.7$ g/mol $203/182.7 = 1.11$ mol $Ni(NO_3)_2$

27) a) 0.967 mol LiCl $\times \dfrac{42.4\ \text{g LiCl}}{1\ \text{mol LiCl}} = 41.0$ g LiCl   b) $17.5 \times 60.0 = 1.05 \times 10^3$ g $HC_2H_3O_2$

   c) $8.60 \times 6.94 = 59.7$ g Li   d) $0.235 \times 400.1 = 94.0$ g $Fe_2(SO_4)_3$ e) $6.28 \times 82.0 = 515$ g $NaC_2H_3O_2$

29) a) Li_2SO_4: $2(6.9) + 32.1 + 4(16.0) = 109.9$ g/mol   As in Problem 27a, $0.973 \times 109.9 = 107$ g Li_2SO_4

   b) $K_2C_2O_4$: $2(39.1) + 2(12.0) + 4(16.0) = 166.2$ g/mol $2.84 \times 166.2 = 472$ g $K_2C_2O_4$

   c) $Pb(NO_3)_2$: $207.2 + 2(14.0) + 6(16.00) = 331.2$ g/mol $0.231 \times 331.2 = 76.5$ g $Pb(NO_3)_2$

31) a) 69.2 g $LiNO_3 \times \dfrac{1\ \text{mol } LiNO_3}{68.9\ \text{g } LiNO_3} \times \dfrac{6.02 \times 10^{23}\ \text{units}}{\text{mol } LiNO_3} = 6.05 \times 10^{23}$ units

   b) $0.515 \times \dfrac{6.02 \times 10^{23}}{45.9} = 6.75 \times 10^{21}$ units   c) $754 \times \dfrac{6.02 \times 10^{23}}{400.1} = 1.13 \times 10^{24}$ units

33) a) I_2: $2(126.9) = 253.8$ g/mol   As in Problem 31a, $0.0320 \times \dfrac{6.02 \times 10^{23}}{253.8} = 7.59 \times 10^{19}$ units

   b) $C_2H_4(OH)_2$: $2(12.0) + 6(1.008) + 2(16.0) = 62.0$ g/mol $411 \times \dfrac{6.02 \times 10^{23}}{62.0} = 3.99 \times 10^{24}$ molecules

   c) $Cr_2(SO_4)_3$: $2(52.0) + 3(32.1) + 12(16.00) = 392.3$ g/mol $1.89 \times \dfrac{6.02 \times 10^{23}}{392.3} = 2.90 \times 10^{21}$ units

35) a) $C_{19}H_{37}COOH$: $20(12.01) + 38(1.008) + 2(16.00) = 310.5$ g/mol

   $3.40 \times 10^{21}\ C_{19}H_{37}COOH$ molecules $\times \dfrac{1\ \text{mol } C_{19}H_{37}COOH}{6.02 \times 10^{23}\ C_{19}H_{37}COOH\ \text{molecules}} \times$

   $\dfrac{310.5\ \text{g } C_{19}H_{37}COOH}{\text{mol } C_{19}H_{37}COOH} = 1.75$ g $C_{19}H_{37}COOH$

   b) F atoms: 19.0 g/mol $7.68 \times 10^{24} \times \dfrac{19.0}{6.02 \times 10^{23}} = 242$ g F atoms

   c) $NiCl_2$: $58.7 + 2(35.5) = 129.7$ g/mol $3.26 \times 10^{23} \times \dfrac{129.7}{6.02 \times 10^{23}} = 70.2$ g $NiCl_2$

37) $\dfrac{£\ 226}{\text{tr oz}} \times \dfrac{\text{tr oz}}{31.1\ \text{g}} \times \dfrac{\$1.76}{£} \times \dfrac{100¢}{\$} \times \dfrac{197.0\ \text{g Au}}{\text{mol Au}} \times \dfrac{\text{mol Au}}{6.02 \times 10^{23}\ \text{Au atoms}} = 4.19 \times 10^{-19}$ ¢/Au atom

39) $C_6H_{12}O_6$: $6(12.01) + 12(1.008) + 6(16.00) = 180.2$ g/mol

$$0.65 \text{ g } C_6H_{12}O_6 \times \frac{1 \text{ mol } C_6H_{12}O_6}{180.2 \text{ g } C_6H_{12}O_6} \times \frac{6.02 \times 10^{23} \text{ molecules}}{\text{mol}} = 2.2 \times 10^{21} \text{ molecules}$$

41) Cl: 35.5 g/mol Cl_2: 71.0 g/mol

a) $$1.44 \text{ g } Cl_2 \times \frac{1 \text{ mol } Cl_2}{71.0 \text{ g } Cl_2} \times \frac{6.02 \times 10^{23} \text{ molecules}}{\text{mol}} = 1.22 \times 10^{22} \text{ molecules}$$

b) $$1.44 \text{ g } Cl_2 \times \frac{1 \text{ mol } Cl_2}{71.0 \text{ g } Cl_2} \times \frac{6.02 \times 10^{23} \text{ molecules}}{\text{mol}} \times \frac{2 \text{ atoms}}{\text{molecule}} = 2.44 \times 10^{22} \text{ atoms}$$

c) $$1.44 \text{ g } Cl \times \frac{1 \text{ mol } Cl}{35.5 \text{ g } Cl} \times \frac{6.02 \times 10^{23} \text{ atoms}}{\text{mol}} = 2.44 \times 10^{22} \text{ atoms}$$

d) $$1.44 \times 10^{23} \text{ atoms} \times \frac{1 \text{ mol atoms}}{6.02 \times 10^{23} \text{ atoms}} \times \frac{35.5 \text{ g } Cl}{\text{mol atoms}} = 8.49 \text{ g } Cl$$

e) $$1.44 \times 10^{23} \text{ molecules} \times \frac{1 \text{ mol molecules}}{6.02 \times 10^{23} \text{ molecules}} \times \frac{71.0 \text{ g } Cl_2}{\text{mol molecules}} = 17.0 \text{ g } Cl_2$$

43) Formula mass numbers are from answer to Question 7.

a) $$\frac{6.9 \text{ g } Li}{42.4 \text{ g } LiCl} \times 100 = 16\% \text{ Li}; \quad \frac{35.5 \text{ g } Cl}{42.4 \text{ g } LiCl} \times 100 = 83.7\% \text{ Cl}$$

b) $$\frac{2(27.0) \text{ g } Al}{102.0 \text{ g } Al_2O_3} \times 100 = 52.9\% \text{ Al}; \quad \frac{3(16.0) \text{ g } O}{102.0 \text{ g } Al_2O_3} \times 100 = 47.1\% \text{ O}$$

c) $$\frac{4(1.01) \text{ g } H}{60.0 \text{ g } HC_2H_3O_2} \times 100 = 6.73\% \text{ H}; \quad \frac{2(12.0) \text{ g } C}{60.0 \text{ g } HC_2H_3O_2} \times 100 = 40.0\% \text{ C}$$

$$\frac{2(16.0) \text{ g } O}{60.0 \text{ g } HC_2H_3O_2} \times 100 = 53.3\% \text{ O}$$

d) $$\frac{24.3 \text{ g } Mg}{148.3 \text{ g } Mg(NO_3)_2} \times 100 = 16.4\% \text{ Mg}; \quad \frac{2(14.0) \text{ g } N}{148.3 \text{ g } Mg(NO_3)_2} \times 100 = 18.9\% \text{ N}; \quad \frac{6(16.00) \text{ g } O}{148.3 \text{ g } Mg(NO_3)_2} \times 100 = 64.73\% \text{ O}$$

e) $$\frac{2(55.9) \text{ g } Fe}{400.1 \text{ g } Fe_2(SO_4)_3} \times 100 = 27.9\% \text{ Fe}; \quad \frac{3(32.1) \text{ g } S}{400.1 \text{ g } Fe_2(SO_4)_3} \times 100 = 24.1\% \text{ S}; \quad \frac{12(16.00) \text{ g } O}{400.1 \text{ g } Fe_2(SO_4)_3} \times 100 = 47.99\% \text{ O}$$

45) LiF: $6.9 + 19.0 = 25.9$ g/mol $688 \text{ g } LiF \times \dfrac{19.0 \text{ g } F}{25.9 \text{ g } LiF} = 505 \text{ g } F$

47) K_2SO_4: $2(39.1) + 32.1 + 4(16.0) = 174.3$ g/mol $311 \text{ g } K \times \dfrac{174.3 \text{ g } K_2SO_4}{2(39.1) \text{ g } K} = 693 \text{ g } K_2SO_4$

49) $Zn(CN)_2$: $65.4 + 2(12.0) + 2(14.0) = 117.4$ g/mol $391 \text{ g } Zn(CN)_2 \times \dfrac{65.4 \text{ g } Zn}{117.4 \text{ g } Zn(CN)_2} = 218 \text{ g } Zn$

51) $PbMoO_4$: $207.2 + 95.9 + 4(16.0) = 367.1$ g/mol $874 \text{ kg } Mo \times \dfrac{367.1 \text{ kg } PbMoO_4}{95.9 \text{ kg } Mo} = 3.35 \times 10^3 \text{ kg } PbMoO_4$

53) CH_3OH: $12.0 + 4(1.0) + 16.0 = 32.0$ g/mol $70.6 \text{ g } CH_3OH \times \dfrac{12.0 \text{ g } C}{32.0 \text{ g } CH_3OH} = 26.5 \text{ g } C$

55) C_5H_{10} cannot be a simplest formula because both 5 and 10 are divisible by 5. The simplest formula is CH_2. C_5H_{10} can be and is a molecular formula of a real substance.

	Element	Grams	Moles	Mole Ratio	Formula Ratio	Simplest Formula	Molecular Formula
57)	C	52.2	4.35	2.00	2		
	H	13.0	13.0	5.96	6		
	O	34.8	2.18	1.00	1	C_2H_6O	

	Element	Grams	Moles	Mole Ratio	Formula Ratio	Simplest Formula	Molecular Formula
59)	Fe	13.51	0.2419	1.000	2		
	O	5.79	0.362	1.50	3	Fe_2O_3	
61)	H	0.54	0.54	2.0	4		
	N	3.75	0.268	1.00	2		
	O	6.44	0.403	1.50	3	$H_4N_2O_3$ (NH_4NO_3)	
63)	C	23.1	1.93	1.00	1		
	H	3.8	3.8	2.0	2		
	F	73.0	3.84	1.99	2	CH_2F_2	
65)	C	40.0	3.33	1.00	1		
	H	6.7	6.7	2.0	2		$\frac{180}{30} = 6$
	O	53.3	3.33	1.00	1	CH_2O	$C_6H_{12}O_6$
67)	C	30.7	2.56	1.00	1		
	H	5.1	5.1	2.0	2		$\frac{60}{30} = 2$
	O	41.0	2.56	1.00	1	CH_2O	$C_2H_4O_2$

70) True: c. False: a, b, d, e, f.

71) No, because 1×10^{23} carbon atoms is about $\frac{1}{6}$ of a mole, or about 2 grams, less than 0.1 ounce.

72) S_8: $8(32.07) = 256.6$ g/mol $\quad 81.4$ g $S_8 \times \dfrac{1 \text{ mol } S_8}{256.6 \text{ g } S_8} \times \dfrac{6.02 \times 10^{23} \text{ } S_8 \text{ molecules}}{\text{mol } S_8} = 1.91 \times 10^{23}$ molecules

73) One mole of NaCl has a mass of 58.5 grams, which is about two ounces. That's the capacity of a typical salt shaker. One mole of $C_{12}H_{22}O_{11}$ has a mass of 342 grams, which is about 12 ounces. That's a little large for a sugar bowl, but not by much.

74) 2.35×10^{22} molecules $\times \dfrac{1 \text{ mol air}}{6.02 \times 10^{23} \text{ molecules}} \times \dfrac{29 \text{ g air}}{\text{mol air}} = 1.1$ g air

75) 4.95×10^{23} molecules $\times \dfrac{1 \text{ mol } C_8H_{18}}{6.02 \times 10^{23} \text{ molecules}} \times \dfrac{114.2 \text{ g } C_8H_{18}}{\text{mol } C_8H_{18}} = 93.9$ g C_8H_{18}

76) CO_2: $12.0 + 2(16.0) = 44.0$ g/mol $\quad H_2O$: $2(1.0) + 16.0 = 18.0$ g/mol

28.4 g $CO_2 \times \dfrac{12.0 \text{ g C}}{44.0 \text{ g } CO_2} = 7.75$ g C $\quad 11.6$ g $H_2O \times \dfrac{2.00 \text{ g H}}{18.0 \text{ g } H_2O} = 1.29$ g H

Element	Grams	Moles	Mole Ratio	Formula Ratio	Simplest Formula
C	7.75	0.646	1.00	1	
H	1.29	1.29	2.00	2	CH_2

Chapter 8   Equation Balancing Exercise

1) $Li_2O + H_2O \longrightarrow 2 LiOH$

2) $2 HgO \longrightarrow 2 Hg + O_2$

3) $CaC_2 + 2 H_2O \longrightarrow C_2H_2 + Ca(OH)_2$

4) $Zn(OH)_2 + H_2SO_4 \longrightarrow ZnSO_4 + 2 H_2O$

5) $2 PbO_2 \longrightarrow 2 PbO + O_2$

6) $2 Al + 6 HCl \longrightarrow 2 AlCl_3 + 3 H_2$

7) $Fe_2(SO_4)_3 + 3 Ba(OH)_2 \longrightarrow 3 BaSO_4 + 2 Fe(OH)_3$

8) $2 Al + 3 CuSO_4 \longrightarrow Al_2(SO_4)_3 + 3 Cu$

9) $3 Mg + N_2 \longrightarrow Mg_3N_2$

10) $3 FeCl_2 + 2 Na_3PO_4 \longrightarrow Fe_3(PO_4)_2 + 6 NaCl$

11) $CaSO_4 \cdot 2 H_2O \longrightarrow CaSO_4 + 2 H_2O$

12) $2 C_3H_7CHO + 11 O_2 \longrightarrow 8 CO_2 + 8 H_2O$

13) $NaHCO_3 + HCl \longrightarrow NaCl + CO_2 + H_2O$

14) $Bi(NO_3)_3 + 3 NaOH \longrightarrow Bi(OH)_3 + 3 NaNO_3$

15) $FeS + 2 HBr \longrightarrow FeBr_2 + H_2S$

16) $P_4O_{10} + 6 H_2O \longrightarrow 4 H_3PO_4$

17) $3 CaCO_3 + 2 H_3PO_4 \longrightarrow Ca_3(PO_4)_2 + 3 H_2O + 3 CO_2$

18) $PCl_5 + 4 H_2O \longrightarrow H_3PO_4 + 5 HCl$

19) $CaI_2 + H_2SO_4 \longrightarrow CaSO_4 + 2 HI$

20) $C_3H_7COOH + 5 O_2 \longrightarrow 4 CO_2 + 4 H_2O$

21) $Mg(CN)_2 + 2 HCl \longrightarrow 2 HCN + MgCl_2$

22) $(NH_4)_2S + 2 HBr \longrightarrow 2 NH_4Br + H_2S$

23) $H_2SO_4 + 2 NaC_2H_3O_2 \longrightarrow Na_2SO_4 + 2 HC_2H_3O_2$

24) $4 Fe + 3 O_2 \longrightarrow 2 Fe_2O_3$

Chapter 8   Questions and Problems

1) $2 Na(s) + O_2(g) \longrightarrow Na_2O_2(s)$

3) $Ba(s) + Br_2(\ell) \longrightarrow BaBr_2(s)$

5) $CaO(s) + H_2O(\ell) \longrightarrow Ca(OH)_2(s)$

7) $2 KCl(s) \longrightarrow 2 K(s) + Cl_2(g)$

9) $H_2CO_3(aq) \longrightarrow H_2O(\ell) + CO_2(g)$

11) $2 C_4H_{10}(g) + 13 O_2(g) \longrightarrow 8 CO_2(g) + 10 H_2O(g)$

13) $2 C_7H_{15}OH(\ell) + 21 O_2(g) \longrightarrow 14 CO_2(g) + 16 H_2O(g)$

15) $2 K(s) + 2 H_2O(\ell) \longrightarrow H_2(g) + 2 KOH(aq)$

17) $Zn(s) + 2 HCl(aq) \longrightarrow H_2(g) + ZnCl_2(aq)$

19) $H_2S(g) + 2 AgNO_3(aq) \longrightarrow Ag_2S(s) + 2 HNO_3(aq)$

21) $BaCl_2(aq) + (NH_4)_2CO_3(aq) \longrightarrow BaCO_3(s) + 2 NH_4Cl(aq)$

23) $NaIO_3(aq) + AgNO_3(aq) \longrightarrow AgIO_3(s) + NaNO_3(aq)$

25) $2 LiOH(s) + H_2SO_4(aq) \longrightarrow Li_2SO_4(aq) + 2 H_2O(\ell)$

27) $KOH(aq) + HBr(aq) \longrightarrow KBr(aq) + H_2O(\ell)$

29) $2 C_3H_8O_3(\ell) + 7 O_2(g) \longrightarrow 6 CO_2(g) + 8 H_2O(g)$

31) $2 P(s) + 3 Br_2(\ell) \longrightarrow 2 PBr_3(s)$

33) $3 Sn(NO_3)_2(aq) + 2 Cr(s) \longrightarrow 2 Cr(NO_3)_3(aq) + 3 Sn(s)$

35) $H_2SO_3(aq) \longrightarrow H_2O(\ell) + SO_2(aq)$

37) $2 Sb(s) + 3 Cl_2(g) \longrightarrow 2 SbCl_3(s)$

39) $2 AgNO_3(aq) + Na_2CrO_4(aq) \longrightarrow Ag_2CrO_4(s) + 2 NaNO_3(aq)$

41) $2 Na_3PO_4(aq) + 3 ZnCl_2(aq) \longrightarrow Zn_3(PO_4)_2(s) + 6 NaCl(aq)$

43) $2 KOH(aq) + ZnCl_2(aq) \longrightarrow 2 KCl(aq) + Zn(OH)_2(s)$

45) $HNH_2SO_3(aq) + KOH(aq) \longrightarrow KNH_2SO_3(aq) + H_2O(\ell)$

47) $C_5H_{12}(\ell) + 8 O_2(g) \longrightarrow 5 CO_2(g) + 6 H_2O(g)$

49) $Ca(OH)_2(s) \longrightarrow CaO(s) + H_2O(\ell)$

51) $2 Cs(s) + 2 H_2O(\ell) \longrightarrow H_2(g) + 2 CsOH(aq)$

53) $SO_3(\ell) + H_2O(\ell) \longrightarrow H_2SO_4(aq)$

55) $4 Al(s) + 3 C(s) \longrightarrow Al_4C_3(s)$

57) $(NH_4)_2S(aq) + Cu(NO_3)_2(aq) \longrightarrow CuS(s) + 2 NH_4NO_3(aq)$

59) $2 AgNO_3(aq) + K_2CO_3(aq) \longrightarrow Ag_2CO_3(s) + 2 KNO_3(aq)$

61) $3 Mn(s) + 2 CrCl_3(aq) \longrightarrow 3 MnCl_2(aq) + 2 Cr(s)$

63) $CO(g) + H_2O(g) \longrightarrow H_2(g) + CO_2(g)$

65) $3 Mg(s) + 2 NH_3(g) \longrightarrow 3 H_2(g) + Mg_3N_2(s)$

67) $4 KMnO_4(aq) + 4 KOH(aq) \longrightarrow 4 K_2MnO_4(aq) + O_2(g) + 2 H_2O(\ell)$

69) $4 Cl_2(g) + 2 CH_4(g) + O_2(g) \longrightarrow 8 HCl(aq) + 2 CO(g)$

Chapter 9

1) a) $8.44 \text{ mol } NH_3 \times \dfrac{5 \text{ mol } O_2}{4 \text{ mol } NH_3} = 10.6 \text{ mol } O_2$

b) $0.528 \text{ mol } O_2 \times \dfrac{6 \text{ mol } H_2O}{5 \text{ mol } O_2} = 0.634 \text{ mol } H_2O$

c) $1.31 \text{ mol } NO \times \dfrac{6 \text{ mol } H_2O}{4 \text{ mol } NO} = 1.97 \text{ mol } NO$

d) $0.738 \text{ mol } NO \times \dfrac{5 \text{ mol } O_2}{4 \text{ mol } NO} \times \dfrac{32.0 \text{ g } O_2}{\text{mol } O_2} = 29.5 \text{ g } O_2$

e) $73.4 \text{ g } NH_3 \times \dfrac{1 \text{ mol } NH_3}{17.0 \text{ g } NH_3} \times \dfrac{4 \text{ mol } NO}{4 \text{ mol } NH_3} = 4.32 \text{ mol } NO$

f) $58.0 \text{ g } NO \times \dfrac{1 \text{ mol } NO}{30.0 \text{ g } NO} \times \dfrac{6 \text{ mol } H_2O}{4 \text{ mol } NO} \times \dfrac{18.0 \text{ g } H_2O}{1 \text{ mol } H_2O} = 52.2 \text{ g } H_2O$

g) $9.36 \text{ g } H_2O \times \dfrac{1 \text{ mol } H_2O}{18.0 \text{ g } H_2O} \times \dfrac{4 \text{ mol } NH_3}{6 \text{ mol } H_2O} \times \dfrac{17.0 \text{ g } NH_3}{\text{mol } NH_3} = 5.89 \text{ g } NH_3$

3) $Na_2CO_3 + CaBr_2 \longrightarrow CaCO_3 + 2 NaBr$

$0.408 \text{ g } Na_2CO_3 \times \dfrac{1 \text{ mol } Na_2CO_3}{106.0 \text{ g } Na_2CO_3} \times \dfrac{1 \text{ mol } CaCO_3}{1 \text{ mol } Na_2CO_3} \times \dfrac{100.1 \text{ g } CaCO_3}{\text{mol } CaCO_3} = 0.385 \text{ g } CaCO_3$

5) $2 KOH + Mg(NO_3)_2 \longrightarrow Mg(OH)_2 + 2 KNO_3$

$6.89 \text{ g } Mg(NO_3)_2 \times \dfrac{1 \text{ mol } Mg(NO_3)_2}{148.3 \text{ g } Mg(NO_3)_2} \times \dfrac{1 \text{ mol } Mg(OH)_2}{1 \text{ mol } Mg(NO_3)_2} \times \dfrac{58.3 \text{ g } Mg(OH)_2}{1 \text{ mol } Mg(OH)_2} = 2.71 \text{ g } Mg(OH)_2$

7) $2 NaOH + H_2SO_4 \longrightarrow Na_2SO_4 + 2 H_2O$

$0.755 \text{ g } Na_2SO_4 \times \dfrac{1 \text{ mol } Na_2SO_4}{142.1 \text{ g } Na_2SO_4} \times \dfrac{2 \text{ mol } NaOH}{1 \text{ mol } Na_2SO_4} \times \dfrac{40.0 \text{ g } NaOH}{1 \text{ mol } NaOH} = 0.425 \text{ g } NaOH$

9) $8.26 \text{ g } Zn \times \dfrac{1 \text{ mol } Zn}{65.4 \text{ g } Zn} \times \dfrac{2 \text{ mol } NH_4Cl}{1 \text{ mol } Zn} \times \dfrac{53.5 \text{ g } NH_4Cl}{1 \text{ mol } NH_4Cl} = 13.5 \text{ g } NH_4Cl$

11) $669 \text{ g } H_2O \times \dfrac{1 \text{ mol } H_2O}{18.0 \text{ g } H_2O} \times \dfrac{4 \text{ mol } C_3H_5(NO_3)_3}{10 \text{ mol } H_2O} \times \dfrac{227.0 \text{ g } C_3H_5(NO_3)_3}{1 \text{ mol } C_3H_5(NO_3)_3} = 3.37 \times 10^3 \text{ g } C_3H_5(NO_3)_3$

13) $739 \text{ g } C_3H_5(C_{17}H_{35}COO)_3 \times \dfrac{1 \text{ mol } C_3H_5(C_{17}H_{35}COO)_3}{891.5 \text{ g } C_3H_5(C_{17}H_{35}COO)_3} \times \dfrac{3 \text{ mol } C_{17}H_{35}COONa}{1 \text{ mol } C_3H_5(C_{17}H_{35}COO)_3} \times$

$\dfrac{306.5 \text{ g } C_{17}H_{35}COONa}{1 \text{ mol } C_{17}H_{35}COONa} = 762 \text{ g } C_{17}H_{35}COONa$

15) $71.7 \text{ g } Na_2CO_3 \times \dfrac{1 \text{ mol } Na_2CO_3}{106.0 \text{ g } Na_2CO_3} \times \dfrac{3 \text{ mol } Na_2S_2O_3}{1 \text{ mol } Na_2CO_3} \times \dfrac{158.2 \text{ g } Na_2S_2O_3}{1 \text{ mol } Na_2S_2O_3} = 321 \text{ g } Na_2S_2O_3$

17) $692 \text{ mg } Ca(C_{18}H_{35}O_2)_2 \times \dfrac{1 \text{ mmol } Ca(C_{18}H_{35}O_2)_2}{607.0 \text{ mg } Ca(C_{18}H_{35}O_2)_2} \times \dfrac{2 \text{ mmol } NaC_{18}H_{35}O_2}{1 \text{ mmol } Ca(C_{18}H_{35}O_2)_2} \times$

$\dfrac{306.5 \text{ mg } NaC_{18}H_{35}O_2}{1 \text{ mmol } NaC_{18}H_{35}O_2} = 699 \text{ mg } NaC_{18}H_{35}O_2$

19) $671 \text{ g } Fe_2O_3 \times \dfrac{1 \text{ mol } Fe_2O_3}{159.8 \text{ g } Fe_2O_3} \times \dfrac{12 \text{ mol } P}{10 \text{ mol } Fe_2O_3} \times \dfrac{31.0 \text{ g } P}{\text{mol } P} = 156 \text{ g } P$

21) $36.3 \text{ g } NaHCO_3 \times \dfrac{1 \text{ mol } NaHCO_3}{84.0 \text{ g } NaHCO_3} \times \dfrac{1 \text{ mol } NH_4HCO_3}{1 \text{ mol } NaHCO_3} \times \dfrac{79.0 \text{ g } NH_4HCO_3}{1 \text{ mol } NH_4HCO_3} = 34.1 \text{ g } NH_4HCO_3$

23) $30.2 \text{ g } NH_3 \times \dfrac{1 \text{ mol } NH_3}{17.0 \text{ g } NH_3} \times \dfrac{1 \text{ mol } Ca(OH)_2}{2 \text{ mol } NH_3} \times \dfrac{74.1 \text{ g } Ca(OH)_2}{1 \text{ mol } Ca(OH)_2} = 65.8 \text{ g } Ca(OH)_2$

25) $53.4 \text{ g } NaOH \times \dfrac{1 \text{ mol } NaOH}{40.0 \text{ g } NaOH} \times \dfrac{1 \text{ mol } H_3PO_4}{2 \text{ mol } NaOH} \times \dfrac{98.0 \text{ g } H_3PO_4}{1 \text{ mol } H_3PO_4} = 65.4 \text{ g } H_3PO_4$

27) $278 \text{ kg } SiC \times \dfrac{1 \text{ kmol } SiC}{40.1 \text{ kg } SiC} \times \dfrac{3 \text{ kmol } C}{1 \text{ kmol } SiC} \times \dfrac{12.0 \text{ kg } C}{1 \text{ kmol } C} \times \dfrac{100 \text{ kg coke}}{84.4 \text{ kg } C} = 296 \text{ kg coke}$

29) a) 50% from Pb and 50% from PbO_2

 b) $75.7 \text{ g } PbO_2 \times \dfrac{1 \text{ mol } PbO_2}{239.2 \text{ g } PbO_2} \times \dfrac{2 \text{ mol } H_2SO_4}{1 \text{ mol } PbO_2} \times \dfrac{98.1 \text{ g } H_2SO_4}{\text{mol } H_2SO_4} = 62.1 \text{ g } H_2SO_4$

31) $5.13 \text{ g } Na_2S_2O_3 \times \dfrac{1 \text{ mol } Na_2S_2O_3}{158.2 \text{ g } Na_2S_2O_3} \times \dfrac{1 \text{ mol } NaBr}{2 \text{ mol } Na_2S_2O_3} \times \dfrac{102.9 \text{ g } NaBr}{\text{mol } NaBr} = 1.67 \text{ g } NaBr \qquad \dfrac{1.58 \text{ g}}{1.67 \text{ g}} \times 100 = 94.6\%$

33) $8.86 \text{ kg } CaCN_2 \times \dfrac{1 \text{ kmol } CaCN_2}{80.1 \text{ kg } CaCN_2} \times \dfrac{2 \text{ kmol } NH_3}{1 \text{ kmol } CaCN_2} \times \dfrac{17.0 \text{ kg } NH_3(\text{theo})}{\text{kmol } NH_3} \times \dfrac{91.6 \text{ kg } NH_3(\text{act})}{100 \text{ kg } NH_3(\text{theo})} = 3.44 \text{ kg } NH_3(\text{act})$

35) $6.00 \times 10^2 \text{ kg } H_2 \times \dfrac{1 \text{ kmol } H_2}{2.02 \text{ kg } H_2} \times \dfrac{2 \text{ kmol } NH_3}{3 \text{ kmol } H_2} \times \dfrac{17.0 \text{ kg } NH_3(\text{theo})}{1 \text{ kmol } NH_3} \times \dfrac{82.9 \text{ kg } NH_3(\text{act})}{100 \text{ kg } NH_3(\text{theo})} = 2.79 \times 10^3 \text{ kg } NH_3(\text{act})$

37) $23.5 \text{ g } CO_2 \times \dfrac{1 \text{ mol } CO_2}{44.0 \text{ g } CO_2} \times \dfrac{1 \text{ mol } C_6H_{12}O_6}{6 \text{ mol } CO_2} \times \dfrac{180.2 \text{ g } C_6H_{12}O_6}{1 \text{ mol } C_6H_{12}O_6} = 16.0 \text{ g } C_6H_{12}O_6 \qquad \dfrac{12.4 \text{ g}}{16.0 \text{ g}} \times 100 = 77.5\%$

39) $168 \text{ kg } CH_3COOC_2H_5(\text{act}) \times \dfrac{100 \text{ kg } CH_3COOC_2H_5(\text{theo})}{70.8 \text{ kg } CH_3COOC_2H_5(\text{act})} \times \dfrac{1 \text{ kmol } CH_3COOC_2H_5}{88.1 \text{ kg } CH_3COOC_2H_5} \times$

$\dfrac{1 \text{ kmol } CH_3COOH}{1 \text{ kmol } CH_3COOC_2H_5} \times \dfrac{60.0 \text{ kg } CH_3COOH}{1 \text{ kmol } CH_3COOH} = 162 \text{ kg } CH_3COOH$

41) $7.8 \text{ t } HCl \times \dfrac{1 \text{ tmol } HCl}{36.5 \text{ t } HCl} \times \dfrac{2 \text{ tmol } Cl_2}{4 \text{ tmol } HCl} \times \dfrac{71.0 \text{ t } Cl_2(\text{theo})}{1 \text{ tmol } Cl_2} \times \dfrac{67 \text{ t } Cl_2(\text{act})}{100 \text{ t } Cl_2(\text{theo})} = 5.1 \text{ t } Cl_2(\text{act})$

43)

	$BaCl_2$	Na_2CrO_4	$BaCrO_4$
g start	1.63	2.40	0
molar mass	208.3	162.0	253.3
mol start	0.00783	0.0148	0.0000
change	−0.00783	−0.00783	+0.00783
mol end	0	0.0070	0.00783
g end	0	1.1	1.98

45)

	$2 NaIO_3$	$5 NaHSO_3$	I_2
g start	6.00	7.33	0
molar mass	197.9	104.1	253.8
mol start	0.0303	0.0704	0.0000
change	−0.0282	−0.0704	+0.0141
mol end	0.0021	0	0.0141
g end	0.42	0	3.58

85) Fluorine and bromine both have ns^2np^5 valence electron configurations, both one short of the number of valence electrons in a noble gas. This gives them similar chemical properties.

87) All four elements have an ns^1 valence electron configuration.

89) Argon, like other noble gases in Group 0 (18), is unreactive and therefore stable at the extremely high temperatures of an electric light bulb.

91) Atomic size is estimated from data obtained by studying crystals, molar volumes, and densities, among other things.

95) Increasing nuclear charge tends to reduce atomic size because outermost electrons are attracted more strongly. This attraction is partially canceled by the ''shielding effect'' of inner electrons. Atoms are larger as the number of occupied energy levels increases.

97) It is easier to remove an electron from boron—it takes less energy to ionize boron—by pulling an electron away from a nucleus with 5 protons than it is to pull the electron away from fluorine, which has 9 protons.

99) Metals form positively charged ions by losing electrons. Nonmetals do not.

103) a) D, A; b) R, T.   105) X, Q, R, Z, M, T.   107) Z, X, Q.

110) True: b, e, g, i, k, m, n, p, q. False: a, c, d, f, h, j, l, o.

111) The quantum and Bohr model explanations of atomic spectra are essentially the same.

112) All species have a single electron. Species with two or more electrons are far more complex.

113) The other lines are outside the visible spectrum, in the infrared or ultraviolet regions.

114) Sc^{3+} is isoelectronic with an argon atom.

115) The highest occupied energy level of a potassium atom at ground state is $n = 4$ ($4s^1$). For a potassium ion it is $n = 3$ ($3d^{10}$). Therefore, potassium atoms are larger than potassium ions.

116) Xenon has the lowest ionization energy of the noble gases and apparently the greatest reactivity. This is characteristic of the more active metals that form ionic compounds.

117) Iron loses two electrons from the $4s$ orbital to form Fe^{2+} and a third from the $3d$ orbital to form Fe^{3+}.

118) a) Ionization energy increases across a period of the periodic table because of increasing nuclear charge.
  b) The breaks in ionization energy trends across periods 2 and 3 occur just after the s orbital is filled and just after the p orbitals are half filled.

119) The ions are listed in order of decreasing size. They are isoelectronic, listed in order of increasing nuclear charge. Size decreases as nuclear charge increases.

120) Zn: [Ar]$3d^{10}4s^2$. The valence electrons of magnesium, calcium, and zinc are all ns^2 electrons. All three elements become monatomic ions by losing their valence electrons, so they have similar chemical properties.

121) H: $1s^1$; He: $1s^2$; halogens: ns^2np^5; noble gases: ns^2np^6. Hydrogen, like each halogen, is one electron short of the configuration of a noble gas. It therefore can become a monatomic ion by gaining one electron, a chemical property of the halogens.

122) (a) Metals form cations with a positive charge; nonmetals form anions with a negative charge. (b) Metals are at the left end of the periodic table; nonmetals are above the stair-step line in the upper right part of the table. Metals have relatively low ionization energies, nonmetals relatively high.

123) LiHe is not a known compound. Helium is able to form few compounds, if any.

Chapter 11

1) K^+, Ca^{2+}, Sc^{3+}, As^{3-}, Se^{2-}, Br^-.   3) Two of K^+, Ca^{2+}, or Sc^{3+}.

5) Two of P^{3-}, S^{2-}, Cl^-, K^+, or Sc^{3+}.   7) a) Argon. b) K^+. c) S^{2-}.

9)
$$\begin{array}{c} K\cdot \searrow \\ \ \ \ \ + \ \ddot{\underset{\cdot\cdot}{S}}: \\ K\cdot \nearrow \end{array} \longrightarrow \begin{array}{c} K^+ \\ \ \\ K^+ \end{array} \left[:\ddot{\underset{\cdot\cdot}{S}}: \right]^{2-} K_2S$$

11) A magnesium atom forms a Mg^{2+} ion by losing two electrons. A chlorine atom can accept one electron to form a Cl^- ion, so it takes two chlorine atoms to accept the two electrons from a single magnesium atom.

15) Sodium is a metal and forms ions by losing one valence electron per atom. Sulfur can form S^{2-} ions by accepting electrons from sodium atoms. Therefore, sodium and sulfur form ionic bonds. The other combinations have too many electrons to form cations, so they form covalent bonds by sharing electrons.

17) $:\ddot{C}l\cdot \ + \ \cdot\ddot{F}: \longrightarrow \ :\ddot{C}l—\ddot{F}:$

19) With one, two, or three valence electrons, a metal atom must receive seven, six, or five electrons from another metal atom to form an anion. No metal atom has that many valence electrons. The maximum number of electrons available for sharing between two metal atoms is six, which is two short of the octet.

23) C—C, nonpolar; N—Cl, essentially nonpolar; S—O, polar; K—Br, ionic; Li—F, ionic.

25) O in S—O; Br in K—Br; F in Li—F.

27) Electronegativity compares attractions of single-bonded atoms for bonding electrons. Noble gases have no electronegativities because they do not form covalent bonds. Initially, noble gases were not assigned electronegativities because they were not believed to form covalent bonds.

29) K—O; Ca—O or Na—O; Al—O; S—O. The group-to-group spread predicts Na—O is more polar than Ca—O, but the period-to-period spread predicts Ca—O is more polar than Na—O.

33) 4; 2.

35) X may share up to four more electrons with other atoms. X may share all four electrons with one atom, forming a double bond. X may share two electrons with each of two other atoms, forming two single bonds. X may share one pair of electrons with one atom and have a lone pair, or bond to no additional atoms and have two lone pairs.

37) The compound must have an even number of valence electrons.

39) Five or six electron pairs may surround a central atom if d orbitals are involved in bonding.

41) Ionization energies are smaller as you go down Group 3A (13), so compounds formed tend to be more ionic as atomic number increases. AlF_3 and BF_3 are a specific example.

44) True: a, b, c, d, e, f, h, k. False: g, i, j.   45) The H^+ ion has no electrons, so it has no electron configuration.

46) c. Mg^{2+} is isoelectronic with neon, not argon.

47) e. Cesium ($Z = 55$) has the lowest electronegativity and fluorine ($Z = 9$) has the highest. The electronegativity difference is therefore greater than any other pair, so the bond is the most ionic.

48) $4p$ from bromine and $2p$ from oxygen.

49) Ionic bonds do not appear in molecular compounds, but covalent bonds exist in polyatomic ions that are present in many ionic compounds. Lewis diagrams of polyatomic ions appear in the next chapter.

50) A bond between identical atoms is completely nonpolar. Their attractions for the bonding electrons are equal. All other bonds have at least a trace of ionic character, as their attractions for bonding electrons are not identical. There are no completely ionic bonds.

51) Electronegativities are highest at the upper right corner of the periodic table and lowest at the lower left-hand corner. Therefore, the electronegativity of A is higher than the electronegativity of B. Because X is higher in the table than Y, the electronegativity of X should be larger than that of Y. But because Y is farther to the right, the electronegativity of X should be smaller than Y. Therefore, no prediction can be made for X and Y.

Chapter 12

1)

3)

5)

7)

9) $C_2H_4Cl_2$:

C_2H_4BrCl:

$C_3H_5F_2I$:

```
   H  :F:  H                    :F:  :F: H                   :F:  H   H
   |  ..  |                     ..   ..  |                    ..  |   |
H—C———C———C—H       or      H—C———C———C—H      or       H—C———C———C—       or
   |  ..  ..                    |    |  ..                    ..  |   ..
   H  :F:  :I:                  H    H  :I:                   :F:  H  :I:
      ..   ..                           ..                    ..       ..
```

```
   H   H  :F:                   H  :F:  :F:                  :F:  H  :F:
   |   |  ..                    |  ..   ..                    ..  |   ..
H—C———C———C—F:      or      H—C———C———C—H      or       H—C———C———C—H
   |   |  ..                    |  ..   |                    |   ..  |
   H  :I:  H                    H  :I:  H                    H  :I: H
       ..                           ..                           ..
```

11) C_4H_{10}:

```
        H  H  H  H                              H
        |  |  |  |                              |
     H—C——C——C——C—H          or             H—C—H
        |  |  |  |                              |
        H  H  H  H                    H         H        H
                                      |         |        |
                                   H—C————C————C—H
                                      |         |        |
                                      H         H        H
```

C_4H_8:

```
     H  H  H                         H  H  H  H
     |  |  |                         |  |  |  |
  H—C=C——C——C—H        or         H—C——C=C——C—H
        |  |  |                      |        |
        H  H  H                      H        H
```

C_2H_4O:

```
     H     ..                       H         H
     |     O:                        \       /
  H—C——C                             C=C
     |     \                        /     \
     H      H                      H       ..
                                           O—H
                                           ..
```

13) C_6H_{14}:

```
     H  H  H  H  H  H                     H  H  H  H  H
     |  |  |  |  |  |                      |  |  |  |  |
  H—C——C——C——C——C——C—H      or          H—C——C——C——C——C—H      or
     |  |  |  |  |  |                      |  |  |  |  |
     H  H  H  H  H  H                      H  H  H     H
                                                 |
                                              H—C—H
                                                 |
                                                 H
```

```
   H  H  H  H  H                          H                           H
   |  |  |  |  |                           |                           |
H—C——C——C——C——C—H       or             H—C—H                       H—C—H
   |  |  |  |  |                           |                           |
   H  H     H  H                  H  H  H  |                  H  H  H  |
         |                         |  |  |  |                  |  |  |  |
      H—C—H                     H—C——C——C——C—H      or      H—C——C——C——C—H
         |                         |  |  |  |                  |  |  |  |
         H                         H  H  H  |                  H     H  H
                                         |  |                     |
                                      H—C—H                    H—C—H
                                         |                        |
                                         H                        H
```

C_3H_8O:

```
     H  H  H                         H  H       H
     |  |  |                         |  |  ..   |
  H—C——C——C—O—H       or          H—C——C——O——C—H
     |  |  |  ..                     |  |       |
     H  H  H                         H  H       H
```

$C_2H_2Cl_2$:

```
   H    :Cl:                     H    :Cl:                    H     H
    \    ..                       \    ..                      \   /
     C=C             or            C=C            or            C=C
    /    \                        /    \                       /   \
   H    :Cl:                   :Cl:    H                   :Cl:   :Cl:
        ..                      ..                          ..     ..
```

15)
```
   H  H   ..
   |  |   O:
H—C——C——C
   |  |   \
   H  H    O—H
          ..
```

17) H—O—O—H   [:O—O:]²⁻
        .. ..

17) $H-\overset{..}{\underset{..}{O}}-\overset{..}{\underset{..}{O}}-H$ $\left[\,:\overset{..}{\underset{..}{O}}-\overset{..}{\underset{..}{O}}:\,\right]^{2-}$

19) BH_3: both trigonal planar. NF_3: tetrahedral, trigonal pyramid. HF: Presumably tetrahedral around F, linear.
21) ClO^-: presumably tetrahedral around both atoms, linear. IO_3^-: tetrahedral, trigonal pyramid. NO_3^-: both planar triangular.
23) Tetrahedral, bent.   25) Both tetrahedral
27) Both atoms, both linear. Each carbon atom forms one single and one triple bond.   29) Both linear.   31) $-\overset{\cdot\cdot}{\underset{|}{X}}-$

33) Trigonal planar and trigonal pyramid. The second molecule has a lone pair and may be polar.
35) Charge is symmetrically distributed in the tetrahedral geometry of CF_4. Charge is unsymmetrically distributed in the trigonal pyramidal structure of NF_3.
37) All the molecules are linear. ClF is the most polar, followed by ICl and BrCl in decreasing polarity. The first element in each formula is the positive end of the molecule. Cl_2 is nonpolar.

39)

Both molecules are polar because of the bent structure around the oxygen atom and the concentration of negative charge near that highly electronegative atom. Carbon chains are nonpolar. The longer the nonpolar carbon chain the less polar the overall molecule. Hence, $C_5H_{11}OH$ is less polar than C_2H_5OH.

41) The only linear molecule is e.   43) Trigonal planar: f. Trigonal pyramid: c.
47) $CH_3(CH_2)_6CH_3$, C_6H_6, and $C_{18}H_8$ are hydrocarbons.
49) The structure of a molecule describes how atoms are bonded to each other. The shape of a molecule describes its geometry in one, two, or three dimensions.
51) Bond angles in C_2H_6 are all tetrahedral, which causes the molecule to be three-dimensional. The geometry around both carbon atoms in C_2H_4 is trigonal planar and the double bond between the carbon atoms keeps all four hydrogen atoms in the same plane.
59) C_3H_7COOH suggests the structure of the molecule. Specifically, it identifies the compound as a carboxylic acid.

62) True: a, d, e, g, h, i, j. False: b, c, f, k, l.   63)

64)   65)

66)

67) All species in the question, including SeO_4^{2-} and CI_4, have 32 valence electrons and four atoms to be distributed about a central atom. Consequently, they all have the same tetrahedral shape.
68) Bond angles between carbon atoms in an alkane are tetrahedral, so the atoms cannot lie in a straight line.
69) (a) Planar with 120° angles around both carbon atoms. (b) Linear. (c) Zig-zag carbon chain, all bond angles tetrahedral (see Question 68).

Chapter 13

1) Gay-Lussac observed that when gases react, the volumes that combine are in the ratio of small whole numbers if the volumes are measured at the same temperature and pressure. This observation is the Law of Combining Volumes. Avogadro interpreted Gay-Lussac's observations as indicating that equal volumes of all gases, measured at the same temperature and pressure, contain the same number of molecules. From this it follows that at constant temperature and pressure, number of molecules is proportional to volume. The interpretation and the proportional relationship are Avogadro's Law.

3) $P = \dfrac{mRT}{V(MM)} = \dfrac{8.33 \text{ g}}{5.00 \text{ L}} \times \dfrac{0.0821 \text{ L} \cdot \text{atm}}{\text{mol} \cdot \text{K}} \times \dfrac{1 \text{ mol}}{44.0 \text{ g}} \times 309 \text{ K} = 0.961 \text{ atm}$

5) $V = \dfrac{nRT}{P} = \dfrac{4.02 \text{ mol}}{18.5 \text{ atm}} \times \dfrac{0.0821 \text{ L} \cdot \text{atm}}{\text{mol} \cdot \text{K}} \times 294 \text{ K} = 5.24 \text{ L}$

7) $n = \dfrac{PV}{RT} = \dfrac{749 \text{ torr}}{294 \text{ K}} \times \dfrac{\text{mol} \cdot \text{K}}{62.4 \text{ L} \cdot \text{torr}} \times 0.844 \text{ L} = 0.0345 \text{ mol}$

9) $T = \dfrac{PV}{nR} = \dfrac{805 \text{ torr}}{0.119 \text{ mol}} \times \dfrac{\text{mol} \cdot \text{K}}{62.4 \text{ L} \cdot \text{torr}} \times 1.97 \text{ L} = 214 \text{ K} = -59°\text{C}$

11) $m = \dfrac{PV(MM)}{RT} = \dfrac{859 \text{ torr}}{304 \text{ K}} \times \dfrac{\text{mol} \cdot \text{K}}{62.4 \text{ L} \cdot \text{torr}} \times \dfrac{44.0 \text{ g}}{\text{mol}} \times 35.0 \text{ L} = 69.7 \text{ g}$

13) $MM = \dfrac{mRT}{PV} = \dfrac{0.681 \text{ g}}{0.629 \text{ atm}} \times \dfrac{0.0821 \text{ L} \cdot \text{atm}}{\text{mol} \cdot \text{K}} \times \dfrac{322 \text{ K}}{0.442 \text{ L}} = 64.8 \text{ g/mol}$

15) $3.37 \text{ g O}_2 \times \dfrac{1 \text{ mol O}_2}{32.0 \text{ g O}_2} \times \dfrac{22.4 \text{ L}}{\text{mol}} = 2.36 \text{ L O}_2$

17) $115 \text{ L F}_2 \times \dfrac{1 \text{ mol F}_2}{22.4 \text{ L F}_2} \times \dfrac{38.0 \text{ g F}_2}{\text{mol F}_2} = 195 \text{ g F}_2$

19)

Element	Grams	Moles	Mole Ratio	Formula Ratio	Simplest Formula	Molecular Formula
C	55.8	4.65	2.00	2		$\dfrac{86.3}{43.0} = 2.01$
H	7.0	7.0	3.0	3		
O	37.2	2.33	1.00	1	C_2H_3O	$C_4H_6O_2$

$MM = \dfrac{mRT}{PV} = \dfrac{3.26 \text{ g}}{0.914 \text{ atm}} \times \dfrac{0.0821 \text{ L} \cdot \text{atm}}{\text{mol} \cdot \text{K}} \times \dfrac{433 \text{ K}}{1.47 \text{ L}} = 86.3 \text{ g/mol}$

21) $0.21 \times 32 \text{ g/mol} + 0.79 \times 28 \text{ g/mol} = 29 \text{ g/mol}$

23) $\dfrac{m}{V} = \dfrac{P(MM)}{RT} = \dfrac{1 \text{ atm}}{273 \text{ K}} \times \dfrac{46.0 \text{ g}}{\text{mol}} \times \dfrac{\text{mol} \cdot \text{K}}{0.0821 \text{ L} \cdot \text{atm}} = 2.05 \text{ g/L}$

25) $MM = \dfrac{mRT}{PV} = \dfrac{2.09 \text{ g}}{\text{L}} \times \dfrac{0.0821 \text{ L} \cdot \text{atm}}{\text{mol} \cdot \text{K}} \times \dfrac{273 \text{ K}}{1 \text{ atm}} = 46.8 \text{ g/mol}$

27) $MM = \dfrac{mRT}{PV} = \dfrac{1.94 \text{ g}}{1.68 \text{ L}} \times \dfrac{0.0821 \text{ L} \cdot \text{atm}}{\text{mol} \cdot \text{K}} \times \dfrac{273 \text{ K}}{1 \text{ atm}} = 25.9 \text{ g/mol}$

29) $\dfrac{m}{V} = \dfrac{P(MM)}{RT} = \dfrac{743 \text{ torr}}{320 \text{ K}} \times \dfrac{\text{mol} \cdot \text{K}}{62.4 \text{ L} \cdot \text{torr}} \times \dfrac{20.2 \text{ g}}{\text{mol}} = 0.752 \text{ g/L}$

31) $\dfrac{m}{V} = \dfrac{P(MM)}{RT} = \dfrac{0.994 \text{ atm}}{293 \text{ K}} \times \dfrac{\text{mol} \cdot \text{K}}{0.0821 \text{ L} \cdot \text{atm}} \times \dfrac{29 \text{ g}}{\text{mol}} = 1.2 \text{ g/L}$

33) $MM = \dfrac{mRT}{PV} = \dfrac{2.74 \text{ g}}{\text{L}} \times \dfrac{62.4 \text{ L} \cdot \text{torr}}{\text{mol} \cdot \text{K}} \times \dfrac{573 \text{ K}}{790 \text{ torr}} = 124 \text{ g/mol @ } 300°\text{C}$

$MM = \dfrac{mRT}{PV} = \dfrac{0.617 \text{ g}}{\text{L}} \times \dfrac{62.4 \text{ L} \cdot \text{torr}}{\text{mol} \cdot \text{K}} \times \dfrac{1273 \text{ K}}{790 \text{ torr}} = 62.0 \text{ g/mol @ } 1000°\text{C}$

At 300°C, $\dfrac{124 \text{ g}}{\text{mol molecules}} \times \dfrac{\text{mol atoms}}{31.0 \text{ g}} = \dfrac{4 \text{ mol atoms}}{\text{mol molecules}}$. Formula: P_4

At 1000°C, $\dfrac{62 \text{ g}}{\text{mol molecules}} \times \dfrac{\text{mol atoms}}{31.0 \text{ g}} = \dfrac{2 \text{ mol atoms}}{\text{mol molecules}}$. Formula: P_2

35) Molar volume is the number of liters occupied by 1 mole of gas. All gases have the same molar volume at the same temperature and pressure because gas volume depends on these two variables.

37) a) $MV = \dfrac{V}{n} = \dfrac{RT}{P} = \dfrac{62.4 \text{ L} \cdot \text{torr}}{\text{mol} \cdot \text{K}} \times \dfrac{322 \text{ K}}{381 \text{ torr}} = 52.7 \text{ L/mol}$

b) $MV = \dfrac{V}{n} = \dfrac{RT}{P} = \dfrac{0.0821 \text{ L} \cdot \text{atm}}{\text{mol} \cdot \text{K}} \times \dfrac{278 \text{ K}}{0.774 \text{ atm}} = 29.5 \text{ L/mol}$

39) $16.2 \text{ L} \times \dfrac{1 \text{ mol}}{45.5 \text{ L}} = 0.356 \text{ mol}$ 41) $6.43 \text{ mol} \times \dfrac{27.3 \text{ L}}{\text{mol}} = 176 \text{ L}$

43) $\text{Density} = \dfrac{m}{V} = \dfrac{P(MM)}{RT} = k \times MM$ $\text{Density} \propto MM$

45) $MV = \dfrac{RT}{P} = \dfrac{0.0821 \text{ L} \cdot \text{atm}}{\text{mol} \cdot \text{K}} \times \dfrac{657 \text{ K}}{1.48 \text{ atm}} = 36.4 \text{ L/mol}$

$423 \text{ L SO}_2 \times \dfrac{1 \text{ mol SO}_2}{36.4 \text{ L SO}_2} \times \dfrac{4 \text{ mol FeS}_2}{8 \text{ mol SO}_2} \times \dfrac{120.1 \text{ g FeS}_2}{1 \text{ mol FeS}_2} = 698 \text{ g FeS}_2$

47) $MV = \dfrac{RT}{P} = \dfrac{0.0821 \text{ L} \cdot \text{atm}}{\text{mol} \cdot \text{K}} \times \dfrac{305 \text{ K}}{0.940 \text{ atm}} = 26.6 \text{ L/mol}$ $2 \text{ H}_2\text{O} \longrightarrow 2 \text{ H}_2 + \text{O}_2$

$12.6 \text{ g H}_2\text{O} \times \dfrac{1 \text{ mol H}_2\text{O}}{18.0 \text{ g H}_2\text{O}} \times \dfrac{2 \text{ mol H}_2}{2 \text{ mol H}_2\text{O}} \times \dfrac{26.6 \text{ L}}{\text{mol}} = 18.6 \text{ L H}_2$

49) $MV = \dfrac{RT}{P} = \dfrac{0.0821 \text{ L} \cdot \text{atm}}{\text{mol} \cdot \text{K}} \times \dfrac{823 \text{ K}}{250 \text{ atm}} = 0.270 \text{ L/mol}$ $\text{N}_2 + 3 \text{ H}_2 \longrightarrow 2 \text{ NH}_3$

$97.0 \text{ L N}_2 \times \dfrac{1 \text{ mol N}_2}{0.270 \text{ L N}_2} \times \dfrac{2 \text{ mol NH}_3}{1 \text{ mol N}_2} \times \dfrac{17.0 \text{ g NH}_3}{1 \text{ mol NH}_3} = 1.22 \times 10^4 \text{ g NH}_3$

51) $MV = \dfrac{RT}{P} = \dfrac{0.0821 \text{ L} \cdot \text{atm}}{\text{mol} \cdot \text{K}} \times \dfrac{1248 \text{ K}}{1.22 \text{ atm}} = 84.0 \text{ L/mol}$

$26.8 \text{ g NH}_4\text{NO}_3 \times \dfrac{1 \text{ mol NH}_4\text{NO}_3}{80.0 \text{ g NH}_4\text{NO}_3} \times \dfrac{3 \text{ mol gas}}{1 \text{ mol NH}_4\text{NO}_3} \times \dfrac{84.0 \text{ L gas}}{1 \text{ mol gas}} = 84.4 \text{ L gas}$

53) $2215 \text{ ft}^3 \text{ O}_2 \times \dfrac{2 \text{ ft}^3 \text{ SO}_2}{1 \text{ ft}^3 \text{ O}_2} = 4.430 \times 10^3 \text{ ft}^3 \text{ SO}_2$

55) $2.09 \text{ L H}_2\text{S} \times \dfrac{354 \text{ K}}{304 \text{ K}} \times \dfrac{0.923 \text{ atm}}{3.52 \text{ atm}} \times \dfrac{3 \text{ L O}_2}{2 \text{ L H}_2\text{S}} = 0.957 \text{ L} = 957 \text{ mL O}_2$

59) $754 - 593 - 149 - 7 = 5 \text{ torr}$ 61) $749 - 25 = 724 \text{ torr}$ 64) True: e. False: a, b, c, d.

65) Volume at start, 350 cm^3. Volume at end, $350 - 309 = 41 \text{ cm}^3$. Compression ratio $= 350/41 = 8.5$.

66) a)   1.0 atm   (b) $1.0 \text{ atm} + 60 \text{ psi} \times \dfrac{1 \text{ atm}}{15 \text{ psi}} = 5.0 \text{ atm}$

c) $n = \dfrac{PV}{RT} = \dfrac{1.0 \text{ atm}}{295 \text{ K}} \times \dfrac{\text{mol} \cdot \text{K}}{0.0821 \text{ L} \cdot \text{atm}} \times 0.39 \text{ L} = 0.016 \text{ mol}$

d) $n = \dfrac{PV}{RT} = \dfrac{1.0 \text{ atm}}{295 \text{ K}} \times \dfrac{\text{mol} \cdot \text{K}}{0.0821 \text{ L} \cdot \text{atm}} \times 1.5 \text{ L} = 0.062 \text{ mol}$

e) $n = \dfrac{PV}{RT} = \dfrac{5.0 \text{ atm}}{295 \text{ K}} \times \dfrac{\text{mol} \cdot \text{K}}{0.0821 \text{ L} \cdot \text{atm}} \times 1.5 \text{ L} = 0.31 \text{ mol}$

f) $0.31 \text{ mol full} - 0.062 \text{ mol empty} = 0.25 \text{ mol to be added}$ $0.25 \text{ mol} \times \dfrac{1 \text{ stroke}}{0.016 \text{ mol}} = 16 \text{ strokes}$

g)   With each stroke you are pumping against a higher pressure.

67) Pressure when full: $1.0 \text{ atm} + 30 \text{ psi} \times \dfrac{1 \text{ atm}}{15 \text{ psi}} = 3.0 \text{ atm}$

$n(\text{full}) = \dfrac{PV}{RT} = \dfrac{3.0 \text{ atm}}{295 \text{ K}} \times \dfrac{\text{mol} \cdot \text{K}}{0.0821 \text{ L} \cdot \text{atm}} \times 41 \text{ L} = 5.1 \text{ mol}$

$n(\text{empty}) = \dfrac{PV}{RT} = \dfrac{1.0 \text{ atm}}{295 \text{ K}} \times \dfrac{\text{mol} \cdot \text{K}}{0.0821 \text{ L} \cdot \text{atm}} \times 41 \text{ L} = 1.7 \text{ mol}$

$5.1 - 1.7 = 3.4 \text{ mol added}$ $3.4 \text{ mol} \times \dfrac{1 \text{ stroke}}{0.016 \text{ mol}} = 213 \text{ strokes}$

Chapter 14

3) Intermolecular attractions are weaker in gases because of the space between the molecules.

9) The ball bearing will reach the bottom of the water cylinder first. Molecules in oil have higher intermolecular attractions than those in water, so the oil is more viscous. Oil molecules resist being pulled apart so the ball bearing can pass through the liquid. This slows the rate of fall.

11) Intermolecular attractions are stronger in mercury than in water. Mercury therefore has a higher surface tension and clings to itself rather than spreading or penetrating paper.

13) The wetting agent reduces surface tension of water, overcoming its intermolecular attractions and allowing it to penetrate the duck's feathers. The duck's buoyancy is reduced, and it sinks.

15) NO_2 has the highest boiling point, which suggests strong intermolecular attractions. It should also have the highest molar heat of vaporization, another property that is associated with strong intermolecular forces.

17) Only N_2O is a liquid at $-90°C$, so it alone has a measurable equilibrium vapor pressure as that term is used in this chapter. NO is a gas, and its vapor pressure is its gas pressure. NO_2, a solid at $-90°C$, probably has a very small vapor pressure.

19) Dispersion forces are likely to be larger than dipole forces when the molecules are very large. With small molecules, dipole forces are stronger than dispersion forces.

21) $NH(CH_3)_2$, hydrogen bonding; CH_2F_2, dipole; C_3H_8 and $BeCl_2$, dispersion.

23) Dipole forces are attractive forces between polar molecules; hydrogen bonds are stronger dipole-like forces between polar molecules in which hydrogen is bonded to a highly electronegative element, usually nitrogen, oxygen, or fluorine.

25) CH_4 has the lower boiling point because it has only weak dispersion forces, versus hydrogen bonds in NH_3.

27) Argon has the lower boiling point because its atoms are smaller than those of krypton.

31) a) Hydrogen bonding. b) Dispersion forces.

33) Dispersion forces are present in all molecules. In addition, dipole forces are present in a) and c), whereas b) and d) have hydrogen bonding.

35) CO_2 should have the lower melting and boiling points because its molecules are smaller than otherwise similar CS_2 molecules.

41) $p = nRT/V$. When T increases, evaporation rate increases until a new equilibrium is reached with a larger number of particles in the vapor state. In other words, n and T both increase while R and V remain constant. Thus, a change in T causes a change in p.

43) Reducing volume increases vapor concentration, which causes an increase in the rate of condensation. Evaporation rate, which depends only on temperature, is not affected.

45) a) L. b) M and S will reach equilibrium vapor pressure, L less than equilibrium pressure. c) All of the liquid will evaporate in L because, even after total evaporation, the vapor concentration will not reach the point at which condensation will begin. In M and S the vapor pressures will be the same, determined by vapor concentration, not box size.

47) $757 + 22.4 = 779$ torr

49) The procedure described measures the *increase* in vapor pressure from its starting point to its value at equilibrium. The increase is equal to the equilibrium vapor pressure only if the starting value is zero, a condition not commonly satisfied by air. Starting with "bone dry air," the system would work.

51) Nothing will happen. Additional liquid will change neither the evaporation nor condensation rates, so the vapor concentration, and hence the vapor pressure, will remain the same.

55) A liquid can be made to boil by reducing the pressure above its surface below the equilibrium vapor pressure at the existing temperature.

57) The water is delivered at very high pressure, pressure greater than the vapor pressure at the temperature of delivery.

59) High-boiling liquids and a high heat of vaporization are both characteristic of relatively strong intermolecular attractions. A liquid with strong attractions would therefore exhibit both properties.

61) M should have both the lower boiling point and lower molar heat of vaporization.

63) Ice is a crystalline solid. Ice crystals have a definite geometric order and ice melts at a definite, constant temperature.

65) A: network solid. B: ionic solid.   67) $\dfrac{44.8 \text{ kJ}}{61.2 \text{ g}} = 0.732$ kJ/g   69) $227 \text{ g} \times \dfrac{4.27 \text{ kJ}}{\text{g}} = 969$ kJ

71) $79.4 \text{ kJ} \times \dfrac{\text{g}}{0.880 \text{ kJ}} = 90.2$ g alcohol   73) $23.8 \text{ g C}_3\text{H}_6\text{O} \times \dfrac{1 \text{ mol C}_3\text{H}_6\text{O}}{58.0 \text{ g C}_3\text{H}_6\text{O}} \times \dfrac{32.0 \text{ kJ}}{1 \text{ mol C}_3\text{H}_6\text{O}} = 13.1$ kJ

75) $3.30 \text{ kg} \times \dfrac{23 \text{ kJ}}{\text{kg}} = 76$ kJ   77) $\dfrac{2.51 \times 10^3 \text{ J}}{36.9 \text{ g}} = 68.0$ J/g   79) $4.45 \times 10^3 \text{ J} \times \dfrac{1 \text{ g Zn}}{100 \text{ J}} = 44.5$ g Zn

81) A. From $q = m \times c \times \Delta T$, when q and m are equal for two objects, c is inversely proportional to ΔT.

83) $467 \text{ g} \times \dfrac{0.38 \text{ J}}{\text{g} \cdot {}^\circ\text{C}} \times (68 - 31)^\circ\text{C} = 6.6 \times 10^3 \text{ J}$ 85) $2.30 \times 10^3 \text{ g} \times \dfrac{0.13 \text{ J}}{\text{g} \cdot {}^\circ\text{C}} \times (88 - 22)^\circ\text{C} = 2.0 \times 10^4 \text{ J} = 2.0 \times 10 \text{ kJ}$

87) $\Delta T = \dfrac{Q}{m \times c} = \dfrac{1.47 \times 10^3 \text{ J}}{144 \text{ g}} \times \dfrac{\text{g} \cdot {}^\circ\text{C}}{0.38 \text{ J}} = 27^\circ\text{C}$ $33 - 27 = 6^\circ\text{C}$

89) $Q = 9.96 \times 10^3 \text{ g} \times \dfrac{4.184 \text{ J}}{\text{g} \cdot {}^\circ\text{C}} \times (28.0 - 17.1)^\circ\text{C} = 4.54 \times 10^5 \text{ J}$

$c = \dfrac{Q}{m \times \Delta T} = \dfrac{4.54 \times 10^5 \text{ J}}{3.62 \times 10^3 \text{ g} \times (92 - 28.0)^\circ\text{C}} = 2.0 \text{ J/g} \cdot {}^\circ\text{C}$

91) P (boiling point), O (freezing point).   93) A, B.   95) G.

97) Gas condenses at boiling point, P; liquid cools from boiling point, P, to freezing point, O.   99) J–K.

101) $Q_1 = 127 \text{ g} \times \dfrac{2.1 \text{ J}}{\text{g} \cdot {}^\circ\text{C}} \times [0 - (-11)]^\circ\text{C} = 2.9 \times 10^3 \text{ J} =$ 2.9 kJ

$Q_2 = 127 \text{ g} \times \dfrac{335 \text{ J}}{\text{g}} = 4.25 \times 10^4 \text{ J} =$ 42.5 kJ

$Q_3 = 127 \text{ g} \times \dfrac{4.18 \text{ J}}{\text{g} \cdot {}^\circ\text{C}} \times (21 - 0)^\circ\text{C} = 1.1 \times 10^4 \text{ J} =$ <u>11 kJ</u>

$Q = Q_1 + Q_2 + Q_3 =$ 56 kJ

103) $Q_1 = 689 \text{ g} \times \dfrac{0.512 \text{ J}}{\text{g} \cdot {}^\circ\text{C}} \times (419 - 552)^\circ\text{C} = -4.69 \times 10^4 \text{ J} =$ −46.9 kJ

$Q_2 = 689 \text{ g} \times -\dfrac{100 \text{ J}}{\text{g}} = -6.89 \times 10^4 \text{ J} =$ −68.9 kJ (Assume 3 SF)

$Q_3 = 689 \text{ g} \times \dfrac{0.38 \text{ J}}{\text{g} \cdot {}^\circ\text{C}} \times (21 - 419)^\circ\text{C} = -1.04 \times 10^5 \text{ J} =$ <u>−104 kJ</u> (2 SF)

$Q = Q_1 + Q_2 + Q_3 = -219.8 \text{ kv} = -2.2 \times 10^2 \text{ kJ}$

105)

$Q \text{ (heat solid)} = 941 \text{ kg} \times \dfrac{0.27 \text{ kJ}}{\text{kg} \cdot {}^\circ\text{C}} \times (264 - 26)^\circ\text{C} = 6.0 \times 10^4 \text{ kJ}$

$Q \text{ (melt solid)} = 941 \text{ kg} \times \dfrac{29 \text{ kJ}}{\text{kg}} = 2.7 \times 10^4 \text{ kJ}$

$Q \text{ (heat liquid)} = 941 \text{ kg} \times \dfrac{0.21 \text{ kJ}}{\text{kg} \cdot {}^\circ\text{C}} \times (339 - 264)^\circ\text{C} = \underline{1.5 \times 10^4 \text{ kJ}}$

Total $Q = 10.2 \times 10^4 \text{ kJ}$

108) True: a, c, f, h, i, o, r. False: b, d, e, g, j, k, l, m, n, p, q.

109) Dissolve the compounds and check for electrical conductivity. The ionic potassium sulfate solute will conduct, whereas the molecular sugar solute will not.

110) Both molecules have dispersion and dipole–dipole forces. CH_3OH has hydrogen bonding and CH_3F does not. The molecules are about the same size. It is reasonable to predict stronger intermolecular forces in CH_3OH, and therefore higher boiling point and lower equilibrium vapor pressure.

111) Without a regular and uniform structure in an amorphous solid, some intermolecular bonds are stronger than others. The weak bonds break at a lower temperature than the strong bonds.

112) Large molecules having strong dispersion forces may have stronger intermolecular attractions than small molecules with hydrogen bonding, and therefore exhibit greater viscosity.

113) As temperature drops, the equilibrium vapor pressure drops below the atmospheric vapor pressure. The air becomes first saturated, then supersaturated, and condensation (dew) begins to form.

114) Heat lost by lemonade = heat gained by ice.   Let M = mass of ice.

$175 \text{ g} \times \dfrac{4.18 \text{ J}}{\text{g} \cdot {}^\circ\text{C}} \times (23 - 5)^\circ\text{C} = M \text{ g} \times \dfrac{2.1 \text{ J}}{\text{g} \cdot {}^\circ\text{C}} \times 8^\circ\text{C} + M \text{ g} \times \dfrac{335 \text{ J}}{\text{g}} + M \text{ g} \times \dfrac{4.18 \text{ J}}{\text{g} \cdot {}^\circ\text{C}} \times 5^\circ\text{C}$

$M = 35 \text{ g}$

Chapter 15

1) The properties of a solution will differ from those of the components. They will also be variable, depending on the solution concentration.

7) A saturated solution holds all the solute it is able to hold at a given temperature. A concentrated solution has a *relatively* high concentration, but it need not be saturated.

9) Solubility is the amount of solute that will dissolve in a given amount of solvent at a given temperature. Any units expressing amount of solute per amount of solvent would satisfy the definition. Examples: mol solute/kg solvent (molality), ounces of solute/gallon of solvent.

11) Acetic acid is soluble in water because it is dispersed uniformly throughout the solution. It is also miscible, a term usually used to express the solubility of liquids in each other.

13) Distillation is one; see Figure 2.3.

21) Solute ions pass from solute to solution faster than from solution to solute in an unsaturated solution.

23) The act of stirring prevents the solution from becoming supersaturated.

29) (a), formic acid; (c), methylamine. Both compounds match the hydrogen bonding found in water. Benzene and tetrafluoromethane are both nonpolar.

31) Glycerine exhibits hydrogen bonding, as does water, so they are miscible. Hexane is nonpolar.

33) Carbon dioxide, CO_2.

35) $\dfrac{2.20 \text{ g solute}}{(2.20 + 57.9)\text{g solution}} \times 100 = 3.66\%$

37) $505 \text{ g solution} \times \dfrac{15.0 \text{ g Na}_2\text{SO}_4}{100 \text{ g solution}} = 75.8 \text{ g Na}_2\text{SO}_4 \qquad 505 \text{ g solution} - 75.8 \text{ g Na}_2\text{SO}_4 = 429 \text{ g H}_2\text{O}$

39) $\dfrac{2.41 \text{ g KI}}{0.0500 \text{ L}} \times \dfrac{1 \text{ mol KI}}{166.0 \text{ g KI}} = 0.290 \text{ M KI}$

41) $\dfrac{18.0 \text{ g NiCl}_2}{0.0900 \text{ L}} \times \dfrac{1 \text{ mol NiCl}_2}{129.7 \text{ g NiCl}_2} = 1.54 \text{ M NiCl}_2 \qquad \dfrac{30.0 \text{ g NiCl}_2 \cdot 6 \text{ H}_2\text{O}}{0.0900 \text{ L}} \times \dfrac{1 \text{ mol NiCl}_2}{237.8 \text{ g NiCl}_2 \cdot 6 \text{ H}_2\text{O}} = 1.40 \text{ M NiCl}_2$

43) $0.250 \text{ L} \times \dfrac{0.058 \text{ mol AgNO}_3}{\text{L}} \times \dfrac{169.9 \text{ g AgNO}_3}{1 \text{ mol AgNO}_3} = 2.5 \text{ g AgNO}_3$

45) $2.50 \text{ L} \times \dfrac{1.40 \text{ mol KOH}}{\text{L}} \times \dfrac{56.1 \text{ g KOH}}{1 \text{ mol KOH}} = 196 \text{ g KOH} \qquad 47) \ 5.19 \text{ mol H}_2\text{SO}_4 \times \dfrac{1 \text{ L}}{18 \text{ mol H}_2\text{SO}_4} = 0.29 \text{ L} = 290 \text{ mL}$

49) $8.33 \text{ g NaCl} \times \dfrac{1 \text{ mol NaCl}}{58.5 \text{ g NaCl}} \times \dfrac{1 \text{ L}}{0.132 \text{ mol NaCl}} = 1.08 \text{ L} \qquad 51) \ 0.0557 \text{ L} \times \dfrac{0.204 \text{ mol AgNO}_3}{1 \text{ L}} = 0.0114 \text{ mol AgNO}_3$

53) $0.0250 \text{ L} \times \dfrac{0.0841 \text{ mol KMnO}_4}{1 \text{ L}} = 2.10 \times 10^{-3} \text{ mol KMnO}_4$

55) $\dfrac{3.30 \text{ mol KNO}_3}{1 \text{ L}} \times \dfrac{1 \text{ L}}{1000 \text{ mL}} \times \dfrac{1 \text{ mL}}{1.15 \text{ g soln}} \times \dfrac{101.1 \text{ g KNO}_3}{1 \text{ mol KNO}_3} \times 100 = 29.0\% \text{ KNO}_3$

57) $\dfrac{44.9 \text{ g C}_{10}\text{H}_8}{0.175 \text{ kg solvent}} \times \dfrac{1 \text{ mol C}_{10}\text{H}_8}{128.2 \text{ g C}_{10}\text{H}_8} = 2.00 \text{ m C}_{10}\text{H}_8$

59) $0.400 \text{ kg eth} \times \dfrac{4.70 \text{ mol (CH}_3\text{CH}_2)_2\text{NH}}{1 \text{ kg eth}} \times \dfrac{73.1 \text{ g (CH}_3\text{CH}_2)_2\text{NH}}{1 \text{ mol (CH}_3\text{CH}_2)_2\text{NH}} = 137 \text{ g (CH}_3\text{CH}_2)_2\text{NH}$

61) $1.00 \times 10^2 \text{ mL B} \times \dfrac{0.879 \text{ g B}}{1 \text{ mL B}} \times \dfrac{1 \text{ kg}}{1000 \text{ g}} \times \dfrac{0.254 \text{ mol C}_4\text{H}_8\text{O}}{1 \text{ kg B}} \times \dfrac{72.1 \text{ g C}_4\text{H}_8\text{O}}{1 \text{ mol C}_4\text{H}_8\text{O}} = 1.61 \text{ g C}_4\text{H}_8\text{O}$

63) Equivalent mass is the mass of a substance that reacts with one mole of hydrogen or hydroxide ions. LiOH has one mole of OH^- ions, so equivalent mass = molar mass. H_2SO_4 can release one or two moles of H^+ ions, so equivalent mass = molar mass or $\frac{1}{2}$ of molar mass.

65) 1 eq/mol HNO_2; 1 eq/mol H_2SeO_4

67) 2 eq/mol $Cu(OH)_2$; 3 eq/mol $Fe(OH)_3$

69) 47.0 g HNO_2/eq; 145.0 g H_2SeO_4/eq

71) 48.8 g $Cu(OH)_2$/eq; 35.6 g $Fe(OH)_3$/eq

73) $\dfrac{2.25 \text{ g KOH}}{0.250 \text{ L}} \times \dfrac{1 \text{ eq KOH}}{56.1 \text{ g KOH}} = 0.160 \text{ N KOH}$

75) $\dfrac{6.69 \text{ g H}_2\text{C}_2\text{O}_4}{0.200 \text{ L}} \times \dfrac{1 \text{ eq}}{90.0 \text{ g H}_2\text{C}_2\text{O}_4} = 0.372 \text{ N H}_2\text{C}_2\text{O}_4$

77) $0.750 \text{ L} \times \dfrac{0.200 \text{ eq NaHSO}_4}{\text{L}} \times \dfrac{120.1 \text{ g NaHSO}_4}{1 \text{ eq NaHSO}_4} = 18.0 \text{ g NaHSO}_4$

79) $0.100 \text{ L} \times \dfrac{0.500 \text{ eq Na}_2\text{CO}_3}{\text{L}} \times \dfrac{286.2 \text{ g Na}_2\text{CO}_3 \cdot 10 \text{ H}_2\text{O}}{2 \text{ eq Na}_2\text{CO}_3} = 7.16 \text{ g Na}_2\text{CO}_3 \cdot 10 \text{ H}_2\text{O}$

81) $0.0731 \text{ L} \times \dfrac{0.834 \text{ eq}}{\text{L}} = 0.0610 \text{ eq}$

83) $0.788 \text{ eq} \times \dfrac{1 \text{ L}}{0.492 \text{ eq}} = 1.60 \text{ L}$

85) $M_d = \dfrac{0.0450 \text{ L}_c \times 17 \text{ mol/L}_c}{1.5 \text{ L}_d} = 0.51 \text{ M HC}_2\text{H}_3\text{O}_2$

87) $L_c = \dfrac{0.750 \text{ L}_d \times 0.69 \text{ mol/L}_d}{16 \text{ mol/L}_c} = .032 \text{ L}_c = 32 \text{ mL}_c$

89) At 2 eq/mol, 18 M H_2SO_4 is 36 N H_2SO_4 $\qquad L_c = \dfrac{3.0 \text{ L}_d \times 2.9 \text{ eq/L}_d}{36 \text{ eq/L}_c} = 0.24 \text{ L}_c$

91) $2 \text{ NaOH} + \text{MgCl}_2 \longrightarrow \text{Mg(OH)}_2 + 2 \text{ NaCl}$

$0.0250 \text{ L} \times \dfrac{0.398 \text{ mol MgCl}_2}{1 \text{ L}} \times \dfrac{1 \text{ mol Mg(OH)}_2}{1 \text{ mol MgCl}_2} \times \dfrac{58.3 \text{ g Mg(OH)}_2}{1 \text{ mol Mg(OH)}_2} = 0.580 \text{ g Mg(OH)}_2$

93) $2 \text{ Fe(NO}_3)_3 \longrightarrow 2 \text{ Fe(OH)}_3 \longrightarrow \text{Fe}_2\text{O}_3$

$0.0350 \text{ L} \times \dfrac{0.516 \text{ mol Fe(NO}_3)_3}{1 \text{ L}} \times \dfrac{1 \text{ mol Fe}_2\text{O}_3}{2 \text{ mol Fe(NO}_3)_3} \times \dfrac{159.8 \text{ g Fe}_2\text{O}_3}{1 \text{ mol Fe}_2\text{O}_3} = 1.44 \text{ g Fe}_2\text{O}_3$

95)

	2 KOH	**+**	**NiCl$_2$**	$\longrightarrow$	**Ni(OH)$_2$ + 2 KCl**
Liters at start	0.0300		0.0250		
Molarity	0.260		0.269		
Moles at start	0.00780		0.00673		
Change in moles	−0.00780		−0.00390		+0.00390
Moles at end	0.0		0.00283		0.00390
Molar mass					92.7
Grams at end					0.362

97) $2.00 \text{ L H}_2 \times \dfrac{\text{mol} \cdot \text{K}}{62.4 \text{ L} \cdot \text{torr}} \times \dfrac{789 \text{ torr}}{295 \text{ K}} \times \dfrac{6 \text{ mol NaOH}}{3 \text{ mol H}_2} \times \dfrac{1000 \text{ mL}}{1.50 \text{ mol NaOH}} = 114 \text{ mL}$

99) $\text{NaOH} + \text{NH}_2\text{SO}_3\text{H} \longrightarrow \text{HOH} + \text{NH}_2\text{SO}_3\text{Na}$

$8.74 \text{ g NH}_2\text{SO}_3\text{H} \times \dfrac{1 \text{ mol NH}_2\text{SO}_3\text{H}}{97.1 \text{ g NH}_2\text{SO}_3\text{H}} \times \dfrac{1 \text{ mol NaOH}}{1 \text{ mol NH}_2\text{SO}_3\text{H}} \times \dfrac{1 \text{ L}}{0.842 \text{ mol NaOH}} = 0.107 \text{ L} = 107 \text{ mL}$

101) $5.038 \text{ g HC}_7\text{H}_5\text{O}_2 \times \dfrac{1 \text{ mol HC}_7\text{H}_5\text{O}_2}{122.1 \text{ g HC}_7\text{H}_5\text{O}_2} \times \dfrac{1 \text{ mol Na}_2\text{CO}_3}{2 \text{ mol HC}_7\text{H}_5\text{O}_2} \times \dfrac{1}{0.05189 \text{ L}} = 0.3976 \text{ M Na}_2\text{CO}_3$

103) $2 \text{ NaOH} + \text{H}_2\text{C}_4\text{H}_2\text{O}_4 \longrightarrow 2 \text{ HOH} + \text{Na}_2\text{C}_4\text{H}_2\text{O}_4$

$0.0500 \text{ L} \times \dfrac{0.500 \text{ mol NaOH}}{1 \text{ L}} \times \dfrac{1 \text{ mol H}_2\text{C}_4\text{H}_2\text{O}_4}{2 \text{ mol NaOH}} \times \dfrac{116 \text{ g H}_2\text{C}_4\text{H}_2\text{O}_4}{1 \text{ mol H}_2\text{C}_4\text{H}_2\text{O}_4} = 1.45 \text{ g H}_2\text{C}_4\text{H}_2\text{O}_4$

105) $\dfrac{17.02 \text{ g NaHCO}_3}{0.5000 \text{ L}} \times \dfrac{1 \text{ mol NaHCO}_3}{84.00 \text{ g NaHCO}_3} = 0.4052 \text{ M NaHCO}_3$

$0.03780 \text{ L} \times \dfrac{0.4052 \text{ mol NaHCO}_3}{1 \text{ L}} \times \dfrac{1 \text{ mol H}_2\text{SO}_4}{2 \text{ mol NaHCO}_3} \times \dfrac{1}{0.02000 \text{ L}} = 0.3829 \text{ M H}_2\text{SO}_4$

107) $\text{H}_2\text{SO}_4 + 2 \text{ OH}^- \longrightarrow 2 \text{ HOH} + \text{SO}_4^{2-}$

$0.01475 \text{ L} \times \dfrac{0.248 \text{ mol H}_2\text{SO}_4}{1 \text{ L}} \times \dfrac{2 \text{ mol OH}^-}{1 \text{ mol H}_2\text{SO}_4} \times \dfrac{1}{0.02000 \text{ L}} = 0.366 \text{ M OH}^-$

109) $19.58 \text{ mL} \times \dfrac{0.201 \text{ mmol NaOH}}{1 \text{ mL}} \times \dfrac{1 \text{ mmol NaH}_2\text{PO}_4}{1 \text{ mmol NaOH}} \times \dfrac{120.0 \text{ mg NaH}_2\text{PO}_4}{1 \text{ mmol NaH}_2\text{PO}_4} = 472 \text{ mg NaH}_2\text{PO}_4$

$\dfrac{472 \text{ mg}}{599 \text{ mg}} \times 100 = 78.8\% \text{ NaH}_2\text{PO}_4, \ 21.2\% \text{ NaH}_2\text{PO}_4$

111) At 2 eq/mol, 0.3976 M Na_2CO_3 = 0.7952 N Na_2CO_3 113) At 2 eq/mol, 0.3829 M H_2SO_4 = 0.7658 N H_2SO_4

115) $\dfrac{39.8 \times 0.405}{25.0} = 0.645 \text{ N Na}_2\text{CO}_3$ 117) $\dfrac{42.2 \times 0.402}{50.0} = 0.339 \text{ N H}_2\text{C}_4\text{H}_4\text{O}_6$

119) $\dfrac{33.4 \times 0.196}{25.0} = 0.262 \text{ N H}_3\text{PO}_4$ 121) $\dfrac{1.21 \text{ g}}{0.0307 \times 0.170 \text{ eq}} = 232 \text{ g/eq}$

123) Partial pressure is a colligative property because it depends on the number of particles and is independent of their identity.

125) $\dfrac{27.2 \text{ g C}_6\text{H}_5\text{NH}_2}{0.120 \text{ kg H}_2\text{O}} \times \dfrac{1 \text{ mol C}_6\text{H}_5\text{NH}_2}{93.1 \text{ g C}_6\text{H}_5\text{NH}_2} = 2.43 \text{ m}$ $\Delta T_b = 0.52 \times 2.43 = 1.3°C$ $T_b = 101.3°C$
      $\Delta T_f = 1.86 \times 2.43 = 4.52°C$ $T_f = -4.52°C$

127) $\dfrac{2.12 \text{ g C}_{10}\text{H}_8}{0.0320 \text{ kg B}} \times \dfrac{1 \text{ mol C}_{10}\text{H}_8}{128.2 \text{ g C}_{10}\text{H}_8} = 0.517 \text{ m}$ $\Delta T_f = 5.10 \times 0.517 = 2.64°C$ $T_f = 5.50 - 2.64 = 2.86°C$

129) $\dfrac{16.6 - 14.1}{3.90} = 0.641 \text{ m}$ 131) $\dfrac{0.28}{0.52} = 0.54 \text{ mol/kg H}_2\text{O}$ $\dfrac{16.1 \text{ g unk}}{0.600 \text{ kg H}_2\text{O}} \times \dfrac{1 \text{ kg H}_2\text{O}}{0.54 \text{ mol unk}} = 5.0 \times 10 \text{ g/mol}$

133) $\dfrac{9.6°C}{3.56°C/m} = 2.7 \text{ m}$ $\dfrac{12.4 \text{ g unk}}{0.0900 \text{ kg P}} \times \dfrac{1 \text{ kg P}}{2.7 \text{ mol unk}} = 51 \text{ g/mol}$

135) $\dfrac{11.4 \text{ g C}_2\text{H}_5\text{OH}}{0.200 \text{ kg}} \times \dfrac{1 \text{ mol C}_2\text{H}_5\text{OH}}{46.0 \text{ g C}_2\text{H}_5\text{OH}} = 1.24 \text{ m}$ $K_f = \dfrac{(28.7 - 22.5)°C}{1.24 \text{ m}} = 5.0°C/m$

138) True: a, d, h, j, m. False: b, c, e, f, g, i, k, l.
139) The bubbles are dissolved air (nitrogen, oxygen) that becomes less soluble at higher temperatures.
140) The boiling point rises.

141) a)

	2 KI	+	Pb(NO₃)₂	⟶	PbI₂	+	2 KNO₃
Volume at start, L	0.0600		0.0200				
Molarity, mol/L	0.322		0.530				
Moles at start	0.0193		0.0106				
Moles used (−), produced (+)	−0.0193		−0.00965		+0.00965		+0.0193
Moles at end	0.0		0.0010		0.00965		0.0193
Molar mass, g/mol					461.0		
Grams at end, g					4.45		

b) Total volume = 0.0600 L + 0.0200 L = 0.0800 L $\dfrac{0.0193 \text{ mol KNO}_3}{0.0800 \text{ L}} \times \dfrac{1 \text{ mol K}^+}{1 \text{ mol KNO}_3} = 0.241 \text{ M K}^+$

c) $\dfrac{0.0010 \text{ mol Pb(NO}_3)_2}{0.0800 \text{ L}} \times \dfrac{1 \text{ mol Pb}^{2+}}{1 \text{ mol Pb(NO}_3)_2} = 0.013 \text{ M Pb}^{2+}$

142) A small sample of pure air is a homogeneous mixture, and is therefore a solution. The "atmosphere," even if it was pure air, is a very tall sample that becomes less dense at higher elevations. The atmosphere is therefore not homogeneous, and consequently it is not a solution.
143) No.
144) The density of a solution must be known to convert concentrations based on mass only (percentage, molality) to those based on volume (molarity, normality).

Chapter 16

5) MnCl_2: Mn^{2+} + 2 Cl^-. $(\text{NH}_4)_2\text{SO}_4$: 2 NH_4^+ + SO_4^{2-}. Na_2S: 2 Na^+ + S^{2-}
7) KNO_2: K^+ + NO_2^-. NiSO_4: Ni^{2+} + SO_4^{2-}. K_3PO_4: 3 K^+ + PO_4^{3-}
9) HBr: H^+ + Br^-. $\text{H}_2\text{C}_4\text{H}_4\text{O}_4$: $\text{H}_2\text{C}_4\text{H}_4\text{O}_4$. HNO_3: H^+ + NO_3^-

11) HF: HF. $HC_2H_3O_2$: $HC_2H_3O_2$. $HClO_4$: $H^+ + ClO_4^-$ 13) NR.   15) $Ba(s) + 2\,H^+(aq) \longrightarrow H_2(g) + Ba^{2+}(aq)$
17) NR.   19) $Pb^{2+}(aq) + 2\,I^-(aq) \longrightarrow PbI_2(s)$ 21) NR.
23) $Ag^+(aq) + Br^-(aq) \longrightarrow AgBr(s)$ 25) $Zn^{2+}(aq) + SO_3^{2-}(aq) \longrightarrow ZnSO_3(s)$
27) $Pb^{2+}(aq) + CO_3^{2-}(aq) \longrightarrow PbCO_3(s)$; $Ca^{2+}(aq) + 2\,OH^-(aq) \longrightarrow Ca(OH)_2(s)$
29) $H^+(aq) + NO_2^-(aq) \longrightarrow HNO_2(aq)$ 31) $H^+(aq) + C_3H_5O_3^-(aq) \longrightarrow HC_3H_5O_3(aq)$
33) $H^+ + OH^-(aq) \longrightarrow H_2O(\ell)$ 35) $2\,H^+(aq) + SO_3^{2-}(aq) \longrightarrow H_2O(\ell) + SO_2(aq)$
37) $2\,H^+(aq) + SO_3^{2-}(aq) \longrightarrow H_2O(\ell) + SO_2(aq)$ 39) $Ba^{2+}(aq) + SO_3^{2-}(aq) \longrightarrow BaSO_3(s)$
41) $Cu^{2+}(aq) + 2\,OH^-(aq) \longrightarrow Cu(OH)_2(s)$ 43) $2\,H^+(aq) + MgCO_3(s) \longrightarrow Mg^{2+}(aq) + CO_2(g) + H_2O(\ell)$
45) $2\,H^+(aq) + Pb(OH)_2(s) \longrightarrow Pb^{2+}(aq) + 2\,H_2O(\ell)$ 47) $H_2C_2O_4(s) + 2\,OH^-(aq) \longrightarrow C_2O_4^{2-}(aq) + 2\,H_2O(\ell)$
49) $Ni(s) + 2\,H^+(aq) \longrightarrow Ni^{2+}(aq) + H_2(g)$ 51) $H^+(aq) + HSO_3^-(aq) \longrightarrow SO_2(aq) + 2\,H_2O(\ell)$ 53) NR.
55) $Mg(s) + 2\,H^+(aq) \longrightarrow Mg^{2+}(aq) + H_2(g)$ 57) $2\,H^+(aq) + Ni(OH)_2(s) \longrightarrow Ni^{2+}(aq) + 2\,H_2O(\ell)$
59) $H^+(aq) + F^-(aq) \longrightarrow HF(aq)$ 61) NR.
63) $2\,Li(s) + 2\,H_2O(\ell) \longrightarrow H_2(g) + 2\,Li^+(aq) + 2\,OH^-(aq)$
65) $2\,Al(s) + 3\,Cu^{2+}(aq) \longrightarrow 2\,Al^{3+}(aq) + 3\,Cu(s)$
67) a) $H^+(aq) + HCO_3^-(aq) \longrightarrow CO_2(g) + H_2O(\ell)$
    b) $2\,H^+(aq) + CO_3^{2-}(aq) \longrightarrow CO_2(g) + H_2O(\ell)$
    c) $H^+(aq) + CO_3^{2-}(aq) \longrightarrow HCO_3^-(aq)$
70) True: b, c, h, i, j, k. False: a, d, e, f, g, l, m.

Chapter 17

3) An Arrhenius base is a source of OH^- ions, whereas a Brönsted–Lowry base is a proton receiver. The two are in agreement, as the OH^- ion is an excellent proton receiver. Other substances, however, can also receive protons, so there are other bases according to the Brönsted–Lowry concept.

5) A Lewis base must have an unshared electron pair to donate to a species that can receive the pair in an empty valence orbital, a Lewis acid.

7) BF_3 is a Lewis acid because the empty valence orbital in the boron atom accepts an electron pair from the oxygen atom in $C_2H_5OC_2H_5$, a Lewis base because it donates the electron pair.

9) F^-; HPO_4^{2-}; HNO_2; H_3PO_4

11) Acids: HSO_4^- (forward) and $HC_2O_4^-$ (reverse); bases: $C_2O_4^{2-}$ (forward) and SO_4^{2-} (reverse).

13) HSO_4^- and SO_4^{2-}; $C_2O_4^{2-}$ and $HC_2O_4^-$ 15) HNO_2 and NO_2^-; $C_3H_5O_2^-$ and $HC_3H_5O_2$

17) NH_4^+ and NH_3; HPO_4^{2-} and $H_2PO_4^-$

19) A strong base has a strong attraction for protons, while a weak base has little attraction for protons. Strong bases are at the bottom of the right column in Table 17.1 and weak bases are at the top.

21) HOH, HClO, $HC_2O_4^-$, H_2SO_3 23) CN^-, ClO^-, HSO_3^-, H_2O, Cl^-

25) $HC_3H_5O_2 + PO_4^{3-} \rightleftharpoons C_3H_5O_2^- + HPO_4^{2-}$ Forward.

27) $HSO_4^- + CO_3^{2-} \rightleftharpoons SO_4^{2-} + HCO_3^-$ Forward.

29) $H_2CO_3 + NO_3^- \rightleftharpoons HCO_3^- + HNO_3$ Reverse.

31) $NO_2^- + H_3O^+ \rightleftharpoons HNO_2 + H_2O$ Forward.

33) $HSO_4^- + HC_2O_4^- \rightleftharpoons SO_4^{2-} + H_2C_2O_4$ Reverse.
    $HSO_4^- + HC_2O_4^- \rightleftharpoons H_2SO_4 + C_2O_4^{2-}$ Reverse.

35) The very small value for K_w indicates that water ionizes to a very small extent.

37) An acidic solution has a higher H^+ concentration than OH^- concentration. The solution is therefore acidic: $10^{-5} > 10^{-9}$.

39) 10^{-12} M

	pH	pOH	[H$^+$]	[OH$^-$]
45)	5	9	10^{-5}	10^{-9}
47)	13	1	10^{-13}	10^{-1}
49)	4	10	10^{-4}	10^{-10}
51)	7	7	10^{-7}	10^{-7}
53)	4.40	9.60	4.0×10^{-5}	2.5×10^{-10}
55)	4.06	9.94	8.7×10^{-5}	1.1×10^{-10}
57)	0.55	13.45	0.28	3.6×10^{-14}
59)	6.60	7.40	2.5×10^{-7}	4.0×10^{-8}

62) True: a, b, c, e, f, g, h. False: d, i, j, k, l.   63) $OH^- + NH_3 \longrightarrow H_2O + NH_2^-$ 65) $pCl = 7.126$

66) When a proton is removed from an H_2X species, the single positive charge is being pulled away from a particle with a single minus charge, HX^-. When a proton is removed from a HX^-, the single positive charge is being pulled away from a particle with a double minus charge, X^{2-}. The loss of the second proton is more difficult, so HX^- is a weaker acid than H_2X.

67) There can be no proton transfer without a proton—an H^+ ion.

68) Carbonate ion is a proton acceptor: $H^+ + CO_3^{2-} \longrightarrow HCO_3^-$.

Chapter 18

3) An electrolytic cell is a cell through which a current driven by an external source passes. A galvanic cell—same as a voltaic cell—causes current to flow through an external circuit by electrochemical action.

7) Oxidation: a, c, and d; reduction: b.

9) a, reduction; b, oxidation.   11)

$$2\,Cr \longrightarrow 2\,Cr^{3+} + 6\,e^-$$
$$\underline{3\,Cl_2 + 6\,e^- \longrightarrow 6\,Cl^-}$$
$$3\,Cl_2 + 2\,Cr \longrightarrow 2\,Cr^{3+} + 6\,Cl^-$$

13) The second equation is the oxidation half-reaction equation. $2\,NiOOH + 2\,H_2O + Cd \longrightarrow 2\,Ni(OH)_2 + Cd(OH)_2$

15) a) $+3, -2, +4, +6$.   b) $+3, +5, +6, +5$.

17) a) Bromine reduced from 0 to -1.   b) Lead oxidized from $+2$ to $+4$.   c) Iodine reduced from $+7$ to -1.

19) a) Oxygen reduced from 0 to -2.   b) Nitrogen oxidized from $+4$ to $+5$.   c) Chromium oxidized from $+3$ to $+6$.

21) Chlorine is the oxidizing agent, and bromide ion is the reducing agent.

23) Pb reduces the lead in PbO_2. PbO_2 oxidizes Pb.

25) From Table 18.2, zinc is a stronger reducer than Fe^{2+}. A strong reducer releases electrons to an oxidizer more readily than a weak reducer releases them.

27) $Br_2, Cu^{2+}, Fe^{2+}, Na^+$.   29) $Br_2 + 2\,I^- \longrightarrow 2\,Br^- + I_2$. Forward reaction favored.

31) $2\,H^+ + 2\,Br^- \longrightarrow H_2 + Br_2$. Reverse reaction favored.

33) $2\,NO + 4\,H_2O + 3\,Fe^{2+} \longrightarrow 2\,NO_3^- + 8\,H^+ + 3\,Fe$. Reverse reaction favored.

35) A strong acid releases protons readily; a strong reducer releases electrons readily. A strong base attracts protons strongly; a strong oxidizer attracts electrons strongly.

37)
$$S_2O_3^{2-} + 5\,H_2O \longrightarrow 2\,SO_4^{2-} + 10\,H^+ + 8\,e^-$$
$$\underline{4\,Cl_2 + 8\,e^- \longrightarrow 8\,Cl^-}$$
$$S_2O_3^{2-} + 5\,H_2O + 4\,Cl_2 \longrightarrow 2\,SO_4^{2-} + 10\,H^+ + 8\,Cl^-$$

39)
$$4\,NO_3^- + 8\,H^+ + 4\,e^- \longrightarrow 4\,NO_2 + 4\,H_2O$$
$$\underline{Sn + 3\,H_2O \longrightarrow H_2SnO_3 + 4\,H^+ + 4\,e^-}$$
$$4\,NO_3^- + 8\,H^+ + Sn \longrightarrow 4\,NO_2 + H_2O + H_2SnO_3$$

41)
$$2\,MnO_4^- + 16\,H^+ + 10\,e^- \longrightarrow 2\,Mn^{2+} + 4\,H_2O$$
$$\underline{5\,C_2O_4^{2-} \longrightarrow 10\,CO_2 + 10\,e^-}$$
$$2\,MnO_4^- + 16\,H^+ + 5\,C_2O_4^{2-} \longrightarrow 2\,Mn^{2+} + 4\,H_2O + 10\,CO_2$$

43) $Cr_2O_7^{2-} + 8\,H^+ + 6\,e^- \longrightarrow Cr_2O_3 + 4\,H_2O$
$$\underline{2\,NH_4^+ \longrightarrow N_2 + 8\,H^+ + 6\,e^-}$$
$$Cr_2O_7^{2-} + 2\,NH_4^+ \longrightarrow Cr_2O_3 + 4\,H_2O + N_2$$

45) $4\,NO_3^- + 16\,H^+ + 12\,e^- \longrightarrow 4\,NO + 8\,H_2O$ 48) True: a, c, g. False: b, d, e, f.
$$\underline{3\,As_2O_3 + 15\,H_2O \longrightarrow 6\,AsO_4^{3-} + 30\,H^+ + 12\,e^-}$$
$$3\,As_2O_3 + 7\,H_2O + 4\,NO_3^- \longrightarrow 6\,AsO_4^{3-} + 14\,H^+ + 4\,NO$$

49) This "property of an acid" is more correctly described as the property of an acid (hydrogen ion) acting as an oxidizing agent. The H^+ ion reacts with only those metals whose ions are weaker oxidizing agents, located below hydrogen in Table 18.2.

50) Water is available in large amounts in any aqueous solution, as is H^+ in an acidic solution.

51) In a simple element $\leftrightarrow$ monatomic ion redox reaction the statement is correct. The *element* oxidized or reduced can always be identified by a change in oxidation number. The oxidizing or reducing agent, however, is a *species,* which may be an element, a monatomic ion, or a polyatomic ion, such as MnO_4^-.

Chapter 19

1) In a dynamic equilibrium, opposing changes continue to occur at equal rates. An equilibrium in which nothing is changing, as a book resting on a table, for example, is called a static equilibrium.

3) Both systems can reach equilibrium. At equilibrium, the salt dissolves at the same rate at which it crystallizes. Whether the container is open or closed is of no importance.

5) The system is not an equilibrium because energy must be supplied constantly to keep it in operation. Also, the water is circulating, not moving reversibly in two opposing directions.

9) ΔE is negative. The reaction is exothermic. $\Delta E = c - b$. Activation energy $= a - b$.

11) Both activation energies are positive. Point a is the highest on the curve; $a > b$ and $a > c$. The activation energy for the reverse reaction is $a - c$.

13) An activated complex is an unstable intermediate species formed during a collision of two reacting particles. The properties of an activated complex cannot be described because the complex decomposes almost as soon as it forms.

15) At a higher temperature, a larger fraction of the molecules has enough kinetic energy to engage in a reaction-producing collision, so reaction rates are higher. Also, collisions are more frequent. At low temperature, a smaller fraction of the collisions produce reactions and there are fewer collisions, so the reaction rate is slower.

19) The reaction rate will increase if either concentration is increased. A higher concentration produces more frequent collisions.

21) See Figure 19.7.   23) The reverse reaction rate reaches its maximum at equilibrium.

25) If O_2 concentration is increased, equilibrium will shift in forward direction, the direction in which more O_2 will be consumed.

27) Additional NH_3 will shift the equilibrium forward, the direction in which some of the added NH_3 will be consumed.

29) The equilibrium will shift in the reverse direction to use up some of the added Cu^{2+}.

31) Gas pressure increases as volume is reduced. The equilibrium will shift to relieve the pressure increase by reducing the number of molecules, which is in the forward direction.

33) It will not shift because there is no change in the total number of molecules.

35) If heat is removed the equilibrium will shift in the direction that produces heat, the forward direction.

37) Heat the system. Heat causes reaction to shift in direction that removes heat, the reverse direction. That lowers the SO_3 yield.

39) $Ca(OH)_2(s) \rightleftharpoons Ca^{2+}(aq) + 2\,OH^-(aq)$ is the equilibrium equation. H^+ ions from the acid combine with OH^- ions to form water molecules. This reduces the OH^- ion concentration and causes a forward shift in the equilibrium. The process continues until all the $Ca(OH)_2$ is dissolved.

41) $K = \dfrac{[CO_2][H_2]}{[CO][H_2O]}$ 43) $K = [Zn^{2+}]^3[PO_4{}^{3-}]^2$ 45) $K = \dfrac{[Cu^{2+}][NH_3]^4}{[Cu(NH_3)_4{}^{2+}]}$

47) $K = \dfrac{[H_2O]}{[CO][H_2]}$ 49) $K = \dfrac{[H_3O^+][NO_2{}^-]}{[HNO_2]}$

51) The equilibrium constant expression for the given equation is $K = \dfrac{[NO_2]^2}{[NO]^2[O_2]}$. If the equation is written in reverse, the equilibrium constant expression is inverted. If different sets of coefficients are used, both the expression and its numerical value change. For example:

$$NO + \tfrac{1}{2}O_2 \rightleftharpoons NO_2 \qquad K_1 = \frac{[NO_2]}{[NO][O_2]^{0.5}} \qquad\qquad 4\,NO + 2\,O_2 \rightleftharpoons 4\,NO_2 \qquad K_2 = \frac{[NO_2]^4}{[NO]^4[O_2]^2}$$

The equilibrium constants are not equal: $K_2 = K^2 = K_1{}^4$.

53) The equilibrium will be favored in the forward direction. If K is very large, at least one factor in the denominator must be very small, indicating that at least one reactant has been almost completely consumed.

55) The equilibrium will be favored in the reverse direction. A very small equilibrium constant results when one species on the right side of the equation is very small.

57) (a) The equilibrium is favored in the forward direction. H_2SO_3 is one of the acids that is unstable, decomposing to H_2O and SO_2, as indicated. K is large for this equilibrium. (b) The equilibrium is favored in the forward direction. $HC_2H_3O_2$ is a weak acid. Nearly all of the ions will combine to form the un-ionized molecule.

59) $[Co^{2+}] = 3.7 \times 10^{-6}$; $[OH^-] = 2 \times 3.7 \times 10^{-6} = 7.4 \times 10^{-6}$ $K_{sp} = (3.7 \times 10^{-6})(7.4 \times 10^{-6})^2 = 2.0 \times 10^{-16}$

61) $\dfrac{8.7\ \text{mg Ag}_2\text{CO}_3}{250\ \text{mL}} \times \dfrac{\text{mmol Ag}_2\text{CO}_3}{276\ \text{mg Ag}_2\text{CO}_3} = 1.26 \times 10^{-4} = [CO_3{}^{2-}]$ (2 SF)

$[Ag^+] = 2 \times 1.26 \times 10^{-4} = 2.52 \times 10^{-4}$ (2 SF)

$K_{sp} = (1.26 \times 10^{-4})(2.52 \times 10^{-4})^2 = 8.0 \times 10^{-12}$ $[CO_3{}^{2-}]$ roundoff to 1.3×10^{-4} yields $K_{sp} = 8.8 \times 10^{-12}$

63) $[Ag^+] = [IO_3^-] = s$ $s^2 = 2.0 \times 10^{-8}$ $s = 1.4 \times 10^{-4}$

$0.10 \text{ L} \times \dfrac{1.4 \times 10^{-4} \text{ mol AgIO}_3}{1 \text{ L}} \times \dfrac{283 \text{ g AgIO}_3}{1 \text{ mol AgIO}_3} = 4.0 \times 10^{-3} \text{ g/100 mL}$

65) $[Mn^{2+}] = s$ $[OH^-] = 2 s$ $4 s^3 = 1.0 \times 10^{-13}$ $s = 2.9 \times 10^{-5}$

67) $K_{sp} = [Ca^{2+}][C_2O_4{}^{2-}] = [Ca^{2+}](0.22) = 2.4 \times 10^{-9}$ $[Ca^{2+}] = 1.1 \times 10^{-8}$

69) $[H^+] = 10^{-2.12} = 7.6 \times 10^{-3}$ $K_a = \dfrac{(7.6 \times 10^{-3})^2}{0.22} = 2.6 \times 10^{-4}$ $\dfrac{7.6 \times 10^{-3}}{0.22} \times 100 = 3.5\% \text{ ionized}$

71) $[H^+] = [(1.8 \times 10^{-5})(0.35)]^{1/2} = 2.5 \times 10^{-3}$ $pH = 2.60$

73) $\dfrac{24.0 \text{ g NaC}_2\text{H}_3\text{O}_2}{0.500 \text{ L}} \times \dfrac{1 \text{ mol C}_2\text{H}_3\text{O}_2{}^-}{82.0 \text{ g NaC}_2\text{H}_3\text{O}_2} = 0.585 \text{ M C}_2\text{H}_3\text{O}_2{}^-$ $[H^+] = 1.8 \times 10^{-5} \times \dfrac{0.12}{0.585} = 3.7 \times 10^{-6}$ $pH = 5.43$

75) $\dfrac{[HC_2H_3O_2]}{[C_2H_3O_2{}^-]} = \dfrac{10^{-4.18}}{1.8 \times 10^{-5}} = 3.7$

77) $[CO] \text{ at start} = \dfrac{0.351}{3.00} = 0.117$ $[Cl_2] \text{ at start} = \dfrac{1.340}{3.00} = 0.447$ $[Cl_2] \text{ at end} = \dfrac{1.050}{3.00} = 0.350$

	$CO(g)$	+	$Cl_2(g)$	$\rightleftharpoons$	$COCl_2(g)$
Initial	0.117		0.447		0
Reacting	−0.097		−0.097		0.097
Equilibrium	0.020		0.350		0.097

$K = \dfrac{0.097}{(0.020)(0.350)} = 14$

80) True: b, d, f, g, i, j, l, m, n. False: a, c, e, h, k, o.

81) Kinetic energies are greater at Time 1 because at Time 2 some of that energy has been converted to potential energy of activated complex.

82) Increase [AB], heat the system, and introduce catalyst.

83) a) High pressure to force reaction to the smaller number of gaseous product molecules.

  b) High temperature, at which all reaction rates are faster.

84) A manufacturer cannot use an equilibrium, which is a closed system from which no product can be removed.

85) The higher temperature is used to speed the reaction rate to an acceptable level. Lower pressure is dictated by limits of mechanical design and safety.

86) a) $Ca(OH)_2 \rightleftharpoons Ca^{2+} + 2 OH^-$

  b) (1) Adding a strong base or soluble calcium compound would increase $[OH^-]$ or $[Ca^{2+}]$, respectively, causing a shift in the reverse direction and reducing solubility of $Ca(OH)_2$. (2) Adding an acid to reduce $[OH^-]$ by forming water; adding a cation that will reduce $[OH^-]$ by precipitation; or adding an anion whose calcium salt is less soluble than $Ca(OH)_2$ would cause a shift in the forward direction, increasing the solubility of $Ca(OH)_2$.

  c) (1) Any anion whose calcium salt is less soluble than $Ca(OH)_2$ will cause a forward shift, increasing $[OH^-]$. (2) An acid that will form water with OH^- or a cation that will precipitate OH^- will reduce $[OH^-]$.

87) Add NO_2: R–I–I–D–I. Reduce temperature: F–D–D–I–I. Add N_2: None. Remove NH_3: R–D–I–D–D. Add catalyst: None.

Chapter 20

1) *Nuclide* is a general term that refers to the nucleus of any atom. An isotope is a specific kind of atom that has a specific nuclear composition.

7) The collision of a radioactive emission with an atom or molecule may rearrange the electrons in the target, possibly ionizing it and causing a potentially harmful chemical change.

9) Radiation can be detected by its ability to expose photographic film, ionize a gas, or cause a substance to emit light.

23) 1/32.

25) a) A–C–B. At the end of one half-life, half of the original A would remain, leaving the other half to be divided between B and C. B disintegrates in one day, so more than half of what was produced in days 1–5 has passed along to C, where it is accumulated.

  b) C–A–B. At the end of two half-lives, A is down to ¼ of the starting amount. Most of the ¾ that disintegrated has passed through B to C.

27) $R = 12.9 \text{ g} \times (\frac{1}{2})^3 = 1.6 \text{ g}$

29) From 1994 to 2050 is 56 years, or two half-lives. Hence, $R = 654 \text{ g} \times (\frac{1}{2})^2 = 1.6 \times 10^2 \text{ g}$
From 1994 to 2070 is 76 years, or $76/28 = 2.7$ half-lives. Hence, $R = 654 \text{ g} \times (\frac{1}{2})^{2.7} = 1.0 \times 10^2 \text{ g.}$

31) $R/S = 1.05 \text{ g}/9.53 \text{ g} = 0.110.$ From graph, this is 3.25 half-lives. $83.2 \text{ hr}/3.25$ half-lives $= 25.6 \text{ hr/half-life} = t_{1/2}$

33) a)　12 min/(3.1 min/half-life) $= 3.9$ half-lives. From graph, $R/S = 0.067.$
　　　　$0.067 \times 84.6 \text{ g} = 5.7 \text{ g remain.}$　　By equation, $R = 84.6 \times (\frac{1}{2})^{3.9} = 5.7 \text{ g remain.}$
　　b)　$R/S = 3.48 \text{ g}/84.6 \text{ g} = 0.0411.$ From graph, this is 4.6 half-lives. 4.6 half-lives $\times 3.1$ min/half-life $= 14$ minutes.

35) $R/S = 0.55.$ From graph, this is 0.86 half-life.　　　0.86 half-life $\times 5730$ yr/half-life $= 4.9 \times 10^3$ years.

39) $^{228}_{89}\text{Ac} \longrightarrow {}^{0}_{-1}\text{e} + {}^{228}_{90}\text{Th}$　　$^{212}_{83}\text{Bi} \longrightarrow {}^{0}_{-1}\text{e} + {}^{212}_{84}\text{Po}$　　41) $^{216}_{84}\text{Po} \longrightarrow {}^{4}_{2}\text{He} + {}^{212}_{82}\text{Pb}$　　$^{234}_{92}\text{U} \longrightarrow {}^{4}_{2}\text{He} + {}^{230}_{90}\text{Th}$

43) "Nuclear chemical properties of lead" is meaningless for two reasons. First, the chemical properties of all isotopes of lead are the same. Second, the nuclear properties of lead isotopes are specific for each individual isotope.

45) Emission of a beta particle would change a calcium atom into a scandium atom: $^{47}_{20}\text{Ca} \longrightarrow {}^{0}_{-1}\text{e} + {}^{47}_{21}\text{Sc}.$

47) The count will remain at 5000/minute. The radioactivity of an element is independent of the form of the element, pure element or in a compound.

49) Radioactivity is spontaneous, while nuclear bombardment reactions are produced by projecting a nuclear particle into a target.

51) A particle accelerator is a device that uses electric fields to build up the kinetic energy of charged particles used in nuclear bombardment to produce radionuclides.

55) $^{44}_{21}\text{Sc}$　　　$^{257}_{103}\text{Lr}$　　　$^{1}_{1}\text{H}$

64) True: a, d, e, i, j, m, n. False: b, c, f, g, h, k, l, p. The answer to o is left to you.

65) The mass values used in the calculations must be masses of the radioactive isotope only, or some masses directly proportional to them. If the radioactive isotope is in a compound, there will be a changing mass of the radioactive isotope, a fixed mass of stable isotopes of the same element, and a fixed mass of other elements in the compound, as well as the mass of the decay products in all mass measurements. The amount of radioactive substance can be measured directly with a Geiger counter in the form of disintegrations per second, or some such quantity, as indicated in Problem 12.

67) Presumably it takes an infinite time for all of a sample of radioactive matter to decay.

68) $1 \text{ lb} \times \dfrac{454 \text{ g}}{1 \text{ lb}} \times \dfrac{1 \text{ mol U}}{238 \text{ g U}} \times \dfrac{2.0 \times 10^{10} \text{ kJ}}{1 \text{ mol U}} \times \dfrac{1 \text{ ton coal}}{2.5 \times 10^7 \text{ kJ}} = 1.5 \times 10^3 \text{ tons coal}$

Chapter 21

1) Defining organic chemistry as the chemistry of carbon compounds includes virtually all the substances containing carbon that are found in living systems. Ironically, carbon dioxide, the end product of respiration in living systems, is not included in this definition.

5) A hydrocarbon is a compound that contains only hydrogen and carbon atoms; C_3H_4, C_8H_{10}, and $CH_3CH_2CH_3$ are hydrocarbons.

7) $C_{11}H_{24}$, $C_{21}H_{44}$ The formulas come from the general alkane formula C_nH_{2n+2}

11) C_7H_{16} is the molecular formula; $CH_3CH_2CH_2CH_2CH_2CH_2CH_3$ is the line formula; $CH_3(CH_2)_5CH_3$ is the condensed formula.

The structural formula is
$$\text{H}-\overset{\displaystyle\underset{|}{\overset{|}{\text{H}}}}{\text{C}}-\overset{\displaystyle\underset{|}{\overset{|}{\text{H}}}}{\text{C}}-\overset{\displaystyle\underset{|}{\overset{|}{\text{H}}}}{\text{C}}-\overset{\displaystyle\underset{|}{\overset{|}{\text{H}}}}{\text{C}}-\overset{\displaystyle\underset{|}{\overset{|}{\text{H}}}}{\text{C}}-\overset{\displaystyle\underset{|}{\overset{|}{\text{H}}}}{\text{C}}-\overset{\displaystyle\underset{|}{\overset{|}{\text{H}}}}{\text{C}}-\text{H}$$

17) 2-methyl-4-ethylhexane

19) $\text{C}-\overset{\displaystyle\overset{\text{C}}{|}}{\text{C}}-\overset{\displaystyle\overset{\text{C}}{|}}{\text{C}}-\text{C}-\text{C}$

21)
$$\text{Cl}-\overset{\displaystyle\overset{\text{Cl}}{|}}{\underset{\underset{\text{Cl}}{|}}{\text{C}}}-\overset{\displaystyle\overset{\text{H}}{|}}{\underset{\underset{\text{H}}{|}}{\text{C}}}-\text{H} \quad (1, 1, 1)$$
$$\text{Cl}-\overset{\displaystyle\overset{\text{Cl}}{|}}{\underset{\underset{\text{H}}{|}}{\text{C}}}-\overset{\displaystyle\overset{\text{H}}{|}}{\underset{\underset{\text{H}}{|}}{\text{C}}}-\text{Cl} \quad (1, 1, 2)$$

25)

27)

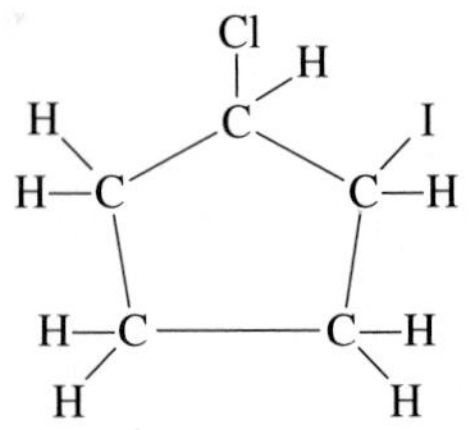

29) 4-ethylheptane   31) 1-bromo-3,3-dichlorobutane   33)

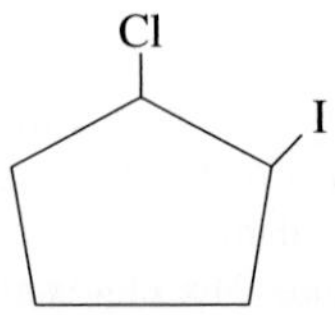

35) 1-chloro-2-ethylcyclohexane

39)

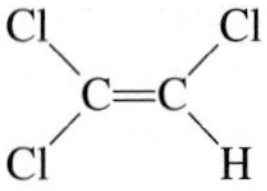

This molecule can have only 2 additional atoms attached to each of its two carbons, so the first 2 chlorines go on one carbon atom. The third chlorine must go on the other carbon atom.

41)

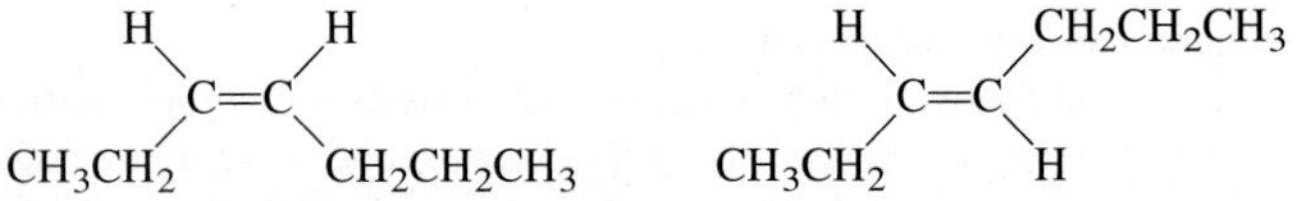

Think of a line drawn in between the two lines that depict a double bond. In *cis*-3-heptene, the hydrogen atoms attached to the double bonded carbons are on the same side of that line. In *trans*-3-heptene, the hydrogen atoms attached to the double bonded carbons are on opposite sides of that line.

43) 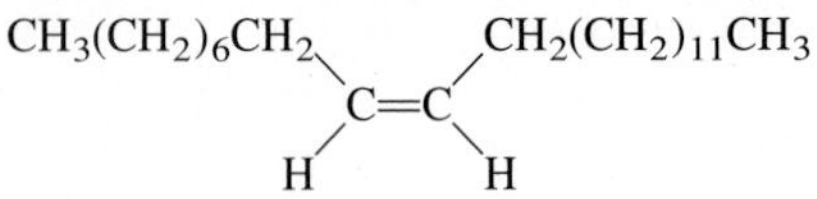 $CH_3(CH_2)_6CH_2$ $CH_2(CH_2)_{11}CH_3$ 45) 2,2-dimethyl-3-heptyne

45) 2,2-dimethyl-3-heptyne

47)

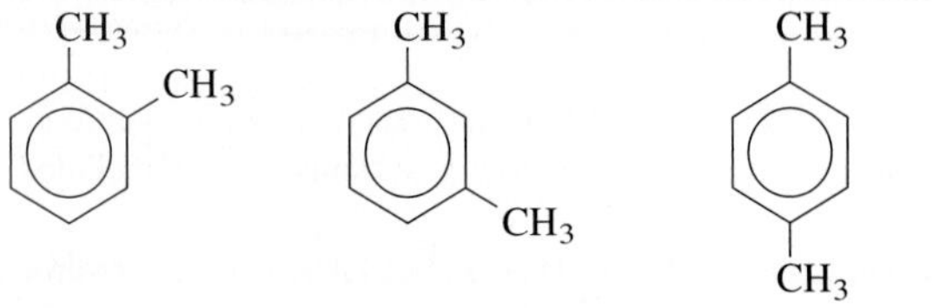

1,2-dimethylbenzene   1,3-dimethylbenzene   1,4-dimethylbenzene

49) 1-bromo-4-chlorobenzene. Because both substituents are halogen atoms, the lower number is given to the first halogen in the alphabet.

51)

1,1-dichloropropane   1,2-dichloropropane   2,2-dichloropropane   1,3-dichloropropane

53)

55) $CH_3—CH{=}CH—CH_3 + H_2 \xrightarrow[\text{catalyst}]{} CH_3—\underset{\underset{H}{|}}{CH}—\underset{\underset{H}{|}}{CH}—CH_3$

Because there is only one straight chain butane molecule, it makes no difference if the starting material is *cis* or *trans*-2-butene.

59) Although ethers have two carbon-oxygen bonds that are polar, the dipoles of these two bonds almost cancel each other by geometry. The forces of attraction in an ether are then weak dipole-dipole. In an alcohol, the hydroxyl proton on one alcohol molecule can hydrogen bond with the lone pair electrons of the oxygen atom on another alcohol molecule. The higher the forces of attraction, the higher the boiling point.

61)
$$H-\overset{\underset{\displaystyle H}{|}}{\underset{\displaystyle H}{C}}-\overset{\underset{\displaystyle H}{|}}{\underset{\displaystyle H}{C}}-\overset{\underset{\displaystyle H}{|}}{\underset{\displaystyle H}{C}}-\overset{\underset{\displaystyle H}{|}}{\underset{\displaystyle H}{C}}-\overset{\underset{\displaystyle H}{|}}{\underset{\displaystyle H}{C}}-\overset{\underset{\displaystyle H}{|}}{\underset{\displaystyle H}{C}}-H$$

(structure with OH on second carbon)

63)
(chain structure: H–C–C–C–C–O–C–C–H with appropriate H substituents)

65)
$$\text{H–C–C–C–OH} + \text{HO–C–C–C–H} \longrightarrow \text{H–C–C–C–O–C–C–C–H} + \text{HOH}$$

67) (two structures: an aldehyde and a ketone)

69)
$$\text{H–C–C–C–H} + \tfrac{1}{2}O_2 \longrightarrow \text{H–C–C–C–H} + H_2O$$

71)
$$\text{H–C–C–C–C–C–C–OH}$$ (carboxylic acid with chain)

73)
$$\text{H–C–C–C–OH} + \text{HO–C–C–H} \longrightarrow \text{H–C–C–C–O–C–C–H} + \text{HOH}$$

The organic reaction product is ethyl propanoate, an ester.

75)
dimethylamine   ethylamine

(structures of dimethylamine and ethylamine)

77) Dimethylamine is a secondary amine; ethylamine is a primary amine.

79)
$$\text{H–C–C–C–OH} + \text{H–N–H} \longrightarrow \text{H–C–C–C–N–H} + \text{HOH}$$

The organic reaction product is propanamide, an amide.

81)
$$\left[\begin{array}{cccccc} \text{H} & \text{H} & \text{H} & \text{H} & \text{H} & \text{H} \\ \text{C–} & \text{C–} & \text{C–} & \text{C–} & \text{C–} & \text{C} \\ \text{H} & \text{Br} & \text{H} & \text{Br} & \text{H} & \text{Br} \end{array} \right]$$

83)
$$\underset{\text{H}\quad\text{CH}_3}{\overset{\text{H}\quad\text{H}}{C=C}}$$

85)
$$\left[\begin{array}{cccccc} \text{Cl} & \text{F} & \text{Cl} & \text{F} & \text{Cl} & \text{F} \\ \text{C–} & \text{C–} & \text{C–} & \text{C–} & \text{C–} & \text{C} \\ \text{F} & \text{F} & \text{F} & \text{F} & \text{F} & \text{F} \end{array} \right]$$

87) (polycarbonate polymer structure)

89) (polyamide / aramid polymer structure)

91)
$$\text{HO–C–}\bigcirc\text{–C–OH} \qquad \text{HO–CH}_2\text{–C–OH}$$ (with CH$_3$)

93)
$$\text{HO–C–CH}_2\text{–CH}_2\text{–CH}_2\text{–CH}_2\text{–CH}_2\text{–NH}_2$$

95) True: b, c, d, g, h, i, j, k, n, o, p, s. False: a, e, f, l, m, q, r.

97) An alkene and a cycloalkane with the same number of carbon atoms have the same molecular formula, C_nH_{2n}. The alkene has a double bond between two carbon atoms and there is no closed loop of carbon atoms. The cycloalkane has only single bonds and the carbon atoms are assembled in a closed ring.

99) Experimentally, there is only one type of carbon-carbon bond in benzene, not two. In addition, benzene does *not* give addition reactions typical of alkenes. The delocalized structure reminds us that benzene is different from both alkanes and alkenes.

101) For an aldehyde to be oxidized to a carboxylic acid, an oxygen atom must be inserted between the carbonyl carbon and the hydrogen bonded to it. If there is no hydrogen atom bonded to the carbonyl atom, as in a ketone, there can be no oxidation. See equation on page 587.

102) The esterification equation is

$$CH_3-\overset{\displaystyle O}{\overset{\|}{C}}-OH + HO\overset{*}{-}CH_3 \longrightarrow CH_3-\overset{\displaystyle O}{\overset{\|}{C}}-O\overset{*}{-}CH_3 + HOH$$

The asterisk on the oxygen atom in methanol identifies the oxygen-18 atom. It is the presence of the radioactive oxygen in the ester product, rather than in the water product, that shows that the water molecule is made up from a hydroxyl from the acid and only a hydrogen atom from the alcohol.

103) Look at the equation describing formation of a peptide linkage.

This condensation reaction requires loss of a water molecule, one hydrogen of which must come from the amine. A tertiary amine has no hydrogens to lose. Trimethylamine (or any tertiary amine) can *not* form an amide.

104) The number of monomers in different polymeric molecules varies, so they have no definite molecular mass.

105) $\{CH{=}CH{-}CH{=}CH\}_n$ $[CH{=}CH{-}CH{=}CH{-}CH{=}CH]_n$ Extend to 6 carbons

106) Because amides are made in an acid-catalyzed condensation reaction, they can be taken apart in the reverse reaction in the presence of an acid catalyst and water. Acid rain gives precisely that combination.

107) 1.8×10^6 g/polymer $\div$ 104 g/molecule = 1.7×10^4 molecules/polymer

Chapter 22

1)

5) V-T-I is valylthreonylisoleucine. I-V-T is isoleucylvalylthreonine. The C terminal amino acid in V-T-I is isoleucine, in I-V-T, threonine.

7)

9) Primary protein structure tells us the order of the atoms, but not their arrangement in space. Secondary protein structure tells us the arrangement in space of a section of a protein chain.

11) The α-helix secondary structure involves hydrogen bonding between the hydrogen attached to the peptide link nitrogen and a peptide link oxygen of an amino acid further down the *same* protein chain.

13) Enzymes are usually proteins.

15) An enzyme substrate is a reactant that the enzyme helps change to product in the enzyme catalyzed reaction.

17) Enzymes, like all catalysts, lower the activation energy of a reaction.

19) When you run a fever, enzyme-catalyzed reactions run faster than at normal body temperature. This may help the body fight off the illness more quickly.

21) Examples of aldose sugars are glucose, ribose, and deoxyribose. Aldose usually refers only to monosaccharides.

23) Find carbon-1 (the only carbon bonded to two oxygen atoms) in both structures. In α-glucose, the OH attached to carbon-1 is vertical, either pointed down or up. In β-glucose, the OH attached to carbon-1 is horizontal, or nearly so.

25) Sucrose is a disaccharide, glycogen a polysaccharide, fructose a monosaccharide.

27) The simple sugars in lactose are galactose and glucose.

29) Glucose, ribose, deoxyribose and lactose would give a positive Benedict's test, because all these sugars can have an aldehyde group in the open chain form.

31)

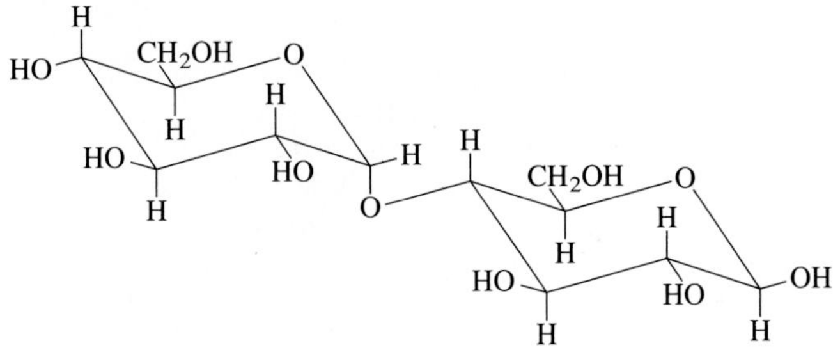

33) Starch has $\alpha 1{\rightarrow}4$ bonds; cellulose has $\beta 1{\rightarrow}4$ bonds. Our enzymes can break the $\alpha 1{\rightarrow}4$ bonds, but not the $\beta 1{\rightarrow}4$ bonds. That's why fibrous vegetables are the dieter's friends. No digestion, no calories.

39)

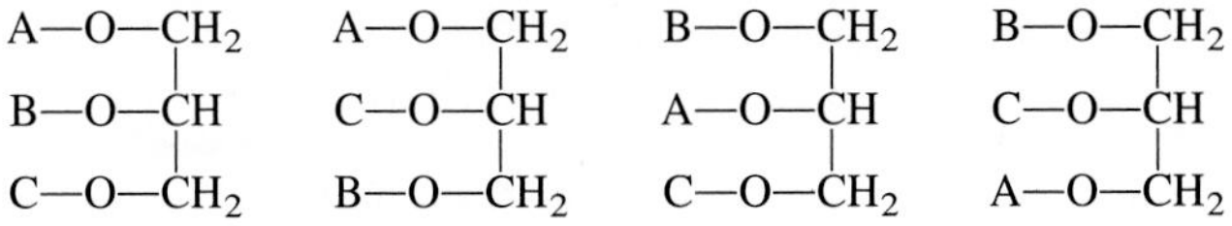

41)

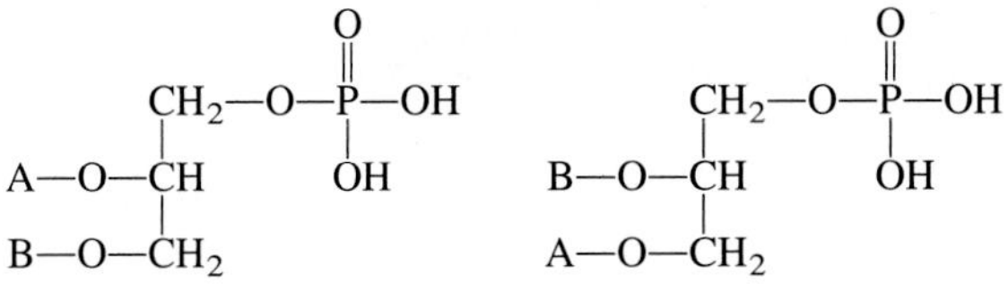

45)

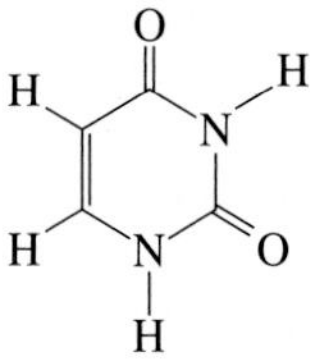

47)

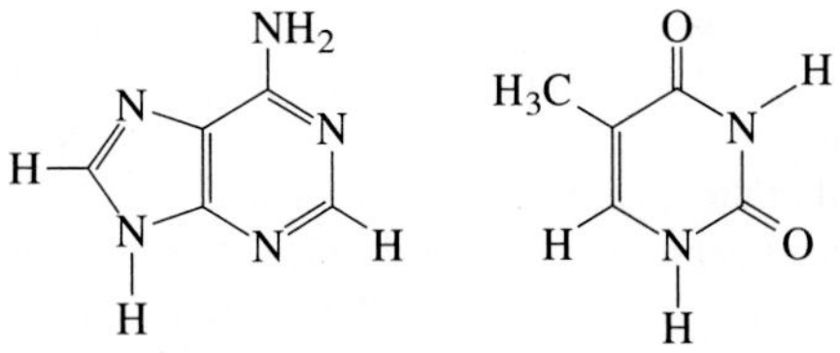

49)

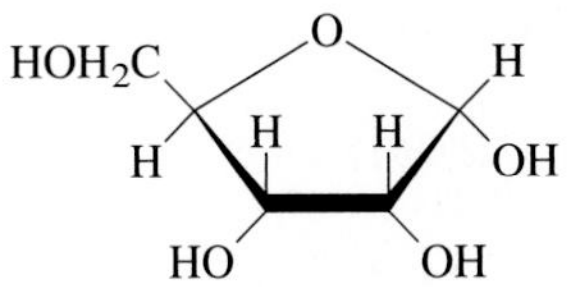

51)

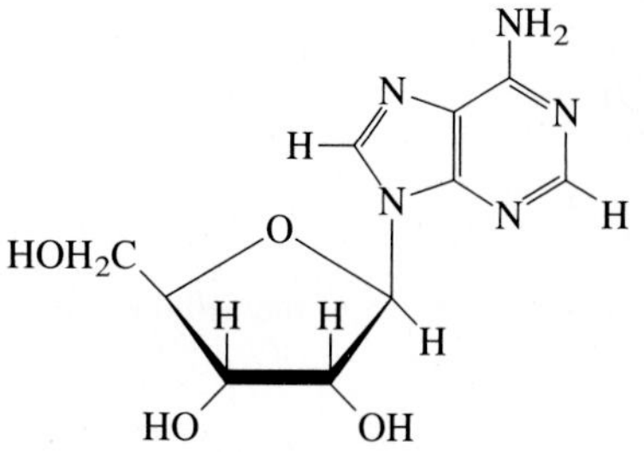

53)

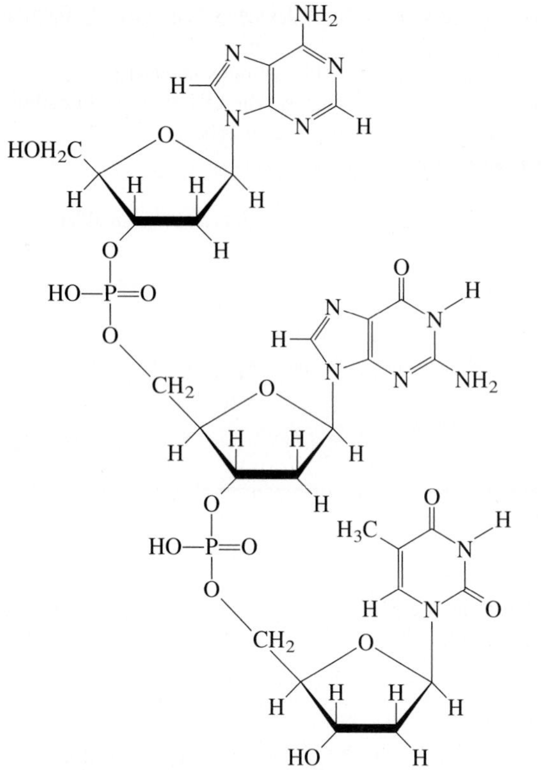

55)

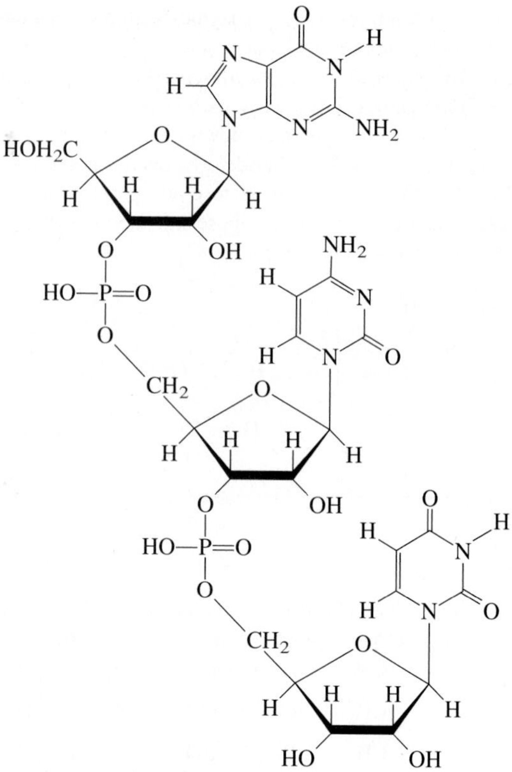

57) Transfer RNA picks up an amino acid molecule and carries it to a protein being synthesized by a ribosome.

60) $\dfrac{64{,}500 \text{ g hemoglobin}}{1 \text{ mole hemoglobin}} \times \dfrac{1 \text{ mole amino acid}}{120 \text{ g amino acid}} = 538 \dfrac{\text{mole amino acid}}{\text{mole hemoglobin}}$

61) 3.8×10^7 g/protein $\div$ 120 g/amino acid $= 3.2 \times 10^5$ amino acids/protein

62) All of these are polymers.

63) Glucose; amino acids; nucleotides (adenine, cytosine, guanine, thymine); glucose; nucleotides (adenine, cytosine, guanine, uracil)

64) Nitrogen is found in all proteins; if you picked sulfur (from cysteine or methionine), that's also true.

65) Phosphorus.

66) The coil of the α-helix, and the strong hydrogen bonding within the protein chains keep water from soaking into the nylon; there is no further hydrogen bonding to be made by the water to the nylon. In cotton, however, there is little hydrogen bonding between the cellulose chains, so water molecules can form hydrogen bonds with (and therefore soak into) the hydroxyl groups on the sugars that make up cellulose.

67)

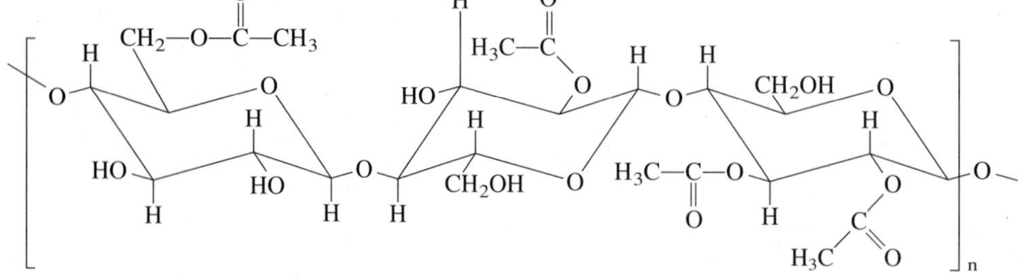

In rayon, the hydroxyl groups of the cellulose react randomly with acetic acid to form rayon.

68) Because guanine and cytosine are complementary base pairs in DNA, there must also be 21% guanine. If guanine + cytosine are 42%, then adenine and thymine must equal 58%. Adenine is then 29%, as is thymine.

69) Because RNA is single stranded, there is no complementary RNA strand. As a result, the percentage guanine has no relationship to the percentage cytosine.

Glossary

...eoretically (absolutely) ...ual to $-273.15°C$ or ...ly stops.

...ydronium) ions in aque- ...stance that donates pro- ...owry definition); a sub- ...epting a pair of electrons

...which the hydrogen-ion ...xide-ion concentration; a ...7.

...03 (Lr).

...olecular species presumed ...ollision) of reacting mole-

...that must be overcome to

...ing of an alkyl group and at ...e general formula ROH.

...a carbonyl group bonded to ...gen, alkyl, or aryl group on ...la RCHO.

...alkene, or alkyne.

...IA of the periodic table.

alkaline basic, ... than 7.

alkaline earth metal a metal from Group 2A of the periodic table.

alkane a saturated hydrocarbon containing only single bonds, in which each carbon atom is bonded to four other atoms.

alkene an unsaturated hydrocarbon containing a double bond, in which each carbon atom that is double bonded is bonded to a maximum of three atoms.

alkyl group an alkane hydrocarbon group lacking one hydrogen atom, having the general formula C_nH_{2n+1}, and frequently symbolized by the letter R.

alkyne an unsaturated hydrocarbon containing a triple bond, in which each carbon atom that is triple bonded is bonded to a total of two atoms.

alpha (α) particle the nucleus of a helium atom, often emitted in nuclear disintegration.

amide a derivative of a carboxylic acid in which the hydroxyl group is replaced by a $-NH_2$ group, and having the general formula $RCONH_2$.

amine an ammonia derivative in which one or more hydrogens are replaced by an alkyl group.

amino acid carboxylic acid containing both an amine group and a variable alkyl group R. Amino acids of the general form $RCH(NH_2)COOH$ can form amide bonds with each other to form peptides and proteins.

amorphous a substance that is without definite structure or shape.

amphiprotic, amphoteric a substance that can act as an acid or a base.

angstrom a length unit equal to 10^{-10} m.

anhydride (anhydrous) a substance that is without water or from which water has been removed.

anion a negatively charged ion.

anode the electrode at which oxidation occurs in an electro-chemical cell.

aqueous pertaining to water.

aromatic hydrocarbon a hydrocarbon containing a benzene ring.

atmosphere (pressure unit) a unit of pressure based on atmospheric pressure at sea level and capable of supporting a mercury column 760 mm high.

atom the smallest particle of an element that can combine with atoms of other elements in forming chemical compounds.

atomic mass the average mass of the atoms of an element compared to an atom of carbon-12 at exactly 12 atomic mass units. Also called *atomic weight*.

atomic mass unit (amu) a unit of mass that is exactly $\frac{1}{12}$ of the mass of an atom of carbon-12.

atomic number (Z) the number of protons in an atom of an element.

atomic weight *see atomic mass.*

Avogadro's number the number of carbon atoms in exactly 12 grams of carbon-12; the number of units in 1 mole (6.02×10^{23}).

barometer laboratory device for measuring atmospheric pressure.

base a substance that yields hydroxide ions in aqueous solution (Arrhenius definition); a substance that accepts protons in chemical reaction (Brönsted–Lowry definition); a substance that forms covalent bonds by donating a pair of electrons (Lewis definition).

basic solution an aqueous solution in which the hydroxide ion concentration is greater than the hydrogen ion concentration; a solution in which the pH is greater than 7.

beta (β) particle a high-energy electron, often emitted in nuclear disintegration.

binary compound a compound consisting of two elements.

boiling point the temperature at which vapor pressure becomes equal to the pressure above a liquid; the temperature at which vapor bubbles form spontaneously anyplace within a liquid.

boiling point elevation the difference between the boiling point of a solution and the boiling point of the pure solvent.

bombardment (nuclear) the striking of a target nucleus by an atomic particle, causing a nuclear change.

bond *see chemical bond.*

bond angle the angle formed by the bonds between two atoms that are bonded to a common central atom.

bonding electrons the electrons transferred or shared in forming chemical bonds; valence electrons.

buffer a solution that resists a change in pH.

calorie a unit of heat equal to 4.184 joules.

calorimeter laboratory device for measuring heat flow.

carbohydrate class of organic compounds consisting mainly of polyhydroxy aldehydes or ketones. Carbohydrates form the supporting tissue of plants and serve as food for animals and people.

carbonyl group an organic functional group, $\diagdown$C$=$O, characteristic of aldehydes and ketones.

carboxyl group an organic functional group. $-$C$\diagup^{\text{O}}_{\diagdown\text{O}-\text{H}}$, characteristic of carboxylic acids.

carboxylic acid an organic acid containing the carboxyl group, having the general formula RCOOH.

catalyst a substance that increases the rate of a chemical reaction by lowering activation energy. The catalyst is either a nonparticipant in the reaction, or it is regenerated. *See inhibitor.*

cathode the negative electrode in a cathode ray tube; the electrode at which reduction occurs in an electrochemical cell.

cation a positively charged ion.

cell, electrolytic a cell in which electrolysis occurs as a result of an externally applied electrical potential.

cell, galvanic *see cell, voltaic.*

cell, voltaic a cell in which an electrical potential is developed by a spontaneous chemical change. Also called a *galvanic cell.*

chain reaction a reaction that has as a product one of its own reactants; that product becomes a reactant, thereby allowing the original reaction to continue.

charge cloud *see electron cloud.*

chemical bond a general term that sometimes includes all of the electrostatic attractions among atoms, molecules, and ions, but more often refers to covalent and ionic bonds. *See covalent bond, ionic bond.*

chemical change a change in which one or more substances disappear and one or more new substances are formed.

chemical family a group of elements having similar chemical properties because of similar valence electron configuration, appearing in the same column of the periodic table.

chemical properties the types of chemical change a substance is able to experience.

cloud chamber a device in which condensation tracks form behind radioactive emissions as they travel through a supersaturated vapor.

colligative properties physical properties of mixtures that depend on concentration of particles irrespective of their identity.

colloid a nonsettling dispersion of aggregated ions or molecules intermediate in size between the particles in a true solution and those in a suspension.

combustion the process of burning.

compound a pure substance that can be broken down into two or more other pure substances by a chemical change.

concentrated adjective for a solution with a relatively large amount of solute per given quantity of solvent or solution.

condense to change from a vapor to a liquid or solid.

condensation the act of condensing.

conjugate acid–base pair a Brönsted–Lowry acid and the base derived from it when it loses a proton; or a Brönsted–Lowry base and the acid developed from it when it accepts a proton.

coordinate covalent bond a bond in which both bonding electrons are furnished by only one of the bonded atoms.

coulomb a unit of electrical charge.

covalent bond the chemical bond between two atoms that share a pair of electrons.

crystalline solid a solid in which the ions and/or molecules are arranged in a definite geometric pattern.

decompose to change chemically into simpler substances.

density the mass of a substance per unit volume.

deoxyribonucleic acid (DNA) large nucleotide *(q.v.)* polymer found in cell nucleus. DNA contains genetic information and controls protein synthesis.

diatomic that which has two atoms.

dilute adjective for a solution with a relatively small amount of solute per given quantity of solvent or solution.

dipole a polar molecule.

diprotic acid an acid capable of yielding two protons per molecule in complete ionization.

dispersion forces weak electrical attractions between molecules, temporarily produced by the shifting of electrons within molecules.

dissolve to pass into solution.

distillation the process of separating components of a mixture by boiling off and condensing the more volatile component.

dynamic equilibrium a state in which opposing changes occur at equal rates, resulting in zero net change over a period of time.

electrode a conductor by which electric charge enters or leaves an electrolyte.

electrolysis the passage of electric charge through an electrolyte.

electrolyte a substance that, when dissolved, yields a solution that conducts electricity; a solution or other medium that conducts electricity by ionic movement.

electrolytic cell *see cell, electrolytic.*

electron subatomic particle carrying a unit negative charge and having a mass of 9.1×10^{-28} gram, or 1/1837 of the mass of a hydrogen nucleus, found outside the nucleus of the atom.

electron (charge) cloud the region of space around or between atoms that is occupied by electrons.

electron configuration the orbital arrangement of electrons in ions or atoms.

electron-dot diagram (structure) *see Lewis diagram.*

electronegativity a scale of the relative ability of an atom of one element to attract the electron pair that forms a single covalent bond with an atom of another element.

electron orbit the circular or elliptical path supposedly followed by an electron around an atomic nucleus, according to the Bohr theory of the atom.

electron orbital a mathematically described region in space within an atom in which there is a high probability that an electron will be found.

electron-pair geometry a description of the distribution of bonding and unshared electron pairs around a bonded atom.

electron-pair repulsion the principle that electron-pair geometry is the result of repulsion between electron pairs around a bonded atom, causing them to be as far apart as possible.

electrostatic force the force of attraction or repulsion between electrically charged objects.

element a pure substance that cannot be decomposed into other pure substances by ordinary chemical means.

empirical formula a formula that represents the lowest integral ratio of atoms of the elements in a compound.

endothermic a change that absorbs energy from the surroundings, having a positive ΔH, an increase in enthalpy.

energy the ability to do work.

enthalpy the heat content of a chemical system.

enthalpy of reaction *see heat of reaction.*

equilibrium *see dynamic equilibrium.*

equilibrium constant with reference to an equilibrium equation, the ratio in which the numerator is the product of concentrations of the species on the right-hand side of the equation, each raised to a power corresponding to its coefficient in the equation, and the denominator is the corresponding product of the species on the left side of the equation; symbol: K, K_c, or K_{eq}.

equilibrium vapor pressure *see vapor pressure.*

equivalent the quantity of an acid (or base) that yields or reacts with one mole of H^+ (or OH^-) in a chemical reaction; the quantity of a substance that gains or loses one mole of electrons in a redox reaction.

ester an organic compound formed by the reaction between a carboxylic acid and an alcohol, having the general formula $R-CO-OR'$.

ether an organic compound in which two alkyl groups are bonded to the same oxygen, having the general formula $R-O-R'$.

excited state the state of an atom in which one or more electrons have absorbed energy—become "excited"—to raise them to energy levels above ground state.

exothermic reaction a reaction that gives off energy to its surroundings.

family *see chemical family.*

fat ester formed from glycerol and three fatty acids. Fats are solids at room temperature. Also called *triacylglycerols or triglycerides.*

fatty acid long-chain carboxylic acid, typically having between 10 and 24 carbon atoms.

fission a nuclear reaction in which a large nucleus splits into two smaller nuclei.

formula, chemical a combination of chemical symbols and subscript numbers that represents the elements in a pure substance and the ratio in which the atoms of the different elements appear.

formula mass (weight) the mass in amu of one formula unit of a substance; the molar mass of formula units of a substance.

formula unit a real (molecular) or hypothetical (ionic) unit particle represented by a chemical formula.

fractional distillation the separation of a mixture into fractions whose components boil over a given temperature range.

freezing point depression the difference between the freezing point of a solution and the freezing point of the pure solvent.

fusion the process of melting; also, a nuclear reaction in which two small nuclei combine to form a larger nucleus.

galvanic cell *see cell, voltaic.*

gamma (γ) ray a high-energy photon emission in radioactive disintegration.

Geiger counter an electrical device for detecting and measuring the intensity of radioactive emission.

ground state the state of an atom in which all electrons occupy the lowest possible energy levels.

group (periodic table) the elements comprising a vertical column in the periodic table.

half-life ($t_{1/2}$) the time required for the disintegration of one half of the radioactive atoms in a sample.

half-reaction the oxidation or reduction half of an oxidation–reduction reaction.

halide ion F^-, Cl^-, Br^-, or I^-.

halogen the name of the chemical family consisting of fluorine, chlorine, bromine, and iodine; any member of the halogen family.

heat of fusion (solidification) the heat flow when one gram of a substance changes between a solid and a liquid at constant pressure and temperature. *See also molar heat of fusion (solidification).*

heat of reaction change of enthalpy in a chemical reaction.

heat of vaporization (condensation) the heat flow when one gram of a substance changes between a liquid and a vapor at constant pressure and temperature.

heterogeneous having a nonuniform composition, usually with visibly different parts or phases.

homogeneous having a uniform appearance and uniform properties throughout.

homologous series a series of compounds in which each member differs from the one next to it by the same structural unit.

hydrate a crystalline solid that contains water of hydration.

hydrocarbon an organic compound consisting of carbon and hydrogen.

hydrogen bond an intermolecular bond (attraction) between a hydrogen atom in one molecule and a highly electronegative atom (fluorine, oxygen, or nitrogen) of another polar molecule; the polar molecule may be of the same substance containing the hydrogen, or of a different substance.

hydronium ion a hydrated hydrogen ion, H_3O^+.

hydroxyl group an organic functional group, —OH, characteristic of alcohols.

ideal gas a hypothetical gas that behaves according to the ideal gas model over all ranges of temperature and pressure.

ideal gas equation the equation $PV = nRT$ that relates quantitatively the pressure, volume, quantity, and temperature of an ideal gas.

immiscible insoluble (usually used only in reference to liquids).

indicator a substance that changes from one color to another, used to signal the end of a titration.

inhibitor a substance added to a chemical reaction to retard its rate; sometimes called a negative catalyst.

ion an atom or group of covalently bonded atoms that is electrically charged because of an excess or deficiency of electrons.

ion-combination reaction when two solutions are combined, the formation of a precipitate or molecular compound by a cation from one solution and an anion from the second solution.

ionic bond the chemical bond arising from the attraction forces between oppositely charged ions in an ionic compound.

ionic compound a compound in which ions are held by ionic bonds.

ionic equation a chemical equation in which dissociated compounds are shown in ionic form.

ionization the formation of an ion from a molecule or atom.

ionization energy the energy required to remove an electron from an atom or ion.

isoelectronic having the same electron configuration.

isomers two compounds having the same molecular formulas but different structural formulas and different physical and chemical properties.

isotopes two or more atoms of the same element that have different atomic masses because of different numbers of neutrons.

IUPAC International Union of Pure and Applied Chemistry.

joule the SI energy unit, defined as a force of one newton applied over a distance of one meter; 1 joule = 0.239 calorie.

K the symbol for the kelvin, the absolute temperature unit; the symbol for an equilibrium constant. K_a is the constant for the ionization of a weak acid; K_{sp} is the constant for the equilibrium between a slightly soluble ionic compound and a saturated solution of its ions; K_w is the constant for the ionization of water.

Kelvin temperature scale an absolute temperature scale on which the degrees are the same size as Celsius degrees, with 0 K at absolute zero, or $-273.15°C$.

ketone a compound consisting of a carbonyl group bonded on each side to an alkyl group, having the general formula R—CO—R′.

kinetic energy energy of motion; translational kinetic energy is equal to $\frac{1}{2}$ mass $\times$ (velocity)^2.

kinetic molecular theory the general theory that all matter consists of particles in constant motion, with different degrees of freedom distinguishing among solids, liquids, and gases.

kinetic theory of gases the portion of the kinetic molecular theory that describes gases and from which the model of an ideal gas is developed.

lanthanides elements 58 (Ce) through 71 (Lu).

Le Chatelier's Principle if an equilibrium system is subjected to a change, processes occur that tend to counteract partially the initial change, thereby bringing the system to a new position of equilibrium.

Lewis diagram, structure, or **symbol** a diagram representing the valence electrons and covalent bonds in an atomic or molecular species.

limiting reagent the reactant first totally consumed in a reaction, thereby determining the maximum yield possible.

line spectrum the spectral lines that appear when light emitted from a sample is analyzed in a spectroscope.

macromolecular crystal a crystal made up of a large but indefinite number of atoms covalently bonded to each other to form a huge molecule. Also called a *network solid*.

manometer a laboratory device for measuring gas pressure.

mass a property reflecting the quantity of matter in a sample.

mass number the total number of protons plus neutrons in the nucleus of an atom.

mass spectrometer a laboratory device in which a flow of gaseous ions may be analyzed in regard to their charge and/or mass.

matter that which occupies space and has mass.

metal a substance that possesses metallic properties, such as luster, ductility, malleability, good conductivity of heat, and electricity; an element that loses electrons to form monatomic cations.

miscible soluble (usually used only in reference to liquids).

mixture a sample of matter containing two or more pure substances.

molality solution concentration expressed in moles of solute per kilogram of solvent.

molar heat of fusion (solidification) the heat flow when one mole of a substance changes between a solid and a liquid at constant temperature and pressure.

molar heat of vaporization (condensation) the heat flow when one mole of a substance changes between a liquid and a vapor at constant temperature and pressure.

molarity solution concentration expressed in moles of solute per liter of solution.

molar mass (weight) the mass of one mole of any substance.

molar volume the volume occupied by one mole, usually of a gas.

mole that quantity of any species that contains the same number of units as the number of atoms in exactly 12 grams of carbon-12.

molecular compound a compound whose fundamental particles are molecules rather than ions.

molecular crystal a molecular solid in which the molecules are arranged according to a definite geometric pattern.

molecular geometry a description of the shape of a molecule.

molecular mass (weight) the number that expresses the average mass of the molecules of a compound compared to the mass of an atom of carbon-12 at a value of exactly 12; the average mass of the molecules of a compound expressed in atomic mass units.

molecule the smallest unit particle of a pure substance that can exist independently and possess the identity of the substance.

monatomic that which has only one atom.

monomer the individual chemical structural unit from which a polymer may be developed.

monoprotic acid an acid capable of yielding one proton per molecule in complete ionization.

negative catalyst *see inhibitor.*

net ionic equation an ionic equation from which all spectators have been removed.

network solid a crystal made up of a large but indefinite number of atoms covalently bonded to each other to form a huge molecule. Also called a *macromolecular crystal.*

neutralization the reaction between an acid and a base to form a salt and water; any reaction between an acid and a base.

neutron an electrically neutral subatomic particle having a mass of 1.7×10^{-24} gram, approximately equal to the mass of a proton, or 1 atomic mass unit, found in the nucleus of the atom.

newton SI unit of force, equal to $1 \text{ kg} \cdot \text{m}^2/\text{sec}^2$.

noble gas the name of the chemical family of relatively unreactive elemental gases appearing in Group 0 of the periodic table.

nonelectrolyte a substance that, when dissolved, yields a solution that is a nonconductor of electricity; a solution or other fluid that does not conduct electricity by ionic movement.

nonpolar pertaining to a bond or molecule having a symmetrical distribution of electric charge.

normal boiling point the temperature at which a substance boils in an open vessel at one atmosphere pressure.

normality solution concentration in equivalents per liter.

nucleotide compound consisting of a nitrogen-containing base, a sugar, and one or more phosphate groups. Nucleotides are the monomers for the polymers DNA and RNA. DNA uses the sugar deoxyribose; RNA uses the sugar ribose.

nucleus the extremely dense central portion of the atom that contains the neutrons and protons that constitute nearly all the mass of the atom and all of the positive charge.

octet rule the general rule that atoms tend to form stable bonds by sharing or transferring electrons until the atom is surrounded by a total of eight electrons.

oil ester formed from glycerol and three fatty acids. Oils are liquids at room temperature. Also called *triacylglycerols* or *triglycerides.*

orbit *see electron orbit.*

orbital *see electron orbital.*

organic chemistry the chemistry of carbon compounds other than carbonates, cyanides, carbon monoxide, and carbon dioxide.

oxidation chemical reaction with oxygen; a chemical change in which the oxidation number (state) of an element is increased; also, the loss of electrons in a redox reaction.

oxidation number a number assigned to each element in a compound, ion, or elemental species by an arbitrary set of rules. Its two main functions are to organize and simplify the study of oxidation–reduction reactions and to serve as a base for one branch of chemical nomenclature.

oxidation state *see oxidation number.*

oxidizer, oxidizing agent the substance that takes electrons from another species, thereby oxidizing it.

oxyacid an acid that contains oxygen.

oxyanion an anion that contains oxygen.

partial pressure the pressure one component of a mixture of gases would exert if it alone occupied the same volume as the mixture at the same temperature.

Pauli exclusion principle the principle that says, in effect, that no more than two electrons can occupy the same orbital.

peptide amino acid polymer typically containing 40 or fewer amino acids. Also called polypeptide.

period (periodic table) a horizontal row of the periodic table.

pH a way of expressing hydrogen-ion concentration; the negative of the logarithm of the hydrogen-ion concentration.

phase a visibly distinct part of a heterogeneous sample of matter.

physical change a change in the physical form of a substance without changing its chemical identity.

physical properties properties of a substance that can be observed and measured without changing the substance chemically.

pOH a way of expressing hydroxide ion concentration; the negative logarithm of the hydroxide ion concentration.

polar pertaining to a bond or molecule having an unsymmetrical distribution of electric charge.

polyatomic pertaining to a species consisting of more than one atom; usually said of polyatomic ions.

polymer a chemical compound formed by bonding two or more monomers. Polymers with molar mass > 5,000 are also called *macromolecules.*

polymerization the reaction in which monomers combine to form polymers.

polyprotic acid an acid capable of yielding more than one proton per molecule on complete ionization.

potential energy energy possessed by a body by virtue of its position in an attractive and/or repulsive force field.

precipitate a solid that forms when two solutions are mixed.

pressure force per unit area.

principal energy level(s) the main energy levels within the electron arrangement in an atom. They are quantized by a set of integers beginning at $n = 1$ for the lowest level, $n = 2$ for the next, and so forth; also called the principal quantum number.

protein amino acid polymer typically containing more than 40 amino acids.

proton a subatomic particle carrying a unit positive charge and having a mass of 1.7×10^{-24} gram, almost the same as the mass of a neutron, found in the nucleus of the atom.

pure substance a sample consisting of only one kind of matter, either compound or element.

quantization of energy the existence of certain discrete electron energy levels within an atom such that electrons may have any one of these energies but no energy between two such levels.

quantum mechanical model of the atom an atomic concept that recognizes four quantum numbers by which electron energy levels may be described.

R a symbol used to designate any alkyl group; the ideal gas constant, having a value of $0.0821 \, L \cdot atm/mol \cdot K$.

radioactivity spontaneous emission of rays and/or particles from an atomic nucleus.

redox a term coined from REDuction–OXidation to refer to oxidation–reduction reactions.

reducer, reducing agent the substance that loses electrons to another species, thereby reducing it.

reduction a chemical change in which the oxidation number (state) of an element is reduced; also, the gain of electrons in a redox reaction.

reversible reaction a chemical reaction in which the products may react to re-form the original reactants.

ribonucleic acid (RNA) nucleotide *(q.v.)* polymer found in the nucleus and other parts of the cell.

salt the product of a neutralization reaction other than water; an ionic compound containing neither the hydrogen ion, H^+, oxide ion, O^{2-}, nor hydroxide ion, OH^-.

saturated hydrocarbon a hydrocarbon that contains only single bonds, in which each carbon atom is bonded to four other atoms.

saturated solution a solution of such concentration that it is or would be in a state of equilibrium with excess solute present.

significant figures the digits in a measurement that are known to be accurate plus one doubtful digit.

SI unit a unit associated with the International System of Units.

solubility the quantity of solute that will dissolve in a given quantity of solvent or in a given quantity of solution, at a specified temperature, to establish an equilibrium between the solution and excess solute; frequently expressed in grams of solute per 100 grams of solvent.

solubility product constant *under K, see K_{sp}.*

soluble a substance that will dissolve in a suitable solvent.

solute the substance dissolved in the solvent; sometimes not clearly distinguishable from the solvent (see below), but usually the lesser of the two.

solution a homogeneous mixture of two or more substances of molecular or ionic particle size, the concentration of which may be varied, usually within certain limits.

solution inventory a precise identification of the chemical species present in a solution, in contrast with the solute from which they may have come; that is, sodium ions and chloride ions, rather than sodium chloride.

solvent the medium in which the solute is dissolved; *see solute.*

specific gravity the ratio of the density of a substance to the density of some standard, usually water at 4°C.

specific heat the quantity of heat required to raise the temperature of one gram of a substance one degree Celsius.

spectator (ion) a species present at the scene of a reaction but not a participant in it.

spectrometer a laboratory instrument used to analyze spectra.

spectrum (*plural:* **spectra**) the result of a dispersion of a beam of light into its component colors; also the result of a dispersion of a beam of gaseous ions into its component particles, distinguished by mass and electric charge.

spontaneous a change that appears to take place by itself without outside influence.

stable that which does not change spontaneously.

standard temperature and pressure (STP) arbitrarily defined conditions of temperature (0°C) and pressure (1 atmosphere) at which gas volumes and quantities are frequently measured and/or compared.

stoichiometry the quantitative relationships between the substances involved in a chemical reaction, established by the equation for the reaction.

STP abbreviation for standard temperature and pressure (see above).

strong acid an acid that ionizes almost completely in aqueous solution; an acid that loses its protons readily.

strong base a base that dissociates almost completely in aqueous solution; a base that has a strong attraction for protons.

strong electrolyte a substance that, when dissolved, yields a solution that is a good conductor of electricity because of nearly complete ionization or dissociation.

strong oxidizer (oxidizing agent) an oxidizer that has a strong attraction for electrons.

strong reducer (reducing agent) a reducer that releases electrons readily.

sublevel the levels into which the principal energy levels are divided according to the quantum mechanical model of the atom; usually specified *s*, *p*, *d*, and *f*.

supersaturated a state of solution concentration that is greater than the equilibrium concentration (solubility) at a given temperature and/or pressure.

suspension a mixture that gradually separates by settling.

tetrahedral related to a tetrahedron; usually used in reference to the orientation of four covalent bonds radiating from a central atom toward the vertices of a tetrahedron, or to the 109°28′ angle formed by any two corners of the tetrahedron and the central atom as its vertex.

tetrahedron a regular four-sided solid, having congruent equilateral triangles as its four faces.

thermal having to do with heat.

thermochemical equation a chemical equation that includes an energy term, or for which ΔH is indicated.

thermochemical stoichiometry stoichiometry expanded to include the energy involved in a chemical reaction, as defined by the thermochemical equation.

titration the controlled and measured addition of one solution into another.

torr a unit of pressure equal to the pressure unit millimeter of mercury.

transition element; transition metal an element from one of the B groups or Group 8 (IUPAC Groups 8–10) of the periodic table.

transmutation conversion of an atom from one element to another by means of a nuclear change.

transuranium elements man-made elements whose atomic numbers are greater than 92.

triprotic acid an acid capable of yielding three protons in complete ionization.

unsaturated hydrocarbon a hydrocarbon that contains one or more multiple bonds.

valence electrons the highest energy *s* and *p* electrons in an atom that determines the bonding characteristics of an element.

van der Waals forces a general term for all kinds of weak intermolecular attractions.

vapor a gas.

vaporize, vaporization changing from a solid or liquid to a gas.

vapor pressure the pressure or partial pressure exerted by a vapor that is in contact with its liquid phase. Often refers to the pressure or partial pressure of a vapor that is in equilibrium with its liquid state at a given temperature.

volatile that which vaporizes easily.

voltaic cell *see cell, voltaic.*

water of crystallization, water of hydration water molecules that are included as structural parts of crystals formed from aqueous solutions.

weak acid an acid that ionizes only slightly in aqueous solution; an acid that does not donate protons readily.

weak base a base that dissociates only slightly in aqueous solution; a base that has a weak attraction for protons.

weak electrolyte a substance that, when dissolved, yields a solution that is a poor conductor of electricity because of limited ionization or dissociation.

weak oxidizer (oxidizing agent) an oxidizer that has a weak attraction for electrons.

weak reducer (reducing agent) a reducer that does not release electrons readily.

weight a measure of the force of gravitational attraction.

yield the amount of product from a chemical reaction.

Z atomic number.

Photo Credits

Chapter 1
Chapter opening photo: Paul Conklin/PhotoEdit
Unnumbered photo on page 4: Bob Daemmrich/The Image Works
Unnumbered photo on page 5: Michael Newman/PhotoEdit
Unnumbered photo on page 6: Hank Morgan/Rainbow
Unnumbered photo on page 7: Yoav Levy/Phototake

Chapter 2
Chapter opening photo: Michael Clay
Unnumbered photo on page 16: Charles D. Winters
Unnumbered photo on page 18: Charles Steele
Fig. 2.4, 2.6 (both), 2.8 (a), 2.8 (b), 2.10: Michael Clay
Unnumbered photo page 21: Leonard Lee Rue/Photo Researchers
Unnumbered photo page 43: Michael Clay

Chapter 3
Chapter opening photo: Michael Clay
Unnumbered photo p. 37 and 48 (top): NASA
Unnumbered photo page 38: (c) CDC/Science Source 1990/Allstock
Fig. 3.1, 3.2, 3.3: Michael Clay
Unnumbered photo p. 57: Robert Brenner/PhotoEdit
Unnumbered photo p. 64: Eunice Harris/Photo Researchers

Chapter 4
Chapter opening photo: Stephen Frink/Allstock
Unnumbered photo page 91: Mark Antman/The Image Works
Unnumbered photo page 101: Charles D. Winters

Chapter 5
Chapter opening photo: Michael Clay

Chapter 6
Chapter opening photo: Michael Clay

Chapter 7
Chapter opening photo, Figs. 7.1a, 7.1b, 7.2: Michael Clay

Chapter 8
Chapter opening photo: Charles D. Winters
Fig. 8.2: UPI/Bettmann
Figs. 8.3, 8.6a, 8.6b: Michael Clay
Fig. 8.7: Charles Steele

Chapter 9
Chapter opening photo: Charles Steele
Unnumbered photo page 244: NASA

Chapter 10
Chapter opening photo, unnumbered photo page 253: Janice A. Peters
Unnumbered photo page 272: Goodyear Tire and Rubber Company
Unnumbered photo page 279 (top and bottom): IBM Corporation, Research Division, Almaden Research Center
Unnumbered photo page 279 (middle): Michael L. Abramson/*Time Magazine*

Chapter 11
Chapter opening photo, Fig. 11.3: Michael Clay
Fig. 11.7: Leon Lewandowski

Chapter 12
Figure 12.2: Charles D. Winters
All photos in Table 12.1, Chapter opening photo, Figs. 12.3, 12.8, 12.10: Michael Clay
Fig. 12.9: Charles Steele

Chapter 13
Chapter opening photo, Fig. 13.3: Michael Clay
Unnumbered photo page 338: Stamp from collection of Professor C.M. Lang, photography by Gary J. Shulfer, University of Wisconsin Stevens Point. Scott Standard Postage Stamp Catalogue Italy #713 (1956).

Chapter 14
Chapter opening photo: Hermann Eisenbeiss/Photo Researchers

Fig. 14.1, 14.16, 14.17, 14.18: Michael Clay
Unnumbered photo page 374: BIOS (V. Bretagnolle)/
Peter Arnold, Inc.
Fig. 14.15: Charles Winters

Chapter 15
Chapter opening photo, unnumbered photo page 398, Fig. 15.5 (all): Michael Clay

Chapter 16
Chapter opening photo, Fig. 16.5: Charles Winters
Fig. 16.1, 16.3, 16.4a, 16.4b, 16.7: Michael Clay
Fig. 16.6: Charles Steele

Chapter 17
Chapter opening photo, Fig. 17.4: Michael Clay
Fig. 17.1, unnumbered photo page 477: Charles Steele

Chapter 18
Chapter opening photo, unnumbered photo page 489 (bottom): Michael Clay
Unnumbered photo page 489 (top, both): UC Irvine
Unnumbered photo page 497 (left): Tom Stack and Associates
Unnumbered photo page 497 (right): Mark Antman/The Image Works

Chapter 19
Chapter opening photo: Charles Steele
Unnumbered photo page 514: Michael Clay
Fig. 19.6: Michael Clay
Fig. 19.8: Marna G. Clarke

Chapter 20
Chapter opening photo: NASA
Unnumbered photo page 540 (top): CNRI/Science Photo Library/Photo Researchers
Fig. 20.2: Michael Clay
Unnumbered photo page 543 (bottom): Tony Freeman/PhotoEdit
Figure 20.5: Sygma
Unnumbered photo page 548: The Granger Collection
Figure 20.10: Michael Clay

Chapter 21
Chapter opening photo, Figures 21.1, 21.2, 21.3, 21.4, 21.5, 21.6, 21.8, 21.9, 21.10, 21.11, 21.12, 21.13, 21.14, 21.15, 21.16, 21.17: Michael Clay

Chapter 22
Chapter Opening photo: Michael Clay
Unnumbered photo page 611 (both), Figure 22.2 (both): Michael Clay
Unnumbered photo page 622: Michael Newman/PhotoEdit
Figure 22.7b: Robert T. Morrison and Robert N. Boyd

Index

Absolute temperature, 66, 93
Accelerators, particle, 551
Acid(s), 143, 328–330, 438–440, 460–480
    (Chapter 17), 587–589, 624–627
  amino, 591, 608
  Arrhenius concept of, 460
  binary, 143
  Brönsted-Lowry concept of, 461–464
  carboxylic, 328–330, 587–589, 592(t)
  equivalents, 409–414
  equivalent mass, 411
  fatty, 620
  ionization of, 143, 329
  Lewis concept of, 464
  names of, 143–150
  nucleic, 624–627
  organic, 328–330, 587–589
  oxyacids, 145–150
  properties of, 460
  reactions of, See Acid-base reactions.
  strong and weak, 438–440, 467–469,
    468(t)
Acid(ic) anions, 151, 152(t)
Acid-base conjugate pairs, 465–467
Acid-base reactions, 214–217, 460–471
  compared to redox reactions, 500
  neutralization, 214–217, 449–451, 460
  predicting, 469–471
  proton transfer, 461–464
Acid constant, K_a, 526–528
Activated complex, 511
Activation energy, 512
Activity series, 442, 443(t), 496
Alcohol(s), 327, 582–584, 592(t)
Aldehyde(s), 585–587, 592(t)
Algebra, how to solve problems by, 66
  review of, A.5–A.8
Aliphatic hydrocarbons, 565, 575
Alkali metal(s), 272
Alkaline earth metal(s), 273
Alkane(s), 326, 566–572
  isomers among, 327, 568
  names of, 566–570
Alkene(s), 572–574
  isomers among, 574
Alkyl group(s), 568
Alkyne(s), 572–574
Alpha rays (particles), 538
Amide(s), 590, 592(t)
Amine(s), 590, 592(t)
Amino acid, 608, 609(t)
Amorphous solid, 375
Anhydrous compound, 159
Anion(s), 140, 144–154, 154(t)

Aromatic hydrocarbons, 575–577
Arrhenius acid-base theory, 460
Arrhenius, Svante, 460
Atmosphere (unit of pressure), 90
Atom(s), 20, 115–123, 254–261
  Bohr model of, 254–257, 259
  Dalton's theory, 115
  mass of, 121–123
  number of in a formula unit, 171
  nuclear model of, 117
  planetary model of, 118
  quantum mechanical model, 258–261
  sizes of, 273–275
Atomic energy, 26, 554–556
Atomic mass, atomic mass unit (amu), 121
Atomic number, Z, 119–121
Atomic size, 273–275
Atomic spectrum, 254–257
Atomic theory, 115
Atomic weight (mass), 121
Avogadro, Amadeo, 338
Avogadro's hypothesis (law), 337, 354
Avogadro's number, 157, 159–162

Barometer, 90
Base(s), 196–198, 432–449, 460–480
  (Chapter 17)
  Arrhenius concept of, 460
  Brönsted-Lowry concept of, 461–464
  Lewis concept of, 464
  properties of, 460
  strong and weak, 467–469, 468(t)
Becquerel, Henri, 538
Benzene ring, 575
Beta rays (particles), 539
Binary acids, 143
Binary compounds, molecular, 138–140
Binding energy, 516
Bohr model of the hydrogen atom, 254–257,
  259
Bohr, Neils, 254
Boiling, 372
Boiling point, 373, 379(t)
  constant, molal, 422
Boiling point elevation, 422
Bonds, 292–301
  covalent, 293, 309
  hydrogen, 366–368
  ionic, 292
  multiple (double, triple), 298
  polar and nonpolar, 295–298
Bond angle, 319
Bonding electrons, 238. See also Valence
  electrons.

Boyle's law, 100–104, 336, 354
Breeder reactor, 554
Brönsted, Johannes N., 462
Brönsted-Lowry acid-base theory, 461–464
Buffer, 527
Burning, 208–210, 217

Calculators, A.1–A.5
Calorie, 242
Carbohydrate(s), 615–620
  monosaccharides, 615
Carbon dating, 546
Carbonyl group, 585, 592(t)
Carboxylic acid(s), 328–330, 587–589,
  592(t)
Carboxyl group, 329, 587, 592(t)
Catalyst, 513
Cation(s), 140–143, 153(t)
Cell, electrolytic, 435, 487
Cell, galvanic, voltaic, 487
Cellulose, 618
Celsius (centigrade) temperature scale, 65,
  93
Chadwick, James, 117
Chain reaction, 553
Charge(s) electrical, 22
Charles' law, 98–100, 336, 354
Chemical bonds. See Bonds
Chemical, common names of, 161(t)
Chemical change, 14
Chemical equations. See Equations,
  chemical.
Chemical equilibrium. See Equilibrium,
  chemical.
Chemical family(ies), 124, 271–273
Chemical formula. See Formulas, chemical.
Chemical group. See Chemical family(ies).
Chemical properties, 13
Chemical symbols of elements, 20
  and the periodic table, 126–128
Colligative properties of solutions, 421–423
Collision theory of chemical reactions, 509–
  512
Combination reactions, 205, 217
Combining volumes, law of, 337
Common ion effect, 525
Composition, percentage, 181–183
Compound(s), chemical, 19–22
  anhydrous, 159
  binary molecular, 138–140
  empirical formula of, 185–190
  formulas of, 20. See also Formulas,
    chemical
  hydrates, 159

Compound(s), chemical (*Continued*)
 ionic, 154–158, 292
  names and formulas of, 154–158
 made up of two nonmetals, 138–140
 mass relationships between elements in, 181–184
 molecular, 138, 293
 names of, 138–140, 154–157
 organic, 325–330, 564. *See also* compound class
 percentage composition, 181–183
 simplest formula of, 185–190
Concentration,
 effect on equilibrium, 516
 effect on reaction rate, 514
 solution. *See* Solution concentration
Condensation, 370
 heat of, 378, 379(t)
Condensation polymerization, 595
Conductivity, solution, 435
Configuration, electron, 262–268
Conjugate acid-base pairs, 465–467
Conservation of energy, 25
Conservation of mass, 24, 115
Constant, equilibrium. *See* Equilibrium constant.
 ideal gas, universal gas, 338, 355
 molal boiling point, 422
 molal freezing point, 422
 proportionality, 95
Constant composition, law of, 21. *See also* Definite composition.
Conversion factor, 42
Coordinate covalent bond, 309
Covalent bonds, 293–298
 polarity of, 295–298
Crookes, William, 117
Critical mass, 553
Crystalline solid(s), 375–377
Crystallization, heat of, 380, 379(t)
 water of, 129
Crystals, 292, 374–377
Curie (unit), 541
Curie, Irene, 551
Curie, Marie and Pierre, 548
Cycloalkanes, 571

Dalton, John, 115
Dalton's atomic theory, 115
Dalton's law of partial pressure, 351–354
Data, interpretation of, 94
Dating, radiocarbon, 546
Decay, radioactive, 539, 547–549
Decomposition reaction, 207, 217, 217(t)
"Defining equation," 68
Definite composition, law of, 21, 115. *See also* Constant composition.
Density, 68–70, 341
Deoxyribonucleic acid (DNA), 616, 624–627
Dilution problems, 414
Dimensional analysis, 41–47
Dipole forces, 365, 368
Disaccharides, 617
 lactose, 618
 sucrose, 617

Dispersion forces, 366, 368
Double replacement equation (reaction), 212–217

Einstein, Albert, 26
Einstein's equation, 26
Electrical charges, 22
Electrode, 409, 487
Electrolysis, 487
Electrolytes, 435, 487
Electrolytic cell, 435, 487
Electromagnetic spectrum, 254
Electron(s), 117
 configuration, 262–268
 energy level diagram, 262
 orbits, 254, 259
 orbitals, 259, 261
 quantized energy levels, 254
 transfer in forming ionic bonds, 292
 transfer in redox reactions, 489
 valence, 268–270
Electron-dot diagram (formula, structure, symbol), 269, 294, 308–315
Electron pair geometry, 316
Electron pair repulsion principle, 316
Electron-transfer reactions. *See* Reactions, oxidation-reduction
Electronegativity, 295–298
Electrostatic force, 24
Element(s), 18–20
 classifications of, 101, 136–138, 141, 512
 formulas of, 137
 Lewis symbols of, 269
 names of, 126–128
 periodic table of, 123–128
 representative, 124
 symbols of, 20, 126–128
 transition, 124
 transuranium, 552
Empirical formula, 185–190
Endothermic change, 23
Energy, 23, 242, 378
 activation, 512
 atomic, 26, 554–556
 and change of state, 378–380, 382–386
 and change of temperature, 380–386
 conservation of, law, 25
 in chemical reactions, 23, 243–245
 ionization, 270
 kinetic, 24, 510–512
  and absolute temperature, 66, 93
  kinetic energy distribution curve, 369, 372, 512
 nuclear, 26, 554–557
 potential, 24, 510–512
 quantization of, 254
 units of, 242
Energy level(s), 254–259
 excited states, 256
 ground states, 256
 principal, 257
 sublevels, 258
English-metric conversion factors, 64(t)
Enthalpy, 243
Enzyme(s), 514, 613

Equation(s), chemical, 23, 199–217 (Chapter 8), 440–453
 balancing, 201–204, 500–503
 combination, 205–207, 217
 decomposition, 207, 217
 double replacement (displacement), 212–217
 half-reaction, 488
 ion combination, 444–453
  molecule formation, 449–451
  neutralization, 449–451
  precipitation, 444–449
 net ionic, 440–453
 neutralization, 214–217, 449–451
 nuclear, 547–550
 oxidation-reduction, 210–212, 217, 441–444, 488–491, 497–503
 precipitation, 212–214, 217, 444–449
 procedure for writing, 205, 440, 500–503
 quantitative interpretation of, 223
 single replacement (displacement), 210–212, 217, 441–444
 summary of, 216, 453
 thermochemical, 243
Equation, ideal gas, 338–344, 354
Equilibrium, 369–372, 400–402, 471, 509–530 (Chapter 19)
 acid-base, 463, 465–471
 chemical, 509–513
  calculations, 523–530
  development of, 515
  Le Chatelier's Principle, 516–520
 effect of temperature on, 519
 gaseous, 528–530
 liquid-vapor, 369–372
 oxidation-reduction, 497–500
 solution, 400–402
 water, 472
Equilibrium constant, 472, 520–523
 acid constant, K_a, 526–528
 solubility product constant, K_{sp}, 523–525
 water constant, K_w, 472, 522
Equivalent, 409–414
Equivalent mass (weight), 411
Ester, 587–589, 592(t)
Ether(s), 327, 582–584, 592(t)
Evaporation, 370
Exclusion principle, 260
Exothermic change, 24
Experimental data, 94
Exponential notation, 36–40
 by calculator, 40, A.4

Fahrenheit temperature scale, 65
Family, chemical, 124, 271–273
Faraday, Michael, 117
Fat(s), 620
Fatty acids, 620
Fission, nuclear, 553–556
 electrical energy from, 554–556
Force(s),
 electrostatic, 22
 intermolecular, 365–369
 and solubility, 402
 dipole, 365, 368

Force(s) (*Continued*)
    dispersion (London), 366, 368
    hydrogen bonds, 366–368
Formula(s), chemical, 20, 137–157, 159, 556–576, 582, 585, 590
    alcohols, 582, 592(t)
    aldehydes, 585, 592(t)
    amides, 590, 592(t)
    amines, 590, 592(t)
    anions, 140–154
    calculations based on, 172–175, 177–184
    compounds made up of two nonmetals, 138–140
    condensed, 566
    elements, 137
    empirical, of a compound, 185–190
    ethers, 582, 592(t)
    hydrates, 159
    hydrocarbons, 566–576
    ketones, 585, 592(t)
    ionic compounds, 154–157
    Lewis, 269, 294, 308–315
    line, 326, 566
    simplest, 185–190
    unit, 137
        number of in a sample, 178–181
Formula mass (weight), 172–175
Formula unit, number of atoms in, 171
    quantitative meaning of, 184
Freezing point constant, 422
Freezing point depression, 422
Fructose, 616
Functional group(s), 582, 592(t). *See also* individual groups
Fusion, heat of. 380, 379(t)
    nuclear, 556

Gamma rays, 539
Gases, 15, 87–106 (Chapter 4), 336–356 (Chapter 13)
    Avogadro's law, 337, 354
    Boyle's law, 100–104, 336, 354
    Charles' law, 98–100, 336, 354
    combined gas laws, 104–106, 336, 354
    Dalton's law of partial pressure, 351–354
    Density of, 341
    Gay-Lussac's law, 96–98, 336, 354
    ideal gas (model), 89
    ideal gas constant, 338, 355
    ideal gas equation, 338–344, 354
    kinetic theory of, 89
    measurable properties of, 90–94
    molar volume of, 342–344
    noble, 272
    partial pressure, Dalton's law of, 351–354
    pressure, 90–94
    properties of, 15, 88, 90–94, 362
    standard temperature and (STP), 105
        molar volume at, 342
    stoichiometry, 344–351
Gas constant, universal, 338, 355
Gas density, 341
Gas equation, ideal, 338–344, 354
Gas phase equilibria, 528
Gas stoichiometry, 344–351

Gay-Lussac's law, 69–98, 336, 354
Geiger-Muller counter, 540
Gene, 626
Geometry, electron pair, 316
Geometry, molecular. *See* Molecular geometry.
"Given quantity," 71
Globular proteins, 613
Glucose, 615
Glycogen, 620
Group(s), functional, 582, 592(t)
Groups in the periodic table, 124

Half-life, 544–547
Half-reaction (equation), 488
Halogen(s), 273
Heat,
    molar, of vaporization, 363
    of condensation, 378, 379(t)
    of fusion, 379(t), 380
    of reaction, 243
    of solidification (crystallization), 379(t), 380
    specific, 380–382, 361(t)
    units of, 242
    of vaporization, 378, 379(t)
Heterogeneous matter, 17
Homogeneous matter, 17
Homologous series, 566
Hydrate(s), 159
Hydrocarbon(s), 326, 565–582. *See also* Alkane(s), Alkene(s), Alkyne(s)
    aliphatic, 565–574, 577(t)
        chemical properties of, 578–580
        sources and preparations, 577
        physical properties of, 570
    aromatic, 565–577
    names and formulas of, 556–570
    saturated, 565
    summary of, 577(t)
    unsaturated, 572–574, 573(t)
    uses of, 580–582
Hydrogen atom, Bohr model of, 254–257, 259
Hydrogen bond, 366–368
Hydrogen ion, 143, 460, 471
    as a proton, 143, 461
    in water, pH, 471–480
Hydronium ion, 462
Hydroxide ion, 471
    in water, pOH, 471–480
Hydroxyl group, 328, 582, 592(t)

Ideal gas. *See* Gases
Ideal gas equation, 338–344, 354
Indicator, 417
Inhibitor, 514, 614
Intermolecular forces. *See* Forces.
Ion(s)
    anion. *See* Anion(s)
    cation. *See* Cation(s)
    formation of, 141–143, 292
    formulas of, 140–154
    hydrogen, 143, 460, 462, 471
    hydronium, 462

Ion(s) (*Continued*)
    hydroxide, 471
    monatomic, 140–143, 289–293
        names and formulas of, 140–143
        with noble gas electron configuration, 289–293
    polyatomic, names and formulas of, 145–154, 154(t)
    sizes of, 274, 276
Ionic bonds, 292
Ionic compound(s). *See* Compound(s), chemical, ionic
Ionic crystals, 292, 374–377
Ionic size, 274, 276
Ionization of acids. *See* Acids, ionization of of water, 471
Ionization energy, 270
Ionization, percentage, 526
Isomer(s), 315, 327, 567, 574
Isotope(s), 119, 122(t)

Joliot, Frederic, 551
Joule, 242

K (equilibrium constant). *See* Equilibrium constant.
Kelvin, 65
Ketone(s), 585–587, 592(t)
Kilogram, 48
Kinetic energy, 24, 93, 369, 510–512
    kinetic energy distribution curve, 369, 372, 512
Kinetic molecular theory, 15, 89
Kinetic theory of gases, 89

Lactose, 618
"Larger/smaller rule," 37, 39, 43
Law(s),
    Avogadro's, 337, 354
    Boyle's, 100–104, 336, 354
    Charles, 98–100, 336, 354
    combined gas law, 104–106, 336, 354
    Dalton's, of partial pressure, 351–354
    Gay-Lussac's law, 96–98, 336, 354
    ideal gas, 89, 338
    of combining volumes, 337
    of conservation of energy and mass, modified, 26
    of conservation of energy, 25
    of conservation of mass, 24, 115
    of definite (constant) composition, 21, 115
    of multiple proportions, 116
Lawrence, E. O., 552
Le Chatelier's Principle, 516–520
Length, measurement of, 49
Lewis acid-base theory, 464
Lewis diagrams (formulas, structures, symbols), 269, 293–295, 308–315
Lewis, G. N., 293
Limiting reactant (reagent) problems, 233–241
Lipid(s), 620–624
    fat(s), 620
    oil(s), 620

Lipid(s) (*Continued*)
  phospholipids, 621
  steroids, 622
  waxes, 622
Liquid(s), 15, 362–373
  and saturated vapor, equilibrium between, 369–372
  intermolecular forces in, 362–369
  miscible and immiscible, 400
  physical properties of, 15, 362–365, 369–372
Liter, 50
Logarithms, 473, 477, A.1, A.8
London (dispersion) forces, 366
Lowry, Thomas M., 462

Macromolecules, 608
Major/minor species, 438
Manometer, 91
Mass, 48
  atomic, 121
  conservation of, Law of, 25, 115
  equivalent, 411
  formula, 172–175
  molar (*also called* molar weight), 177
  molecular, 172–175
  unit of, 48
Mass number, 119–121
Mathematics, review of, A.5–A.9
  algebra and arithmetic, A.5–A.8
  exponential (scientific) notation, 36–40, A.4
  logarithms, A.8
  percent, 181
  proportionalities, 42, 94
    proportionality constant, 95
  significant figures, 54–63, 478
Matter, 15–17, 22. *See also* Gases; Liquids; Solids
  electrical character of, 22
  and energy, 13–26 (Chapter 2)
  heterogeneous and homogeneous, 17
  kinetic molecular theory of, 15, 89
  properties of, 13, 15–17
  pure or mixture, 17
  states of, 15
Measurement, 36
  length, 49
  gases, 90–94
  mass, 48
  metric prefixes, 48, 49(t)
  metric system, 3, 48–54, 77
  SI system, 36, A.11
  temperature, 65–68. *See also* Absolute temperature; Celsius temperature; Fahrenheit temperature scale; Kelvin
Melting point, 379(t)
Mendeleev, Dimitri, 123
Metal(s), 124, 138, 277
  transition, 124
Metallic bonds, crystals, 376
Metalloid, 277
Meter, 49
Metric prefixes, 48, 49(t)
Metric–English conversion factors, 63–65, 64(t)

Meyer, Lothar, 123
Mixtures, 17
Molal boiling point constant, 422
Molal freezing point constant, 422
Molality, 408, 414
Molarity, 405–408, 414
Molar volume, 342–344
Molar mass (weight), 177
Mole, 175–177
  and mass, conversion between, 178–181
  relationships in a chemical equation, 223
Molecular compound(s), 293
Molecular formulas, 185, 189
Molecular geometry, 319–323
  bond angle, 319
  electron pair, 316
  electron pair repulsion principle, 316
  of the multiple bond, 322
  of water, 367, 373
Molecular polarity, 323–325, 365
Molecular structure of organic compounds, 325–330, 564, 565(t), 592(t)
Molecular mass (weight), 172–175
Molecule(s), 15, 137–140, 293
  diatomic, elements that form, 137
  number of in a sample, 178–181, 184
  polarity of, 323–325, 365
Monatomic ions. *See* Ions, monatomic
Monomer, 593
Monosaccharides, 615–617
Multiple bonds, 298
  geometry of, 322
Multiple proportions, law of, 116

Names of chemicals
  acids, 143–150
  anions, 141, 144–154, 154(t)
  compounds of two nonmetals, 138–140
  cations, 141–143, 153(t)
  common names, 161
  elements, 137
  ionic compounds, 157
  monatomic ions, 140–143
  organic compounds. *See* compound class
Net ionic equations, 440–453, 453(t)
Network solid, 376
Neutralization, 214–217, 449–451
Neutron, 117
Noble gases, 272, 289
Nomenclature, 137–161 (Chapter 6), 556–570, 571, 573–576, 582, 585, 590. *See also* Formulas, chemical; Names of chemicals.
Nonelectrolyte, 435
Nonmetals, 124, 138, 277
Normal boiling point, 373
Normality, 409–414
Nuclear chemistry, 537–557 (Chapter 20)
Nuclear energy, 26, 554–557
Nuclear equations, 547–550
Nuclear fission, 553–556
  electrical energy from, 554–557
Nuclear fusion, 556
Nuclear reactions, 547–552, 553, 556
  bombardment, 550–552
  chain, 553

Nuclear reactions (*Continued*)
  compared to ordinary chemical reactions, 550
  equations for, 547–549
  fission, 553
  fusion, 556
Nuclear symbol, 120, 539
Nucleic acids, 624–627
Nucleotide(s), 624
Nucleus (atomic), 118
Nuclide(s), 538

Octet, 295
Octet rule, 295
  exceptions to, 300
Oil(s), 20
Orbit(s), electron, 254, 259
Orbital, electron, 259, 261
Organic chemistry, 325–330, 554–597 (Chapter 21)
Oxidation,
  burning, 208–210, 217
  in redox reactions, 488, 494
Oxidation number (state), 142, 491–495
Oxidation-reductions reactions. *See* Reactions, oxidation-reduction
Oxidizing agents (oxidizers), 495–497, 496(t)
Oxyacids, 145–150
Oxyanions, 151–152, 154(t)

Partial pressure, 351–354, 402
Particles, subatomic, 117
  accelerator, 551
Pascal (unit of pressure), 90
Pauli exclusion principle, 260
Peptide linkage, 591, 610
Percent, percentage
  composition of a compound, 181–184
  concentration of a solution, by weight, 403–405
  ionization, 526
  yield, 229–233
"Per" expression, 41
Periodic table of the elements, 123–128
  and charges on monatomic ions, 140
  and electron configurations, 264
  group in, 124
  period in, 124
  trends in, 270–277
pH and pOH, 473–480, 527
  of common fluids, 475(t)
Phospholipids, 621
Phosphorescence, 540
Physical changes, properties, 13
"Plan" (how to solve a problem), 72
Planetary model of the atom, 118
pOH. *See* pH.
Polarity,
  of covalent bonds, 295–298
  and intermolecular attractions, 365
  of molecules, 323–325, 365
Polymer, 593
Polymerization, 593, 595

Polysaccharides, 618–620
cellulose, 618
starch, 618
glycogen, 620
Positron, 551
Potential energy, 24
Precipitation reactions, 212, 217, 444–449
Prefixes, metric, 48, 49(t)
numerical, 139
Pressure, 90–93
gauge, 91
measurement of, 90–94
partial, Dalton's law of, 351–354
standard, 105, 355
vapor, *See* vapor pressure
water vapor, 352, 372–374
units of, 90
Principal energy levels, 254, 257–261
Principal quantum number, 254
Problem solving, 70–72
by algebra, 66, A.7
by dimensional analysis, 47
summary, 71
Properties, chemical and physical, 13
Proportionality, 42, 94
proportionality constant, 95
Protein, 608,
amino acids found in, 609(t)
globular, 613
primary structure, 608
secondary structure, 610–612
tertiary structure, 612
Proton, 117, 143, 461
Proton transfer reaction. *See* Acid-base
reactions.
Pure substance, 17

Quanta, quantum, 254
Quantized energy levels, 254
Quantum mechanical model of the atom,
258–262
Quantum number, principal, 257

Rad (unit), 542
Radioactive decay, 539, 547–549
Radioactivity, 538–549
detection and measurement of, 539–541
effects of on living systems, 541–543
Radiocarbon dating, 546
Radionuclides (radioisotopes), 538, 552, 557
Ray(s), alpha, beta, gamma, 538
Reaction rate,
effect of a catalyst on, 513
effect of concentration on, 514
effect of temperature on, 512
Reaction(s), chemical, (*See also* Equations,
chemical
acid-base. *See* Acid-base reactions.
addition, 578
addition polymerization, 593
burning of compounds of carbon,
hydrogen and oxygen, 208–210, 217
chain, 553
collision theory of, 509–512
combination, 205–207, 217
condensation polymerization, 595

Reaction(s), chemical (*Continued*)
decomposition, 207, 217
double replacement (displacement), 212–
217, 444
electron-transfer. *See* Reactions, oxidation-
reduction
enthalpy of, 243
endothermic, 24
exothermic, 23
in water solutions. *See* Solutions, reactions
in.
ion combination, 444–453
molecule formation, 449–451
precipitation, 212, 217, 444–449
neutralization, 214–217, 449–451
nuclear. *See* Nuclear reaction(s)
oxidation-reduction, 210–212, 217, 441–
444, 487–503 (Chapter 18)
compared to acid-base reactions, 500
oxidation numbers and, 491
predicting, 441–444, 497–500
"single replacement," 210–212, 441–
444
writing equations for, 441–444, 500–
503
polymerization, 593, 595
precipitation, 212, 217, 444–449
rate of. *See* Reaction rate.
redox. *See* Reaction(s), oxidation-
reduction
reversible, 371
single replacement (displacement), 210–
212, 217, 441–444
substitution, 579
synthesis, 205
Redox reactions, *See* Reaction(s), oxidation-
reduction
Reducing agents (reducers), 495–497, 496(t)
Reduction, 488, 494
Rem (unit), 542
Representative element(s), 124
Resonance, 311
Reversible reaction (change), 207, 371, 401,
509
Ribonucleic acid (RNA), 616, 624–627
Ribose, 616
Roentgen (unit), 541
Roentgen, Wilhelm, 538
Rounding off (numbers), 58
Rule of eight, 295
Rutherford, Ernest, 117
Rutherford scattering experiment, 118

Salt, 214
Scattering experiment, Rutherford's, 118
Schrodinger, Irwin, 258
Scientific method, 4
Scientific notation. *See* Exponential notation.
Scintillation counter, 540
SI system of measurement, 36, A.11
Significant figures, 54–63, 478
Simplest formula, 185–190
Single replacement (redox) reactions, 210–
212, 217, 441–444
Sizes, of atoms, 273–275
of ions, 274, 276

Solids, 15, 374–377
Solidification, heat of, 379(t), 380
Solubility, 399, 402, 446, 447(t)
rules, table, 447
Solubility product constant, 523–525
Solute, 399
Solution conductivity, 409
"Solution inventory," 436
Solution stoichiometry, 415–417
Solutions, 398–423 (Chapter 15)
colligative properties of, 421–423
concentration of, 403–414
molality, 408, 414
molarity, 405–408, 414
normality, 409–414
percentage by weight, 405–408, 414
conductivity, 435
electrolytes, 487
strong, weak, and nonelectrolytes, 436
formation of, 400–402
"inventory," 436
miscible and immiscible, 400
rate of dissolving, 401
reactions in, 441–453
molecule formation, 449–453
neutralization, 214–217, 449–451
oxidation-reduction, single replacement,
210–212, 217, 441–444
precipitation, 212, 217, 444–449
unstable product, 451
stoichiometry, 415–417
Solvent, 399
Specific heat, 380–386, 381(t)
Spectator ion, 440
Spectrum, 122, 254–257
atomic, 254–257
electromagnetic, 254
mass (Fig. 5.5), 122
Standard temperature and pressure, STP, 105
Starch, 618
State(s) of matter, 15
change of, 378–380, 382–386
State symbols, 199
Steroids, 622
Stoichiometry, 223–241, 244, 344–351, 415–
417
gases, 344–351
limiting reactant (reagent), 233–241
pattern for solving problems, 225, 344,
415
percent yield, 229–233
solution, 415–417
thermochemical, 244
STP. (standard temperature and pressure),
105
Structure. *See* Molecular geometry
Subatomic particles, 117
Sublevel(s), 258, 261
Substance, pure, 17
Substitution reaction, 579
Sucrose, 617
Sugar(s), 615–618
fructose, 616
glucose, 615
ribose, 616
Surface tension, 364

Symbol(s) of elements, 20, 126
   Lewis (electron dot), 269
   nuclear, 120, 539
Synthesis, 205

Temperature, 65–68, 93
   absolute, 66, 93
   and kinetic energy, 93, 369, 512
   Celsius, 65
   effect of, on equilibrium, 519
     on reaction rate, 512
     on vapor pressure, 371
   Fahrenheit, 65
   Kelvin, 65, 93
   standard, 105
Tetrahedral angle, 317, 564
Tetrahedron, 319, 564
Theory,
   atomic, 115
   collision, of chemical reactions, 509–512
   kinetic, of gases, 89
   kinetic molecular, 15, 89
Thermochemical equations, 243
Thermochemical stoichiometry, 244
Thomson, J. J., 117
Titration, 417

Torr, 90
Transition metal (element), 124
Transuranium elements, 552

Unit(s),
   atomic mass, 121
   cancellation of, 42
   energy, 242
   length, 49
   mass, 48
   metric-English conversion, 63–65, 64(t)
   metric prefixes, 48, 49(t)
   pressure, 90–93
   SI system of, 36, A.11
   volume, 49
   temperature, 65–68
Unit path, 42
Universal gas constant, 338, 355

Valence electrons, 268–270
Valence shell electron pair repulsion
     (VSEPR), 316
Vapor pressure, 352, 362, 371
   effect of temperature on, 371
   equilibrium, 369–372

Vaporization, 363, 378
   heat of, 378, 379(t)
   molar heat of, 363, 363(t)
Viscosity, 364
Volume, 49
   effect of, on equilibrium, 517–519
   molar, of a gas, 342–344
   units of, 49

Water, 373, 471
   of crystallization (hydration), 159
   equilibrium, 471
   equilibrium constant, K_W, of 472
   hydrogen bonding in, 366–368
   ionization of, 471, 522
   molecular structure of, 373
   properties of, 374, 379(t), 381(t)
   specific heat of, 379(t)
   vapor pressure, 352, 372 (Fig. 14.10)
Wave properties, 254
Wax(es), 622
Weight, 48. *See also* Mass.

Yield, percent, 229–233

Z (symbol for atomic number), 119